Industrial Pollution and Its Control

Volume I

Industrial Pollution and Its Control

Volume I

S. C. Bhatia
BE (Chemical), BITS Pilani & MBA

Edited by

Pramod Kumar
Assistant Professor & Head of Department – Biotechnology
Hindustan College of Science and Technology, (Mathura – UP)

and

Sarvesh Devraj
B.Tech (Mechanical), UPTU
M.Tech (Renewable Energy Engineering & Management), TERI University,
(Research Associate – TERI, New Delhi)

Published by Woodhead Publishing India Pvt. Ltd.
Woodhead Publishing India Pvt. Ltd.,
303, Vardaan House, 7/28, Ansari Road,
Daryaganj, New Delhi - 110002, India
www.woodheadpublishingindia.com

First published 2017, Woodhead Publishing India Pvt. Ltd.

Woodhead Publishing India Pvt. Ltd. ISBN: 978-93-85059-24-7
Woodhead Publishing India Pvt. Ltd. e-ISBN: 978-93-85059-82-7

Typeset by Asian Enterprises, Vasant Kunj, New Delhi
Printed and bound India by Replika Press Pvt. Ltd.

Contents

Volume I

Preface

Industries contributes various kinds of pollutants to the environment. Different countries in the world are facing different types of industrial pollution problems. Industry produces both traditional pollutants such as organic substances, sulphur dioxide, particulates and nutrients, etc., and newly-recognised pollutants such as dioxin and other specific toxic substances. The pollutants are mainly in gas, water, and solid forms that can cause serious damage to the biosystems. Industrial pollution has attracted a lot of attention. Great efforts have been made to solve the problems. The problems of industrial pollution and their control measures are briefly discussed in various chapters of this book. In recent years, the sustainable development concept has been widely recognised, which has promoted the implementation of integrated management of industrial production. The development of 'industrial ecology' aims to provide theories and methods to harmonise the industrial sectors with the biosphere, that may bring solutions of sustainable development to the industry and society. Industry plays an important role in the process of economic development in the world. It enhances the economic welfare of citizens and supplies the material goods they consume. The way in which society will develop in the future is largely dependent on how the growth which industry generates is distributed. The industrial sector generates both traditional pollutants (e.g., organic substances, sulphur dioxide, particulates and nutrients) and newly-recognised pollutants (e.g., specific toxic substances). The industrial sector includes a number of diverse activities. As a result, there is a wide range of different resource and environmental impacts created by industry.

Thus, industry has particular environmental responsibilities in terms of such factors as plant location and design, environmental pollution, vibration and noise controls, waste disposal, occupational health and safety aspects, and long-range planning.

Industrial pollution control has been paid a lot of attention. Increasing efforts have been made to protect the environment, both in terms of reducing point-source emissions, risk management during chemical use and handling of hazardous wastes. New legislation, more stringent emission standards, stricter controls and growing consumer demands for environmentally-sound products have been promoting the implementation of environmental friendly technologies and integrated pollution management strategies. Recent practices in pollution control tend to move towards the core of industrial operation. Nevertheless, end-pipe treatments played and still play an important role in industrial pollution control.

The industrial pollution problems faced by different countries worldwide are different. Generally speaking, in developed countries, the pressures created by industrial activates (i.e., the emission of traditional pollutants from iron and steel, metal fabrication and petrochemicals) has grown slowly in recent decades. Other types of environmental problems, e.g., contamination of soil and buildings at closed sites, with subsequent high costs for remedial treatments have received growing attention. In developing countries, the environmental pressure coming from the traditional pollutants created by industries is still very heavy. For both developed and developing countries, the growing technology-based industries, created new problems due to the use of toxic material in their production processes, which can cause soil and water contamination. Industrial growth is commonly regarded as being accompanied by an increase in consumption of energy and raw materials. However, industrial experiences in many countries show that the opposite situation can prevail. Industrial growth may favour environmental protection work and govern research and development, thereby promoting new technologies in industry to further minimise environmental risks. It also provides the necessary financial conditions under which large investments in new technology, necessary for further reducing environmental effects, can be made. As a result, the prerequisites are created for a sustainable industrial development of products with lower requirements for natural resources, and enhanced waste recycling and minimisation.

This book on industrial pollution and its control summarises various aspects of pollution prevention and waste minimisation aspects in various industries and is divided in two volumes. Volume I contains 1 to 22 chapters and Volume II has 23 to 50 chapters.

Section I discusses general concepts and engineering considerations. Chapter 1 is devoted to industrial pollution: An overview. Industrial pollution deteriorates the environment in a range of ways and it has negative impact on human lives and health. Chapter 2 focuses on emissions from particulate matter and its control. Particulates are microscopic solid or liquid matter suspended in earth's atmosphere. Chapter 3 concentrates on control of gaseous pollutants. The chapter discusses some of the important gaseous pollutants such as– carbon oxides, nitrogen oxides, sulphur oxides, ozone and hydrocarbons. Chapter 4 deals with odour pollution and its control. The chapter discusses odour spread by various industries and methods of prevention of odour. Chapter 5 focuses on industrial wastewater and its treatment. Industrial wastewater contains a diversity of impurities and therefore, its treatment constitutes a special task.

Chapter 6 is devoted to industrial solid wastes and its treatment. Solid wastes is the term now used the world over to describe all non-liquid waste material arising from industrial, agricultural, chemical and mining activities. Chapter 7 deals with management of hazardous wastes. Hazardous wastes is any unwanted material the disposal of which poses a threat to the environment, i.e., it is explosive, flammable, oxidising, poisonous/infectious, radioactive, corrosive and/or toxic/ecotoxic. Chapter 8 concentrates on nuclear pollution and radioactive wastes. Nuclear pollution comes from nuclear processes such as nuclear fission and radioactive mining. Chapter 9 focuses on process integration for pollution prevention and energy saving. Process integration is a holistic approach to process design, retrofitting and operation of industrial plants, the applications of which are focused on resource conservation, pollution prevention and energy management. Chapter 10 is devoted to noise pollution and its control. Sources and various control methods for reducing noise are discussed. Chapter 11 deals with common effluent treatment plant (CETP). The CETP has been accepted as a solution for collecting, conveying, treating and disposing of the effluents from the industrial estates. This CETP concept helps small and medium scale industries to dispose of their effluents.

Chapter 12 concentrates on environmental auditing. Environmental audits can help companies identify ways to reduce waste, save money, improve overall efficiency and minimise liability risks. Chapter 13 deals with environmental impact assessment. Environmental impact assessment is a process analysing the positive and negative effects of a proposed project, plan or activity on the environment. Chapter 14 focuses on industrial biotechnology and climate change. Industrial biotechnology has a great potential in transformation and enhancement of resource base, reduction of energy consumption, pollution abatement, enhancement of biomass production and conservation of biomass etc. Some of these applications have indirect beneficial effects on wider problems such as global warming, acid rain and climate changes.

Section II discusses pollution control in mechanical and electrical industries. Chapter 15 is devoted to pollution control in mechanical industry. Metal fabricating plants generate lot of liquid wastes during metal removal and metal finishing. Chapter 16 concentrates on pollution control in semiconductors and electronic industry and describes the various pollution prevention techniques including process modification and resources recovery and recycling.

Chapter 17 deals with environmental aspects for e-wastes. E-waste is any refuse created by discarded electronic devices and components as well as substances involved in their manufacture or use. Chapter 18 is devoted to pollution control in thermal power plants. Thermal or heat pollution is common in the vicinity of thermal plants generating electricity. The water used for cooling the working steam is lead back to the sources at an appreciable higher temperature, affecting the aquatic organism. The thermal power industry has an important role in the development of economy, but a thermal power station involves large scale combustion which results in the hazard of flue gases and other pollutants.

Section III discusses pollution control in metallurgical and mining industry. Chapter 19 deals with pollution control in iron and steel industry and describes the sources and treatment of air, water and solid wastes pollutants and stressing the need to modify or alter technological processes. Chapter 20 concentrates in pollution control in non-ferrous industry and discusses variety of gaseous, solid and liquid wastes for disposal in non-ferrous industry such as– aluminium, copper, lead and zinc. Chapter 21 focuses on pollution control in foundry industry and discusses to control particulate matter such as dust, fly ash and gaseous pollutants emitted from foundry industry. Chapter 22 deals pollution control in mining industry and discusses mining activities along with environmental problems due to mining such as air pollution due to release of non-toxic gases and generation of dust, land degradation due to large scale excavation, disposal of mine wastes and tailings and subsidence, noise and vibration due to blasting, etc.

Section IV discusses pollution control in cement, ceramic and glass industry. Chapter 23 is devoted to pollution control in cement industry which contributes to air pollution in the form of suspended particulate matter and gases such as oxides of nitrogen and sulphur dioxide. Chapter 24 focuses on pollution control in ceramic industry. Various sources of air pollution, particulate matter emissions, water pollution and prevention techniques are discussed. Chapter 25 deals with pollution control in glass industry. Emission from glass industry include particulate matter, oxides of nitrogen, sulphur, chlorides and fluorides, etc.

Section V discusses pollution control in food and agriculture industries. Chapter 26 concentrates on pollution control in food processing industry: A review. The key environmental issues for the food industry include wastewater, solid wastes, etc. Chapter 27 is devoted to pollution control in wheat, pulse and rice mills. Environmental issues related to this industry are dust emissions, air pollution, water pollution and noise pollution. Chapter 28 deals with pollution control in vegetable and palm oil industry. The key environmental issues related to this industry are solid wastes and by-products, greenhouse gas emissions, etc.

Chapter 29 focuses on pollution control in dairy industry which is a major sources of water pollution, the special characteristics of dairy waste being a high percentage of dissolved organic matter responsible for the high BOD of dairy effluents. Various methods of recovery and reuse of milk by-products and prevention of dairy wastes are also discussed. Chapter 30 is devoted to pollution control in bakery industry. Even though bakery industry is a low polluting industry but for the ovens, boilers, hot water generators, DG sets they all come under the air pollution act due to emission of flue gases through stack. Chapter 31 deals with pollution control in poultry industry and discusses technical options to mitigate environmental impacts, such as improvements to farm management, animal waste management and nutrition management, along with options to reduce the impacts of intensive feed production.

Chapter 32 focuses on pollution control in meat industry which has the potential for generating large quantities of solid wastes and wastewater with a BOD of 600. Chapter 33 concentrates on pollution control in seafood food industry. Environmental issues in fish processing industries primarily include– water consumption and wastewater generation, solid waste generation and by-products production, emission to air and energy consumption.

Section VI discusses pollution control in chemical process industries. Chapter 34 deals with pollution control in chemical process industries: A review. The environmental impact of an existing process can be reduced by using clean technologies. Chapter 35 concentrates on pollution control in fertiliser industry. The fertiliser industry implements wet scrubbers, condensate strippers and recovery and recycle methods to control various emissions.

Chapter 36 focuses on pollution control in petroleum refinery which includes extracting useful petroleum products. The refineries, because of their complex nature of the chemical and physical processes involved in their working, produce solid, liquid and gaseous effluents, all of which require special treatment and handling. Various types of hydrocarbons, harmful gases and tars comprise the effluent streams. Chapter 37 deals with pollution control in petrochemical industry. The characteristics of various effluents including treatment of wastewater and solid wastes are discussed in detail.

Chapter 38 is devoted pollution control in chlor-alkali industry. Emissions and waste generation from chlorine, brine circuit, cell plants processing are discussed in detail. Chapter 39 concentrates on pollution control in paint industry and describes various types of wastes generated in paint industry along with methods of prevention and treatment of wastes. Chapter 40 focuses on pollution control in dyes and dye intermediates industry which generates air emission, liquid and solid wastes. Various methods to prevent pollution and cleaner production are discussed in detail.

Chapter 41 is devoted to pollution control in pesticides industry which has been identified as one of the highly polluting industries needing pollution control on top priority. The possible sources of pollutants are raw materials used in pesticides synthesis in excess of their stoichiometric requirements and solvents used as an extractive medium. Methods of recycling and reuse of wastewater within the industry has been discussed which helps in minimising fresh water requirements.

Section VII discusses pollution control in other important industries. Chapter 42 deals with pollution control in pulp and paper industry. The main polluting constituents in pulp and paper mill wastewater are suspended solids, colour, foam, inorganics such as sodium carbonate, bicarbonates, chlorides and sulphates, toxic chemicals such as mercaptans and inorganic sulphides, mercury if caustic chlorine plant forms a part of pulp and paper mill. Chapter 43 focuses on pollution control in sugar industry. Sugar factory effluent produces obnoxious odour and unpleasant colour when released into the environment without proper treatment. Chapter 44 concentrates on pollution control in distillery industry. Distillery wastewater, generally known as stillage, slops, vinnasse or dunder, constitutes a high value, high strength acidic waste that presents significant disposal or treatment problems.

Chapter 45 is devoted to pollution control in rubber industry. Pollution abatement involves– in-plant control of waste and end-process treatment of wastewater. Chapter 46 deals with pollution control in plastic industry. Various recycling methods such as– mechanical recycling, chemical or feedstock recycling, biodegradation aspects of plastics, etc., are discussed in detail.

Chapter 47 focuses on pollution control in drug and pharmaceutical industry. Gaseous effluents like hydrogen chloride, hydrogen cyanide, sulphur dioxide, chlorine and solvent vapours are emitted from various processing units manufacturing basic drugs. Chapter 48 concentrates on pollution control in textile industry which represents a range of industries with operations and processes as diverse as it products. Various physical, chemical and biological treatment methods of wastewater are discussed. Chapter 49 deals with pollution control in tannery industry and explains the processes used in tanneries, the characteristics and effects of wastes generated and methods of pollution control and waste disposal. Chapter 50 is devoted to nanotechnology for industrial pollution prevention. Pollution results from resource production and consumption, which in their current state are very wasteful. Nanotechnology plays a vital role in air and water pollution control. Nanofabrication holds much potential for effective pollution control. Air pollution can be remediated using nanotechnology in several ways. One is through the use of nano-catalysts with increased surface area for gaseous reactions. Harmful pollutants in water can be converted into harmless chemicals through chemical reactions. Nanotechnology eases the water cleansing process because inserting nanoparticles into underground water sources is cheaper and more efficient than pumping water for treatment.

Such wide coverage makes this book a treatise on the subject. Diagrams, figures, tables and index supplement the text. All topics have been covered in a cogent and lucid style to help the reader grasp the information quickly and easily.

This book could not have been completed without the help of Professor Pramod Kumar (Assistant Professor and Head of Department- Biotechnology Hindustan College of Science and Technology) and Mr. Sarvesh Devraj (Research Associate – TERI, New Delhi), who worked hard in locating and organising the material and spent many hours checking the manuscript. Appreciations are also extended to Mr Harinder singh, senior DTP operator, who did type setting and drew labelled the flow diagrams and worked long hours to bring the book on time. I am also thankful to the editorial team of Woodhead Publishing India Pvt. for their wholehearted cooperation in bringing out the book in time.

It may not be wrong to hold that this book on Industrial pollution and its control is essential reading for professionals and students pursuing engineering courses. Besides students, this book will prove useful to industrialists and consultants in the respective fields.

It has been prepared with meticulous care, aiming at making the book error-free. Constructive suggestions are always welcome from users of this book.

S. C. Bhatia

Section I

General concepts and engineering considerations

1

Industrial pollution: An overview

1.1 Introduction

Industrialisation is the process of social and economic change that transforms a human from a pre-industrial society into and industrial one. Industrial pollution deteriotes the environment in a range of ways and it has negative impact on human lives and health.

Pollutants can kill animals and plants, imbalance ecosystems, degrade air quality, damage buildings and adversely affect quality of life.

Any form of pollution that can trace its immediate source to industrial practices is known as industrial pollution. Most of the pollution on the planet can be traced back to industries of some kind. In fact, the issue of industrial pollution has taken on grave importance for agencies trying to fight against environmental degradation. Countries facing sudden and rapid growth of such industries are finding it to be a serious problem which has to be brought under control immediately.

Industrial pollution takes on many faces. It contaminates many sources of drinking water, releases unwanted toxins into the air and reduces the quality of soil all over the world. Major environmental disasters have been caused due to industrial mishaps, which have yet to be brought under control. Below are few of the causes of industrial pollution that have resulted in environment degradation.

1.2 Causes of industrial pollution

1. Lack of policies to control pollution: Lack of effective policies and poor enforcement drive allowed many industries to bypass laws made by pollution control board which resulted in mass scale pollution that affected lives of many people.
2. Unplanned industrial growth: In most industrial townships, unplanned growth took place wherein those companies flouted rules and norms and polluted the environment with both air and water pollution.
3. Use of outdated technologies: Most industries still rely on old technologies to produce products that generate large amount of waste. To avoid high cost and expenditure, many companies still make use of traditional technologies to produce high end products.

4. Presence of large number of small scale industries: Many small scale industries and factories that don't have enough capital and rely on government grants to run their day-to-day businesses often escape environment regulations and release large amount of toxic gases in the atmosphere.
5. Inefficient waste disposal: Water pollution and soil pollution are often caused directly due to inefficiency in disposal of waste. Long term exposure to polluted air and water causes chronic health problems, making the issue of industrial pollution into a severe one. It also lowers the air quality in surrounding areas which causes many respiratory disorders.
6. Leaching of resources from our natural world: Industries do require large amount of raw material to make them into finished products. This requires extraction of minerals from beneath the earth. The extracted minerals can cause soil pollution when spilled on the earth. Leaks from vessels can cause oil spills that may prove harmful for marine life.

1.3 Effects of industrial pollution

1. Water pollution: The effects of industrial pollution are far reaching and liable to affect the ecosystem for many years to come. Most industries require large amounts of water for their work. When involved in a series of processes, the water comes into contact with heavy metals, harmful chemicals, radioactive waste and even organic sludge. These are either dumped into open oceans or rivers. As a result, many of our water sources have high amount of industrial waste in them which seriously impacts the health of our ecosystem. The same water is then used by farmers for irrigation purpose which affects the quality of food that is produced. Water pollution has already rendered many ground water resources useless for humans and wildlife. It can at best be recycled for further usage in industries.
2. Soil pollution: Soil pollution is creating problems in agriculture and destroying local vegetation. It also causes chronic health issues to the people that come in contact with such soil on a daily basis.
3. Air pollution: Air pollution has led to a steep increase in various illnesses and it continues to affect us on a daily basis. With so many small, mid and large scale industries coming up, air pollution has taken toll on the health of the people and the environment.
4. Wildlife extinction: By and large, the issue of industrial pollution shows us that it causes natural rhythms and patterns to fail, meaning that the wildlife is getting affected in a severe manner. Habitats are being lost, species are becoming extinct and it is harder for the environment to recover from each natural disaster. Major industrial accidents like oil

spills, fires, leak of radioactive material and damage to property are harder to cleanup as they have a higher impact in a shorter span of time.

5. Global warming: With the rise in industrial pollution, global warming has been increasing at a steady pace. Smoke and greenhouse gases are being released by industries into the air which causes increase in global warming. Melting of glaciers, extinction of polar beers, floods, tsunamis, hurricanes are few of the effects of global warming. The issue of industrial pollution concerns every nation on the planet. As a result, many steps have been taken to seek permanent solutions to the problem. Better technology is being developed for disposal of waste and recycling as much polluted water in the industries as possible. Organic methods are being used to clean the water and soil, such as using microbes that naturally uses heavy metals and waste as feed. Policies are being pushed into place to prevent further misuse of land. However, industrial pollution is still rampant and will take many years to be brought under control.

1.4 Pollution and pollutants

Human activities directly or indirectly affect the environment adversely. A stone crusher adds a lot of suspended particulate matter and noise into the atmosphere. Automobiles emit from their tail pipes oxides of nitrogen, sulphur dioxide, carbon dioxide, carbon monoxide and a complex mixture of unburnt hydrocarbons and black soot which pollute the atmosphere. Domestic sewage and run off from agricultural fields, laden with pesticides and fertilisers, pollute water bodies. Effluents from tanneries contain many harmful chemicals and emit foul smell. These are only a few examples which show how human activities pollute the environment. Pollution may be defined as addition of undesirable material into the environment as a result of human activities. The agents which cause environmental pollution are called pollutants. A pollutants may be defined as a physical, chemical or biological substance unintentionally released into the environment which is directly or indirectly harmful to humans and other living organisms.

1.5 Types of industrial pollution

Pollution may be of the following types:

1. Air pollution.
2. Water pollution.
3. Thermal pollution.
4. Radiation pollution.
5. Soil pollution.
6. Noise pollution.

1.5.1 Air pollution

Air pollution is a result of industrial and certain domestic activity. An ever increasing use of fossil fuels in power plants, industries, transportation, mining, construction of buildings, stone quarries had led to air pollution. Air pollution may be defined as the presence of any solid, liquid or gaseous substance including noise and radioactive radiation in the atmosphere in such concentration that may be directly and indirectly injurious to humans or other living organisms, plants, property or interferes with the normal environmental processes. Air pollutants are of two types: (i) suspended particulate matter and (ii) gaseous pollutants like carbon dioxide (CO_2), NO_x, etc. Some of the major air pollutants, their sources and effects are given in Table 1.1.

Table 1.1: Particulate air pollutants, their sources and effects.

Pollutant	*Sources*	*Effects*
Suspended particulate matter/dust	Smoke from domestic, industrial and vehicular soot	Depends on specific composition Reduces sunlight and visibility, increases corrosion, pneumoconiosis, asthma, cancer and other lung diseases
Fly ash	Part of smoke released from chimneys of factories and power plants	Settles down on vegetation, houses. Adds to the suspended particulate matter (SPM) in the air. Leachates contain harmful material

Among these pollutants, carbon monoxide (CO), nitrogen oxide (NO), sulphur dioxide, (SO_2), ozone (O_3), particulate matter and lead oxide damage the environment. These are discussed briefly in chapter and for detail discussion ref. chapters 2 and 3 on pollution from particulate matter and gaseous pollutants.

Carbon monoxide

Carbon monoxide (CO) is a colourless, odourless and highly toxic gas, generated when carbon-containing material burns with insufficient oxygen. It is discharged from gasoline engines and burning of coal. It occurs naturally in the atmosphere in trace amounts. Another source of CO is the burning of firewood and wood-waste. CO is also produced in large quantities in thermal power plants as a result of burning of coal. Approximately 250 million tonnes of carbon monoxide is poisoning the global atmosphere every year.

Carbon dioxide

Human activities are responsible for the increasing amount of carbon dioxide in the atmosphere. It is not a pollutant itself and it comes from burning of fossil fuels and the oxidation of carbon stored in trees and soil humus. It is a natural component of the earth's atmosphere and plays a vital role in global carbon circulation through the environment. It is the main feedstock for the photosynthesis process in green vegetation.

Nitrogen oxides

All the oxides of nitrogen, for example, nitric oxide (NO) and nitrogen dioxide (NO_2) are represented by NO_x. NO_x emission comes about in equal parts from the transport sector and from energy and industrial sectors. The major sources of oxides of nitrogen are motor vehicle exhausts, soft coal burning and acid manufacturing. On a global level, about 60% of NO_x is produced from natural resources, including biomass burning, fixation by lighting, inflow from the stratosphere, chemical conversion from ammonia in the troposphere and loss of gaseous nitric oxide from the soils.

Sulphur dioxide

Various oxides of sulphur have harmful effects on the environment, among which sulphur dioxide (SO_2) is the most damaging. SO_2 is a colourless gas with a suffocating and strong pungent odour. Industrial processes and burning of sulphur-containing coals or heavy oil, particularly in thermal power plants, are responsible for release of SO_2 in the air. In the atmosphere, sulphur dioxide combines with oxygen and water to form sulphuric acid (H_2SO_4), the major component of acidic rain. On a global scale, about half of the ambient SO_2 originates from the oxidation of hydrogen sulphide (H_2S) given off by decaying organic matter.

Particulate matter

Particulate matter includes those air pollutants which may be in the form of solid particles or liquid droplets, including fumes, smoke, fog, dust, pollen, bacteria, viruses and aerosols. This category includes about 5% of the weight of all pollutants present in the atmosphere. Natural dust forms about half of the total mass of particulate matter in the air. Dust has a relatively small impact because it tends to be coarse. Fine particulate matter (< 2.5 μg dia) is more harmful to human health than the coarse particulate matter (> 2.5 μg dia). Particulates remain suspended in the air and are transported over a long distance by strong wind. Small particulates with a size of less than 2 μg have an important impact on health although they contribute only about 1% of the particulate load in the atmosphere.

Particulate pollutants

Particulate matter suspended in air are dust and soot released from the industrial chimneys. Their size ranges from 0.001 to 500 μm in diameter. Particles less than 10 μm float and move freely with the air current. Particles which are more than 10 μm in diameter settle down. Particles less than 0.02 μm form persisent aerosols. Major source of SPM (suspended particulate matter) are vehicles, power plants, construction activities, oil refinery, railway yard, market place, industries, etc.

Fly ash: Fly ash is ejected mostly by thermal power plants as by-products of coal burning operations. Fly ash pollutes air and water and may cause heavy metal pollution in water bodies. Fly ash affects vegetation as a result of its direct deposition on leaf surfaces or indirectly through its deposition on soil. Fly ash is now being used for making bricks and as a land fill material.

Lead and other metals particles: Tetraethyl lead (TEL) is used as an anti-knock agent in petrol for smooth and easy running of vehicles. The lead particles coming out from the exhaust pipes of vehicles is mixed with air. If inhaled it produces injurious effects on kidney and liver and interferes with development of red blood cells. Lead mixed with water and food can create cumulative poisoning. It has long term effects on children as it lowers intelligence. Oxides of iron, aluminium, manganese, magnesium, zinc and other metals have adverse effect due to deposition of dust on plants during mining operations and metallurgical processes. They create physiological, biochemical and developmental disorders in plants and also contribute towards reproductive failure in plants.

Gaseous pollutants

Power plants, industries, different types of vehicles–both private and commercial use petrol, diesel as fuel and release gaseous pollutants such as carbon dioxide, oxides of nitrogen and sulphur dioxide along with particulate matter in the form of smoke. All of these have harmful effects on plants and humans.

Table 1.2 lists some of these pollutants, their sources and harmful effects.

Table 1.2: Gaseous air pollutants: their sources and effects.

Pollutant	*Source*	*Harmful effect*
Carbon compound (CO and CO_2)	Automobile exhaust burning of wood and coal	• Respiratory problems • Green house effect
Sulphur compounds (SO_2 and H_2S)	Power plants and refineries volcanic eruptions	• Respiratory problems in humans • Loss of chlorophyll in plants (chlorosis) • Acid rain
Nitrogen compound (NO and N_2O)	Motor vehicle exhaust atmospheric reaction	• Irritation in eyes and lungs • Low productivity in plants • Acid rain damages material (metals and stone)

(Cont'd...)

Pollutant	*Source*	*Harmful effect*
Hydrocarbons (benzene, ethylene)	Automobiles and petroleum industries	• Respiratory problem • Cancer causing properties
SPM (suspended particulate matter) (any solid and liquid) particles suspended in the air, (flush, dust, lead)	Thermal power plants, Construction activities, metalurgical processes and automobiles	• Poor visibility, breathing problems • Lead interfers with the development of red blood diseases and cancer. • Smog formation leads to poor visibility and aggravates asthma in patients
Fibres (cotton, wool)	Textiles and carpet weaving industries	• Lung disorders

Prevention and control of air pollution

Indoor air pollution: Poor ventilation due to faulty design of buildings leads to pollution of the confined space. Paints, carpets, furniture, etc., in rooms may give out volatile organic compounds (VOCs). Use of disinfectants, fumigants, etc., may release hazardous gases. In hospitals, pathogens present in waste remain in the air in the form of spores. This can result in hospital acquired infections and is an occupational health hazard. In congested areas, slums and rural areas burning of firewood and biomass results in lot of smoke. Children and ladies exposed to smoke may suffer from acute respiratory problems which include running nose, cough, sore throat, lung infection, asthama, difficulty in breathing, noisy respiration and wheezing.

Prevention and control of indoor air pollution: Use of wood and dung cakes should be replaced by cleaner fuels such as biogas, kerosene or electricity. But supply of electricity is limited. Similarly kerosene is also limited. Improved stoves for looking like smokeless chullahs have high thermal efficiency and reduced emission of pollutants including smoke. The house designs should incorporate a well ventilated kitchen. Use of biogas and CNG (Compressed Natural Gas) need to be encouraged. Those species of trees such as baval (*Acacia nilotica*) which are least smoky should be planted and used. Charcoal is a comparatively cleaner fuel. Indoor pollution due to decay of exposed kitchen waste can be reduced by covering the waste properly. Segregation of waste, pre-treatment at source, sterilisation of rooms will help in checking indoor air pollution.

Prevention and control of industrial pollution: Industrial pollution can be greatly reduced by:

1. Use of cleaner fuels such as liquefied natural gas (LNG) in power plants, fertiliser plants, etc., which is cheaper in addition to being environmentally friendly.

2. Employing environment friendly industrial processes so that emission of pollutants and hazardous waste is minimised.
3. Installing devices which reduce release of pollutants. Devices like filters, electrostatic precipitators, inertial collectors, scrubbers, gravel bed filters or dry scrubbers are described below:
   (a) Filters: Filters remove particulate matter from the gas stream. The medium of a filter may be made of fibrous materials like cloth, granular material like sand, a rigid material like screen, or any mat like felt pad. Baghouse filtration system is the most common one and is made of cotton or synthetic fibres (for low temperatures) or glass cloth fabrics (for higher temperature up to 290°C).
   (b) Electrostatic precipitators (ESP): The emanating dust is charged with ions and the ionised particulate matter is collected on an oppositely charged surface. The particles are removed from the collection surface by occasional shaking or by rapping the surface. ESPs are used in boilers, furnaces and many other units of thermal power plants, cement factories, steel plants, etc.
   (c) Inertial collectors: It works on the principle that inertia of SPM in a gas is higher than its solvent and as inertia is a function of the mass of the particulate matter this device collects heavier particles more efficiently. 'Cyclone' is a common inertial collector used in gas cleaning plants.
   (d) Scrubbers: Scrubbers are wet collectors. They remove aerosols from a stream of gas either by collecting wet particles on a surface followed by their removal, or else the particles are wetted by a scrubbing liquid. The particles get trapped as they travel from supporting gaseous medium across the interface to the liquid scrubbing medium.

Gaseous pollutants can be removed by absorption in a liquid using a wet scrubber and depends on the type of the gas to be removed, e.g., for removal of sulphur dioxide alkaline solution is needed as it dissolves sulphur dioxide. Gaseous pollutants may be absorbed on an activated solid surface like silica gel, alumina, carbon, etc. Silica gel can remove water vapour. Condensation allows the recovery of many by-products in coal and petroleum processing industries from their liquid effluents.

Apart from the use of above mentioned devices, other control measures are: (i) increasing the height of chimneys, (ii) closing industries which pollute the environment, (iii) shifting of polluting industries away from cities and heavily populated areas and (iv) development and maintenance of green belt of adequate width.

1.5.2 Photochemical smog

Photochemical smog is initiated by the photochemical dissociation of NO_2 and the consequent secondary reactions involving unsaturated hydrocarbons, other organic compounds and free radicals, leading to the formation of organic peroxides and ozone. This phenomenon takes place during sunny days with low winds and low level inversion. Photochemical smog and the consequent formation of aerosols reduce visibility, cause irritation to eyes and damage plants and rubber goods. The oxidation of SO_2 can also take place by interaction with the free radical HO· present in photochemical smog:

$$SO_2 + HO\cdot \rightarrow HOSO_2\cdot$$

$$HOSO_2\cdot + O_2 \rightarrow HOSO_2\,O_2\cdot$$

$$HOSO_2\,O_2\cdot \rightarrow \underset{\text{(sulphate)}}{HOSO_2\,O\cdot + NO_2\cdot}$$

Chemical oxidation of SO_2 may also take place in water droplets, present in aerosols. This reaction is accelerated in the presence of NH_3 and catalysts, e.g., oxides of Mn, Fe, Cu, Ni.

Solid particles, such as soot, bring about catalytic oxidation of SO_2 by providing a heterogeneous phase for contact. Soot is formed during combustion of solid and liquid fuels in domestic and industrial operations and automobile emissions. Sulphur dioxide is a pollutant responsible for smog formation, acid rains and corrosion of metals and alloys.

Oxidation of organic compounds

Organic compounds such as hydrocarbons, aldehydes and ketones absorb solar radiation and undergo various photochemical and chemical reactions involving free radicals. Some of these reactions are catalysed by particulate matter such as soot and metal oxides. Some of the resultant intermediates and final products contribute to photochemical smog formation.

1.5.3 Greenhouse effect

The earth is heated by sunlight and some of the heat that is absorbed by the earth is radiated back into space. However, some of the gases in the lower atmosphere, acting like glass in a greenhouse, allow solar radiations (in the range 300 to 2500 nm, i.e., near UV, visible and near infrared region, while filtering the dangerous UV radiations, i.e., < 300 nm) but do not allow the earth to re-radiate the heat into space. In other words, these gases in the atmosphere are transparent to the sunlight coming in, but they strongly absorb infrared radiation, which the earth sends back as heat. A part of the heat so trapped in these atmospheric gases is re-emitted to the earth's surface. The net result is the heating of the earth's surface by this phenomenon, called the 'greenhouse effect'. The gases that are responsible for this greenhouse effect are CO_2, water vapour, CH_4 and man-made chlorofluorocarbons (CFCs). Water vapour strongly absorbs infrared radiations in the range 4,000 to 8,000 nm and CO_2 in the range 12,000 to 16,300 nm. The radiations in the range 8,000 to 12,000 nm escape unabsorbed and this is known as the region of atmospheric window.

Carbon dioxide is released by volcanoes, oceans, decaying plants as well as human activities, such as deforestation and combustion of fossil fuels. Automobile exhausts account for 30% of CO_2 emissions in developed countries. Methane is released from coal mines, decomposition of organic matter in swamps, rice paddy cultivation, guts of termites in forest debris and stomachs of ruminants.

Chlorofluorocarbons (CFCs) are used as coolants in refrigerators, propellants in aerosol sprays, plastic foam materials like 'thermocoles' or 'styrofoam' and in automobile air-conditioners.

In fact, the 'greenhouse gases' (particularly CO_2 and water vapour) are responsible for keeping our planet warm and thus sustaining life on the earth. If the greenhouse gases were very less or totally absent then the average temperature on the earth would have been at subzero levels. But, however, if the concentration of greenhouse gases increases, they may trap too much of heat, which may threaten the very existence of life on earth. For instance, the CO_2 present in the atmosphere of the planet *Venus*, is about 60,000 times more than that on earth. Hence, the average temperature of *Venus* is about 425°C, making the existence of life impossible there.

Oceans and biomass are the major sinks for atmospheric CO_2. Oceans convert CO_2 into soluble bicarbonates. The photosynthetic activity in the green plants increases with the increase in CO_2 level in the atmosphere. Forests are the places where lot of photosynthetic activity occurs. They also act as vast reservoirs of fixed but readily oxidisable carbon in the form of vegetation, wood and humus. Hence, forests maintain a balance in the atmospheric CO_2 level and deforestation upsets this balance and increases the atmospheric CO_2 level. It is estimated that the atmospheric CO_2 content has increased by 25% during the last two centuries. This is mostly attributed to the industrial revolution and is one of the reasons for the slight increase in the global temperature (about 0.5°C). Since the concentrations of greenhouse gases have been continuously increasing because of deforestation, industrialisation, increased burning of fossil fuels, mining, exhausts from increasing number of automobiles and other anthropogenic activities, there is an increasing concern about the possible 'global warming'. Some scientists fear that if proper precautions are not taken, the concentration of greenhouse gases in the atmosphere may double within the next 50–100 years. If this happens, the average global temperature may increase by 4–5°C. This will increase the evaporation of surface waters, which may influence climatic changes depending upon the pattern of cloud formation. For instance, low-level dense clouds may exert cooling effect whereas high level thin cloud formation may exert heating effect due to increased greenhouse effect.

The projections from computer modelling regarding the climatic changes that could be triggered off due to 'global warming' reveal alarming scenarios. Even a 1.5°C rise in surface temperature can adversely affect food production in the world. Thus, the wheat growing zones in the northern latitude may be shifted from the USSR and Canada to the polar regions, i.e., from fertile soils to poor soils near the North Pole. The biological productivity of the ocean would also decrease due to warming of the earth's surface layer, which in turn, may reduce the transport of nutrients from deeper layers to the surface by vertical circulation. Computer modelling also indicates the following effects due to 'global warming': melting of the polar ice caps; dry areas becoming drier; humid areas like the Amazon suffering more intense tropical storms; drastic drop in food production, particularly in lands within 35 degrees north and south of the Equator; increased breeding of pests and diseases due to more humid conditions; shorter, wetter and warmer winters and longer, hotter and drier summers, particularly in mid-continental areas. Global warming may also trigger increased thermal expansion of oceans and melting of glaciers, which may result in an increase in the sea-level by 20 cm to 1.5 metres by the latter part of the 21st century. Thus, cities like Mumbai, Miami, London, Venice, Bangkok and Leningrad may become extremely vulnerable. Defences against the rising sea-levels and expanding oceans are very difficult and expensive, which many nations cannot afford. Further, a global temperature rise, is likely to cause more floods, hurricanes and tornadoes.

There are differences of opinion among experts regarding the dynamics and effects of 'global warming' due to the complexity of natural phenomena that might be operating simultaneously. More accurate future climatic projections will be possible with better super-computer models, based on greater understanding of the complex natural climatic forces involved. But until that time, the possible devastating effect due to 'global warming' by the 'greenhouse effect' cannot be underestimated. Some of the steps suggested to minimise the 'greenhouse effect' include reduction in the use of fossil fuels, encouraging the use of alternative sources of energy (e.g., solar, geothermal, wind, biogas, etc.), conservation of forests, extensive afforestation, encouraging community forestry, reduction in the use of automobiles, research in the development of more efficient automobile engines, ban on CFCs and nuclear explosions, development of environmentally compatible technologies with the help of intensive interdisciplinary research, effective check on the growth of population and imparting of non-formal and formal environmental education.

1.5.4 Formation and depletion of ozone in the stratosphere

Ozone is an important chemical species present in the stratosphere. At an altitude of about 30 km, its concentration is about 10 ppm. The ozone layer present in the stratosphere acts as a protective shield for life on earth. It strongly absorbs ultraviolet radiations from the sun in the region 220–330 nm and thereby protects life on earth from severe radiation damage, such as DNA mutation and skin cancer. Thus only a small fraction of UV radiation reaches the lower atmosphere and the earth's surface.

Ozone is formed in the stratosphere by photochemical reaction:

$$O_2 + h\nu \text{ (242 nm)} \rightarrow O + O$$

$$O + O_2 + M \text{ (third body, such as } N_2 \text{ or } O_2) \rightarrow O_3 + M$$

The third body absorbs the excess energy liberated by the above reaction and thereby the ozone molecule is stabilised. Thus, ozone is constantly formed in the stratosphere. However, it is also destroyed by chlorine, released due to volcanic activity and also by reaction with: (i) nitric oxide, (ii) atomic oxygen and (iii) reactive hydroxyl radical, which are also present in the atmosphere.

In the atmosphere, nitrogen oxide (NO) comes from chemical and photochemical reactions, supersonic jets, nuclear explosions, etc., Cl_2 comes from CFC's and volcanoes and OH comes from biomass burning and from natural water systems by the following reactions:

$$O_3 + NO \rightarrow NO_2 + O_2$$

$$O_3 + O \rightarrow O_2 + O_2$$

$$O_3 + HO\cdot \rightarrow HO_2 + HOO\cdot$$

$$HOO\cdot + O \rightarrow HO\cdot + O_2$$

Ozone, in the stratosphere, is also destroyed by man-made chlorofluorocarbons (CFCs), which are used as coolants in refrigerators, air-conditioners, propellants in aerosol sprays and in plastic foams, such as 'thermocole' or 'styrofoam'. The CFC molecules, escaping into the atmosphere, decompose to release chlorine in the ozone layer (by photo-dissociation) and each atom of chlorine, thus liberated is capable of attacking several ozone molecules.

$$Cl + O_3 \rightarrow ClO + O_2$$

This reaction is followed by:

$$ClO + O \rightarrow Cl + O_2$$

which regenerates Cl atoms, so that a long chain process is involved, which conserves Cl atoms. The environmental hazards of CFCs were recognised as early as 1970. In fact, temporary thinning in the stratospheric ozone layer, leading to the formation of 'Ozone hole' was actually detected over the Antarctica during September to November 1985. Reported increase in cases of skin cancer in South Australia are also attributed to UV radiations reaching the earth, due to depletion of the ozone layer.

The detection of the 'Ozone hole' over Antarctica in 1985 attracted the attention of global scientific community. The U.S., immediately banned the use of CFCs in spray cans. Further, in the year 1987, 24 nations signed the montreal protocol, which aimed at 35% reduction in the global production of the CFCs by the year 1999. Simultaneously, efforts to produce chlorine-free substitutes have also started. In fact, synthesis of a product called HFC-134a has already been reported as an effective substitute for CFC. The use of hydrofluorocarbons (HFCs), hydrochlorofluorocarbons (HCFCs) and methyl cyclohexane (MCH) as substitutes for CFCs is envisaged for several applications.

Almost all the sulphur present in liquid and gaseous fuels and about 80% of sulphur present in solid fuels appears as SO_2 in the flue gases. Depending on the sulphur content of the fuel burnt and the conditions of combustion (e.g., percentage of excess air used), the concentration of SO_x in flue gases varies from 0.05 to 0.4%. However, in metallurgical operations such as smelting of sulphide ores, the SO_2 concentration in stack gases may be 5 to 10%.

SO_2 is oxidised to SO_3 in atmospheric air by photolytic and catalytic processes involving ozone, NO_x and hydrocarbons, giving rise to the formation of photochemical smog. Oxidation of SO_2 can take place in presence of catalysts such as NO_x, metal oxides, soot and dust. Under normal humid conditions, SO_3 reacts with water vapour to produce droplets of H_2SO_4 aerosol which gives rise to the so-called 'acid rain'. The sulphuric acid and sulphate aerosols present in urban air are smaller than 2 μ and hence can easily reach the pulmonary region of lungs, causing serious respiratory problems, particularly in older people.

$$SO_2 + O_3 \rightarrow SO_3 + O_2$$

$$SO_2 + \tfrac{1}{2}O_2 + H_2O \xrightarrow[\text{metal oxide, soot, etc.}]{\text{Catalyst such as}} H_2SO_4 \xrightarrow[\text{Aerosol}]{\text{Aerosol}} (H_2SO_4)_n$$

Control of SO_x emissions from the anthropogenic activities is contemplated along the following lines:

1. Removing SO_x from flue gases before letting them out into the atmosphere: Chemical scrubbers such as: (i) lime stone and (ii) citric acid are suggested to absorb SO_2 from the flue gases.

(a) $2CaCO_3 + 2SO_2 + O_2 \rightarrow 2CaSO_4 + CO_2$

(b) $SO_2 + H_2O \rightarrow HSO_3^- + H^+$

$HSO_3^- + H_2\,cit^- \rightarrow (HSO_3\,.\,H_2\,cit)^{-2}$

2. Removing sulphur from the fuels used for combustion: Pyritic sulphur in coal can be removed by grinding and washing in coal washeries. However, organically bound sulphur cannot be easily removed from coals. Research is in progress to synthesise special type of micro-organisms using biotechnology, which are capable of converting organically bound sulphur into soluble form.
3. Utilising low-sulphur fuels.
4. Generation of power by alternative energy sources and discouraging fossil-fuel based thermal power-plants.

1.5.5 Acid rain

Rain has always been valued by mankind, because good crops and abundant water supplies are possible only due to timely and plentiful rainfall. Summer rains refresh people. Spring rains recharge the aquifers and cleanse the groundwater. Autumn rains and winter snow help clean the air. Rain, in general, brings with it a sense of hope, vitality and a promise for the future.

Over the last few decades, simple rainfall has taken on a threatening complexity in some parts of the world. In these locales, rain must pass through an atmosphere polluted with oxides of sulphur (SO_x) and of nitrogen (NO_x). The falling rain and snow often react with these oxide pollutants to produce often a mixture of sulphuric acid, nitric acid and water. This is known as acid precipitation or acid rain.

Rain tends to be naturally acidic with a pH of 5.6 to 5.7 due to the reaction of atmospheric CO_2 with water to produce carbonic acid. This small amount of acidity is sufficient to dissolve minerals the earth's crust and make them available to plant and animal life, but it is not acidic enough to inflict any major damage. Other atmospheric substances from volcanic eruptions, forest fires and other similar natural phenomena also contribute to the acidity in rain. Thus, even with the enormous amounts of acids created by nature annually, normal rainfall is able to assimilate them to the point where they cause little, if any, known damage.

But, it is the contributions of SO_x, NO_x, etc., from anthropogenic activities that disturb this acid balance and convert natural and mildly acidic rain into precipitation with far-reaching environmental consequences.

Acid rain represents one of the major consequences of air pollution, because of large SO_x and NO_x emissions from big industrial areas. The longer the SO_x and NO_x remain in the atmosphere, the greater are the chances of their oxidation to H_2SO_4 and HNO_3 due to photochemical and catalytic chemical reactions. Acid rains may cause extensive damage to materials and terrestrial ecosystems, such as water, fish, vegetation, stone, steel, paint, soil and mankind.

The only practical approach to counter the problem of acid rain is to reduce SO_x and NO_x emissions. The following three general options are considered for this purpose:

1. Energy conservation resulting in reduced fuel consumption and hence slower emissions of SO_x and NO_x. Conservation via more efficient fuel use and through improved thermal insulation is also being studied.
2. Desulphurisation and denitrification of fuels of stack gases and increased use of fuels naturally low in sulphur content or use of technologies that reduce SO_x and NO_x emissions. Desulphurisation and use of low NO_x-producing technologies are the only viable control options today and will perhaps continue to be so for some more time.
3. Substitutions for fossil fuels by other alternative energy forms may offer future solutions to this problem.

Reduction of SO_x emissions can be accomplished by: (i) removing the sulphur content before the fuel is burnt with the help of techniques such as coal cleaning, coal gasification and desulphurisation of liquid fuels, (ii) removing the sulphur content during combustion, as in fluidised-bed combustion and (iii) removal of sulphur emissions after combustion, as in stack or flue gas desulphurisation systems or scrubbers. The future of SO_x control from traditional fuel sources lies in the perfection of these techniques.

Reduction of NO_x emissions from stationary combustion sources can be achieved by modification of furnace and burner design, and/or modification of operating conditions. The combustion modification techniques available now include using two-stage combustion, precisely controlling air, injecting water during combustion, recirculating flue gases, and/or by altering design of firing chambers. Reductions in NO_x emissions from mobile combustion sources may be achieved by lowering the combustion temperatures in the engine and catalytic removal of NO_x from exhaust gases using devices such as a three-way system that simultaneously reduces carbon monoxide, hydrocarbons and NO_x.

1.5.6 Water pollution

Addition or presence of undesirable substances in water is called water pollution. Water pollution is one of the most serious environmental problems. Water pollution is caused by a variety of human activities such as industrial, agricultural and domestic. Agricultural run off laden with excess fertilisers and pesticides, industrial effluents with toxic substances and sewage water with human and animal wastes pollute our water thoroughly. Natural sources of pollution of water are soil erosion, leaching of minerals from rocks and decaying of organic matter. Rivers, lakes, seas, oceans, estuaries and ground water sources may be polluted by point or non-point sources. When pollutants are discharged from a specific location such as a drain pipe carrying industrial effluents discharged directly into a water body it represents point source pollution. In contrast non-point sources include discharge of pollutants from diffused sources or from a larger area such as run off from agricultural fields, grazing lands, construction sites, abandoned mines and pits, roads and streets.

Sources of water pollution

Water pollution is the major source of water born diseases and other health problems. Sediments brought by runoff water from agricultural fields and discharge of untreated or partially treated sewage and industrial effluents, disposal of fly ash or solid waste into or close to a water body cause severe problems of water pollution. Increased turbidity of water because of sediments reduces penetration of light in water that reduces photosynthesis by aquatic plants.

Pollution due to pesticides and inorganic chemicals:

1. Pesticides like DDT and others used in agriculture may contaminate water bodies. Aquatic organisms take up pesticides from water get into the food chain (aquatic in this case) and move up the food chain. At higher trophic level they get concentrated and may reach the upper end of the food chain.
2. Metals like lead, zinc, arsenic, copper, mercury and cadmium in industrial wastewaters adversely affect humans and other animals. Arsenic pollution of ground water has been reported from the various countries of the world. Consumption of such arsenic polluted water leads to accumulation of arsenic in the body parts like blood, nails and hairs causing skin lesions, rough skin, dry and thickening of skin and ultimately skin cancer.
3. Pollution of water bodies by mercury causes Minamata disease in humans and dropsy in fishes. Lead causes displexia, cadmium poisoning causes Itai – Itai disease, etc.
4. Oil pollution of sea occurs from leakage from ships, oil tankers, rigs and pipelines. Accidents of oil tankers spill large quantity of oil in seas which kills marine birds and adversely affects other marine life and beaches.

Ground water pollution

Lot of people around the world depend on ground water for drinking, domestic, industrial and agricultural uses. Generally groundwater is a clean source of water. However, human activities such as improper sewage disposal, dumping of farm yard manures and agricultural chemicals, industrial effluents are causing pollution of ground water.

Methods for control of water pollution and water recycling

Control water pollution: Wastewater from domestic or industrial sources or from garbage dumps is generally known as sewage. It may also contain rain water and surface runoff. The sewage water can be treated to make it safe for disposal into water bodies like rivers, lakes, etc. The treatment involves three stages: primary, secondary and tertiary.

This includes: (i) sedimentation, (ii) coagulation/flocculation, (iii) filtration, (iv) disinfection, (v) softening and (vi) aeration.

The first four steps are of primary treatment. The first three steps are involved in primary treatment remove suspended particulate matter. Secondary treatment removes organic solids, left out after primary treatment, through their microbial decomposition. Effluents after secondary treatment may be clean but contain large amounts of nitrogen, in form of ammonia, nitrates and phosphorous which can cause problem of eutrophication upon their discharge into a receiving water body such as river, lake or pond. The tertiary treatment is meant to remove nutrients, disinfect for removing pathogenic bacteria and aeration removes hydrogen sulphide and reduce the amount of carbon dioxide and make water healthy and fit for aquatic organisms. This treatment of wastewater or sewage is carried out in effluent treatment plants especially built for this purpose. The residue obtained from primary treatment one known as sludge.

Water recycling: With increasing population the requirement for water is increasing rapidly. However, the availability of water is limited but an ever increasing water withdrawal from different sources such as rivers, lakes and ground water is depleting these sources and deteriorating their water quality. Therefore, it is essential to utilise the available water with maximum economy. This involves recycling of wastewater for certain uses with or without treatment. Recycling refers to the use of wastewater by the original user prior to the discharge either to a treatment system or to a receiving water body. Thus the wastewater is recovered and repetitively recycled with or without treatment by the same user.

1.5.7 Thermal pollution

Power plants- thermal and nuclear, chemical and other industries use lot of water (about 30% of all abstracted water) for cooling purposes and the used hot water is discharged into rivers, streams or oceans. The waste heat from the boilers and heating processes increases the temperature of the cooling water. Discharge of hot water may increase the temperature of the receiving water by 10 to 15°C above the ambient water temperature. This is called thermal pollution. Increase in water temperature decreases dissolved oxygen in water which adversely affects aquatic life. Unlike terrestrial ecosystems, the temperature of water bodies remain steady and does not change very much. Accordingly, aquatic organisms are adopted to a uniform steady temperature of environment and any fluctuation in water temperature severely affects aquatic plants and animals. Hence discharge of hot water from power plants adversely affects aquatic organisms. Aquatic plants and animals in the warm tropical water live dangerously close to their upper limit of temperature, particularly during the warm summer months. It requires only a slight deviation from this limit to cause a thermal stress to these organisms. Discharge of hot water in water body affects feeding in fishes, increases their metabolism and affects their growth. Their swimming efficiency declines. Running away from predators or chasing prey becomes difficult. Their resistance to diseases and parasites decreases.

Due to thermal pollution biological diversity is reduced. One of the best methods of reducing thermal pollution is to store the hot water in cooling ponds, allow the water to cool before releasing into any receiving water body. Thermal pollution is discussed in detail in chapter 18.

1.5.8 Nuclear and radioactive pollution

One of the major sources of environmental concern nowadays is the production of nuclear power which releases radioactive substances in the environment. Radioactivity is toxic in the sense that it causes harm to living organisms. Accidents at three-mile island in the United States and Chernobyl in the Soviet Union, both of which released radioactive rays, have increased public concern worldwide about the safety of nuclear power. The management and disposal of the accumulating radioactive wastes remain a key long-term problem facing the nuclear power sector. Wastes related to nuclear power generation occur in a variety of physical and chemical forms and in different class levels based on the relative radioactivity of the radio nuclides that they contain.

Radioactive wastes can be divided into two categories: high-level wastes and low-level wastes. Both these wastes are temporarily stored at product sites awaiting disposal. Nuclear pollution and radioactive wastes is discussed in detail in chapter 8.

High-level wastes

These wastes are predominantly used as fuel. Uranium oxide is fabricated into small pellets and sealed in metal tubes, which are then assembled in bundles. After about 18 months in the reactors, the fuel bundles need to be replaced. They are intensely radioactive and some elements in them emit radiation for tens of thousands of years (e.g., plutonium-239 has a half-life of 24,000 years). Extreme caution must be exercised to ensure that such high-level wastes do not contaminate the air, water and food materials. High-level wastes are stored mainly in deep, water-filled pools at the reactor sites to cool them and shield their radiation and a small fraction is stored dry in concrete containers. Although on-site storage is an adequate means of dealing with these wastes over the short term, eventually a long-term solution is required. For the disposal of nuclear fuel wastes in the long-term, deep geological formations are required. The buried radioactive material would have to be encased in non-corrosive (titanium or copper) containers to ensure safe and secure disposal so that they cannot enter the groundwater.

Low-level wastes

These are related to nuclear energy production and the wastes arising from uranium refineries and fuel fabrication plants. In low-level mines, radioactive wastes are generated from the extraction of uranium from ores. To check low-level radioactive wastes, monitoring of temporary sites is necessary to ensure safety of the surrounding public.

1.5.9 Soil pollution

The contamination of soil (or) land with acid rain, excess of fertilisers, wrong fertilisers, insecticides and herbicides is termed as 'soil pollution'.

Sources of soil pollution

The following are some of the important sources which pollute the soil:

1. Repeated and excess use of fertilisers and pesticides at random cause soil pollution. For example, due to excess use of $(NH_4)_2SO_4$ chemical fertiliser for several times, the SO_4^{-2} ion present in it gets accumulated into the soil. The soil then becomes infertile to plant growth due to acidity. Similarly, due to repeated and excess use of fertilisers containing KNO_3 (or) $NaNO_3$, plant growth will be retarded due to the accumulation of Na^+, K^+ ions in the soil which makes it alkaline.
2. Soil pollution arises due to the application of defective methods in the cultivation processes.
3. Soil gets polluted due to the release of the cysts from antameba, ascaris, pigworm, etc., which enters into the human body through food chain.
4. Acid rain is another source of soil pollution. Sulphur and Nitrogen oxides in the air undergo photo, chemical and catalytic oxidation followed by the interaction with rainy water or moisture to form H_2SO_4 and HNO_3.

$$2SO_2 + O_2 \rightarrow 2SO_3$$

$$SO_3 + H_2O \rightarrow H_2SO_4$$

$$4NO_2 + 2H_2O + O_2 \rightarrow 4HNO_3$$

$$2NO_2 + H_2O + O_3 \rightarrow 2HNO_3 + O_2$$

When rain water containing H_2SO_4 and HNO_3 is absorbed by the soil, its fertility is damaged, making it unfit for plant growth. Due to the acid rain, the pH of the soil becomes 3.0 to 4.0. This soil causes health hazards to the living organisms on the earth. It reduces agricultural yield and leads to soil cracks. The toxic metallic compounds present in the soil will flow along with the acid rain and enter into the rivers, lakes, tanks and sea, endangering aquatic life. They also enter the human body through food chain causing health hazards.

5. Serious cadmium pollution of soil occurred in Japan. Over a period of 20 years around 100 people (mostly older woman) living near the Jintsu river died of a syndrome called Itai-Itai (ouch-ouch) disease due to bone damage.
6. Spraying vegetables and fruit plants with insecticides and herbicides leads to soil pollution. When the standing vegetable and fruit plants are sprayed with insecticides and herbicides to protect them, they accumulate in their growing tissues. Such pesticides enter the human body by contaminated grains, fruit and vegetables as food material, causing health hazards that may damage even our heart. Hence, it is advised that before eating food material, they should be properly washed with sufficient quantity of good water. Chemicals like $CaCl_2$ used for ripening of fruit like banana and mangoes can easily be eliminated by washing with water.
7. Heat from molecular reactors and industries also cause soil pollution.
8. Sewage is the dirty water which contains human and animal excretions. The discharge of large quantities of sewage into rivers, lakes and water tanks and discharge of industrial wastes, sewage from household activities, dumping of synthetic, plastic waste material on the earth cause severe soil pollution.
9. Deforestation also leads to soil pollution. Due to deforestation, the concentration of CO_2 in the atmosphere increases, leading to an increase in the atmospheric (or) earth's temperature with the result that water source in the soil is lowered.

Methods of control of soil pollution

1. The use of fertilisers and pesticides should be minimised as far as possible.
2. Waste material should be covered in earth layers with soil.
3. Waste gases from various automobiles and industries should be purified before they reach the earth surface.
4. Before dumping sewage into rivers, lakes, tanks and sea, it should be treated chemically to remove harmful organic material.

5. The polluted water should be purified by installing purification machinery.
6. Deforestation (indiscriminate cutting of plants and trees) should be minimised and plantation carried out as a universal standard project work under a clean and green programme.
7. The usage of machines using biogas as fuel should be encouraged, including for electricity generation.
8. Various operations must be carried out to neutralise the acidity of acid rain.
9. The digging of larger pits or grooves in view of granite mines, aqua culture, addition of various chemicals to grow water bodies, etc., must be avoided.

1.5.10 Noise pollution

Noise is one of the most pervasive pollutant. A musical clock may be nice to listen during the day, but may be an irritant during sleep at night. Noise by definition is 'sound without value' or 'any noise that is unwanted by the recipient'. Noise in industries such as stone cutting and crushing, steel forgings, loudspeakers, shouting by hawkers selling their wares, movement of heavy transport vehicles, railways and airports leads to irritation and an increased blood pressure, loss of temper, decrease in work efficiency, loss of hearing which may be first temporary but can become permanent in the noise stress continues. It is therefore of utmost importance that excessive noise is controlled. Noise level is measured in terms of decibels (dB). World Health Organisation (WHO) has prescribed optimum noise level as 45 dB by day and 35 dB by night. Anything above 80 dB is hazardous. The Table 1.3 gives the noise intensity in some of the common activities.

Table 1.3: Sources of some noises and their intensity.

Source	*Intensity*	*Source*	*Intensity*
Quiet conversation	20–30 dB	Radio music	50–60 dB
Loud conversation	60 dB	Traffic noise	60–90 dB
Lawn mower	60–80 dB	Heavy truck	90–100 dB
Aircraft noise	90–120 dB	Space vehicle	140–179 dB
Beat music	120 dB	Launch	
Motor cycle	105 dB	Jet engine	140 dB

Sources of noise pollution

Noise pollution is a growing problem. All human activities contribute to noise pollution to varying extent. Sources of noise pollution are many and may be located indoors or outdoors.

Indoor sources include noise produced by radio, television, generators, electric fans, air coolers, air conditioners, different home appliances and family conflict. Noise pollution is more in cities due to a higher concentration of population and industries and activities such as transportation.

Noise like other pollutants is a by-product of industrialisation, urbanisation and modern civilisation. Noise pollution and its control is discussed in detail in chapter 10.

2

Emissions from particulate matter and its control

2.1 Introduction

Particulate matter (PM) is the term for particles and liquid droplets suspended in the air. Particles emitted directly into the air are known as 'direct' or 'primary' PM. Other particles are formed indirectly in the atmosphere from the chemical reaction of gaseous pollutants known as 'precursors.' Sources of direct PM and PM precursors include factories, power plants, vehicles, construction activity and natural sources such as fires and wind blown dust.

2.2 Particulate matter

Particulates present in air in the form of aerosols, dusts, mist, smoke, smog, smaze or cloud pose a potential pollution hazard and the worsening situation demands an immediate relief by means of a proper particulate control technology. As particulates are more clearly seen, their removal from flue gases has assumed greater significance. Airborne small, solid particles and liquid droplets are commonly known as particulates. They are present in air in excess and pose serious pollution problems.

Size: It is the most important physical property of particulates. They vary in size ranging from a diameter of 0.0002 μ to 500 μ (μ or micron is equal to 10^{-6} metres). Particles having a diameter of 20 μ are relatively large while particles with the least diameter exist in aerosol. The life period of particulates vary from a few seconds to several months, which depends on their setting rate, size, density and turbulence of air.

Concentration of particulates: The density of particulates in the atmosphere vary from several hundred per cm^3 in clean air to more than 100,000 per cm^3 in densely polluted air. However, their concentration may range from 60 μg to 2000 μg per cm^3 in an urban area. Generally widely spread particulates in the air are found in the size range of 0.1 μ to 10 μ. In remote areas, particulate content may be as low as 10 μg/m^3.

Particulates possess large surface areas and hence appear to form attractive sites for sorption of various organic and inorganic matter.

PM_{10} and $PM_{2.5}$: Particles come in a wide variety of sizes and have been historically assessed based on size, typically measured by the diameter of the particle in micrometers. PM_{10} refers to particles that are 10 micrometers in diameter or less. $PM_{2.5}$, or fine PM, refers to particles that are 2.5 micrometers in diameter or less. (Note: a human hair is about 70 micrometers in diameter and a grain of sand is about 90 micrometers in diameter). Areas of the country are designated nonattainment or attainment separately for the PM_{10} and $PM_{2.5}$ standards.

2.3 Types of particulates

The particulate pollutants may be classified according to their optical properties, nature and size as follows:

1. Smoke: Smoke particulates consist of solid and liquid particles ranging in size from 0.05 to 1.0 micron, which are formed during incomplete combustion of carbonaceous materials. This includes smoke of gaseous pollutants like oxides of sulphur, oxides of nitrogen, carbon monoxide and hydrocarbons, etc.
2. Dust: Dust is composed of fine solid particles and their size ranges from 1 micron to 100 microns. Even dust particles of the size of 0.1 micron have been observed in the atmosphere. Dust particles are formed by: (i) entrained materials used in mechanical processing, e.g., sand from sand blasting, (ii) the particles obtained directly from material when it undergoes a mechanical operation, e.g., saw dust from wood works and (iii) directly from the material being processed, e.g., coal.
3. Mist: Mist or liquid particles are formed by condensation of a vapour, having a size less than 10 microns.
4. Spray: Spray constitutes liquid particles obtained from the parent liquid by the process of mechanical disintegration like atomisation.
5. Fumes: Fumes are generally obtained by the condensation of vapours by sublimation, distillation, boiling, calcination and by several other chemical reactions. Generally organic solvents, metals and metallic oxides form fume particles having a size less than 1 micron.
6. Fly ash: Fly ash (i.e., smaller ash particles which enter furnace flues and emerge from the stack system in the absence of collector devices) is another major pollutant. The bulk of mineral particulate matter in a highly polluted atmosphere exists in the form of oxides of metals and other compounds resulting from the combustion of high-ash fossil fuel. The mineral matter like coal, wood or lignite present in fossil fuel is converted during combustion to a fused glassy bottom ash that settles down on the earth and creates little air pollution problems. However, a larger percentage of fly ash escapes from the industrial combustion processes and enters the atmosphere. These fine particles cause sufficient damage to human, animal and plant life.

Carbon particles, ash, asbestos, oil, grease and acids form particulates which are widely distributed in air.

Composition of particulates: Particulates include organic and inorganic matter, nitrogen compounds, sulphur compounds, polycyclic aromatic hydrocarbons, several metals and radio nuclides.

An estimate indicates that particulate contaminants contain nearly 22 metallic elements. The most abundant elements are calcium, sodium, silicon, aluminium and iron. Considerable quantities of zinc, lead, copper, magnesium and manganese are also present in the air. The concentration of these particulate pollutants depends upon the nature of industrial emission. For instance, the level of lead (Pb) increases with an increase in traffic density, since lead tetra ethyl is added as an anti-knock to petroleum fuel.

Spectrographic studies of the organic fraction conclude that a variety of aliphatic and aromatic hydrocarbons, organic bases, phenols, cresols, organic acids, carcinogenic compounds and various types of complex compounds occur in the atmosphere as particulate pollutants.

2.4 Sources of particulates

Natural processes inject 800–2000 million tonnes of particulate matter each year into the atmosphere. These processes include volcanic eruptions, blowing of dust and soil by the wind, spraying of salt and various other solid particles by the seas and oceans, etc.

Human activities, however, emit 450 million tonnes of particulates every year. For example, particulates in the form of dust and asbestoes are formed during construction, fly ash rich in particulates is given out from power plants, smelters, mining process and smoke from incomplete combustion processes.

According to an estimate: (i) stationary combustion sources (coal, wood, fuel, oil, natural gas), (ii) industrial process and (iii) miscellaneous sources (coal refuse burning, agricultural burning, forest fires, structural fires, etc.), contribute equally (nearly one-third) of the total particulate emission (450 mn tonnes) by human activities.

2.4.1 Physical methods involved in particulate formation

Some of the important physical methods involved in particulate formation are:

1. Dispersion process mainly yields dusts which are solid dispersion aerosols. Particulates of nearly 1 μ in size are formed by the disintegration of larger particles.
2. Adhesion of smaller particles by chemical process yields particulates of size ranging from 10 μ to 20 μ.

3. Natural sources also produce dispersed aerosols as from sea sprays, wind blown dust during cultivation and volcanic dust, etc.
4. By coagulation, extremely small particles form larger aggregates; sorption, absorption, adsorption and chemisorption also results in the formation of particulate matter.

2.4.2 Formation of inorganic particulate matter

Inorganic particulates generally originate from metallic oxides, sulphides, carbonates, etc. These are produced when fuels containing metals are burned. For instance:

1. The particulate Fe_3O_4 is formed during the combustion of pyrites containing coal.

$$3FeS_2 + 8O_2 \rightarrow Fe_3O_4 + 6SO_2$$

2. A part of calcium carbonate in the ash fraction of coal gets converted to calcium oxide, which is emitted into the atmosphere through stack.

$$CaCO_3 \xrightarrow{\text{Heat}} CaO + CO_2$$

3. Organic vanadium in residual fuel oil is converted to particulate vanadium oxide (V_2O_3).
4. Lead halides are generated by the combustion of leaded gasoline. Tetra ethyl lead, $Pb(C_2H_5)_4$ in leaded gasoline combines with oxygen and halogenated scavengers, dibromo ethane, dichloro-ethane, etc. These lead halides emerge through the exhaust system and condense to form particulates, entering the atmosphere.

$$Pb(C_2H_5)_4 + O_2 + C_2H_4Cl_2 + C_2H_4Br_2 \rightarrow CO_2 + H_2O + PbCl_2 + PbBr_2 + PbBrCl$$

This resultant, lead pollution can be minimised by using unleaded gasoline.

5. Aerosol mists are formed by the oxidation of atmospheric sulphur dioxide to sulphuric acid:

$$2SO_2 + O_2 + 2H_2O \rightarrow 2H_2SO_4$$

This forms salts with basic air pollutants such as ammonia or calcium oxide:

$$\underset{\text{(Droplet)}}{H_2SO_4} + \underset{\text{(Gas)}}{2NH_3} \rightarrow \underset{\text{(Droplet)}}{(NH_4)_2SO_4} \qquad \underset{\text{(Droplet)}}{H_2SO_4} + \underset{\text{(Particle)}}{CaO} \rightarrow \underset{\text{(Droplet)}}{CaSO_4 + H_2O}$$

Organic particulate matter (OPM)

Organic particulate matter originates mainly from combustion of fuels, automobiles and vegetation. Polycyclic aromatic hydrocarbons (PAH) like chrysene, benzofluoranthene, benzo α-pyrene, benzidine, etc., are some of the constituents of organic particulate matter of carcinogenic nature. PAH compounds mostly occur in urban atmosphere at levels of ~20 $\mu g/m^3$.

Aldehydes, ketones, peroxides, epoxides, esters, quinones and lactones are found among the oxygenated neutral organic compounds.

Organic acids present in OPM include lauric, palmitic, stearic, myristic, oleic, linoleic, behemic acids, etc. Particulate paraffins are pyrolysed to yield $C_{10}H_{22}$ which again disintegrates into fine particles.

Oxidised polymerised hydrocarbons and nitrogenous azoheterocyclic compounds are released into the atmosphere from automobile exhaust.

PAH compounds remain absorbed in soot particles. Soot itself is a highly condensed product of these compounds. A soot particle is composed of several thousand interconnected crystallites, i.e., graphitic platelets, each having about 100 condensed aromatic rings. It consists of 1–3% hydrogen and 5–10% oxygen due to partial oxidation.

Soot particles act as a disease carrier containing toxic trace metals like Be, Cr, Mn, Ni, Cd, Fe and poisonous organics such as benzo-(α–) pyrene.

2.4.3 Effects of particulate pollutants

Some of the effects of particulate pollutants are discussed below.

Effects on plants

1. Plants are adversely affected by gaseous pollutants and deposition of particulates on soil. This deposition of toxic metals on the soil makes it unsuitable for plant growth.
2. Several particulate pollutants fall on the soil by acid rain which tend to lower its pH, making it more acidic and infertile.
3. Particulates such as dust, fog, soot, deposited on plant leaves block the stomata of plants, thus inhibiting the rate of transpiration of minerals from the soil.
4. Deposited particulates restrict the absorption of CO_2, thereby reducing the rate of photosynthesis, retarding plant growth and crop production.
5. Dust mixed with mist or light rain forms a thick crust on the upper leaf surfaces, which shields the bright sunlight necessary for carbon assimilation.
6. Some plants are very sensitive to the traces of toxic metals, where particulates inhibit the action of their enzyme system.
7. Arsenic is a cumulative, potent, protoplasmic poison which inhibits SH-group in enzymes. This is present in almost all types of soils in minute quantities and affects plant growth.

Effects on humans

The effects of particulate pollutants are largely dependent on the particle size. Airborne particles, i.e., dust, soot, fumes and mists, are potentially dangerous for human health.

1. Particulate pollutants have a bearing on the penetration of particles beyond the respiratory passages into the lungs. Nasal passage prevents coarser particulates bigger than 5 microns from entering into the respiratory system. Particles with a size of about 1 micron enter into lungs easily and rapidly. Actually, aerosols less than 1 micron may reach the alveoli of lungs and damage lung tissues.
2. Soluble aerosols will be absorbed into the blood from the alveoli while insoluble aerosols are carried to the lymphatic stream and get deposited in pulmonary lymphatic depot points or in the lymph glands, where they create toxicity on the respiratory system.
3. Workers exposed to pollutant asbestos mostly develop cancer called mesothelioma, which occurs in the tissue lining the abdomen. They also have a greatly elevated risk of lung cancer.
4. Insoluble particulates which can not be phagocytised by white blood corpuscles (WBCs) pass through the alveolar walls into lymph channels. They also accumulate in various specific organs and their increased concentration exerts actions on lungs.
5. Lead, the most serious pollutant released from automobile exhaust, is reported to have detrimental effect on children's brains.
6. Lead interferes with the development and maturation of red blood cells.
7. Workers exposed to lead excrete porphyrins, the precursors of haemoglobin, in their urine.
8. It has been reported that smokers can more easily develop symptoms of asthma, which is also due to excess concentration of lead, than non-smokers.
9. Silcosis, a chronic disease of lungs, is caused by inhalation of dust containing free silica, SiO_2.
10. Acid particulates and aldehydes cause eye, nose and throat irritation.
11. Formaldehyde is extremely toxic to human health. Acrolein causes irritation to mucus membranes and also poses bronchioconstriction.
12. Lead and asbestos act as cumulative poison and are dangerous to children, causing brain damage and cancer. Small fibres of asbestos irritate lung tissues causing asbestosis, a condition characterised by lung fibrosis.
13. Black lung disease is common among coal miners, while white lung disease occurs frequently among textile workers.

14. Carbon, in the form of soot, deposits in the nose, throat and respiratory tract.
15. Particulates such as silica, asbestos and different forms of carbon are capable of exerting a noxious or fibrotic local action in the interstitial areas of the lungs and in the lymphatic tissues.
16. Beryllium compounds like $BeSO_4$ and $BeCl_2$ cause acute inflammation of the lungs.
17. Small exposure of cadmium causes cardiovascular diseases and hypertension. It also interferes with copper and zinc metabolism.
18. Traces of mercury cause nerve damage and death, while arsenic creates acute and chronic cancer.
19. Fine particulates of less than 2 μ size are the worst causes of lung damage while the larger particles (3 μ) are trapped in the nose and throat. These particles create various breathing troubles due to nasal tract blockage and irritation of lung capillaries.
20. Arsenic is absorbed through the lungs and skin and causes diarrhoea, peripheral neuritis, conjunctivitis, hyperketosis, lung and skin cancer. Chronic exposure to arsenic leads to the so-called 'black foot' disease.

Effects on materials

1. Particulates affect a variety of materials in various ways. They damage buildings, paints, furniture, etc. Painted surfaces are very susceptible to damage in wet conditions. Particulate fumes and mists react directly with any painted surfaces and cause cracks in them.
2. Particulates accelerate corrosion of metals, mainly in urban and industrial areas.
3. Particles, including fumes, dust, soot, mists and aerosols, can cause severe damage to soil, buildings, sculpture and monuments.
4. As a result of the 1984 Bhopal disaster, soil within 160 km, radius was coated with thick dust due to methyl isocyanate leakage and soil fertility was lost for at least the next ten years.
5. Particulates cause extensive damage when they themselves are corrosive and carry toxic harmful chemicals along with them.
6. Particulate pollutants are responsible for polluted cloud formation, rain and snow in which the particles act as the nuclei on which water droplets condense.
7. Particulates accumulate on soil surfaces causing erosion. The particles are generally sticky, tarry and acidic, so they adhere to surfaces and act as acid reservoirs.

8. Precipitation effect over localities or suburbs is due to particulate pollutants. They also influence the formation of rain, dew and snow.

Effects on solar radiation

1. Particulates reduce visibility by absorption and scattering of solar radiation. Decreased visibility creates chronic and annoying problems.
2. They cause illumination problems by reducing sunlight by nearly one-third.
3. In winter, increased heating requires more power which releases more particulate particles in the atmosphere.
4. They also upset the delicate heat balance of the earth's atmosphere. Increased content of CO_2 (10%) causes global warming but the particulates (i.e., aerosols) have the ability to reject more solar radiations thus compensating the climatic effect of increased carbon dioxide concentration.

2.5 Control of particulates

There are three broad approaches to the control of particulates: (i) dilution in the atmosphere, (ii) control at source and (iii) control by using pollution control equipment.

2.5.1 Dilution in the atmosphere

Dilution of particulates and gases can be accomplished by the use of tall stacks. Pollutants released from taller stacks disperse easily and hence low ground level concentrations are observed. Tall stacks penetrate the inversion layer and disperse the contaminants easily so that the ground level concentrations are less harmful.

However, dilution is only a short-term control measure and has highly undesirable long-range effects.

Industries have to ensure a minimum stack height of 30 metres. Theoretically, the minimum height of stack, H, required for effective dispersion of particulates is given by $H = 74\ Q^{0.27}$ where Q is particulate emission rate in tonnes per hour. This often demands stack heights in excess of 400 m, especially in cement industries and thermal power plants and therefore may prove more uneconomical when compared to particulate control by treatment.

2.5.2 Control at source

The most effective means of dealing with the problem of air pollution is to prevent emission at the source itself. In the case of industrial pollutants, this can often be achieved by investigating various approaches at the early stage of process design and development and selecting those methods which do not contribute to air pollution (or) have the minimum air pollution potential. It may be difficult to implement these methods in existing plants, but still some of the methods can be applied without upsetting the economy of the operation.

Control of particulates at the source can be accomplished in several ways through raw material changes, operational changes, modification or replacement of process equipment and by more effective operation of existing equipment. Source locations through allocation of land usage, i.e., proper planning and zoning of industrial areas, is also an important means of control of pollutants. Some of the source correction methods are discussed briefly as follows.

Raw material changes

Some raw materials are primarily responsible for causing air pollution. Use of pure grade of raw material is often beneficial and may reduce the formation of undesirable impurities and by-products and may even eliminate troublesome effluents. Ore handling operations usually result in the emission of large quantities of dust into the atmosphere. In the steel industry, replacement of raw ore with briquetted or pelleted sintered ore has greatly reduced dust emissions during ore handling and also helped reduce blast furnace 'slips,' which result in emission of enormous amounts of uncleaned, blast furnace gas when the safety dampers open, by-passing the dust collectors.

Process changes

Changing the processing methods is still another important method of controlling emissions at their source. For example, petroleum/chemical industries have undergone radical changes in processing methods which emphasise continuous automatic operations, often computer controlled and completely enclosed systems that minimise the release of materials into the atmosphere. The volatile substances from storage tanks, etc., are recovered by condensation and the non-condensable gases are recycled for additional reactions such as polymerisation and alkylation of gaseous hydrocarbons to produce gasoline. Hydrogen sulphide, which was once flared in refineries, is now recycled using the Clauss process to recover elemental sulphur.

Replacing open hearth furnaces with controlled basic oxygen furnaces or electric furnaces can reduce smoke, carbon monoxide and metal fumes while at the same time conserve energy. Such changes coupled with various combinations of gas cleaning devices can be very effective in reducing air pollution.

In cement plants, rotary kilns are the major sources of dust generation. This can be reduced to some degree by adjusting operating conditions like reduction of the gas velocities within the kiln, modification of the rate and location of feed, introduction and employment of a dense curtain of light weight chain at the discharge end of the kiln, etc.

Emissions of sulphurous materials from smelting and paper industries are highly objectionable. These emissions are being curtailed by major process changes such as hydro-metallurgical separations of ores and avoiding the use of sulphides in paper making.

In the steel industry, a radically different process has been proposed to lower sulphurous emissions during combustion. In this process, the sulphur bearing fuel, limestone and air are injected into a molten bath. The combustibles in the fuel are partially oxidised to carbon monoxide within the molten iron bath. Here the gaseous CO comes off at the top of the molten iron and is burnt efficiently in a conventional manner, the sulphur is retained in the iron bath and forms a slag with the limestone, which is removed.

Other examples involving process changes include:

1. Reduction of the formation of nitric oxides in combustion chambers by low excess air combustion in two stages, fuel gas recirculation and water injection.
2. Washing the coal before pulverisation to reduce fly ash emissions.
3. Substitution of bauxite flux for fluorine-containing fluorspar in the open hearth furnace method.
4. The use of liquid and gaseous fertiliser chemicals like anhydrous ammonia, applied by injection into the earth instead of being spread across the surface as finely divided powder.
5. Reduction in oxidation of SO_2 to SO_3, by reducing excess air from 20% to less than 1% when burning fossil fuels, has eliminated sulphuric acid emissions. However, care should be taken as absence of excess air tends to result in greater soot production.

Equipment modification (or) replacement

Another method of control of pollutants at the source involves the proper use of existing equipment and modification and replacement of equipment. For example, the unburnt carbon monoxide and hydrocarbons in the cylinders of an automobile engine, which are otherwise emitted into the atmosphere through the tail pipe, can be burnt by injecting air into the hot exhaust manifold of the engine. Similar results can be obtained by suitable modification in the carburetion and ignition systems.

Hydrocarbon vapours from petroleum refineries are released into the atmosphere from storage tanks due to temperature changes, direct evaporation and displacement during filling. These losses can be minimised by designing the tanks with floating roof covers or by pressurising them.

In addition to the above mentioned source correction methods, air pollutants emitted from industrial operations can be reduced by proper equipment maintenance, handkeeping and cleanliness in the facilities and premises. Often, changes in the design of local exhaust hood and proper installation can minimise the emission of pollutants.

Chemical process plants often have excessive leakage around ducts, piping, valves and pumps. Many such leaks can be prevented by routine checking of the seals and gaskets. Floors, decks, storage bins and silos, loading areas and material transfer conveyors must also be kept clean to reduce dust pollution.

2.5.3 Particulate control by using equipment

The most effective methods of particulate control are reduction at the source by the application of control equipment and process control. If air pollution problems are properly considered in an industry prior to its design, real economy can be effected. But unfortunately in most cases, air pollution control is an after-thought and the ways and means to control it are derived and designed hurriedly at the eleventh hour.

To remove particulate matter from gas streams, various types of control equipment are available. But to select the required equipment, certain basic data must be available. The required data are:

1. Quantity of gas to be treated and its variation with time.
2. Nature and concentration of the particulate matter to be removed.
3. Temperature and pressure of the gas stream.
4. Nature of the gas phase (for solubility and corrosive effects).
5. Desired quality of the treated effluent, i.e., efficiency of removal of particulates required.

Objectives of using control equipment

1. Prevention of nuisance.
2. Prevention of physical damage to property.
3. Elimination of health hazards to plant personnel and the general population.
4. Recovery of valuable waste products.
5. Minimisation of economic losses through reduction of plant maintenance.
6. Improvement of product quality.

As pollutants, originating from a variety of sources, primarily from industrial processes, airborne particulates exert a significant influence on atmospheric phenomena, plants, property and on animals, including man. Many techniques for the control of particulate emission have been developed. These are discussed below.

Gravity settling chamber

It consists of a simple horizontal rectangular gravity apparatus for the removal of particulates from gas streams. Effluent gases are led into this chamber which allows larger particles (250 μ) to settle down. However, the method is unsuitable for finer particles which have longer settling time.

Cyclone collector

A gas containing of particulates flowing in a tight circular spiral creates a centrifugal force on suspended particles forcing them to move away from the gas stream to a wall where they are collected. This inertial separator due to centrifugal force exerts greater inertial effects on the dispersoids to effect particulate and gas separation. It is one of the least expensive and efficient techniques by which 95% of particulates in the range of 5 to 20 μ can be separated.

Cyclonic separators and trajectory separators

These have been extensively employed for industrial dust collection. They are also used for the control of gas-borne dispersoids in industrial operations, e.g., manufacture of cement, paper and pulp industries, mineral processing, feed and grain processing and wood working factories.

Filters

Filters are now widely used to collect extremely fine particulates. Solid dispersoids can be removed from the carrier gas by filtration of the effluents through porous cloth or fibre. Next to cyclonic separators, filters constitute the most effective devices of controlling particulate pollutants.

Scrubbers

Scrubbers may be classified as wet washers, cyclonic scrubbers, gravity spray towers, impingement scrubbers, disintegrator scrubbers and spray de-dusters.

Wet scrubbers: These utilise water to remove solid, liquid or gaseous contaminants. The effectiveness of the device depends upon the degree of contact and interaction between the liquid phase and the contaminants to be removed. Spray chambers or towers are used, where the liquid is introduced into the gas stream as a fine spray.

Another recent technique makes the liquid percolate downward through a bed of materials, while the gas stream moves upward and gets separated.

2.6 Particulate emission control by electrostatic precipitators

The electrostatic precipitator (ESP) is one of the most widely used collection devices for particulates. An electrostatic precipitator (ESP) is a particulate collection device that removes particles from a flowing gaseous stream (such as air) using the force of an induced electrostatic charge.

ESP can be operated at high temperature and pressures and its power requirement is low. For these reasons the electrostatic precipitation is often the preferred method of collection where high efficiency is required with small particles. ESP are highly efficient filtration devices that minimally impede the flow of gases through the device and can easily remove fine particulate matter such as dust and smoke from the air stream. In the electrostatic precipitation process the basic force which acts to separate the particles from the gas is electrostatic attraction. The particles are given an electrical charge by forcing them to pass through a corona, a region in which gaseous ions flow. The electrical field that forces the charged particles to the walls comes from electrodes maintained at high voltage in the centre of the flow lane.

Control of emissions from the industrial sources has served the threefold purpose of:

1. Recovery of the emission for economic reason.
2. Removal of abrasive dusts to reduce wear of fan component.
3. Removal of objectionable natter from gases being discharged into the atmosphere.

2.6.1 Requirement of electrostatic precipitation process

1. Source of high voltage.
2. Discharge and collecting electrode.
3. Inlet and outlet for gas.
4. A means for disposal of collected material.
5. Cleaning system, outer casing.

2.6.2 Steps in electrostatic precipitation

1. Generation of electric field high voltage direct current 20–80 kv.
2. Generation of electric charges.
3. Transfer of electric charge to a dust particle.
4. Movement of the charge dust particle in an electric field to the collection electrodes.
5. Adhesion of the charge dust particle to the surface of the collection electrode.
6. Dislodging of dust layer from collection electrode.
7. Collection of dust layer in a hopper.
8. Removal of the dust from the hopper.

Electrostatic precipitator is shown in Fig. 2.1.

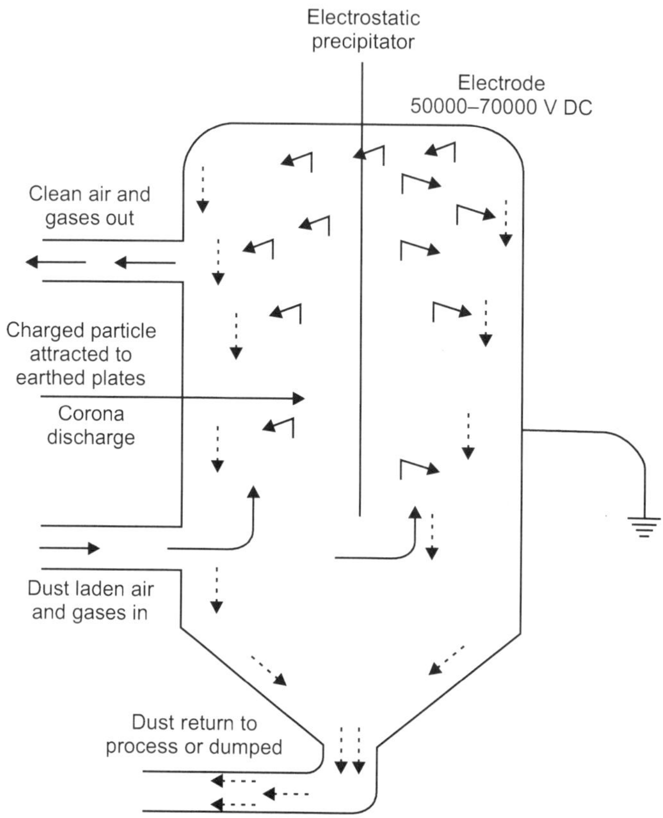

Figure 2.1: Movement of dust and air in ESP.

2.6.3 Principle of ESP

Principle of ESP has four distinct phases as follows:

1. Ionisation or corona generation: When the potential difference between the wire and electrode increases, a voltage is reached where an electrical breakdown of the gas occurs near the wire. This electrical break down or ion discharge is known as corona formation and thereby gas is transformed from insulating to conducting state.

   Two types of corona discharge can be generated which are:

   (a) Negative corona: In negative corona, discharge electrode is of negative polarity and the process of electron generation occurs at narrow region.

   (b) Positive corona: When positive voltage is applied to discharge electrodes in the same way as negative corona, large number of free electron and positive ions are generated. Or large number of positive ions produced move towards collecting electrode and thus transfer charge to dust particles upon collision.

Negative coronas are more commonly used in industrial application, while for cleaning air in inhabited space positive coronas are used. Due to ozone generation in negative corona its application for air cleaning in inhabited area is avoided. Variation of field strength between wire and plate electrodes is shown in Fig. 2.2.

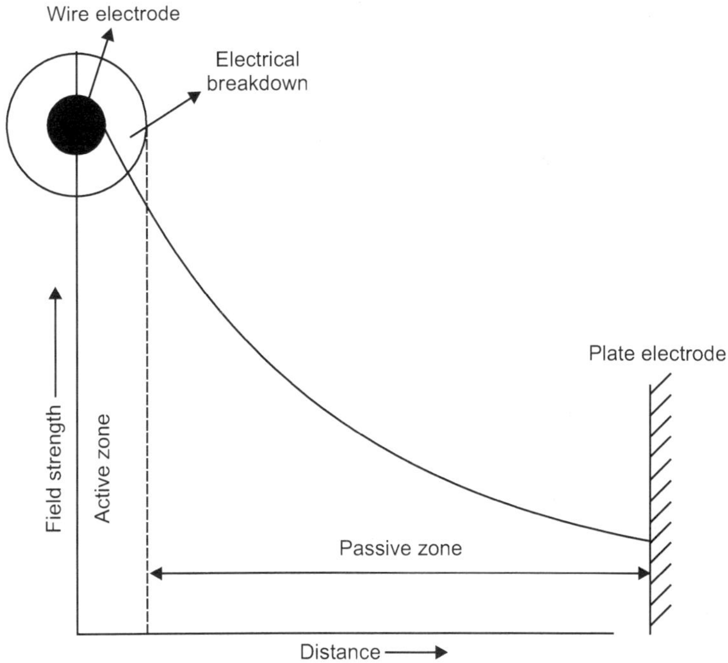

Figure 2.2: Variation of field strength between wire and plate electrodes.

2. Charging of particles: Particle charging takes place in region between the boundary of corona glow and the collection electrode, where particles are subjected to the rain of negative ions from the corona process. Mainly two mechanisms are responsible for particle charging. Each mechanism becomes significant according to particle size ranges. For particles having diameter greater than 1 μm, field charging is dominant force and for particle size less than 0.2 μm diffusion charging predominates.
3. Migration and precipitation of particle.
4. Removal of deposited dust: Once collected, particle can be removed by coalescing and draining, in the case of liquid aerosols and by periodic

impact or rapping, in case of solid material. In case of solid material, a sufficiently thick layer of dust must be collected so that it falls into the hopper or bin in coherent masses to prevent excessive re-entrainment of the material into the gas system.

2.6.4 Types of electrostatic precipitators

ESPs are configured in several ways. Some of these configurations have been developed for special control action and others have evolved for economic reasons.

Single stage precipitators

1. Plate-wire precipitators:
   (a) In a plate-wire ESP, gas flows between parallel plates of sheet metal and high-voltage electrodes.
   (b) These electrodes are long wires weighted and hanging between the plates or are supported there by mast-like structures (rigid frames).
   (c) Within each flow path, gas flow must pass each wire in sequence as flows through the unit.
   (d) Plate-wire ESPs are used in a wide variety of industrial applications, including coal-fired boilers, cement kilns, solid waste incinerators, paper mill recovery boilers, petroleum refining catalytic cracking units, sinter plants, basic oxygen furnaces, open hearth furnaces, electric arc furnaces, coke oven batteries and glass furnaces.
2. Flat plate precipitators:
   (a) A significant number of smaller precipitators [100,000 to 200,000 actual cubic feet per minute (acfm)] use flat plates instead of wires for the high-voltage electrodes.
   (b) A flat plate ESP operates with little or no corona current flowing through the collected dust, except directly under the corona needles or wires.
   (c) Flat plate ESPs seem to have wide application for high-resistivity particles with small (1 to 2 μm) mass median diameters.
   (d) Fly ash has been successfully collected with this type of ESP, but low-flow velocity appears to be critical for avoiding high rapping losses.

Tubular precipitators

1. The original ESPs were tubular like the smoke stacks they were placed on, with the high-voltage electrode running along the axis of the tube.

2. Tubular precipitators have typical applications in sulphuric acid plants, coke oven by-product gas cleaning (tar removal) and recently, iron and steel sinter plants.

Wet precipitators

1. Any of the precipitator configurations discussed above may be operated with wet walls instead of dry.
2. The water flow may be applied intermittently or continuously to wash the collected particles into a sump for disposal.
3. The advantage of the wet wall precipitator is that it has no problems with rapping re-entrainment or with back coronas.
4. The disadvantage is the increased complexity of the wash and the fact that the collected slurry must be handled more carefully than a dry product, adding to the expense of disposal.

Two-stage precipitators

1. The previously described precipitators are all parallel in nature, i.e., the discharge and collecting electrodes are side by side.
2. Two-stage precipitators are considered to be separate and distinct types of devices compared to large, high-gas-volume, single-stage ESPs.
3. The two-stage precipitator invented by Penney is a series device with the discharge electrode, or ioniser, preceding the collector electrodes.
4. Advantages of this configuration include more time for particle charging, less propensity for back corona and economical construction for small sizes.

2.6.5 Operational issues in ESP

1. Pre-scrubbing.
2. Wash-down sprays and wires.
3. Wet/dry interface.
4. Current suspension.
5. Sparking.
6. Mist elimination.

2.6.6 Advantages and disadvantages of electrostatic precipitators

Advantages

1. High collection efficiency.
2. Particles as small as 0.1 micron can be removed.

3. Low maintenance and operating costs.
4. Low pressure drop (0.25 to 1.25 cm of water).
5. Satisfactory handling of large quantities of high temperature gas.
6. Treatment time is negligible (0.1 to 10 seconds).
7. Cleaning is easy by removing units of the precipitator from operation.
8. There is no limit to solid, liquid or corrosive chemical usage.

Disadvantages

1. High initial cost.
2. Space requirement is more because of the large size of the equipment.
3. Possible explosion hazards during collection of combustible gases or particulates. Well trained personnel are needed.
4. The poisonous gas, ozone is produced by the negatively charged discharge electrodes during gas ionisation.
5. Precautions are necessary to maintain safety during operation (i.e., proper gas flow distribution, gas resistivity, particulate conductivity, etc.).
6. Gases cannot be removed by ESPs.

2.7 Applications of electrostatic precipitators

The important applications of electrostatic precipitators in industries are given in Table 2.1.

Table 2.1: Important applications of electrostatic precipitators.

Industry	*Application*
Cement factories	Cleaning of flue gas from cement kilns
	Recovery of cement dust from kilns
Pulp and paper	Soda fume recovery in kraft pulp mills
Steel plants	Cleaning blast furnace gas
	Removing tars from coke oven gases
	Cleaning open hearth and electric furnace gases
Chemical industries	Collection of SO_x, phosphoric acid mist
	Cleaning various types of gases, i.e., hydrogen, CO_2, SO_2.
	Removing dust from elemental phosphorus in the vapour state
Petroleum industry	Recovery of catalyst dust
Carbon black industry	Agglomeration and collection of carbon black
Thermal power plants	Collecting fly ash from coal fired boilers

3

Control of gaseous pollutants

3.1 Introduction

Combustion processes generate various primary and secondary air pollutants such as carbon oxides (mainly CO), nitrogen oxides (NO_x), sulphur oxides (SO_x), ozone, along with organic acid, inorganic acid, petrochemical oxidant gas and hydrocarbons (HC). For controlling these gases, the mechanisms adopted are chemical engineering unit operations, which include adsorption, absorption, condensation and combustion. In addition the concept of mass transfer is also used in the unit operations. Gaseous pollutant streams are like chemical process streams. The major exception is that the gaseous pollutant concentrations are usually lower than the chemical process stream concentrations. These lower concentrations make it possible to consider gaseous pollutants as ideal gases. Also, the carrier gas (air), may be treated as an ideal gas in the operating range for most air pollution control equipment. Gaseous pollutants can be controlled by a wide variety of devices and choosing the most cost effective, most efficient units requires careful attention to the particular case for which the control devices are intended.

3.2 Control of gaseous pollutants

Emission of gaseous pollutants can be controlled by: (i) absorption, (ii) adsorption, (iii) combustion and (iv) fugitive dust control measures.

Different treatment processes are applied for controlling these and other gaseous emissions. These processes include adsorption and absorption. Selection of appropriate technique depends in part, physical and chemical characteristic of specific gas and the vapour phase compounds present in the gas stream. For stationary air pollution sources, we can select single or combined air pollution control technique. Variety of devices are used for controlling gaseous pollutant and choosing most cost effective, most cost efficient unit requires careful attention to the particular operation for which control devices are intended. In order to control the emissions within given standard emissions it is necessary to monitor emissions carefully after selecting best control technique.

Properties of gas stream for selection of a control system: The selection and design of a gaseous contaminant control system is done based on some specific information concerning the gas stream to be treated.

Following are some factors considered during selection of a process:

1. Gas stream particulate matter characteristics.
2. Gas stream average and peak flow rates.
3. Gas stream average and peak temperatures.
4. Gas stream particulate matter average and peak concentrations.
5. Gas stream minimum, average and maximum oxygen concentrations.
6. Contaminant average and peak concentrations.
7. Contaminant ignition characteristics.

3.2.1 Absorption

Absorption is a process where transfer of a gaseous component from gas phase to liquid phase takes place. More specifically in air pollution control, absorption involves the removal of objectionable gaseous contaminant from a process stream by dissolving them in liquid. Common terms used in absorption process are as follows:

1. Absorbent: The liquid, usually water, into which contaminant is absorbed.
2. Absorbate or solute: The gaseous contaminant being absorbed, such as SO_2, H_2S, etc.
3. Carrier gas: The inert portion of gas stream, usually air, from which the contaminant is to be removed.
4. Interface: The area where the gas phase and the absorbent contact each other.
5. Solubility: The capability of the gas to be dissolved in a liquid.

Absorption equipment used to remove gaseous contaminants are referred to as absorber or wet scrubber. Wet scrubbers usually cannot be operated to optimise simultaneous removal of both gases and particulate matter. In designing absorber from gaseous emissions, optimum mass transfer can be accomplished by:

1. Providing a large interfacial contact area.
2. Providing good mixing between gas and liquid phases.
3. Allowing sufficient residence or contact time between the phases.
4. Ensuring a high degree of solubility of the contaminant in the absorbent.

Mechanism of absorption

The gaseous contaminant are removed in absorption process by passing (contacting) a contaminated laden gas through a liquid. The following three steps occur during this process:

Step1: The pollutant diffuses from bulk area of the gas phases to the gas liquid interface.

Step2: Gaseous pollutant transfer across the interface to the liquid phase. This second step is extremely rapid.

Step3: The pollutant diffuses bulk area of the liquid, making room for additional gas molecule to absorb.

The rate of mass transfer (absorption) is dependent on the diffusion rate in either the gas phase or liquid phase. The diffusion rate of gaseous pollutant molecule through a gas is always faster than its diffusion rate through the liquid because molecules in the gas are further apart than are molecules in the liquid. However the mass transfer rate depends primarily upon the solubility of the pollutant in the liquid.

Solubility: Solubility of contaminant affects the amount of contaminant that can be adsorbed. It is a function of both the temperature and, to a lesser extent pressure of a system. As we increase the temperature of the system the amount of gas that can be absorbed by liquid decreases, while as with increasing the pressure generally absorption increases.

Absorption unit

Spray tower: Spray towers are useful (Fig. 3.1) for large volume handling with relatively low pressure drop and high efficiency. In general smaller the droplet size the greater the turbulence, the more chance for absorption of the gas. Absorbing liquid usually water is sprayed through the contaminated gas and the absorbent contaminant solution falls downwards for removal while clean gas exits through an outlet valve in the top of unit. Moisture eliminator reduces the amount of moisture in the gases being released. Spray tower has less gas liquid interfacial area so they are less effective in removal of gaseous contaminant.

Plate or tray tower: This type of tray contains horizontal trays or plate that provides large gas liquid interfacial areas. The polluted air is introduced from one side of the bottom of the column, rise up through the opening in each tray and the rising gas prevents the liquid from draining through the opening. Due to repeated contact of gas and liquid the contaminant are removed and the clean air emerges from the top.

In bubble cap tray column (Fig. 3.2) the contaminated gas moves upward until they strike the cap, at which point they are diverted downward and discharged as small bubbles from slots at the bottom of the caps. Since gas continues flow in upward direction so repeated interaction takes place and contaminated gas is removed and clean gas emerged from the top. The contaminant laden liquid flows to the bottom and is drawn off.

Packed tower: In packed tower (Fig. 3.3) the contact time between vapour and liquid is increased by introducing packing. The packing material has a large surface to volume ratio and a large void ratio that offers minimum

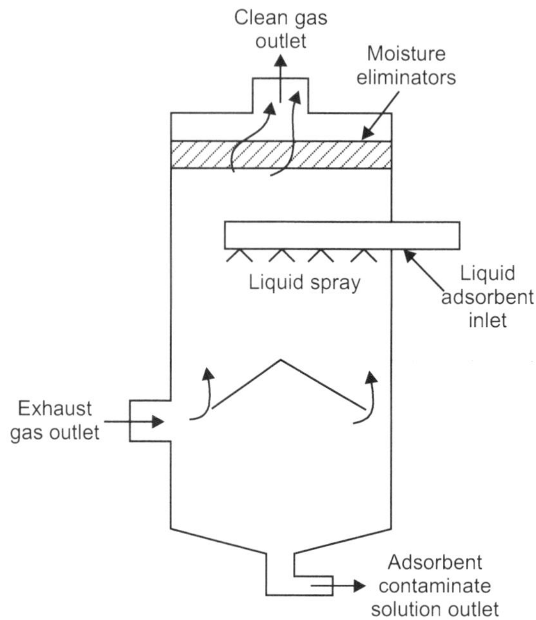

Figure 3.1: Spray tower.

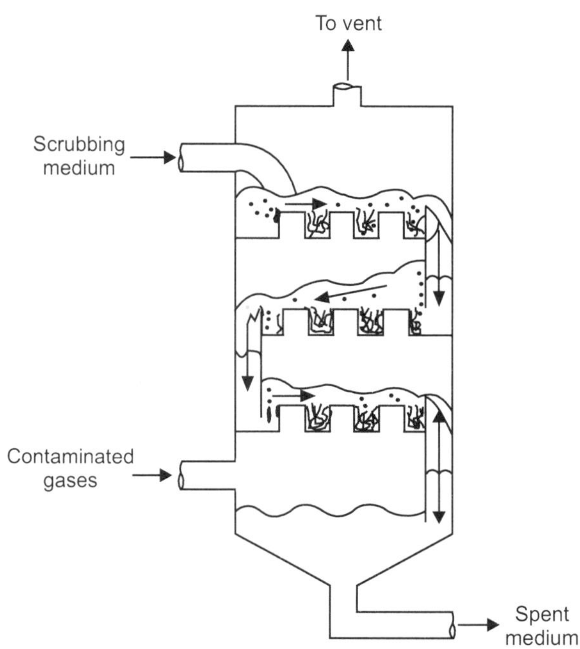

Figure 3.2: Bubble plate tower.

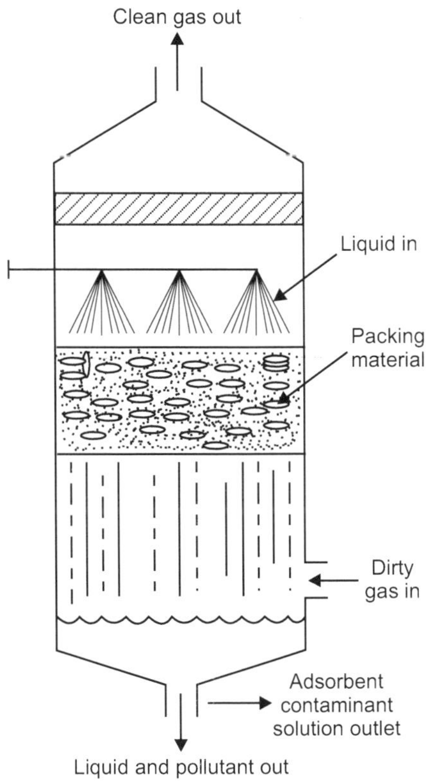

Figure 3.3: Counter current flow packed tower.

resistance to gas flow. Generally packed tower are operated counter currently, with gas entering at the bottom of tower and liquid entering from the top. Liquid flows over the surface of the packing in a thin film causing continuous contact with the gases. Packed towers are highly efficient but they become easily clogged when gas with high particulate loads are introduced.

3.2.2 Adsorption

In adsorption process the contaminant removal is done by passing a stream of effluent gas through a pours solid material (adsorbent) contained in adsorption bed. The surface of porous solid material attracts and holds the gas (the adsorbate) by either by physical or chemical adsorption. The basic difference between physical and chemical adsorption is the manner in which the gas molecule is bonded to the adsorbent. Difference between physical and chemical adsorption are shown in Table 3.1 and Table 3.2 shows physical properties of major type of adsorbents.

Table 3.1: Difference between physical and chemical adsorption.

Physical adsorption	*Chemical adsorption*
Gas or vapour molecule is weakly held to the solid surface by intermolecular attractive forces	Gas contaminant and is held strongly to the solid surface by valence forces
It is accompanied by capillary condensation within the pores	Chemical reaction occurs between the adsorbent and the gaseous contaminant
Physical adsorption is easily reversed by the application of heat or by reducing the pressure	Chemical reaction is usually irreversible
Commonly used for the capture and concentration of organic compounds	It is frequently used for the control of acid gases. Chemical adsorption is also used for the control of mercury vapour
Gas temperature is usually maintained at levels less than approximately 120°F	Chemical adsorption can be conducted at higher temperatures (100°F to 400°F)
Higher the boiling point greater will be adsorption	Much slower than the physical adsorption. Amount of gas adsorb depends upon temperature and pressure
It is directly proportional to surface area available and multilayer adsorption can take place	Liberates greater amount of heat and hence requires much energy

Table 3.2: Physical properties of major type of adsorbents.

Adsorbent	*Internal porosity (%)*	*Surface area (m^2/gm)*	*Pore volume (cm^3/gm)*	*Bulk dry density (gm/cm^3)*	*Mean pore diameter (Å)*
Activated carbon	55–75	600–1600	0.8–1.2	0.35–0.50	1500–2000
Activated alumina	30–40	200–300	0.29–0.37	0.90–1.00	1800–2000
Zeolite (molecular sieves)	40–55	600–700	0.27–0.38	0.80	300–900
Synthetic polymers	–	1080–1100	0.94–1.16	0.34–0.40	–

Salient feature of adsorption process

1. Adsorption processes are used extensively on large-scale applications having solvent vapour concentrations in the range of 10 to 10,000 ppm.
2. Prior to becoming saturated with the solvents, the adsorbent is isolated from the gas stream and treated to drive the solvent compounds out of the solid adsorbent and into a small volume, high concentration gas stream.
3. The desorbed gas stream is then treated to recover and reuse the solvents.
4. The adsorbent is cooled (if necessary) and returned to adsorption service.

5. Because the adsorbent is treated and placed back in service, these adsorption processes are termed regenerative.
6. Adsorption processes usually operate at efficiencies of 90% to 98% over long time periods.

Steps in adsorption process

Adsorption occurs in three steps:

Step 1: The contaminant diffuses from the bulk gas stream to the external surface of the adsorbent material.

Step 2: The contaminant molecule migrate external surface to the macropores, transitional pores and micropores within each adsorbent.

Step 3: The contaminant molecule adheres to the surface in the pore.

Steps 1 and 2 are diffusional processes that occur because of the concentration difference between the bulk gas stream passing through the adsorbent and the gas near the surface of the adsorbent. Step 3 is the actual physical bonding between the molecule and the adsorbent surface. This step occurs more rapidly than steps 1 and 2.

Adsorbent regeneration methods

After a long period of operation and when adsorption bed becomes saturated, replacement or regeneration of the adsorbent bed is necessary in order to maintain continuous operation. When the adsorbate concentration is high and/or the cycle time is short (less than 12 hours), replacement of the adsorbent is not feasible and *in situ* regeneration is required. Regeneration is accomplished by reversing the adsorption process, usually increasing the temperature or decreasing the pressure.

Following four main methods used commercially for regeneration.

1. Thermal swing: The bed is heated so that the adsorption capacity is reduced to a lower level. The adsorbate leaves the surface of the carbon and is removed from the vessel by a stream of purge gas. Cooling must be provided before the subsequent adsorption cycle begins.
2. Pressure swing: The pressure is lowered at a constant temperature to reduce the adsorbent capacity.
3. Inert purge gas stripping: The stripping action is caused by an inert gas that reduces the partial pressure of the contaminant in the gas phase, reversing the concentration gradient. Molecules migrate from the surface into the gas stream.
4. Displacement cycle: The adsorbates are displaced by a preferentially adsorbed material. This method is usually a last resort for situations in

which the adsorbate is both valuable and heat sensitive and in which pressure swing regeneration is ineffective.

Factors affecting the performance of adsorption system

Temperature: For physical adsorption processes, the capacity of an adsorbent decreases as the temperature of the system increases. With increase in the temperature, the vapour pressure of the adsorbate increases, raising the energy level of the adsorbed molecules. Adsorbed molecules now have sufficient energy to overcome the van der Waals' attraction and migrate back to the gas phase. Molecules already in the gas phase tend to stay there due to their high vapour pressure.

Pressure: Adsorption capacity increases with an increase in the partial pressure of the vapour. The partial pressure of a vapour is proportional to the total pressure of the system. Any increase in pressure will increase the adsorption capacity of a system. The increase in capacity occurs because of a decrease in the mean free path of vapour at higher pressures.

Gas velocity: The gas determines the contact or residence time between the contaminant stream and adsorbent. The slower the contaminant stream flows through the adsorbent bed, the greater the probability of a contaminant molecule reaching an available site. In order to achieve 90% or more capture efficiency, most carbon adsorption systems are designed for a maximum airflow velocity of 100 ft/min (30 m/min) through the adsorber. A lower limit of at least 20 ft/min (6 m/min) is maintained to avoid flow problems such as channeling. Gas velocity through the adsorber is a function of the cross-sectional area of the adsorber for a given volume of contaminant gas.

Humidity: Activated carbon has more affinity towards nonpolar hydrocarbons over polar water vapour. The water vapour molecules in the exhaust stream exhibit strong attractions for each other rather than the adsorbent. At high relative humidity, over 50%, the number of water molecules increases to the extent that they begin to compete with the hydrocarbon molecules for active adsorption sites. This reduces the capacity and the efficiency of the adsorption system.

Bed depth: Providing a sufficient depth of adsorbent is very important in achieving efficient gas removal due to the rate that VOC compounds are adsorbed in the bed. There are practical minimum and maximum limits to the bed depth.

3.2.3 Combustion or incineration

This is used when the pollutants in the gas stream are oxidisable to an inert gas. Pollutants like hydrocarbons and carbon monoxide can be easily burned, oxidised and removed from the combustion equipment. This is achieved by: (i) direct flame incineration and (ii) catalytic incineration.

Direct flame incineration

Direct flame incineration is a control technique for combustible organic air pollutants (Fig. 3.4). This is accomplished in the presence of a flame and sufficient oxygen by raising the temperature of the gases above their ignition temperature and then maintaining this temperature until the oxidation reactions are complete.

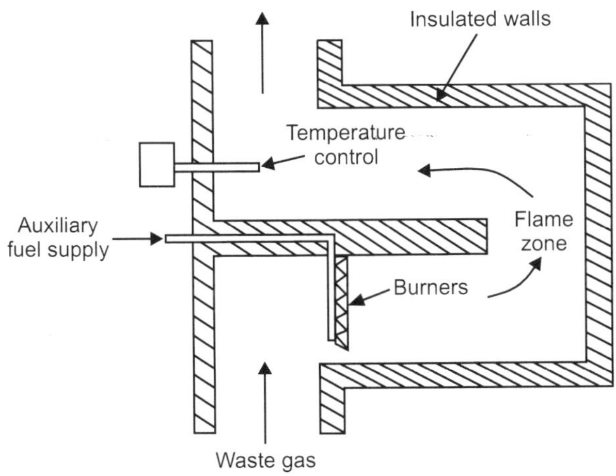

Figure 3.4: Direct flame incinerator.

Catalytic incineration

In catalytic incineration (Fig. 3.5), a mixture of dilute organic gases and oxygen is exposed to a catalytic surface at a temperature which is high enough for oxidation to occur and for a length of time sufficient for the oxidation to be completed. Catalysts are usually solids that are neither reactants nor products of a reaction, yet they alter the rate of chemical reactions. The effect of the catalyst is to reduce the temperature required to oxidise the organic compounds and hence the inlet gases need not be heated to ignition temperature. This requires less fuel than would be needed for direct flame incineration. The catalytic oxidation reaction occurs in a very rapid sequence at the catalytic surface. Hence, the overall residence time required is very much less than that in a direct flame incinerator. Due to this, the size of the unit is reduced.

3.2.4 Fugitive dust control measures

Sources of fugitive dust emissions

1. Coal yards.
2. Construction sites.

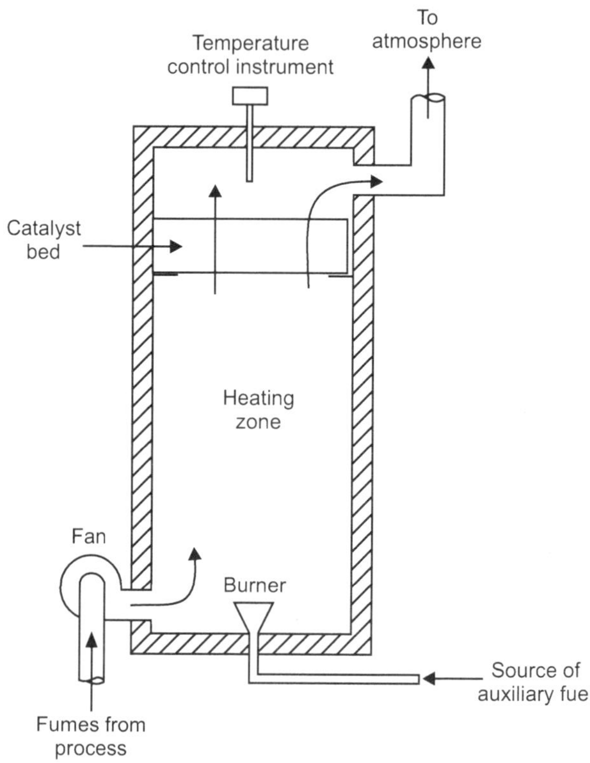

Figure 3.5: Catalytic incinerator.

3. Unpaved road surfaces.
4. Raw material storage.
5. Demolition of building.
6. Renovation of building.

Methodology of reducing fugitive dust emissions

1. Wet suppression:
   (a) Spray of water for dust suppression.
   (b) Only temporary and must be repeated at regular intervals.
2. Chemical stabilisation:
   (a) Salt (CaCl, MgCl): Absorb and retain moisture in the surface layer.
   (b) Wetting agents and surfactants: Lower the surface tension of water, rapid penetration into the surface layer.
   (c) Dust suppressants: Bind fines to large particles in the surface layer.

3. Physical stabilisation: Place cover on exposed surfaces to prevent particles from becoming airborne due to wind or machinery action.
4. Vegetative stabilisation: Deploy vegetation to control erosion.
5. Specialised techniques:
   (a) Vehicle speed reduction – less turbulence.
   (b) Surface cleaning – reduces re-entrainment.
   (c) Traffic control – reduces vehicle access.
   (d) Windbreaks – lessens surface wind speed.
   (e) Good operating practices.

3.3 Sulphur compounds (SO_x) as pollutants

Oxides of sulphur, i.e., sulphur dioxide (SO_2) and sulphur trioxide (SO_3), represented as SO_x, hydrogen sulphide (H_2S), carbonyl sulphide (COS), carbon disulphide (CS_2), dimethyl sulphide [$(CH_3)_2S$] and sulphates (SO_4) are the most serious air pollutants.

3.3.1 Oxides of sulphur (SO_x)

SO_2 is the second most important contribution of all air pollutants as it accounts for about 29% of the total weight of all pollutants. Sulphur in low concentration is essential for the metabolism of both animals and plants, but it becomes injurious when its concentration increases. There are two sources of SO_2: (i) natural and (ii) anthropogenic or man-made.

Natural sources (e.g., volcanoes) provide about 67% of the SO_x pollution all over the globe, while man-made sources contribute about 33% of SO_x pollution, which is mainly localised in some urban areas. Among man-made sources, fossil fuel combustion (coal, etc.), accounts for 74%, industries 22% and transportation 2% of the total SO_x emission. This clearly indicates that coal fired power stations are mainly responsible for the SO_x pollution, followed by industrial plants.

The burning of fossil fuels in thermal power plants, manufacture of sulphuric acid and fertilisers, smelting industries and other processes like electric power plants accounts for 75% of total SO_2 emission, while automobiles and refineries contribute to the remaining 25%.

3.3.2 Reaction of SO_2 in the atmosphere

SO_2 undergoes several chemical reactions in air, forming particulate matter and aerosols, etc., which are scavenged from the atmosphere. SO_2 and SO_3 gases are washed down from the atmosphere in the form of sulphuric acid (H_2SO_4). The presence of increasing concentrations of H_2SO_4 in the troposphere is evident from the increasing and more widespread occurrence of acid rain. A number of factors, such as temperature, light intensity, humidity, air traffic, suspended particulate matter (SPM) may enter into these reactions. However, SO_2 reacts through several ways in the atmosphere. These are:

1. Photochemical reactions.
2. Chemical reactions in the presence of NO_x and hydrocarbons.
3. Chemical reactions in water droplets and solid particles.

Effects of SO_2 on man

SO_2 affects human health in various ways:

1. It causes intense irritation to eyes and respiratory tract even at 2.5 ppm levels.
2. SO_2 is absorbed by the nasal system, leading to swelling and stimulated mucus secretion. It severely affects the aged and chronically ill persons.
3. Lung cancer is known to result from increased levels of SO_2 in the atmosphere.
4. SO_2 inhalation causes bronchitis, emphysema and other lung diseases. The intensity increases with the increased atmospheric concentration of SO_2.
5. London or sulphurous acid smog, called the killer smog of January 1956, was formed when SO_2 concentration rose to 0.40 ppm. As a result, the mortality rate among the aged increased from 130 to 180 per day.
6. Moisture and fog enhance SO_2 danger due to the formation of H_2SO_3 and H_2SO_4 and also sulphates. H_2SO_4 is nearly 5 to 20 times as irritant as SO_2..
7. SO_2 is considered as the most serious single air pollutant causing health hazard, obstructing breathing.
8. Oxides of sulphur are the major contributors to lung diseases, cough and choking. Their increased concentration causes acute and chronic asthma.
9. SO_2 is a severe allergenic agent.

Effects of SO_2 on plants

Plants are relatively more sensitive to SO_2 than humans and animals. The threshold levels of SO_2 injury in plants are quite low as compared to man and animals. Some of the effects of SO_2 on plants are:

1. SO_2 damages vegetable crops and affects plant growth and nutrient quality of plant products.
2. Acute exposure to high levels of SO_2 kills leaf tissues causing leaf necrosis. The edges and area between leaf veins are severely damaged.

3. Chronic SO_2 exposure to plants causes bleaching of leaf pigments due to conversion of chlorophylls to phecophytin—reducing plant productivity.
4. Concentration of SO_2 even as much as 1.00 ppm. is injurious to trees causing chlorosis and dwarfing.
5. Susceptible species like cucumber, oats and spinach may be damaged by exposure to air containing 0.05 to 0.5 ppm. of SO_2 for 8 hours. Cotton, wheat, barley and apple are most sensitive to SO_2.
6. Plants are mostly injured from SO_2 when their stomata are open during the day time.

Effects of SO_2 on materials

1. SO_2 also rapidly attacks marble, limestone, roofing, slate, electrical contacts, paper, textiles and buildings. It can even dissolve nylon. Some textile fibres obtained from vegetable sources lose strength when exposed to H_2SO_4. However, wool is somewhat more resistant to SO_2.
2. Paper also absorbs SO_2 which is oxidised to H_2SO_4, causing it to become brittle and fragile.
3. Leather too has much affinity towards SO_2, which affects its strength and causes it to disintegrate.
4. SO_2 is also involved in erosion of building materials such as marble, mortar and in the deterioration of statues.
5. Pollutant emissions of SO_2 from the nearby railway marshalling yard, thermal power stations and about 300 foundries are accelerating the deterioration of world famous Taj Mahal in Agra. Pollutants from Mathura refinery may also cause serious damage to the Taj Mahal.
6. Petroleum refineries, craft paper mills, smelters and industries liberating SO_2 adversely affect historic monuments.
7. Acid rain produced by the oxidation of SO_2 corrodes metals, attacks fibres and washes out basic materials like lime from the soil. The rapid attack of H_2SO_4 on marble is known as 'stone leprosy'.
8. Long exposure to SO_2 increases the drying and hardening time of paints. It affects durability in paint films. For example, exposure of linseed oil paint films 1–2 ppm SO_2 increased drying times by 50–100%.
9. SO_2 polluted air accelerates corrosion rates of metals such as Fe, steel, Zn and Cu. High humidity, particulate matter and temperature also enhance the corrosion of metals.

3.3.3 Effects of hydrogen sulphide (H_2S) and organic sulphides

1. Mostly sulphides cause odour nuisance when present even in minute concentrations.
2. H_2S causes headache, nausea, collapse, coma and death even at 1–3 ppm.
3. H_2S at 5 ppm affects the digestive system destroying appetite.
4. An exposure to 150 ppm of H_2S gives rise to conjunctivitis and irritation of the mucos membrane.
5. H_2S gas rapidly passes through the alveolar membrane of lungs and penetrates blood. It causes death due to respiratory failure.
6. Even a short exposure, for 10–30 minutes at 500 ppm, of H_2S causes colic diarrhoea and bronchial pneumonia.
7. H_2S reacts with lead paints to form lead sulphide, thereby producing brown to black discolouration.

3.4 Oxides of carbon as pollutants

CO, a significant contaminant is produced in the atmosphere by natural processes, i.e., forest fires, natural gas emission, marsh gas production and volcanic actions contribute to form CO. However, human activities, mainly automobile exhausts, contribute significantly (about 80%) to CO emission. Its concentration varies depending upon the density of vehicular traffic.

3.4.1 Sinks of carbon monoxide (CO)

1. The possible sinks which oxidise CO into CO_2 in the atmosphere are atomic oxygen, hydroxyl radical, NO_2, N_2O, O_3 and excited oxygen molecules.
2. The major CO sinks are some soil micro-organisms.
3. Plants are world's natural pollutant sink of CO and CO_2. They fix and metabolise CO also with the help of chlorophyll in the presence of light. This is known as photosynthesis and even in the dark, plants can carry on the fixation process non-photosynthetically.

CO is removed by converting it into CO_2 which then metabolises. Thus, the fixation of CO by plants is of immense importance as green plants are a major global sink of CO. This CO absorption by plants increases linearly with the increase of CO concentration.

Therefore, in cities where CO concentration is higher, the rate of CO absorption by plants may be greater by a factor of 10 to 100. Thus, plants play a vital role in the global CO sink.

Distribution and concentration of CO

Although CO sinks absorb five times the annual discharge of CO into the atmosphere, considerable concentration of CO still exists in the atmosphere. The reason is that neither CO nor the soil sinks are distributed uniformly. Even the largest CO producing areas have the least amount of soil sink available. Higher CO levels are generally observed in areas of heavy vehicular traffic, as the automobiles constitute the single largest (about 59.2%) CO producing source.

Effects of CO pollutants

1. On plants: CO has some detrimental effects on plants, when they are exposed to CO for long duration. For example, it inhibits the nitrogen fixation ability of bacteria when they are exposed to CO levels of 2000 ppm for 33–38 hours. Similarly, nitrogen fixing ability of bacteria living in clover roots is inhibited when exposed to a 100 ppm CO concentration for 35 days. Since the CO levels in the atmosphere normally does not reach 100 ppm it creates no significant effects on plants and vegetation. However, CO concentration from 100–10,000 ppm affects leaf drop, leaf curling, reduction in leaf size, synthesis and chlorophyll with premature ageing, etc.
2. On humans: All gaseous pollutants cause severe damage to the respiratory system. But the adverse effects of CO in the human body are unique. On inhalation, it passes through the lungs into the blood stream. It actually reacts with the haemoglobin (Hb) of the red blood corpuscles (RBC) forming a stable co-ordinated complex, called carboxy-haemoglobin, which restricts transport of oxygen from lungs to cells. A relatively low concentration of CO, because of its high affinity to haemoglobin, is able to displace considerable amount of oxygen forming carboxy haemoglobin complex (COHb).

Effects of carbon dioxide (CO_2)

Carbon dioxide is non-toxic, therefore it is not harmful to human health, unlike CO. CO_2 is utilised by green plants to prepare starch during photosynthesis. But today its increasing concentration in the atmosphere (10%) has long-term effects. It produces adverse physiological effects only at very high levels. The increased amount of CO_2 in air is mainly responsible for global warming. CO_2 molecules absorb heat energy and tend to prevent the long wave infrared heat radiation from earth from escaping into space and it deflects these radiations back to earth. The phenomena called 'atmospheric effect', 'green house effect' or 'global warming' has become a serious threat to global food production. The increased temperature may also lead to melting of polar ice caps and hence flooding of low lakes a caged bird into lying areas. A person may die in an atmosphere of CO_2, not because it is poisonous, but because of lack of oxygen. Birds and animals are more sensitive to CO_2. If a person takes a caged bird into a long deep cave, which is suspected to contain CO_2 in dangerous proportion, the bird will die first. This serves as a warning to the person entering the cave.

Mixed with oxides of sulphur and carbon monoxide, CO_2 gas becomes poisonous, pungent, suffocating and causes diseases like chronic asthma and bronchitis. Mixed with CO, it causes headache, dizziness, vomitting and difficulty in breathing.

3.5 Hydrocarbons as pollutants

Among a variety of hydrocarbons involved in air pollution, 56 have been clearly identified by making use of the technique of gas chromatography.

Natural sources, particularly trees, emit huge quantities of hydrocarbons in air. Methane has mean residence time of 3 to 7 years in the atmosphere. Human activities contribute nearly 20% of the hydrocarbons emitted to the atmosphere every year and animals contribute about 80–85 million tonnes of methane in the atmosphere every year.

About 45% of reactive hydrocarbons in the atmosphere originate from the combustion of gasoline while 15% are produced by other pollutant sources. Emissions are reported to occur from the crank-case of an automobile also. Automobile exhausts emit maximum hydrocarbons in the atmosphere.

Industrial processes like processing, storage and transfer of products, etc., constitute the next largest source of hydrocarbons. The evaporation of organic solvents leads to about 10% hydrocarbons in the air during industrial operation. These solvents are ingredients of paints, varnishes, lacquers, undercoatings and other products. During their manufacture, a large quantity of reactive hydrocarbons are emitted into the atmosphere.

Gasoline hydrocarbons evaporate and get mixed with air, increasing their concentration in the atmosphere. Hydrocarbons in the presence of nitrogen oxides (NO_x) are mainly responsible for the formation of photochemical smog. Photochemical reactions initiate most of the chemical processes, culminating in the removal of hydrocarbons from the atmosphere. The chief culprit in photochemical smog is NO_2.

3.5.1 Effects of hydrocarbons

Effects on human beings

1. Hydrocarbons at high concentration (500–1000 ppm) have carcinogenic effects on lungs, mainly causing swelling when they enter the lungs.

2. Aromatic hydrocarbons like benzene, toluene; etc., are more dangerous than acyclic and alicyclic hydrocarbons. The inhalation of their vapours causes much irritation to the mucous membrane. However, their different levels create different types symptoms in the body.
3. Secondary pollutants (PAN) produced by hydrocarbon and NO_x results in the formation of photochemical smog which causes irritation of eyes, nose, throat and respiratory distress.
4. Excess of hydrocarbon increases mucous secretion as a result of which respiratory tracts are blocked and a person coughs continuously, exerting pressure on the trachea of lungs due to which the lining membrane of alveoli bursts. So very little area is left for exchange of oxygen and carbon dioxide.
5. Benzpyrene, which is present as trace amounts in tobacco, charcoal, boiler stacks and gasoline exhausts, etc., is a dangerous cancer inducing hydrocarbon pollutant.
6. Methane (marsh gas) is a severe gaseous pollutant and occurs in air by volume to the extent of 0.0002%. Its higher levels in the absence of oxygen create narcotic effects on human beings.
7. A group of hydrocarbons, especially carcinogenic hydrocarbons, cause cancer in man and animals affecting DNA and cell growth. The effects of carcinogens on animals have been studied by US Occupational Safety and Health Administration.

Effects on plants

1. Hydrocarbons and photochemical oxidants are injurious to plants. Exposure to high levels of ozone to plants causes chlorosis, i.e., yellowing of green portion of leaves.
2. Ozone enhances plant injury creating light flecks or stipples (clusters of dead cells) on the upper leaf surface inhibiting photosynthetic activity.
3. Ethylene even at 1 ppm concentration shows adverse effects on vegetation.
4. Acetylene and propylene at 50–500 ppm show extreme toxicity towards plants damaging their growth.
5. Ethylene hydrocarbons inhibits plant growth and damages leaf tissues and flowering plants.

Effects on materials

1. Even low levels of ozone induces chemical alteration in natural synthetic textiles, paper, rubber and polymers. Higher the number of carbon-to-carbon double bonds in the material, greater is the susceptibility of their attack.

2. Hydrocarbon pollutants damage long chains of carbon atoms in a polymer, which loses its tensile strength.
3. Ozone forms new carbon chain links between parallel carbon chains so that the material becomes less elastic and more brittle.

Control of hydrocarbons

Hydrocarbons and NO_x produce PAN and O_3, etc., which are chronic secondary pollutants. So their control ultimately depends on the control of the primary precursors, i.e., hydrocarbons and NO_x which are the main culprit of air pollution. (Refer to NO_x control). Hydrocarbons from auto-exhaust emissions can be controlled by applying the techniques like incineration, absorption, adsorption and condensation, etc. By adopting these methods, all the three pollutants (hydrocarbons, NO_x and CO) can be converted to less harmful end products.

3.6 Control and removal of SO_x pollutants

There are six procedures for controlling SO_x emissions. They are either in-plant control measures or effluent treatment methods.

1. Natural dispersion by dilution.
2. Using alternate fuels.
3. Removal of sulphur from fuels (desulphurisation).
4. Process modifications.
5. Control of SO_x in the combustion process.
6. Treatment of flue gas emissions.

These different control methods are described further.

3.6.1 Natural dispersion by dilution

The control method is based on natural dispersion at high elevation so that the ground-level concentrations are acceptable at all times. In India, minimum stack heights of 30 m are recommended.

It is also a common practice to stop discharging the effluents into the atmosphere during adverse meteorological conditions. However, this technique is not possible in large-scale power plants, etc.

3.6.2 Using alternate fuels

A switch to natural gas from the conventional high-sulphur fuels like coal and petroleum to lessen SO_x emissions is an available alternative. Liquefied natural gas, LNG, also is an effective alternative. However, for utility use, its cost will be much higher than that of other alternatives. Low sulphur coal is another alternative, but obtaining low sulphur coal from the ground is neither quick nor cheap.

3.6.3 Removal of sulphur from fuels

The process of removing sulphur prior to combustion is theoretically attractive but practically ineffective. Coal consists of sulphur in both organic and inorganic forms. The inorganic form of sulphur is iron disulphide (FeS_2) mostly available in the forms of pyrites and marcasites. Apparently, washing seems to be an effective process with more than 30% of sulphur being removed. But this results in a loss of combustible material and may increase the requirement of coal and thus the cost. Organic sulphur is present in the form of crystine, thiols, sulphides and some other cyclic compounds, which can only be removed by chemical processing.

Hydro-desulphurisation of coal using a solvent extraction process can remove both organic sulphur as well as inorganic sulphur. In this process finely ground coal is mixed with anthracene oil to form slurry, is heated at a very high temperature of 450°C. The ash residue consisting of both organic and inorganic forms of sulphur is eliminated by pressure filteration. To avoid repolymerisation a small amount of hydrogen is also added. The coal solution filterate is sent into a chamber to remove lighter fractions. The hot liquid residue is cooled to a brittle solid fuel which can be pulverised. The product will be liquefied at about 450°C. It has a higher heating value than raw coal and contains less than 1% sulphur.

Coal gasification is another process widely used all over the world. There are several processes to convert coal into the gaseous form. The earlier processes like water-gas and producer-gas processes produced gas of lower quality. The new processes give a better quality, besides removing both organic and inorganic types of sulphur.

Catalytic hydrogenation of coal suspended in tar at 100–250 bars at 450°C can achieve 75% desulphurisation with the consumption of 20 kg hydrogen per tonne of feed coal. About 5% of this hydrogen is transformed to hydrogen sulphide, an unwanted by-product. Similarly, impure natural gas can be freed of hydrogen sulphide by scrubbing with monoethanolamine or other amines. The sulphur can be recovered by dry catalytic conversion when molten sulphur is obtained. Usually 0.5 to 5% of sulphur is present in heavy fuel oils. The hydro-desulphurisation process is used for desulphurisation of fuel oil. In this process, temperatures are maintained between 400°C to 550°C. The fuel oil is treated with hydrogen in the presence of a catalyst. The residue is eliminated by pressure filtration, with pressures ranging from 35 to 70 atmospheres (atm). The sulphur is recovered by sending it into a flash evaporator where lighter fractions are removed. Almost all coals dissolve in solvent oils at high temperatures (550°C) and pressures (75 bars) in the presence of hydrogen which prevents polymerisation and helps in the removal of organic sulphur.

Coal gasification appears to be promising for the abundant coal reserves in India. Oil gas can be scrubbed free of sulphur and can be used in gas-burning devices.

Process changes: Process changes involve new or modified techniques lowering atmospheric pollutant emissions.

Control of SO_x in the combustion process

Here, finely powdered limestone is injected directly into the conventional combustion chamber. The limestone is calcinated to CaO and it reacts with SO_2 contained in the flue gas to form sulphites and sulphates. The unreacted materials and flyash are removed by dry collectors.

$$CaCO_3 \rightarrow CaO + CO_2$$

$$CaO + SO_2 \rightarrow CaSO_3$$

$$CaO + SO_2 + \tfrac{1}{2}O_2 \rightarrow CaSO_4$$

The formation of $CaSO_4$ is most favoured at temperatures above 1000°C. Due to smaller residence time in combustion zone, limestone does not completely react with SO_2 to produce a stoichiometric yield of $CaSO_4$. At temperatures above 1200°C, $CaSO_3$ is unstable and SO_2 removal by sulphite formation is inhibited. As a result, the process removes less than half of the sulphur oxides. Due to this, the dry limestone injection process, although relatively simple and easy to operate, is not very attractive from the point of view of control technology. In the widely used fluid bed combustion process, limestone and crushed coal together form the fluidised bed and oil is used as the fluidising medium.

The operating temperatures are of the order of 700–1000°C, since the fluidised beds are capable of transferring high heat. This process proves to be quite effective as it removes more than 90% of sulphur. In addition to high SO_2 removal efficiencies, the fluidised bed combustion process prevents the onset of ash fusion and as a result, the fouling and corrosion of boiler tubes associated with the molten slag are considerably reduced. Also, the formation of nitrogen oxides by the nitrogen fixation reaction are reduced. The method, however, has some drawbacks. It requires design modification for boilers and also additional installations for the preparation of limestone. If the limestone is ground to the same size as coal, then a practically inseparable mixture of ash and lime is produced. The fine ash can be elutriated and partially sulphated lime is regenerated.

$$2CaSO_4 + C \rightarrow 2CaO + 2SO_2 + CO_2$$

This regeneration, which requires temperatures in excess of 1000°C will substantially suppress the consumption of limestone.

Treatment of sulphur from flue gas emissions

It seems unlikely that a single desulphurisation method will be developed that is capable of controlling effluents from all types of sources. The control techniques to be employed depend upon such factors as boiler size, configuration, load pattern, geographical location and the like. Nearly 50 flue gas desulphurisation processes have been proposed, but as such no ideal process exists. The general classification of these processes may be wet and dry.

Dry methods

The dry methods can be divided, broadly, into two types:

1. Oxidation/reduction methods.
2. Metal oxides usage.

Oxidation/Reduction

CatOx process. This method of oxidation produces sulphuric acid. In this process, fly ash is first removed from the flue gas by a high temperature electrostatic precipitator (Fig. 3.6). SO_2 is then catalytically oxidised to SO_3 and recovered as sulphuric acid. If the exit gases are at a lower temperature, they may be heated. V_2O_5 at 400–500°C is used as a catalyst for good conversion efficiency. In a modified oxidation process, SO_2 and oxygen present in the stack gas are absorbed on the surface of an active carbon catalyst, which catalyses the oxidation of SO_2 to SO_3. SO_3 reacts with the moisture present to form H_2SO_4 in the pores of active carbon.

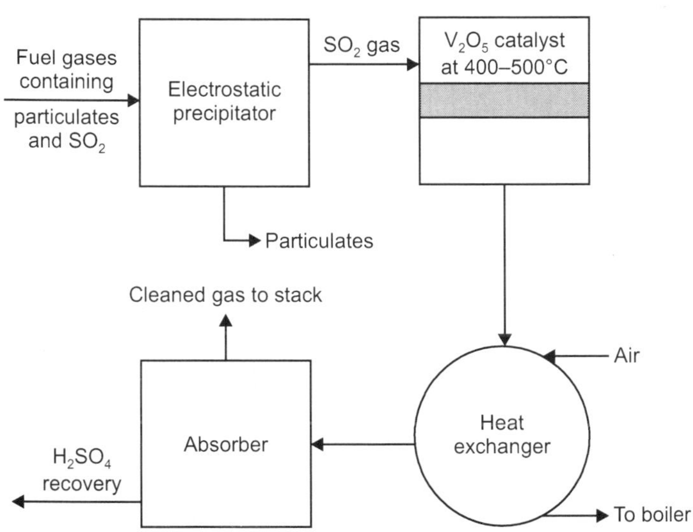

Figure 3.6: CatOx process of removal of SO_x.

The combined effect of absorption and catalysis by the active carbon leads to a complete conversion of SO_2.

Westvaco process. This process also involves oxidation/reduction. This process is unique in that it utilises fluidised beds of high efficiency activated carbon and converts H_2SO_4 to sulphur. Flue gas is contacted with activated carbon in the absorber unit and SO_2 is oxidised to SO_3.

The carbon which is used as a catalyst is fed to a sulphur generator where it is contacted with H_2S to form sulphur.

$$H_2SO_4 + 3H_2S \rightarrow 4S + 4H_2$$

A fraction of the sulphur is recovered by vapourisation and is recondensed as a molten product. The remaining fraction of sulphur reacts with hydrogen in a hydrogen sulphide generator to form H_2S.

SCOT process: In the SCOT (Shell Clauss Off-gas Treatment) process, developed in the Netherlands, almost complete removal of sulphur or its compounds is possible. The gases are reduced while sulphur and its compounds are converted to H_2S. In this process, SO_2 reacts with copper oxide at a temperature of about 400°C to form copper sulphate.

$$SO_2 + \tfrac{1}{2}O_2 + CuO \rightarrow CuSO_4$$

Copper sulphate is then reduced at the same temperature in a hydrogen rich gas:

$$CuSO_4 + 2H_2 \rightarrow Cu + SO_2 + 2H_2O$$

The concentrated SO_2 is then sent to a Clauss sulphur recovery plant. The Clauss off-gas is heated with the addition of a reducing gas (H_2,CO) and passed through a reactor containing a cobalt-molybdenum catalyst. The absorber used for removal of H_2S is often a diisopropanolamine (DIPA) solution. SO_2 concentrations of less than 250 ppm can be obtained with this process.

Use of metal oxides

In this process, sodium aluminate ($Na_2O.Al_2O_3$) is used to remove SO_2 in a fluidised bed. The dust free flue gas is fed to a reactor wherein the absorbent, a porous form of sodium aluminate ($Na_2O.Al_2O_3$), adsorbs SO_2 at a temperature of 300°C.

$$Na_2O.Al_2O_3 + SO_2 + \tfrac{1}{2}O_2 \rightarrow Na_2SO_4 + Al_2O_3$$

The product from the above reaction is contacted with a reducing gas such as hydrogen in a regenerator at about 680°C to form hydrogen sulphide.

$$Na_2SO_4 + Al_2O_3 + 4H_2 \rightarrow Na_2O.Al_2O_3 + H_2S + 3H_2O$$

From this, the sodium aluminate is recycled and hence the process is called a cyclic adsorption process.

Sulphur can also be produced as a by-product from the H_2S gas by sending it into a Clauss unit.

$$2H_2S + SO_2 \rightarrow 3S + 2H_2O$$

However, maintenance of the granular strength of sorbent is the main problem relating to this process. The rigorous temperature and chemical cycling to which the sorbent is subjected, deteriorate the sorbent by causing crystalline growth and loss of surface area.

Wet processes

The main unit responsible in a wet process is the scrubber, a spray tower, a cyclone scrubber, a venturi scrubber or a packed tower. Most of the currently available wet flue gas desulphurisation methods use slurries of compounds of calcium, magnesium and sodium. In the wet process, the treated gases are kept at low temperatures in the range of 25–50°C. This creates a problem in the dispersion of the flue gas. Hence, reheating is required which consumes heat in the range of 3% of the heat of fuel used in the combustion chamber. Some of the wet processes are described below:

Calsox process (lime and limestone scrubbing): Here, the flue gas is scrubbed with a 5 to 15% slurry of calcium sulphite/sulphate salts which also contains some amounts of lime (CaO) and limestone ($CaCO_3$). The SO_2 reacts with the slurry to form additional sulphite and sulphate salts. The solids are continuously separated from the slurry and discharged into a settling pond. The remaining liquor, at a pH of 6 to 8, is recycled to the scrubbing tower after fresh lime or limestone has been added. A schematic representation of the process is shown in Fig. 3.7.

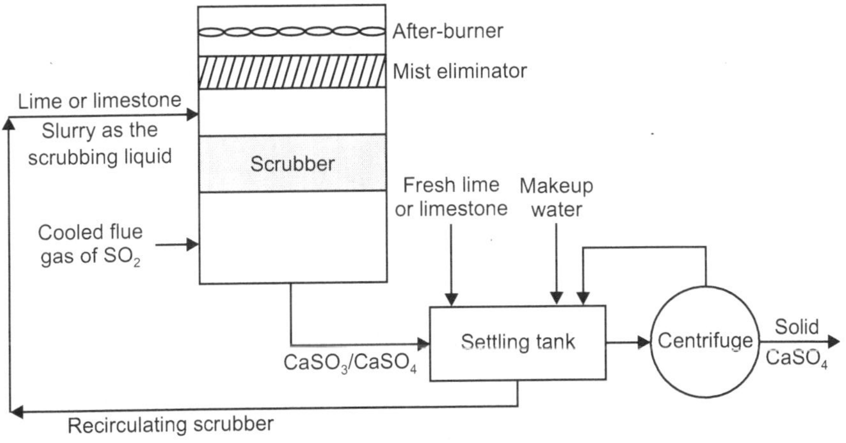

Figure 3.7: SO_x controlled by lime/limestone scrubbing

The problems in this process include scaling, corrosion, erosion and solid waste disposal. A sizeable disposal area adjacent to plant is required. Another drawback of the process is the necessity for reheating the cleaned flue gas. Reheating is accomplished by installing a gas cooler before the scrubber and a gas stack heater after it. Thus, two additional units must be provided to the gas flow system.

Chemico process (magnesium oxide scrubbing): In this process, known as the 'Chemico Process', magnesium oxide acts as a venturi scrubber by absorbing SO_2 and generating magnesium sulphite and sulphate. The mixed sulphite/sulphate slurry along with unreacted MgO is separated from the liquid phase in a centrifuge. This process is a regenerative one as the mother liquor is regenerated. Concentrated SO_2 (10 to 15%) removed from the flue gas is used in making elemental sulphur or reprocessed to manufacture H_2SO_4.

Welman lord (single alkali) process: This process removes sulphur dioxide by washing fuel gases with an aqueous solution of sodium sulphite. It is quite a common practice in chemical industries, in which 90% of desulphurisation is possible. In this process, sulphite is converted to bisulphite as the SO_2 from flue gases is absorbed by a saturated solution of sodium sulphite. The sodium bisulphite formed is led to a double effect evaporator cum crystalliser where it decomposes into sodium sulphite at a temperature of around 100°C. This results in the ejection of SO_2 and steam.

Clear solutions of either sodium or ammonia are excellent absorbers of SO_2. The regeneration step can be carried out at a relatively low temperature in a liquid system. The one advantage that sodium scrubbing has over ammonia is that the cation is non-volatile. Fume development is another problem in almost all ammonia scrubbers. Both processes produce an unavoidable side product, sodium sulphate in one case and ammonium sulphate in the other.

CuO/CuSO$_4$ process: This process removes NO_x and SO_x simultaneously by using copper oxide (CuO) supported on stabilised alumina placed in two or more parallel passage reactors.

Flue gas is introduced at about 400°C into one of the reactors where SO_2 reacts with CuO to form copper sulphate ($CuSO_4$). The $CuSO_4$ and to a lesser extent, the CuO act as catalysts in the reduction of NO_x with NH_3. When the reactor is saturated with $CuSO_4$ flue gas is switched to a fresh reactor for acceptance of the flue gases and the spent reactor is regenerated. In the regeneration cycle, hydrogen (H_2) is used to reduce $CuSO_4$ to copper (Cu), yielding an SO_2 stream of sufficient concentration for conversion to sulphur or sulphuric acid.

American smelting and refining company (ASARCO) process: In this process, emissions of SO_x are controlled by absorption of flue gases in dimethylaniline. The desulphurised gas is further sent through sodium carbonate solution and then to diluted H_2SO_4, where the gas is scrubbed for traces of sulphur and dimethylaniline, if any. The SO_2 absorbed dimethylaniline solution is passed through a steam distillation column where dimethylaniline is treated and recycled. The tail gas left in the stack after this scrubbing process consists of SO_2 of about 500 ppm concentration. This process can be used for SO_2 gas concentrations of more than 3%.

Consolidated mining and smelting company (COMINCO) process: In this process, SO_2 is removed by using aqueous ammonia solution in zinc plants. This process can be used for gas concentrations of 5% to as low as 0.3%. In this process, a 30% aqueous ammonia solution absorbs SO_2, producing ammonium bisulphite (NH_4HSO_4). Then in a stripping column, the NH_4HSO_4 solution is added to 95% H_2SO_4 which strips the solution and ammonium sulphite is produced. This process has a removal efficiency of about 95% in zinc roasters.

3.7 Oxides of nitrogen as pollutants

Of the six or seven oxides of nitrogen known, only three—nitrous oxide (N_2O), nitric oxide (NO) and nitrogen oxide (NO_2)—are formed in any appreciable quantities in the atmosphere. Often NO and NO_2 are analysed together in air and are referred to as NO_x. The various reasons for the increase of oxides of nitrogen in the ambient atmosphere may be the increased emission of NO_x from industries and automobiles and various other sources. Well over 90% of all man-made nitrogen oxides that enter our atmosphere are produced by the combustion of various fuels.

The real danger posed by NO_x at the concentration found in metropolitan areas lies in its role in photochemical reactions leading to smog formation. These atmospheric reactions lead to the formation of chemical compounds that have a direct adverse effect on human beings and plants. In some situations, NO_x may be present in a high enough concentration, yet not react to form smog because other necessary conditions for the reaction are absent. However, nearly every major city in India experiences the effects induced by the presence of NO_x.

The actual quantities of NO_x produced by any given industry can be quite large. For example, a 750-MW gas or coal-fired power plant produces around 75 to 100 tonnes of NO_x per day. The type of fuel used can change the amount of NO_x released significantly.

About 60% of NO_x is contributed by fuel combustion in the stationary sources and about 40% by transportation. The stable gaseous oxides of nitrogen include N_2O (nitrous oxide), NO (nitric oxide), N_2O_3 (nitrogen trioxide), NO_2 (nitrogen dioxide) and N_2O_5 (nitrogen pentoxide). An unstable form, NO_3

also exists. Of all these oxides of nitrogen, the only ones present in the atmosphere in any significant amount are N_2O, NO and NO_2. These three are the potential contributors to air pollution. The main chemical reactions in the formation of nitrogen oxides are:

$$\tfrac{1}{2}N_2 + \tfrac{1}{2}O_2 \rightarrow NO \text{ and } NO + \tfrac{1}{2}O_2 \rightarrow NO_2$$

Nitric oxide (NO) is a colourless gas and its ambient concentration is usually far less than 0.5 ppm. At this concentration, its biological toxicity in terms of human health is insignificant. N_2O is an inert gas with an anaesthetic characteristic. Its ambient air concentration is 5 ppm. It has a balanced environmental cycle which is independent of other oxides of nitrogen.

However, nitric oxide is a precursor to the formation of nitrogen dioxide and is an active compound in photochemical reactions leading to air pollution. NO_2 (nitrogen dioxide) is a reddish brown gas and is quite visible in sufficient amounts. NO_2 directly affects human health only if it is present in extremely high concentrations. Exposure to 15 ppm of NO_2 causes eye and nose irritation and pulmonary, discomfort is noted at 25 ppm for exposure of less than 1 hour. NO_2 concentration of 1 ppm can be detected by the eye. The threshold limit in ambient air is considered to be 5 ppm for daily exposure. In general, the concentrations of oxides of nitrogen are expressed under the formula NO_x (or on an 'equivalent NO_2' basis).

3.7.1 Occurrence of nitrogen oxides in the atmosphere

These oxides occur in atmosphere as follows:

1. NO: It is the main product of combustion of nitrogen and automobile exhaust produced by the combustion of gasoline. It is slowly oxidised to NO_2 by O_2 but rapidly by ozone.
2. N_2O: Present in air at concentration level of 0.25 ppm. Maximum level is 0.5 ppm. It is not a product of combustion.
3. NO_2: In atmosphere NO_2 levels are about 0.001 ppm. It is a strong absorber of ultraviolet light and is the chief constituent of photochemical smog. It initiates photochemical reactions in the troposphere.
4. N_2O_3: It reacts with water vapour to form HNO_3, which combines with ammonia to form ammonium nitrate.
5. N_2O_5: Forms HNO_3 with water and thus reduces the pH of rain water.

3.7.2 Sources of NO_x pollution

Natural stratospheric NO_x are also produced by the action of cosmic rays in the upper atmosphere. Man-made sources of NO_x varies depending upon global areas. NO_x are 10 to 100 times greater in urban atmosphere as compared to rural areas. Major man-made activities include combustion of coal, oil, natural gas and gasoline, etc., which produce up to 50 ppm. of nitrogen.

NO_x are also produced as by-products of some chemical industries like nitric acid and sulphuric acid industry. They are also formed during the manufacture of nylon intermediates.

Another potential serious source of man-made stratospheric NO_x involves rising fire balls associated with atmospheric nuclear explosion. Abundant fluxes of nuclear produced NO_x are immediately transported to high altitudes by aerodynamic forces created by the explosion itself.

The average residence time of NO and NO_2 in the atmosphere are four days and three days, respectively. This indicates that natural processes, including photochemical reactions, take care of NO_x, the product being HNO_3. The latter is precipitated as nitrate salts in either rain or as dust.

$$O_3 + NO_2 \rightarrow NO_3 + O_2$$

$$NO_3 + NO_2 \rightarrow N_2O_5$$

$$N_2O_5 + H_2O \rightarrow 2HNO_3$$

The end product of NO_x, however, is HNO_3. The latter reacts rapidly to form various particulate nitrates.

Effects of nitrogen oxides (NO_x)

Mostly the oxides of nitrogen are not so dangerous, but the role they play in the formation of photochemical oxidants, etc., constitute the most harmful effect. These NO_x affect plants, human health and the atmosphere.

Effects of NO_x on plants

1. Higher concentrations of NO_2 damage leaves of plants, retard the photosynthetic activity and cause chlorosis.
2. Plants exposed to 100 ppm of NO_2 suffer from leaf spotting and breakdown of tissues.
3. Exposure to 10 ppm of NO checks the metabolic activities in plant tissues, e.g., bean and tomato plants on fumigation showed decreased activity of CO_2 absorption and photosynthetic rate.
4. Damage to vegetation probably results from the production of secondary pollutants such as O_3, PAN and smog, etc.
5. NO_2 is highly injurious to plants.

Effects of NO_x on human health

1. Acute toxicity has, however, not been found in man due to NO_x, NO is biochemically inert and not extremely toxic.

2. The health effects of NO_2 vary with the degree of exposure. An exposure to 50 to 100 ppm of NO_2 for 30 to 50 minutes for a period of 5 to 8 weeks causes inflammation of lung tissues.
3. NO_2 has irritating effects on mucous membrane.
4. Higher doses of NO_2 cause bronchitis and respiratory problems.
5. An exposure to 500–600 ppm of NO_2 for 2 to 10 days results in death of the victim.
6. Death often results due to irritation caused by gases containing NO_x from burning celluloid and nitro cellulose films.
7. Cigarette smokers readily develop lung diseases as cigarettes contain 330–1500 ppm of NO_x.
8. NO_2 lowers the resistance to influenza and irritates eyes.
9. Nitric, nitrous acid and several nitrates cause respiratory, digestive and nervous ailments.
10. NO gets attached to haemoglobin and reduces oxygen transport efficiency of blood.
11. Higher levels of NO_x cause gum inflammtion, internal bleeding, pneumonia, oxygen deficiency and lung cancer, etc.
12. Both NO and NO_2 are fairly toxic at low concentrations.
13. Oxides of nitrogen are the second-most abundant atmospheric pollutants. These are extremely dangerous to human health. Their acute effects are sometimes more severe than that of carbon monoxide.

Effects of NO_x on materials

1. NO_x fades away a number of textile materials like cotton, rayon and acetates. It is found that NO_x level reaches to 1 to 2 ppm, during the combustion of natural gas which is used to heat dryers.
2. However, damaging effects of NO_x on various textile fibres is not much pronounced. Higher levels of NO_x cause 10% loss of fibre strength in cotton and rayon.
3. NO_2 produces aerosols which damage nylon fibres.
4. Abnormally high levels of NO_2 present in ambient air, during dynamite blast operations in New York had adverse effects on nylon stockings and cotton.
5. NO_2 along with hydrocarbons produces peroxides, which combine with ozone and cause cracks in rubber.
6. Oxides of nitrogen are involved in the generation of serious air pollutants such as ozone, PAN and peroxy organic compounds, which are toxic at

0.1 ppm levels and responsible for photochemical smog. Hydrocarbons and NO_x are the key factors in multiplying photochemical smog problems.

7. Higher concentrations of particulate nitrates in airborne dust accumulate adjacent to cracked areas of metals like nickel and brass.

NO_x and acid rain

Oxides of nitrogen (NO, NO_2 and N_2O) through photochemical chain reactions, produce irritating gases which are toxic and corrosive. These gases produce nitirc acid which is washed down and contributes to acid rain. Acidity of rain water increases when the amount of oxides of nitrogen and sulphur increases in the atmosphere to produce more nitric and sulphuric acid. Acid rain reduces fertility of the soil. It is also affecting the Taj Mahal in Agra. NO_2 is a part of the photolytic NO_2 cycle.

Effect of NO_x on the stratospheric atmosphere

In the stratosphere, SST (super sonic transport), airplanes produce NO_x which catalyses the stratospheric ozone layer depletion. Ozone absorbs harmful ultraviolet (UV) radiations near 250–300 nm and disintegrates into oxygen and atomic oxygen:

$$O_3 \xrightarrow{hv} O_2 + O \quad O_2 \xrightarrow{hv} O + O$$

atomic oxygen reacts with O_2 in presence of energy absorbing bodies (M):

$$O_2 + O + M \rightarrow O_3 + M$$

Ozone reacts with NO_x in the stratosphere

$$NO + O_3 \rightarrow NO_2 + O_2 \quad NO_2 + O \rightarrow NO + O_2$$

Thus, a chain reaction decreases the concentration of O_3 allowing the chronic uv radiation to reach the earth's surface.

$$O + O_3 \rightarrow O_2 + O_2$$

The cycle, alongwith other air pollutants, is responsible for several adverse effects in the atmosphere.

3.7.3 Control of NO_x pollution

NO_x emissions from flue gases can be reduced by: (i) dilution in the atmosphere by increasing stack height or by discharging effluents into the atmosphere only during favourable meteorological conditions, (ii) modification of operating and design conditions and (iii) treatment of flue gases. The method of control of NO_x by dilution is similar to that of SO_x, particulates and other pollutants.

NO_x control by modification of operating and design conditions

On the basis of the thermodynamics and kinetics involved in the formation of NO_x, the following in-plant measures can be adopted to reduce NO_x emissions.

These are based on the principles that NO_x emissions can be reduced by: (i) reducing the peak temperature, (ii) reducing the residence time at the peak temperature and (iii) reducing the availability of oxygen, i.e., by maintaining higher N/O ratios. The in-plant control measures are:

Low excess air combustion: The presence of excess air affects both the temperature and oxygen concentration of gases in the post-combustion zone. Normal boiler units operate in the presence of 10–20% excess air to ensure complete combustion of the fuel, but this excess air provides enough oxygen for reacting with nitrogen.

Decreasing combustion air temperature: In many industries, waste heat is available to help pre-heat the air entering a combustion process. Although this process leads to appreciable energy savings, the added energy increases the flame temperature. Thus, NO_x emissions increase. Data from full-sized boiler tests indicate a three-fold increase in NO_x emissions when combustion air is pre-heated from 25° to 300°C. Since temperature has a major influence on nitric oxide formation, an effective method of control is cooling of primary flame zone by heat transfer to surrounding surfaces.

Two stage combustion: This is one of the most effective control methods for the suppression of the formation of nitrogen oxides. In the first stage, the fuel and air are burned at near—stoichiometric conditions. All the fuel is fired with only 85–95% of total air requirements at the bottom of the furnace followed by secondary air injection higher up in the furnace to complete the combustion. Thus, in the primary zone, incomplete combustion of the fuel takes place resulting in reduced concentrations of NO_x in furnace gases. Heat removal and gas dilution between zones cause the temperature of the gases after the primary zone to decrease so that the final stage of combustion process occurs at a lower temperature. Reduction in the emission of NO_x by 38% for coal and oil firing and by 50% for natural gas combustion has been observed under these conditions.

Flue gas recirculation: The recirculation of coal flue gas into the combustion chamber is found to be quite effective in reducing NO_x formation from stationary sources. A portion of the cooled flue-gas is injected back into the combustion zone. This additional gas acts as a thermal sink and reduces the overall combustion temperature. In addition, the oxygen concentration is lowered as the flue gases going out have a higher N/O ratio, i.e., lower oxygen availability. Both these effects favour a reduction in NO_x emissions. One disadvantage is the increased cost of duct work, since large volumes of gas are mixed with the primary air prior to combustion.

Injection of water and steam: Injection of water or steam into the combustion zone is also effective in reducing emissions of nitrogen oxides. This method is similar to that in flue gas circulation. Injecting water to the extent of 10–15% of the total combustion air can reduce the temperature level in the furnace to some extent.

Modification of furnace burner configuration: The two basic furnace designs are the tangential and the horizontal methods of firing. In the tangential method of firing, the flame and the combustion products rotate in an upward spiral around the walls of the furnace and the furnace itself is used as the burner. This results in low peak flame temperature and consequently reduction in NO_x emissions. In the horizontal firing, the flame is at right angles to the walls of the fire box. This tends to concentrate the hot gases, leading to higher flame temperatures and more NO_x emissions.

3.7.4 NO_x control by treatment

The modification of design and combustion techniques, in general, is to prevent or to suppress the emission of nitrogen oxides in the combustion chambers. However, if the measures are not enough or too costly, stack gas treatment is necessary. Stack gas treatment is an efficient method for the control of NO_x. The methods are: (i) absorption by liquids, (ii) adsorption by solids, (iii) catalytic reduction—selective and non-selective and (iv) electron beam irradiation.

Absorption by liquids

The oxides of nitrogen can be absorbed by water, hydroxide and carbonate solutions, sulphuric acid, organic solutions and molten alkali carbonates and hydroxides. Several scrubbing techniques were developed which initially used solutions of sodium and calcium hydroxide. Some of the commonly adopted methods are:

1. Absorption by alkaline solutions: NO_x absorption by using aqueous alkaline solutions like NaOH and MgOH also yields good results. This is a common method adopted during desulphurisation of power plant emissions by such alkaline solutions. In the desulphurisation process, about 10% of NO is oxidised to NO_2 before the flue gas reaches the scrubber. The scrubber then removes about 20% of the total NO_x in equal parts of NO and NO_2.
2. Absorption by lime: Aqueous suspension of calcium hydroxide can be used as the scrubber to reduce NO_x levels to 200 ppm. The calcium nitrite in the solution can further be converted to more valuable calcium nitrate by treating with sulphuric acid.

   $$2H_2SO_4 + 3Ca\,(NO_2)_2 \rightarrow 2CaSO_4 + 4NO + Ca(NO_3)_2 + 2H_2O$$

   The NO evolved may be recycled to the nitric acid plant and calcium nitrate can be used as a fertiliser. Thus, the process reduces NO_x, recovers NO and gives a valuable fertiliser simultaneously.

3. NO_x absorption by magnesium hydroxide: In this process, the oxides of nitrogen are absorbed by hydroxide liquor in an absorption tower. The magnesium nitrite solution leaving the absorber is taken to a pressure reactor where the nitrite is converted to nitrate. The by-product NO is oxidised to NO_2. The liquid leaving the pressure reactor, consisting of Mg $(NO_3)_2$/Mg $(OH)_2$ is sent to a settling chamber where nitrate is separated from the hydroxide. Part of the NO_2 from the oxidiser is sent to the absorber to maintain equimolar concentrations of NO and NO_2, while the rest of NO_2 is used for nitric acid production.

Adsorption by solids

The adsorbents that show some capacity for oxidising NO and NO_2 and for adsorbing nitrogen dioxide are activated carbon, silica gel, molecular sieves, ion-exchange resins and certain metal oxides, particularly manganese and alkalised ferric oxides. The use of activated carbon to adsorb oxides of nitrogen has been studied extensively. Activated carbon has a high adsorption rate and capacity compared to other materials. However, regeneration may be a problem.

A potential fire and explosion hazard may be another difficulty with this material since O_2 is usually present in most stack gases. Thus, the efficiency of activated carbon decreases with quantity of O_2 present in flue gas. Manganese oxides and alkalised ferric oxides show technical potential. However, sorbent attrition is a major technical stumbling block. The most suitable adsorbent for NO_x is the one which can be regenerated and at the same time, which does not preferentially react with water vapour or with CO_2 in the flue gas. The most promising adsorbent is ferrous salt. Molecular sieves also can be used for NO_x control, particularly for NO_x from nitric acid plants. In this process, two beds operate batch wise—one adsorbs NO_x from the tail gas while the other is regenerated.

Catalytic reduction

Selective catalytic reduction (SCR): Selective catalytic reduction refers to a process that chemically reduces NO_x with NH_3 over a heterogeneous catalyst in the presence of O_2. The process is termed selective because the reducing agent NH_3 preferentially attacks NO_x rather than O_2. However, the O_2 enhances the reaction and is indeed a necessary part of the reaction scheme.

Non-selective catalytic reduction: Reducing agents other than NH_3, such as CO, H_2 or CH_4, completely reduce NO at temperatures of 300–400°C. Non-noble metal catalysts such as supported copper oxides may also be used for NO_x control. These reductants act non-selectively, reacting with O_2 and SO_x present in the flue gas, in addition to NO_x and thus huge amounts of reductants must be added. For non-selective catalytic reduction of NO_x in stationary combustion sources, stoichiometric or sub-stoichiometric air must be used in the primary combustion zone. This reduces the cost of the reducing agent. NO_x removal by CH_4 is as follows: When hydrocarbons like methane are used for removing NO_x, higher oxides of nitrogen are converted to NO in the first stage. In the second stage, NO is reduced to nitrogen. Oxygen in the tail gases react simultaneously with the hydrocarbons to produce water vapour and carbon dioxide.

Electron beam irradiation

This process simultaneously removes 90% of NO_x and SO_x with NH_3 activated by electron beam irradiation, without the need for catalysts. The process first requires removal of fly ash from the flue gas. Then ammonia is added and the gas enters the electron beam reactors. There, ammonia in the presence of electrons converts NO_x and SO_x into a dry powder of ammonium sulphate [$(NH_4)_2SO_4$] and ammonium nitrate sulphate [$(NH_4)_2SO_4$, 2 NH_4NO_3)]. After the dry powder is removed by a second particulate collection device, the treated flue gas leaves the system hot enough to be exhausted through the stack without reheat. The waste generated also is a potential fertiliser feedstock.

4

Odour pollution and its control

4.1 Introduction

Odour can be defined as the 'perception of smell' or in scientific terms as 'a sensation resulting from the reception of stimulus by the olfactory sensory system'. Whether pleasant or unpleasant, odours are induced by inhaling airborne volatile organics or inorganics.

Controlling odours is an important consideration for protecting the environment and our community amenity. It is possible to detect odours scientifically and measure their impact on the environment. Once this is done there are different methods of control that can be implemented, depending on the source of the odour and various other factors. Officers should call in expert consultants if they are at all unsure when assessing development approvals where odours may be a problem.

4.2 Odour nuisances

4.2.1 Odour from process plants

Odorants often appear in connection with chemical processes and form part of the process discharge, unless they are cleaned before emission from the plant. But the most thorough cleaning of discharges cannot prevent odour from a plant in which process equipment is badly maintained, with leakages through which odorous gas emitted in the atmosphere. Gas may also escape in connection with emptying/filling of tanks or with repair and cleaning of manufacturing equipment.

4.2.2 Odour sources

It is not possible to make a complete list of odour-producing plants. Below are given some examples of plants, processes and activities which are well-known emitters of odorous compounds and, thus, sources of offensive smell: (i) oil refinery, (ii) foundry, (iii) paper mill, (iv) plastics processing, (v) fertiliser production, (vi) oil seed factory, (vii) pharmaceuticals manufacture, (viii) detergent manufacture, (ix) tobacco factory, (x) sugar mill, (xi) waste incineration, (xi) refuse disposal, (xii) transport infrastructure, (xiii) rubber manufacture (vulcanising), (xiv) sewage and sludge treatment, (xv) meat meal and bone meal manufacture, (xvi) fishing port (unloading of trash fish),

(xvii) fishing industry (fish meal, fish oil, fish products), (xviii) food processing (coffee roasting, onion roasting, canned products and (xix) manufacture, smokehouse, slaughter house)

4.2.3 Odour generation

Odours are generated in many different situations, in industry sometimes in connection with processing of natural products, e.g., oil distillation or synthesis of chemical compounds. In some plants smells are generated by microbial decomposition of organic compounds, e.g., hydrogen sulphide, mercaptans, but also nitrogen compounds in the form of ammonia, amines and skatole may cause malodour.

Smell generating reactions often occurs where organic substances are kept in containers without oxygen or in anaerobic water, for instance sewers, wells, settling tanks, etc. Some of the problems may be solved by adequate circulation or supply or oxygen or oxidising substances.

Problems relating to odour control

Odour nuisances are sometimes very difficult to prevent, because even negligible amounts of some odorants are sensed by humans. Hydrogen sulphide and mercaptans for instance can be sensed in concentrations far below concentrations which are harmful to man. Due to our ability to detect odorants in very small concentrations, otherwise insignificant air pollution sources (valve and joint leakages, small spills, evaporation from wells, sewers or open basins) may give rise to complaints from neighbours.

4.3 Measuring odours

An odour is defined as a sensation resulting from the reception of a stimulus by the olfactory sensory system. The way the human response to an odour is evaluated depends on the particular sensory property that is being measured, including the intensity, detectability, character and hedonic tone of the odour. The combined effect of these properties is related to the annoyance that may be caused by the odour.

Odour intensity: Odour intensity is the strength of the perceived odour sensation. It is related to the odorant concentration, which is an entirely different type of measurement.

Odour detectability: Odour detectability or threshold is a sensory property referring to the minimum concentration that produces an olfactory response or sensation. This threshold is usually determined by an odour panel consisting of a specified number of people. The numerical result is typically expressed as occurring when 50% of the panel correctly detect the odour.

Odour character: Odour character or quality is that property that identifies an odour and differentiates it from another odour of equal intensity. The odour character is described by a method known as multidimensional scaling or profiling. In this method, the odour is characterised by either the degree of its similarity to a set of reference odours or the degree to which it matches a scale of various descriptor terms. The result is an odour profile.

Hedonic tone: Hedonic tone is a property of an odour relating to its pleasantness or unpleasantness. A distinction should be made between the acceptability and the hedonic tone of an odour. When an odour is evaluated in the laboratory for its hedonic tone in the neutral context of an olfactometric presentation, the panellist is exposed to a controlled stimulus in terms of intensity and duration. The degree of pleasantness or unpleasantness is determined by each panellist's experience and emotional associations.

4.3.1 Odour air pollutants

Odorous air pollutants are often judged important primarily for their nuisance value and the number of complaints they generate. In only a few cases are there adverse health effects documented in measurable physiological terms. However, odours detected from biological processes may indicate contamination of the air by pathogens.

4.4 Assessing odour impact

In recent times attention has been devoted to better and more repeatable measurement techniques for assessing odour levels. This, coupled with more realistic modelling of odour dispersion, can provide a more reliable assessment of odour impact.

Mixtures of compounds require dynamic olfactometry for assessment of odour level. This involves exposing a selected and controlled panel of observers to precise variations in the concentrations in a controlled sequence, to determine the point at which only half the panel can detect the odour. This point is called the odour threshold or one odour unit. The number of odour units is the concentration of a sample divided by the odour threshold.

There are several odour thresholds that can be determined using olfactometry. They are:

1. The detection threshold.
2. The recognition threshold.
3. The description threshold.

The detection threshold is defined as the lowest concentration that will elicit a response without reference to odour quality.

The recognition threshold is defined as the minimum concentration that is recognised as having a characteristic odour quality.

The description threshold is the point at which the panellist is asked to distinguish the odour.

Odours that are produced from large areas (such as sewage treatment works, cattle feedlots and the spreading of effluent or manure) are classified as odours from area sources. Odours emitted from a process through a chimney or flue are classified as odours from process industries.

4.5 Controlling odours from area sources

For large area sources like sewage treatment farms, cattle feedlots, composting, household or industrial tips and manure spreading, there are only two proven methods that can be used to reduce odour complaints. These are:

1. Excluding development close to the site.
2. Ensuring that the operation is carried out under best management practice.

If development close to the site is to be excluded, a reasonable 'buffer zone' around the area source has to be determined. The actual size of this zone will depend upon a number of factors, including the size of the area from which the odour emanates, the intensity of the odours being emitted, the duration and frequency of the odour emissions, the actual process being undertaken, the topography of the site, the weather conditions that prevail at the site and the neighbours' perception of offensiveness of the odours being produced.

For example, a rural community may not regard low intensity feedlot odours as offensive, but the same odours close to an urban community may generate many complaints.

Best management practices (BMPs) will vary according to the industry producing the odour. However, for all new developments, BMPs will start with the site selection and the building of the facility. For example, composting in Holland is being conducted in specially designed, fully enclosed process chambers ('tunnels') instead of in the open.

In some cases these strategies reduce the production of odours because there is closer process control; if the odours are still produced it is often easier to control them. There are a large number of chemicals and proprietary products that are claimed to reduce odour when they are applied to area sources. Be careful if you are considering using these products. To reduce odours they would have to be applied over very large area sources; the cost of materials and labour would be very high. The large quantities of these compounds required could themselves cause pollution. However, in certain industries it may be feasible to apply biologically active agents to change the odours being emitted to less odorous compounds.

4.6 Controlling odours from process industries

The gas stream emitted from a chimney or flue can be collected and made available for treatment. The choice of method or combination of methods to be used for controlling odours in gas streams will be influenced by the following factors:

1. The volume of gas (or vapour) being produced and its flow rate.
2. The chemical composition of the mixture causing the odour.
3. The temperature.
4. The water content of the stream.

While most odours are caused by gases, problems may also result from aerosols in the fumes. Odorous air streams frequently contain high concentrations of moisture. If these vapour discharges can be cooled to less than 40°C, a substantial quantity of the water vapour will be condensed and so reduce the volume of gases to be incinerated or otherwise treated. Mist filters can be used, also, to remove solids and liquids from the gas stream; if the odour is caused by these particles the odour will be reduced. The methods available to control odours in a gas stream are:

1. Dispersion.
2. Scrubbing.
3. Incineration.
4. Adsorption on to a solid.
5. Biofiltration.

For complete control of odours it may be necessary to use more than one of these techniques (for example, scrubbing may be needed before adsorption or biofiltration). If the temperature of the gas stream is high, it may also be necessary to cool the gases before they are treated. This may be needed before the chemical scrubbing or the activated carbon adsorption method is used.

4.6.1 Dispersion

Odorous gases require special consideration and treatment. The human nose is capable of detecting wafts of odorous gases lasting as little as one or two seconds. For such very short periods the concentration of gases in a dispersing plume may be up to ten times higher than the design values averaged over a three minute period. Dispersion of odorous emissions via a chimney is not recommended unless the emission source is properly quantified by an odour panel test.

4.6.2 Incineration

Incineration is the oxidation of the odour into carbon dioxide and water by the combustion of the odour with a fuel and air. In some cases other compounds may be formed, depending on the mixture of fuel and air used, the flame temperature and the composition of the odour. These compounds could include carbon monoxide, oxides of nitrogen and sulphur oxides.

It is important to reduce the moisture content of any gas stream requiring incineration in order to reduce fuel consumption.

An afterburner is basically a refractory-lined furnace fitted with one or more burners. The furnace should be fitted with a temperature indicator-controller and an independent high temperature alarm. Explosion protection systems should be installed if the odorous gas is capable of forming an explosive mixture or if gas is used as the fuel.

The furnace will normally consist of two chambers: a mixing section in which the odorous gases are mixed with auxiliary fuel and ignited and a combustion section in which combustion is completed. The gas velocity in the mixing section is normally 8 to 15 metres per second to ensure adequate turbulence, but reduces to between 6 and 12 metres per second in the combustion section. The afterburner temperature should be measured where the gases leave the combustion chamber with the sensor shielded from furnace radiation. The critical part of the afterburner is the contact between the odorous gases and the flame and, ideally, all odorous air should be used as primary or secondary combustion air. The efficiency of combustion will decrease with decreasing contact between the odours and the flame.

Catalysts can be used to allow incineration at lower temperatures with consequent fuel savings. However, these are susceptible to catalyst poisoning and plugging by solid or viscous particles, which makes the system ineffective. Designing the afterburner to allow for alternative thermal incineration in such events removes any capital cost advantages of a catalytic incinerator. Proponents of such systems would need to provide clear evidence of successful extended use in similar applications.

A chimney of adequate height should be installed on the after burner to ensure proper dispersion of combustion products. A reduction in the efficiency of combustion may arise because of a change in volume or composition of the odorous gas stream. Problems may occur with the burner or fuel supply and this could reduce the combustion efficiency. A tall chimney will help disperse any residual odours.

A boiler or furnace may be used as an after burner, provided it is operating at a reasonable load when it is required to act as an afterburner. A liquid or gas-fired boiler, which may be on low fire for considerable periods, is unsatisfactory as an afterburner, unless a separate afterburner is installed. Such a system requires a changeover valve actuated by the boiler to divert the odorous gas to the afterburner when the boiler is on low fire. Non-condensed gases should be admitted to a boiler as primary burner air.

A coal-fired boiler with a chain-grate stoker is suitable as an afterburner because the odorous gases can be admitted below the grate as underfire air. Such coal-fired boilers normally have high residence times. The chimney height should be increased, if practicable, when it is known that a boiler will also be used as an afterburner. The additional stack height will improve dispersion and possibly prevent complaints if the odours are not completely incinerated.

4.6.3 Liquid scrubbing

Liquid scrubbing of gases to remove odours involves either absorption in a suitable solvent or chemical treatment with a suitable reagent.

Scrubbing brings the odorous gas stream into intimate contact with the scrubbing liquid. Unless the odorous substances are readily soluble in the liquid, it is imperative that a large liquid surface is exposed to the gas.

It has been suggested that liquid scrubbing becomes economically attractive compared with incineration and adsorption on activated carbon when the volume of odorous gas to be treated is greater than 5000 cubic metres per hour. It is important that hot, moist streams are cooled before they contact scrubbing solutions. If this is not done the scrubbing solution will be heated and less efficient and the scrubbing medium will become diluted from condensation of water vapour.

If a hypochlorite solution is used there is the chance that chlorine will be lost and the cost of replenishment can be high. There is also the chance that odours will be released from a hot scrubbing solution.

Condensation of moisture from the air stream can be achieved with either a direct or indirect condenser. An indirect water-cooled condenser separates the condensed water or condensate for the cooling water. In a direct condenser, cooling water is sprayed into the air stream and the cooling water will be contaminated with odorous condensate. If hot, contaminated cooling water is circulated through a cooling tower it is likely that odours will be released to the atmosphere. With an indirect condenser, the smelly condensate is segregated from the cooling water and may be discharged to sewer or a water treatment plant. The use of indirect condensers is preferred.

The principal types of gas absorption equipment are:

1. Packed towers.
2. Plate or tray towers.
3. Spray towers.
4. Venturi scrubbers.

If the gases contain hydrogen sulphide a solution of sodium hydroxide may be used. When the odour is caused by the presence of unsaturated organic compounds, it may be necessary to use an oxidising agent such as chlorine, sodium hypochlorite, potassium permanganate, ozone or hydrogen peroxide. If chlorine, sodium hydroxide, sodium hypochlorite or ozone is used you may need to monitor exit gas streams.

Satisfactory results have been achieved using a sequence of chlorine gas, diluted sulphuric acid and sodium hydroxide to treat odours. The concentration of reagent in the scrubbing solutions must be maintained either by the use of a metering pump or by regular additions of reagent. (For example, you may specify 4% free caustic soda in a sodium hydroxide reagent and this should be kept consistent.)

Make provision to sample the exit gases from the scrubber to determine if the scrubbing is effective. Exit gases should be discharged through a stack, which should be higher than nearby buildings. This will avoid problems with building downwash, which may cause odour complaints, particularly if the scrubbing efficiency decreases.

A stack will safeguard against process changes or equipment malfunctions. Some additional instrumentation will be needed on the control equipment in order to monitor scrubber pressure drop, liquid flow, pump pressure, temperature and reagent concentration of the scrubbing solution.

4.6.4 Adsorption on to activated solids

A method that is suitable for controlling odorous substances, even at low concentrations, is adsorption on to activated carbon. To be effective, the contaminated air stream must be free of substances (such as dust) that might clog the carbon particles. The cost of replacing the carbon can be high, as simple systems use the carbon once only. More complex and expensive systems allow regeneration of the carbon for reuse. Regeneration can produce either a wastewater, which will require further treatment before disposal, or a concentrated vapour stream, which can be incinerated more cheaply than the original air stream.

One proprietary system uses activated alumina impregnated with potassium permanganate. The alumina adsorbs the odorous substances so that the permanganate can oxidise them, usually to carbon dioxide, water, nitrogen and sulphur dioxide, depending on their composition. The alumina bed is replaced progressively as the permanganate is exhausted. This has an advantage over carbon because no further treatment is needed; this may offset the cost of the alumina. Proponents wanting to use an adsorption system should provide evidence of successful long-term application of the particular process.

4.6.5 Biofiltration

This method is becoming an acceptable and successful way of reducing odours from biological processes. For odours from process industries, where the odour is caused by one or a simple mixture of known chemicals, techniques are becoming available to increase the scope of biofiltration.

The procedure is similar to chemical scrubbing, except that in biofiltration the odour is removed by bacterial action. The bacteria grow on inert supports, allowing intimate contact between the odorous gases and the bacteria. The process is self-sustaining. In biological industries (for example, rendering works) it is usual to place the biofilter after the condenser.

Biofilters require careful attention to ensure continued operation. The bed may have to be replaced regularly because of mechanical failure.

Another type of biofilter is the soil-bed filter. Here the odorous gas stream is allowed to flow through a porous soil with a typical depth of 60 cm. The bacteria in the soil are responsible for the destruction of the odorous compounds. Typical reductions of odour by 99.9% are claimed, with no decrease after a year's operation.

4.7 Odour pollution from some important industries

This section discusses odour pollution from paper industry, petroleum refinery and foundry industry.

4.7.1 Odour pollution from paper industry

In papermaking, odorous substances impair product quality and may result in complaints and reduced revenue. Odour emissions may lead to grievances from neighbours, complaints from authorities and even call the industrial location into question. Occupational safety may even be endangered in some extreme cases. Odour problems are not a singular problem, but are rather widespread throughout the pulp and paper industry.

Odours in paper and paperboard may be caused in principle by the following factors:

1. Anaerobic microbiological processes in the water and pulp systems.
2. Odours arising from additives and coating binders.
3. The oxidation of wood extractives.
4. Odour arising from printing inks (when they interact with the paper).

One frequent cause of odour problems is anaerobic metabolic processes of microbes in the water circuit, in effluents and sludges which give rise to hydrogen sulphide, volatile organic acids and other compounds, some of which are truly ill smelling. Odour problems as a result of microbial processes are usually linked with increasing closure of water circuits and rising recovered paper utilisation rates.

Microbial odour formation is related to a number of preconditions:

1. Pre-microbial growth with odour forming micro-organisms.

2. An oxygen-deficient or oxygen-free (anaerobic) environment.
3. Sufficient supply of nutrients.
4. Sufficient dwell time.
5. Additional factors that support environmental conditions such as specific pH and temperature ranges.

Odorous substances form microbiologically in the absence of air, whereas emission requires contact with air. Odorous substances therefore inevitably pass from the water phase to the air phase at a location other than where they form. Odorous substances, once they have formed, are therefore emitted after transport and a change of environment, at locations of high turbulence by stripping or at large boundary layers between water and air.

It is possible to obtain a meaningful inventory using the measurands redox potential, strippable hydrogen sulphide, pH and organic acids. Additional parameters confirm results and help to interpret measurements. What is important is that the water circuitry and process control be evaluated at the same time the analytical parameters are being determined. Conducting and interpreting such a process analysis should be carried out by trained personnel in order to guarantee success.

Determination of the content of low molecular-weight organic acids by ion chromatography

Paper can contain intense smelling contents of low-molecular-weight fatty acids such as acetic acid, propionic acid, butyric acid or valeric acid. Greater amounts of these substances impart to the paper an acidic, rancid smell ranging from moldy to sewerage-like. These acids are formed as metabolic products by anaerobic and facultative anaerobic micro-organisms in water and pulp systems.

One possible method of determination is ion chromatography. In this procedure, the aqueous sample is passed through a suitable separation column where the dissolved organic acids are adsorbed for different lengths of time, thus separating them. The presence of the organic acids is verified by conductivity. The low-molecular-weight organic acids contained in the paper are previously dissolved by hot water extraction.

Organoleptic analysis

Organoleptic analysis can provide helpful evidence to explain the causes of an off-odour coming from a product. For this purpose, samples are subjected to an odour test in accordance with DIN EN 1230-1, a sample that is found to be unobjectionable and odourless being used as a standard reference sample. The additives and raw materials used during production can be employed as reference samples with the product odour being compared with them.

4.7.2 Odour pollution from petroleum refinery

Odours can be created in the petroleum refinery. Odours in a petroleum refinery are mainly created by sulphur compounds (e.g., hydrogen sulphide, mercaptans, sulphides, disulphides); nitrogen compounds (e.g., ammonia, amines); and hydrocarbons (e.g., aromatics). The main sources of odour in refineries are storage (such as sour crudes), bitumen production, desalter water, sewers, uncovered DAF, oil/water/solid separation and biological treatment units and flaring. Different approaches can be applied to refinery odour problems such as instrumental measurements (e.g., GC) and GC-MS), sensory methods (e.g., olfactometry (dilution technique)), odour panels and hybrid instrumentation. Generally, odour measurement is necessary for odour regulation and control. Odour detectability or threshold or concentration, odour Intensity, odour persistence, hedonic tone, odour character or quality and annoyance are the terms associated with odour measurement. The use of nitrate-based products in septic water areas; reduction of VOCs and odour generation by covering some units of wastewater treatment plant (e.g., CPI and API separators) with closed sealed covers; reduction of odours from water buffer tanks by maintaining the smallest possible surface area of oil and water in contact with air via using a fixed roof tank or a floating roof tank; reduction and control of fugitive emissions; control of flares and prevention or reduction of emissions from them; control of fuel quality; the use of scrubbing systems for odorous gases; and the use of incineration systems for odorous gases may be used to reduce or control the odour generation in the petroleum refinery.

Refinery odours and their possible sources are shown in Table 4.1.

Table 4.1: Refinery odours and their possible sources.

Type of smell	*Odour compounds*	*Sources*
Bad eggs	H_2S + trace of disulphides	Crude storage, distillation of gases, sulphur removal, flare stacks (cold flare)
Sewer smell	Dimethyl sulphide, ethyl and methyl mercaptans	Effluent water, biological treatment plants, LPG odourising spent caustic loading and transfer
Burnt oil	Unsaturated hydrocarbons	Catalytic cracking unit, coking asphalt blowing, asphalt storage
Gasoline	Hydrocarbons	Product storage, American Petroleum Institute (API) and corrugated plate interceptor (CPI) separators
Aromatics (benzene)	Benzene , toluene	Aromatic plants, Naphtha reformers
Hot tar	H_2S, mercaptans, hydrocarbons	Asphalt storage

Control of odours in the petroleum refinery

The techniques which can be used to reduce or control the odour generation in the petroleum refinery are:

1. The use of nitrate-based products in septic water areas (e.g., storage tanks, sewage systems, oil/water separators) in order to replace bacteria feedstock and to favour the development of denitrificative bacteria, which will both reduce added nitrates in nitrogen and existing hydrogen sulphide in sulphates.
2. Reduction of Volatile Organic Compounds (VOCs) and odour generation by covering some units of Wastewater Treatment Plant (WWTP) (e.g., CPI and API separators) with closed sealed covers.
3. Reduction of odours from water buffer tanks by maintaining the smallest possible surface area of oil and water in contact with air via using a fixed roof tank or a floating roof tank.
4. Reduction and control of fugitive emissions.
5. Control of flares and prevention or reduction of emissions from them.
6. Control of fuel quality.
7. The use of scrubbing systems for odorous gases.
8. The use of incineration systems for odorous gases.

4.7.3 Odour pollution from foundry

The main operations conducted at foundries that have the potential to generate odorous emissions can be summarised as follows:

1. Smelting. The potential for odour generation is generally dependent upon the type of metals utilised and can vary widely from site to site. Processing of recycled metals tends to pose the highest risk from this perspective.
2. Casting. Generation of odour during the addition of the molten metal to sand cores is the most common source of odour. Odours are primarily associated with the release of amines and aromatic odorants, which are produced from the phenolic resin binders used to manufacture the cores. Similar odours also tend to be released during the subsequent recycling of used cores.

Control of odour emissions from foundries can be achieved to some extent through adoption of good practice and use of resins with low phenol content. However, where foundries are located in close proximity to sensitive receptors such as residential housing, maintaining effective odour control will generally necessitate the use of some form of enclosure and odour treatment system (e.g., chemical scrubber).

It is very important to note that, when measuring odour from foundries for the purposes of assessing odour impact or designing odour abatement plans, the odour samples should be analysed as soon as possible, since they tend to be unstable and can change substantially in odour concentration and character during storage.

Some odour control methods are given in Table 4.2.

Table 4.2: Odour control methods.

Operation type	*Emissions*	*Odour control*
Boiler	Sulphur dioxide	A
Acid plant	Acid gases	A, B, E
Fertiliser	Fluorides, fertiliser odour	A, B, F
Rendering works	Decomposing flesh, amines	A, C, D, G
Coffee	Aldehydes, amines	A, C, D
Chicken feathers	Amines	B, D
Fish meal	Amines, aerosols	B, C, D, E
Garbage	Decomposing organics, sulphides	C, G
Ammonia	Ammonia	A, B
Detergent and soap	Soapy	A, B, I
Oil refinery	Hydrocarbons, sulphides (complex)	A,C,G
Chemical plant	various (complex)	A, B, C
Rockwool	Burnt oil, aldehydes	A, B, C
Varnish and paint	Burnt oil, hydrocarbons, aldehydes	A, C
Solvent storage	Various solvents	D, A, G
Animal incineration	Amines, aldehydes	A, C, F, I
Grease trap waste	Amines	B, D, G, I
Fermentation	Yeast, fermenting	A, D, G
Non-ferrous foundry	Burnt core odour, aldehydes	A, C
Ferrous foundry		A
Laminated plastic	Phenols, formaldehyde	A, C
Fibreglass	Styrene, acrylates	A, D

Key

A. Dispersion: Tall chimney required, moderate capital cost, low running costs.

B. Wet scrubbing: Absorption: moderate to high costs, three stages often required. Needs careful selection of scrubber liquor and usually trials. Not always successful, requires regular maintenance and daily tests of active agent and pH control in some cases.

C. Afterburner (direct): Temperatures between 600°C and 1000°C with residence time of 0.3 to 1 second. Doubling of residence time may enable afterburner temperature to be reduced 20°C to 100°C. Requires careful design to reduce air volume to a minimum. Capital cost high and cost of running is high if air volume large. Needs further control if sulphur or chlorine present in exhaust gases.

   Afterburner (catalytic). Temperature 500°C. Lower temperature operation than direct method, but the catalyst may be destroyed if not operated and maintained correctly. This is frequently a problem.

D. Carbon adsorption: Needs regenerating at regular intervals. Can be effective but expensive for large volumes. Small operations are reasonably inexpensive to deal with.

E. Mist filter: Mist filter of plastic or metal, self cleaning, inexpensive.

F. Best management practices: Cleanliness and avoiding spills requires human effort but is relatively inexpensive.

G. Masking odour: Deodorant disguises the problem. Usually not effective but can help in marginal cases or with accidental releases. 'Inexpensive'.

H. Biological filtration: Has been successful for rendering works.

I. Condensers: Condensable liquids removed from the exhaust gas stream, reduces the amount of odorous gas to be treated.

5

Industrial wastewater and its treatment

5.1 Introduction

A large amount of water is discharged back after domestic and industrial usage. This is contaminated with domestic waste and industrial effluents. When this contamination reaches beyond certain allowed concentrations, it is called pollution and the contaminants are called the pollutants. Water pollution may be defined as the contamination of streams, lakes, seas, underground water or oceans by substances, which are harmful for living beings. If the concentration of substances naturally present in water increases then also the water is said to be polluted. Water may be called polluted when the following parameters stated below reach beyond a specified concentration in water.

Physical parameters: Colour, odour, turbidity, taste, temperature and electrical conductivity constitute the physical parameters and are good indicators of contamination. For instance, colour and turbidity are visible evidences of polluted water while an offensive odour or a bitter and difference than normal taste also makes water unfit for drinking.

Chemical parameters: These include the amount of carbonates, sulphates, chlorides, fluorides, nitrates and metal ions. These chemicals form the total dissolved solids, present in water.

Biological parameters: The biological parameters include matter like algae, fungi, viruses, protozoa and bacteria. The life forms present in water are affected to a good extent by the presence of pollutants. The pollutants in water may cause a reduction in the population of both lower and higher plant and animal lives. Thus, the biological parameters give an indirect indication of the amount of pollution in water.

5.2 Water pollutants

Water pollutants refer to the substances which are capable of making any physical, chemical or biological change in the water body. These have undesirable effect on living organisms. As mentioned earlier, the water used for domestic, agricultural and industrial purposes is discharged with some undesirable impurities in it. You have read the various sources from where pollutants enter the water bodies. Let us now learn about the various types of pollutants arising out of these sources.

These can be broadly put under the following types:

1. Sewage pollutants (domestic and municipal waste).
2. Industrial pollutants.
3. Agricultural pollutants.
4. Radioactive and thermal pollutants.

Domestic and municipal pollutants: The sewage contains garbage, soaps, detergents, waste food and human excreta and is the single largest sources of water pollution. Pathogenic (disease causing) micro-ogranisms (bacteria, fungi, protozoa, algae) enter the water system through sewage making it infected. Typhoid, chloera, gastroenteritis and dysentery are commonly caused by drinking infected water. Water polluted by sewage may carry certain other bacteria and viruses which cannot grow by themselves, but reproduce in the cells of host organisms. They cause a number of diseases, such as, polio, viral hepatitis and may be cancer which are resistant to like the organic matter are oxygen demanding substances. They are responsible for deoxygenation of water-bodies which is harmful for aquatic life.

Other ingedients which enter the various water bodies are the plant nutrients, i.e., nitrates and phosphates. They support growth of algae, commonly called algal bloom (blue-green species). This process is called eutrophication and is discussed in details in the next section.

Industrial pollutants: Many industries are located near rivers or fresh water streams. These are responsible for discharging their untreated effluents into rivers like highly toxic heavy metals such as chromium, arsenic, lead, mercury, etc., along with hazardous organic and inorganic wastes (e.g., acids, alkalies, cyanides, chlorides, etc.). River Ganges receives wastes from textile, sugar, paper and pulp mills, tanneries, rubber and pesticide industries. Most of these pollutants are resistant to breakdown by micro-organisms (called non-biodegradable), therefore damage the growth of crops and the polluted water is unsafe for drinking purposes. Factories manufacturing plastic, caustic soda and some fungicides and pesticides release mercury (a heavy metal) along with other effluents in nearby water body. Mercury enters the food chain through bacteria, algae, fish and finally into the human body. The toxicity of mercury became evident by the Minamata Bay tragedy in Japan during the period 1953–1960. Fish died due to mercury consumption and those who ate fish were affected by mercury poisoning and quite a few died. The milder symptoms of mercury poisoning are depression and irritability but acute toxic effects can cause paralysis, blindness, insanity, birth defects and even death. The high concentration of mercury in water and in fish tissues results from formation of soluble monomethylmercury ion, (CH_3, Hg^+) and volatile dimethylmercury [$(CH_3)_2$ Hg] by anaerobic bacteria in sediments.

Agricultural waste: Manure, fertilisers, pesticides, wastes from farms, slaughterhouse, poultry farms, salts and silt are drained as run-off from agricultural lands. The water body receiving large quantities of fertilisers (phosphates and nitrates or manures becomes rich in nutrients which leads to eutrophication and consequent depletion of dissolved oxygen. Consumption of water rich in nitrates is bad for human health especially for small children.

Pesticides (DDT, dieldrin, aldrin, malathion, carbaryl, etc.), are used to kill insect and rodent pests. Toxic pesticide residues enter the human body through drinking water or through food chain (biomagnification). These compounds have low solubility in water but are highly soluble in fats. For example, the concentration of DDT in river water may be very low but some fish over a period of time accumulate so much of DDT that they become unfit for human consumption. The use of pesticides in our country is increasing very rapidly.

Some of these chemicals which are highly toxic become metabolised by animals that graze on fields. Therefore, these poisonous chemicals have been often observed in the human food chain. The presence of these chemicals in humans even in minute amounts can cause hormonal imbalance and may lead to cancer.

5.3 Water pollution and some biological effects

The natural source of water in the form of precipitation or rain is the purest form available in nature. However after reaching the surface and then underground it gets contaminated by a number of pollutants. There are some biological factors also mentioned earlier responsible for spoiling the quality of water. These include the lower plants like algae and bacteria which are the causes of nutrient accumulation in aquatic systems. This nutrient accumulation gives rise to a condition called eutrophication explained below.

5.3.1 Eutrophication

Eutrophication is a process by which a water body slowly becomes rich in plant nutrients such as nitrates and phosphates due to soil erosion and run off from the surrounding land. Let us try to understand this phenomenon. A water system like a lake or any reservoir may get a large inflow of organic matter from domestic wastes and run off from the surrounding land. Increasing human population, intensive agriculture and rapid industrial growth have led to an increasing release of domestic waste, agricultural residues, industrial wastes and land runoff into various water bodies. Nutrients are released from organic waste by aerobic (oxygen requiring) bacteria which start decomposing it. Dissolved oxygen is consumed in this process. As more and more organic matter enters a water body, more is the deoxygenation of the water body and larger is the production of nutrients. These nutrients fertilise an abnormal growth of algae and other large water plants such as duckweed. As more plants grow, some of them die also due to larger oxygen demand and therefore oxygen deficiency in the water body (i.e., deoxygenation of the water body).

Eutrophication of a water body results due to the release of large amount of nutrients by the action of aerobic bacteria on organic wastes entering a water body naturally or by human activity.

The above discussion leads us to a concept called biological oxygen demand (BOD). Let us try to understand by the description given below.

5.3.2 Biological oxygen demand (BOD)

The quality of oxygen used up by micro-organisms at 27°C and in darkness during 3 days in breaking down organic wastes in a water body is called its biological oxygen demand (BOD).

It can be explained in the following manner.

We know that there are many organic compounds or waste present in a water body. The micro-organisms present in the system act upon this waste for their own consumption and growth. In the process the metabolic activity requires oxygen which is met by the dissolved oxygen present in water. It is this amount of oxygen which is defined as biological oxygen demand (BOD). The BOD value of an aquatic system depends upon:

1. The type and amount of organic waste.
2. The organisms acting on it.
3. Temperature and pH.

The greater the amount of organic waste in the water body, the greater is the amount of oxygen required to break it down biologically and therefore higher is the BOD value of water. This value is a good measure in evaluating the degree of pollution in a water body. The less polluted water shows comparatively low value of BOD. Its value is used as a criterion for managing water pollution of a water body. An evaluation is made by determining oxygen concentration in water before and after incubation at 20°C in dark for 5 days.

5.3.3 Biomagnification

A variety of toxic chemicals move through food chains. Toxic pesticides may be sprayed for controlling insect pests, fungi, herbs, but they concentrate in the food chain and harm to other (non target) organisms. For example, DDT was sprayed in the U.S. to control mosquitoes at a concentration expected to be harmless to non target organisms like fish and birds. DDT accumulated in the marshes and planktons. Planktons were eaten by fish and the fish had a higher concentration of DDT in its body. Further, when birds ate the fish, they accumulated still higher concentration. This increase in concentration of accumulated toxic chemicals as one goes higher in the food chain is termed biomagnification. Biomagnification has at times threatened the reproduction and survival of carnivores (secondary consumers) who occupy the highest level of the food chain.

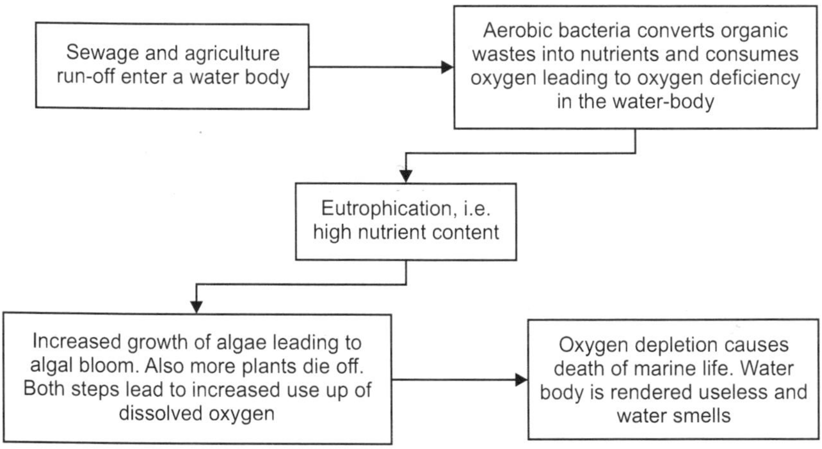

5.4 Water pollution – some control measures

Wastewater generated by household activity, industries or garbage landfills is called sewage which is classified as the municipal water pollution. Sewage contains solid matters in the form of suspended colloidal and dissolved organic matter, detergent, mineral matter, nutrients and gases.

Sewage is one of the major causes of water borne diseases and therefore the treatment of sewage is one of the important tasks. For a long time treatment of municipal waste in the form of sewage involved mainly of the removal of suspended solids, oxygen demanding materials and harmful bacteria. Now the disposal of the solid residue from sewage has been improved by applying municipal treatment processes.

5.5 Pre-treatment of wastewater

5.5.1 Equalisation

Among the most effective waste management procedures is equalisation of the wastewater. Equalisation can be of two types: flow equalisation and contaminant equalisation (COD, TSS, pH, etc.).

The use of flow equalisation tanks following the API separators provides a relatively constant flow rate and contaminant loading to the biological treatment system.

With equalisation, we can provide a more uniform effluent to downstream processes. This becomes particularly beneficial (and important) when biological treatment follows. And here again, we find in wastewater, as in life, that size really does matter, particularly so when it comes to EQ tanks. The equalisation tank(s) must be large enough to absorb and dampen concentrated batch dumps or spills within the process.

In complex chemical and petrochemical plants there are numerous waste streams that enter the wastewater treatment plant. Some of these streams are a readily biodegradable source of substrate (food) for the bacteria in the biological treatment unit. Other streams are inhibitory while still others might have a toxic impact to the bacteria. The EQ tank dampens or neutralises the impact of inhibitory or toxic compounds in the wastewater.

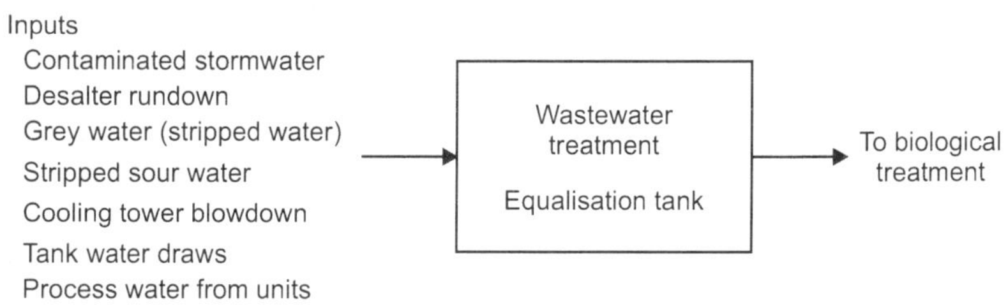

Goal of an EQ tank

1. Equalise the flow rate.
2. Equalise or balance the organic concentration.
3. Dilute or dampen the impact from inhibitory compounds.
4. Neutralise the pH.
5. Temperature fluctuations.
6. Minimise chemical usage (e.g., downstream pH adjustment).

Some aspects to the design and operation of an EQ tank are as follows:

1. The equalisation tank should be completely mixed.
2. Can be operated at constant volume which results in variable outflow.
3. Can be operated at variable volume which results in constant outflow.
4. Highly variable waste streams favour variable volume/constant outflow operation.

Operator input for an EQ is quite basic and involves the following:

1. Sample points on the side of the EQ tank should be checked regularly to confirm the surface oil level in the tank.
2. Oil skimming should occur as needed.
3. Bottom solids should be removed from the tank periodically.

5.5.2 Neutralisation

Neutralisation involves adjusting the pH of a liquid to approach the 'neutral' pH of 7.0 (neither acid nor base). Generally, neutralisation involves the use of an acid (pH less than 7) to lower the pH of a tank of basic (or alkaline) liquid (pH greater than 7), or the use of a base (or alkali) to raise the pH of a tank of acidic liquid.

In the water treatment industry, wastewater is generated from the regeneration of the resins used in cation, anion and mixed bed ion exchange systems. The chemicals used to regenerate the resins have extreme pH levels. For example, sulphuric acid, with a pH of 1–2, is used for regeneration of cation resin and sodium hydroxide, with a pH of 13–14, is used for regeneration of the anion resin. After the regeneration process, the water containing these chemicals must be disposed of. Due to its extreme pH, the wastewater cannot be sent directly to drain, as this will typically violate local regulations regarding the pH of wastewater discharge. For example, the NPDES (National Pollutant Discharge Elimination System) standard for stream discharge pH is 6–9. The acceptable pH of the discharge varies by location and the destination of the discharge (such as a stream, a well or a sewer leading to a municipal waste treatment plant).

To neutralise the highly acidic wastewater from a cation resin regeneration, a supply of alkaline liquid is needed. An anion resin regeneration is usually performed at the same time and its wastewater is alkaline. When these two wastewater streams are combined, the pH is 'neutralised' and approaches 7. If the resulting pH of the combined liquids is not within the acceptable range for discharge, an additional amount of either acid or base must be added to shift the pH into the acceptable range. Generally, the chemicals used to adjust the wastewater are the same chemicals used to regenerate the resins in the ion exchange systems.

5.5.3 Oil and grease

Oil and grease (O&G) are a common occurrence in wastewater. Removing oil and grease from wastewater is relatively simple and cheap. Process industries should take the lead in addressing these questions, so that they are ready when the law takes effect. Furthermore, if they design their system such that the wastewater can be recycled, they actually can use this new law to save costs and lessen public concern about discharge of contaminated wastewater.

Effective removal of O&G requires an understanding of emulsions, mechanical versus chemical. Furthermore, the operator must know how to break emulsions and how to test effective treatment methods in the laboratory. He then must know how to remove the now mechanically emulsified oil most economically (i.e., how to coalesce the oil droplets effectively and how to reduce the O&G content to non-detect) so that recycling of the wastewater is feasible. This means being familiar with post-polishing techniques, particularly the use of organically modified clays (organoclays). Organoclays remove oil and grease from water at seven times the rate of activated carbon.

Oil: Oil appears in wastewater in a number of different forms including free oil and grease (FOG), mechanically emulsified oil, oil wet solids, chemically emulsified oil and dissolved oil.

Free oil and grease: Free oil rises rapidly to the surface of the water tank under calm conditions. The droplet size is 150 microns. This oil can be removed by an overflow weir in the tank and a skimmer. The traces can be removed by passing the wastewater through an adsorber tank (such as a carbon tank) filled with organoclay.

Mechanically emulsified oil: These oil droplets range in size from 20–150 microns. Mechanically emulsified oil is stabilised by electrical charges and other forces that result in the coating of suspended solids. Such oils mix with water due to shear that can result from the wastewater travelling through a pump, wastewater splashing into a tank and anything that will break up and disperse larger oil droplets.

Oil wet solids: This category includes oil that adheres to sediments and other particulate matter such as metal chafings that are common in industrial wastewater. Such oil solids are removed with sand filters, bag filters, flocculants and oil/water separators. These processes are followed by polishing with organoclay.

Chemically emulsified oil: Oil is chemically emulsified in water when emulsifiers such as surfactants or soaps are present. The surfactants have hydrocarbon chains. The simplest ones are sodium laurel sulphate or stearic acid, which have a hydrophilic (water loving) and a lipophilic (oil loving) end. The lipophilic end enters the oil droplet, while the hydrophilic end remains in the water. Since this creates a charge on the otherwise neutral oil droplet, the droplets will repel each other and disperse. This is called a chemically stabilised emulsion. The droplets are less then 20 microns, while the colour of the water is white. The white colour is an indicator that the emulsion must be split to allow removal of the oil. The source of such oils is metal working fluids, coolants, lubricants, motor oil, hydraulic fluids, etc.

Dissolved oil: These are oils from the light end of the oil spectrum such as benzene, toluene and xylene. The molecules are less than five microns in size. They are removed very effectively by activated carbon.

Emulsion description: If oil is not emulsified, it floats on top of the water as a sheen. Oil can be either mechanically or chemically emulsified. This means that the oil is broken up into droplets that disperse in the water. The smaller the droplets, the more stable the emulsion. When the drops contact each other they tend to coalesce and rise to the surface.

In chemical emulsions, an emulsifier (usually a surfactant, detergent or soap) is present. Surfactants consist of a hydrophilic/oleophobic end and a hydrophobic/oleophilic/lyophylic end. They act as a coupling agent between the oil/water phase. Because the emulsifier is polar on one end (i.e., it has a charge) and is non-polar at the other end, it prevents the oil from approaching and coalescing.

Thus, an emulsion is a heterogeneous system that consists of at least one immiscible liquid intimately dispersed in another liquid in the form of droplets, whose diameter generally exceeds 0.1 micron. Surfactants and finely divided solids increase the stability of the emulsion.

The treatment of this wastewater is carried out in the following three stages:

1. Primary treatment.
2. Secondary treatment.
3. Tertiary treatment.

5.6 Primary treatment of wastewater

Wastewater contains large solids and grit that can interfere with treatment processes or cause undue mechanical wear and increased maintenance on wastewater treatment equipment. To minimise potential problems, these materials require separate handling. Preliminary treatment removes these constituents from the influent wastewater.

5.6.1 Screening

Screening is the first unit operation used at wastewater treatment plants (WWTPs). Screening removes objects such as rags, paper, plastics and metals to prevent damage and clogging of downstream equipment, piping and appurtenances. Some modern wastewater treatment plants use both coarse screens and fine screens.

Coarse screens: Coarse screens remove large solids, rags and debris from wastewater and typically have openings of 6 mm (0.25 in) or larger. Types of coarse screens include mechanically and manually cleaned bar screens, including trash racks.

Fine screens: Fine screens are typically used to remove material that may create operation and maintenance problems in downstream processes, particularly in systems that lack primary treatment. Typical opening sizes for fine screens are 1.5 to 6 mm (0.06 to 0.25 in). Very fine screens with openings of 0.2 to 1.5 mm (0.01 to 0.06 in) placed after coarse or fine screens can reduce suspended solids to levels near those achieved by primary clarification.

5.6.2 Comminutors and grinders

Processing coarse solids reduces their size so that they can be removed during downstream treatment operations, such as primary clarification, where both floating and settleable solids are removed. Comminuting and grinding devices are installed in the wastewater flow channel to grind and shred material up to 6 to 19 mm (0.25 to 0.75 in) in size. Comminutors consist of a rotating slotted cylinder through which wastewater flow passes. Solids that are too large to pass through the slots are cut by blades as the cylinder rotates, reducing their size until they pass through the slot openings.

Grinders consist of two sets of counter rotating, intermeshing cutters that trap and shear wastewater solids into a consistent particle size, typically 6 mm (0.25 in). The cutters are mounted on two drive shafts with intermediate spacers. The shafts counter rotate at different speeds to clean the cutters.

The chopping action of the grinder reduces the formation of rag 'balls' and rag 'ropes' (an inherent problem with comminutors). Wastewaters that contain large quantities of rags and solids, such as prison wastewaters, utilise grinders downstream from coarse screens to help prevent frequent jamming and excessive wear.

5.6.3 Grit removal

Grit includes sand, gravel, cinder, or other heavy solid materials that are 'heavier' (higher specific gravity) than the organic biodegradable solids in the wastewater. Grit also includes eggshells, bone chips, seeds, coffee grounds and large organic particles, such as food waste. Removal of grit prevents unnecessary abrasion and wear of mechanical equipment, grit deposition in pipelines and channels and accumulation of grit in anaerobic digesters and aeration basins. Grit removal facilities typically precede primary clarification and follow screening and comminution. This prevents large solids from interfering with grit handling equipment. In secondary treatment plants without primary clarification, grit removal should precede aeration.

Many types of grit removal systems exist, including aerated grit chambers, vortex-type (paddle or jet induced vortex) grit removal systems, detritus tanks (short-term sedimentation basins), horizontal flow grit chambers (velocity-controlled channel) and hydrocyclones (cyclonic inertial separation). Various factors must be taken into consideration when selecting a grit removal process, including the quantity and characteristics of grit, potential adverse effects on downstream processes, head loss requirements, space requirements, removal efficiency, organic content and cost. The type of grit removal system chosen for a specific facility should be the one that best balances these different considerations.

Aerated grit chamber

In aerated grit chambers, grit is removed by causing the wastewater to flow in a spiral pattern, as shown in Fig. 5.1. Air is introduced in the grit chamber along one side, causing a perpendicular spiral velocity pattern to flow through the tank. Heavier particles are accelerated and diverge from the streamlines, dropping to the bottom of the tank, while lighter organic particles are suspended and eventually carried out of the tank.

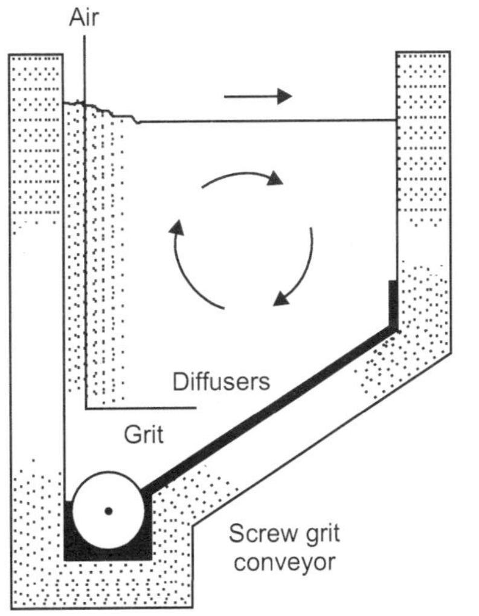

Figure 5.1: Aerated grit chamber.

5.6.4 Advantages and disadvantages

Advantages

Screening: Manually cleaned screens require little or no equipment maintenance and provide a good alternative for smaller plants with few screenings. Mechanically cleaned screens tend to have lower labour costs than manually cleaned screens and offer the advantages of improved flow conditions and screening capture over manually cleaned screens.

Communitors and grinders: A major advantage of using communitors and grinders is that removal of grit reduces damage and maintenance to downstream processes. Comminutors and grinders also eliminate screenings handling and disposal, which may improve the aesthetics of the plant, reducing odours, flies and the unsightliness associated with screenings. Some recently developed grinders can chop, remove, wash and compact the screenings. The use of comminutors in cold weather eliminates the need to prevent collected screenings from freezing. Comminutors and grinders typically have a lower profile than screens, so cost savings can be significant when the units must be enclosed.

Grit removal: Some advantages of aerated grit chambers include:

1. Consistent removal efficiency over a wide flow range.
2. A relatively low putrescible organic content may be removed with a well controlled rate of aeration.
3. Performance of downstream units may be improved by using preaeration to reduce septic conditions in incoming wastewater.
4. Aerated grit chambers are versatile, allowing for chemical addition, mixing, preaeration and flocculation.

Disadvantages

Screening: Manually cleaned screens require frequent raking to avoid clogging and high backwater levels that cause buildup of a solids mat on the screen. The increased raking frequency increases labour costs. Removal of this mat during cleaning may also cause flow surges that can reduce the solids-capture efficiency of downstream units. Mechanically cleaned screens are not subject to this problem, but they have high equipment maintenance costs.

Communitors and grinders: Comminutors and grinders can create problems for downstream processes, such as increasing plastics buildup in digestion tanks or rag accumulation on air diffusers. In addition, solids from comminutors and grinders will not decompose during the digestion process. If these synthetic solids are not removed, they may cause biosolids to be rejected for reuse as a soil amendment.

Grit removal: Grit removal systems increase the headloss through a wastewater treatment plant, which could be problematic if headloss is an issue. This could require additional pumping to compensate for the headloss.

The following section describe the specific disadvantages of different types of grit removal systems.

Aerated grit chamber: Potentially harmful volatile organics and odours may be released from the aerated grit chamber. Aerated grit chambers also require more power than other grit removal processes and maintenance and control of the aeration system requires additional labour.

5.6.5 Gravity sedimentation

Slowly settling particles are removed with gravity sedimentation tanks. For the most part, these tanks are designed on the basis of retention time, surface overflow rate and minimum depth. A sedimentation tank can be rectangular or circular. The important factor affecting its removal efficiency is the hydraulic flow pattern through the tank.

The energy contained in the incoming wastewater flow must be dissipated before the solids can settle. The wastewater flow must be distributed properly through the sedimentation volume for maximum settling efficiency. After the solids have settled, the settled effluent should be collected without creating serious hydraulic currents that can adversely affect the sedimentation process. Effluent weirs are placed at the end of rectangular sedimentation tanks and around the periphery of circular sedimentation tanks to ensure uniform outflow. Once the solids have settled, they must be removed from the sedimentation tank floor by scraping and hydraulic flow.

Conventional sedimentation tanks have sludge hoppers to collect the concentrated sludge and to prevent removal of excess volumes of water with the settled solids. Cross-sectional diagrams of conventional sedimentation tanks are shown in Figs 5.2 and 5.3.

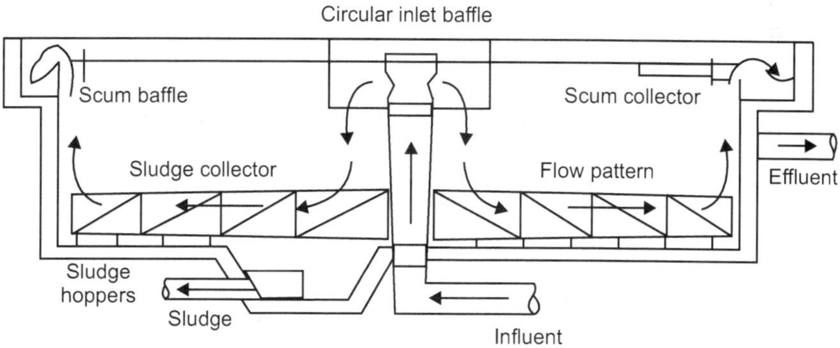

Figure 5.2: Schematic diagram of an aerated grit chamber.

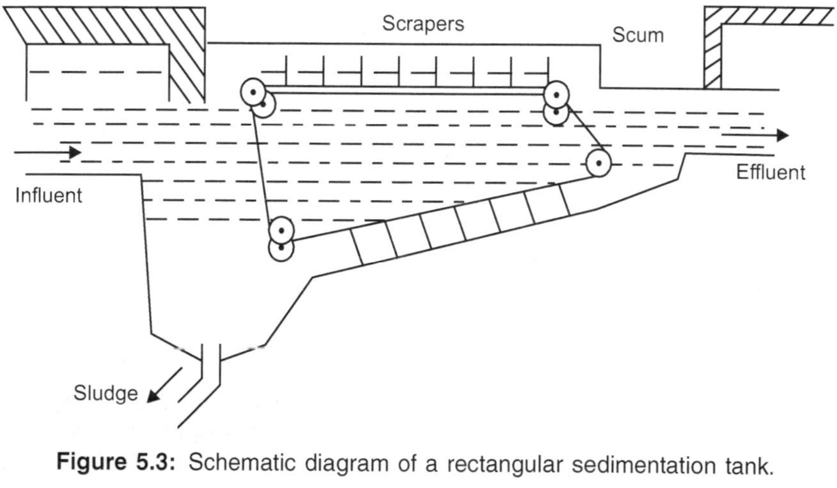

Figure 5.3: Schematic diagram of a rectangular sedimentation tank.

Design criteria for gravity sedimentation tanks normally provide for 2-h retention based on average flow, with longer retention periods used for light solids or inert solids that do not change during their retention. Care should be taken that sedimentation time is not too long; otherwise, the solids will compact too densely and affect collection and removal. Organic solids generally will not compact to more than 5 to 10%. Inorganic solids will compact up to 20 or 30%. Centrifugal sludge pumps can handle solids that compact up to 5 or 6%, while positive-displacement sludge pumps can handle solids up to 10%. With solids above 10%, the sludge tends to lose fluid properties and must be handled as a semi-solid rather than a fluid. Circular sedimentation tanks have steel truss boxes with angled sludge scrapers on the lower side. As the sludge scrapers rotate, the solids are pushed toward the sludge hopper for removal on a continuous or semi-continuous basis. The rectangular sedimentation tanks employ chain-and-flight sludge collectors or rail-mounted sludge collectors. When floating solids can occur in primary sedimentation tanks, surface skimmers are mounted on the sludge scrapers so that these are removed at regular intervals.

The surface overflow rate (SOR) for primary sedimentation is normally held close to 40.74 $m^3/(m^2.day)$ [1000 $gal/(ft^2.day)$] for average flow rates, depending upon the solids characteristics. Lowering the SOR below 40.74 $m^3/(m^2.day)$ does not produce improved effluent quality in proportion to the reduction in SOR. Generally, the minimum depth of sedimentation tanks is 3.0 m (10 ft), with circular sedimentation tanks having a minimum diameter of 6.0 m (20 ft) and rectangular sedimentation tanks having length-to-width ratios of 5:1. Chain-and-flight limitations generally keep the width of rectangular sedimentation tanks to increments of 6.0 m (20 ft) or less. While hydraulic overflow rates have been limited on the effluent weirs, operating experience has indicated that the recommended limit of 186 $m^3/(m.day)$ [15,000 gal/(ft.day)] is lower than that necessary for good operation. A circular sedimentation tank with single-edge weir provides adequate weir length and is easier to adjust than one with a double-sided weir. More problems appear to be created from improper adjustment of the effluent weirs than from their improper length.

5.6.6 Chemical precipitation

Chemical precipitation is the most common method for removing dissolved metals from wastewater solution containing toxic metals. To convert the dissolved metals into solid particle form, a precipitation reagent is added to the mixture. A chemical reaction, triggered by the reagent, causes the dissolved metals to form solid particles. Filtration can then be used to remove the particles from the mixture. How well the process works is dependent upon the kind of metal present, the concentration of the metal and the kind of reagent used. In hydroxide precipitation, a commonly used chemical precipitation process, calcium or sodium hydroxide is used as the reagent to create solid metal hydroxides. However, it can be difficult to create hydroxides from dissolved metal particles in wastewater because many wastewater solutions contain mixed metals.

Chemical precipitation

Lightweight suspended solids and colloidal solids can be removed by chemical precipitation and gravity sedimentation. In effect, the chemical precipitate is used to agglomerate the tiny particles into large particles that settle rapidly in normal sedimentation tanks. Aluminium sulphate, ferric chloride, ferrous sulphate, lime and polyelectrolytes have been used as coagulants. The choice of the coagulant depends upon the chemical characteristics of the particles being removed, the pH of the wastewaters and the cost and availability of the precipitants. While the precipitation reaction results in removal of the suspended solids, it increases the amount of sludge to be handled. The chemical sludge must be considered along with the characteristics of the original suspended solids in evaluating sludge-processing systems.

Normally, chemical precipitation requires a rapid mixing system and a flocculation system ahead of the sedimentation tank. However, in a circular sedimentation tank, the rapid-mixer and flocculation units are built into the tank. Schematic diagrams of chemical treatment systems are shown in Figs 5.4 and 5.5. Rapid mixers are designed to provide 30-s retention at average flow with sufficient turbulence to mix the chemicals with the incoming wastewaters. The flocculation units are designed for slow mixing at 20-min retention. These units are designed to cause the particles to collide and increase in size without excessive shearing. Care must be taken to move the flocculated mixture to the sedimentation unit without disrupting the large floc particles.

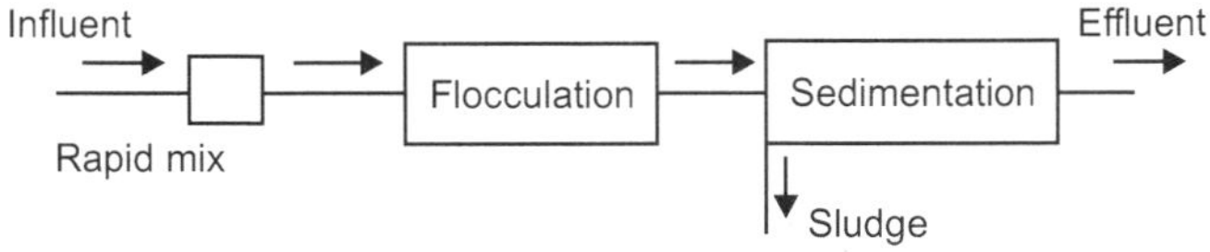

Figure 5.4: Schematic diagram of a chemical precipitation system for rectangular sedimentation tanks.

5.7 Secondary treatment of wastewater

Secondary treatment is designed to remove colloidal and soluble organics through microbial metabolism. Biological treatment systems convert the biodegradable organics in the solution into suspended organics which flocculate

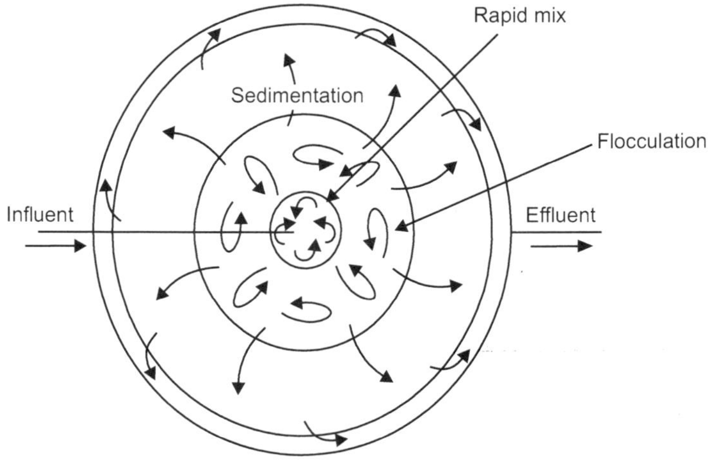

Figure 5.5: Schematic diagram of a chemical precipitation system for circular sedimentation tanks.

and are removed by gravity sedimentation. Colloidal solids tend to be adsorbed onto the microbial flocs.

Biological wastewater treatment systems employ bacteria as the primary micro-organisms responsible for removing excess bacteria and producing a clarified effluent. The bacteria average 0.5 to 1.0 μm in diameter and 1.0 to 2.0 μm in length. Protozoa range from 10 μm to several hundred micrometers in size. The bacteria cannot metabolise solid organics but instead convert solid particles to soluble organics prior to metabolism. Fortunately, the bacteria have enzymes on their cell surface which can hydrolyse the complex organics to simple organic molecules. Bacteria are widely distributed throughout the environment with different biochemical characteristics. Soil is the best source of concentrated microbes, since the bacteria in the soil are responsible for the stabilisation of waste materials. Normally, the desired bacteria are extracted from the soil by growing them in the desired wastewaters until sufficient numbers have been obtained for the desired reactions. While specialised bacteria are available for purchase, it is not normally necessary to do so as soil will supply the desired micro-organisms more economically.

In order to grow, bacteria must have a suitable environment with all the proper nutrients. The environment must provide good mixing for adequate contact between the micro-organisms and the pollutants being metabolised. It can be either aerobic with excess dissolved oxygen or anaerobic without any dissolved oxygen. The pH should be maintained between 6 and 9. There should be sufficient nitrogen, phosphorus, iron and trace metals for good growth, but there should not be too high a concentration of heavy metals to be toxic. The temperature should be maintained between 5 and 35°C, although it is possible to operate thermophilic systems up to 65°C. If the environment is not balanced, improper microbial growth and inadequate treatment can occur. One of the major problems lies with filamentous microbes, which adversely affect flocculation and sedimentation. In the absence of adequate oxygen and iron, filamentous bacteria can predominate over normal bacteria. Under low-pH conditions and low-nitrogen or high-carbohydrate wastes, filamentous fungi predominate over bacteria.

The best quality effluent is produced under aerobic conditions. The bacteria metabolise the biodegradable organics, using dissolved oxygen as their terminal electron acceptor. Approximately one-third of the organics metabolised are oxidised to carbon dioxide and water to obtain the energy to convert the remaining two-thirds of the organics to microbial protoplasm. Microbial protoplasm is similar regardless of the bacteria species or the organics being metabolised. Chemical analysis indicates that microbial protoplasm contains 90% organics and 10% inorganics. The organic fraction contains approximately 50% C, 7% H, 31% O and 12% N by weight. The major problem in aerobic treatment is the large volume of microbial solids to be processed. Microbial protoplasm is not stable but rather continues to be degraded over time in a process known as 'endogenous respiration'. Approximately 80% of microbial protoplasm can be oxidised, but 20% of the organic fraction is stable in aerobic treatment systems and must be processed further.

In the absence of dissolved oxygen, the bacteria must find another electron acceptor. Chemically bound oxygen is the primary electron acceptor under anaerobic conditions. NO_3, NO_2, SO_4, CO_2 and organic oxygen compounds are used as electron acceptors. Nitrates are reduced to nitrites and to various intermediates before being reduced to nitrogen gas. It is important to recognise that denitrification results in the ultimate production of nitrogen gas that is insoluble in the wastewaters rather than in the production of ammonia or ammonium ions. If the nitrates are needed as a source of nitrogen for the bacteria, they will be reduced to ammonium ions in proportion to the nitrogen needs for protoplasm. A special group of bacteria use sulphates as their electron acceptor and can produce thiosulphates, free sulphur, or hydrogen sulphide. The methane bacteria can reduce carbon dioxide with the production of methane, which is relatively insoluble in water and can be used as a source of energy. Organic oxygen compounds such as carbohydrates are broken down to simple organic acids, aldehydes, ketones and alcohols, which can be metabolised by the methane bacteria to methane and carbon dioxide.

Since it takes the same amount of energy to produce a unit of microbial protoplasm aerobically and anaerobically, less microbial cell mass is produced per unit of organic pollutants metabolised anaerobically rather than aerobically.

The anaerobic reactions do not yield as much energy to the microbes as the aerobic reactions and their end products contain considerable energy that is not available to the microbes. Current research into anaerobic wastewater treatment systems is being stimulated by the production of methane, which can be used as an energy source and by the lower microbial solids production. Approximately 80 to 95% of the organics can be converted to methane and carbon dioxide, with 5 to 20% being converted into microbial solids.

Time is a critical variable in sizing wastewater treatment systems. A definite time period is required to metabolise a given amount of organic matter by a unit of cell mass. By retaining the microbes in the treatment system, the treatment time per unit of organic matter is reduced. Unfortunately, the time for aerobic treatment is controlled by oxygen transfer. Under anaerobic conditions, contact between the microbes and the organic pollutants controls the total reaction time. With proper design, more organic matter can be treated anaerobically than aerobically in the same time period. Unfortunately, the anaerobic system cannot produce as high-quality effluent as the aerobic system. The net result is that anaerobic systems are often used as the first stage of biological treatment, with aerobic systems being used as the second stage.

5.7.1 Facultative lagoons

These lagoons have been designed to use both aerobic and anaerobic reactions. Normally, facultative lagoons consist of two or more cells in series. The settleable solids tend to settle out in the first cell and undergo anaerobic metabolism with the production of organic acids and methane gas, which bubbles out into the atmosphere. Algae at the surface of the lagoon utilise sunlight for their energy in converting carbon dioxide, water and ammonium ions into algal protoplasm with the release of oxygen as a waste product. Aerobic bacteria utilise the oxygen released by the algae to stabilise the soluble and colloidal organics. Thus, the bacteria and algae form a symbiotic relationship as shown in Fig. 5.6. The interesting aspect of facultative lagoons is that the organic matter in the incoming wastewaters is not stabilised but rather is converted to microbial protoplasm, which has a slower rate of oxygen demand. In fact, in some facultative lagoons, inorganic compounds in the wastewaters are converted into organic compounds with a total increase in organics within the lagoon system.

Facultative lagoons are designed on the basis of organic load in relationship to the potential sunlight availability. The facultative lagoons can be designed on the basis of 2.2g/(m^2.day) [20 lb BOD5/(acre.day)]. In some cases, the organic load can be increased to 3.4 to 4.5 g/(m^2. day) [30 to 40 lb BOD5/(acre.day)], or 6.7 g/(m^2.day)[60 lb BOD5/(acre.day)]. The depth of lagoons is normally maintained between 1.0 and 1.7 m (3 and 5 ft). A depth less than

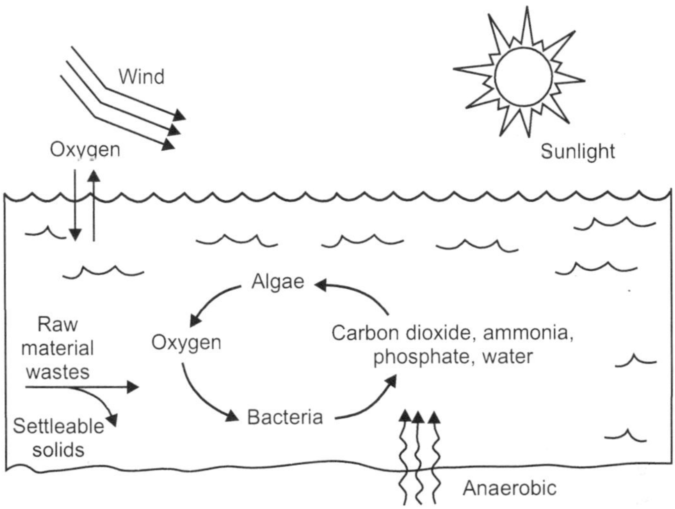

Figure 5.6: Processes in the facultative pond.

1.0 m (3 ft) encourages the growth of aquatic weeds and permits mosquito breeding. In dry areas, the maximum depth may be increased above 1.7 m (5 ft) depending upon evaporation. Most facultative lagoons depend upon natural wind action for mixing and should not be placed in screened areas where wind action is blocked.

Effluent quality from facultative lagoons is related primarily to the suspended solids created by living and dead microbes. The long retention period in the lagoons allows the microbes to die, leaving a small particle that settles slowly. The release of nutrients from the dead microbes permits the algae to survive by recycling the nutrients. Thus, the algae determine the ultimate effluent quality. Use of a series ponds with well-designed transfer structures of permits maximum retention of algae within the ponds and the best-quality effluent. Normally, the soluble BOD5 is under 5 or 10 mg/L with a total effluent BOD5 under 30 mg/L. The effluent suspended solids vary widely in different seasons, being a maximum of 70 to 100 mg/L in the summer months and a minimum of 10 to 20 mg/L in the winter months. If suspended-solids removal is essential, chemical precipitation is currently the best method available. Slow sand filters and rock filters have been studied for suspended-solids removal; they work well as long as the effluent suspended solids are relatively low, 40 to 70 mg/L.

5.7.2 Aerated lagoon

Aerated lagoons are one of the aerobic suspended growth processes. An aerated lagoon is a basin in which wastewater is treated either on a flow through basis or with solids recycle. Oxygen is usually supplied by means of surface aerators on floats or on fixed platforms or diffused air aeration units instead of photosynthetic oxygen yield as in case of oxidation pond. The action of the aerators and that of the rising air bubbles from the diffuser are used to keep the contents of the basin in suspension.

Aerated lagoons have the advantages such as ease of operation and maintenance, equalisation of wastewater and a high capacity of heat dissipation when required. The disadvantages of aerated lagoons are large area requirement, difficulty in process modification, high effluent suspended solids concentration, sensitivity of process efficiency to variation in ambient air temperature.

Anaerobic lagoons

These lagoons were developed when a major fraction of the organic contaminants consisted of suspended solids that could be removed easily by gravity sedimentation. The anaerobic lagoons relatively are deep [3.0 to 6.0 m (10 to 20 ft)], with a short fluid-retention time (3 to 5 days) and a high BOD5 loading rate, up to 3.2 kg/(m^3.day) [200 lb/1000 ft^3.day)]. Microbial metabolism in the settled-solids layer produces methane and carbon dioxide, which quickly rise to the surface, carrying some of the suspended solids. A scum layer that retards oxygen transfer and release of obnoxious gases is quickly produced in anaerobic lagoons. Mixing with a grinder pump can provide a better environment for metabolism of the suspended solids. The key for anaerobic lagoons is adequate buffer to keep the pH between 6.5 and 8.0. Protein wastes have proved to be the best pollutants to be treated by anaerobic lagoons, with the ammonium ions reacting with carbon dioxide and water to form ammonium bicarbonate as the primary buffer. High-carbohydrate wastes are poor in anaerobic lagoons since they produce organic acids without adequate buffer, making it difficult to maintain a suitable pH for good microbial growth.

Anaerobic lagoons do not produce a high-quality effluent, but are able to reduce the BOD load by 80 to 90% with a minimum effort. Since anaerobic lagoons work best on strong organic wastes, their effluent must be treated by either aerated lagoons or facultative lagoons. An anaerobic lagoon is simply the first stage in the treatment of strong organic wastewaters.

5.7.3 Trickling filters

For years, trickling filters were the mainstay of biological wastewater treatment systems because of their simplicity of design and operation. Trickling filters were displaced as the primary biological treatment system by activated sludge because of the better effluent quality. Tricking filters are simply fixed-medium biological reactors, with the wastewaters being spread over the surface of a solid medium where the microbes are growing. The microbes remove the organics from the wastewaters flowing over the fixed medium. Oxygen from the air permits aerobic reactions to occur on the surface of the microbial layer, but anaerobic metabolism occurs at the bottom of the microbial layer where oxygen does not penetrate.

5.7.4 Rotating biological contactors (RBC)

The newest form of trickling filter is the rotating biological contactor with a series of circular plastic disks, 3.0 to 3.6 m (10 to 12 ft) in diameter, immersed to approximately 40% diameter in a shaped contact tank. The RBC disks rotate at 2 to 5 r/min. As the disks travel though the wastewaters, a small layer adheres to them. As the disks travel into the air, the microbes on the disk surface oxidise the organics. Thus, only a small amount of energy is required to supply the required oxygen for wastewater treatment. As the microbes build up on the plastic disks, the shearing velocity that is created by the movement of the disks through the water causes the excess microbes to be removed and discharged to the final sedimentation tank.

Rotating biological contactors have been very popular in treating industrial wastes because of their relatively small size and their low energy requirements. Unfortunately, a number of problems must be recognised prior to using RBCs. Strong industrial wastes tend to create excessive microbial growths which are not easily sheared off and which create high oxygen-demand rates with the production of hydrogen sulphide and other obnoxious odours. The heavy microbial growths can damage some of the disks and cause shaft failures. The disks are now being covered with plastic shells to prevent nuisance odours from occurring. Air must be forced through the covered RBC systems and be chemically treated before being discharged back into the environment. Recirculation of wastewater flow around the RBC units can distribute the load over all the units and reduce the heavy initial microbial growths. RBC units also work best under uniform organic loads, requiring surge tanks for many industrial wastes. The net result is that the cost of RBC units is similar to that of other treatment units in terms of organic matter stabilised.

5.7.5 Activated sludge

Activated sludge is the most widespread biological wastewater treatment system because of its high quality effluent. The high energy requirements for activated sludge have, however, caused some questions to be raised about the continued use of such systems. There is no doubt that these systems are energy intensive, but there does not appear to be any way to provide treatment without the expenditure of a definite amount of energy. Care should be taken with regard to activated sludge systems providing a high degree of treatment with a minimum expenditure of energy.

Modified activated sludge systems offer considerable choice in processes. Complete-mixing of activated sludge is the most popular system for industrial wastes because of its ability to absorb shock loads better than other modifications. Contact stabilisation is a modification of activated sludge that is best suited to wastewaters having high suspended solids and low soluble organics. Contact stabilisation employs a short-term mixing tank to adsorb the suspended solids and metabolise the soluble organics, a sedimentation tank for solids separation and a reaeration tank for stabilisation of the suspended organics. Extended aeration systems are actually long-term aeration, completely mixed activated sludge systems. They employ 24 to 48 hour aeration periods and high mixed liquor suspended solids to provide complete stabilisation of the organics and aerobic digestion of the activated sludge in the same aeration tank. The oxidation ditch is a popular form of the extended aeration system employing mechanical aeration. Pure oxygen systems are designed to treat strong industrial waste in series, completely mixed units having relatively short contact periods. One of the latest modifications of activated sludge system employs powdered activated carbon to adsorb complex organics and assist in solids separation. Another modification employs a redwood-medium trickling filter ahead of a short-term aeration tank, with mixed liquor recycled over the redwood-medium tower to provide heavy microbial growth on the redwood as well as in the aeration tank.

Each activated sludge process employs an aeration tank for waste stabilisation, with conversion of the organics to new microbial solids and a sedimentation tank for solids separation.

Aeration systems

These systems control the design of aeration tanks. Aeration equipment has two major functions: mixing and oxygen transfer. Diffused aeration equipment employs either a fixed speed positive displacement blower or a high-speed turbine blower for readily adjustable air volumes. Air diffusers can be located along one side of the aeration tank or spread over the entire bottom of the tank. They can be either fine-bubble or coarse-bubble diffusers. Fine bubble diffusers are more efficient in oxygen transfer, but require more extensive air cleaning equipment to prevent them from clogging as a result of dirty air. Mechanical surface aeration equipment is more efficient but is not as flexible. Economics has dictated the use of large power aerators, but tank configuration has tended to favour the use of more lower-power aerators. Mixing is a critical problem with mechanical surface aerators since they are a point source pump of limited capacity. Experience shows that bearings are a serious problem with mechanical-aeration equipment. Wave action generated within the aeration tank tends to produce lateral stresses on the bearings and results in failures and increased maintenance costs. Slow-speed mechanical surface-aeration units present fewer problems than the high-speed mechanical surface-aeration units. Deep tanks, greater than 3.0 m (10 ft), require draft tubes to ensure proper hydraulic flow. Short-circuiting is one of the major problems associated with mechanical aeration equipment. Combined mechanical and diffused aeration systems enjoy some popularity for industrial waste systems that treat variable organic loads. The mechanical mixers provide the fluid mixing with the diffused aeration varied for different oxygen transfer rates.

Sedimentation tanks

These tanks are an integral part of any activated sludge system. It is essential to separate the suspended solids from the treated liquid if a high quality effluent is to be produced. Circular sedimentation tanks with various types of hydraulic sludge collectors have become the standard secondary sedimentation systems. Square tanks have been used with common wall construction for compact design with multiple tanks.

Most secondary sedimentation tanks use centre feed inlets and peripheral weir outlets. Recently, efforts have been made to employ peripheral inlets with submerged orifice flow controllers and either centre weir outlets or peripheral weir outlets adjacent to the peripheral inlet channel.

5.8 Tertiary treatment of wastewater

Tertiary treatment is the final treatment, meant for 'polishing' the effluents from the secondary treatment processes, to improve its quality further. The major objectives of tertiary treatment are:

1. Removal of fine suspended solids.
2. Removal of bacteria.
3. Removal of dissolved inorganic solids.
4. Removal of final traces of organics, if it is felt necessary.

Removal of finely divided suspended solids can be achieved with the help of micro-strainers and sand filters. Removal of bacteria, particularly of faecal origin, can be achieved by retaining the effluents from secondary biological treatment plants in maturation ponds or lagoons for specified periods of time. Three or four lagoons arranged in series give an excellent final effluent with very low BOD and low suspended solids. The final effluent is chlorinated if necessary. Removal of dissolved inorganic solids is a major problem with wastewaters from industries such as fertilisers, textile processing, tannery and electroplating. Depending upon the required quality of the final effluent and the cost of treatment that can be afforded in a given situation, any of the following treatment methods can be employed.

5.8.1 Reverse osmosis

Reverse osmosis (RO) systems can often improve the quality of water. The reverse osmosis water treatment method has been used extensively to convert brackish or seawater to drinking water, to clean up wastewater and to recover dissolved salts from industrial processes. It is becoming more popular in the home market as home owners are increasingly concerned about contaminants that affect their health, as well as about non-hazardous chemicals that affect the taste, odour, or colour of their drinking water.

Reducing contaminants through RO

Reverse osmosis treatment reduces the concentration of dissolved solids, including a variety of ions and metals and very fine suspended particles such as asbestos that may be found in water. An RO device may be installed following a water softener to reduce the concentration of sodium ions exchanged for hardness ions. RO also removes certain organic contaminants, some detergents and specific pesticides.

Although RO membranes can remove virtually all micro-organisms, it is currently recommended that only microbiologically safe (i.e., coliform negative) water be fed into RO systems. Reverse osmosis will not remove all contaminants from water. Dissolved gases such as oxygen and carbon dioxide pass through RO membranes into the treated water. Unfortunately, hydrogen sulphide gas, with its notorious odour of rotten eggs, also passes through the RO membrane. RO in general is not a very effective treatment for trihalomethanes (THMs), some pesticides, solvents and other volatile organic chemicals (VOCs). If the water source is severely polluted or untreatable, a public water supply or a reliable private water source must be used.

Reverse osmosis process: In the reverse osmosis process a cellophane-like membrane separates purified water from contaminated water. An understanding of osmosis is needed before further describing RO. Osmosis occurs when two solutions containing different quantities of dissolved chemicals are separated by a semi permeable membrane (allowing only some compounds to pass through). Osmotic pressure of the dissolved chemical causes pure water to pass through the membrane from the dilute to the more concentrated solution (Fig. 5.7) There is a natural tendency for chemicals to reach equal concentrations on both sides of the membrane.

In reverse osmosis, water pressure applied to the concentrated side forces the process of osmosis into reverse. Under enough pressure, pure water is 'squeezed' through the membrane from the concentrated to the dilute side (Fig. 5.8). Salts dissolved in water as charged ions are repelled by the RO membrane. Treated water is collected in a storage container. The rejected impurities on the concentrated side of the membrane are washed away in a

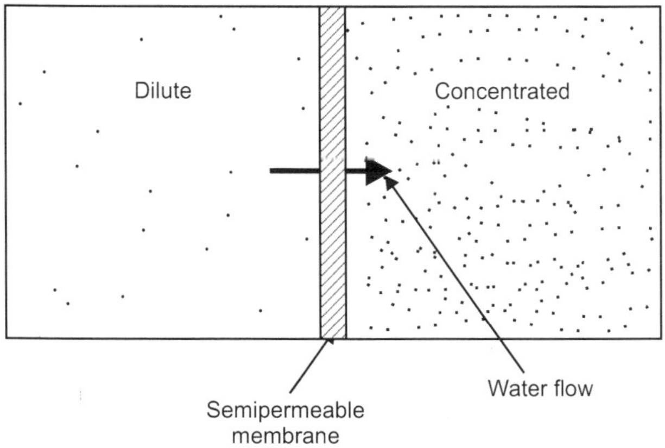

Figure 5.7: In osmosis, water across the membrane from the dilute to the concentrated solution.

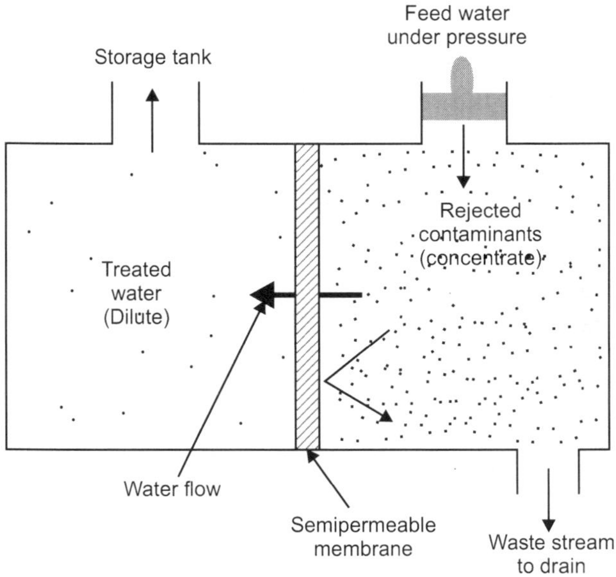

Figure 5.8: The reverse osmosis, pressure is applied to the concentrated solution reversing the natural direction of flow, forcing water across the membrane from the concentrated solution into the more dilute solution.

stream of wastewater, not accumulated as on a traditional filter. The RO membrane also functions as an ultrafiltration device, screening out particles, including micro-organisms that are physically too large to pass through the membrane's pores. RO membranes can remove compounds in the 0.0001 to 0.1 micron size range (thousands of times smaller than a human hair).

Design of an RO system

Although the reverse osmosis process is simple, a complete water treatment system is often complex, depending on the quality of the incoming water before treatment and the consumer's needs. Most home RO systems are point-of-use (POU) units placed beneath the kitchen sink to treat water used for cooking and drinking. Point-of-entry (POE) systems that treat all the water entering the household are more expensive to purchase and operate than POU systems.

A typical home reverse osmosis system consists of pre-treatment and post-treatment filters as well as the RO membrane, flow regulator, storage container for the treated water and dispensing faucet (Fig. 5.9). The pressure for RO is usually supplied by the feed line pressure of the water system in the home, but a booster pump may be needed to produce an adequate volume of treated water. A sediment pre-filter is essential for removing relatively large sand grains and silt that may tear or clog the RO membrane or clog a pump or flow regulator. Water softeners are used in advance of the RO system when household water is excessively hard. If the water is chlorinated or contains other oxidising chemicals such as bromine, an activated carbon pre-filter is needed to protect membranes sensitive to these chemicals.

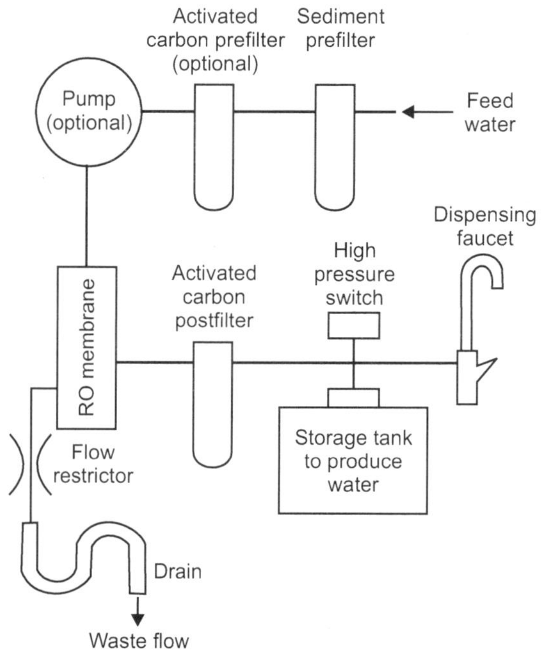

Figure 5.9: A schematic of a typical RO system.

To remove certain pesticides and organic solvents, an activated carbon (AC) post-treatment must be included in the system. A standard AC filter positioned after the storage tank removes compounds that cause unpleasant taste and odours, including those from the tank or plastic tubing, just before water is dispensed. To remove high levels of organic chemicals such as trihalomethanes, volatile organic chemicals and chloramines, an additional prolonged contact carbon filter (PCCF) is placed between the RO membrane and the storage tank. Combining an activated carbonfilter with RO expands the range of chemicals the system can remove.

Furthermore, AC treatment is improved because RO removes compounds that adversely affect AC adsorption.

The storage tank, tubing and dispensing faucet should be made of plastic, stainless steel, or other nontoxic materials. The low pH and mineral content of RO-treated water may corrode copper pipes and allow lead to leach into the drinking water from brass components.

RO membrane materials: The most common RO membrane materials are polyamide thin film composites (TFC) or cellulosic types (cellulose acetate [CA], cellulose triacetate [CTA], or blends). Very thin membranes are made from these synthetic fibres. Membrane material can be spiral wound around a tube, or hollow fibres can be bundled together, providing a tremendous surface area for water treatment inside a compact cylindrical element (Fig. 5.10).

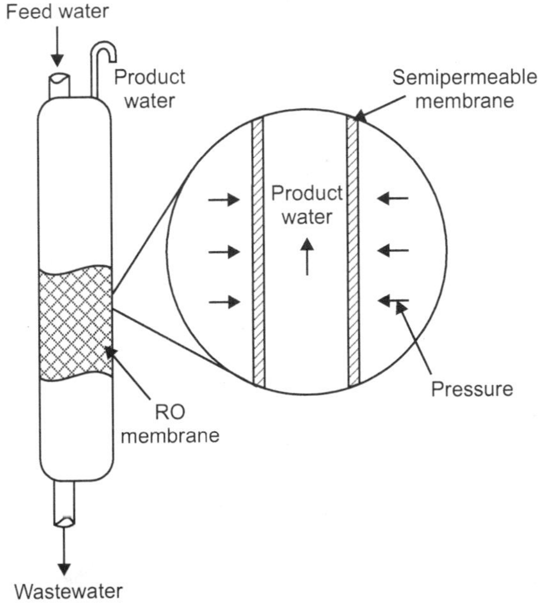

Figure 5.10: A typical osmosis device.

Hollow fibre membranes have greater surface area (and therefore greater capacity) but are more easily clogged than the spiral wound membranes commonly used in home RO systems. The flux, or capacity, of the RO membrane indicates how much treated water it can produce per day. Typically, RO membranes for home systems are rated in the range of 24 to 100 gallons per day. Thus, under standard operating conditions it could take from two to six hours to fill a two and-a-half-gallon storage tank. CA/CTA membranes have adequate capacity for most households, but TFC membranes should be used if large volumes of treated water are needed.

RO membranes are rated for their ability to reject compounds from the contaminated water. A rejection rate (% rejection) is calculated for each specific ion or contaminant as well as for reduction of total dissolved solids (TDS). It is important that consumers know their specific requirements for water quality when buying a system. For example, high rejection rates are essential when high nitrates or lead concentrations in the water must be brought below the EPA maximum contaminant or action levels.

Although thin film composite membranes are initially more expensive, they have superior strength and durability, as well as higher TDS rejection rates (>95%), than cellulosic membranes (88–94%). TFC membranes are more resistant to microbial attack, stand up better to high pH (greater than 9) and are better able to handle higher levels of total dissolved solids (1500–2000 ppm) than cellulosic (CA/CTA) membranes. The advantages of cellulosic membranes are their lower cost and ability to tolerate chlorine, which curtails the growth of micro-organisms in the system. Thin film composites deteriorate in chlorinated water but perform well with an activated carbon prefilter to remove chlorine.

Membranes made of sulphonated polysulphone (SPS) are chlorine tolerant (like CTA) and can withstand higher pH levels (like TFC), but they do not match the low costs of CTA or the performance of TFC. SPS membranes are best used in RO systems when the feed water is soft and high in pH or when high nitrate levels are a primary concern.

Nanofiltration membranes (also referred to as loose RO or softening membranes) have much higher flow rates than other membranes. They tend to reject negatively charged ions such as sulphates very well but do not perform well in removing total dissolved solids.

Efficiency of RO systems

The performance of an RO system depends on membrane type, flow control, feed water quality (e.g., turbidity, TDS and pH), temperature and pressure. The standard at which manufacturers rate RO system performance is 77°F, 60 pounds per square inch (psi) and TDS at 500 parts per million (ppm). Only part of the water that flows into an RO system comes out as treated water. Part of the water fed into the system is used to wash away the rejected compounds and goes down the drain as waste.

The recovery rate, or efficiency, of the system is calculated by dividing the volume of treated water produced by the volume of water fed into the system:

$$\% \text{ Recovery} = \frac{\text{Volume of treated water produced}}{\text{Volume of feed water used}}$$

If not properly designed, RO systems can use large quantities of water to produce relatively little treated water. Most home RO systems are designed for 20 to 30% recovery (i.e., 2–3 gallons of treated water are produced for every 10 gallons put into the system). Home RO systems can operate at higher recovery rates, but doing so may shorten membrane life. The flow regulator on the reject stream must be properly adjusted. If the flow is slow, the recovery rate is high, but RO membranes are easily fouled if concentrated impurities are not washed away quickly enough. If the flow is too fast, the recovery rate is low and too much water goes down the drain.

Overall water quality affects the efficiency of an RO system and its ability to remove specific contaminants. The higher the TDS, the lower the recovery rate of treated water. The amount of treated water produced decreases 1 to 2% for every degree below the standard temperature of 77°F. An RO system supplied with well water at a temperature of 60°F produces only three-quarters of the volume it would produce at 77°F.

For an RO system to function properly, there must be enough water pressure. Although most home RO systems are rated at 60 pounds per square inch, the incoming feed fine pressure of many private water systems is less than 40 psi. The RO system must work against back pressure created in the storage tank as it fills with water and compresses the air in the tank. The RO device must also overcome osmotic pressure, the bonding between water molecules and the dissolved impurities; the higher the TDS level, the greater the osmotic pressure. The net water pressure at the RO membrane can be calculated by subtracting back pressure and osmotic pressure from the feed line pressure. If the net water pressure at the membrane is lower than 15 psi, treated water production is less efficient and contaminant rejection rates are lower. Auxiliary pumps can be added to the treatment system to boost pressure and improve the quality and quantity of water produced. High-quality RO systems tank pressure reaches two-thirds of the feed pressure; at that point, low net water pressure can result in low rejection rates. In some systems, once the storage tank is filled, surplus treated water is discarded; water loss from such units is frequently excessive. A system that automatically shuts off when the pressure on the tank reaches a given level saves water.

Maintenance of an RO system

An RO system must be well maintained to ensure reliable performance. Clogged RO membranes, filters, or flow controls will decrease water flow and the system's performance. If fouling is detected in the early stages the membrane can often be cleaned and regenerated. The cleaning procedure varies depending on the type of membrane and fouling. Completely clogged or torn RO membranes must be replaced. In addition, pre- or post-filters must be replaced once a year or more often, depending on the volume of water fed through the system and the quality of the feed water.

Damage to RO membranes cannot be seen easily. The treated water must be analysed periodically to determine whether the membrane is intact and doing its job. Many systems now have a built-in continuous monitor that indicates a high TDS level, a sign that the system is not operating properly. It may also be necessary to test regularly for specific health-related contaminants such as nitrates or lead.

Micro-organisms, dead or alive, can clog RO membranes. To prevent bio-fouling, RO units must be disinfected periodically with chlorine or other biocides provided by the manufacturer. Continuous chlorination can be used with cellulose membranes to protect the system from bio-fouling and eliminate the particle trapping slime that worsens other forms of fouling such as scaling.

Chlorine and other oxidising disinfectants are harmful to thin film composite membranes. If the feed water is chlorinated, an activated carbon unit must be in place to remove the oxidising chemicals before they reach the TFC membrane. Activated carbon (AC) prefilters should not be used on nonchlorinated water supplies because they provide a place for micro-organisms to multiply and lead to increased bio-fouling of the RO membrane surface. It is important to replace AC filters periodically following the manufacturer's instructions, especially after an extended shutdown period during which micro-organisms can flourish.

5.8.2 Ion exchange process

Softening of water by the lon exchange process involves the exchange substitution of the hardness minerals chiefly calcium and magnesium for sodium minerals. The exchange is made possible because the minerals are ionic in nature often called ionised impurities means they have an electrical charge. The ion exchange process is based on the fact that like charges repel one another and unlike charges attract. Calcium and magnesium ions in water are actually dissolved rock. They have been dissolved by water the universal solvent tricking down through strata rock and soil dissolving the calcium and magnesium deposits as it goes. This dissolved rock eventually finds its way into an underground aquifer and when water from the aquifer is pumped to the surface it contains the dissolved hardness minerals of calcium and magnesium and is said to be hard water. The water softening and recharge process is shown in Fig. 5.11.

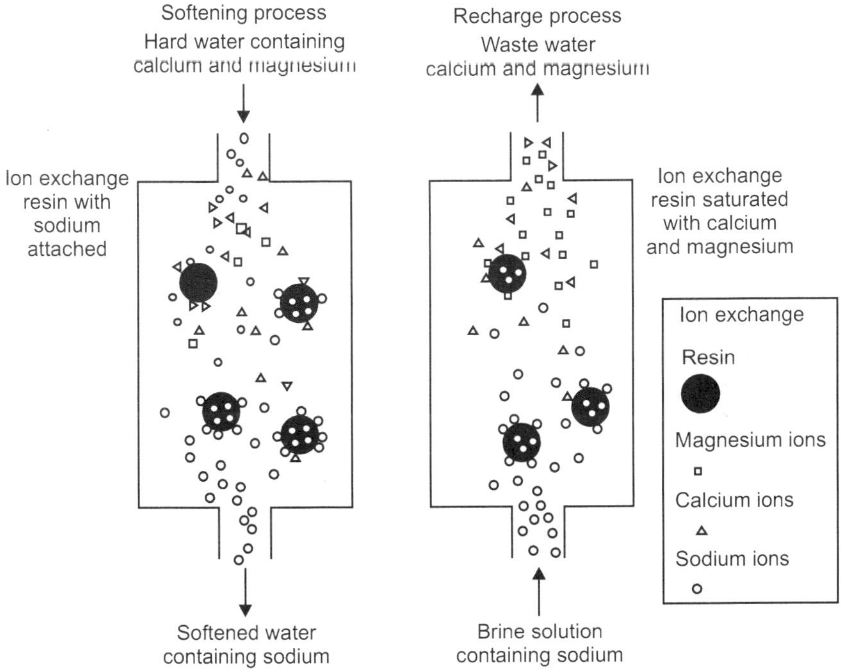

Figure 5.11: The water softening and recharge process.

An ion exchange softener exchanges the hardness minerals calcium and magnesium dissolved in water for the soft mineral sodium contained on the softener resin which does far less damage to objects it contacts. Principally because it does not build up on surfaces as scale deposits.

All three minerals are positively charged ions called cations. The exchange takes place by passing water containing hardness minerals over a man-made ion exchange resin in a suitable pressure vessel (tank). The resin polystyrene divinyl benzene in most modern softeners consists of millions of tiny plastic balls (beads) all of which contain many negatively charged exchange sites attracting positive cations when the resin is in the regenerated state these negatively charged exchange sites hold positively charged sodium cations.

As the calcium and magnesium contact the resin beads in their travel through the resin tank they displace the sodium ions from the exchange sites during the ion exchange process relatively small amounts of other strongly charged cations, such as ion and manganese are also removed along with the calcium and magnesium.

Ion exchange is possible for two reasons:

1. All cations do not have the same strength of positive charge.
2. The resin prefers the more strongly charged cations calcium and magnesium than it does the weaker sodium cations.

The displaced sodium cations then pass downward through the resin 'bed' and out the softener outlet thus the softener delivers soft water. Eventually all of the resin exchange sites are occupied by calcium and magnesium and no further exchange can take place The resin is solid to be exhausted and must be regenerated. The resin of the softener is regenerated with a dilute solution of sodium chloride (common salt) and water-brine. During regeneration the flow of service water from the softener is first stopped. Brine is drawn from the brine tank mixing with a separate stream of water. The brine solution flows downward through the resin contacting the resin beads loaded with calcium and magnesium ions. Even though the calcium and magnesium are more strongly charged than the sodium the concentrated brine solution contains literally billions of the more weakly charged sodium ions which have the power to displace the smaller number of calcium and magnesium ions when the calcium and magnesium ions are displaced (exchanged) the positive sodium ions are then attracted to the negative exchange sites eventually all sites are taken up by sodium ions and the resin is said to be regenerated and ready for the next softening cycle.

5.8.3 Electrodialysis

The principle behind electrodialysis is that electrical potential gradients will make charged molecules diffuse in a given medium at rates far greater than attainable by chemical potentials between two liquids as in conventional dialysis. When a DC electric current is transmitted through a saline solution, the cations migrate toward the negative terminal or cathode and the anions toward the positive terminal, the anode. By adjusting the potential between the terminals or plates, the electric current and therefore, the flow of ions transported between the plates can be varied.

Electrodialysis can be applied to the continuous-flow type of operation needed in industry. Multi-membrane stacks can be built by alternately spacing anionic and cationic-selective membranes. Among the technical problems associated with the electrodialysis process, concentration polarisation is perhaps the most serious. Other problems in practical applications include membrane scaling by inorganics in feed solutions as well as membrane fouling by organics. Efficient separation or pre-treatment in the influent streams can include activated carbon absorption to reduce or prevent such problems.

Principal applications of electrodialysis include: (i) recovery of materials from liquid effluents, such as processes related to conservation, cleanup, concentration and separation of desirable fractions from undesirable ones, (ii) purification of water sources and (iii) effluent water renovation for re-use or to meet point source disposal standards required to maintain suitable water quality in the receptor streams.

Treatment of brackish water in the production of potable supplies has been the largest application of electrodialysis. Costs associated with electrodialysis processes depend on such factors as the total dissolved solids (TDS) in the feed, the level of removal of TDS (per cent rejection) and the size of the plant.

A combination of electrodialysis with conventional evaporation, for example, may be substantially cheaper than evaporation alone for the production of dry salt from saline solutions. Competing technologies include reverse osmosis and crystallisation.

Added control of the movement of the ions can be obtained by placing sheet-type membranes of cation or anion-exchange material between the outer plates, as shown diagrammatically in Fig. 5.12. These sheets of cation-selective resins and anion-selective resins permit the passage of the respective ions in the solution. Under an applied DC field, the cations and anions will collect on one side of each membrane through which they are transported and vacate the other side. Thus, if a NaCl solution is supplied to the central zone of the cell shown in Fig. 5.12, the Na^+ ions will migrate through membrane A, depleting the central zone (termed the diluting or product feed stream) of the salt ions. The two outer zones where the ions collect are commonly known as the concentrating or brine streams.

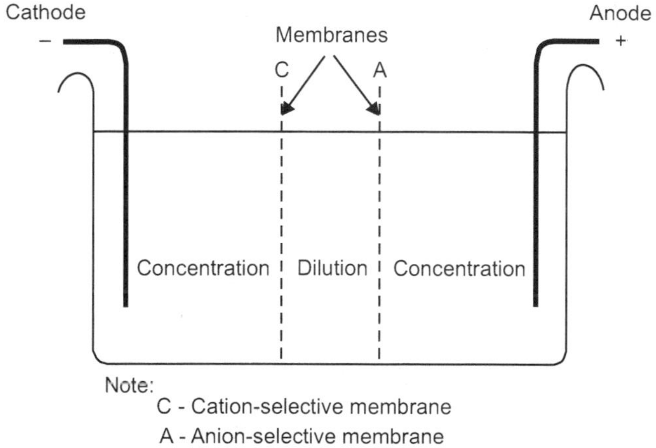

Figure 5.12: Electrodialysis cell diagram.

5.8.4 Evaporation

This is an energy intensive and hence an expensive process. It is used only when the recovered solids or the concentrated solutions are reused, e.g., some electroplating wastes. This method is also used when the volume of the wastewater to be treated is less. This method is also employed for concentrating radioactive liquid wastes. It permits the recovery of a wide variets of process chemicals. It is applicable to remove or concentrate chemicals which cannot be accomplished by any other means.

5.8.5 Adsorption

Adsorption by activated carbon is advantageous to remove small quantities of organic contaminants from wastewater. Special adsorbents are commercially available for the removal and retrieval of toxic heavy metal ions from industrial wastewater. Activated carbon treatment is particularly useful for the removal of pesticides (e.g., DDT) and carbamate insecticides.

5.9 Sludge treatment technologies

Sewage sludge is generated as a result of treating municipal wastewater to remove organic and inorganic impurities from dilute solutions. The treatment results in concentrating the impurities into a smaller volume of liquid, called sludge. Normally sewage sludge is a mixture of primary sludge from primary clarifier and biological sludge from biological treatment units. If the treatment process includes tertiary treatment, then sewage sludge may also include tertiary sludge. Thus, sewage sludge is the concentrated form of impurities extracted from domestic wastewater, in the attempt to improve the quality of the effluent. The proper management of sludge during its disposal holds the key for the success of any wastewater treatment operation.

Dumping has been the most common disposal method for sewage sludge. However, this practice is limited due to lack of dumping sites, problems in handling, increasing costs and government regulations. These have provided the impetus for new methods in sewage sludge management. Whereas in the past locally organised disposal practices have been sufficient, requirements for future sludge management will demand more regional management concepts. Depending on the specific local and regional situations, sewage sludge management and disposal options can be broadly categorised into the following:

1. Disposal without any treatment.
2. Disposal after treatment.
3. Utilisation for recycle and reuse of nutrients and other resources, with or without treatment.

The disposal of sewage sludge without any treatment is more suited to land fill disposal. There are plenty of treatment options available, which depend mainly on the ultimate disposal goal. By far, the most desirable option for management of sewage sludge is to utilise it as a resource so that the nutrient and other resources, which are the constituents of sludge, will be recycled. The most important treatment processes are sludge stabilisation, dewatering, incineration, composting and drying.

5.9.1 Sludge stabilisation

The major objectives of stabilisation are to make the sludge less odorous and putrescible and to reduce the pathogenic organisms. Primary processes of stabilisation are aerobic digestion, anaerobic digestion, thermo-chemical treatment and chemical treatment. Application of thermo-chemical stabilisation is very limited. The application of aerobic digestion, which uses extended aeration process for stabilisation until now, has also been limited to small and medium size treatment plants, due to its energy-intensive nature.

Wet oxidation also is one of the aerobic sludge digestion process in which the sludge is oxidised in a deep shaft reactor. In this process, organic and partly inorganic compounds get oxidised in a 1250 cm deep tube reactor at temperatures up to 280°C and a pressure of 8.5–11 MPa under injection of pure oxygen. The products remaining from the oxidation process are the process gas and a suspension with water and solids. The process gas undergoes catalytic post-oxidation for treatment for carbon monoxide and hydrocarbons. The suspension gets dewatered in a chamber filter press with the process water being treated in a multiple stage wastewater treatment plant i.e., biological wastewater treatment for nitrification/denitrification.

The sludge cake, which contains very low concentrations of organic matter at 10–25% gets disposed off to landfill sites. The disadvantages in this process are the geo-technical reliability and the operation and maintenance security with limited access to the process.

The most practiced process for stabilisation has been anaerobic digestion. During anaerobic digestion, the sludge is stabilised through biological degradation of complex organic substances in the absence of free oxygen. Typically, 25–45% of the raw sludge solids are destroyed during anaerobic digestion through conversion to methane, carbon dioxide, water and soluble organic material. Most commonly, anaerobic digesters are operated between the mesophilic temperature range of 32–35°C. Thermophilic digestion operating in the range of 50–60°C has high efficiency, but require a high-energy input.

The recommended average detention time for conventional anaerobic mesophilic digesters is 20–30 days and for high rate digesters, it ranges from

15–20 days. The detention time for thermophilic digesters, on the other hand, is 3–5 days. The major advantages of anaerobic digestion process include:

1. A 25–45% reduction in the volume of sludge.
2. Generation of methane rich biogas, which can be used as a fuel.
3. Inactivation of pathogens.
4. The disadvantages are.
5. High capital cost requirement.
6. Sensitivity of anaerobic micro-organisms to environmental conditions.

Aerobic composting is a method of sludge stabilisation in which sludge organics are decomposed by the micro-organisms in the presence of oxygen. The result of sludge composting is a humus like product that can be used, for example, as a soil amendment, for erosion control, as mulch, or other soil like products. The types of sludge that can be composted include both digested and undigested primary and secondary sludge.

Composting can be practiced only for sludge or a mixture of sludge with municipal solid waste. Aerobic composting depends on several operational parameters, which include oxygen availability within the compost, moisture content, temperature and biodegradable volatile solid content of the compose. For adequate air supply, composting systems use methods of forced aeration, addition of bulking agents such as rice straw, wood chips, saw dust, etc., or turning of pile. The composting of sludge can be carried out within a reactor (in-vessel system) or in the open (non-reactor system). During decomposition of volatile organics the temperature produced within the compost ranging from 50–70°C. At this high temperature, moisture is evaporated and pathogenic organisms in the compost are inactivated. The temperature must be maintained for a minimum duration of time in order to be effective in destroying pathogens.

5.9.2 Sludge dewatering

The mechanical dewatering of sludge plays a crucial role, both operationally and economically, particularly when the methods of disposal involve landfill disposal or heat treatment combined with the subsequent disposal of the residues. The dewatering process significantly increases the solid content and reduces the volume of sludge. The makes the sludge treatment process more competitive as a reduction in volume leads to a smaller amount for any subsequent treatment processes and transportation. This is particularly so when thermal treatment is part of the sludge treatment process. Factors affecting dewaterability of sludge are:

1. Catchment area and general sewerage treatment scheme.
2. Efficiency of screening and grit removal processes during treatment.

3. Sludge load and carbon:nitrogen:phosphorus ratio.
4. Detention time in secondary clarifier.
5. Type and efficiency of the sludge stabilisation process adopted.
6. Dry solids (DS) content of sludge.

DS content of sludge is basically used as the indicator for efficiency of dewatering processes. DS is measured as solids concentration on dry weight basis. General classification of dewaterability based on DS is presented in Table 5.1.

Table 5.1: Sludge dewaterability classification.

Dewaterability	*Percentage of dry solids*
Good	26–30
Sufficient to medium	22–26
Bad	8–22

There are different types of dewatering processes such as:

1. Centrifugal dewatering.
2. Centripress.
3. Centridry.
4. Belt filter dewatering.
5. Chamber filter press.
6. Membrane chamber filter press

Centrifugal dewatering is a process by which centrifugal force is applied to bring about the separation of solids from the liquid in the sludge. Dewatering is accomplished through clarification and solids compaction.

In the centripress process, clarification and compaction of sediments occur simultaneously. This system has been developed with the aim of combining the well-known principle of continuously operated centrifuges with the dewatering performance achieved by filter presses. It can achieve dewatered sludge DS concentration of between 20–40%.

Centridry is the combination of several dewatering steps in one continuous, compact, enclosed process and is an improvement over the centripress technology. Here, the dewatered sludge is further dried by heated air resulting in DS concentration of 60–90%.

The heated air is generated using a number of alternative fuels depending on what is available at a particular site.

Belt filters employ single or double moving belts made of woven synthetic fibre to dewater sludge continuously. The belt passes over and between rollers to exert increasing pressure on the sludge as it moves between the belts.

A high pressure belt filters can achieve a DS concentration of 20–35%, by application of pressure upto 1500 KPa.

The chamber filter press consists of series of parallel plates, each fitted with filter cloth and rigidly held together in a structural frame. The sludge is pumped at high pressure in between the plate packs and the plates are compressed by a hydraulic closing device, thereby dewatering the sludge.

In membrane chamber filter presses, sludge holding chambers are formed by filter cloth on one side and a membrane plate on the other side. The filter presses can achieve a DS efficiency of 25–40%.

6

Industrial solid wastes and its treatment

6.1 Introduction

Industrial solid waste refers to solid waste generated in production activities such as industry, traffic and resource development. These include solid wastes, semi-solid wastes and liquid and gaseous wastes in vessels that are not permitted to discharge into the environment. Industrial solid wastes are classified into organic wastes and inorganic wastes based on their components; into solid wastes, semi-solid wastes and liquid (gaseous) wastes based on their species; into hazardous wastes and common wastes based on pollution characteristics. Because many industrial solid wastes hold hazardous characteristics, they usually receive special attention.

Industrial solid waste pollution has become an increasingly serious problem in the world. The developing countries have to specially confront it. Every year, large quantities of industrial solid wastes are generated from the growing industries. However, there are no adequate treatment and disposal facilities and qualified personnel in these developing countries. All these have seriously hindered the development of industries in these countries and done harm to the human being health and the environment. Solid wastes is the term now used the world over to describe all non-liquid waste materials arising from domestic, trade, commercial, industrial, agricultural and mining activities, as also from public services. The term 'refuse' is often used interchangeably with 'solid waste'. Industry is the principal source with regard to the amount and the level of toxicity of several types of waste, such as production and processing residues and packing materials. The total amount of waste generated is related to the level of industrial production of the country concerned.

Industrial wastes can be broken down into a number of categories, including raw materials, process wastes, packing wastes and products that have not been manufactured to the required standard. In general, the amount and quality of the refuse generated in any processing operation will depend upon the mechanics of that particular unit. No exact statement can be made about the type, quantity and quality of the wastes produced by any particular industry. The wastes produced by a steel unit will definitely differ from those produced in the chemical or food processing industry. But the waste materials generated from paper and plastic industries are similar to the packing of paper and plastic materials to be found in household rubbish. On the other side, the metal processing factory will obviously generate metallic wastes, but in addition, it will produce large quantities of slag, processing chemicals and other residues. Hazardous wastes produced by chemical industries and other specialised industries will depend upon the particular product of the manufacturing process. Flow diagram 6.1 shows solid wastes from various industrial sectors.

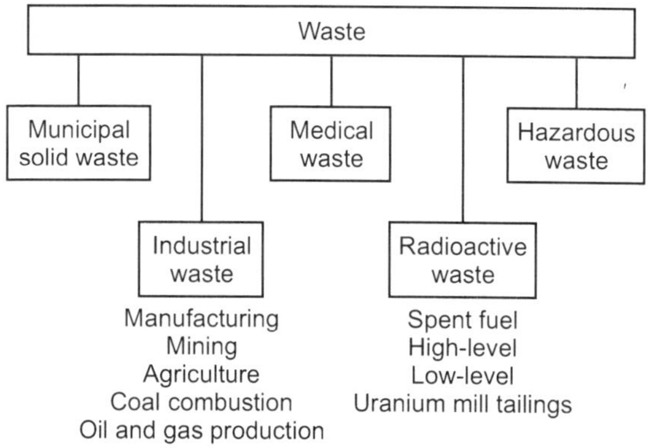

Figure 6.1: Solid wastes from various industrial sectors.

However, very often the waste of one industry may get recycled or reused in the same industry, or may be a source of raw material in another industry. Such reuse or recycling reduces pollution and conserves energy and resources. However, very often the distance between the respective industries and the possible higher cost of transportation come in the way to such recycling and reuse. Each industrial unit as such should be studied carefully and methods for disposal or recycling recommended. Some of the wastes can be grouped as biodegradable, non-biodegradable and hazardous.

Biodegradable wastes. Biodegradable wastes are generated in fruit processing industries, slaughter houses, cotton grinning (separation of cotton from seeds), textile mills, etc. Most of the wastes from these industries are reused; those which cannot be reused; those need to be processed as they are putrescible. Some of the wastes are amenable to composting, while others can also be pyrolysed to produce char.

Non-biodegradable wastes. Coal, stone, timber, metal scraps, sludge, etc., are the inert wastes generated from colliery operations. Refineries produce inert dry solids, combustible dry solids and varieties of sludge containing oil. Fly ash is the major solid waste from thermal power plants. Depending upon the activities, steel plants produce blast furnace slag, blast furnace flue dust, mill scale, etc. The problem of each plant should be considered separately; most of the wastes normally can be recycled or reused. The wastes not suitable for reuse or recycled are often found to be suitable for landfilling.

Hazardous solid wastes. Chemical, biological, explosive or radioactive wastes that pose a substantial danger, immediately or over time, to human, plant or animal life are classified as hazardous. Radioactive wastes are regulated in India by the atomic energy commission and separately disposed of. But all others must also be handled and disposed of with great care and caution. Chemical wastes are generated by various activities and include many poisonous substances and synthetic organics; explosives may come from ordnance factories and flammable substances may exist along with organic solvents, oils, plasticisers and organic sludge. Biological wastes normally come from hospitals as well as research facilities and include malignant tissues, bandages, contaminated hypodermic needles, etc. When an industry is not able to obtain any useful product from its waste, normally it opts for land disposal. Very often, the toxic substances then leach out and pollute both ground water and surface water. Some of the toxic wastes may also cause fire if they are inflammable. Sometimes the wastes release poisonous gases and may also cause explosions. On occasions, the dumped wastes may degrade the soil and make it unfit for cultivation. Hence, utmost care should be taken in the disposal of such wastes. Essentially the handling and disposal of toxic and hazardous wastes demands specific operating standards, which should include proper safety measures, development of emergency procedures, monitoring and training of employees, etc.

6.2 Types of industrial solid wastes

In general, industrial solid wastes are classified into the following major types:

1. Solid wastes from the mining industry: This kind of waste includes waste stones generated during mining and tailings. Waste stones mean wall rocks peeled off from major ore during the mining of metal and non-metal mines. Tailings mean the residue slag after distilling refining mines during mill run.
2. Industrial solid wastes from the metallurgical industry: This kind of waste includes varies of slag from the metallurgical procedure and processing of metals and non-metals. Some industrial solid wastes from the metallurgical industry are listed in Table 6.1.

Table 6.1: Industrial solid wastes from metallurgical industry.

Slag	*Sources*
Blast furnace slag	Generated in blast furnace iron smelting
Steel slag	Generated in steel smelting by level furnace, converter and electric stove
Non-ferrous metal slag	Generated in smelting processing of the non-ferrous metals, such as copper, nickel, lead and zinc
Red mud	Generated in abstracting aluminium oxide

3. Solid wastes from the power industry: This kind of waste includes coal fly ash, coal slag and flue ash from power plants in which coals are used as fuel. It also includes gangue generated from coal excavating and coal-washing.
4. Solid wastes from the chemical industry: This kind of wastes includes inferior products (semi-finished products), outgrowth, disabled catalysts, waste additives, raw materials that have not reacted and impurity in raw materials discharged from chemical reaction during production processes, such as chemical combination, decomposition and synthesis. They also include wastes discharged from refining, separating and washing procedures and from devices. Furthermore, they include the pyritic slag, acidic slag, alkali slag, salt mud, mud from kettle, residues of refining or distillation, pharmaceutical wastes, waste medicines from the producing and processing sectors in the chemical industry and waste pesticides from medicine and insecticide production. In addition, they include dust from air pollution control facilities, sludge from wastewater treatment facilities, solid wastes from equipment examination and repairing, equipment scraps, vessels and industrial refuses.
5. Solid wastes from the oil chemical industry: This kind of wastes includes oil mud, tar shale slag, waste catalysts and waste organic solvent in oil processing.
6. Solid wastes from light industry: This kind of wastes includes sludge, animal residues, waste acid, waste alkali and other wastes from the processing procedure in light industries, such as food industry, paper making and printing industry, spinning and dye-printing industry and leather industry.
7. Other industrial solid wastes: These kinds of waste mainly include metal dross from mechanical processing, plating sludge, construction wastes and slag from processing in other industries.

In the U.S., solid wastes means any garbage, refuse, sludge from a wastewater treatment plant, a water supply treatment plant, or an air pollution control facility and other discarded materials including solid, liquid, semi-solid, or contained gaseous materials arising from industrial, commercial, mining and agricultural operations and from community activities; but does not include solid or dissolved materials in domestic sewage, or solid or dissolved materials in irrigation return flows or industrial discharges.

6.2.1 Agricultural wastes

An agricultural waste management system consists of six basic functions (Fig. 6.2).

1. Production.
2. Collection.
3. Storage.
4. Treatment.
5. Transfer.
6. Utilisation.

For a specific system these functions may be combined, repeated, eliminated, or arranged as necessary.

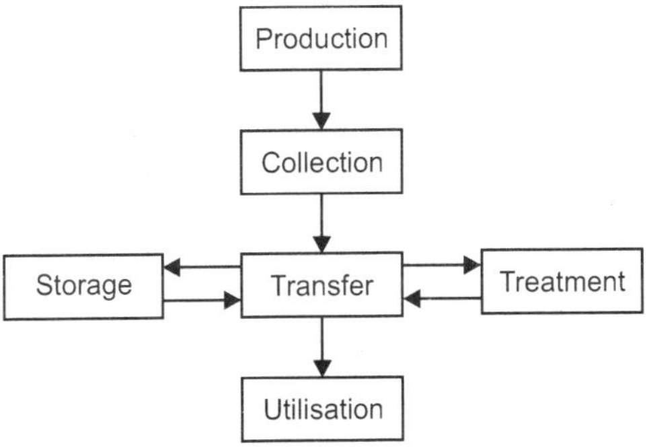

Figure 6.2: Waste management functions.

Production

Production is the function of the amount and nature of agricultural waste generated by an agricultural enterprise. The waste requires management if quantities produced are sufficient enough to become a resource concern. A complete analysis of production includes the kind, consistency, volume, location and timing of the waste produced.

The waste management system may need to accommodate seasonal variations in the rate of production. The production of unnecessary waste should be kept to a minimum. For example, a large part of the waste associated with many livestock operations includes contaminated runoff from open holding areas. The runoff can be reduced by restricting the size of open holding areas, roofing part of the holding area and installing gutters and diversions to direct uncontaminated water away from the waste. A proverb to remember is, 'Keep the clean water clean.' Leaking watering facilities and spilled feed contribute to the production of waste. These problems can be reduced by careful management and maintenance of feeders, watering facilities and associated equipment. A record should be kept of the data, assumptions and calculations used to determine the kind, consistency, volume, location and timing of the waste produced. The production estimates should include future expansion.

Collection

This refers to the initial capture and gathering of the waste from the point of origin or deposition to a collection point. The effective plan should identify the method of collection, location of the collection points, scheduling of the collection, labour requirements, necessary equipment or structural facilities, management and installation costs of the components and the impact that collection has on the consistency of the waste.

Storage

Storage is the temporary containment of the waste. The storage facility of a waste management system is the tool that gives the manager control over the scheduling and timing of the system functions. For example, with adequate storage the manager has the flexibility to schedule the land application of the waste when the spreading operations do not interfere with other necessary tasks, when weather and field conditions are suitable and when the nutrients in the waste can best be used by the crop. The storage period should be determined by the utilisation schedule.

The waste management system should identify the storage period; the required storage volume; the type, estimated size, location and installation cost of the storage facility; the management cost of the storage process; and the impact of the storage on the consistency of the waste.

Treatment

Treatment is any function designed to reduce the pollution potential of the waste, including physical, biological and chemical treatment. It includes activities that are sometimes considered pre-treatment, such as the separation of solids. The plan should include an analysis of the characteristics of the waste before treatment; a determination of the desired characteristics of the waste following treatment; the selection of the type, estimated size, location and the installation cost of the treatment facility; and the management cost of the treatment process.

Transfer

This refers to the movement and transportation of the waste throughout the system. It includes the transfer of the waste from the collection point to the storage facility, to the treatment facility and to the utilisation site. The waste may require transfer as a solid, liquid, or slurry, depending on the total solids concentration. The system plan should include an analysis of the consistency of the waste to be moved, method of transportation, distance between points, frequency and scheduling, necessary equipment and the installation and management costs of the transfer system.

Utilisation

Utilisation includes recycling reusable waste products and reintroducing nonreusable waste products into the environment. Agricultural wastes may be used as a source of energy, bedding, animal feed, mulch, organic matter or plant nutrients. Properly treated, they can be marketable. A common practice is to recycle the nutrients in the waste through land application. A complete analysis of utilisation through land application includes selecting the fields; scheduling applications; designing the distribution system; selecting necessary equipment and determining application rates and volumes, value of the recycled products and installation and management costs associated with the utilisation process.

6.3 Processing and resource recovery

The techniques and methods used to recover materials, conversion products and energy from solid wastes include (i) processing techniques, (ii) processing techniques for hazardous wastes, (iii) materials recovery systems, (iv) recovery of biological conversion products, (v) recovery of chemical conversion products and (vi) recovery of energy from conversion products.

Because many of the techniques, especially those associated with the recovery of materials and energy and the processing of solid hazardous wastes, are in a state of flux with respect to application and design criteria, the objective here is only to introduce them to the reader. If these techniques are to be considered in the development of waste management systems, current engineering design and performance data must be obtained from the records of operating installations, from field tests, from equipment manufacturers and from literature.

6.3.1 Processing techniques

Processing techniques are used in solid waste management systems to: (i) improve the efficiency of the systems, (ii) to recover resources (usable materials) and (iii) to prepare materials for recovery of conversion products and energy. The more important techniques used for processing solid wastes are summarised in Table 6.2.

Manual component separation

The manual separation of solid waste components can be accomplished at the source, at a transfer station, at a centralised processing station, or at the disposal site. Manual sorting at the source of generation is the most positive way to achieve the recovery and reuse of materials. The number and types of components salvaged or sorted (e.g., cardboard and high quality paper, metals and wood) depend on the location, the opportunities for recycling and the resale market.

Table 6.2: Summary of techniques used for processing solids wastes.

Processing techniques	*Functions*	*Representative equipment and/or facilities and applications*
Manual component separation	Separation of recoverable materials, usually at point of generation	
Storage and transfer	Storage and transfer of wastes to be processed	Open storage pits for unprocessed wastes, storage bins and silos for processed wastes; transfer equipment, including metal and rubber belt conveyors, vibratory conveyors with unprocessed wastes, pneumatic conveyors and screw conveyors with processed wastes.
Mechanical volume	Reduction of solid waste volume, alteration of shape of solid waste components; all modern collection vehicles essentially equipped with compaction equipment	Hydraulic piston type compactors for collection reduction vehicles, onsite compactors and transfer station compactors; roll crushers used to fracture brittle materials and to crush tin and aluminium cans and other ductile materials
Chemical volume reduction	Reduction of volume of solid wastes through burning (incineration)	Mass fired incinerators, with and without heat recovery, for unprocessed wastes
Mechanical size and shape alteration	Alteration of size and shape of solid waste components	Equipment used to reduce the size of solid waste, including hammer mills, shredders, roll crushers, grinders, chippers, jaw crushers, rasp mills and hydropulpers
Mechanical component separation	Separation of recoverable materials, usually at a processing facility	
Magnetic and electromechanical separation	Separation of ferrous and nonferrous materials from processed solid wastes	Magnetic separation for ferrous materials, eddy current separation for aluminium; electrostatic separation for glass from wastes free of ferrous and aluminium scrap; magnetic fluid separation for nonferrous materials from processed wastes
Drying and dewatering	Removal of moisture from solid wastes	Convection, conduction and radiation dryers used for solid wastes and sludge; centrifugation and filtration used to dewater treatment plant sludge

Storage and transfer

When solid wastes are to be processed for materials recovery, storage and transfer facilities should be considered an essential part of the processing operation. Important factors in the design of such facilities include: (i) the size of the material before and after processing, (ii) the density of the material, (iii) the angle of repose before and after processing, (iv) the abrasive characteristics of the material and (v) the moisture content.

Mechanical volume reduction

Mechanical volume reduction is perhaps the most important factor in the development and operation of solid water management systems. Vehicles equipped with compaction mechanisms are used for the collection of most industrial solid wastes. To increase the useful life of landfills, wastes are compacted. Paper for recycling is baled for shipping to processing centres.

When compacting industrial solid wastes, it has been found that the final density (typically about 1100 kg/m^3) is essentially the same regardless of the starting density and applied pressure. This fact is important in evaluating manufacturer's claims.

Chemical volume reduction

Incineration has been the method most commonly used to reduce the volume of wastes chemically. One of the most attractive features of the incineration process is that it can be used to reduce the original volume of combustible solid wastes by 80 to 90%.

6.4 Methods of solid wastes disposal

1. Incineration.
2. Composting.
3. Pyrolysis.
4. Sanitary landfill.

6.4.1 Incineration

The term incinerates means to burn something until nothing is left but ashes. An incinerator is a unit or facility used to burn trash and other types of waste until it is reduced to ash. An incinerator is constructed of heavy, well-insulated materials, so that it does not give off extreme amounts of external heat.

The high levels of heat are kept inside the furnace or unit so that the waste is burned quickly and efficiently. If the heat were allowed to escape, the waste would not burn as completely or as rapidly. Incineration is a disposal method in which solid organic wastes are subjected to combustion so as to convert them into residue and gaseous products. This method is useful for disposal of residue of both solid waste management and solid residue from wastewater management. This process reduces the volumes of solid waste to 20 to 30% of the original volume.

Incineration and other high temperature waste treatment systems are sometimes described as 'thermal treatment'. Incinerators convert waste materials into heat, gas, steam and ash. Incineration is carried out both on a small scale by individuals and on a large scale by industry. It is used to dispose of solid, liquid and gaseous waste. It is recognised as a practical method of disposing of certain hazardous waste materials. Incineration is a controversial method of waste disposal, due to issues such as emission of gaseous pollutants.

Incineration with heat recovery

Heat contained in the gases produced from the incineration of solid wastes can be recovered by conversion to steam. The low-level heat remaining in the gases after heat recovery can also be used to pre-heat the combustion air, boiler make-up water, or solid waste fuel.

1. In existing incinerators: With existing incinerators, waste heat boilers can be installed to extract heat from the combustion gases without introducing excess amounts of air or moisture. Typically, incinerator gases will be cooled from a range of 1250 to 1375 K (1800 to 2000°F) to a range from 500 to 800 K (600 to 1000°F) before being discharged into the atmosphere. Apart from the production of steam, the use of a boiler system is beneficial in reducing the volume of gas to be processed in the air pollution control equipment.
2. In water wall incinerators: In these incinerators, the internal walls of the combustion chamber are lined with boiler tubes that are arranged vertically and welded together in continuous sections. When water walls are employed in place of refractory materials, they are not only useful for the recovery of steam, but also extremely effective in controlling furnace temperature without introducing excess air; however, they are subject to corrosion by hydrochloric acid produced from the burning of some plastic compounds.

Gasification

The gasification process involves the partial combustion of a carbonaceous fuel to generate a combustible fuel gas rich in carbon monoxide and hydrogen. A gasifier is basically an incinerator operating under reducing conditions.

Heat to sustain the process is derived from exothermic reactions, while the combustible components of the low energy gas are primarily generated by endothermic reactions. The reaction kinetics of the gasification process are quite complex and still the subject of considerable debate. When a gasifer is operated at atmospheric pressure with air as the oxidant, the end products of the gasification process are a low energy gas typically containing (by volume) 10% CO_2, 20% CO, 15% H_2 and 2% CH_4, with the balance being N_2 and a carbon rich char. Because of the diluting effect of the nitrogen in the input air, the low energy gas has an energy content in the range of 5.2 to 6.0 MJ/m^3 (140 to 160 Btu/ft^3). When pure oxygen is used as the oxidant, a medium energy gas with an energy content in the range of 12.9 to 13.8 MJ/m^3 (345 to 370 Btu/ft^3) is produced.

6.4.2 Composting

Due to shortage of space for landfill in bigger cities, the biodegradable yard waste (kept separate from the municipal waste) is allowed to degrade or decompose in a medium. A good quality nutrient rich and environmental friendly manure is formed which improves the soil conditions and fertility.

Organic matter constitutes 35%–40% of the municipal solid waste generated in India. This waste can be recycled by the method of composting, one of the oldest forms of disposal. It is the natural process of decomposition of organic waste that yields manure or compost, which is very rich in nutrients.

Composting is a biological process in which micro-organisms, mainly fungi and bacteria, convert degradable organic waste into humus like substance. This finished product, which looks like soil, is high in carbon and nitrogen and is an excellent medium for growing plants.

The process of composting ensures the waste that is produced in the kitchens is not carelessly thrown and left to rot. It recycles the nutrients and returns them to the soil as nutrients. Apart from being clean, cheap and safe, composting can significantly reduce the amount of disposable garbage.

The organic fertiliser can be used instead of chemical fertilisers and is better specially when used for vegetables. It increases the soil's ability to hold water and makes the soil easier to cultivate. It can help the soil retain more of the plant nutrients.

Vermicomposting has become very popular in the last few years. In this method, worms are added to the compost. These help to break the waste and the added excreta of the worms makes the compost very rich in nutrients.

To make a compost pit, you have to select a cool, shaded corner of the garden or the school compound and dig a pit, which ideally should be 3 feet deep. This depth is convenient for aerobic composting as the compost has to be turned at regular intervals in this process.

Preferably the pit should be lined with granite or brick to prevent nitrite pollution of the subsoil water, which is known to be highly toxic. Each time organic matter is added to the pit it should be covered with a layer of dried leaves or a thin layer of soil which allows air to enter the pit thereby preventing bad odour. At the end of 45 days, the rich pure organic matter is ready to be used.

Some benefits of composting are given below:

1. Compost allows the soil to retain more plant nutrients over a longer period.
2. It supplies part of the 16 essential elements needed by the plants.
3. It helps reduce the adverse effects of excessive alkalinity, acidity, or the excessive use of chemical fertiliser.
4. It makes soil easier to cultivate.
5. It helps keep the soil cool in summer and warm in winter.
6. It aids in preventing soil erosion by keeping the soil covered.
7. It helps in controlling the growth of weeds in the garden.

Anaerobic digestion

Anaerobic digestion or anaerobic fermentation, as it is often called, is the process used for the production of methane from solid wastes. In most processes in which methane is to be produced from solid wastes by anaerobic digestion, three basic steps are involved. The first step involves preparation of the organic fraction of the solid wastes for anaerobic digestion and usually includes receiving, sorting, separation and size reduction. The second step involves the addition of moisture and nutrients, blending, pH adjustment to about 6.7, heating of the slurry to between 327 and 333K (130 and 140°F) and anaerobic digestion in a reactor with continuous flow in which the contents are well mixed for a period of time varying from 8 to 15 days. The third step involves capture, storage and if necessary, separation of the gas components evolved during the digestion process. The disposal of the digested sludge is an additional task that must be accomplished.

6.4.3 Pyrolysis

Pyrolysis is a form of incineration that chemically decomposes organic materials by heat in the absence of oxygen. Pyrolysis typically occurs under pressure and at operating temperatures above 430°C (800°F).

In practice, it is not possible to achieve a completely oxygen-free atmosphere. Because some oxygen is present in any pyrolysis system, a small amount of oxidation occurs. If volatile or semi-volatile materials are present in the waste, thermal desorption will also occur.

Organic materials are transformed into gases, small quantities of liquid and a solid residue containing carbon and ash. The off-gases may also be treated in a secondary thermal oxidation unit. Particulate removal equipment is also required. Several types of pyrolysis units are available, including the rotary kiln, rotary hearth furnace and fluidised bed furnace. These units are similar to incinerators except that they operate at lower temperatures and with less air supply.

Limitations of pyrolysis

1. The technology requires drying of soil prior to treatment.
2. Limited performance data are available for systems treating hazardous wastes containing polychlorinated biphenyls (PCBs), dioxins and other organics. There is concern that systems that destroy chlorinated organic molecules by heat have the potential to create products of incomplete combustion, including dioxins and furans. These compounds are extremely toxic in the parts per trillion ranges. The MSO process reportedly does not produce dioxins and furans.
3. The molten salt is usually recycled in the reactor chamber. However, depending on the waste treated (especially inorganics) and the amount of ash, spent molten salt may be hazardous and require special care in disposal.
4. Pyrolysis is not effective in either destroying or physically separating in organics from the contaminated medium. Volatile metals may be removed as a result of the higher temperatures associated with the process, but they are not destroyed. By-products containing heavy metals may require stabilisation before final disposal.
5. When the off-gases are cooled, liquids condense, producing an oil/tar residue and contaminated water. These oils and tars may be hazardous wastes, requiring proper treatment, storage and disposal.

6.4.4 Sanitary land filling

In a sanitary landfill, garbage is spread out in thin layers, compacted and covered with clay or plastic foam. In the modern landfills the bottom is covered with an impermeable liner, usually several layers of clay, thick plastic and sand. The liner protects the ground water from being contaminated due to percolation of leachate.

Leachate from bottom is pumped and sent for treatment. When landfill is full it is covered with clay, sand, gravel and top soil to prevent seepage of water. Several wells are drilled near the landfill site to monitor if any leakage is contaminating ground water. Methane produced by anaerobic decomposition is collected and burnt to produce electricity or heat. Guidelines for sanitary landfills site selection are given below:

1. Should be above the water table, to minimise interaction with groundwater.
2. Preferably located in clay or silt.

3. Do not want to place in a rock quarry, as water can leach through the cracks inherent in rocks into a water fracture system.
4. Do not want to locate in sand or gravel pits, as these have high leaching. Unfortunately, most of long island is sand or gravel and many landfills are located in gravel pits, after they were no longer being used.
5. Do not want to locate in a flood plain. Most garbage tends to be less dense than water, so if the area of the landfill floods, the garbage will float to the top and wash away downstream.

A large number of adverse impacts may occur from landfill operations. These impacts can vary:

1. Fatal accidents (e.g., scavengers buried under waste piles).
2. Infrastructure damage (e.g., damage to access roads by heavy vehicles).
3. Pollution of the local environment (such as contamination of groundwater and/or aquifers by leakage and residual soil contamination during landfill usage, as well as after landfill closure).
4. Off gassing of methane generated by decaying organic wastes (methane is a greenhouse gas many times more potent than carbon dioxide and can itself be a danger to inhabitants of an area).
5. Harbouring of disease vectors such as rats and flies, particularly from improperly operated landfills.

Landfilling methods and operations

To use the available area at a landfill site effectively, a plan of operation for the placement of solid wastes must be prepared. Various operational methods have been developed primarily on the basis of field experience. The principal methods used for landfilling dry areas may be classified as: (i) area, (ii) trench and (iii) depression.

1. Area method: The area method is used when the terrain is unsuitable for the excavation of trenches in which to place the solid wastes. The filling operation is usually started by building an earthen levee against which wastes are placed in thin layers and compacted (Fig. 6.3). Each layer is compacted as the filling progresses until the thickness of the compacted wastes reaches a height varying from 2 to 3 m (6 to 10 ft). At that time and at the end of each day's operation, a 150 to 300 mm (6 to 12 in) layer of cover material is placed over the completed fill. The cover material must be hauled in by truck or earth moving equipment from adjacent land or from borrow pit areas. In some newer landfill operations, the daily cover material is omitted. A completed lift, including the cover material, is called a 'cell'. Successive lifts are placed on top of one another until the final grade called for in the ultimate development plan is reached.

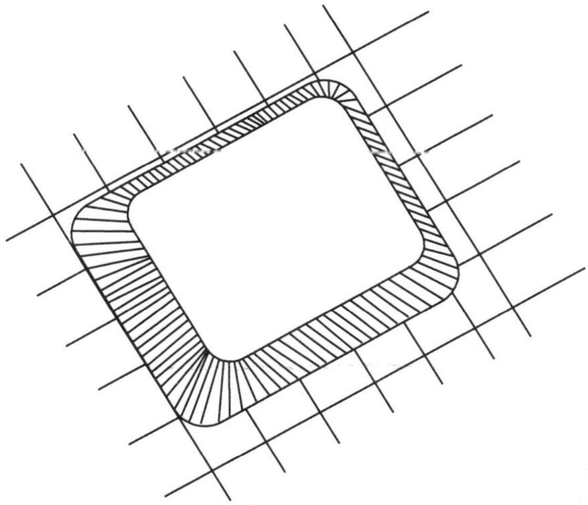

(a)

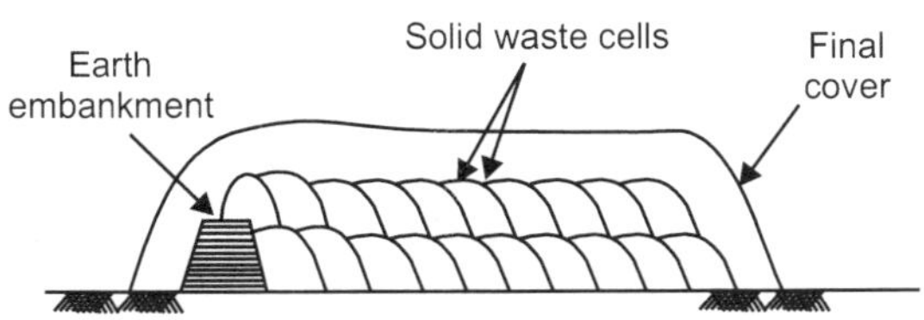

(b)

Figure 6.3: Area method for landfilling, solid-wastes: (a) pictorial view of completed landfill and (b) section through landfill.

A final layer of cover material is used when the fill reaches the final design height.

2. Trench method: The trench method (Fig. 6.4) of landfilling is ideally suited to areas where an adequate depth of cover material is available at the site and where the water table is well below the surface. To start the process, a portion of the trench is dug and the dirt is stockpiled to form an embankment behind the first trench. Wastes are then placed in the trench, spread into thin layers and compacted. The operation continues until the desired height is reached. Cover material is obtained by excavating an adjacent trench or continuing the trench that is being filled.
3. Depression method: At locations where natural or artificial depressions exist, it is often possible to use them effectively for landfilling operations. Canyons, ravines, dry burrow pits and quarries have all been used for this purpose. The techniques to place and compact solid wastes in

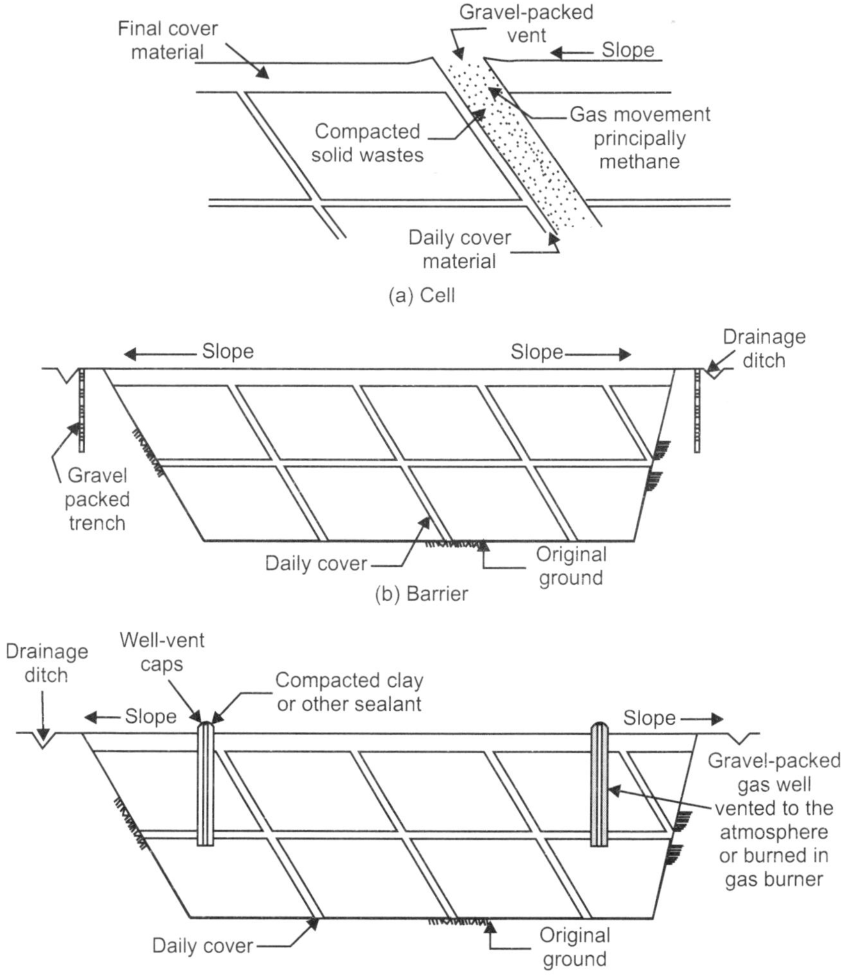

Figure 6.4: Vents used to control the lateral movement of gases in landfills: (a) cell, (b) barrier and (c) well.

depression landfills vary with the geometry of the site, the characteristics of the cover material, the hydrology and geology of the site and access to the site.

In a canyon, site-filling starts at the head end of the canyon and ends at the mouth. This practice prevents the accumulation of water behind the landfill. Wastes are usually deposited on the canyon floor and from there are pushed up against the canyon face at a slope of about 2 to 1. In this way, a high degree of compaction can be achieved.

4. *Landfills in wet areas:* Because of the problems associated with contamination of local groundwater, the development of odours and structural stability, landfills are seldom used in wet areas. If wet areas such as swamps and marshes, tidal areas and ponds, pits or quarries must be used as landfill sites, special provisions must be made to contain or eliminate the movement of leachate and gases from completed cells. Usually, this is accomplished by first draining the site and then lining the bottom with a clay liner or other appropriate sealants. If a clay liner is used, it is important to continue operation of the drainage facility until the site is filled to avoid the creation of uplift pressures that can cause the liner to rupture from heaving.

6.4.5 Occurrence of gases and leachate in landfills

The following biological, physical and chemical events occur when solid wastes are placed in a sanitary landfill: (i) biological decay of organic materials, either aerobically or anaerobically, with the evolution of gases and liquids, (ii) chemical oxidation of waste materials, (iii) escape of gases from the fill, (iv) movement of liquids caused by differential heads, (v) dissolving and leaching of organic and inorganic materials by water and leachate moving through the fill, (vi) movement of dissolved material by concentration gradients and osmosis and (vii) uneven settlement caused by consolidation of material into voids.

With respect to biological decay, bacterial decomposition initially occurs under aerobic conditions because a certain amount of air is trapped within the landfill. However, the oxygen in the trapped air is exhausted within days and long-term decomposition occurs under anaerobic conditions.

1. Gases in landfills: Gases found in landfills include air, ammonia, carbon dioxide, carbon monoxide, hydrogen, hydrogen sulphide, methane, nitrogen and oxygen. Carbon dioxide and methane are the principal gases produced from the anaerobic decomposition of the organic solid waste components. The anaerobic conversion of organic compounds is thought to occur in three steps: the first involves the enzyme mediated transformation (hydrolysis) of higher weight molecular compounds into compounds suitable for use as a source of energy and cell carbon; the second is associated with the bacterial conversion of the compounds resulting from the first step into identifiable lower molecular weight intermediate compounds; and the third step involves the bacterial conversion of the intermediate compounds into simpler end products, such as carbon dioxide (CO_2) and methane (CH_4). The rate of decomposition in unmanaged landfills, as measured by gas production, reaches a peak within the first two years and then slowly tapers off, continuing in many cases for periods up to 25 years or more.

2. Leachate in landfills: Leachate may be defined as the liquid that has percolated through solid waste and has extracted dissolved or suspended materials from it. In most landfills, the liquid portion of the leachate is composed of the liquid produced from the decomposition of the wastes and liquid that has entered the landfill from external sources, such as surface drainage, rainfall, groundwater and water from underground springs.

Gas and leachate movement and control

Under ideal conditions, the gases generated from a landfill should be either vented into the atmosphere or, in larger landfills, collected for the production of energy. The leachate should either be contained within the landfill or removed for treatment.

1. Gas movement: In most cases, over 90% of the gas volume produced from the decomposition of solid wastes consists of methane and carbon dioxide. Although most of the methane escapes into the atmosphere, both methane and carbon dioxide have been found in concentrations of up to 40% at lateral distances of up to 120 m (400 ft) from the edges of landfills. If vented into the atmosphere in an uncontrolled manner, methane can accumulate (because its specific gravity is less than that of air) below buildings or in other enclosed spaces on or close to a sanitary landfill. With proper-venting, methane should not pose a problem.
   Because carbon dioxide is about 1.5 times as dense as air and 2.8 times as dense as methane, it tends to move toward the bottom of the landfill. As a result, the concentration of carbon dioxide in the lower portions of landfill may be high for years. Ultimately, because of its density, carbon dioxide will also move downward through the underlying formation until it reaches the groundwater. Because carbon dioxide is readily soluble in water, it usually lowers the pH, which, in turn, can increase the hardness and mineral content of the groundwater through the solubilisation of calcium and magnesium carbonates.
2. Control of gas movement: The lateral movement of gases produced in a landfill can be controlled by installing vents made of materials that are more permeable than the surrounding soil. Typically, as shown in Fig. 6.4a, gas vents are constructed of gravel. The spacing of cell vents depends on the width of the waste cell but usually varies from 18 to 60 m (60 to 200 ft). The thickness of the gravel layer should be such that it will remain continuous even though there may be differential settling; 300 to 450 mm (12 to 18 in) is recommended. Barrier or well vents can also be used to control the lateral movement of gases. Well vents are often used in conjunction with lateral surface vents buried below grade in a

gravel trench. Control of the downward movement of gases can be accomplished by installing perforated pipes in the gravel layer at the bottom of the landfill. If the gases cannot be vented laterally, it may be necessary to install gas wells and vent the pumped gas into the atmosphere. The movement of landfill gases through adjacent soil formations can be controlled by constructing barriers of material that are more impermeable than the soil.

3. Control of gas movement by recovery: The movement of gas in landfills can also be controlled by installing gas recovery wells in completed landfills. Clay and other liners are used when landfill gas is to recovered. In some gas recovery systems, leachate is collected and recycled to the top of the landfills and reinjected through perforated lines located in drainage trenches. Typically, the rate of gas production is greater in leachate recirculation systems. Although gas recovery systems have been installed in some large municipal landfills, the economics of such operations are, at present, not well defined. The cost of the gas clean-up and processing equipment may limit the recovery of landfill gases, especially from small landfills.
4. Leachate movement: Under normal conditions, leachate is found at the bottom of landfills. From there, it moves through the underlying strata, although some lateral movement may also occur, depending on the charateristiscs of the surrounding material. The rate of seepage of leachate from the bottom of a landfill can be estimated by Darcy's law by assuming that the material below the landfill to the top of the water table is saturated and that a small layer of leachate exists at the bottom of the fill. Under these conditions, the leachate discharge rate per unit area is equal to the value of the coefficient of permeability K, expressed in metres per day. The computed value represents the maximum amount of seepage that would be expected and this value should be used for design purposes. Under normal conditions, the actual rate will be less than this value, because the soil column below the landfill will not be saturated.
5. Control of leachate movement: As leachate percolates through the underlying strata, many of the chemical and biological constituents originally contained in it will be removed by the filtering and adsorptive action of the material composing the strata. In general, the extent of this action depends on the characteristics of the soil, especially the clay content. Because of the potential risk involved in allowing leachate to percolate to the groundwater, best practice calls for its elimination or containment. Ultimately, it may be necessary to collect and treat the leachate. The use of clay has been the favoured method of reducing or eliminating the percolation of leachate. Membrane liners have also been

used, but they are expensive and require care so that they are not be damaged during filling operations. Equally important in controlling the movement of leachate is the elimination of surface water infiltration, which is the major contributor to the total volume of leachate. With the use of an impermeable clay layer, an appropriate surface slope (1 to 2%) and adequate drainage, surface infiltration can be controlled effectively.

6. Settlement and structural characteristics of landfills: The settlement of landfills depends on the initial compaction, characteristics of wastes, degree of decomposition and effects of consolidation when the leachate and gases are formed. The height of the completed fill will also influence the initial compaction and degree of consolidation. The degree of consolidation can be modelled with a first order equation.

7

Management of hazardous wastes

7.1 Introduction

Hazardous waste is any unwanted material the disposal of which poses a threat to the environment, i.e., it is explosive, flammable, oxidising, poisonous/infectious, radioactive, corrosive and/or toxic/ecotoxic.

Sources of hazardous waste in include waste from chemical industries, petrol storage, metal finishing, paint manufacture, iron and steel industries, tanneries, wastes from agriculture/horticulture and electricity distribution.

The waste can be treated chemically (i.e., by neutralisation, oxidation, reduction, hydrolysis, precipitation), physically (encapsulation, separation), biologically (using micro-organisms) or thermally (incineration). Most treated waste is then deposited in landfills. To prevent landfills themselves being environmental hazards, conditions of resource consents granted under the Resource Management Act 1991 generally control the types of waste that can be deposited in them and the landfill design.

Many of the materials used or produced in chemical processes possess hazardous properties. As such they require appropriate management throughout their life cycle so as to minimise adverse effects on public health and safety or to the environment generally. For the purposes of this discussion the life cycle of a hazardous substance can be regarded as covering all stages from its recovery from natural resources through to its final disposal as a waste.

Examples of hazardous wastes from various sectors are given in Table 7.1.

Table 7.1: Examples of hazardous wastes arising from various sectors.

Origin	*Waste type*	*Constituents*	*Hazardous characteristic*
Chemical and allied industries	Difficult to treat	DMT, aromatic	Toxic
Petrol storage	Sludges/liquids	Hydrocarbons,	Flammable, toxic
Metal finishing	Sludges/liquids	Sulphuric and/or Hydrochloric acids	Corrosive
		Cyanides	Toxic
		Cadmium, copper chromium, etc.	Toxic
Paint manufacture	Sludges/liquids	Organic solvents	Flammable, toxic

(Cont'd...)

Origin	*Waste type*	*Constituents*	*Hazardous characteristic*
Tanneries	Sludges/liquids	Chromium, sulphides, etc.	Toxic
Electricity distribution	Oils from capacitors, transformers, etc.	Polychlorinated PCBs	Toxic, ecotoxic
Agriculture/ horticulture	Obsolete herbicides/ pesticides	Organophosphorus pesticides DDT, Aldrin, Dieldrin, chlordane,	Toxic/Ecotoxic

7.2 Managing hazardous wastes

7.2.1 Waste management hierarchy

In deciding on the best method for managing any waste there is a hierarchy for decision making which addresses issues such as sustainability, cleaner production, health, safety and environmental protection. It is applied to existing or proposed practices, examining and testing these at each level, starting at the top of the hierarchy.

For hazardous waste the hierarchy is as follows:

1. Eliminate the production of hazardous waste.
2. Where elimination is not possible apply methods to reduce the quantity or hazard involved.
3. Minimise amount of waste for disposal by recycling, reuse and/or recovery. This includes the recovery of energy which may be available from the waste.
4. Treat waste to stabilise, immobilise, contain or destroy hazardous properties.
5. Dispose of residues with a minimum of environmental impact.
6. Appropriately contain, isolate and store hazardous waste for which no acceptable treatment or disposal option is currently available.

7.2.2 Cleaner production

Cleaner production refers to a precautionary approach which includes the goal of preventing the generation of hazardous waste. Minimising the amount of hazardous waste produced would be one of the objectives of a cleaner production programme. In many cases the introduction of cleaner production measures brings economic benefits in addition to savings in waste disposal costs.

7.2.3 Waste minimisation

Examples of ways in which waste minimisation can be achieved include:

1. Substituting a hazardous material used in a process with a non-hazardous material.
2. Process changes.
3. Reducing the amount of hazardous materials used.
4. Recovering and reusing materials.

Table 7.2 gives examples indicating how these have been incorporated producing hazardous waste.

Table 7.2: Examples of waste minimisation practices.

Practice	*Industry*	*Activity*
Substitution	Leather production	Replacement of ammonium salts by carbon dioxide in dehairing operations
Process change	Manufacture of plastic containers	Labels molded into lids eliminating need for glues containing organic solvents
Reduction of waste quantity by recycling/	Wire manufacturing	Recycling by sulphuric acid use for pickling mild steel
Recovery and reuse	Timber treatment	Recover of copper, chromium and arsenic and/or boron from sludges and reuse in wood treatment process
	Dry cleaning	Improved recovery of perchlorethylene
	Vehicle servicing	Recovery of used lubricating oils for • Refining and reuse • Use as a supplementary source of fuel in cement kilns

7.3 Treatment of hazardous wastes

7.3.1 Methods available

The purpose of treating hazardous waste is to convert it into nonhazardous substances or to stabilise or encapsulate the waste so that it will not migrate and present a hazard when released into the environment. Stabilisation or encapsulating techniques are particularly necessary for inorganic wastes such as those containing toxic heavy metals. Treatment technologies exist for most if not all hazardous wastes.

7.3.2 Chemical methods

Neutralisation

Waste acid with an alkali, e.g., sulphuric acid with sodium carbonate:

$$H_2SO_4 + CO_3^{2-} \rightarrow SO_4^{2-} + CO_2 + H_2O$$

Oxidation

Using common oxidising substances such as hydrogen peroxide or calcium hypochlorite, e.g., cyanide waste with calcium hypochlorite:

$$CN^- + OCl^- \rightarrow OCN^- + Cl^- \qquad OCN^- + H_3O^+ \rightarrow CO_2 + NH_3$$

Reduction

Used to convert inorganic substances to a less mobile and toxic form, e.g., reducing Cr(VI) to Cr(III) by the use of ferrous sulphate:

$$14H^+ + Cr_2O_7^{2-} + 6Fe^{2+} \rightarrow 6Fe^{3+} + 2Cr^{3+} + 7H_2O$$

Hydrolysis

Decomposition of hazardous organic substances, e.g., decomposing certain organophosphorus pesticides with sodium hydroxide.

Precipitation

Particularly useful for converting hazardous heavy metals to a less mobile, insoluble form prior to disposal to a landfill, e.g., precipitation of cadmium as its hydroxide by the use of sodium hydroxide:

$$Cd^{2+}(aq) + 2OH^- \rightarrow Cd(OH)_2(s)$$

7.3.3 Physical methods

Encapsulation

Immobilising hazardous materials by stabilisation and incorporation within a solid matrix such as cement concrete or proprietary organic polymers prior to and filling, e.g., encapsulating beryllium in concrete.

Filtration/centrifuging/separation

Physically separating phases containing hazardous substances from other nonhazardous constituents, e.g., separation of oils from ship bilge waters.

7.3.4 Biological methods

These involve the use of micro-organisms under optimised conditions to mineralise hazardous organic substances, e.g., the use of pseudomonas under aerobic conditions break down phenols.

7.3.5 Thermal methods

These are the treatment processes which involve the application of heat to convert the waste into less hazardous forms. It also reduces the volume and allows opportunities for the recovery of energy from the waste.

7.3.6 High temperature incineration

In North America and Europe the treatment method most commonly used to destroy hazardous organic wastes, including organochlorines such as polychlorinated biphenyls (PCBs), is high temperature incineration. Figure 7.1 schematically illustrates the processes involved. Currently in New Zealand incineration is only available for hospital, clinical and quarantine wastes. There is a high temperature incinerator for the destruction chemical wastes in New Plymouth but its use is dedicated solely to those generated on site.

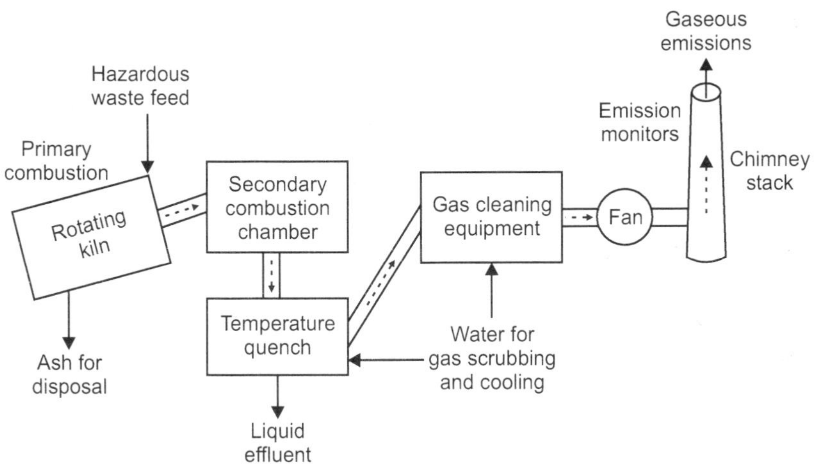

Figure 7.1: High temperature incineration.

Incineration is the controlled combustion process which can be used to degrade organic substances. For a simple hydrocarbon, involving complete combustion, for example the chemical reaction can be illustrated as follows:

$$C_xH_y + (x + \frac{1}{4}y)O_2 \rightarrow xCO_2 + \frac{1}{2}y\,H_2O$$

In practice, complete combustion is difficult if not impossible to achieve but for hazardous waste 99.99% or greater destruction or removal is required for the process to be generally acceptable.

Combustion parameters

Incinerators for the treatment of hazardous waste must be carefully designed and operated if they are to achieve the efficiency of destruction required.

Combustion of organics occur in two stages. In the primary stage, volatile matter is driven off leaving the remainder to burn to ash. The volatiles are combusted in the secondary stage. Incinerators are designed accordingly. High temperatures are required, for most wastes 800–900°C is sufficient but for materials with high thermal stability 1100°C or higher is required. This temperature must be maintained for sufficient time to allow for complete combustion. For example, two seconds at 1200°C would be suitable for most organic waste. As well as temperature and time, sufficient air must be provided to supply the oxygen required for combustion. The process should be designed to ensure that the air is provided in appropriate locations and in a manner so as to promote the turbulence necessary to achieve effective mixing with the combustible materials.

Toxic combustion by-products

Public concern relating to the use of incineration for the disposal of hazardous waste relates particularly to emissions of potentially toxic combustion products from the process. Simple examples of such emissions are:

1. Carbon monoxide and hydrocarbons resulting from incomplete combustion of organic waste.
2. Sulphur dioxide resulting from the combustion of wastes containing sulphur.
3. Hydrogen chloride from the combustion of wastes containing chlorinated compounds.
4. Heavy metal fume and particulates resulting from the incineration of organic wastes which also contain heavy metals such as lead, cadmium or chromium.

Polychlorinated dibenzodioxins - Dioxins: Polychlorinated dibenzodioxins (PCDDs) are often associated with emissions from waste incinerators. While these are often referred to as dioxin they are in fact a group of 75 compounds with different numbers of chlorine atoms and structures. These are very persistent in the environment and some are established carcinogens or cancer promoters. PCDDs and PCDFs are formed in trace quantities whenever combustion involving compounds containing chlorine occurs. As such they have been found to be wide spread in the environment but in very low concentrations, normally in nanogram or picogram amounts.

Polycyclic aromatic hydrocarbons - PAHs: Another group of compounds that can be found in trace amounts in emissions from the combustion of organic wastes are the polycyclic aromatic hydrocarbons (PAHS). Many of these are established carcinogens.

Control of gaseous emissions: Emissions of hazardous pollutants resulting from incomplete combustion of wastes can be minimised by good incinerator design and efficient combustion practices. In the case of PCDDs and PCDFs rapid reduction of flue gas temperature immediately following combustion is necessary to prevent the reformation of these compounds. Gas scrubbers using alkaline liquors are used to control acid gases such as sulphur dioxide and hydrochloric acid.

Particulate emissions, including heavy metals in the form of particles, are controlled by the use of bag filters (both wet and dry), high energy scrubbers or, less frequently, electrostatic precipitators.

High temperature incineration of organic hazardous wastes in properly designed and operated facilities can be performed in a manner that complies with stringent standards regulating the emission of gaseous pollutants, such as those enacted in North America and the European Community.

Other potential environmental impacts

The ash resulting from incineration of hazardous waste may itself possess hazardous properties. This is likely to be the case when toxic heavy metals are involved. The ash must therefore be constantly monitored and may require stabilisation and encapsulation before disposal to landfills.

Liquid effluent results where water is used for temperature reduction of gases by quenching and/or, where wet scrubbers are used for emission control. Some recycling may be incorporated after cooling and chemical treatment, but a quantity of liquid effluent will need to be discharged after appropriate treatment.

Use of cement kilns

Kilns used for the production of cement clinker in the manufacturing of Portland cements are designed and operated in a manner that achieves the required parameters for the destruction of hazardous waste, such as time at high temperatures, as discussed under incineration above. In many developed countries overseas such kilns are licensed by authorities for the destruction of appropriate hazardous wastes. Such wastes include PCBs and recently in Japan, waste chlorofluorocarbons (CFCs). An additional advantage from the use of cement clinker kilns is that the alkaline particulates involved act to neutralise acidic combustion products.

The general use of cement kilns for hazardous waste destruction in New Zealand has not gained favour. Substantial recovery of energy is available however by the use of surplus waste lubricating oils as supplementary fuel for these kilns.

Plasma arc

A thermal process developed for commercial application in Australia uses the very high temperatures, in excess of 10,000K, which can be attained in arcs formed across high voltage electrodes. This is particularly useful for the destruction of difficult hazardous liquids and gases such as some of the halogenated organics. This process is particularly applicable for the destruction of waste halons and CFCs.

7.4 Other developed technologies

7.4.1 Problem wastes

The problem, or intractable, wastes which cannot be treated are organochlorines such as PCBs and chlorinated pesticides. In the case of PCBs legislative controls have resulted in its withdrawal from use. Contaminated oils containing PCBs have been collected regionally and most exported overseas for destruction in high temperature incinerators. Other organochlorine wastes must be appropriately stored until treatment options become available.

Two different processes for organochlorine wastes are now commercially available in Australia. Both are based on chemical methods of treatment.

Base catalysed dechlorination (BCD) and the hydrogen reduction processes

The first, referred to as the BCD process, involves catalysed substitution of chlorine by hydrogen atoms under alkaline conditions. The second, operated commercially as the eco logic process, is based on complete reduction of organochlorines by hydrogen to carbon monoxide, methane and hydrogen chloride. The water shift reaction:

$$CH_4 + H_2O \rightarrow CO + 3H_2$$

is used to convert the CH_4 to H_2 thus providing the hydrogen for the process. The CO is used as a supplementary fuel for the process furnaces. Hydrogen chloride is recovered and sold from the process as hydrochloric acid.

7.5 Landfills as a disposal method

7.5.1 Requirement for landfills

As discussed above some of the treatment processes discussed above result in residues that themselves require disposal. This disposal is best carried out in properly designed and operated landfills. Controlled quantities of specific hazardous wastes may be broken down to nonhazardous substances, immobilised or adequately diluted by the physical, chemical and biochemical processes which occur in landfills accepting predominantly normal municipal refuse. Such a practice is known as codisposal and entails a degree of management and monitoring usually restricted to the modern larger regional landfills.

7.5.2 Types of landfills

Most refuse dumps or landfills can be classified into two categories according to the manner in which they have been designed, sited and installed, particularly with regard to leachate management.

Dedicated landfills

A further type of landfill is that dedicated solely for the containment of hazardous waste over a long period of time. The hazardous waste may or may not be encapsulated before placement in the landfill. Extra precautions are taken to prevent the formation of leachate from rainfall migration and to isolate and contain the wastes.

This is done by the incorporation of at least a double liner and capping system. At least one of these liners is a thick synthetic material, such as high density polyethylene (HDPE), which must be compatible with the retained waste. Monitor drains underneath and around the landfill are installed so as to illustrate the integrity of containment over time.

A system of solution mining allowing for surface treatment by aerobic biodegradation has been incorporated to breakdown these wastes over time.

The use of landfills dedicated solely for the containment of untreated hazardous wastes is not now widely accepted internationally. There is insufficient experience to allow acceptance of indefinite integrity of clay or synthetic liners. It should be considered to be in the ground long term storage of hazardous waste for which re-excavation and treatment will be required some time in the future.

Some of the waste currently produced by industrial chemical processes possess hazardous properties and require special attention in respect to disposal. Concern relating to these wastes has resulted in the development of international conventions aimed at controlling their intercountry movement. These conventions also assist in the management of these waste by providing listings that can be used as the basis for a identification and classification scheme.

Modern refuse landfills have a role in the disposal of residuals from treatment processes and possibly for limited codisposal of hazardous waste with normal refuse. This requires a level of management of these landfills considerably in excess of that practised at refuse dumps in the past.

7.6 Electrochemical methods for treatment of hazardous organic wastes

As already discussed hazardous organic wastes are produced as spent industrial solvents, effluents from textile processing units, tanneries, plastics and metal processing units, pulp and paper industries, chemical plants and food-processing units. The daily discharges from hospitals and municipal bodies also contribute considerably to these organic wastes. Even though some of the municipal wastes are disposed off as landfills in secured areas, the leachates contain a very high concentration of organic compounds, contaminating the surrounding soils and the underground water. Lastly, even the discharges from the chemical laboratories contain generous quantities of organic compounds and need treatment before disposal.

The conversion of such diverse and difficult wastes to harmless molecules is a recognised problem. The common methods for treatment of such effluents are: (i) biological treatment by bacteria, (ii) adsorption on suitable substrates such as activated charcoal, clays, etc., (iii) coagulation and filtration and (vi) thermal methods such as incineration at high temperatures.

The first three methods are age-old proven technologies, but are time consuming and generate a high volume of sludge, which often needs further treatment before disposal. Shrinking landfill areas pose a problem for such disposal methods. The adsorption and coagulation methods actually separate the harmful molecules into a different medium, but mostly do not destroy the molecules. The bacteria destroy the molecules, but all effluents are not conducive to the growth and sustenance of bacteria and hence cannot be treated biologically. Plus, the wide variations in the effluent characteristics even from the same unit and in the same stream due to process variations often make a single scheme of treatment ineffective and need multiple treatments followed by secondary and tertiary processes.

Thus, the thermal method (incineration) appears to be a much better option and is still considered to be the Best Demonstrated Available Technology (BDAT) for the destruction of most organic wastes, particularly when these are available in solid or highly concentrated form. Incineration is largely immune to variations in feed characteristics and a large volume of wastes can be managed within a short time. Incineration reduces the total solid volume to just 5–6% of its original volume. This residue is mostly siliceous matter and metals and can be safely disposed off in secured landfills in small areas. Most important, a majority of the organic compounds are themselves highly combustible and evolve considerable heat during combustion, which partly can be recovered and used to generate electricity. Therefore, extensive researches have been done on the incineration methods and a host of advanced integrated waste management technologies with emphasis on energy recovery are available today commercially.

But in recent years, the concept of incineration itself is being increasingly questioned, because some fractions of these organic wastes are often converted to highly poisonous dioxins and furans. Apart from causing cancer, all dioxins have many serious adverse effects on human health.

They cause birth defects, decrease fertility, prematurely terminate pregnancy, reduce sperm and testosterone levels, suppress the immune system and cause skin disorders, lung problems, diabetes, etc. There are now numerous reports on the formation of dioxins and other toxic substances in spite of extreme care during incineration.

Therefore, many countries are now seriously considering banning incineration of organic wastes altogether. Thus, there is a growing opinion that incineration, in spite of its many advancement, does not have any future and may be considered today a dying technology.

Therefore, many alternative technologies have been developed over the years. Some of these new technologies are novel processes and based on completely different concepts altogether.

Broadly, these new technologies may be classified into two types: (i) thermal methods and (ii) electrochemical methods or methods that use electricity.

7.6.1 Thermal methods

These methods are primarily based on a parallel school of thought that, although incineration invariably produces dioxins, it is possible to convert more than 99.9% of carbons into CO and H_2O thermally without forming the dioxins or furans. Several technologies have been developed in recent years towards that goal and some of them are briefly discussed below.

7.6.2 Plasma processing

In plasma processing, soils with toxic organic residues (such as superfund sites) are struck by plasma. The very high temperature of about 10,000°C of the plasma breaks down all organics to atoms and radicals, while the solid is melted to a black glassy lava-like mass. This is called hot plasma technology. In the cold plasma technology the electrons and radicals of the plasma react with organic compounds changing them into non-toxic and non-hazardous substances, which can be suitably scrubbed in an absorber. Today many commercial plasma systems are available for waste incineration.

7.6.3 Thermal blanketing

This method also called as *in situ* thermal decomposition, which was originally developed by Shell in 1980s for extracting more oil from the almost exhausted underground strata, the soil is covered with a thermal blanket, which raises the soil temperature locally to 600°–700°C, while vacuum is applied. This leads to a temperature-assisted desorption of the contaminants from the soil surface and then their thermal decomposition into CO and H_2O. Any organic molecule that escapes the decomposition is trapped by activated carbon. The halogens of the organic compounds react with the soil forming halides. It is claimed that the soil becomes free from all contaminants, including mercury, arsenic, lead, hydrocarbons and other toxins, which volatilise at 1000°C or so. This thermal blanketing technology also is .available commercially and has been applied to a large number of superfund sites in U.S.

7.6.4 Gasification

Another thermal method is the gasification of carbonaceous material. Gasification is an age-old technology with many commercial scale processes being operated all over the world with different feedstock. But specific attempts have been done to develop commercial scale methods to destroy the oxygenated and halogenated organic compounds by using coal and gasifying the mixture. In one such process, pulverized coal is mixed with water, ammonia and the organic waste matter and the slurry is pumped into the reactor vessel and heated to high temperatures of 1200°C–1650°C under pressures of more than 300 psi for a retention time of 1.5–4 seconds.

Under these conditions the coal is gasified and the product gases react with the organic molecules to break them into CO, CO_2, H_2 and acid gases (such as H_2S, HCN, HCl, etc.).

The process has been claimed to have advantages over other thermal waste destruction processes in terms of (i) energy and materials conservation, (ii) fewer process control parameters, (iii) is applicable for all types and quantities of wastes, (iv) limited exposure to the environment of any by-products, (v) ability to recycle any incomplete destruction products, (vi) most metals entering the system are fixated in a silicate slag and (vii) can handle waste not suitable to conventional incineration.

7.6.5 Molten salt oxidation (MSO)

Another thermal technology is the molten salt oxidation (MSO) process, in which the wastes containing hazardous organic materials are injected with an excess of air into a pool of molten carbonate salts at temperatures between 700–950°C. Flameless oxidation takes place within the salt bath converting the organic waste into CO_2 and water. The product gas leaving the vessel is treated to remove any entrained salt particulates and water vapour before being discharged to the environment. Halogens and sulphur are converted into acid gases, which are then 'scrubbed' and trapped as NaCl and Na_2SO_4.

Other non-oxidisable inorganic constituents, including heavy metals, are held captive in the salt, either as metals or oxides and most are easily separated for disposal. This technology was originally developed by Rockwell for coal gasification in 1970s. Later in 1990s this has been modified by Lawrence Livermore National Laboratory (LLNL) for treating hazardous wastes and subsequently it has been extensively tested at integrated pilot scale.

This technology appears to be suitable for the destruction of the huge stock pile of chemical weapons, which are mainly halogen, sulphur and phosphorous containing organic molecules and are to be converted to CO and H_2O before disposal.

It is a fact that these various thermal technologies are quite effective and energy-efficient in destroying many difficult organic wastes, but some of these technologies such as plasma processing and thermal blanketing are not energy-efficient or suitable for large volumes of aqueous solutions containing varieties of organic compounds. Moreover, It is necessary that effluent treatment technology should be relatively simple and adaptable to various scale operations so that it is economically useable by most small and medium scale industries. The electrochemical methods precisely provide such technology, which is suitable to all types of organic molecules and can be used in the industry at all scales. Some of the electrochemical technologies are discussed below:

7.6.6 Electrochemical methods

In an electrochemical process, the organic pollutants in an aqueous phase are used as electrolyte. Since many organic molecules are non-conductors of electricity, often suitable acids, salts, etc., and even ionic liquids are added to enhance the electrical conductivity and the mixture is electrolysed between two electrodes by applying a small DC potential. The cathode is usually a stainless steel plate or graphite, whereas the anodes are various materials, but mostly titanium, coated with suitable electro-catalysts like TiO_2, Pt, RuO_2, etc.

The main purpose of the electro-catalysts is to promote the desired oxidation reactions and at the same time to increase the over voltage for H_2 and O_2 formation so as to prevent the energy loss by decomposition of water:

$$H_2O \rightarrow 2\,H^+ + \tfrac{1}{2}\,O_2 + 2\,e$$

which occurs at an applied voltage of 1.23 V and above. In the process of direct electrolysis, as the electrical current passes through the solution, the organic molecules are converted to radicals and ions, which subsequently break down completely forming CO_2 and H_2O. But in view of the conductivity problem, there are processes, where the organic molecules are not electrolysed directly. Instead these are oxidised completely at first in a separate reactor by a strong oxidising agent like Ce^{4+}, Ag^{2+}, Co^{3+}, etc.

In the next step, the reduced species (Ce^{3+}, Ag^{+} or Co^{2+}) is sent to an electrolysis cell, where these are re-oxidised to their original state. This process is called Mediated Electrochemical Oxidation or simply, as MEO. Thus, the electrochemical methods may be broadly classified into four types: (i) Mediated Electrochemical Oxidation (MEO), (ii) Direct Oxidation by Electrolysis, (iii) Electro-assisted Advanced Oxidation Processes and (vi) Indirect Electrochemical Methods such as Electrocoagulation, which do not electrolyse, but use electricity as the main driver of the process.

Mediated electrochemical oxidation (MEO)

It is claimed that this process is capable of destroying any type of organic wastes and even the chemical weapon such as nerve gas or sarin. However, it cannot handle the wastes, which are sparsely dispersed in another media.

The mediated electrochemical oxidation process as a concept is very popular and has been studied widely because it circumvents the electrical conductivity problem of the organic compounds by carrying out the oxidations chemically and regenerates the main reactant through electrolysis. Thus, in the ultimate analysis, the main driver of the process is electricity and it has been reported that the total CO_2 generated by this process (including that produced during electricity generation) per molecule of the organic pollutant destroyed is about 25% less than the amount of CO_2, which would have been generated, had the organic pollutant been destroyed by a thermal method. Therefore, the mediated electrochemical oxidation processes have been studied extensively and many commercial electrolytic cells are available today for treating organic wastes by this process.

Direct electrolysis

Most of fine chemicals production are characterised by relatively slow, multi-step batch operations in multipurpose plants and generation of a very high volume of organic waste per unit volume of the final product. These wastes contain all sorts of organic aliphatic and aromatic compounds with a high dose of ethanol and methanol as solvents. The average COD is about 6000 kg/L and a typical TOC/COD load is about 0.27. The typical electrical conductivity is also very high (~ 25 ms/cm) due to the presence of a large quantity of chlorides. In recent years, boron-doped diamond (BDD) electrodes have opened a new horizon for the treatment of such wastes. Diamond is an electrical insulator due to extensive covalent bonding. But a doping of boron to the extent of 0.01% and 0.5% by weight changes the electronic properties of the diamond drastically, making it electrically conducting. This, in combination with the extraordinary hardness of chemical resistance of diamond, has made it an ideal electrode for many electrochemical applications. A major benefit of these electrodes is that like platinum-coated titanium electrodes, these also show high over voltages for both H_2 and O_2 evolution in aqueous solutions. Today a large number of processes have been developed to make such conducting boron-doped diamond electrodes cheaply' and a range of electrochemical processes has been studied and developed to oxidise organic wastes. These processes are known as conducting-diamond electrochemical oxidation (CDEO) processes.

Electro-assisted advanced oxidation processes

In recent years, a process has been developed to produce H_2O_2 electrolytically *in situ* and this H_2O_2 then reacts with ferrous sulphate to produce the hydroxyl radical. This process is called as Electro-Fenton process. There have been numerous studies on the Electro-Fenton process, particularly in the treatment of wastewaters containing organic pollutants. In the Electro-Fenton process, the electrolysis is carried out with conventional electrodes, but the conditions (pH, applied voltage and current density) are so maintained that oxygen is reduced at the cathode to form H_2O_2. Then the addition of Fe^{2+} leads to the formation of hydroxyl radical possibly by the following reactions:

$$O_2 + 2H^+ + 2e \rightarrow H_2O_2$$

$$Fe^{2+} + H_2O_2 \rightarrow Fe^{3+} + OH^- \text{ (Radical)}$$

$$Fe^{3+} + e \rightarrow Fe^{2+}$$

Electro-Fenton process shows generally more efficiency than Fenton process because of the *in situ* formation of H_2O_2 and the regeneration of Fe^{2+} at the cathode.

Indirect electro-assisted processes

Electro-coagulation: In a different type of process, which uses electricity, but one that does not electrolyse the substances, an electrode of pure copper or iron is used as anode, which is slowly oxidised to form Al^{3+} or Fe^{3+} ions, respectively. These ions then slowly hydrolyse to form hydroxyl ions.

The process is called electro-coagulation and the reactions are as follows:

At iron anode: $Fe \rightarrow Fe^{3+} + 3\ e$

At cathode: $H_2O + e \rightarrow \frac{1}{2} H_2 + OH^-$

Then the hydrolysis of Fe^{3+} and dimerisation and polymerisation of the various iron-hydroxy species take place, ending finally in the formation of very fine flocs of iron hydroxide with strong affinity to adsorb the organic molecules.

$$Fe^{3+} + H_2O \rightarrow Fe(OH)^{2+} + H^+$$

$$Fe(OH)^{2+} + H_2O \rightarrow Fe(OH)_2^{+} + H^+$$

$$2Fe(OH)^{2+} \rightarrow Fe_2(OH)_2^{4+}$$

$$2\ Fe(OH)_2^{+} \rightarrow Fe_2(OH)_4^{2+}$$

$$Fe(OH)^{2+} + H_2O \rightarrow Fe(OH)_3 + H^+$$

$$Fe(OH)_2^{+} + OH^- \rightarrow Fe(OH)_3$$

The same type of reactions occurs with aluminium. It has been observed that both iron and aluminium are equally active and efficient for removal of the organic impurities, but apparently iron is slightly more active and effective than aluminium, particularly as its range of operation (pH range) is wider. This process has been used widely to treat the wastewater from textile processing units.

Electro-mediation

Another process, which uses electricity, but does not electrolyse the solution, is the Electro-mediation of soils. In this process, a direct current is passed into the moist contaminated soil at various points through a number of electrodes typically under a potential gradient of 20–200 V/m and/or a constant current density of 0.025–5 A/m^2. The current is kept low so that there is no significant heating effect. The current is passed from a few hours to several days or weeks. The heavy metal ions such as Ni, Cd, Pb, Hg, Sn, Zn, Cu, Co, etc., which are present in the soil (contaminating the soil) migrate through the soil under the action of the potential gradient and accumulate around the cathode over a period of time. In case of Cr or As, which form oxyanions, these are accumulated around the anode. The final extent of removal of all these ions is 75–95%, depending upon the soil condition such as pH, porosity and texture, moisture content, hydrophobicity/hydrophilicity, zeta potential, iso-electric point, etc. This technology has been in use for corrosion protection for many years. But since 1990s there have been many attempts to use this technology at lab scale to remediate soils contaminated with organic substances and it has been found that it is possible to remove various organic molecules from the soil by combined effects of electro-osmosis, i.e., osmosis under an applied electric field, electrophoresis and diffusion. As it has been cited by page and others many workers have observed successful removal of acetic acid, phenol, nitro-compounds and various types of chlorinated hydrocarbons from contaminated soils by this method.

A major problem, which has been observed during field trials, is that stones, hard objects, metal pieces, concrete embedded in the earth disturb the variance in conductivity, affecting the outcome of the trial.

Use of 3-D electrodes

The main disadvantage of any electrochemical process is that its rate is determined only by the magnitude of current flow and thus is quite slow in comparison with the thermal or incineration methods. Thus, whatever may be the process conditions, 96,500 coulombs of electricity leads to the destruction of only molecules (Faradays laws of electrolysis), where number of electrons involved in the oxidation of one molecule. So, prima facie, an increase in the current may increase the output, but there is a limit to this increase, as a very high current density and/or applied voltage may promote secondary reactions. Therefore, today the main emphasis in the electrochemical technology is towards: (i) use of electro-catalysis and/or other coating materials, which increase the over voltage of the electrode and discourage secondary reactions like water electrolysis and (ii) use of higher surface area of the electrodes. This is possible using three-dimensional electrodes, which are packed bed of conducting particles through which the electrolyte flows.

In recent times, electrodes made of activated carbon fibre (ACF) has been extensively used for wastewater treatment applications. In a typical electrolytic system, iron is used as anode and ACF is used as cathode and the typical electrode reactions are:

At anode $Fe \rightarrow Fe^{2+} + 2e$

At cathode $2\,H_2O + 2e \rightarrow H_2(g) + 2OH^-$

Thus, overall reaction is $Fe + 2H_2O \rightarrow Fe(OH)_2 + H_2$

This type of system has been used in the destruction of dyes in wastewater. For example, Shen and others have reported positive results for 29 dyes of different classes in an electrolysis system using activated carbon fibre electrodes. It has been found that almost all dyes are degraded rapidly, but some dyes having –C=O, –NH–, aromatic and hydrophobic groups, tend to be adsorbed. For dyes with $-SO_3-$, –COOH and –OH groups, there is also a tendency for adsorption and flocculation, particularly if the molecules linearly spread in solution and have considerable hydrogen bonding within the molecules. With an applied voltage of 15 V, electrolyte concentration 0.05 M Na_2SO_4 solution, volumetric current density 110 mA/dm^3, it has been reported that up to 85% of colour is removed and the total organic carbon (TOC) is removed up to 70%. The efficiency of the process is higher in case of vat dyes and dispersed dyes than in case of acid dyes, reactive dyes and neutral dyes.

Thus, the electrolysis, being done under near ambient conditions is an inexpensive and relatively simple operation. Only major expenditure is the initial capital cost towards the cell and the special type of electrodes. The main disadvantage is that the electrolysis, being governed by Faraday's laws of electrolysis and Coulomb's law, is relatively slow in comparison with the conventional chemical processes. Therefore, world-over efforts are being put in towards the development of this form of technology. It must be admitted that in spite of a huge volume of work all over the world, not many of these technologies are available as because of the unfavourable economics. However, a number of electrochemical methods are available commercially. The correct infrastructure; systems and process should be identified. The locations should be for away from residential facilities and any water/natural body resource.

7.7 Storage and distribution of hazardous chemicals

The encouraging growth of chemical industry has naturally brought the issues concerning hazardous chemical storage and distribution into sharp focus. The bottlenecks pertaining to infrastructure, suitable storage facility, equipment and technology is posing serious challenges for the growth of the sector. There is a lack of continuous engagement at the strategical and tactical level between the chemical industry and logistics service providers. Most of the engagement takes place at an operational level, which doesn't enable any major investments on a long term basis and thus does not help in adding value to the entire supply chain. Today, there is a lack of education at both the service provider and the service takers end. The discussions in various countries unfortunately do not focus on deliverables and expectations, but mainly on commercials. It is important to 'know your customer' and the chemical very well before storage and distribution of the same.

7.7.1 Logistics sector: Moving on the right track

However, with the government providing a few sops in the areas of storage facility, equipment, etc., the logistics sector seems to be moving on the right track. Hazardous materials storage and distribution risk management involves establishing, organising, planning, executing and monitoring a set of operations that aims to decrease the probability of accidents and reduce the relevant potential consequences. This risk management process pertains to preventive and repressive safety measures.

7.7.2 Preventive and repressive safety measures

The preventive measures aim to eliminate the potential causes of accidents, i.e., the roadway system defects, mishandling by drivers, mechanical errors in vehicles, etc. Besides, driver training, policies related to driving hours, container specifications, vehicle and mechanical conditions monitoring equipment and routing through safe roadway segments are other important elements. The laws related to storage and distribution of hazardous chemicals are convoluted and there are no clear guidelines. The chemical manufacturing companies are complying with laws in order to get 'certifications' and to be compliant legally rather than with the aim of creating a world class infrastructure and bring in efficiencies in the supply chain.

7.7.3 Right storage location

It is imperative to identify the right storage location for any kind of chemical. The civil design of the warehouse is such that it has provision for fire water retention, sump for storage of water for fire fighting and fire resistant walls, depending on the type of products that need storage.

The usage of Personal Protective Equipment (PPE) is a must for all operations carried out in a warehouse. The man-power deployed is trained on effective usage of PPE's.

Intensive training sessions for manpower constitute of trainings related to:

1. Material (MSDS for chemicals, spill control, material handling training).
2. Labour (fire fighting and first aid, safety regulations, emergency response).
3. Driving – MHE operations (safe driving, driving behaviour, importance of PPE).
4. Basic operations and handling (mock drills, fire safety and life support).

High degree of HSE (health, safety and environment) aspects are considered. Compliance to all statutory requirements, necessary licenses and NOCs required for storage and handling of such products are obtained. There are safety boards and signage's displayed across the warehouse. There is a defined area for the movement of people and visitors.

The parking areas should be designated and the speed limits within the warehouse should be restricted and well defined. Bar code enabled inventory and system driven process management leads to higher productivity and effective operations at warehouses.

In various countries today, most of the operations are 'drum in drum out'. There should be unitisation by palletisation and adequate infrastructure should be created at both the service provider and the customers end. Multiple manual handling of hazardous cargo should be avoided and mechanisation should be introduced.

7.7.4 Need for collaborative logistics

Today, there is a need for collaborative logistics, where the service provider should be ready to invest if the chemical industry cumulatively takes advantage of such an investment. The service provider should get involved at the strategic level.

8

Nuclear pollution and radioactive wastes

8.1 Introduction

Radioactivity is the term used to describe the natural process by which some atoms spontaneously disintegrate, emitting both particles and energy as they transform into different, more stable atoms. This process, also called radioactive decay, occurs because unstable isotopes tend to transform into a more stable state. Radioactivity is measured in terms of disintegrations, or decays, per unit time. Common units of radioactivity are the Becquerel, equal to 1 decay per second and the Curie, equal to 37 billion decays per second.

Radiation: Radiation refers to the particles or energy released during radioactive decay. The radiation emitted may be in the form of particles, such as neutrons, alpha particles and beta particles or waves of pure energy, such as gamma and X-rays.

Sources of radiation: Everyone is exposed to radiation on a daily basis, primarily from naturally occurring cosmic rays, radioactive elements in the soil and radioactive elements incorporated in the body. Man-made sources of radiation, such as medical X-rays or fallout from historical nuclear weapons testing, also contribute, but to a lesser extent.

Uranium and radioactivity: All isotopes of uranium are radioactive, with most having extremely long half-lives. Half-life is a measure of the time it takes for one half of the atoms of a particular radionuclide to disintegrate (or decay) into another nuclear form. Each radionuclide has a characteristic half-life. Half-lives vary from millionths of a second to billions of years. Because radioactivity is a measure of the rate at which a radionuclide decays (for example, decays per second), the longer the half-life of a radionuclide, the less radioactive it is for a given mass. The half live of uranium-238 is about 4.5 billion years, uranium-235 about 700 million years and uranium-234 about 25 thousand years.

Decay products of uranium: Uranium atoms decay into other atoms or radionuclides, that are also radioactive and commonly called 'decay products.' Uranium and its decay products primarily emit alpha radiation, however, lower levels of both beta and gamma radiation are also emitted.

8.2 Radioactive waste management

All parts of the nuclear fuel cycle produce some radioactive waste (radwaste) and the relatively modest cost of managing and disposing of this is part of the electricity cost, i.e., it is internalised and paid for by the electricity consumers. At each stage of the fuel cycle there are proven technologies to dispose of the radioactive wastes safely. For low- and intermediate-level wastes these are mostly being implemented. For high-level wastes some countries await the accumulation of enough of it to warrant building geological repositories.

Unlike other industrial wastes, the level of hazard of all nuclear waste – its radioactivity – diminishes with time. Each radionuclidea contained in the waste has a half-life – the time taken for half of its atoms to decay and thus for it to lose half of its radioactivity. Radionuclides with long half-lives tend to be alpha and beta emitters – making their handling easier – while those with short half-lives tend to emit the more penetrating gamma rays. Eventually all radioactive wastes decay into non-radioactive elements. The more radioactive an isotope is, the faster it decays.

The main objective in managing and disposing of radioactive (or other) waste is to protect people and the environment. This means isolating or diluting the waste so that the rate or concentration of any radionuclides returned to the biosphere is harmless. To achieve this, practically all wastes are contained and managed – some clearly need deep and permanent burial. From nuclear power generation, none is allowed to cause harmful pollution.

All toxic wastes need to be dealt with safely, not just radioactive wastes. In countries with nuclear power, radioactive wastes comprise less than 1% of total industrial toxic wastes (the balance of which remains hazardous indefinitely).

8.3 Types of radioactive wastes

8.3.1 Exempt waste and very low level wastes

Exempt waste and very low level waste (VLLW) contains radioactive materials at a level which is not considered harmful to people or the surrounding environment. It consists mainly of demolished material (such as concrete, plaster, bricks, metal, valves, piping, etc.), produced during rehabilitation or dismantling operations on nuclear industrial sites. Other industries, such as food processing, chemical, steel, etc., also produce VLLW as a result of the concentration of natural radioactivity present in certain minerals used in their manufacturing processes. The waste is therefore disposed of with domestic refuse, although countries such as France are currently developing facilities to store VLLW in specifically designed VLLW disposal facilities.

8.3.2 Low-level wastes

Low-level waste (LLW) is generated from hospitals and industry, as well as the nuclear fuel cycle. It comprises paper, rags, tools, clothing, filters, etc., which contain small amounts of mostly short-lived radioactivity. It does not require shielding during handling and transport and is suitable for shallow land burial. To reduce its volume, it is often compacted or incinerated before disposal. It comprises some 90% of the volume but only 1% of the radioactivity of all radioactive waste.

8.3.3 Intermediate-level wastes

Intermediate-level waste (ILW) contains higher amounts of radioactivity and some requires shielding. It typically comprises resins, chemical sludges and metal fuel cladding, as well as contaminated materials from reactor decommissioning. Smaller items and any non-solids may be solidified in concrete or bitumen for disposal. It makes up some 7% of the volume and has 4% of the radioactivity of all radwaste. Its radioactive decay generates heat of less than about 2 kW/m^3 so does not require heating to be taken into account in design of storage or disposal facilities.

8.3.4 High-level wastes

High-level waste (HLW) arises from the 'burning' of uranium fuel in a nuclear reactor. HLW contains the fission products and transuranic elements generated in the reactor core. It is highly radioactive and hot due to decay heat, so requires cooling and shielding. It has thermal power above about 2 kW/m^3 and can be considered as the 'ash' from 'burning' uranium. HLW accounts for over 95% of the total radioactivity produced in the process of electricity generation. There are two distinct kinds of HLW:

1. Used fuel itself.
2. Separated waste from reprocessing the used fuel.

HLW has both long-lived and short-lived components, depending on the length of time it will take for the radioactivity of particular radionuclides to decrease to levels that are considered no longer hazardous for people and the surrounding environment. If generally short-lived fission products can be separated from long-lived actinides, this distinction becomes important in management and disposal of HLW. HLW is a major focus of attention regarding nuclear power and the industry has proposed that any option for the management of used nuclear fuel is sustainable if:

1. It covers all the steps of used fuel management until final disposal, in accordance with an acceptable, practical plan.
2. It proves to be feasible with an acceptable impact level by meeting defined key criteria.
3. It includes a realistic and balanced financing plan.
4. It does not impose undue burdens on future generations.

8.3.5 Mining and milling

Traditional uranium mining generates fine sandy tailings, which contain virtually all the naturally occurring radioactive elements naturally found in uranium ore. These are collected in engineered tailings dams and finally covered with a layer of clay and rock to inhibit the leakage of radon gas and ensure long-term stability. In the short term, the tailings material is often covered with water. After a few months, the tailings material contains about 75% of the radioactivity of the original ore. Strictly speaking these are not classified as radioactive wastes.

8.3.6 Conversion, enrichment, fuel fabrication

Uranium oxide concentrate from mining, essentially 'yellowcake' (U_3O_8), is not significantly radioactive – barely more so than the granite used in buildings. It is refined then converted to uranium hexafluoride gas (UF_6). As a gas, it undergoes enrichment to increase the U-235 content from 0.7% to about 3.5%. It is then turned into a hard ceramic oxide (UO_2) for assembly as reactor fuel elements.

The main by-product of enrichment is depleted uranium (DU), principally the U-238 isotope, which is stored either as UF_6 or U_3O_8. About 1.2 million tonnes of DU is now stored. Some is used in applications where its extremely high density makes it valuable, such as the keels of yachts and military projectiles. It is also used (with reprocessed plutonium) for making mixed oxide fuel and to dilute highly-enriched uranium from dismantled weapons which is now being used for reactor fuel.

8.4 Electricity generation

In terms of radioactivity, high-level waste (HLW) is the major issue arising from the use of nuclear reactors to generate electricity. Highly radioactive fission products and also transuranic elements are produced from uranium and plutonium during reactor operations and are contained within the used fuel. Where countries have adopted a closed cycle and utilised reprocessing to recycle material from used fuel, the fission products and minor actinides are separated from uranium and plutonium and treated as HLW (uranium and plutonium is then reused as fuel in reactors). In countries where used fuel is not reprocessed, the used fuel itself is considered a waste and therefore classified as HLW.

Low- and intermediate-level waste is produced as a result of operations, such as the cleaning of reactor cooling systems and fuel storage ponds, the decontamination of equipment, filters and metal components that have become radioactive as a result of their use in or near the reactor.

8.5 Quantity of wastes produced

As already noted, the volume of nuclear waste produced by the nuclear industry is very small compared with other wastes generated. Each year, nuclear power generation facilities worldwide produce about 200,000 m^3 of low- and intermediate-level radioactive waste and about 10,000 m^3/12,000 T of high-level waste including used fuel designated as waste. In the OECD countries, some 300 million tonnes of toxic wastes are produced each year, but conditioned radioactive wastes amount to only 81,000 m^3 per year.

A typical 1000 MWe light water reactor will generate (directly and indirectly) 200–350 m^3 low- and intermediate-level waste per year. It will also discharge about 20 m^3 (27 T) of used fuel per year, which corresponds to a 75 m^3 disposal volume following encapsulation if it is treated as waste. Where that used fuel is reprocessed, only 3 m^3 of vitrified waste (glass) is produced, which is equivalent to a 28 m^3 disposal volume following placement in a disposal canister.

This compares with an average 400,000 T of ash produced from a coal-fired plant of the same power capacity. Today, volume reduction techniques and abatement technologies as well as continuing good practice within the work force all contribute to continuing minimisation of waste produced, a key principle of waste management policy in the nuclear industry. Whilst the volumes of nuclear wastes produced are very small, the most important issue for the nuclear industry is managing their toxic nature in a way that is environmentally sound and presents no hazard to both workers and the general public.

8.5.1 Managing high-level wastes

Used fuel gives rise to high-level waste (HLW) which may be either the used fuel itself in fuel rods, or the separated waste arising from reprocessing this. In either case, the amount is modest – as noted above, a typical reactor generates about 27 T of used fuel which may be reduced to 3 m^3 per year of vitrified waste. Both can be effectively and economically isolated and have been handled and stored safely since nuclear power began.

Storage of used fuel is mostly in ponds associated with individual reactors, or in a common pool at multi-reactor sites, or occasionally at a central site.

If the used fuel is reprocessed, as is that from U.K., French, German, Japanese and Russian reactors, HLW comprises highly-radioactive fission products and some transuranic elements with long-lived radioactivity. These are separated from the used fuel, enabling the uranium and plutonium to be recycled. Liquid HLW from reprocessing must be solidified. The HLW also generates a considerable amount of heat and requires cooling. It is vitrified into borosilicate (Pyrex) glass, encapsulated into heavy stainless steel cylinders about 1.3 metres high and stored for eventual disposal deep underground. This material has no conceivable future use and is unequivocally waste. The hulls and end-fittings of the reprocessed fuel assemblies are compacted, to reduce volume and usually incorporated into cement prior to disposal as ILW. France has two commercial plants to vitrify HLW left over from reprocessing oxide fuel and there are also plants in the U.K., and Belgium. The capacity of these Western European plants is 2500 canisters (1000 T) a year and some have been operating for three decades.

If used reactor fuel is not reprocessed, it will still contain all the highly radioactive isotopes and then the entire fuel assembly is treated as HLW for direct disposal. It too generates a lot of heat and requires cooling. However, since it largely consists of uranium (with a little plutonium), it represents a potentially valuable resource and there is an increasing reluctance to dispose of it irretrievably.

Either way, after 40–50 years the heat and radioactivity have fallen to one thousandth of the level at removal. This provides a technical incentive to delay further action with HLW until the radioactivity has reduced to about 0.1% of its original level.

After storage for about 40 years the used fuel assemblies are ready for encapsulation or loading into casks ready for indefinite storage or permanent disposal underground.

Direct disposal of used fuel has been chosen by the U.S., and Sweden among others, although evolving concepts lean towards making it recoverable if future generations see it as a resource. This means allowing for a period of management and oversight before a repository is closed.

Used fuel is subject to international safeguards due to its uranium and plutonium content. Separated (and vitrified) HLW is not subject to safeguards, which is another factor in easier handling.

Radioactive waste handling system is shown in Fig. 8.1.

8.6 Recycling of used fuel

Any used fuel will still contain some of the original U-235 as well as various plutonium isotopes which have been formed inside the reactor core and the U-238. In total these account for some 96% of the original uranium and over half of the original energy content (ignoring U-238). Reprocessing, undertaken in Europe and Russia, separates this uranium and plutonium from the wastes so that they can be recycled for reuse in a nuclear reactor. Plutonium arising from reprocessing is recycled through a MOX fuel fabrication plant where it is mixed with depleted uranium oxide to make fresh fuel. European reactors currently use over 5 T of plutonium a year in fresh MOX fuel.

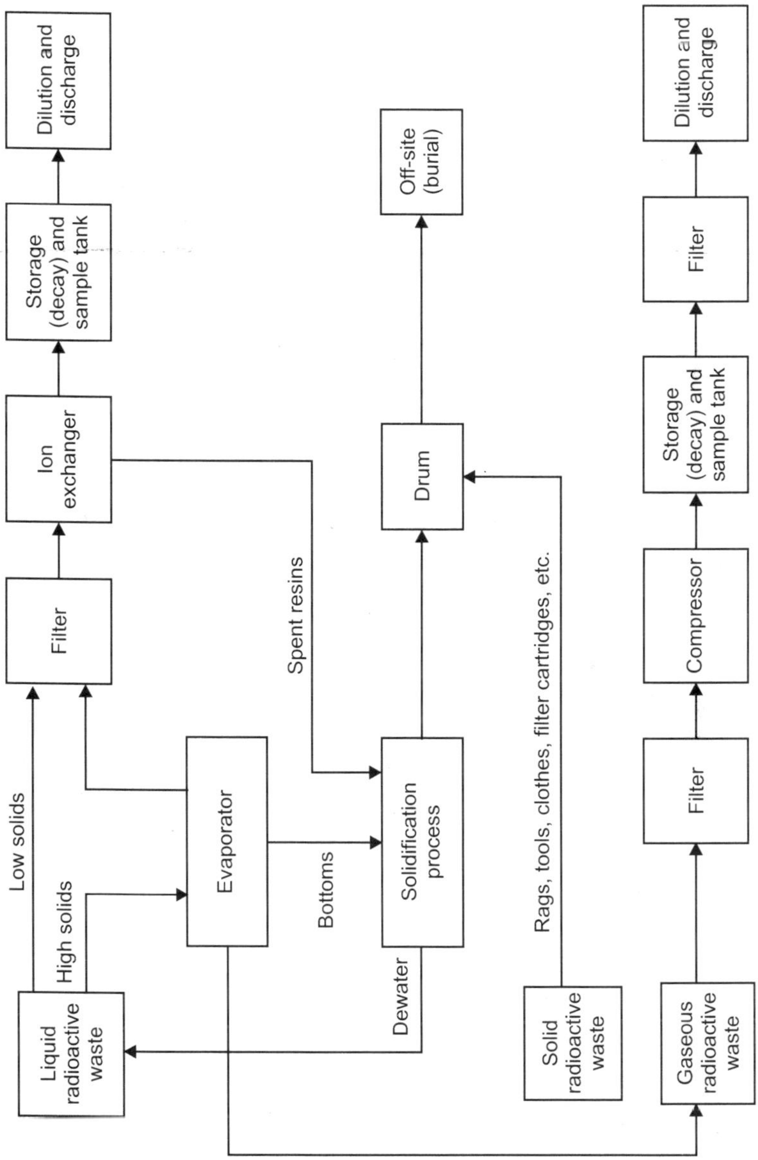

Figure 8.1: Radioactive wastes handling system.

Major commercial reprocessing plants operate in France, U.K., and Russia with a capacity of some 5000 T per year and cumulative civilian experience of 80,000 T over 50 years. A new reprocessing plant with an 800 T per year capacity at Rokkasho in Japan is undergoing commissioning. France and UK also undertake reprocessing for utilities in other countries, notably Japan, which has made over 140 shipments of used fuel to Europe since 1979. Until now most Japanese used fuel has been reprocessed in Europe, with the vitrified waste and the recovered uranium and plutonium (as MOX fuel) being returned to Japan to be used in fresh fuel. Russia also reprocesses some used fuel from Soviet-designed reactors in other countries.

There are several proposed developments of reprocessing technologies. One technology under development would separate plutonium along with the minor actinides as one product. This however cannot be simply put into MOX fuel and recycled in conventional reactors; it requires fast neutron reactors which are as yet few and far between. On the other hand, it will make disposal of high-level wastes easier.

So the options for used fuel are:

1. Direct disposal (after storage) to a geological repository. The material has very long-lived radioactivity and will take about 300,000 years to reach the same level as the original ore.
2. Aqueous reprocessing to remove only uranium and plutonium. The material then only takes about 9000 years to reach the same level of radioactivity as the original ore.
3. Advanced electrometallurgical reprocessing which removes uranium, plutonium and minor actinides together for recycling in a fast reactor. The wastes then only need 300 years to reach the same level of radioactivity as the original ore. This is not yet operational on any commercial scale.

8.7 Storage and disposal of used fuel and separated HLW

There is about 240,000 T (As on 2014) of used fuel in storage, much of it at reactor sites. About 90% of this is in storage ponds (smaller versions of that illustrated above), the balance in dry storage. Much of the world's used fuel is stored thus and some of it has been there for decades. Annual arisings of used fuel are about 7000 T (6000 T from LWRs) and up to 3000 T of this are intended for reprocessing. Final disposal is not urgent in any logistical sense.

Storage ponds at reactors and those at centralised facilities such as CLAB (CLAB means central interim storage facility for spent nuclear fuel) in Sweden, are 7–12 metres deep, to allow several metres of water over the used fuel comprising racked fuel assemblies typically about 4 m long and standing on end. The multiple racks are made of metal with neutron absorbers incorporated in it. The circulating water both shields and cools the fuel.

These pools are robust constructions made of thick reinforced concrete with steel liners. Ponds at reactors are often designed to hold all the used fuel for the life of the reactor.

Some storage of fuel assemblies which have been cooling in ponds for at least five years is in dry casks, or vaults with air circulation inside concrete shielding. One common system is for sealed steel casks or multi-purpose canisters (MPCs) each holding about 80 fuel assemblies with inert gas. Casks/ MPCs may be used also for transporting and eventual disposal of the used fuel. For storage, each is enclosed in a ventilated storage module made of concrete and steel. These are commonly standing on the surface, about 6m high, cooled by air convection, or they may be below grade, with just the tops showing. The modules are robust and provide full shielding. Each cask has up to 45 kW heat load.

A collection of casks or modules comprises an Independent Spent Fuel Storage Installation (ISFSI), which in the U.S., is licensed separately from any associated power plant and is for interim storage only. About one quarter of U.S., used fuel is stored thus.

For disposal, to ensure that no significant environmental releases occur over tens of thousands of years, 'multiple barrier' geological disposal is planned. This immobilises the radioactive elements in HLW and some ILW and isolates them from the biosphere. The main barriers are:

1. Immobilise waste in an insoluble matrix such as borosilicate glass or synthetic rock (fuel pellets are already a very stable ceramic: UO_2).
2. Seal it inside a corrosion-resistant container, such as stainless steel.
3. Locate it deep underground in a stable rock structure.
4. Surround containers with an impermeable backfill such as bentonite clay if the repository is wet.

HLW from reprocessing must be solidified. France has two commercial plants to vitrify HLW left over from reprocessing oxide fuel and there are also significant plants in the U.K., and Belgium. The capacity of these western European plants is 2500 canisters (1000 T) a year and some have been operating for three decades. By mid-2009, the U.K., sellafield vitrification plant had produced it 5000th canister of vitrified HLW, representing 3000 m^3 of liquor reduced to 750 m^3 of glass. The plant fills about 400 canisters per year.

The Australian Synroc (synthetic rock) system is a more sophisticated way to immobilise such waste and this process may eventually come into commercial use for civil wastes.

To date there has been no practical need for final HLW repositories, as surface storage for 40–50 years is first required so that heat and radioactivity can decay to levels which make handling and storage easier.

The process of selecting appropriate deep geological repositories is now underway in several countries. Finland and Sweden are well advanced with plans for direct disposal of used fuel in mined repositories, since their parliaments decided to proceed on the basis that it was safe, using existing technology. Both countries have selected sites, in Sweden, after competition between two municipalities. The U.S., has opted for a final repository at Yucca Mountain in Nevada, though this is now stalled due to political decision. There have also been proposals for international HLW repositories in optimum geology.

A current question is whether wastes should be emplaced so that they are readily retrievable from repositories. There are sound reasons for keeping such options open–in particular, it is possible that future generations might consider the buried waste to be a valuable resource. On the other hand, permanent closure might increase long-term security of the facility. After being buried for about 1000 years most of the radioactivity will have decayed. The amount of radioactivity then remaining would be similar to that of the naturally-occurring uranium ore from which it originated, though it would be more concentrated. In mined repositories, which represent the main concept being pursued, retrievability can be straightforward, but any deep borehole disposal is permanent. Deep boreholes are more appropriate for smaller amounts of wastes than national programmes involving direct disposal of used fuel and hence are more likely to be used for smaller volumes of shorter-lived wastes arising from Generation IV fuel cycles.

France's 2006 waste law says that HLW disposal must be 'reversible', which was clarified in a 2015 amendment to mean guaranteeing long-term flexibility in disposal policy, while 'retrievable' referred to short-term practicality. France, Switzerland, Canada, Japan and the U.S., require retrievability and that is policy also in most other countries, though this presupposes that in the long-term, the repository would be sealed to satisfy safety requirements.

8.8 Waste management for used fuel and HLW from nuclear power reactors

Waste management for used fuel and HLW from nuclear power reactors is shown in Table 8.1.

Table 8.1: Waste management for used fuel and HLW from nuclear power reactors.

Country	*Policy*	*Facilities and progress towards final repositories*
Belgium	Reprocessing	• Central waste storage at Dessel • Underground laboratory established 1984 at Mol • Construction of repository to begin about 2035
Canada	Direct disposal	• Nuclear waste management organisation set up 2002 • Deep geological repository confirmed as policy, retrievable • Repository site search from 2009, planned for use 2025
China	Reprocessing	• Central used fuel storage at LanZhou • Repository site selection to be completed by 2020 • Underground research laboratory from 2020, disposal from 2050
Finland	Direct disposal	• Programme start 1983, two used fuel storages in operation • Posiva Oy set up 1995 to implement deep geological disposal • Underground research laboratory Onkalo under construction • Repository planned from this, near Olkiluoto, open in 2023
France	Reprocessing	• Underground rock laboratories in clay and granite • Parliamentary confirmation in 2006 of deep geological disposal, containers to be retrievable and policy 'reversible' • Bure clay deposit is likely repository site to be licensed 2015, operating 2025
Germany	Reprocessing but moving to direct disposal	• Repository planning started 1973 • Used fuel storage at Ahaus and Gorleben salt dome • Geological repository may be operational at Gorleben after 2025
India	Reprocessing	• Research on deep geological disposal for HLW
Japan	Reprocessing	• Underground laboratory at Mizunami in granite since 1996 • Used fuel and HLW storage facility at Rokkasho since 1995 • Used fuel storage under construction at Mutsu, start up 2013 • NUMO set up 2000, site selection for deep geological repository under way to 2025, operation from 2035, retrievable
Russia	Reprocessing	• Underground laboratory in granite or gneiss in Krasnoyarsk region from 2015, may evolve into repository • Sites for final repository under investigation on Kola peninsula

(Cont'd...)

Country	Policy	Facilities and progress towards final repositories
		• Pool storage for used VVER-1000 fuel at Zheleznogorsk since 1985 • Dry storage for used RBMK and other fuel at Zheleznogorsk from 2012 • Various interim storage facilities in operation
South Korea	Direct disposal, wants to change	• Waste programme confirmed 1998, KRWM seet up 2009 • Central interim storage planned from 2016
Spain	Direct disposal	• ENRESA established 1984, its plan accepted 1999 • Central interim storage at Villar de Canas from 2016 (volunteered location) • Research on deep geological disposal, decision after 2010
Sweden	Direct disposal	• Central used fuel storage facility – CLAB – in operation since 1985 • Underground research laboratory at Aspo for HLW repository • Osthammar site selected for repository (volunteered location), likely to open in 2028
Switzerland	Reprocessing	• Central interim storage for HLW and used fuel at ZZL Wurenlingen since 2001 • Smaller used fuel storage at Beznau • Underground research laboratory for high-level waste repository at Grimsel since 1983 • Deep repository by 2020, containers to be retrievable
United Kingdom	Reprocessing	• Low-level waste repository in operation since 1959 • HLW from reprocessing is vitrified and stored at Sellafield • Repository location to be on basis of community agreement • New NDA subsidiary to progress geological disposal
U.S.	Direct disposal but reconsidering	• DoE responsible for used fuel from 1998, accumulated $32 billion waste fund • Considerable research and development on repository in welded tuffs at Yucca Mountain, Nevada • The 2002 Congress decision that geological repository be at Yucca Mountain was countered politically in 2009 • Central interim storage for used fuel now likely

Note: in most countries repositories or at least storage facilities for low-level wastes and intermediate-level wastes are operating.

8.9 Wastes from decommissioning nuclear plants

In the case of nuclear reactors, about 99% of the radioactivity is associated with the fuel. Apart from any surface contamination of plant, the remaining radioactivity comes from 'activation products' such as steel components which have long been exposed to neutron irradiation. Their atoms are changed into different isotopes such as iron-55, cobalt-60, nickel-63 and carbon-14. The first two are highly radioactive, emitting gamma rays, but with correspondingly short half-lives so that after 50 years from final shutdown their hazard is much diminished. Some caesium-137 may also be in decommissioning wastes.

Some scrap material from decommissioning may be recycled, but for uses outside the industry very low clearance levels are applied, so most is buried and some is recycled within the industry. Generally, short-lived intermediate-level wastes (mainly from decommissioning reactors) are buried, while long-lived intermediate-level wastes (from fuel reprocessing) will be disposed of deep underground. Low-level wastes are disposed of in shallow burial sites.

8.10 Disposal of other radioactive wastes

Some low-level liquid wastes from reprocessing plants are discharged to the sea. These include radionuclides which are distinctive, notably technetium-99 (sometimes used as a tracer in environmental studies) and this can be discerned many hundred kilometres away. However, such discharges are regulated and controlled and the maximum radiation dose anyone receives from them is a small fraction of natural background radiation.

Nuclear power stations and reprocessing plants release small quantities of radioactive gases (e.g., krypton-85 and xenon-133) and trace amounts of iodine-131 to the atmosphere. However, they have short half-lives and the radioactivity in the emissions is diminished by delaying their release. Also the first two are chemically inert. The net effect is too small to warrant consideration in any life-cycle analysis. A little tritium is also produced but regulators do not consider its release to be significant.

The U.S., Nuclear Regulatory Commission classifies low-level wastes into four categories based on radioactivity corresponding to management and disposal requirements: Class A waste has the lowest radioactivity level and decays to background level after about 100 years. It accounts for about 99% of the volume of LLW generated in the U.S., and includes slightly contaminated paper products, clothing, rags, mops, equipment and tools, as well as depleted uranium. Class B and C wastes include filters, resins, irradiated hardware with activation products and longer-lived radioisotopes that decay after 300 and 500 years, respectively. Greater-than-Class C LLW has radionuclide concentration limits greater than those specified for Class C waste.

Following the Fukushima accident, large areas were contaminated mainly with caesium fallout. In 2016 the government announced that material with less than 8 kBq/kg caesium would no longer be subject to restriction regarding disposal or use for embankments. This is about six times the average radioactivity of the Earth's crust and about three times the average radioactivity of Australian coal ash. It is noteworthy that coal burning produces some 280 million tonnes of ash per year, most of it containing low levels of natural radionuclides. Some of this could be classified as LLW. It is simply buried.

8.11 Costs of radioactive wastes management

Financial provisions are made for managing all kinds of civilian radioactive waste. The cost of managing and disposing of nuclear power plant wastes represents about 5% of the total cost of the electricity generated.

The actual arrangements for paying for waste management and decommissioning also vary. The key objective is however always the same: to ensure that sufficient funds are available when they are needed. There are three main approaches.

8.11.1 Provisions on the balance sheet

Sums to cover the anticipated costs of waste management and decommissioning are included on the generating company's balance sheet as a liability. As waste management and decommissioning work proceeds, the company has to ensure that it has sufficient investments and cashflow to meet the required payments.

8.11.2 Internal fund

Payments are made over the life of the nuclear facility into a special fund that is held and administered within the company. The rules for the management of the fund vary, but many countries allow the fund to be reinvested in the assets of the company, subject to adequate securities and investment returns.

8.11.3 External fund

Payments are made into a fund that is held outside the company, often within government or administered by a group of independent trustees. Again, rules for the management of the fund vary. Some countries only allow the fund to be used for waste management and decommissioning purposes, others allow companies to borrow a percentage of the fund to reinvest in their business.

8.11.4 Legacy wastes

In addition to the routine wastes from current nuclear power generation there are other radioactive wastes referred to as 'legacy wastes'. These wastes exist in several countries which pioneered nuclear power and especially where power programmes were developed out of military programmes. These are sometimes voluminous and difficult and arose in the course of those countries getting to a position where nuclear technology is a commercial proposition for power generation.

8.11.5 Regulation

The nuclear and radioactive waste management industries work to well-established safety standards for the management of radioactive waste. International and regional organisations such as the International Atomic Energy Agency (IAEA), the Nuclear Energy Agency (NEA) of the Organisation for Economic Co-operation and Development (OECD), the European Commission (EC) and the International Commission on Radiological Protection (ICRP) develop standards, guidelines and recommendations under a framework of co-operation to assist countries in establishing and maintaining national standards. National policies, legislation and regulations are all developed from these internationally agreed standards, guidelines and recommendations. Amongst others, these standards aim to ensure the protection of the public and the environment, both now and into the future. Other international conventions and directives seek to provide for inter alia, the safe transportation of radioactive material, protection of the environment (including the marine environment) from radioactive waste and the control of imports and exports of radioactive waste and transboundary movements.

8.12 International atomic energy agency

The IAEA is the international organisation that advises on the safe and peaceful uses of nuclear technology. It is an agency of the United Nations, based in Vienna, Austria founded in 1957 and it currently has 134 member states from countries with and without nuclear energy programmes.

The IAEA develops safety standards, guidelines and recommendations and inter alia provides technical guidance to member states on radioactive waste principles. Member states use the standards and guidelines in developing their own legislation, regulatory documents and guidelines. It also verifies through a safeguards inspection programme compliance with the Nuclear Non-Proliferation Treaty (NPT).

The IAEA's Waste and Environmental Safety Section works to develop internationally agreed standards on the safety of radioactive waste. The Radioactive Waste Safety Standards Programme (RADWASS) provides guidance to member states to produce their own policies and regulations for the safe management of radioactive waste, including disposal. In addition, the IAEA helps member states by providing technical assistance with services, equipment and training and by conducting radiological assessments.

8.12.1 Nuclear energy agency

The nuclear energy agency of the OECD is based in Paris, France. It has a variety of waste management programmes involving its 28 member states.

The organisation aims to assist these states in developing safe waste disposal strategies and policies for spent nuclear fuel, HLW and waste from decommissioning nuclear facilities. It also works closely with the IAEA on nuclear safety standards and other technical activities. The NEA has a project aimed at preserving records, knowledge and management (RK&M) of long-lived nuclear waste disposal for future generations.

8.12.2 European commission

For several years, the European Commission (EC) has attempted to pass Directives aimed at ensuring a common approach to nuclear safety and radioactive waste management. The so-called 'nuclear package' of directives on nuclear safety and waste management was a top-down approach which met with considerable opposition from several member states and was revised on several occasions leading to the 2011 adoption of a scaled-back version.

In July 2011 the European Union adopted a directive for the disposal of used nuclear fuel and radioactive wastes which required member countries to develop national waste management plans for European Commission review by 2015. The plans must include firm timetables for the construction of disposal facilities, descriptions of needed implementation activities, cost assessments and financing schemes. Safety standards promulgated by the IAEA would become legally binding within the EU-wide policy framework. International peer reviews should be invited at least every ten years.

The agreement allows two or more member nations to develop joint disposal facilities and allows transport of used fuel and radioactive wastes within the EU. Exports outside the EU will only be possible to countries that already have a repository in operation that meets IAEA standards. For overseas reprocessing, ultimate wastes must be returned to the originating EU country. The directive acknowledges that no country currently operates such a repository and projects that a minimum of 40 years would be required to develop one. The shipment of used fuel and radioactive wastes to African, Pacific and Caribbean countries and to Antarctica is explicitly banned. Plans are expected to use a step-by-step approach to geologic disposal based on the voluntary involvement of potential host communities. Two routes are acknowledged: one to dispose of used nuclear fuel as waste; the other to reprocess the fuel and recycle the uranium and plutonium while disposing of the remainder as waste. The directive became effective in August 2011 and national governments, which retain ultimate responsibility for wastes, had two years to bring their nuclear waste legislation into line with it. There are 143 nuclear energy facilities generating used fuel in 14 of the EU's 27 member nations. The remaining nations possess radioactive waste requiring disposal that has been produced by research, medicine and industry.

8.12.3 International commission on radiological protection

The International Commission on Radiological Protection (ICRP) is an independent registered charity that issues recommendations for protection against all sources of radiation. The IAEA interprets these recommendations into international safety standards and guidelines for radiological protection. National regulators may also adopt the recommendations by the ICRP for their own radiation protection standards.

Perspective

Nuclear wastes are a significant part of the nuclear power picture and need to be managed and disposed of properly. However in more than 50 decades of civil nuclear power experience they have not caused any serious health or environmental problems, nor posed any real risks to people.

Alternatives for power generation are not without challenges and for a variety of reasons they – particularly those from coal combustion – have not always been well controlled. Both flyash and bottom ash are often loaded with heavy metals (including uranium and thorium). Flyash is mostly retained for land disposal today and bottom ash is normally buried also, but not always securely and without effects on groundwater. Groundwater pollution with arsenic, boron, cobalt and mercury is not unusual and the U.S. EPA in 2011 listed 181 U.S., coal ash ponds which posed a significant hazard, 47 of these a high hazard and threat to life. Burning any fossil fuels gives rise to carbon dioxide emissions and gaseous pollutants such as nitrogen oxides and often sulphur oxides.

9

Process integration for pollution prevention and energy saving

9.1 Introduction

Process integration is a holistic approach to process design and operation that emphasises the unity of the process. It is based on fundamental chemical engineering and systems principles and therefore provides a set of generally-applicable tools. This is an active research area that promises to lead to significant contributions on the engineering principles of integrated systems. It can be broadly categorised into mass integration and energy integration. Mass integration is a holistic approach to the generation, separation and routing of species and streams throughout the process. It has been developed and applied to identify global insights, synthesise strategies and address the root causes of the environmental and mass processing problems at the heart of the process. Mass integration is a systematic methodology that provides a fundamental understanding of the global flow of mass within the process and employs this understanding in identifying performance targets and optimising the allocation, separation and generation of streams.

In the environmental context, the development of the methodologies for waste reduction (mass integration) has been driven by the promulgation of more stringent environmental regulations coupled with the desire to improve industrial competitiveness. In the past, for waste reduction tasks, the industrial goal was to identify a recovery system that would effectively allow the recycle and reuse of certain wastes. This goal was generally accomplished by postulating a variety of process system configurations and operating conditions and then individually screening these alternatives to evaluate their overall economic impact to the company (operating cost, capital investment, etc.).

Recently, significant progress has been made toward developing systematic design methodologies that not only identify a system that accomplishes the waste reduction task, but also a system that represents the most cost-effective approach. The primary focuses of these efforts have been towards the development of systematic design methodologies for identifying cost-effective wastewater minimisation systems, end-of the-pipe separation and recycle systems and in-plant separation systems. The other important category of process integration is energy integration that deals with the global allocation, generation and exchange of energy throughout the process. The development of the methodologies for energy conservation (*energy integration*) has been driven by the increasing demand for expensive utilities within chemical industries. A review of some of the process integration design tools for addressing energy conservation and waste reduction is provided in the following sections.

9.2 Applications of process integration

Generally speaking Process Integration (PI) is concerned to the advanced management of material, energy and information flows in a production plant and the surrounding community based on the multi criteria optimisation of the processing systems. PI can be used to:

1. Minimise energy consumption (energy efficiency), improve raw material utilisation (improved heat recovery will allow increased recycling in the process) and reduce equipment (investments), by identifying the optimal trade off between operating cost (raw materials and energy) and investment cost (equipment).
2. Increase production volume (plant capacity) for plant debottlenecking.
3. Reduce operating problems (operability, start-up and shut-down).
4. Increase plant controllability (which is an inherent feature of the design itself, not the control system) by selecting plant topology and equipment parameters in a way that makes the control easier, irrespective of the actual control system.
5. Enabled the plant flexibility with respect to planned (new feedstock and/or products, adjusted production volume, etc.), and undesired (such as heat exchanger fouling and/or catalyst deactivation) changes in operating conditions.
6. Minimise undesirable plant emissions, by reduced use of fossil fuels, switching to alternative energy sources, closing the processes, (environment).
7. Add to the efforts in the process industries and society for a sustainable development.

9.2.1 Process integration and process design

Process integration is a holistic approach to process design, retrofitting and operation of industrial plants, the applications of which are focused on resource conservation, pollution prevention and energy management.

In order to cover the areas of PI application the onion diagram (Fig. 9.1) is commonly used which indicates the hierarchy of most processes, represented symbolically by its layers.

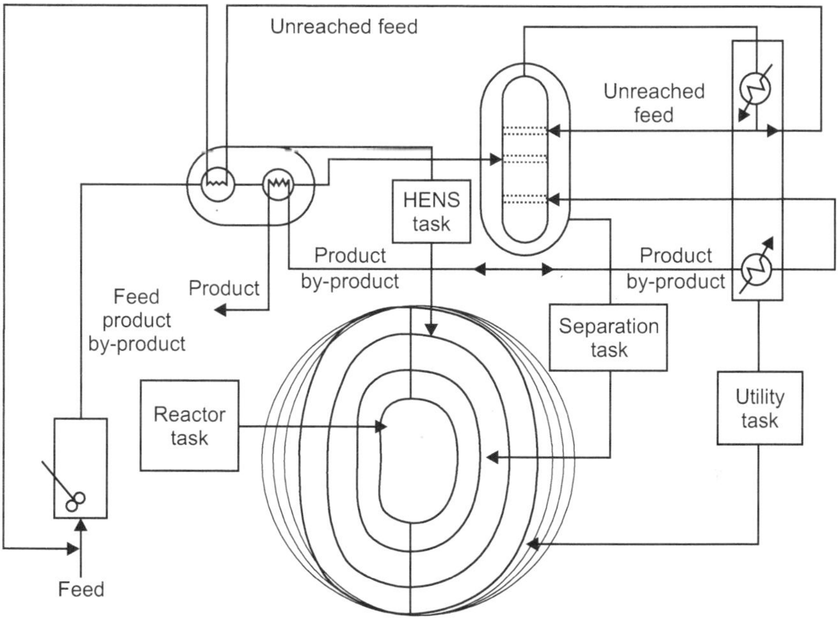

Figure 9.1: Onion diagram in process integration.

9.2.2 Process integration methods

Process integration includes a number of methods that allow engineers to evaluate entire processes or sites, rather than focusing on individual unit operations. These methods are particularly significant for large and complex industrial facilities, because the more complex the process becomes, the harder it is to identify the best design opportunities. Classification of PI methods can be expressed according to the following characteristics:

1. Algorithmic (the use of mathematical models) vs. Heuristic (use design experience often in the form of rules of thumb).
2. Simultaneous (address all design decisions at the same time) vs. Sequential (broke the design problem to a number of subproblems).
3. Quantitative vs. qualitative.
4. Automatic vs. interactive or manual.
5. Analysis (targeting) vs. design (synthesis) vs. optimisation.
6. Use of graphical diagrams and representations.
7. Use of thermodynamics (first and second law).

The large number of structural alternatives can be significantly reduced by the three major features of process integration methods heuristics, thermodynamics and optimisation. This features enabled a possible classification of PI methods, based on the two-dimensional diagram shown in Fig. 9.2.

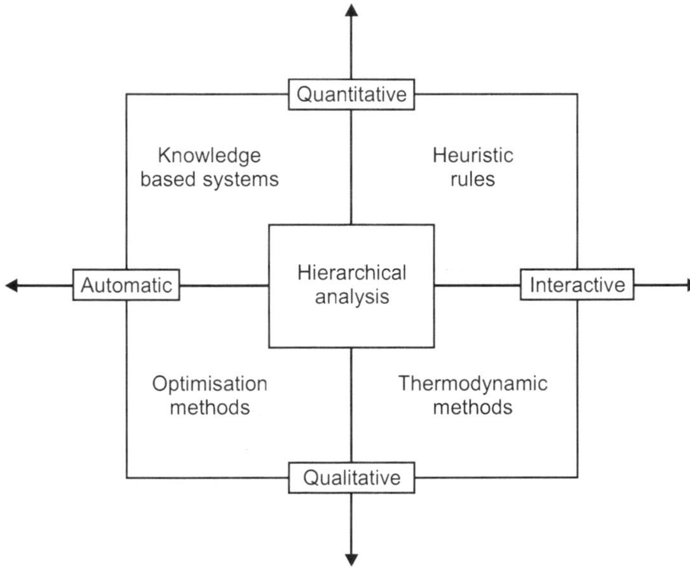

Figure 9.2: Possible classification of PI methods.

9.3 Process integration tools for designing energy conservation systems

During the past decade, rising energy costs have required operating companies to look for ways to improve energy conservation. Energy integration deals with all forms of energy such as heating, cooling, power generation/consumption, pressurisation/depressurisation and fuel. Much of the effort in this area has been directed toward increasing heat recovery in chemical processes. Industrial heat exchange networks, 'HENs', are of particular importance because of their role in recovering process heat. An HEN is a network consisting of one or more heat exchangers that collectively satisfy the energy conservation task. Therefore, in most chemical process industries it is necessary to synthesise cost-effective HENs that can transfer heat among the hot and cold streams (temperature conditions refer to the initial stream state). During the design stage, temperature specifications for the hot and cold streams must be met and a decision must be made regarding the use of a process stream or an external utility (e.g., steam, cooling water, etc.), to accomplish the required heat duty. Even in relatively simple situations, the problem of pairing and sequencing of exchanger streams becomes a large one and the use of systematic techniques is necessary. Figure 9.3 shows as a general representation of the HEN synthesis task.

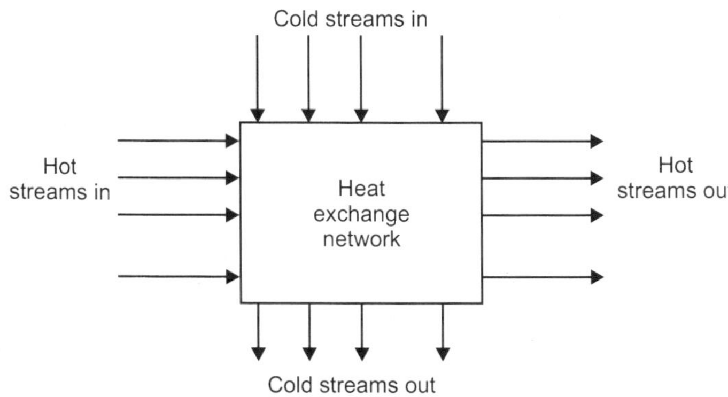

Figure 9.3: Heat exchange network (HEN) synthesis.

For a given system, the synthesis of HENs entails answering several questions including:

1. Which heating/cooling utilities should be employed?
2. What is the optimal heat load to be removed/added by each utility?
3. How should the hot and cold streams be matched (i.e., stream pairings)?
4. What is the optimal system configuration (e.g., how should the heat exchangers be arranged?, is there any stream splitting and mixing?, etc.)?

9.4 Process integration tools for designing wastewater reduction and water conservation systems

Wastewater reduction and water conservation are becoming increasingly more important issues in process industries. More stringent environmental regulations, concerns over long-term health effects on humans and nature and the future availability of 'clean' water resources are just a few of the factors that are driving efforts toward improvements in water conservation and wastewater reduction in manufacturing processes. As these issues continue to receive intense government scrutiny and heightened concern from community-organised environmental groups, the ability of industry to address these issues may soon impact their right-to-operate within these communities and their sustainability of future operations.

These critical concerns have refocused efforts over the past decade towards identifying cost-effective wastewater reduction and water conservation process designs, involving direct recycle and reuse of water, that can be implemented within a variety of process industries. Several recent research efforts have focused on the development of process design methodologies and tools that are generic and can be utilised across a wide variety of industries.

These industries include petroleum process industries, pulp and paper manufacturing industries, plastics production industries, food products industries, pharmaceutical industries, specialty chemical industries, industrial laundries and fibre dying industries, to name a few.

The design methodologies and approaches cover a variety of techniques ranging from graphical based approaches, including 'water pinch' analysis the source-sink graphical methodology to mathematical optimisation-based approaches.

For a given system, the synthesis of wastewater reduction and water conservation networks entails answering several questions including:

1. Which wastewater streams should be recycled or reused?
2. What is the optimal load of each wastewater stream to be recycled or reused?
3. What is the optimal allocation of wastewater streams that are to be routed to process water users?
4. What is the optimal system configuration (e.g., how should the water allocation system be arranged, is there any stream splitting and mixing, etc.)?

9.5 Process integration tools for designing end-of-the-pipe separation and recycle waste minimisation systems

Separation systems (a combination of unit operations) designed to allow the recycle/reuse of waste streams, or certain constituents in waste streams, are generally envisioned as an end-of-the-pipe process system. The initial thrust to identify the most cost-effective waste separation system from a large group of process options (multiple technologies and/or separating agents for the separation task) resulted in the notion of synthesising a mass-exchange network, 'MEN'. An MEN is a network consisting of one or more mass exchangers that collectively satisfy the waste recovery task. A single mass exchanger is defined as a direct-contact, countercurrent unit that employs a mass-separating agent, 'MSA', to effect the transfer of the pollutant from the waste stream to the MSA and a general schematic of a single mass exchanger is shown in Fig. 9.4. Examples of mass exchangers would be absorption columns, adsorbents, extraction units, strippers and ion-exchange units and examples of MSAs would be liquid absorbents, adsorbents and extractants such as solvents, activated carbon, liquid extractants, etc. The design task of synthesising an MEN is to systematically identify a cost-effective network of mass exchangers for the

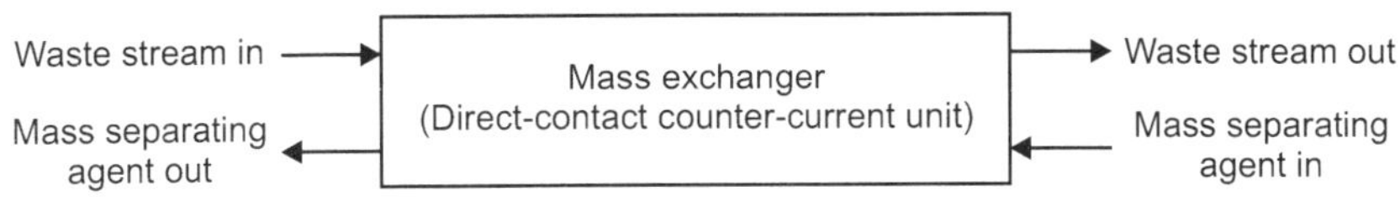

Figure 9.4: Schematic of a single mass exchanger for environmental process design.

selective transfer of a certain undesirable species from a set of 'rich' (waste) streams to a set of 'lean' (MSA) streams. This undesirable species generally represents a pollutant if it is discharged to the environment, but may be a valuable raw material if it can be recovered for reuse within the plant. Figure 9.5 is shown as a general representation of the MEN synthesis task for end-of-the-pipe waste minimisation process design.

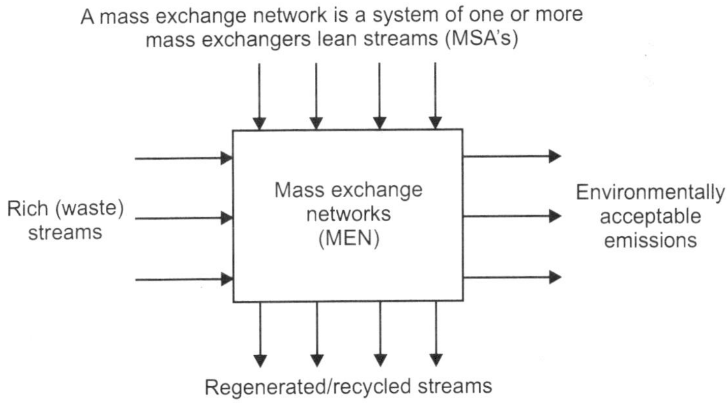

Figure 9.5: Mass exchange network (MEN) synthesis.

For a given system, the synthesis of MENs entails answering several questions including:

1. Which mass separating agents should be employed?
2. What is the optimal mass load to be removed/added by each MSA?
3. How should the waste and MSA streams be matched (i.e., stream pairings)?
4. What is the optimal system configuration (e.g., how should the mass exchangers be arranged, is there any stream splitting and mixing, etc.)?

Over the past few years, several important categories of the MEN synthesis task have been identified and addressed:

1. MENs for multiple component systems.
2. MENs with regeneration systems.
3. MENs with chemical reactions.

4. MENs with heat integration.
5. MENs via a structural based approach.
6. MENs for wastewater reduction.
7. MENs with flexibility.
8. MENs for fixed load removal.
9. MENs with controllability.

Synthesis techniques have also been developed for other separations systems that are traditionally used for waste minimisation via end-of-the-pipe recycle/reuse systems. For instance, there is a wide class of separation systems that employ energy separating agents, 'ESAs' (hot and cold process streams and/or utilities such as steam and cooling water), to separate species from a waste stream via a phase change. These include unit operations such as condensers, evaporators, dryers and crystallisers and they are collectively grouped in the category of heat-induced separators. A general schematic of a single heat-induced separator is shown in Fig. 9.6.

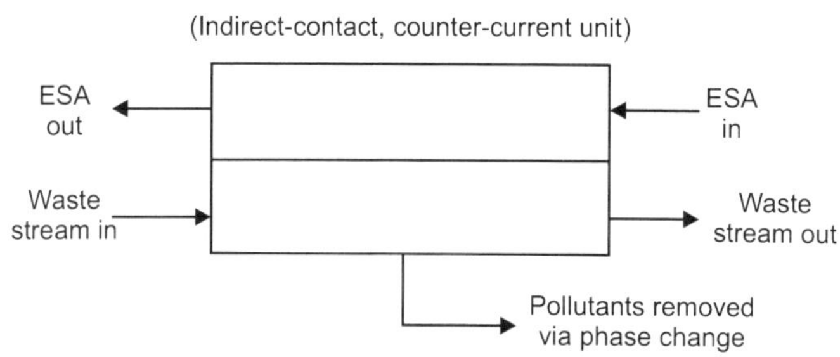

Figure 9.6: Schematic of a single heat-induced separator for environmental process design.

The notion of synthesising heat-induced separation networks (HISENs) for recycle/reuse waste minimisation process design encompasses the task of identifying a cost-effective system of heat-induced separators and heat exchangers that can achieve a specified waste reduction task (single component or multiple component waste streams) by heating/cooling the streams to produce a phase separation.

Figure 9.7 shows as a general representation of the HISEN synthesis task for end-of-the-pipe waste minimisation process design.

More recently, the HISEN synthesis methodology has been generalised to include pressurisation/depressurisation equipment in conjunction with heat-induced separation specifically to address gaseous emissions containing volatile organic compounds, 'VOCs' and the resulting recovery system was referred to as an energy-induced separation network, 'EISEN' One of the simplest

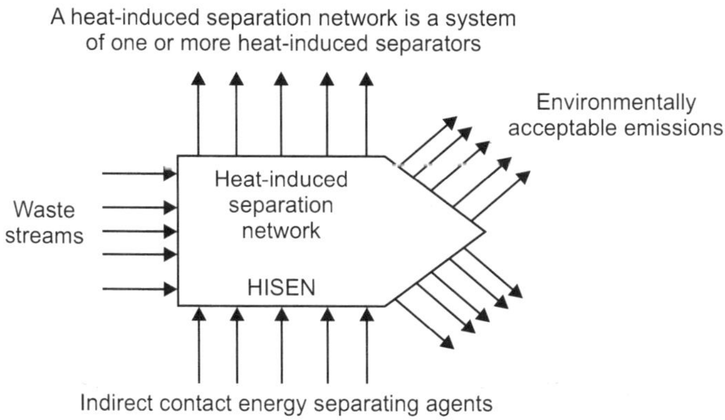

Figure 9.7: Heat-induced separation network synthesis.

techniques for VOC recovery is to use heat-induced separation networks to affect condensation via cooling, however, VOC condensation is, in general, a function of both temperature and pressure. Hence, a more general condensation design task should capitalise on the synergism between two energy forms: cooling and pressurisation/depressurisation. This synergism was addressed by the design methodology of Dunn and coworkers whose objective was to create a cost-effective network of heat-induced separators, heat exchangers and pressurisation/depressurisation devices, which can separate one or more species from a set of waste streams via phase change.

For a given system, the synthesis of HISENs/EISENs entails answering several questions including:

1. Which energy separating agents should be employed?
2. Should stream pressurisation or depressurisation be employed and, if so, to what level?
3. What is the optimal mass and heat load to be removed/added by each ESA?
4. How should the waste and ESA streams be matched (i.e., stream pairings)?
5. What is the optimal system configuration (e.g., how should the heat-induced separators, heat exchangers and compressors/turbines be arranged, is there any stream splitting and mixing, etc.)?

Over the past few years, several important categories of the HISEN and EISEN synthesis task have been identified and addressed as summarised below:

1. HISENs for single component VOC condensation systems.
2. A shortcut graphical approach for HISENs for single component VOC condensation systems.

3. HISENs for multiple component VOC condensation systems.
4. HISENs for fixed load removal.
5. Hybrid HISEN and membrane systems.
6. A spread sheet-based approach for identifying cost-effective HISENs and EISENs for condensation hybrid processes.
7. HISENs for crystallisation systems.
8. HISENs for infinite component VOC condensation systems using clusters.

In addition to mass exchange operations and heat and energy-induced separation systems, systematic design techniques have also be developed for end of-the-pipe pressure-based membrane systems.

Also, the application of MEN, EISEN and membrane synthesis techniques has been illustrated via the design of cost effective VOC recovery systems.

9.6 Process integration tools for designing in-plant separation systems for waste minimisation

Although end-of-the-pipe recycle/reuse systems provide an attractive approach for waste minimisation, there may be a greater economic incentive to address pollution prevention from a source reduction perspective. In fact, reducing/eliminating the generation/use of the undesirable species at the source may significantly reduce/eliminate costs associated with end-of-the-pipe separation and recycle. The following four strategies are commonly pursued to address source reduction of pollutants:

1. If the undesirable species is generated in the process (via reaction), find an alternative, environmentally acceptable, reaction path. This reaction path should eliminate or minimise the use of raw materials or the production of by-products that are not environmentally friendly.
2. If the undesirable species is not generated in the process (no reaction), substitute it with a more environmentally benign material. An example of this type of approach could be the substitution of a more environmentally acceptable solvent for use in a chemical process.
3. If the emissions are associated with in-plant utility systems (e.g., blow downs, greenhouse gas emissions, etc.), implement in-plant heat integration modifications to minimise the amount of emissions from thermal pollution or to reduce the amount of emissions associated with utility systems.
4. If the undesirable species is not generated in the process and cannot be replaced with a more environmentally benign material, implement in-plant modifications to recycle streams with the undesirable species. This design approach has resulted in the novel concepts of the design of wastewater minimisation systems, waste interception and allocation networks,

'WINs', heat-induced waste minimisation networks, 'HIWAMINs' and energy-induced waste minimisation networks, 'EIWAMINs'. The HIWAMIN and EIWAMIN design approaches simultaneously address waste reduction and energy conservation process design to further improve economics.

Reaction-path modifications and/or material/solvent substitutions generally have a greater potential to tackle waste reduction at the source and are important design approaches to pursue during the overall process identification/conception phase; however, in-plant modifications may be a more readily acceptable option for an existing process. In-plant modifications are generally less capital intensive and easier to retrofit within an existing process than the modifications generally required by reaction/material substitution.

Furthermore, in a manufacturing environment, reaction/material substitution usually involves several years of research and development and, in some cases, could also depend on an invention. Thus, source reduction via in-plant modifications is an attractive approach towards tackling waste minimisation for existing process plants.

Recently, there have been several systematic methodologies developed for the design for cost-effective waste reduction systems based on in-plant modifications, such as the synthesis of waste interception and allocation networks, 'WINs' the synthesis of heat-induced waste minimisation networks, 'HIWAMINs' and the synthesis of energy-induced waste minimisation networks, 'EIWAMINs'. The WIN synthesis methodology is based on tracking streams containing an undesirable species within a process and identifying the optimal location(s) to intercept one or more in-plant streams with mass exchangers to achieve a specified waste reduction task. The WIN design technique features the use of direct-contact *mass separating agents* to intercept the undesirable species. Figure 9.8 shows representation of the WIN design methodology. The HIWAMIN methodology is also based on tracking streams containing an undesirable species within a process and identifying the optimal location(s) to intercept one or more streams with heat induced separators and heat exchangers to achieve a specified waste reduction and heat integration task. However, the HIWAMIN design technique features the use of indirect-contact, *energy separating agents* to intercept the undesirable species. Furthermore, the HIWAMIN design approach simultaneously addresses the waste minimisation and process heat integration (heat exchange network) design tasks. The EIWAMIN design technique extends the HIWAMIN approach to include the use of pressurisation and/or depressurisation devices in addition to heat-induced separators and heat exchangers to further improve separation efficiencies and the cost effectiveness of the final design. Figure 9.9 shows as a schematic representation of the HIWAMIN and EIWAMIN design techniques.

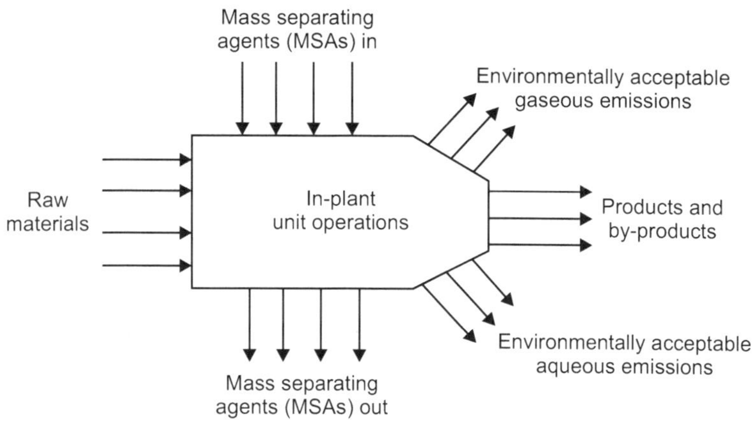

Figure 9.8: WIN synthesis representation.

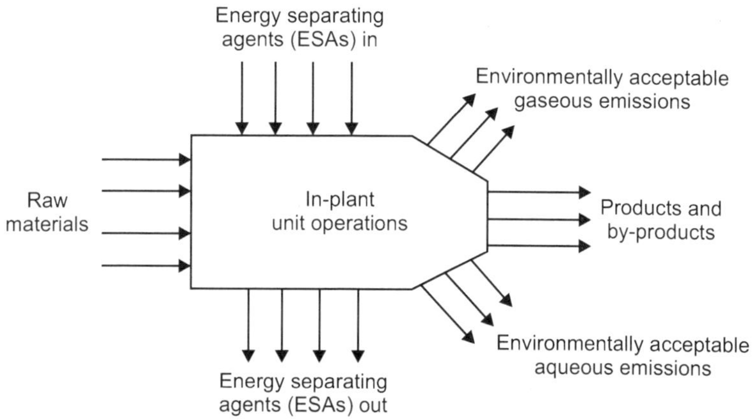

Figure 9.9: HIWAMIN and EIWAMIN synthesis representation.

9.7 Environmental friendly chemicals

So far, the previous strategies have focused on the process aspects of pollution prevention. Another important element of mass-integration strategies is based on material substitution and chemistry and chemical changes. Examples include the selection of environmentally benign chemical reactions, raw materials, solvents and products. Over the past few years, significant progress has been made in this area. This section provides a brief overview of the recent advances in synthesising 'green' reactions and species.

In many cases, it is possible to replace environmentally hazardous chemicals with more benign species without compromising the technical and economic performance of the process. Examples include alternative solvents, polymers and refrigerants. Group contribution methods have been commonly used in predicting physical and chemical properties of synthesised materials. Two main frameworks have been employed to synthesise alternative materials: knowledge base and computer-aided optimisation. Knowledge-based approaches depend on understanding the criteria of the materials to be replaced along with general rules and algorithms that link properties with structure. Furthermore, software can be used to screen solvents based on their properties and performance. An example of this approach is the PARIS (Programme for Assisting in the Replacement of Industrial Solvents). Computer-aided optimisation approaches are based on formulating the molecular design problem as an optimisation programme that seeks to maximise a performance function or minimise deviation from desired properties subject to various constraints including structural feasibility, property-structure correlations and environmental criteria. Examples of this approach include synthesis of solvents, polymers and refrigerants.

Table 9.1 discusses summary of the design methodologies.

Table 9.1: Summary of some methodologies for environmental process design.

Design methodology	*Description*	*Example technologies targeted*
Heat integration systems	The identification of heat recovery devices that minimise environmental emissions resulting from utility generation systems.	• Heat exchangers • Heat pumps • Boilers/cooling towers
Wastewater minimisation systems	A design strategy for reuse, regeneration reuse, and regeneration recycling of wastewater streams that minimises water usage and minimises wastewater discharge.	• Direct recycle opportunities • Regeneration reuse and recycling opportunities
Mass exchange networks (MENs) and reactive mass exchange networks (REAMENs)	A network of process units that removes pollutant(s) from end-of-pipe streams via the use of physical or chemical, direct-contact, mass separating agents (MSAs).	• Adsorption • Absorption • Liquid–liquid extraction • Ion exchange
Heat-induced separation networks (HISENs) and energy-induced separation networks	A network of process units that removes pollutant(s) from end-of-pipe streams via the use of indirect-	• Condensation • Evaporation • Drying • Crystallisation

(Cont'd...)

Design methodology	*Description*	*Example technologies targeted*
(EISENs)	contact energy separating agents (ESAs), including stream pressurisation and/or depressurisation.	• Compressors • Vacuum pumps
Membrane separation networks	A network of process units that removes pollutant(s) from end-of-pipe streams via the use of membranes and stream pressurisation and/or depressurisation.	• Reverse osmosis • Pervapouration
Environmentally acceptable reactions (EARs)	The development of an alternative reaction path that will no longer result in the generation of the pollutant.	• Reactions
Solvent selection systems	The identification of alternative solvents and/or solvent combinations that can be used to provide a desired effect (solubility, material transport, separating agent, etc.).	• Substitute refrigerants • Substitute coating constituents • Simultaneous solvent and separation unit selection
In-plant separation design via waste interception and allocation networks (WINs)	A network of process units that removes pollutant(s) from in-plant streams via the use of physical or reactive direct-contact mass separating agents (MSAs) and/or rerouting of in-plant process streams.	• Direct recycle opportunities • Adsorption • Absorption • Liquid–liquid extraction • Ion exchange
In-plant separation design via heat-induced waste minimisation networks (HIWAMINs) and energy-induced waste minimisation networks (EIWAMINs)	A network of process units that removes pollutant(s) from in-plant streams via the use of indirect-contact energy separating agents (ESAs) with stream pressurisation and/or depressurisation and/or rerouting of in-plant process streams. Full site heat integration is simultaneously addressed by this technique.	• Direct recycle opportunities • Heat exchange/heat integration • Condensation • Evaporation • Drying • Crystallisation • Compressors • Vacuum pumps • Heat pumps

9.8 Developing sustainable process integration

The foregoing discussion has focused on process integration tools. These are key elements for identifying opportunities and developing sound strategies. Nonetheless, tools alone will not deliver solutions; people, work processes and working cultures and environments will.

The following are key building blocks:

1. Articulate a clear vision of the broad goals of the company.
2. Perform a preliminary targeting analysis to determine priority areas of work.
3. Develop a preliminary framework of tasks needed and the required human, technical and financial resources.
4. Establish realistic expectation and targets of what process integration can deliver and what resources it will take to perform the tasks.
5. Describe anticipated constraints, corporate unique aspects and challenges.
6. Get enthusiastic support from senior management.
7. Recruit local champions from among the process experts and the stakeholders.
8. Form task-driven teams.
9. Encourage an open environment which fosters creativity and out-of-the-box integrated thinking where the dominating culture is 'how do we make it happen?' instead of 'why it won't work'.
10. Measure, analyse, use process integration tools to develop, improve, synthesise, feedback, refine and sustain projects and strategies.
11. Present the proposed changes in an way which focuses on gained insights, is easy to follow and highlights the key characteristics of the findings.
12. Give focus to important issues.
13. Consult with relevant individuals all along to capture process know-how, ensure that appropriate details are included and hurdles are overcome.
14. Start now!

Another important aspect is to take initiative and not be discouraged by resisting responses.

9.9 Observations on industrial application of process integration technology

In applying process integration methodologics in the process industries has led to the following insights:

1. These methodologies are well suited for a variety of chemical processes including, but not limited to, the chemical, polymer, petroleum, pulp and paper, food products and pharmaceutical industries.

2. These methodologies work equally well for small, medium and large size chemical processes. Obviously application to larger processes often results in a greater magnitude of savings and a greater magnitude of capital cost for the implementation of identified designs.
3. These methodologies are applicable for both continuous and batch processes; however, batch processes are more difficult to analyse and to implement the identified solutions as they typically involve dynamic performance and require an additional layer of scheduling solutions.
4. These methodologies are well suited for both retrofit and grassroots process designs. Utilisation of these tools during grassroots plant design can allow the added flexibility of locating equipment to be 'integrated' within a close proximity of each other. In addition, personal experience has shown that even relatively 'new' plants that are analysed for retrofit process integration designs can offer substantial cost savings and waste minimisation opportunities.
5. Companies involved in applying process integration methodologies must be prepared to pay for the implementation of identified solutions. The process designs may often be low cost but almost none are free. It is true that 'it takes money to make money'. If no capital funds are available, the potential results of the process integration study can be severely limited. The good news, however, is that most process integration projects enjoy a very attractive payback period (typically less than two years) and can 'pay for themselves'.
6. Implementation of identified process designs can be sequenced as a part of a master-plan investment strategy for the company. Implementation may start with the low-cost projects and proceeds to the high-cost projects as the financial resources become available. However, process integration provides an overarching framework for sequencing the implementation.
7. The nature of the problems calls for the use of specific tools within the process integration methodology. A basic knowledge of the entire methodology and set of tools referenced in this paper is recommended.
8. Process integration tools are systematic and the application of these methodologies often identifies process designs that are not intuitively obvious even by experienced engineers employing heuristics and brainstorming. Equally important to the tools is the overall integration approach and mindset.

10

Noise pollution and its control

10.1 Introduction

The rapid growth of industrialisation, urbanisation, communication and transport has led to noise pollution in big cities, affecting the common man in offices, institutions and houses. Noise pollution is proportionately related to developmental and growth-oriented activities. It may also be caused due to unmindful activities of man. When any activity performed by a person or machine or any change in nature produces a sound that exceeds a certain level and which is not liked by others, we call it noise. For example, the listening of TV programmes at a high volume may give joy to a few listeners but it may create a problem of restlessness in the neighbourhood. Similarly, one noisy truck or auto rickshaw crossing the streets can disturb the sleep of many people in the locality. In simple words, noise can be defined as a wrong sound, at a wrong place and at a wrong time. In other words, noise may be defined as an 'unwanted sound' and noise pollution as unwanted sound dumped into the atmosphere without any regard to its adverse affects. The term 'noise' in the electronic communication system is defined as perturbations that interfere with communication.

10.2 Noise sources

Industrial machinery and processes are composed of various noise sources such as rotors, stators, gears, fans, vibrating panels, turbulent fluid flow, impact processes, electrical machines, internal combustion engines, etc.

The mechanisms of noise generation depends on the particularly noisy operations and equipment including crushing, riveting, blasting (quarries and mines), shake-out (foundries), punch presses, drop forges, drilling, lathes, pneumatic equipment (e.g., jack hammers, chipping hammers, etc.), tumbling barrels, plasma jets, cutting torches, sandblasting, electric furnaces, boiler making, machine tools for forming, dividing and metal cutting, such as punching, pressing and shearing, lathes, milling machines and grinders, as well as textile machines, beverage filling machines and print machines, pumps and compressors, drive units, hand-guided machines, self-propelled working machines, in-plant conveying systems and transport vehicles. On top of this there are the information technology devices which are being encountered more and more in all areas.

Noise is therefore a common occupational hazard in a large number of workplaces such as the iron and steel industry, foundries, saw mills, textile mills, airports and aircraft maintenance shops, crushing mills, among many others. In many countries, noise-induced hearing loss is one of the most prevalent occupational diseases. A cross-section analysis in Germany of working equipment and processes in operational noise areas with a hearing impairment hazard has shown that 80% of the - several million - sound sources can be attributed to machine operations, conveying systems, control and regulation devices and turbo machines, while 20% are accounted for by manual working and conveying operations. About three quarters of the machine operations can be attributed to machine tools. The main concern of noise control is therefore the development, production and preferred use of low-noise working equipment and processes.

10.3 Industrial noise sources

In this section, the fundamental mechanisms of noise sources are discussed, as well as some examples of the most common machines used in the work environment. The sound pressure level generated depends on the type of the noise source, distance from the source to the receiver and the nature of the working environment. For a given machine, the sound pressure levels depend on the part of the total mechanical or electrical energy that is transformed into acoustical energy. Sound fields in the workplace are usually complex, due to the participation of many sources: propagation through air (air-borne noise), propagation through solids (structure-borne noise), diffraction at the machinery boundaries, reflection from the floor, wall, ceiling and machinery surface, absorption on the surfaces, etc. Therefore any noise control measure should be carried out after a source ranking study, using identification and quantification techniques. The basic mechanism of noise generation can be due to mechanical noise, fluid noise and/or electromagnetic noise.

The driving force for economic development is mainly the endeavour to produce consumer goods ever more cost-effectively. From the point of view of the machine manufacturer, this generally means offering products with a low space, material, energy and production time requirement (smaller, lighter, more economical and more productive). At the same time account is being taken increasingly of resource conservation and environmental friendliness, although the rise in noise levels which frequently goes along with increased output and productivity is often overlooked. Personnel are then exposed to higher noise levels than before, despite noise-reducing measures taken in the machine's design. This is because the noise emission rises non-linearly because of higher rotary and travelling speeds in machine parts. For example, for every doubling of the rotary speed the noise emission for rotating print machines rises by about 7 dB, for warp knitting looms 12 dB, for diesel engines 9 dB, for petrol engines 15 dB and for fans is between 18 to 24 dB. For the purpose of comparison: the doubling of sound power produces an increase in emission of 3 dB only.

Mechanical noise: Mechanical noise is generally created by a vibrating panel making intermittent contact. Care must be taken to reduce the vibrating area and/or reduce the vibration velocity. Reducing the vibrating area can be carried out by separating a large area into small areas, using a flexible joint. Reduction of the vibration velocity can be carried out by using damping materials at resonance frequencies and/or blocking the induced forced vibration.

Fluid noise: Air turbulence and vortices generate noise, especially at high air flow velocities. Turbulence can be generated by a moving or rotating solid object, such as the blade tip of a ventilator fan, by changing high pressure discharge fluid to low (or atmospheric) pressure, such as a cleaning air jet or by introducing an obstacle into a high speed fluid flow.

10.3.1 Examples of machinery noise sources

In this section, noise sources are presented for the most common machines used in industrial installations. For each case, the mechanism of noise generation is discussed.

Industrial gas jets

Industrial jet noise probably ranks third as a major cause of hearing damage after that of impact and material handling noise. Air jets are used extensively for cleaning, for drying and ejecting parts, for power tools, for blowing off compressed air, for steam valves, pneumatic discharge vents, gas and oil burners, etc. Typical sound pressure levels at 1 m from a blow-off nozzle can reach 105 dB(A).

Ventilator and exhaust fans

It is rare not to find one or more ventilators or exhaust fans in each department of an industrial or manufacturing complex. Fan and blower noise is the easiest and most straightforward noise problem to solve, using an absorptive type silencer. Fans are used to move a large volume of air for ventilation, by bringing in fresh air from the outside, blowing out dust, vapour or oil mist from an industrial environment and for a drying or cooling operation, etc. Industrial fans are usually low speed, low-static-pressure and have a large volume flow rate. Ideally, fans should operate at the maximum efficiency point on the pressure flow curve characteristic. Therefore, the choice between axial or centrifugal fans is made by the manufacturer to satisfy maximum efficiency at a certain static pressure/flow rate.

Three basic noise sources are:

1. Broadband aerodynamic noise generated by the turbulent flow.
2. Discrete tones at the blade passing frequency.
3. Mechanical noise due to mounting, bearing, balancing, etc.

Compressors

Compressors are usually very noisy machines with high pressure. There are several types of compressor: rotary positive displacement, gear or screw compressors, reciprocating compressors and liquid ring compressors are the most common.

The basic noise sources are caused by trapping a definite volume of fluid and carrying it around the case to the outlet with higher pressure. The pressure pulses from compressors are quite severe and equivalent sound pressure levels can exceed 105 dB(A). The noise generated from compressors is periodic with discrete tones and harmonics present in the noise spectrum.

Electric motors

Noise from electrical equipment such as motors and generators is generally a discrete low frequency, superimposed on a broadband cooling system noise. The electric motor converts electrical energy to magnetic and then mechanical energy with the output of a useful torque at the motor shaft. Part of the energy transformation is converted to heat, causing a rise in rotor, stator and casing temperature, therefore an electric motor must be supplied with a cooling fan system. The cooling fan can be incorporated inside as in the case of an 'OPEN' motor or outside as in the case of a 'Totally Enclosed Fan Cooled (TEFC)' motor. TEFC motors are more widely used, due to their robust construction which can withstand a dirty environment.

OPEN motors are less used due to possible contamination by the environment. An OPEN motor is sometimes (but not always) less noisy than a TEFC motor since the noisy fans are incorporated inside.

There are three basic sources involved in the noise generated by electric motors:

1. Broadband aerodynamic noise generated from the end flow at the inlet/outlet of the cooling fan. The cooling fan is usually the dominant noise source.
2. Discrete frequency components caused by the blade passing frequencies of the fan.
3. Mechanical noise caused by bearing, casing vibration, motor balancing shaft misalignment and/or motor mounting. Thus careful attention should be given to the vibration isolation, mounting and maintenance.

Noise generated by the motor fan is the dominant motor noise source, especially for TEFC motors. A sharp increase in noise occurs as the shaft rotational speed increases from 1800 to 3600 rpm. For large motors in the range of 1000 kW, 3600 rpm, a sound pressure level of as high as 106 dB(A) occurs. Measurements carried out in the laboratory for a range of TEFC motors from 25 to 2500 HP, no load, with and without the straight blade motor fan, show a difference of up to 50 dB(A) in the total sound pressure level. This large distribution of the fan noise is due to the fan shape. Motor fan blades are usually straight, so that the motor cooling is independent of rotation direction. Straight blade fans are very noisy, due to the large aerodynamic turbulent sound generated. Noise reduction in electric motors can be achieved by the use of an absorptive silencer or by redesign of the cooling fan, e.g., with irregular spacing of straight blades.

Pneumatic tools

Compressed air-powered, hand-held tools such as drills, grinders, rivetting guns, chipping hammers, impact guns, pavement breakers, etc., are widely used within a broad spectrum of different industries. There are three basic types of sources that dominate the noise generated:

1. Noise produced by contact between the machine and the working surface. The vibration transmitted from the tool tends to vibrate the working surface and work bench, generating high radiation noise, especially at mid and high frequencies.
2. Exhaust air noise caused by the turbulent flow generated as the compressed air passes the motor and by the aerodynamic noise generated in the air exhaust.
3. Sound radiation from tool vibration caused by air flow inside the tool.

The noise level of hand held tools can reach as high as 110 dB(A) at the operator's ear.

10.4 Control of noise pollution

Noise pollution is a serious threat to the life of man, wildlife and non-living things. Noise pollution has become a major problem not only in advanced countries but also in developing countries. There is a need to prevent and control the increasing noise pollution because its effects have far reaching consequences in destroying the eco system and deteriorating the qualities of environment. Though, it is impossible to control such pollution completely at this stage, but with certain measures and strict actions, we can reduce noise pollution to its minimum. The ways and means to check and control noise pollution may be general as well as specific.

10.4.1 General measures of control

The following general measures can be adopted to control and prevent the occurrence of noise pollution:

1. General awareness: General awareness among the masses is an important step to control noise pollution. The people must be educated through films, newspapers, radio, television and other suitable media about the hazards of noise pollution. The users of loudspeakers and other sources of entertainment must realise the common feelings of the public in general to avoid hinderances due to noise.
2. Noise reduction at source: The haphazard layouts of plants and machinery in factories unnecessarily produce a lot of noise. Such noise can be reduced throughout proper planning and methods of layout. Machines should not be put into operation when there is no work. The noise level can also be decreased if repair and maintenance is made a regular feature. Efforts should be made to substitute the noisy process and install noise-free equipment. Hence, effective steps can be taken at source to control noise pollution.
3. Screen the noise source: Noise pollution can be reduced if the noise source is screened by thick walls and other absorbing materials. For example, if the machinery has been installed in the open and the unit is located near residential areas, the same can be moved to the other side of the factory or the machine can be operated in a covered shed. Similarly, other rubberised material and anti-vibration pads, etc., can reduce noise level considerably.
4. Screen the receiver: Noise in industries has become a matter of routine, adversely affecting workers, in particular and public, in general. Workers should be provided with ear muffs and protectors. Their working hours should be adjusted to ensure that they are not exposed to noise beyond the acceptable limit. Audiometric tests must be carried out regularly. Restrictions should be imposed while constructing residential buildings near the roadside. Plantation is another source that can absorb sound in order to reduce the noise level.
5. Remove noise sources: The noise level can also be reduced if the source is removed to areas where the noise created is less likely to affect the surroundings. For example, aerodromes should not be near a populated area, but outside the city. Similarly, if a factory is situated near a locality, it can be shifted to another area.
6. Co-ordination: In order to check and control noise pollution, co-ordination between the planners and environmental authorities is essential. The authorities responsible for the development of cities must

consult the Pollution Control Board (PCB) authorities before planning and executing various schemes. This cooperation can prevent an increase in noise levels. Clearance from the PCB must be made mandatory. New devices to control noise pollution of traffic and construction work must be developed and implemented.

7. Engineering approach:
   - (a) Use of silencers, improvement in design, better installation for those motor vehicles and industries, which create noise pollution above the low limit.
   - (b) Workers should be provided with ear muffs and protectors. It is the responsibility of the employer that the degree of protection is both adequate and comfortable.
   - (c) Noise produced by vehicles can be reduced by banning pressure horns and streamlining traffic flow.
   - (d) Utilisation of loudspeakers of low frequency.
   - (e) Construction of walls in highly noise-polluted areas.
   - (f) Plantation of trees and shrubs along the roads and railway lines and around industrial complexes.
8. Legislative approach:
   - (a) Coordination of the noise research and control efforts of all concerned departments.
   - (b) Establishment of the criteria for concern about noise, effects of noise on health and welfare.
   - (c) Identification of major sources of noise.
   - (d) Need for understanding, cooperation with political, social, administrative and religious leaders.
   - (e) Separate noise pollution control act where standards for noise emission should be enforced.

10.5 Techniques for reducing machinery noise

Total elimination of noise can be very difficult, but machinery noise can be reduced to safe, livable limits. Sheet lead, lead-impregnated vinyl, sheet lead sandwiched between two layers of polyurethane foam, lead-loaded epoxy, sheet lead laminated to a variety of substrates and other specialty lead products offer excellent sound isolation properties and can be matched perfectly for specific tasks of controlling unwanted noise.

In approaching a noise control problem, the type of machine must be considered. Fully automatic machines may be completely enclosed in a sound-insulating structure. Doors may be built into the enclosure to permit necessary manual adjustments or tooling changes. Manually-fed equipment, however, must provide for operator access so it is not possible to cover the machine completely. In such cases, efficiently constructed, partial enclosures, are capable of reducing noise levels to safe limits.

10.5.1 Noise control in weaving rooms of textile industry

Noise represents in various departments and above all in weaving mills a problem of primary importance, especially if there is not enough room available and no adequate soundproofing intervention on the machine and on the rooms have been carried out. In such cases the alternative is the use of individual safety devices. A high noise level can entail a reduction in the functions and other secondary collateral effects.

In the textile industry, the noise problem in the various working departments is a cause of serious concern. The highest noise levels are to be found in the weaving rooms, where the operators are exposed to levels of 94 to 100 dBa. The needs of having the possibility to control fabric quality prevent any casing or partial shielding of the weaving machines, it is however possible to correct the acoustics of the working room. The mostly used materials are:

1. Glass wool baffles put in a glass fabric envelope and hang up on the ceiling.
2. Glass fibre panels with an interspace between panel and adjoining wall.

These measures, unfortunately, are not very effective, so that the personnel is anyway compelled to use the devices for individual protection. In fact these measures reduce the noise level only by 1 to 1.5 dBa between the weaving machines and by 2 to 3 dBa between the beams and in the department passageways. The above mentioned modest results, typical of the weaving rooms, are due to the preponderance of the direct waves (coming from the noise sources) over the waves which are reflected by other bodies and to the distance of the sound absorbent material from the noise sources.

The devices for individual protection which the workers have to use against noise are of various types and give different results with the variation of the frequency. There are devices which protect better at high frequency values (1000–8000 Hz) and others which are more efficient at low to medium frequency 125–1000 Hz).

Noise origin and problems in weaving machines

Noise is caused by the vibration of the mechanical parts of the machine. These parts can be either in motion (various kinematic motions) or standing (structural parts, boxes, casings). The moving parts are the main origin of vibrations, which are then transmitted to the other parts of the machine. The vibrations are the higher, the more intense are the load variations to which the moving elements are submitted: sley, heald frames, weft inserting elements.

These movements are alternative motions and have rather high operation frequency levels; as such motions generate the maximum load variation values on the involved mechanical elements, it is easily understandable that the resulting vibrations and the pertaining noise, can attain very high values. The noise of a machine depends therefore to a very large extent on the operating speed but also on the machine equipment viz., on its composition, as this entails a different quantity and typology of the mechanical units, each with different vibration mode. For this reason the weaving machine manufacturers are very particular about two well known basic lines in their production:

1. Noise reduction already at the designing stage.
2. Reduction of the noise reaching the operator by means of physical barriers between the noise sources and the subject (casings).

A further possibility could be, as previously indicated, the modification of the mill acoustics. In fact, although a great deal of progress has been accomplished to reduce noise in the weaving rooms, there is still a long way to go. We need only to consider that the noise emitted by a modern rapier machine is about 90 dBa (maximum level of acoustic pressure in 8 hours per day for a single person at 1 meter distance from the machine surface) when the machine turns at 500 strokes per minute without screenings, viz., the same noise level emitted by an old shuttle loom running at 180–200 strokes/minute. Thanks to the technological development, the weaving speed in the last 20 years has more than doubled, however without increasing the level of acoustic pressure. The attention which most of the industrial countries give today to the issue of environment pollution is more than justified. The noise is not only annoying, but can be harmful to health and at the end increase the social costs.

The EEC guideline machines 89/392 draws the attention to this problem and invites the manufacturers to design machines in such a way, that the risks due to noise emission are reduced to a minimum, in consideration of the technical progress and of the technical means available to reduce the emissions at their source. This guideline obliges the manufacturers to declare the noise levels emitted by their machines. The noise evaluation of a single weaving machine is anyway not sufficient; in the textile mills dozens, not to say hundreds, work simultaneously in one and the same weaving room and the sound level increases in proportion to the number of looms, even exceeding the threshold of 90 dBa indicated by the present Italian legislation. The Fig. 10.1 shows the noise increase in relation to the variation in the intensity of the sound produced by a certain number of sources positioned side by side.

We can note that, with 8 noise sources at 89 dBa, the noise level on a central measurement point is equal to 89 dBa; in the case for instance of two noise sources, by increasing the noise level of each source by 5 dBa, we get a

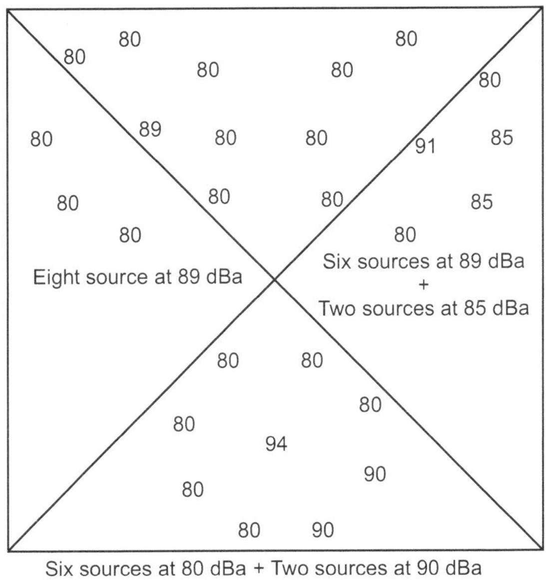

Figure 10.1: Noise increases in variation to sound produced by sources.

variation in the central point of 2 dBa. In the third graphic, if we bring the same noise sources to 90 dBa, we get a central point at a level of 94 dBa. This variation in the value of the central detection point in relation to the change of the sound level of the two noise sources follows a logarithmic trend.

10.5.2 Lead for noise reduction

Sheet lead for use in sound insulation applications is normally specified in pounds per square foot. Sheets can be obtained in weights ranging from 1/2 lb (1/128 in. thick) to 8 lb (1/8 in. thick). In most cases, weights normally used for noise reduction applications are 1/2 lb and 1 lb. Sheet lead can be cut with ordinary scissors, formed by hand and applied to surface with elastic type adhesives. Lead can be readily laminated to many substrates, including steel and aluminium sheet. These lead laminates can be formed, drawn or otherwise shaped.

Leaded vinyl sheet is another useful material for sound attenuation. One manufacturer's adaptation of this product comprises two sheets of lead loaded vinyl laminated to a core of glass fibre cloth to give the material more supporting strength. Sheet lead and polyurethane sandwich material can be used for sound conditioning areas, such as the inside of existing machine shrouds or guards. For structure-borne noise, lead-loaded epoxies make ideal damping compounds. The compounds are made by mixing powdered lead into epoxies. The material is usually applied with a trowel and can be used on resonating structures, such as machinery guards to reduce vibration and the resultant noise. The various lead materials for noise insulation are given in Table 10.1.

Table 10.1: Lead materials for noise insulation.

Material	*Description*	*Uses*
Sheet lead	Usual weight 1/4 lb to 4 lb	Used alone or laminated to substrates
Lead/foam	1/2 lb and 1 lb sheet lead sandwiched between layers of polyurethane foam	Laminated to enclosures
Leaded plastic Sheets	Lead-loaded sheet vinyl or neoprene with or without fabric reinforcement	As a curtain or to line enclosures
Damping tile	Lead-loaded epoxy or urethane tiles	Damping heavy machinery
Casting compounds	Lead-loaded epoxy	Potting, filling complex voids
Trowelling and damping compounds	Lead-load epoxy, neoprene and urethane	Damping enclosures, surfaces, resonate members and rattling panels

10.5.3 Quieting machinery

Reducing noise from machinery is still an emerging technique. In this 'cut and try' period, much must be borrowed and adapted from the manner in which noise problems have been solved by others.

10.5.4 Mechanical power presses

The obvious solution to the problem of noise emanating from an automatic machine is to enclose the unit in a sound-attenuating structure. A large maker of die sets for mechanical power presses usually adopts this approach to quiet the presses in its shop. One of the presses, a straight side, 20 T model fitted with a slide feed, normally operates at 600 rpm.

To prevent excessive heat build-up louvers are cut into the top of the press enclosure. Flat panels are located beneath the louvers to act as baffles. Exact locations of the baffles are determined by trial and error. Slight shifting of the baffles makes a great difference in the reduction of noise. A small exhaust fan is mounted in one of the louvers to further reduce heat build-up.

This press, running under high-speed conditions, now has a noise level in front of the press of 86 dBA scale. In the rear, the reading is now 85 dBA. This reduction brings the press noise well within occupational health and safety act requirements.

Power presses with automatic feed devices can also be enclosed in noise barriers made from leaded vinyl. These floor-to-ceiling curtains can be used to contain single presses. A press attendant can go in and out of the area to load additional stock in feed devices and to adjust press malfunctions. Total exposure time within the area can be adjusted to the time permitted by the noise level. If fail-safe electronic sensors are used on the presses, they can be adapted to activate alarms when a press needs attention for a malfunction or loading. Most noise produced by a mechanical power press is from the impact produced by the die set closing on a piece part. In such cases, a bellows constructed of leaded vinyl attached to the press slide can be timed to enclose the impact area when the press is stroked. The balance of the upper part of the press, with attendant gears or belts and motor, can be enclosed with a plywood or lead or lead/foam structure. If such an arrangement does not reduce the noise level to permissible limits, a second structure can be built around the lower section of the press.

Pneumatic tools

Pneumatic shop tools produce impact noise. By enclosing the barrel of the tool in a leaded vinyl sheet and rock wool, some quieting can be achieved by muffling exhaust noises.

High impact machines

High tonnage presses, drop hammers and other high impact machines create vibrations in addition to airborne noise. These vibrations, unless short-circuited, can be transmitted throughout an entire plant and even into office areas, where they may be translated into audible noise. To prevent this, machines must be isolated from their foundations. One highly, efficient method of accomplishing this isolation is through the use of lead anti-vibration pads.

10.5.5 Lathes and automatics

Most noise generated by a lathe or automatic screw machine comes from the gears and drive train. Noise emanating from the tool-chip interface should not be of concern in most cases. The solution to a gear noise problem is in the use of partial enclosures. One-pound sheet lead or lead/foam material laminated to any material used for drywall construction can be a successful sound barrier built around gear cases. Hinged doors or ports should be incorporated in the design of the enclosure to provide access to speed and feed controls. The access doors should be reasonably air tight to be an effective noise seal. The hinge side of the door should be gasketed with leaded vinyl sheet attached to the door and the jamb. The top, bottom and the side opposite the hinge should also be gasketed with a lead-loaded material. Smaller gear boxes can generally be quieted through the use of pillows made of lead-loaded vinyl and acoustical wool attached to their top. Heavy gears such as those on hoists and crane equipment generate high noise levels. The resonance of these huge gears can be substantially reduced by applying damping materials such as lead-loaded epoxies or lead-loaded vinyl sheets to the gear face.

Machinery guards

In some instances, large machinery guards of sheet metal have a tendency to set up a high-pitched ringing noise if stiffening ribs have not been added. This problem may be solved by laminating pieces of sheet lead to the guard by the trial and error method until the ringing is reduced or eliminated. Self adhesive leaded vinyl sheet may also be used to line the guard with the same result.

High pressure pipes and valves

High pressure transmission lines and valves for liquids and gases may be silenced through the application of sheet lead or leaded plastics. By first wrapping the pipe with absorbent material such as glass fibre or mineral wool to act as a thermal insulator and then covering the material with sheet lead or leaded vinyl sheet, noise levels will be significantly reduced. The final installation step should be the sealing of all joints on the line or valve with a self-adhesive lead tape.

In actual practice, noise levels of 110 to 130 dB at a gas pressure reducing station on a pipeline can be successfully muffled to 100 dB by wrapping the line with a 2-in. thickness of fine yarn glass fibre and covering it with a single blanket of reinforced leaded vinyl sheet. Greater attenuation is attained if a second layer of glass fibre and leaded vinyl sheet are used.

Air intakes and exhausts

Noise generated by the movement of great volumes of air can reach painful and dangerous levels. A leaded vinyl bellows installed between the intake of a large air compressor and the outside of a building will reduce the noise of the air movement within the enclosure. Air ducts, which serve as pathways for the intakes and exhausts, may be enclosed in glass fibre and sheet lead to further decrease transmission noise.

10.5.6 Air compressors

Air compressors, a source of irritating noise in textile plant, can be silenced by enclosing them in a shroud consisting of sheet lead or leaded vinyl laminated to any material intended for drywall construction. Louvered openings with baffles should be provided for the dissipation of heat build-up within the enclosure. Lead/foam material may also be used as a lining for the enclosure.

Vibratory feeds

A source of highly distracting noise in production machines devoted to automatic assembly of spark plugs can be created by air-vibrated feeds on each of the units. The noise, caused when the steel spark plug shells are vibrated into position on the metal feed track, is like machine gun fire. This unavoidable, metal-on-metal noise is a nerve-wracking distraction for workers. Because of the construction of the feed devices, it is impossible to completely enclose them. To solve the problem, the 2 ft × 4 ft area around each vibrating feeder is surrounded with curtains made from reinforced, lead-loaded vinyl. Even though the curtained area is left open at top and bottom, the noise at the individual feeds is reduced from 93 dB to 86 dB.

Transformers

The hum of transformers located at a substation can be a troublesome noise. When faced with such a noise problem, a West Coast utility used lead adhesive-bonded to steel as the noise barrier because of its weight and because it did not tend to resonate. The barrier is made by laminating 4-lb (1/16-in.-thick) sheet lead to 10-gauge steel with epoxy adhesive. The enclosure and its frame are built to be free standing and isolated from the transformer. Within this sheet, a lining of a 2-in.-thick blanket of glass fibre is spaced about 1 in. from the inner face. Pipes and fittings piercing the barrier are isolated from the walls by neoprene gaskets. After the transformer radiators are remounted, the units can show a noise reduction of 17 dB.

10.6 Lead-loaded fabrics

Lead-loaded fabrics keep outside noise out: When production facilities are expanded in a plant, additional office space is often necessary for the resultant increased clerical work. To keep office workers convenient to the production operation, yet keep noise problems to a minimum, offices are usually constructed at one end of the production area. Since production area ceilings are high, office walls are often constructed of brick or acoustical partitions. The ceiling is then suspended from the underside of the plant roof.

The office walls will usually prove to be good sound barriers. However, the sound-deadening capacity of the ceiling will be far below what is required for conversation, even when acoustical ceiling tile is used in the T-bar suspension systems. The ceiling will control sound originating within the office, but will not serve as a barrier to external noise. Noise from production equipment can reflect off the underside of the plant roof and focus on the acoustical tile ceiling of the office. As a result of this unique focusing problem, noise levels within the office can be more distracting than those in the plant. Constructing full-height insulated office walls to the underside of the plant roof is usually uneconomical in high-ceiling areas. High cost of materials and labour, interruption to production and modifications to fire control sprinkler systems usually make high-wall office systems prohibitive in cost.

However, using a lead-powder-loaded, vinyl-coated fabric can help reduce outside noise substantially. The material is limp, dense and non-porous and is easily cut with a knife or scissors. It weighs about 0.9 lb per sq ft and is approximately 1/16 in. thick. Most important, the material can be installed easily over one office in an entire group and its effectiveness evaluated before making a commitment for a total installation.

The lead-loaded fabric can be installed by plant personnel, quickly and easily, usually needing only oral instructions to explain the installation technique. Rolls of vinyl are unrolled on the floor, cut to length and laid directly over the T-bar suspension of the ceiling. No additional support is necessary. Material is cut to go around suspension wires and edges of the fabric are overlapped to prevent sound waves from passing through the joints. Edges can be sealed with adhesive tape; access to the sprinkler system is maintained by not sealing them. Noise levels ranging from 95–100 dB in offices have been reduced after installation of lead-loaded fabrics to 75–80 dB.

10.6.1 Using glass for noise reduction

A glass barrier offers the potential for controlling noise while providing visual communication. For the same thickness, it is a better sound barrier than most brick, tile or plaster materials. An average noise insulation of 35 dB can be obtained by using a single-leaf wall weighing about 6 lb/ft^2. However single-leaf wall construction is not efficient, if an average insulation of 40 dB or more is required. It is better then to use double-leaf type construction.

Sound transmission mechanism

A single-leaf partition is one which points on opposite sides of the structure move in the same direction at the same time at all frequencies. The simplest sound insulation treatment of this type of partition is known as the mass law.

10.6.2 Workable solutions to common machinery noise problems

The first inclination that a pollution engineer may have is to build a rigid closure or install a flexible sound curtain or a permanent wall around the equipment. In many cases, these approaches constitute the only possible solution and are the most efficient, expedient and economical. Also, when the machine can be treated without dramatically altering the production cycle, this is generally the most desirable approach.

Plastic grinders or granulators

Existing situation: Noise levels of 105–115 dBA were encountered in grinding poly-carbonates and other tough plastics. Normal frequency was 600 Hz, with intermittent impact peaks of 2000 Hz.

Noise sources: Noise sources are the grinding action itself, hopper impact in the loading and whipping action of the part, electric motor, belt gear drive and impact at the chipper and material chute. No attempt was made to determine the individual levels of each source, but it was obvious that hopper impact and grinding action were the major contributors.

Considerations

Solutions considered were a flexible curtain closure and a supplemental rigid plywood noise-treated closure. This method could have worked; however, operator accessibility, material flow and light and ventilation would have been sacrificed. The design of the equipment made it a closure in itself. Also, by treating the machine, the original equipment manufacturer could have reduced the noise levels substantially at the time of manufacture, providing a guaranteed noise-treated grinder at lower cost than a remedial retrofit.

Solutions

While the following solutions were developed for the OEM supplier, the same techniques can be used by pollution engineers. Mount the machine on vibration isolation mounts (fibre glass pre-compressed pads). Damp the impact noise areas with a viscoelastic spray-on damping compound. Make the existing steel housing a sound barrier by adding a noise control composite (1/2-in. polyurethane foam-1/24 lb/sq ft).

The outside of the metal hopper, pellet tray and chute should be sprayed with a viscoelastic damping compound to a thickness of 1½ times the thickness of the metal. Composite material should be applied to all other metal surfaces using a brush-on, solvent type thermosetting compound with room temperature cure properties.

Metal components should be treated before the machine is assembled. After assembly, 1/2-lb/sq ft 2-in. tape of loaded vinyl film with the same thermosetting adhesive should be applied to seal corners and mating joints, except at access doors and other removable panels.

10.6.3 Forge presses

Existing situation

Noise levels of 100 dB were generated from 36 presses arranged in two rows. The peak frequency was 300 to 600 Hz. The primary noise was generated by high-pressure, water-cooled hydraulic pumps and the hydraulic lines. The noise generated by the forging operation itself was less than 85 dB, low frequency and intermittent.

Considerations

The machines were already isolated on fibre glass pre-compressed pads to protect against structure-borne noise. A complete curtain closure was considered around each machine, but was ruled out when the problem was pinpointed to the pumps and piping.

Solutions

The most practical and economical approach was to cover the pumps with 0.87-lb loaded vinyl with 1/4-in. foam. For ease of installation, the composite was adhered to the pump housing with pressure-sensitive adhesive on the foam side of the composite. Joints were taped with a 2-in. wide tape of 0.87 lb/sq ft loaded vinyl. A hydraulic piping on the machine, from the pump to the hydraulic cylinder, was also treated with the same composite and adhesive.

Results

The sound level of the hydraulic system was reduced to below the 85 dB being generated by the press operation.

Observations

Pressure sensitive adhesives are more expensive than brush-on or spray-on rubber-based, contact adhesives with thermosetting properties. However, they are ideal for prototype work, save labour and make installation easier for remedial treatment.

Pressure-sensitive adhesives should not be used directly on lead vinyl or loaded vinyl, because the plasticiser in the vinyl will soften and weaken or destroy the adhesive bond. Also, these adhesives will not provide good adhesion to vinyl surfaces.

Lead-loaded and loaded vinyls are available with a barrier coat when pressure sensitive adhesives are required. In all other cases, the pressure sensitive adhesive must be applied on the foam side.

The sound control material in this installation must be removed for pump maintenance. An alternative would have been to vacuum-form clamshell halves of unreinforced loaded vinyl and foam and join the halves with zippers or catches. This system will provide easier access and replacement. However, in this particular case, when a pump requires maintenance, the unit is removed and replaced by a rebuilt pump. This factor makes the prescribed solution more economical than the vacuum-formed halves.

10.6.4 Gear housings

Existing situation

Noise levels generated were approximately 92 to 105 dBA, with peak frequencies of 250 to 4000 Hz. In all cases, the equipment or machine performing its operation was below 90 dBA. The excessive noise was caused by structure-borne noise, causing the gear housing or guards to resonate and vibrate.

Considerations

The machines were damped to ensure that structure-borne noise would not reach the floor or building. However, reducing the resonance vibration requires isolation within the machine. Generally, this entails a whole new design of the equipment, in that soft isolators inside the machine can change the work function. Gear noise can be reduced by having closer tolerances on gears, using different materials and incorporating less backlash in the design. However, in a remedial problem, complete gear redesign can be costly and time-consuming. If the problem is housing resonance, even redesign of the gears may not help.

Solutions

If the noise is housing resonance, spray a viscoelastic damping compound on the inside or outside—either way is effective. If the problem is gear noise being transmitted through the housing as airborne noise, use a composite barrier on the exterior. If both housing resonance and gear noise are problems, then use a combination of the damping compound and the composite.

10.6.5 Turret lathes

Existing situation

Noise levels of 100 to 110 dBA, with peak frequencies of 2000 to 4000 Hz are common in turret lathe operations. The primary noise is impact-created at automatic stock feed tubes.

Considerations

Total curtain enclosures and fibre glass sound baffles close to the tool holder have been tried to solve this noise problem. The curtain system has worked well. Fibre glass has had marginal results, because it becomes oil soaked and loses efficiency, as well as becoming a potential fire hazard.

Solutions

The recommended solution involves a cut-and-try approach, depending on the design of the turret lathe and severity of the noise problem.

The following measures can be taken:

1. Install isolation mounts on the lathe.
2. Apply a viscoelastic damping compound on the outside of the stock feed tubes to reduce ringing impact noise.
3. Check the gear housing and machine panels for vibration, if vibration can be felt, spray-on damping materials.
4. If noise is still above acceptable levels, apply composite panels (noise barrier and absorber) around the stock feed tubes. (This is not a total enclosure which will cause material handling and productivity problems.)
5. If excessive noise persists, use noise control composite panels around the tool holder case, in a manner that will not impair operator movement. Make sure the absorbing surface of the noise composite is supplied with an impervious film to eliminate oil absorption.

10.7 Control of noise in pumps and valves

10.7.1 Noise control in pumps

While there are numerous reasons why a pump may be generating an abnormal degree of noise, the greatest number of questions and concerns are usually about pumps involved in cooling tower applications or problems resulting from poor suction piping design.

In either instance, the following is typically experienced:

1. The noise is very similar, if not identical, to classical cavitation (growling or pumping gravel).
2. The phenomenon can occur with either a forced draft or induced draft-cooling tower. The resulting noise tends to be more prevalent on negative suction pressure systems but will occur on positive suction pressure as well.
3. Unlike classic cavitation, throttling the pump discharge to a lower capacity typically has little impact on the overall noise level.

Well, in addition to being very irritating, noisy pumps often mean there is diminished overall performance that can severely reduce the life cycle of the critical bearing and seal components in the pump.

There are really only two primary sources of pump noise: liquid and mechanical. Both sources produce acoustic pressure fluctuations that can be transmitted as audible noise. For centrifugal pumps, mechanical noise is generally the result of component imbalance (impeller and/or coupler), coupler misalignment, rubbing components or improper installation of the base plate and/or motor. These mechanical mechanisms generate distinct frequencies equal to rotational speed and/or multiples of rotational speed. If a noise spectra analysis is completed and it does not reveal distinct frequencies, then the noise is not being mechanically generated. The second mechanism for generating noise is liquid sources. Liquid noise is directly produced by water movement and is fluid dynamic in character. Turbulence, flow separation (vortex), cavitation, water hammer, flashing and impeller interactions with the volute cut water are all examples of fluid dynamic noise sources.

Pulsation sources

There are generally four types of pulsation sources in pumps that are the result of liquid noise:

1. Discrete frequency components generated by the pump impeller involving impeller vane passing.
2. Broadband turbulent energy resulting from high flow velocities.
3. Impact noise consisting of intermittent bursts of broad-band noise caused by cavitation and water hammer.
4. Flow induced pulsations caused by periodic vortex formation when flow is past obstructions and side branches in the piping system.

Techniques of controlling noise

Several successful control techniques have been employed in the past to mitigate excessive noise.

1. Increase or decrease the pump speed to avoid system resonances of the mechanical or liquid systems.
2. Increase liquid pressures (NPSHA, etc.) to avoid cavitation or flashing; decrease suction lift. This could include raising the tower, lowering the pump or straightening the suction piping to reduce friction losses.
3. Modify the pump so that the clearance between the impeller diameter and casing cut water (tongue) or diffuser vanes is increased.
4. Inject a small quantity of air into the suction of a centrifugal pump to reduce cavitation noises by providing a shock absorbing cushion to minimise the impact of recondensation of water vapour within the pump impeller.

For an existing facility, the first three techniques may not be practical, as they would be extremely difficult, expensive and time consuming to accomplish. However, injecting air into the suction end has proven itself to be a very viable and cost-effective solution in many instances to cushion and minimise the impact of the collapse of the vapour bubbles. The injection of small amounts of air can usually be accomplished quickly and easily in the field with minimal expense. If the pump is pulling a vacuum, the addition of a small vacuum breaker on the suction flange works perfectly.

Alternatively, for positive pressure situations, the connection of a compressed air line to the pump suction flange and adjusted to allow a small quantity of air into the pump, is equally acceptable. How much is enough? Use your ears! Attach the compressed air line to the pump with the air valve closed. Crack open the valve allowing a small amount of air to enter the pump. Continue to slowly open the valve until you hear the crackling cavitation noise subside or stop. Very small additions of entrained air introduced at the pump will not cause any problem to the pump or to the cooling tower/condenser circuit. This method is very desirable and has been proven on numerous occasions to be a viable solution to eliminating excessive noise. At a minimum, this approach is also an excellent analytical tool that can be utilised to better ascertain the severity of the overall noise problem.

Additional factors for reducing noise

In addition to the techniques outlined previously to reduce or eliminate noise, attention must also be given to two other factors that can exacerbate the situation: vortexing of the liquid in the tower pan -which is the most common source of air within a pump - and the suction piping arrangement.

Excessive vortexing: The amount of air entrainment contained in the pump fluid as a result of the vortexing is heavily dependent on several variables: the vortex size and the submergence level of the pump suction pipe below the water level of the pan. The most common method of eliminating vortexing in the tower pan is by the inclusion of baffle assemblies to abolish the formation of vortexes. Raising the fluid level in the pan to a sufficient depth can sometimes resolve this problem.

Suction piping: Improper pump suction piping is one of the most significant contributors to noise, as well as to diminished overall performance of a pump. Unwittingly, many centrifugal pump troubles are the result of poor suction conditions. Coupled with the vortexing phenomenon described above, the improper layout of the pump suction piping can be a significant contributor to the generation of pump noise.

Avoiding problems

There are several field-proven best practices that should be followed to avoid these types of problems from developing which are as under:

1. Suction pipe velocities should never exceed the 5 to 8 ft/sec range. Otherwise, higher velocities will result in an increase in friction loss and can result in air or vapour separation.
2. Suction headers should be sized for 5 ft/sec of total system demand.
3. All suction piping pressure drops must be equal for pumps operating in parallel.

4. The suction pipe should never be smaller than the suction connection of the pump and in almost all cases it should be at least one size larger.
5. The suction lines should be as straight as possible and no less than 5 pipe diameters in length going into the pump.
6. The addition of short radius elbows and/or tees next to the pump suction nozzle will result in an uneven flow pattern to the impeller resulting in a hydraulically unbalanced distribution of water into the dual impeller eyes. The resulting hydraulic imbalance leads to possible cavitation, vibration and excessive shaft deflection - all of which can severely reduce the overall life cycle of the critical bearing and seal components in the pump.
7. Use tapering Y-branches for multiple pumps off a header, versus straight branch headers.
8. The suction piping should be arranged with as few twists and turns as possible. When bends become necessary, a long radius elbow must be utilised.

10.7.2 Control of noise in valves

Control valves are generally present whenever fluid flow regulation is required, therefore they are used in a wide variety of industrial applications. Control valves regulate flow by increasing or decreasing the fluid pressure drop across the element. These pressure drop adjustments are usually accompanied by noise generation.

Sources of valve noise

Noise generated by control valves can come from multiple sources that contribute to the overall noise level. Mechanical noise sources include piping and plug instability noise. More prevalent is aerodynamic noise that is generated by the release of energy from the pressure drop of the fluid, flow turbulence, flow path obstructions, rapid expansions or decelerations of the flow, or directional changes.

To effectively combat aerodynamic noise, it is important to understand its two main sources. The first is noise contributed by the turbulent energy released as the pressure of the fluid drops at the trim. This noise is reduced by using abatement trim that combats the noise through pressure staging, frequency shifting and jet interaction prevention.

The second source of aerodynamic noise occurs in the expansion area downstream of the trim, which can arise when the valve outlet is smaller than the downstream pipe diameter. Here, the velocity-related turbulence generated by the expanded valve outlet acts as an independent noise source that can exceed that of the trim. In effect, two noise sources exist in series.

Proper valve sizing is important as a means to combat expansion noise. When noise exceeds customer acceptable levels, or 110 dBa, valves with larger expansion areas may be needed. An alternative is to use a diffuser downstream of the valve to increase back pressure, which effectively reduces the required outlet area and slows the fluid velocity to less than 0.3 Mach.

The noise is typically due to both, the mechanical vibration of valve components due to flow and the flow itself, both aerodynamic and hydrodynamic.

Mechanical noise produced by control valves is a result of random pressure fluctuations within the valve body and/or the fluid impingement of valve internals that come into mechanical contact with the fluid flow. The sound that is produced by this type of vibration is normally in the frequency rang of 1500 Hz and perceived as metallic rattling. Though the physical damage that is typically associated with this type of vibration is generally more of a concern than the noise that is emitted. Noise as a result of such vibrations is in general not predictable and can only be eliminated by improving the valve or fluid network design.

Control valves handling liquid flows can cause significant amounts of noise as a result of the flow of the liquid. The noise resulting from hydrodynamic flows can be generally classified into three distinct classifications:

1. Non-cavitating.
2. Cavitating.
3. Flashing.

Noise resulting from non-cavitating flow is not typically of sufficient intensity to be problematic. The generally accepted mechanism for noise production as a result of non-cavitating flow is turbulent velocity fluctuations in the fluid stream which are usually referred to as ‘Reynolds stresses’. Such velocity fluctuations are present in control valves as a result of the large decrease in the magnitude of the linear velocity downstream of the vena contracta. Noise resulting from cavitation is more likely to be significant. Cavitation is a two-stage process. First bubbles of vapour form as the liquid pressure drops below its vapour pressure. When the pressure recovers and rises above the vapour pressure, these bubbles collapse, or implode. Cavitation affects the capacity of a valve, causes noise, vibration and erosion of valve components.

Cavitation may cause severe damage to metal valve components due to mechanical stress.

Flashing is a condition that occurs when the downstream pressure of a control valve is lower than the upstream vapour pressure, causing a partial liquid vapourisation with the vapour remaining in the flow. The flow is then chocked and therefore independent of the downstream pressure. Noise as a result of a valve that is handling such a flow is due to the deceleration and expansion of the two-phase flow.

Finally, aerodynamic noise, another major contributor of valve noise, is due to the direct conversion of the energy of a turbulent gas stream. The principle source of aerodynamic noise from control valves is the recovery region, which is downstream of the vena contracta. The shear forces in this location create flow patterns that are discontinuous and chaotic.

The ability to predict control valve noise has become a critical task for manufacturers of valves. The demand for such information stems partly from increased government regulations, requiring sound levels to be within acceptable limits. According to the Occupational Safety and Health Administration (OSHA), 15% of workers exposed to noise levels of 85 dBA or higher will develop material hearing impairment. Persons should not be exposed to noise above 115 dBA. The National Institute for Occupational Safety and Health (NIOSH) and the American Conference of Governmental (ACGIH) both believed that 85 dBA is the recommended exposure limit. Diffusers offer another method for noise attenuation. A diffuser is a pressure-reducing device that is installed downstream from the control valve. The total system pressure drop is divided across the valve and diffuser. Up to 40 dBA noise attenuation is possible with properly sized and selected valve and diffusers.

10.8 Management of noise pollution control

The above only touches the highlights of the broad effort to halt the increase of noise in our environment and to roll it back to more acceptable levels. The number of central, state and local regulations clearly indicates that it is now public policy to prevent infringement on the rights of those exposed to unnecessary noise not of their making. No one in commerce, business or government can ignore the accumulative effect of this large-scale effort. Therefore, it is necessary to become informed as to the regulations on noise and to devise programmes for meeting the requirements of the law.

10.8.1 Noise pollution control programmes

Noise pollution control programmes contain some or all of the following elements: (i) planning, (ii) measurements and (iii) implementation.

Qualified engineering assistance is required in formulating and carrying out the tasks that make up these major elements. As in other forms of pollution control, such as air and water, the decisions that have to be made must take into account psychological and physiological factors as well as the purely physical. The exact noise level that will cause annoyance, physical damage to structures, or even hearing damage cannot be determined because of the wide variability of tolerance. Scientists trying to determine proper noise levels have resorted to statistics in the development of criteria. The use of statistical measures can be expected to increase because of the availability of computers to handle a vast amount of data. The attempt to find the exact amount of noise reduction that is desirable should not obscure the fact that in many situations, the noise levels far exceed any reasonable criteria.

Planning

The first step in planning for the reduction or control of noise is to become familiar with all the regulations and ordinances that might now, or in the future, set the criteria that must be met. Where specific levels are shown, they can be included in the plan. Where there are only qualitative allusions to requirements for noise control, the intent and thrust of the statues must be interpreted in terms of actual maximum levels.

Since these decisions should be made by competent acoustical engineers, it is essential that the initial planning stage contain provisions for obtaining such help through outside consultation or hiring of an in-house staff. In either case, it will be necessary to appoint a programme manager. For large communities, large industrial establishments and transportation planning offices, a full-time programme manager will be required. Smaller programmes can be managed as a part of the public health and safety office, industrial safety department or environmental planning office. Failure to determine all of the requirements for noise control and to plan a programme adequate for meeting these requirements can only result in an inadequate programme.

Measurements

Whether maximum noise levels are stated in the statutes or determined from reference to non-specific ordinances, measurements of noise are necessary to determine the ambient level and to monitor the performance of the control measures. These measurements can be made by in-house engineers, using equipment owned or rented by the control agency or by contract to outside engineering firms. The simplest measurement is by direct reading of a sound level meter set on the 'A-weighting' network. Except where diagnostic data is required, as in machinery noise control, this is sufficient.

If engineering decisions are to be made on the basis of diagnostic measurements, more sophisticated instrumentation is required. An octave band analyser, which divides the sound spectrum into frequency bands an octave wide, can be used. This is required when it is necessary to detect the presence of discrete frequencies such as those from fans, gears and blowers. The analysers are generally battery operated instruments with a series of band pass filters.

The octave filter bands are centered on standard preferred frequencies of 31.5, 63, 125, 250, 500, 1000, 2000, 4000, 8000 and 16000 Hz. Frequently, a third octave analyser is used to obtain more detail.

If more detail than that provided by the one-third octave band analyser is needed, a narrow band analyser having the capability of separating pure tones close in frequency should be used. In using these analysers, it is more convenient to record the sound on a quality tape recorder and then reduce the data in a laboratory. The standard sound level meter and analysers are sufficient for continuous or slowly varying noise. If the noise to be measured varies rapidly or is 'impulsive', a meter capable of following the rapidly changing signal and storing its peak is required.

As previously stated, there is a trend towards using statistical measures to describe the noise levels in industrial and community studies. For measuring and computing these measures, some form of distribution analyser is required. These analysers determine the time during which the noise is at a particular level. For example, a community noise study may show that the noise exceeded 75 dBA for a total of two hours. In other types of equipment, the total energy reaching the microphone over the period of measurement is summed up and an average value given. One form of this instrument is the dosimeter used for occupational noise exposure.

However, it must be stated that the accuracy of the result depends as much on the training and skill of the operator as it does on the instrument. The cost of noise control programmes is directly related to the number and magnitude of the noise sources and is, therefore, dependent on the accuracy of the data acquisition and reduction.

Implementation

After a well-planned and staffed programme has been initiated and sufficient measurements made to determine the areas that require control measures, design and implementation of these measures is the next step. If it is a community noise problem, this effort falls on those who are in violation. For industrial noise control, implementation will be a continuing function of the in-house group or the consulting engineer. Here, the first serious steps are taken to reduce the noise. Established criteria are compared with the measured noise level and the degree of reduction determined.

For a punch press having a noise level of 106 dBA at the operator's ear, it will be necessary to reduce this level to 85 or 90 dBA. The acoustical engineer will utilise those materials and techniques which he feels confident will reduce the noise by 16 to 21 dBA. Where there are a number of sources in an industrial plant, the cumulative effect of these sources should be taken into account. It is important to realise that sound from a number of sources in a factory area is additive and that while each machine is below the required level, the total may be as much as 10 to 15 dB above the desired level. At this point in the programme, it becomes possible to make cost and time estimates for achieving the necessary criteria. A pollution control manager should encourage the development of alternative plans with the acoustical design engineers to determine which has the least impact on cost and productivity.

10.9 Case study: Noise control in textile industry

Ahmedabad Textile Industry's Research Association (A TIRA) conducted noise pollution surveys in the Indian textile mills over a period of 15 years. The results (Table 10.2) indicate that the noise level in spinning department is between 80 dBA and 90 dBA, of which the lowest is in blow room and the highest in ring spinning. Noise level in weaving preparatory is low. Excessive noise level of 94–99 dBA is in loom shed, depending upon the design, type, erection and number of looms used, condition of machines, fabric structure, building type, building size, etc.

Table 10.2: Noise level in different departments of a textile mill.

Section	*Noise level, dBA*
Blow room	80–83
Carding	84–89
Draw frames	84–88
Inter frames	82–86
Ring frames	86–90
Winding	82–86
Warping	80–86
Sizing	73–86
Loom shed (Non-auto)	94–99
Loom shed (Auto)	95–97

Talukdarl evaluated quantitatively the noise level of conventional automatic looms and observed that the noise during weaving is mostly impulsive and periodic in character. When the loom is run with a shuttle, noise level increases by about 1.5–2.0 dBA. The maximum noise level is observed at the front of the machine because of the movement of sley, which causes air turbulence. As the 100 m speed increases, the noise level also increases significantly due to the higher impact forces acting on the different parts of a loom. Spectrum analysis shows that the peak noise level occurs at the frequencies between 1.4 kHz and 5.0 kHz and is mainly due to the impact between different parts of the picking and checking mechanisms.

Of the shuttleless weaving machines, noise level of water-jet weaving machine is the lowest (85 dBA) followed by that of air-jet and rapier weaving machines (91 dBA) and projectile weaving machine (92 dBA).

10.9.1 Effect of noise on human beings

The effect of noise on human health has been a subject of research for some time. Nevertheless, in India, no systematic study has been conducted so far. Damage risk criteria of workers exposed to sound level of ~ 85 dBA at the frequencies 500, 1000 and 1500 Hz are given in Table 10.3. Balachew and Berhane carried out a detailed study in a textile factory of Ethiopia during October-December 2009. They found the prevalence of and risk factors for, noise-induced hearing loss (NIHL) in the textile mill. A sample of 630 from 5900 workers was selected from the factory rosters by means of systematic sampling technique. Data were collected through interview, otology examination and pure tone audiogram measurement.

Table 10.3: Damage risk criteria.

Equivalent control sound level, dBA	*Risk percentage*[a]			
	10[b]	*20*[b]	*30*[b]	*40*[b]
85	3	6	8	10
90	10	16	18	21
95	17	28	31	29
100	29	42	44	41

[a] A 40-hour week with 50 weeks per year

[b] Years of exposure = Age - 18 years

Environmental noise survey and personal dosimeter data were also collected to determine noise exposure levels at every section of the mill. The highest noise level in area samples was observed in the weaving section (99.5±3.2 dBA). History of hearing loss was reported by 51 (8.1 %) participants, while 57(9.0%) reported a history of ear-related diseases. More than one third (34.3%) complained of current ear problems and 186 (29.5%) complained of ear pain. On physical examination, 154 (24.4%) were found to have detectable ear problems, of which 66 (42.9%) had otitis, while the remainder had ceremonious occlusion of the auditory canal.

There was no significant difference by gender in the prevalence of detectable ear problems. Audiometric tests, carried out at a frequency of 4000 Hz, revealed a 34% overall prevalence of NIHL (hearing threshold level exceeding 25 dBA) with the highest prevalence of 71.1 % observed among the weavers. Preventive measures were generally absent, with no employee reporting use of personal protective devices (PPDs). Textile factories are among the many occupational settings that pose the risk of noise induced hearing loss. Apart from damage to hearing, there is evidence that noise also affects the people in the following other ways:

1. Reduced performance, e.g., reduced ability to concentrate.
2. Disturbance of sleep.
3. Annoyance (oral communication is difficult or impossible), which manifests itself primarily in emotional responses.
4. Excitement (activation) of the central and vegetative nervous system, e.g., increased blood pressure, higher heart frequency and effect on metabolism.

10.9.2 Methods of reducing noise and vibration

In approaching the possibilities of reducing noise and vibration, a fundamental distinction must be drawn between the active and passive measures. Active measures are all primary measures, which prevent noise and vibration to occur in the first place. Passive measures are all secondary actions aimed at reducing the radiation of existing noise and vibration.

Active measures: Despite the enormous increase in speed of the textile machines, the machinery manufacturers have succeeded in keeping the noise level reasonably low. For example, the speed of the projectile weaving machines has gone up from 550 to 1500 m/min during the last two decades, but the sound level of the machine on the emission side (effect on human beings) reduced from 92 to 86 dBA. Some of the active measures taken by the textile machine manufacturers, especially weaving, to control noise and vibration are briefly discussed below:

Reducing inertia forces: Inertia forces that are produced in a machine are responsible for noise and vibration. If noise and vibration are to be reduced, this presupposes a reduction in the inertia forces. Reducing the mass and/or reducing the acceleration can achieve this. With this in mind, the textile machinery manufacturers have repeatedly optimised the most important components of the textile machinery, especially in weaving and, therefore, there is very little scope to reduce the mass further. However, the reduction in mass can also be achieved through substitution with another material like carbon. Some progress has been made in this direction . Since the introduction of new material usually means redesigning of entire section of the system, the modifications of this kind are generally not transferable to older machines. They are, therefore, restricted to newer generation machines. The simplest method of reducing acceleration is to reduce the rotational speed and/or to increase the distance, but both are not feasible for well-known reasons. A reduction of 10% in speed of a weaving machine, for example, would give a noise reduction of 2 dBA. Another possibility of reducing noise is the optimum design of the sequence of motion at a given speed and a given stroke. Today, cam gears are replaced by crank gears since the former give acoustic and vibration problems which arise from the choice of long standstill times and discontinuous transmission functions, while the latter give harmonious transmission functions, resulting in a steep drop in the excitation spectrum of forces to give acoustically problem free machine.

General machine components (toothed gears, motor, fans, etc.), are also a major source of noise on all machines. Noise control is extremely important, for example, in toothed gears.

Faulty gear teeth cause uneven transmission and bumpy loading of the teeth. Mass balancing is another way of reducing noise.

Active noise control: Active noise control (ANC) is the state-of-the-art technique which is most successfully demonstrated for controlling noise in enclosed spaces such as ducts, vehicle cabins, exhaust pipes and headphones. However, the most demonstrations have not yet made the transition into successful commercial products, probably due to the high capital cost. In 2010, Yigone undertook a project to use active noise control technique on weaving machines. Initial laboratory study indicated that newly designed active noise control is able to work in reverberant field as the looms' hall. The basic principle of ANC technique is outlined as follows:

ANC is sound field modifications, particularly sound field cancellation, by electro-acoustical means. In simplest form, a control system drives a speaker to produce a sound field that is an exact mirror image of the offending sound (disturbance). The speaker thus cancels the disturbance and the net result is no sound at all. In practice, of course, active control is somewhat more complicated. The idea of active noise control was actually conceived in the 1930's and more developments were done in the 1950's. However, it was not until the advent of modern digital computers that active control became truly practical. ANC works best when the wavelength is long compared to the dimensions of its surroundings, i.e., low frequencies.

The four major parts of an active control system are:

1. The plant is the physical system to be controlled – typical examples are a headphone and the air inside it and the air travelling through an air-conditioning duct.
2. Sensors are the microphones, accelerometers or other devices that sense the disturbance and monitor how well the control system is performing.
3. Actuators are the devices that physically do the work of altering the plant response; usually these are electromechanical devices such as speakers or vibration generators.

4. The controller is a signal processor (usually digital) that tells the actuators what to do; the controller bases its commands on sensor signals and, usually, on some knowledge of how the plant responds to the actuators.

Passive measures: These measures should be looked at keeping following three basic solutions in mind:

1. Blocking airborne sound.
2. Absorption of airborne sound.
3. Vibration damping.

For most applications, a solution will consist of one to all of these categories.

Blocking airborne sound: Individual ear protectors are by far the most effective and cheapest means of reducing airborne noise emissions. They are available virtually everywhere in any forms. However, they must be worn and, therefore, require the exercise of a minimum of discipline.

Normally, all machines including textiles have enclosures. The enclosure is basically made up of a metal or plastic sheet and its primary function is for cosmetic purposes or as a safety feature to protect the work force from a possible hazard. Whatever the case may be, it is convenient when an enclosure, of any sort, can be utilised for noise abatement.

At frequencies from 250 Hz to 750 Hz, a barrier can be extremely effective if a complete enclosure exists. The major stumbling block associated with barrier materials is that they are best utilised with complete enclosures. A small amount of open area, even 1%, will significantly reduce the barrier's performance. In general, a complete enclosure gives substantial noise reduction, but the method has considerable drawbacks in respect of space requirements, monitoring, accessibility, operation and maintenance, material flow, cleaning and cost. Partial enclosure has rather fewer disadvantages as regards space requirements, monitoring and accessibility.

However, the effect achieved is much less compared to that achieved with a total enclosure. When the absorber is coupled with a barrier, significant reduction can be realised. The acoustical energy, which is not dissipated by the absorber, is reflected, in part, by the barrier and is again dissipated by the absorber. The total energy within the enclosure is reduced, which minimises the problem of acoustical energy that escapes out of the open area of partial enclosures.

Absorption of airborne sound: If the major noise problem lies in the medium frequency range (500–4000 Hz), the acoustical absorption can be a good starting point. By lining the interior walls of the enclosure with an absorber, a significant amount of the acoustical energy incident on the walls can be absorbed and dissipated. This fabric utilises controlled permeability to optimise the absorption characteristics. Another advantage of membrane is the impervious film facings available that prevent absorption of dirt, oil, grease and moisture, which will degrade the sound attenuating properties and shorten the life of the foam. Since the most industrial applications have somewhat of an adverse environment, a film facing is a necessity.

As sound waves, which are pressure waves travelling in air, pass through a porous open cell of membranes, the movement of the air molecules through the openings across the strands and membranes generates heat through friction which takes energy away from the sound wave. The most significant attribute affecting an absorber's acoustical performance is the permeability performance of the product.

Rakshit and others studied the noise absorption of 300 gsm needle-punched nonwoven fabrics using a number of blend compositions of wool and polypropylene. They concluded that wool waste/polypropylene blend could be used effectively for noise reduction. Sound absorbing covers as well as machine and accessory components made of special materials or with special coatings may be used in modern spinning and weaving machines. These along with other measures can reduce the sound power level of these machines. For example, the sound level of projectile weaving machines at maximum rotational speed is reduced from 105 dBA to 102 dBA. This corresponds on the emission side (effect on human beings) to a reduction of the sound pressure level from 89 dBA to 86 dBA. This noise reduction of 3 dBA is equivalent to a halving of the noise exposure. However, in the areas where low frequency airborne sound is predominant, an absorber is limited in its efficiency. Because the wavelength of sound waves is inversely proportional to the frequency, the lower is the frequency, the longer is the wavelength. A sound wave at 100 Hz has a wavelength of approximately 3 nm.

In an enclosure or partial enclosure application, a sound absorber cannot reduce the sound energy below what is being produced. The ideal situation is to absorb enough energy so that the reflected sound waves do not reinforce the sound energy being produced. To illustrate this point, if a noise source produces 90 dBA of sound energy and a partial or full bare metal enclosure is put over the noise source, the reflected sound will reinforce the sound produced and the sound level within the enclosure may increase to 100 dBA . By lining the enclosure with a properly selected absorption product, the energy of the reflected sound waves does not reinforce the incident sound waves. The sound level within the enclosure can be reduced to 90 dBA, but not below this level since this is what is being produced.

Vibration damping: Constantly increasing textile machine speeds produce not only steadily higher noise levels but also higher vibration emissions. In the process, the forces increase proportionally to the square of the speed (a 10% increase in speed means a 20% increase in forces). Damping pads are used to dissipate mechanical vibration. For high speed weaving machines, plastic spring elements were used to reduce vibration emission. Subsequently, air spring absorbers were developed to enhance the performance of vibration isolation. Today, steel spring/laminated base plate elements are replacing air spring absorbers since the former is just effective as the latter as regards isolation effect, but it is virtually maintenance free. These elements can be retrofitted. By damping an enclosure or panel, the panel's resonance will be reduced. This may reduce the sound level if the resonance of the panel is significant enough to produce airborne sound. Another benefit of damping a panel is the reduction of natural frequency vibration and resonance, which reduces the effectiveness of the panel to block noise due to its mass.

Weaving machine widths: Increased output on a weaving machine can be achieved by increasing the speed or by using greater machine widths. The noise level for a 10% increase in weft insertion rate thus changes as follows:

1. With a 10% increase in rotational speed and unchanged machine width, the noise level rises by about 2 dBA.
2. With a 10% increase in width and the same speed, the noise level rises by only 0.5–0.7 dBA. Wide weaving machines thus record lower noise levels than the narrow weaving machines at the same weft insertion rate.

11

Common effluent treatment plant

11.1 Introduction

The Common Effluent Treatment Plant (CETP) has been accepted as a solution for collecting, conveying, treating and disposing of the effluents from the industrial estates. The effluent include industrial wastewaters and domestic sewage generated from the estate. This CETP concept helps small and medium scale industries to dispose of their effluents. Otherwise it may not be economical for these industries to treat their wastewaters or there may be space constraints. Some of these industries may require to give preliminary treatment (for removal of solids) so that the receiving sewers can be maintained free flowing. It may be required to correct pH or removal of specific pollutant before the industry discharges in CETP.

11.2 Concept of common treatment

The concept of effluent treatment, by means, of a collective effort, has assumed reasonable gravity by being especially purposeful for cluster of small scale industrial units. Common effluent treatment plant not only help the industries in easier way to control the pollution, but also act as a step towards cleaner environment and service to the society at large.

The concerted approach of joint or common effluent treatment provisions has many advantages. Wastewater of individual industries often contain significant concentration of pollutants; and to reduce them by individual treatment upto the desired concentration, become techno-economically difficult. The, combined treatment provides a better and, economical option because of the equalisation and neutralisation taking place in the CETP.

Other important issues for the merit of common treatment include, scarcity of land at the industry's level and a comparatively easier availability of professional and trained staff for the operation of CETP, which can otherwise be difficult, at the individual industry level. For the regulatory authorities also, common treatment facility offers a comparatively easier means of ensuring compliance of stipulated norms. The handling and disposal of solid- waste also becomes increasingly easier as the infrastructure is created in the project itself. The concept of common treatment, based on feasibility, should be part of the new industrial estates as essential component of infrastructure, In fact, the location of industries should always be such that units with compatible nature of activity are located in a cluster which in-turn can facilitate in providing common treatment. Typical effluent treatment scheme for common effluent treatment plant is shown in Fig. 11.1.

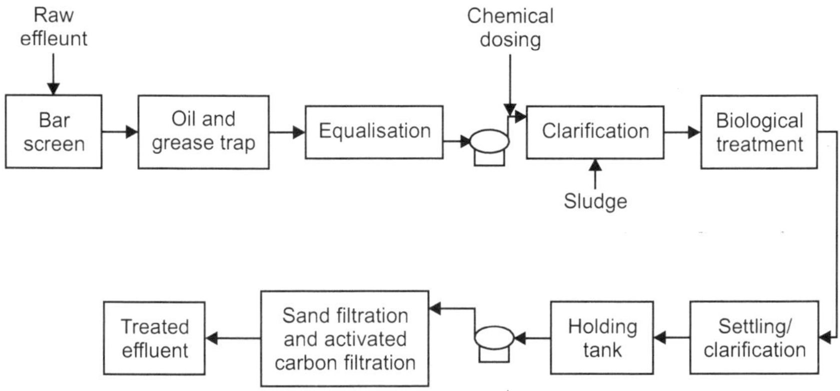

Figure 11.1: Typical effluent treatment scheme for common effluent treatment plant.

11.3 Measures for optimum efficiency

Operation of CETP being a participatory mechanism, the primary requirement is hence to define the ultimate responsibility for the proper functioning of the plant after it is commissioned. The important issues which merit consideration are discussed below.

11.3.1 Aspect of ownership

Various ownership alternatives include- the plant owned by government, consortium of industries or by an independent body. Whatever be the case, the primary emphasis should be on responsiveness in terms of effective and optimal operation of the plant and accountability. The member industries should also be made to realise that they are equally responsible for the sustenance of the plant.

11.3.2 Conveyance system

Different from the discharge characteristics of an integrated (big) industrial unit, small scale units usually generate higher proportion of floating or suspended particles in their effluent streams. At times their job operation result in high corrosive effluent. In either of these cases , to effectively convey their effluent to the CETP, it becomes necessary for individual units to setup a 'pre-treatment' device. It is also necessary that the conveyance network be so designed as to ensure their periodic desludging. Care should also be taken to minimise on cost of operation by facilitating conveyance through gravity flow, instead of multistage pumping. In fact the location of CETP should be selected after a careful topographical survey of the drainage area to keep the conveyance route as short as possible. The chances of flooding in monsoon and accidental surface run-off into the conveyance route should also be looked into. Conveyance by tankers is another option, provided the chances of leakage are effectively checked and their transit is strictly monitored in accordance to a properly laid down system.

11.3.3 Cost of treatment

The cost effective treatment supported with a system of regular collection/ payment of treatment charges by each member unit, while maintaining its effluent quality within acceptable norms are some of the prerequisites. The system of payment should be legally supported to provide a check for non-payment of dues and to take steps against defaulters.

11.3.4 Criteria for cost

The cost sharing should be decided in such a way that volume of effluent becomes an important norm, but its share in the total cost should not be such as to encourage by-passing of dilute streams and conveying highly toxic/non-biodegradable waste to CETP. The treatability factor should also be given due consideration in cost estimation. An effort by the industry to segregate toxic, highly acidic/highly basic, or toxic metal bearing waste be made to explore the possibility to de-toxify/neutralise or to attempt the recovery of metals by installing recovery plants, which are feasible and economically viable on account of their payback potential.

11.3.5 Plant design

The approach to provide treatment at low cost, an important factor in common treatment, depends on appropriate design of CETP. In keeping with the diverse nature and scale of operations, typical of small scale units, low capital investment and lower operation and maintenance cost incurred on treatment is a prime factor. In such a situation mechanical and chemical processes are advantageous over biological systems. And the least preferred are conventional anaerobic processes on account of huge space requirements and least flexibility. Though the advanced Upflow Anaerobic Sludge Blanket (UASB) technique with less hydraulic retention and space requirement being significantly low, anaerobic system is also a possible option. In order to obviate the need of excessive civil work at CETP in making huge equalisation and settling units, the member units should also provide settling and neutralisation of their individual waste. In order to minimise on the electrical cost, the possibility of substituting bioenergy should be explored to the extent possible. Proper management of sludge with its nutritive value would mobilise resources to substitute the operational cost. While designing the plant it would be of additional advantage to keep manpower requirement as low as possible but high in technical skills to reduce down-time for maintenance.

11.4 Advantages of common treatment

1. Saving in capital and operating cost of treatment plant. Combined treatment is always cheaper than small scattered treatment units.
2. Availability of land which is difficult to be ensured by all individual units in the event they go for individual treatment plants. This is particularly important in case of existing old industries which simply do not have any space.
3. Contribution of nutrient and diluting potential, making the complex industrial waste more amenable to degradation.
4. The neutralisation and equalisation of heterogeneous waste makes its treatment techno-economically viable.
5. Professional and trained staff can be made available for operation of CETP which is not possible in case of individual plants.
6. Disposal of treated wastewater and sludge becomes more organised.
7. Reduced burden of various regulatory authorities in ensuring pollution control requirement.

The pre-requisite for the concept of common treatment is the treatment at low cost. With more and more improvisation and stress on in-house measures, it is now established that the nature of job in small scale industries has immense potential of not only adopting certain measure to control pollutants at source thereby reducing pollution load and cost of treatment but also to reap rich benefits. Benefits of CETP in various industries are given in Table 11.1.

Table 11.1: Benefits of common effluent treatment plant in various industries.

Industry	*Measures*	*Benefits*
Hosiery	Manufacturing liquid detergent from spent liquor	Reduction in pollution load
Electroplating	Reactive rinsing by use of rinsewater after alkaline cleaning for rinsing components after pickling	Reduced water consumption and wastewater generation Effective cleaning of components
	Modification in dragout tray arrangements (hang the workpiece vertically instead of dragging out horizontally)	Better rinsing Reduced chemical consumption

(Cont'd...)

Industry	*Measures*	*Benefits*
	Installation of double plating line instead of Single line in plating bath	Enhanced production capacity
Cotton dyeing	Substituting soda ash by caustic soda in dyeing process	Reduced chemical consumption
Tannery	Installation of flow meter	Exact measurement of water/chemical
	Scrapping salt from hides before rinsing	Salt recovery, low total dissolved solid (TDS) in effluent
	Measurement of chemicals by weight and not by volume	Reduced wastage of chemicals
	Substitution of Benzidine based dye with non- Benzidine based dye stuff	Elimination of toxicity due to dye stuff
	Recycling of spent alkaline bath after screening for next batch	Reduction in chemical consumption and pollution load
Small pulp and paper	Increase in wire mesh size in de-pither	Increased pith recovery Decrease in chemical consumption

11.5 Cost sharing system

Various systems for cost sharing as exercised by common treatment provisions operating in India as well as in other countries are discussed below. The concerned management body can opt the most suited and viable system based on the local situation and feasibility to implement. The system exercised in most of CETPs operational in India is partly based on quality-quantity method with case specific modifications.

11.5.1 Quantity method

1. Simplest method and applicable for joint treatment of sewage and low strength industrial waste.
2. Operation cost and the cost for debt service are divided by total volume (quantity) of waste in a given time.
3. Advantageous where nature of waste from industries are generally uniform.

11.5.2 Quantity quality method

Charges are proportional to the benefits from the treatment plant. The computation of treatment cost can be summarised in following steps:

Step-1: Identify the critical design - parameters for the treatment plant. Let us assume they are BOD, SS and volume.

Step-2: Work-out fixed cost (depreciation, loan repayment, etc.), and operational cost on annual basis, for all major components like conveyance system, pumps, clarifiers, digesters, etc.

Step-3: Identify the variables (BOD, SS, volume) and extent of dependence the variable is considered for design specification for the given component, e.g., conveyance system and digesters are designed based primarily on volume and SS respectively, whereas for clarifier, volume and SS are equally considered.

Step-4: Work-out apportionment of the total treatment cost (fixed and operational), based on the function of the variables.

Step-5: Depending upon design specification of the plant, against critical parameters, work out the cost component for unit volume treated, kg BOD removed and kg SS removed.

Step- 6: Calculate the cost of treatment for effluent from any industry based on unit cost component as in Step-4 and effluent characteristics in terms of quantity (volume) and quality (BOD, SS, etc.), for the given industry.

11.5.3 Malz formulation

Based on 'polluter pays' principle. The factors considered in cost computation are volume, degree of noxiousness 'S' of wastewater and dilution factor 'V' to ensure survival of fish (48 hr. exposure).

$$V = A/A_0 + B/B_0 + TDS/\ TDS_0 + F\ 1$$

where,

A	=	Settleable matter, mg/L
B	=	COD of settled sample, mg/L
TDS	=	Total dissolved solids mg/L
F	=	Fish Toxicity
A_0, B_0, TDS_0	=	Standards

Noxiousness degree 'S' is estimated from dilution factor expressed in appropriate slabs.

Annual treatment cost apportioned to each polluter is calculated as follows:

Cost share	=	S.Q.Z./(S.Q.)
Q	=	Annual flow of an individual industry.
Z	=	Total cost of collection/conveyance and treatment.

11.5.4 Fukashiba formulation

Japanese system adopted for CETP catering to petrochemical industries. Charges levied are calculated differently on the basis of quantity and differently for quality.

While quantity rate is mutually decided and is binding to all member industries, the quality rate is determined based on BOD, COD, suspended solids and concentration of *N*-hexane extract in effluent from individual industry

11.5.5 Roman formulation

Based on the idea that:

1. Construction cost of CETP should correspond to the volume and strength of effluent from individual industry.
2. Conveyance cost should be proportional to the volume by each industry irrespective of its distance from CETP.

Provision of penalties in terms of transgression of 'discharge volume' regulated for each industry.

Incorporates Roman concept that economic aspect of pollution control should never be considered as the main factor influencing economic decision.

Different factors considered in cost computation are:

1. Average flow of effluent from industry to CETP.
2. Total flow of effluent reaching CETP.
3. Construction cost of CETP.
4. Construction cost of all additional facilities used for conveyance of effluent.
5. Average strength of effluent of the given industry expressed as BOD_5 in kg O_2/m^3.
6. Average strength of total effluent expressed as BOD_5 in kg O_2/m^3.
7. Cost of treatment plant units depending on the pollution load to construction cost of the whole treatment plant.

11.5.6 Chemtech formulation

This formulation is based on the proportion of the CETP usage.

Industries pay according to their volume of effluent and the influence of chemical, physical and biological conditions of the waste on capital and operational cost of CETP.

Cost of neutralisation at CETP is not considered as this is considered to be a part of pre-treatment by the industry.

11.5.7 Graduated payment formulation

Based on concept of graduated payment, whereby:

1. Large industries pay marginally higher cost.
2. Cost paid by smaller industries is fixed as minimal amount thus eliminating intricate accounting procedures.

3. The cost for large or medium industries can be calculated based on factors which include effluent generation of individual industry, total quantum of effluent reaching CETP and pollution factor 'P' as follows:

$$P = 0.5 \text{ (BODi + 200) } 600 + 0.5 \text{ (CODi + 500) } 1500$$

where, BODi and CODi are average BOD and COD in effluent from individual industry.

11.5.8 Flecksedar methodology

Cost apportioning can be attempted based on following system:

Cost be adjudged separately for designed and actual treatment capacity of CETP The type of treatment and the cost incurred on it, be levied from the industry(s) responsible for particular pollutant.

11.6 Design criteria for CETP

1. Inventory of industries.
2. Qualitative and quantitative characterisation of wastewater from industries.
3. Classification of industries based on wastewater generation.
4. Classification of wastewater based on bio-degradability.
5. Site-specific, effective and easy-to-maintain design of conveyance system.
6. Bench scale and pilot scale treatability study.
7. Segregation of wastewater.
8. Pre-treatment of wastewater.
9. Assessment of appropriate treatment technology.
10. Waste minimisation and resource recovery.
11. Disposal mechanism of treated effluent and sludge.
12. Estimation of treatment cost.
13. Cost benefit analysis.
14. Selection of best suited cost sharing pattern.
15. Stress on cleaner technologies.

11.7 Common problems and causes of CETP

11.7.1 Common problems

The common problems are given below:

1. Frequent disruption in the conveyance system resulting in overflow of effluent.
2. Stock loading and overloading of the CETP.

3. Corrosion of machinery and equipment, break-downs, below-rated-capacity performance of pumps, aerators, etc.
4. Inefficient performance of various treatment units of a CETP—much below the designed capacity.
5. Inconsistent results in the effluent discharged after treatment.
6. Difficulty in management and disposal of sludge generated.
7. Higher cost of treatment than anticipated.

11.7.2 Common causes

A CETP generally consists of the following units—pre-treatment systems, collection and conveyance system, screens, receiving sump, screens and grit remover, equalisation tank, chemical dosing (flash mixer), primary clarifier, anaerobic/aerobic treatment or two stage aerobic treatment, secondary clarifier, sludge thickener, sludge drying (mechanical or solar drying beds), discharge of effluent.

The efficiency of a CETP depends on the efficiency of each individual unit. If any of these units does not work properly it will be reflected in the operation of other units and the overall efficiency of the plant.

12

Environmental auditing

12.1 Introduction

Environmental auditing originated in the United States in the 1970s. At first reactive in focus, environmental considerations were dealt with by 'end of pipe' solutions. Control measures were heavily influenced by the need to reduce remediation costs and fines which might stem from industrial accidents and from the need to manage environmental liabilities.

Environmental auditing was introduced to the U.K., and elsewhere mainly by multi-national companies who began to apply the audit procedures corporately and via subsidiaries (horizontally). It is now being encouraged down the supply chain when large companies demand that their suppliers have green credentials (vertically). Environmental auditing has also become more proactive as organisations have recognised potential market and stakeholder benefits, efficiency gains, financial savings and the importance of improved public relations.

An environmental audit is a tool that enables companies to assess:

1. Their compliance with the environmental laws and the conditions of their environmental approvals or permits.
2. The effectiveness of their environmental plans, policies, practices, controls and records.

Environmental audits can also help companies identify ways to reduce waste, save money, improve overall efficiency and minimise liability risks. But the environmental laws don't require companies to conduct environmental audits. So many companies don't—and as a result, they miss out on many valuable opportunities.

Private sector environmental auditing has been defined as: A management tool comprising a systematic, documented, periodic and objective evaluation of the performance of the organisation, management system and processes designed to protect the environment with the aim of: (i) facilitating management control of practices which may have impact on the environment and (ii) assessing compliance with company policies' the systematic examination of the interaction between any business operation and its surrounding. This includes all emissions to air, land and water legal constraints; the effects on the neighbouring community, landscape and ecology and the public's perception of the operating company in the local area.

Many types of audit have been carried out by companies:

1. Compliance audit: The most common type of audit consisting of checks against environmental legislation and company policy.
2. Issues audit: An evaluation of how a company's activities relate to an environmental issue or (e.g., global pollution, energy use) or an evaluation of a specific issue (e.g., buildings, supplies).
3. Health and safety audit: An assessment of risks and contingency planning (sometimes merged with environmental auditing because of the interconnected impacts of industrial processes and hazards).
4. Site audit: An audit of a particular site to examine actual or potential environmental problems.
5. Corporate audit: An audit of the whole company and its polices, structures, procedures and practices.
6. Due diligence audit: An assessment of potential environmental and financial risks and liabilities carried out before a company merger or site acquisition or divestiture (e.g., contaminated land remediation costs).
7. Activity or operational audit: An assessment of activities that may cross company departments or units (e.g., energy or waste management).
8. Product or life cycle audit: An analysis of environmental impacts of a product throughout all stages of its design, production, use and disposal, including its reuse and recycling (cradle to grave).

The above have been described as partial environmental audits to distinguish them from the more specific environmental audit and, in particular, the periodic audit that forms a crucial step in Environmental Management Systems (EMS).

External audit: An assessment of the condition of the local environment, usually resulting in a State of the Environment Report (SOER).

Internal audit: Consisting of two areas:

1. Policy impact assessment: A review of the activities (objectives, services, practices and policies) of the authority.
2. Management audit: A review of the procedures and structures by which environmental policies are managed by the authority.

Strategic Environmental Assessment (SEA) has emerged in recent years as a way of appraising the environmental impacts of polices, programmes and plans. As such it is a method for policy impact assessment. Many local authorities in England and Wales have undertaken or are currently preparing one. In Scotland SEA was piloted in Gordon District in 1994, followed by Clackmannan District Council in 1995. SEA is also being explored as a way of identifying the potential environmental impacts of some of the structural fund programmes.

12.2 Environmental auditing practice and procedures

The more specific type of environmental audit involves the collection, collation, analysis, interpretation and presentation of information which is used to:

1. Assess performance against a set of requirements or targets, related to specific issues.
2. Evaluate compliance with environmental legislation and corporate policies.
3. Measure performance against the requirements of an environmental management system standard.

The systematic, periodic, documented and objective aspects of environmental auditing are fundamental to effectiveness. It is fast developing as an important and powerful tool in the corporate environmental assessment and management toolkit. The requirement periodically to repeat audits ensures that there is an ongoing commitment and a systematic process to improve environmental performance. The scope of repeat audits can also broaden to become more comprehensive as experience and expertise are accrued or as new issues or legislation emerge.

Sometimes the terms assessment, appraisal, monitoring or review have been used interchangeably with audit. Audit implies detailed statistical verification with a periodic cycle between audits. An assessment or review is usually a one-off event which is carried out in less detail and with less direct checking of data. Environmental reviews provide a baseline overview of current environmental effects or impacts, relevant environmental legislation and a statement of existing environmental performance. The reviews provide a basis for establishing a management action plan. They can become part of an environmental management system to help implement the plan. When they are undertaken as the first of a series of periodic environmental audits they have been referred to as a 'baseline environmental audit'. Environmental audits should be appropriate to the particular circumstances. As environmental auditing draws upon various methodologies, each organisation will define its own system depending upon its size, its activities and its corporate culture.

12.2.1 Pre-audit stage

1. Full management commitment.
2. Setting overall goals, objectives, scope and priorities.
3. Selecting a team to ensure objectivity and professional competence.

12.2.2 Audit stage

1. On-site audit, well defined and systematic using protocols or checklists.
2. Review of documents and records.

3. Review of policies.
4. Interviews.
5. Site inspection.

12.2.3 Post-audit stage

1. Evaluation of findings.
2. Reporting with recommendations.
3. Preparation of an action plan.
4. Follow-up.

There is an increasing demand for the results of auditing to be disclosed. Recent European initiatives on access to environmental information and the requirement of the eco-management and audit scheme for participants to publish environmental statements confirm the importance of this.

12.2.4 Need for conducting environmental audit

In general, no environmental law in Canada requires companies to conduct environmental audits as a matter of course. There are two jurisdictions in which a company may be required to conduct an environmental audit in certain circumstances. Under the federal CEPA, a court may require a company to conduct an environmental audit after being convicted of an environmental violation. And under Alberta's Environmental Enhancement and Protection Act, an environmental protection order may require a company to conduct an environmental audit.

Avoid liability: Some environmental laws encourage companies to conduct environmental audits by providing incentives.

Even if your company isn't in one of these four jurisdictions, an environmental audit will help uncover violations and give the company a chance to address them before the government gets involved and decides to prosecute.

Prove due diligence: Due diligence requires companies to prove they took all reasonable efforts to address known risks, prevent violations and ensure compliance with the environmental laws. Environmental audits are ideal for helping companies meet that burden. After all, one of the purposes of an audit is to measure the company's compliance with the environmental laws and the effectiveness of its EHS programme. So if your company ever gets prosecuted, an audit programme will be a major boost to a due diligence defence. More importantly, if done correctly, environmental audits help prevent prosecutions to begin with by enabling you to ferret out and correct violations and improve the company's environmental policies and programmes.

Save money: Lastly, environmental audits can help identify costly problems, such as inefficiencies in the company's operations. Fix these problems and the company will benefit financially.

ISO 14001

Many companies have implemented environmental management systems (EMSs) and had them certified as compliant with ISO 14001, a standard for EMSs developed by the International Organisation for Standardisation, or informally adopted the ISO 14001 principles without getting formal certification.

12.3 Environmental auditing and environmental management systems (EMS)

An environmental management system is part of an enterprise's overall management system. It includes the organisational structure, planning activities, responsibilities, practices, procedures, processes and resources for implementing and maintaining environmental management. It includes those aspects of management that plan, develop, implement, achieve, review, maintain and improve the enterprise's environmental policy, objectives and targets.

12.3.1 Role of environmental management system (EMS)

Role of an environmental management system is intended to help an enterprise:

1. Identify and control environmental aspects, impacts and risks relevant to the organisation.
2. Achieve its environmental policy, objectives and targets, including compliance with environmental legislation.
3. Define a basic set of principles that guide the organisation's approach to its environmental responsibilities in the future.
4. Establish short, medium- and long-term goals for environmental performance, making sure to balance costs and benefits for the organisation and for its various shareholders and stakeholders.
5. Determine what resources are needed to achieve these goals, assign responsibility for them and commit the necessary resources.
6. Define and document specific tasks, responsibilities, authorities and procedures to ensure that every employee acts in the course of his daily work to help minimise or eliminate the enterprise's negative impact on the environment.
7. Communicate these throughout the organisation and train people to effectively fulfil their responsibilities.
8. Measure performance against pre-agreed standards and goals and modify the approach as necessary.

The integration of environmental management into the overall management function of an organisation is critical because environment is one among a number of external issues impacting the enterprise. A stand alone environmental management system will, therefore, not be effective.

EMS can be described as a planned and co-ordinated set of management actions, operating procedures, documentation and record-keeping, implemented by a specific structure with defined responsibilities, accountabilities and resources and aimed at the prevention of adverse environmental effects as well as the promotion of action and activities that preserve and/or enhance environmental quality.

An environmental management system follows the well-known quality management approach of 'plan, do, check, improve'. It is a problem identification and problem-solving tool, which can be implemented in an organisation in many different ways, depending on the precise sector of activity and the needs perceived by the management. The specific system implemented depends entirely on the needs and objectives of the organisation.

12.3.2 Origin of EMS

Enterprises have adopted a range of responses to environmental challenges from doing nothing, to crisis response, to integration of environmental management into the overall management of the enterprise via a well-defined environmental management system. Environmental management systems have benefited from the development of and the experience with two separate management tools over the past 15 years:

1. Increased environmental awareness led companies in North America and Europe to develop environmental auditing as a management tool to identify environmental problems and to monitor the company's environmental performance similar to the way a financial audit is used to measure financial performance. The first goal was to ensure the company's compliance with environmental laws and regulations. Later, the scope was extended to cover the monitoring of 'best management practices' for environmental vulnerabilities.
2. 'Total Quality Management' (TQM) concepts, although originally aimed at reducing and eventually eliminating defects (non-compliance with specifications) in manufacturing and at improving the efficiency of business processes, have increasingly been applied to managing environmental issues.

Environmental auditing was developed by companies such as Westinghouse, Philips, ICI and others in response to the mounting cost of not complying with environmental health and safety legislation in highly regulated countries such as the U.S., Canada and in Western Europe. A 'compliance audit' monitors compliance with laws and regulations. A 'due diligence' or 'pre-acquisition' audit is used by enterprises to identify environmental vulnerabilities and problems of a site or enterprise before any investment is made. The results of the pre-acquisition audit may affect the level of investment, the purchase price, future investments and operating costs connected with the site, or even the ultimate decision about the investment.

Today, the term environmental audit has expanded and is sometimes used to mean either the initial environmental review to identify an enterprise's environmental impacts and issues, or the environmental management system audit. The environment review is performed to identify risks, problems and environmental opportunities when no environmental management system is in place; it is similar to the pre-acquisition audit. The EMS audit is a tool to determine whether the environmental management system conforms to the organisation's planned arrangements and whether it has been properly implemented and maintained. The origins of TQM derive from industrial efficiency, experts who systematically began investigating the working methods. It has been applied to industrial redevelopment and to environmental management throughout the world. Historically, enterprises have managed environmental issues in three ways:

1. No action: They do not recognise environmental opportunities or threats until it is too late.
2. Reactively: Management waits for events and issues to be decided externally before taking steps to deal with them. These enterprises may benefit in the short term, but they will never be certain of approaching problem areas and are ill-equipped when problems occur.
3. Proactively: These enterprises monitor environmental issues and concerns as a part of their daily business and formulate responses before a situation becomes critical. This policy does not mean that these companies escape difficulty, only that they are better prepared to deal with crisis situations.

12.3.3 Purpose of EMS

An environmental review or an audit has at least one serious drawback: although it is a very valuable management tool, it can only describe the environmental situation of the enterprise at the time the review or audit is carried out. The follow-up of non-conformities is not a guarantee that after some time the same or other deficiencies in the organisation's control of environmental aspects will not occur. Some form of management system is, therefore, needed to ensure that the enterprise's environmental targets and objectives are pursued effectively. Furthermore, because relationship between and even within, different organisational functional units are often complex, there is a need to have a system of co-ordination.

An EMS, therefore, is intended to link these different processes through a network of management actions, procedures, documentation and records with the aim to:

1. Identify and control significant environmental vulnerabilities and effects.
2. Identify significant environmental opportunities.
3. Identify relevant environmental regulatory requirements.
4. Establish a sound environmental policy and basis for environmental management.
5. Establish priorities, determine objectives and work towards their achievement.
6. Monitor performance and evaluate the effectiveness of the system, including promotion of system improvements and adaptation to meet new and changing conditions and demands.

The communication of an enterprise's environmental objectives and results to external audiences is also gaining importance in environmental management, but is normally not considered to be a core element in an EMS.

12.4 Public sector environmental auditing

Increasingly, public sector bodies and local authorities are adapting auditing methods to establish baselines of environmental performance. These then inform management action.

Interest and action was stimulated in the late 1980s by the Friends of the Earth. A number of authorities prepared environmental charters, follow-up environmental strategies and action plans, which are generally referred to as Green Plans. It was not long before leading authorities also realised the greater corporate performance and environmental benefits of the broader and deeper approaches of the internal and external auditing.

In 1989, the first local authority environmental audit was undertaken by Kirklees District Council with the assistance of Friends of the Earth. Since then a number of Scottish local authorities have produced environmental audits, notably Fife and Grampian Regional Councils, Ross and Cromarty, Gordon, Falkirk, Clackmannan and Dundee District Councils. The scope of public sector audits is different from that of the industrial sector in that the effects of service provision are considered as well as the direct effects of the activities of the local authority:

1. Direct effects: Environmental impacts that result from the way in which day to day activities are undertaken. Direct effects are covered by internal management audit.

2. Service effects: Environmental impacts that result from the organisations efforts in implementing environmental policies and objectives. They would be covered by the policy impact assessment type of audit.

Additionally, audit techniques have been adapted to prepare State of the Environment Reports. After several authorities, including Fife Regional Council and Ross and Cromarty District Council, piloted the approach, the DoE published a guide to implementing environmental management systems within local authorities. It is referred to as UK-EMAS, as it was derived from the European eco-management and audit scheme.

Typically public sector audits cover a number of target areas such as:

1. Energy use.
2. Recycling.
3. Hardwoods from sustainable sources.
4. Environmental education.
5. Habitat conservation and creation.
6. Green purchasing.
7. Traffic calming.

Inevitably there are issues which may have been overlooked or might be misinterpreted and further and clearer guidance will be necessary. Scottish Natural Heritage (SNH) along with the Countryside Commission and the Countryside Council for Wales have commissioned guidance on the treatment of countryside and conservation issues within state of the environment reports.

12.5 Benefits of auditing

While environmental audits are designed to identify environmental problems, there may be widely differing reasons for undertaking them: compliance with legislation, pressure from suppliers and customers, requirements from insurers or for capital projects, or to demonstrate environmental activities to the public. The benefits of environmental auditing include:

1. Ensuring compliance, not only with laws, regulations and standards, but also with company policies and the requirements of an Environmental

Management system (EMS) standard:

1. Enabling environmental problems and risks to be anticipated and responses planned.
2. To demonstrate that an organisation is aware of its impact upon the environment through providing feedback.
3. Increased awareness amongst stakeholders.
4. More efficient resource use and financial savings.

12.6 Trends/future developments

Audit programmes are becoming a standard environmental management tool and pressures for the disclosure of audit results are increasing. Public statements of environmental information with external validation are required by those participating in the European or local authority eco-management and audit scheme.

The utility of environmental audits vary from organisation to organisation. It is likely that audits will be used increasingly to:

1. Provide baseline information to enable organisations to evaluate and manage environmental change, threat and risk.
2. Form the basis for initiating and monitoring the performance of environmental management systems.
3. Contribute to environmental management approaches which become integrated with environmental impact assessment and the management of predicted impacts, mitigation and monitoring measures.
4. Support the implementation and management of integrated pollution control procedures and assist in the definition of 'best practicable environmental options' (BPEO).
5. Tackle external off-site impacts which consider the broader environmental footprint of an organisation's activities.
6. Pass environmentally responsible approaches down the supply chain.

Environmental audits have traditionally dealt with the environmental effects of industrial processes and to a lesser extent, with resource consumption. Guided by the legislation and compliance procedures, the environment has usually been considered in terms of air land and water. Considerable conservation benefits could be achieved by broadening the focus of auditing to include natural heritage features and objectives. This would include natural heritage legislation and by the application of audit techniques to habitats and land use, such as farm units, forest management units, or sporting estates. Generic approaches could contribute to the development of conservation management plans.

13

Environmental impact assessment

13.1 Introduction

Environmental impact assessment is a process analysing the positive and negative effects of a proposed project, plan or activity on the environment. The specific purpose of the assessment is to provide the decision makers with information allowing them to introduce environmental protection considerations in the decision-making process leading to the approval, rejection or modification of the project, plan or activity under examination.

Thus, the Environmental Impact Assessment (EIA) process is an interdisciplinary and multistep procedure to ensure that environmental considerations are included in decisions regarding projects that may impact the environment. Simply defined, the EIA process helps identify the possible environmental effects of a proposed activity and how those impacts can be mitigated.

The requirements for environmental impact assessments of projects or plans which may have adverse effects on the human environment are today embodied in the national legislation of many countries and in a number of international agreements. However, the procedures used for preparation of such assessments vary considerably from country to country and at present there is no agreement, on a global or regional level, on procedures which may lead to comparable results.

The purpose of the EIA process is to inform decision-makers and the public of the environmental consequences of implementing a proposed project. The EIA document itself is a technical tool that identifies, predicts and analyses impacts on the physical environment, as well as social, cultural and health impacts. If the EIA process is successful, it identifies alternatives and mitigation measures to reduce the environmental impact of a proposed project. The EIA process also serves an important procedural role in the overall decision-making process by promoting transparency and public involvement. It is important to note that the EIA process does not guarantee that a project will be modified or rejected if the process reveals that there will be serious environmental impacts.

In some countries, a decision-maker may, in fact, choose the most environmentally-harmful alternative, as long as the consequences are disclosed in the EIA. In other words, the EIA process ensures an informed decision, but not necessarily an environmentally beneficial decision.

13.2 Benefits of the EIA process

1. Potentially screens out environmentally-unsound projects.
2. Proposes modified designs to reduce environmental impacts.
3. Identifies feasible alternatives.
4. Predicts significant adverse impacts.
5. Identifies mitigation measures to reduce, offset or eliminate major impacts.
6. Engages and informs potentially affected communities and individuals.
7. Influences decision-making and the development of terms and conditions.

13.3 Preparing an EIA

Depending on the EIA system, responsiblity for producing an EIA will be assigned to one of two parties: (i) the government agency or ministry and (ii) the project proponent. If EIA laws permit, either party may opt to hire a consultant to prepare the EIA or handle specific portions of the EIA process, such as public participation or technical studies.

Some EIA laws recognise the inherent conflict of interest produced when a mining company or other project proponent hires a consultant to prepare an EIA. Using a consultant carries the risk that the document will be biased in favour of proceeding with the project. If a consultant is hired by the mining company, conflicts may arise if the consultant believes it will receive future work if the project is approved, or even indirect benefits from related activities (e.g., consulting work for a port where ore will be exported). Some laws require consultants to be registered with the government and/or professionally accredited in EIA preparation. In some instances, a consultant may be required to file a statement disclosing any financial or other interest in the outcome of the project. The basic elements of good EIA practice are shown in Fig. 13.1.

13.4 Stages of the EIA process

The EIA process, while not uniform from country to country, generally consists of a set of procedural steps culminating in a written impact assessment report that will inform the decision-maker whether to approve or reject a proposed project. The flowchart below depicts the basic elements of good EIA practice.

Identifying and defining the project or activity: Although this step may seem relatively simple, defining a 'project' for the purposes of an EIA can become complex and even controversial if a mining project is large, has several phases, or involves multiple sites. The goal of this step is to define the project with enough specificity to accurately determine the zone of possible impacts and to include activities that are closely connected with the proposal so that the entire scope of environmental impacts is evaluated.

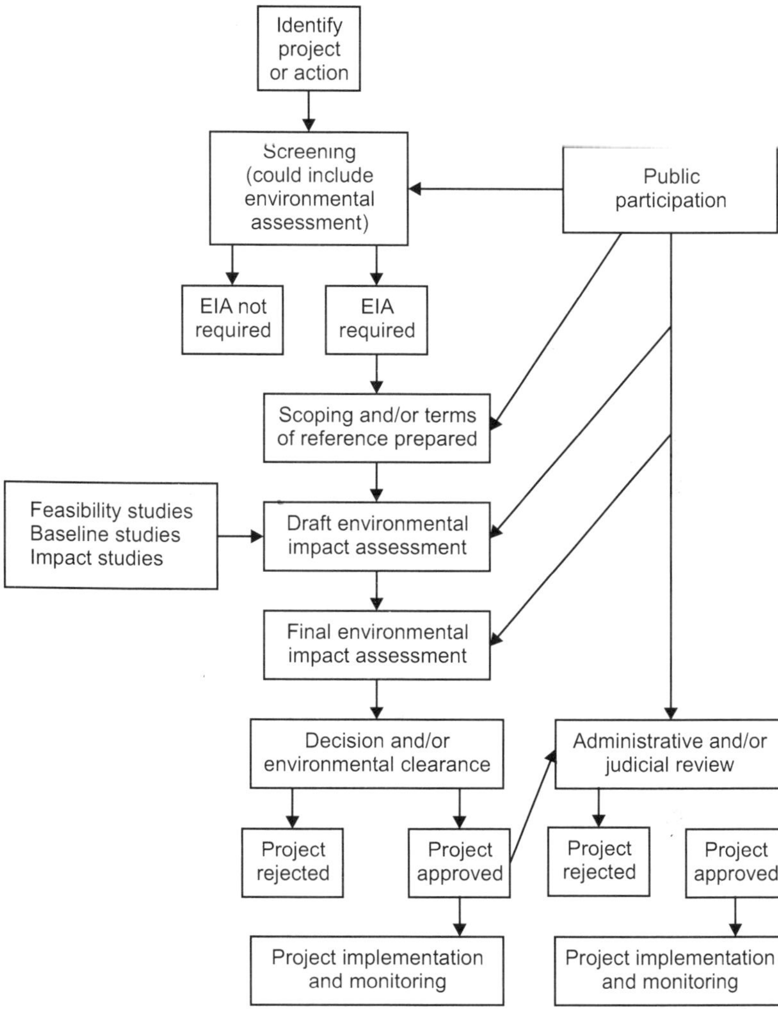

Figure 13.1: The basic elements of good EIA practice.

Screening: The screening process determines whether a particular project warrants preparation of an EIA. The threshold requirements for an EIA vary from country to country – some laws provide a list of the types of activities or projects that will require an EIA, others require an EIA for any project that may have a significant impact on the environment or for projects that exceed a certain monetary value.

In some cases, particularly if the possible impacts of a project are not known, a preliminary environmental assessment will be prepared to determine whether the project warrants an EIA.

Scoping: Scoping is a stage, usually involving the public and other interested parties, that identifies the key environmental issues that should be addressed in an EIA. This step provides one of the first opportunities for members of the public or NGOs to learn about a proposed project and to voice their opinions. Scoping may also reveal similar or connected activities that may be occurring in the vicinity of a project, or identify problems that need to be mitigated or that may cause the project to be cancelled.

Preparing terms of reference: The terms of reference serve as a roadmap for EIA preparation and should ideally encompass the issues and impacts that have been identified during the scoping process. A draft terms of reference may be made available for public review and comment. Public review at this early stage of the process provides a key opportunity to ensure that the EIA is properly framed and will address issues of community concern.

Preparing draft EIA: A draft EIA is prepared in accordance with the terms of reference and/or the range of issues identified during the scoping process. The draft EIA must also meet the content requirements of the overarching EIA law or regulations. This step will ideally engage a wide range of technical specialists to evaluate baseline conditions, predict the likely impacts of the project and design mitigation measures.

Public participation: Best EIA practice involves and engages the public at numerous points throughout the process with a two-way exchange of information and views. Public participation may consist of informational meetings, public hearings and opportunities to provide written comments about a proposed project. However, there are no consistent rules for public participation among current EIA systems. Even within a particular country, there can be variations in the quality and extent of public involvement in the EIA process, depending on the type of project being considered, the communities that may be affected, or government agencies that are overseeing the project.

Preparing final EIA: This step produces a final impact assessment report that addresses the viewpoints and comments of the parties that reviewed the draft EIA. These comments may prompt revisions or additions to the text of the draft EIA. In some cases, the final EIA will contain an appendix summarising all of the comments received from the public and other interested parties and provide responses to those comments.

Decision: A decision to approve or reject a mining project is generally based on the final EIA, but in some instances, an environmental clearance may be just one step in the mine permitting process. The decision may be accompanied by certain conditions that must be fulfilled, such as posting a reclamation bond or filing an environmental management plan.

Administrative or judicial review: Depending on the jurisdiction, there may be opportunities for a party to seek administrative and/or judicial review of the final decision and the EIA process. An appeal may address procedural flaws in the EIA process, such as a failure to hold any required public hearings, or may point to substantive issues that the decision-maker failed to consider. A country's judicial review or administrative procedure act, or sometimes the EIA law itself, will usually identify the kinds of issues that can be raised in an appeal and the type of relief that may be granted.

Project implementation: Provided all regulatory requirements are met and permits are obtained, mine development will proceed following the project decision and once opportunities for administrative and/or judicial review are exhausted.

Monitoring: Monitoring is an important part of project implementation. Monitoring serves three purposes: (i) ensuring that required mitigation measures are being implemented, (ii) evaluating whether mitigation measures are working effectively and (iii) validating the accuracy of models or projections that were used during the impact assessment process.

13.5 Goals and principles of EIA

EIA means an examination, analysis and assessment of planned activities with a view to ensuring environmentally sound and sustainable development.

The EIA goals and principles set out below are necessarily general in nature and may be further refined when fulfilling EIA tasks at the national, regional and international levels.

13.5.1 Goals

1. To establish that before decisions are taken by the competent authority or authorities to undertake or to authorise activities that are likely to significantly affect the environment, the environmental effects of those activities should be taken fully into account.
2. To promote the implementation of appropriate procedures in all countries consistent with national laws and decision-making processes, through which the foregoing goal may be realised.
3. To encourage the development of reciprocal procedures for information exchange, notification and consultation between States when proposed activities are likely to have significant transboundary effects on the environment of those states.

13.5.2 Principles

Principle 1: States (including their competent authorities) should not undertake or authorise activities without prior consideration, at an early stage, of their environmental effects. Where the extent, nature or location of a proposed activity is such that it is likely to significantly affect the environment, a comprehensive environmental impact assessment (EIA) should be undertaken in accordance with the following principles.

Principle 2: The criteria and procedures for determining whether an activity is likely to significantly affect the environment and is therefore subject to an EIA, should be defined clearly by legislation, regulation, or other means, so that subject activities can be quickly and surely identified and EIA can be applied as the activity is being planned.

For instance, this principle may be implemented through a variety of mechanisms, including:

1. List of categories of activities that by their nature are, or are not, likely to have significant effects.
2. Lists of areas that are of special importance or sensitivity (such as national parks or wetland areas), so that any activity affecting such areas is likely to have significant effects.
3. Lists of categories of resources (such as water, tropical rain forests, etc.), or environmental problems (such as increased soil erosion, desertification, deforestation) which are of special concern, so that any diminution of such resources or exacerbation of such problems is likely to be 'significant'.
4. An 'initial environmental evaluation', a quick and informal assessment of the proposed activity to determine whether its effects are likely to be significant.
5. Criteria to guide determinations whether the effects of a proposed activity are likely to be significant.

If a listing system is used, it is recommended that states reserve the discretion to require the preparation of an EIA on an adhoc basis, to ensure that they have the flexibility needed to respond to unanticipated cases:

Principle 3: In the EIA process the relevant significant environmental issues should be identified and studied. Where appropriate, all efforts should be made to identify these issues at an early stage in the process.

Principle 4: An EIA should include, at a minimum:

1. A description of the proposed activity.
2. A description of the potentially affected environment, including specific information necessary for identifying and assessing the environmental effects of the proposed activity.
3. A description of practical alternatives, as appropriate.
4. An assessment of the likely or potential environmental impacts of the proposed activity and alternatives, including the direct, indirect, cumulative, short-term and long-term effects.

5. An identification and description of measures available to mitigate adverse environmental impacts of the proposed activity and alternatives and an assessment of those measures.
6. An indication of gaps in knowledge and uncertainties which may be encountered in compiling the required information.
7. An indication of whether the environment of any other state or areas beyond national jurisdiction is likely to be affected by the proposed activity or alternatives.
8. A brief, non-technical summary of the information provided under the above headings.

Principle 5: The environmental effects in an EIA should be assessed with a degree of detail commensurate with their likely environmental significance.

Principle 6: The information provided as part of EIA should be examined impartially prior to the decision.

Principle 7: Before a decision is made on an activity, government agencies, members of the public, experts in relevant disciplines and interested groups should be allowed appropriate opportunity to comment on the EIA.

Principle 8: A decision as to whether a proposed activity should be authorised or undertaken should not be taken until an appropriate period has elapsed to consider comments pursuant to principles 7 and 12.

Principle 9: The decision on any proposed activity subject to an EIA should be in writing, state the reasons therefore and include the provisions, if any, to prevent, reduce or mitigate damage to the environment. This decision should be made available to interested persons or groups.

Principle 10: Where it is justified, following a decision on an activity which has been subject to an EIA, the activity and its effects on the environment or the provisions (pursuant to principle 9) of the decision on this activity should be subject to appropriate supervision.

Principle 11: States should endeavour to conclude bilateral, regional or multilateral arrangements, as appropriate, so as to provide, on the basis of reciprocity, notification, exchange of information and agreed-upon consultation on the potential environmental effects of activities under their control or jurisdiction which are likely to significantly affect other States or areas beyond national jurisdiction.

Principle 12: When information provided as part of an EIA indicates that the environment within another state is likely to be significantly affected by a proposed activity, the state in which the activity is being planned should, to the extent possible:

1. Notify the potentially affected state of the proposed activity.

2. Transmit to the potentially affected state any relevant information from the EIA, the transmission of which is not prohibited by national laws or regulations.
3. When it is agreed between the states concerned, enter into timely consultations.

Principle 13: Appropriate measures should be established to ensure implementation of EIA procedures.

13.6 Advantages and limitations of environmental impact assessment

Public discussion and participation: One major strength and outstanding feature of EIA in many countries is increased public discussion and participation. However, in practice a number of limitations occur. In some instances where public participation is required by EIA legislation, such participation could be limited in practice. For example, the law in India required that a public hearing is conducted; however, NGOs often considered the public hearing as a staged process that appeared to involve citizens when the decision had already been made.

In response, the 2006 EIA notification changed the requirement from public hearing to public consultation in an attempt to force project proponents to proactively seek the views of affected communities. In other instances, where projects are only discussed openly for their potential environmental merits and shortcomings, but not for social or economic ones, public discussion could focus on allegedly environmental problems when there are underlying social or economic reasons for delaying or stopping a project.

Over-reliance on EIA to achieve environmental management objectives: EIA should be used in conjunction with other policy tools and should not be over-emphasised for achieving environmental management objectives. For example, in El Salvador, the National Environment Law introduced nine instruments of environmental policy including, among others, environmental zoning, environmental evaluation, environmental information and public participation and economic and financial incentives.

However, the implementation of the environmental policy focused on the EIA and this led to overburdening the tool. Further development of other instruments was recommended to complement a comprehensive environmental policy. South Africa, where EIA is applied within the framework of an integrated environmental management system, the strategic plan for the environment sector has called for developing a broader toolkit‘ of environmental impact management approaches and instruments in addition to EIA in order to address the issue of over-reliance on EIA.

Time-intensive process: An effective screening system is important to ensure that the EIA does not become a prerequisite for too many activities, thus limiting the responsible authority's ability to adequately assess important projects and weakening the legitimacy of EIA. For example, in Colombia the lack of a screening procedure resulted in an EIA being required for any project, regardless of the intensity and magnitude of its potential impacts, location, the sensitivity of the surrounding area, or public opinion regarding the project. The need for vast numbers of EIAs coupled with an absence of baseline environmental data resulted in mass production of EIAs of poor quality and little value.

Lack of screening has led to a situation where the EIA regulations become a hurdle to projects and add little value to the environmental planning and management processes for which they were intended. Similarly, in El Salvador, broad screening procedures have led to a situation where 300–400 EIAs were prepared annually. The number was beyond the capacity of the responsible reviewing authority (MARN) and resulted in a backlog of 2500 EIAs and led EIA to becoming a bottleneck to development.

Potential for rent seeking: Where the EIA is linked to the licensing process, it may become a tool for rent seeking and could be a source of illegal influence on the responsible authorities. For example, in Bangladesh, the Environment Conservation Act required environmental clearance for development projects, but it also reserved the right to the government to waive the clearance requirement. This provided the incentive for project proponents to exert influence to avoid this requirement. Because public consultation and public participation were limited, these two forces for were precluded from countering the illegal influence and they were blocked from providing an incentive for the effective identification and mitigation of potential environmental impacts.

14

Industrial biotechnology and climate change

14.1 Introduction

Carbon dioxide (CO_2) has the second-largest direct contribution to the greenhouse effect of all gases. Since the beginning of the Industrial Revolution, the burning of fossil fuels has contributed to the increase in carbon dioxide in the atmosphere. Moreover, national CO_2 emissions are strongly linked to wealth and a global strategic goal should be to prevent growth at the expense of sustainability. A large proportion of CO_2 production arises from industrial activities. Therefore, meeting market demands whilst reducing the impact on climate change is critically important for industry.

To meet the challenge of Greenhouse Gas (GHG) emissions reductions, industry has responded in most countries and as a result GHG emissions from industry have substantially decreased in recent years. However, the targets for reductions are very ambitious and more remains to be done. Thus the idea of 'sustainable growth' has taken hold. New technologies are being developed to counter GHG emissions and the new industrial biotechnology should be seen as part of the technology toolkit.

Industrial biotechnology, based on renewable resources, can save energy and significantly reduce CO_2 emissions. It is an embryonic industry, but has already proven its worth in climate change mitigation. It holds much greater promise for the future by avoiding the use of fossil raw materials. It involves the use of enzymes and micro-organisms to make biobased products in a diverse variety of industry sectors. The feedstocks are agricultural biomass and organic waste materials, even wastewaters.

Many drivers of industrial biotechnology are clearly linked to the global challenges of climate change, energy security and the financial crisis, yet there are still many barriers to its growth and optimal uptake across industry sectors. Technical barriers are not as daunting as they once were, as biology research is maturing rapidly. Progress in DNA sequencing, strides in proteomics and the emergence of the field of synthetic biology are leading to unprecedented progress in the biological sciences.

The new industrial biotechnology arose from international interest in the production of biofuels from a variety of feedstocks. Many countries now have bioenergy strategies and targets. In short order, many supportive policies were developed for the production and utilisation of biofuels, particularly bioethanol.

However, very quickly the biofuels boom courted controversy, especially the food versus fuel debate. Now there is a shift in policy towards second generation biofuels using non-food crops as feedstocks.

Besides biofuels, industrial biotechnology can contribute to climate change mitigation through diverse products in the plastics and chemicals sectors. These products are less controversial, some are closer to market than second generation biofuels and yet do not enjoy the wealth of supportive supply and demand policies as seen with biofuels.

Several sources predict that biobased chemicals could in the near future occupy a much larger market share than at present. As befits the remarkable biodiversity of micro-organisms, the diversity of potential biobased chemicals is wide. In many cases, organic chemistry has no feasible replacement; in other cases, biobased chemicals can replace their fossil-based counterparts with significant GHG emissions reductions. Climate change is challenging almost all human endeavours, including the future ways in which energy will be generated and consumed. It could adversely affect water supplies and agricultural productivity and the need to cut CO_2 emissions to avoid harmful environmental degradation has made the transition from conventional fossil fuels to alternative and renewable resources a global priority.

Meeting market demands whilst at the same time reducing the impact on the climate is of critical importance to industry. Although past discussions have centred around the causes of climate change, society, industry, governments and other stakeholders are now searching for solutions to mitigate the impacts on health and the environment. In most countries, industrial greenhouse gas (GHG) emissions have substantially decreased over recent years. Both regulators and industry have responded to the need to work towards a sustainable future and have focused on developing and innovating technological improvements for the benefit of the environment. However, in spite of these endeavours, current practices and technologies will most likely be insufficient or inadequate to achieve the ambitious objectives set by countries to tackle climate change. Industries should take bold steps towards establishing more sustainable growth and should be given the opportunity to implement new technologies including the use of biotechnology to reduce the carbon footprint.

14.2 Industrial biotechnology as part of the solution to climate change

Since the industrial revolution, economic growth has been inextricably linked with accelerating negative environmental impact. Industrial biotechnology challenges this pattern and has the potential to break the cycle of resource consumption by allowing for a rethinking of traditional industrial processes. Industrial biotechnology, by providing a range of options for competitive industrial performance in selected sectors, could enhance economic growth, while at the same time save water, energy, raw materials and reduce waste production.

Industrial biotechnology, based on renewable resources, can save energy in production processes and significantly reduce CO_2 emissions. The impacts of biotechnology on industry are confirmed by scientific studies on the application of biotechnology to industrial sustainability and, more recently, by a EuropaBio report on the potential of industrial biotechnology to help mitigate climate change. It has been claimed that the full climate change mitigation potential of biotechnology processes and biobased products ranges from between 1 billion and 2.5 billion tons CO_2 equivalent per year by 2030.

14.3 Role of industrial biotechnology in the biobased economy

Industrial biotechnology, also known as white biotechnology, uses enzymes and micro-organisms to make biobased products in sectors as diverse as chemicals, food and feed, healthcare, detergents, paper and pulp, textiles and energy. Agricultural products, biomass and organic waste, including food processing waste and effluents (also referred to as renewable raw materials) are transformed into other substances, in the same way as crude oil is used as a feedstock in the production of chemicals.

Existing energy infrastructure and production processes are largely based on fossil fuels, which result in high levels of GHG emissions. In a biobased economy society is no longer wholly dependent on fossil fuels and industrial raw materials.

By contrast, industrial biotechnology holds promise by avoiding the use of limited fossil resources as starting materials, but in some instances it competes with edible feedstocks. This important issue, specially raised in the case of biofuels, can be solved by the introduction of second generation biofuels using non-edible biomass as a sole feedstock. Besides biofuels, the wide variety of intermediate products that may enter at different stages in different value chains introduces complexity when analysing biotechnological products.

Industrial biotechnology applications are given below:

1. Improved enzymes for a growing range of applications in the chemical sector.
2. Improved micro-organisms that can produce an increasing number of chemical products in one step, some of which build on genes identified through bioprospecting.
3. Biosensors for real-time monitoring of environmental pollutants and biometrics for identifying people.

4. High energy-density biofuels produced from sugar cane and cellulosic sources of biomass.
5. Greater market share for biomaterials such as bioplastics, especially in niche areas where they provide some advantage.

In addition, significant growth in the biobased polymers sector will result from the development of new polymers with new properties, greater incentives to reduce costs through the use of renewable materials and increasing regulatory pressure to reduce carbon footprint (for example for packaging applications). In addition, enzymes will be used increasingly in a growing range of applications, due to improvements and advantages, particularly in the food, cosmetic and textile industries, in line with customer requirements and stricter environmental regulations.

14.4 Industrial biotechnology: Greening the economy

14.4.1 Current trends in technology applications

Industrial biotechnology has been established for decades in the production of biofuels, biochemicals for the pharmaceutical markets, food and feed, fine chemicals, detergents and hygiene products. Bioethanol production is increasing for biofuel, with worldwide annual growth rates of more than 10%. Biopolymers (e.g., poly-lactic acid, PLA) are emerging products with diversifying applications. Most of these established products are available only through biotechnological processes because chemical synthesis offers no alternative.

14.4.2 Emerging trends

Chemical or biochemical products obtained using biotechnological processes need not only to be developed but their quality, production capacity and share of the market have to be assured and improved. Certain building block (platform) chemicals derived from sugar sources may potentially replace petrochemical building blocks; this means the development of biotechnological platform intermediates based on renewable carbon sources.

Future synthesis of polyesters and biobased biodegradable polymers may be achieved by using building blocks such as fumaric, succinic or malic acid, among others, which are used at present in the food industry.

Ethanol and succinate are good examples of precursors which, after they are fermented, isolated and purified, may enter into synthetic processes to be transformed into ethylene and subsequently polyethylene products and polyesters, polyurethanes and polyamides, respectively. Thus the combination of biotechnological and chemical technologies may be critical in the future.

When and if these biotechnology processes and applications become more economically and environmentally sustainable, the industrial relevance of biotechnology will change dramatically. This development opens not only a potentially broad field of new applications, but would also require a combination of biology and clean chemistry processes. For example, one area of great interest currently is the field of biopolymers.

It can be predicted that the growing sugar consumption for the production of (bio)-chemicals will compete with the food and increasingly the biofuel industry. The industrial biotechnology industry should avoid competing with biomass intended for food production. Lignocellulose as carbon source from biomass, plant breeding for industrial purposes and algae production may represent ways to prevent this competition. The concept of biorefineries as a combination of integrated plants addressing:

1. Processing and fractionation of renewable raw materials.
2. Transforming feedstocks to various products.
3. Recycling and integrated valorisation of the products after use where possible is becoming a promising strategy.

Finally, synthetic biology represents a very important step forward, since it allows designing chemicals that would not occur by natural pathways. In this sense the tailor-made synthesis of products by modified micro-organisms opens new possibilities for the production of chemicals for widely differing purposes. There are compelling reasons to believe that synthetic biology will strongly influence the bioresearch agenda in the 21st century and in fact may be the vehicle that moves biotechnology into the economic mainstream.

14.4.3 From incremental to radical innovation

During the last decade there has been a tangible shift from incremental innovation in the industrial biotechnology field, where some specific steps in existing technologies were replaced to save energy or to make the overall production process more efficient, towards more radical innovation with the development of new production processes entirely based on renewable resources. This makes industrial biotechnology today one of the rare technologies which might help to radically change production chains in terms of sustainability along the whole product life cycle.

Industrial biotechnology techniques were first introduced within traditional industries, such as the food industry, where enzymes and yeasts are used in many food processing techniques, resulting in an increased efficiency and reduced environmental and GHG impact. The experiences developed in the food industry have found application in a number of traditional industries, such as the pulp and paper, leather and textile sectors. Enzymes and micro-organisms can carry out processes that traditional manufacturing systems would perform with much higher quantities of energy, using aggressive chemicals and producing significant amounts of potentially hazardous waste products. In addition, the biotechnological alternatives can deliver productivity improvements and can help to reduce the impact of GHG emissions.

The feedstock processing and fermentation technologies developed in traditional industries were essential tools for the development of industrial biotechnology solutions for the transformation of agricultural feedstock into biofuels such as bioethanol. The largest share of ethanol production is obtained from sugar cane and corn, which are the crops most utilised by the world's largest ethanol producers: Brazil and the United States. Today, new technologies are being developed to extract sugars from lignocellulosic biomass. The so-called cellulosic ethanol is bioethanol made from lignocellulosic feedstock, meaning agricultural residues such as straw or corn stover.

This has significant economic and environmental potential, as biotechnological processes using highly specific biocatalysts and fermentation systems can produce cellulosic ethanol in high yield with GHG emission reductions of almost 95% compared to fossil fuels. In addition to the transport sector cellulosic ethanol can play an important role as a platform compound in the production of green chemicals. For example, ethanol can be converted into ethylene which is the source for one of the most commonly used plastics polyethylene (PE). In Brazil, one of the leading ethanol producing countries, green polyethylene is already being made from bioethanol made from sugar cane. Much of the biofuel growth in the market is driven by supporting policies and measures, such as the establishment of some targets for biofuel use.

Although currently the market share of biotechnology-produced biobased chemicals is relatively small, the importance could grow very quickly depending on the substitution potential that biobased materials have compared with their petrochemical counterparts. Some industrial biotechnology processes create the same molecules that are produced petrochemically. In these instances, the substitution could be complete, given that they are price-competitive. Other biotechnological processes lead to the creation of different compounds that have similar functionalities to petrochemical products and in this case the biotechnology-derived molecules may not be suitable for a complete substitution of their petrochemical counterpart. They may nevertheless occupy or even create a niche within the overall market.

Other factors driving the uptake of industrial biotechnology products are: the relative price of feedstocks compared to the price of crude oil; the technological progress and the relevant policies and measures put in place.

Insightful companies have already discovered the potential of biotechnology to cut GHGs emissions. However, as with most technologies, the potential to achieve sustainability goals does not automatically translate into such goals being realised. There are still numerous challenges for industrial biotechnology

—including technical, commercial and sustainability challenges—inhibiting large scale diffusion and commercialisation. The impact of industrial biotechnology on global GHG emissions will depend on the framework conditions in which this technology is applied. In the short term industrial biotechnology will mainly increase process efficiency and reduce emissions. In the longer term however, the impact of industrial biotechnology on GHG reductions will not be determined by the development of the technology alone, but by the supporting socioeconomic environment and policy context and proactive effort from both industry and policy makers, as well as by the main actors of the biobased value chain.

14.4.4 Industrial biotechnology and biorefineries

Biorefineries are similar to petroleum refineries in concept. However, biorefineries use biological matter (as opposed to petroleum or other fossil sources) to produce transportation fuels, chemicals and heat and power. Plastics and ethanol were among the first chemical products to use biorefining methods for production. As the technology advanced, methods using bioconversion entered the arena and advanced methodologies were developed to produce a variety of biochemicals. The latest enabling technologies include synthetic biology and systems biology which bring new production methods to biofuels, renewable chemicals, specialty chemicals and other bioproducts. The discovery tools of metagenomics and quantitative proteomics are now reaching maturity and are starting to be rolled out from large research institutions to smaller universities and research organisations.

A crucial step in developing this industry is to establish integrated biorefineries capable of efficiently converting a broad range of industrial biomass feedstocks simultaneously into affordable biofuels, energy and a wide range of biochemicals and biomaterials. Integrated biorefineries are similar to conventional petrochemical refineries in that they produce a range of products in order to optimise use of the feedstock and improve process economics. However, for integrated biorefineries to utilise a range of feedstocks efficiently, significant technology development and financial risk is implied, which increases the capital costs considerably, with the ultimate objective of lowering the operating costs.

A recent report by the World Economic Forum (WEF) identified industrial biorefineries as one possible solution that may help mitigate the threat of climate change and the burgeoning demand for energy, fuels, chemicals and materials. The report also concludes that there are still numerous large technical, commercial and sustainability challenges inhibiting – including both technical and commercial as well as sustainability challenges – large scale commercialisation.

14.5 Challenges and recommendations

14.5.1 Feedstock competition and land availability

Rapid development of first generation biofuels was accompanied by significant and public controversy as the feedstock is the same raw material (e.g., sugars and plant oils) as for food production.

In addition biofuels received some of the blame for the food price increase mid-2008, resulting in the 'biomass: food versus fuel' slogan. And although agriculture has supplied biomass and crops, which have provided products for non-food applications for a very long time (e.g., cotton or linen for clothing, wood for building, paper and energy, rubber), producing biobased products in huge quantities is a relatively recent trend. A generally accepted priority list for the different applications of biomass has recently been defined, known as the 5 F-cascade:

1. Food and feed.
2. Fine and bulk chemicals, pharmaceutical ingredients.
3. Fibres and biomaterials (wood, pulp and paper).
4. Fuels and bioenergy.
5. Fertiliser and soil conditioners (e.g., compost).

Although it is crucial to secure a sustainable supply of feedstock for industrial uses, there are concerns about resource availability in the long term: both on the potential of biomass feedstock to deliver sufficient raw material for all future applications (chemicals, biomaterials, bioenergy), as well as on its long term price level. The Nova Institute in Germany estimates that 500 million hectares of land are available worldwide (and this without cutting down forest or using artificial irrigation) to produce biomass in a sustainable way in response to growing food, chemicals, materials and energy needs. Securing a sustainable supply of feedstock for the biobased economy requires further research into methods of improving feedstock yields and/or the composition of biomass for optimal conversion efficiency.

This research will involve plant genomics and new breeding programmes and also research into efficient crop rotation, land management and land-use change issues. In addition feedstocks must be produced according to broadly supported sustainability criteria.

Although climate change mitigation is one of the main drivers of industrial biotechnology and the biobased economy, it should not be forgotten that climate change will also impact upon productivity, yields, water and land availability. While yields in forests, for example, might increase in parts of the world such as Northern Europe, the factors destabilising forest ecosystems (insect attacks, changing weather conditions) will have growing impacts. In agriculture,

International Food Policy Research Institute (IFPRI) noted that yields, especially in developing countries, will likely decrease and South Asia will be especially hard hit.

To avoid competition with food, the use of agro-food by-products and wastes should be strongly encouraged. They represent huge amounts of biomass already available and at low costs. However investments in developing and optimising infrastructures and logistics capacities are crucial to ensure that all the biomass that can be mobilised in a sustainable way (both from environmental and economic points of view) is actually used. Incentives are also needed for farmers to collect agricultural residues.

Another important discussion, although mainly related to biofuels, is the inclusion of GHG emissions-related land use change, both direct (dLUC) and indirect (iLUC), as land conversion can cause large changes in carbon content of the soil and the upper soil vegetation. The principle of iLUC is as follows: when biofuels are grown on existing arable land, current demand for food and animal feed may push these production activities into new areas such as forests or grasslands. Conversion of forest or grassland to agricultural land can lead to very significant releases of carbon to the atmosphere. When this indirect change is also allocated to biofuels, this can have an important effect on the overall GHG balance, when compared to fossil fuels.

However the methodology for calculating iLUC emissions is still under discussion as most of the current calculation methods base their reference on a status-quo (e.g., poor agricultural policy in developing countries, lack of protection of valuable nature areas, inefficient use of food and feed products). If real improvements in agricultural policy and subsequent yield increases in the developing world, or a stricter protection of valuable natural areas and rainforest (for all applications of biomass) are taken into account, this could significantly reduce these iLUC effects, e.g., CO_2 fixation by bacteria or algae as well as CO_2 fixation by micro-organisms in soils.

In order to develop and secure a supply of feedstock for industrial biotechnology and the biobased economy, without hindering the production of food for a fast growing world population and without negatively impacting the environment and biodiversity, the following measures are interesting additional pathways by which biotechnology can help mitigate climate change:

1. Optimise production per hectare of land through increased crop productivity by developing the crops themselves (traditional crops as well as 'energy' or lignocellulosic crops), the cropping system and new biorefining technologies to make better use of the biomass.

Develop and support a reliable upstream supply chain able to mobilise a sufficient level of feedstock for conversion, by investing in local and regional infrastructures and logistical capabilities to allow all biomass, including agricultural, forestry and waste-based raw material, to be utilised.

1. Introduce specific financial incentives to improve the logistical capabilities to collect and stabilise biomass residues (e.g., agricultural residues and forestry waste).
2. Develop policy instruments which could secure access to sustainable renewable feedstock, well balanced between bioenergy and biobased products. A higher focus should be put on resource efficiency and climate protection regarding the use of land and biomass flow. Cascading utilisation (the sequential utilisation of renewable raw materials for material and energy uses) could be one option for future support.

Research and innovation

For industrial biotechnology to meet its promise in climate change mitigation requires large-scale public and private investment in R&D. There is a serious mismatch between the level of private sector investment in industrial biotechnology R&D and the potential market opportunities for the sector. Therefore there is a pressing need to boost research in industrial biotechnologies by increasing public research investment, reducing regulatory burdens to bringing products to market and by encouraging public-private partnerships.

Industrial biotechnology is still a relatively new field and therefore immature; there are major areas of knowledge still to be explored. This presents a bottleneck to greater exploitation, but also offers tremendous opportunities for further research and breakthrough innovation. Both basic and applied sciences are essential: basic science to develop our fundamental knowledge base and applied science to introduce innovative products and processes based on this knowledge. In addition, industrial biotechnology is by nature a multidisciplinary area, comprising biology, microbiology, plant sciences, biochemistry, molecular biotechnology, chemistry, bioinformatics, engineering and fermentation technology.

Special attention should also be placed on specific key areas, such as:

1. Synthetic biology and metabolic pathway engineering are examples of emerging technologies that will significantly increase the diversity of biotechnological processes and products, driving the development of innovative products.
2. Some biobased products will require further chemical processing and unless these chemical processes are made available there will be no market for these precursors. Therefore dedicated research on the combination of technologies such as biochemical and chemical processes should also be given special attention.

3. Another important field is the development of efficient and robust enzymes, particularly for the conversion of lignocellulosic material. This should enable the conversion of a variety of feedstock.
4. Specific research is also needed to improve feedstock yield and/or the composition of biomass involving both plant genomics and new breeding programmes and also incorporating further research into efficient crop rotation, land management and land-use change issues.

One of industry's remaining major challenges is to translate research to products, including the development of new product applications. Setting up public-private partnerships would result in a pooling of resources, thus allowing more ambitious goals to be set in terms of reducing the time-to-market. This would also enable industry to adopt longer-term investment plans in the field of the bioeconomy, taking into account the market perspective.

Lessons learned from existing biorefineries show that the construction of demonstration facilities is a crucial step towards developing a fully-fledged biorefining industry. Demonstration activities are able to close a critical gap between scientific feasibility and industrial application. They enable the measurement of actual operating costs and specific strengths and weaknesses of technological processes before costly, large-scale facilities are built. They dramatically reduce the risk of introducing new technology on an industrial scale and therefore make a biorefinery venture much less risky for investors. Stimulating the construction of demonstrators via public-private partnerships is therefore one of the most important measures that can be taken in the development of the bioeconomy.

The initial construction of biorefinery demonstration plants is not only a costly undertaking but it also involves bringing together market actors along new and highly complex value chains. This includes the diverse suppliers of biomass raw materials (farmers, forest owners, wood and paper producers, biological waste suppliers, producers of macro- and microalgae), the industrial plants that convert the raw materials and industries providing them with the necessary technologies and the various end users of intermediate or final products. Countries like the United States, Brazil, China, the Netherlands and others are increasing investment into research, technology development and innovation and are supporting large scale demonstrators.

In order to make a fast and efficient shift towards more integrated and sustainable production and processing systems, the following measures are proposed.

1. Support technological progress and increase public and private R&D investment in industrial biotechnology and related technologies, through multidisciplinary research programmes at both national and international level.

2. Build networks between industry and academia to overcome the competence hurdle and knowledge gap.
3. Mobilise sufficient resources to support large coordinated research initiatives, focusing on the market perspective covering the whole value chain.
4. Stimulate and support technology transfer in the area of industrial biotechnology and green technologies.
5. Establish programmes to accelerate and alleviate risks of transforming knowledge into commercial products, by financially supporting access to pilot and demonstration facilities and by integrating production processes.

14.5.2 Dedicated policies and incentives to promote the use of biobased products

Specific policies for the development of biobased products are more extensive for bioenergy (including liquid biofuel use and solid biomass applications) than for biochemicals or biomaterials. Worldwide, many governments support their emerging biofuel industries far more than other sectors via subsidies, mandates, adjustments to fuel taxes and incentives for the use of flex-fuel vehicles (e.g., United States, European Union, Brazil, China) and several countries have started to develop extra incentives for the so-called second generation biofuels.

Communication and education

Biobased products comprise a great variety of innovative products in many applications. However most of these products cannot be easily recognised by users and consumers, due to the fact that no external, perceptible characteristics differentiate biobased products from traditional products made with petrochemicals. Thus the specific features of biobased products are mostly invisible to the purchaser. This lack of awareness and knowledge of the potential of industrial biotechnology, in manufacturing industry and amongst policy makers, consumers and even investors represents a major obstacle to increased market uptake of biobased products. Sustainability should be made more visible as the core value driving the development process. As environmental awareness grows, the demand for products from sustainable manufacturing processes, including biobased products, is likely to grow.

Consumers are increasingly making purchasing decisions based on environmental considerations. Such buying patterns also give signals to retailers and manufacturers that factors in addition to quality and price are important. However in order for the consumer to make an informed choice, communication with the public has to provide simple and clear information. Consumers have limited knowledge of biobased products in the market place. If the aim is to create a market-pull effect by information and specific campaigns, then brand owners and retailers are a very important target group. Such business to business (B2B) information should be based on standards and certification schemes. Meaningful information campaigns can increase consumer awareness and understanding.

The challenge in communicating about biobased products is their diversity. Though they are all created at least partially from biomass or the transformation of biomass components, they fall into many product categories, ranging from pharmaceutical ingredients to consumer products, from bulk materials to chemicals. The main characteristic of biobased products is their biobased carbon or biomass content, while other important characteristics related to climate change are environmental performance and carbon footprint.

The public needs simple and easy-to-recognise information such as labels. Labels are frequently used to inform consumers about characteristics of the products, options for disposal (recyclable, biodegradable), or provide a quality indicator. However the efficiency of a label is limited by two factors:

1. First, with the multiplicity of labels and information tags on products, consumers tend to be overwhelmed by information. Thus it is important to review the efficiency and coherence of the labelling system and evaluate the need for consumers to have labels and their actual impact on consumer behaviour.
2. Second, in driving a value chain towards increased sustainability, a label is only as efficient as its adoption by the market—especially if the label is not compulsory—and its recognition and understanding by consumers.

Labels can be used for improving information and transparency for consumers regarding the renewable origin of products and promoting these products.

If such tools are not well developed or improperly used, they can lead to mistrust in the public and damage the reputation of the industry.

14.6 Conversion towards low-carbon renewable-based production systems

One of the main questions is how we can economically valorise GHG emission savings or avoidance. Producing chemicals through biochemical routes is currently still more expensive compared to traditional production routes. In addition, existing production facilities for chemical syntheses cannot be converted to biotechnological production without massive new investments and in many cases, there are clear economic restrictions in biotechnological production processes due to higher operating costs and higher levels of R&D costs and investments. When competing against an existing process based on fossil carbon sources, the alternative bioprocess must also show competitiveness in scenarios with high energy- and feedstock-cost volatility.

Investments required for building a new bio-industrial facility, especially if it competes with the conventional one, might present a significant barrier. In addition, it has become even more difficult to obtain financing for investing in building new, full-scale commercial plants and infrastructure as result of the recent worldwide financial crisis. Governments tend only to provide financial support and incentives on a relatively short-term basis, while the pathway to success for many enterprises is a long term proposition.

Governments interested in supporting biorefineries for reasons of environmental protection, energy security and innovation leadership need to support market growth, in order to trigger private sector investments.

In the United States—at federal as well as at state-level—several programmes have been set up to stimulate the construction of new plants (producing biofuels and biobased products) and/or new biorefineries. The United States is currently evaluating the feasibility of bioenergy and biofuel tax policies, programmes and carbon legislation for other biobased chemicals and products.

14.7 Need for an integrated and strategic approach

In order to achieve a competitive bioeconomy, broad approaches, such as creating and maintaining markets for environmentally sustainable products, funding basic and applied research and investing in multipurpose infrastructure and education will be necessary. In addition, these will need to be combined with shorter term policies such as the application of biotechnology for improving plant and animal varieties, improving access to technologies for use in a wider range of plants, fostering public dialogue and increasing support for the adoption and use of internationally accepted standards for life cycle analysis together with a range of other incentives designed to reward environmentally sustainable technologies. EuropaBio and the Biotechnology Industry Organisation in the United States, recently published policy guides setting out similar requirements and called for a more integrated and strategic approach, with supportive policies in the areas of climate change, energy security, renewable feedstock supplies, research and innovation, the environment and trade.

It can be assumed that the net GHG impact of industrial biotechnology will be strongly influenced by the overall socioeconomic environment and the policy landscape promoting the dissemination of these technologies. In many instances, current policies and private sector strategies are neither strong nor efficient enough to fully develop and harvest the GHG emissions reduction potential of industrial biotechnologies. A significant effort is therefore required to achieve the socioeconomic environment and policy landscape that would nurture such potential.

At the heart of the climate change debate is GHG emissions. There is a strong linkage between economic growth and increasing CO_2 production. The problem then becomes one of how to break this linkage, i.e., to continue economic growth but in a sustainable manner that leads to GHG emissions reductions. Industrial biotechnology is one part of the solution to the problem.

It is incumbent upon any new environmentally competitive technology to be also performant and cost-competitive. In its early years industrial biotechnology must be supported financially at a very high level as it competes head-to-head with the oil and petrochemical industries. These are very well established industries that benefit from large economies of scale that would make industrial biotechnology processes and products uncompetitive. The level of support required needs to come from both government and private industry in partnership.

To date the United States first generation bioethanol industry provides a good example of how policy measures can change the landscape for a promising technology. Bioethanol has been massively supported by regulation, volume and GHG mandates, public procurement and private industry has responded positively. For biofuels to continue to be a success requires, amongst other challenges, that the technical challenges associated with the use of agricultural waste materials as feedstock be overcome. The maturation of the integrated biorefinery will go a long way to addressing economic issues. Biobased chemicals and bioplastics lack the national and international policy frameworks at present that have made first generation bioethanol successful.

It is assumed that producing chemicals, polymers and liquid fuels by biological processes rather than petrochemical ones, or by including biobased ingredients, should lead to a reduction in GHG emissions as:

1. Biological processes are more benign than most petrochemical processes (lower temperature and pressure, low water and energy consumption as well as lower CO_2 and waste production).
2. The raw materials for biological processes are renewable and therefore the products are more likely to be carbon-neutral.

The assumption needs to be proven on a case-by-case basis, using techniques such as life cycle analysis and carbon footprint testing. To produce such proof requires standardisation of techniques used. To create a globalised industrial biotechnology industry requires harmonisation of those techniques to prevent trade barriers developing. Best practices are an essential prerequisite. Great care is required in determining what are the most useful sustainability indicators and then making sure that there is great clarity in their application to prevent confusion. For example, when biodegradable plastics started entering the market, there were a series of misconceptions about the term degradable and biodegradable.

Industrial biotechnology has suffered a lack of investment at all levels e.g., R&D, demonstrator plants, full-scale process development, feedstock development, land use issues. There exists a serious mismatch between future expectations of this industry (high) and the level of investment (low). Part of the problem is the lack of visibility of a biobased product compared to, say a pharmaceutical product. It is difficult for industrialists, investors and the public to visualise industrial biotechnology products and therefore to understand their benefits. Part of the policy mix therefore is communication and education. Designing appropriate labels, by consequence of such misconceptions, is therefore a complex problem.

The stakes involved environmentally and financially are very high indeed. Industrial biotechnology offers the promise to be a force for good, not a panacea, in the ongoing climate change struggle. Policy intervention is required across the three broad criteria—social/environmental, industrial performance and economical. To make this happen requires policy to be both national and international if industrial biotechnology is to fulfil its promise in a rapidly globalising world.

14.8 Impact of industrial biotechnology in different industrial sectors

Several studies outline the potential of industrial biotechnology and biobased products to lower the impact of industry on the environment, with particular regard to lowering GHG emissions. This section discusses some specific industrial sectors and it should be pointed out that the product specific claims on impacts were measured via transparent life cycle analyses (LCA). The studies indicate that—in most cases—bio based products consume less energy and emit less carbon dioxide (CO_2) than products from fossil resources. This is mainly due to the fact that fermentation or bioconversion processes occur at or near room temperature and atmospheric pressures, resulting in lower combustion energy requirements and fewer associated emissions than in conventional chemical processing.

According to Ademe, biobased products can create 40–80% savings in non-renewable energy consumption and at least 50% lower CO_2 emissions.

14.8.1 Textiles and detergents

Laundry and dish washing are among the most energy intensive household activities, requiring the heating of large quantities of water and the consumption of electricity to drive the wash process. Washing machines are one of the biggest consumers of household electricity and 80% of the electricity for washing laundry is used to heat the water. So reducing the temperature required to effectively clean soiled clothes or dishes can result in significant energy savings and reduced CO_2 emissions. Detergent enzymes represent in fact one of the largest and most successful applications of modern industrial biotechnology as one of the main environmental advantages of enzymes is that clothes can be washed at a lower temperature.

With the new generation of cold water enzymes, washing temperatures can be reduced to 30°C, without sacrificing cleanliness. Studies show that CO_2 emissions can be reduced by 100 g per wash by washing at 30°C rather than 40°C. By washing at 30°C rather than 40°C or 60°C, the CO_2 savings potential in the United States and Europe alone is around 32 million tons annually, equivalent to the emissions of 8 million cars. Natural fabrics such as cotton are normally bleached with hydrogen peroxide before dying. In the classical process, bleaching was followed by at least two rinses in hot water (80–95°C). With the use of an enzyme that degrades residual peroxide during the second post-bleach rinse, water heated to 30–40°C can be used resulting in lower non-renewable energy consumption. Another application of enzymes in the textile industry is in the treatment of cotton fibres. Scouring is a cleaning process that removes pectins and thereby assists with the removal of impurities such as waxes and mineral salts from cotton yarns and fabric before dying. Traditionally, scouring involves a number of high-temperature steps with a large consumption of chemicals. Enzymes enable a faster and gentler scouring process with lower energy and chemical consumption. Compared to the traditional chemical process, the enzymatic process reduces temperature from 95°C to 55°C, meaning that there are important energy savings.

14.8.2 Plastics production

Plastics in general are important materials contributing significantly to environmental protection: due to their tailor-made properties (e.g., light weight, excellent insulation ability, tuneable properties for optimum food protection, etc.) they reduce energy use by 26% and GHG emissions (GHG) by 56% across a variety of applications compared to alternatives. Besides crude oil, natural gas and coal, biomass is an additional raw material source for plastics (biobased plastics). Depending on the type of biobased plastic, they may contribute up to a 50% decrease in terms of energy consumption and up to 67% savings of CO_2 emissions during the production process. Depending on the extent to which policies and measures supporting biobased plastics are implemented, the potential GHG emission reductions range between 3.0 and 8.5 million T CO_2 eq. by 2020.

Polylactic acid or PLA, produced from corn starch as a renewable feedstock, is being used in fibres, films, food and beverage containers, textiles, coated papers and many other applications. LCA data indicate that from cradle to polymer factory gate, PLA requires 30–50% less non-renewable energy and releases 30–60% less GHG to the atmosphere than the most common traditional plastics.

14.8.3 Pulp and paper

Converting wood into paper is an energy, water and chemical intensive process. The conventional chemical process requires boiling wood chips at around 160°C before bleaching the pulp with chlorine dioxide. With the application of new biotechnology processes, it is now possible to reduce the energy consumption by 32% used during the bleaching process.

14.8.4 Chemicals

Biotechnology can be used to produce various bulk and fine chemicals that are currently produced from fossil fuel based feedstocks. Biobased substances can also act as building blocks for many other materials provided that they are cost-competitive. Recently a company has developed a patented process to manufacture 1,3 propanediol, the so-called bio-PDO™, from renewable resources instead of petrochemicals. Bio-PDO™ is one of the first commercial-scale industrial applications of metabolic engineering designed to make a 100% renewable resources based material from corn starch and is the key element for many materials in use every day. The production of Bio-PDO™ has a much smaller impact on the environment than its petroleum-based equivalent. The production of Bio-PDO™ consumes 40% less energy and reduces GHG emissions by 20% versus petroleum-based propanediol. It has a broad range of applications in fibres, films, thermoplastics, detergents, cosmetics products, ink and several other industrial areas.

Chemical synthesis

There are a number of chemicals. like acrylamide, vitamin C, monosodium glutamate, gluconic acid, lactic acid and L-malic acid, produced by bio-transformation using enzymes. Even the conversion of glucose to fructose, penicillin-G to 6-APA, glycerides for emulsifiers is a major application of industrial enzymes. Also, enzymes are now being used in diagnostics, biosensors and medical research.

14.8.5 Drug and pharmaceutical

Vitamin B2, also known as riboflavin, plays a very important role in maintaining human and animal health. Vitamin B2 is central to energy metabolism and is required for the metabolism of fats, carbohydrates and proteins. Industrial biotechnology replaced a laborious and costly traditional multi-step chemical synthesis and purification process by a one-step fermentation process whereby a micro-organism directly transforms vegetable oil into the required vitamin. Evaluation of the benefits of the one-step process shows a reduction of CO_2 emissions of 30%. Industrial biotechnology made it possible to replace the purely chemical production process of cephalosporin, a commonly prescribed antibiotic, by an enzyme-catalysed manufacturing process. Biotechnological production of 7-aminodeacetoxycephalosporanic acid, a key intermediate for the synthesis of cephalosporin, resulted in 37% less electricity use and CO_2 reductions of 75%.

14.8.6 Food

Enzymes have been used in food manufacturing for hundreds of years, mainly based on fermentation by micro-organisms. Today, novel enzyme applications are also being implemented in baking, fruit and vegetable processing, brewing, wine-making, processing of vegetable oils, cheese manufacturing and meat- and fish-processing. Many enzymes applications in the food industry are advantageous due to their impact on processing conditions in food manufacturing plants, where enzyme use may result in savings of energy and chemicals.

Recently a special amylase has been commercialised that diminishes the crystallisation of starch, reducing waste of bread by allowing bread to stay fresh and moist longer. This effect has provided industrial bakeries with new opportunities for changing their production and delivery setup in order to produce at larger, centralised bakeries and make fewer deliveries to retailers. The industry can save both money and energy, while less waste bread also means more efficient use of agricultural raw materials. There are significant environmental gains due to better utilisation of agricultural raw materials. A reduction of GHG emissions of up to 54 T per million loaves of bread sold can be obtained. The major contribution to the reduction of CO_2 emissions is also, in this case, the agricultural load. Some 65% of the reduction stems from savings in the production of wheat, including agricultural emissions, production of fertiliser and traction. Some 15% of the reduction comes from savings in energy consumption during milling and baking and 15% comes from reduced transportation.

Malting barley for use in brewing is an energy and water intensive process. The barley has to be soaked in water to allow it to germinate and then has to be heated for drying. Brewers and biotechnologists developed a process removing this entire step with the use of an enzyme. By removing the malting process, the application of this enzyme to brewing is estimated to save 350,000

T of CO_2 for every 10% of global beer production that is converted to unmalted barley—a CO_2 savings equivalent to the emissions of 85,000 cars.

Enzyme usage in starch

Starch conversion will be the major application of enzymes. The major enzymes used in this sector are alpha amylase, glucoamylase and glucose isomerase. Starch hydrolysis has gradually become a wholly enzymatic one in the last couple of years and a range of starch degrading enzymes are now available in sufficient quantities at competitive rates. Growth of this sector is related to the growth in demand for high-fructose corn syrup used by the soft drink industry. Soft drinks are yet very less explored in third world countries by major soft drink manufacturers and this can enhance the demand for enzymes.

Animal feed

Enzymes are relatively a new class of animal feed additive. Their function is to act as catalysts to enable more efficient utilisation of animal feeds. They act as bio-catalysts and assist in digestion process for the total utilisation of nutrients and minerals, which are excreted unused.

Solid state fermentation of agricultural wastes into vitamin rich products as well as conversion of such waste into pro-biotics or enriching poultry and animal feed are also picking up fast.

Dairy

Milk coagulating enzymes have the dominant market in this sector. India is the largest manufacturer of milk in the world. In recent years, microbial rennets and recombinant chymosin have very good market for cheese making from milk. The by-product of cheese manufacturing process (whey) can be enzymatically converted into glucose and galactose, which have an application as a sweetener.

14.8.7 Tanning and leather

Enzymes have been used in the tanning industry for centuries because they are efficient in degrading protein and fat. In early times, the enzymes were derived from animal excrement and later on from the pancreas of cattle. Today, the enzymes are almost entirely produced by microbial fermentation. Soaking enzymes reduces the required soaking time, leading to electricity savings in turning the drum where the hides rest. Enzymes that remove hair during the tanning process reduce the sulphide requirements for the process. The environmental impacts of producing and delivering the enzymes to the tannery on the one hand and the savings in chemicals and electricity on the other have been evaluated via a LCA study. This comparison of conventional and enzyme-assisted bovine soaking and de-hairing/liming processes indicates that the application of enzymes in the tanning industry is justified by considerable energy savings and considerable reductions in the contribution of the processes to global warming. Assuming that the environmental improvements by switching from conventional to enzyme-assisted soaking and de-hairing/liming are applicable worldwide, the global saving potential is in the order of 8 million GJ of energy and 0.7 million tonnes of CO_2 per year.

14.8.8 Dyes

Bioprocesses to produce biobased colourants have recently been developed as an alternative to traditional chemical synthesis. While the creation of chemical-based dyes requires temperatures of up to 70–90°C in harsh conditions, the enzymatic synthesis of these colourants can be obtained at ambient temperature, under mild conditions. LCAs have illustrated that on an industrial scale, enzymatic processes could help to reduce CO_2 emissions and toxicity towards the environment.

Enzymatic treatment of coloured wastewater cannot only cut human toxicity in half, but a LCA study showed that, as compared to classical chemical sludge, enzymes can reduce by 10 times the impact on global warming, reduce by a factor of two the impact on the ozone layer, reduce by a factor of three the impact on abiotic (non-living) components in the environment and decrease by a factor of three the impact on marine toxicity.

14.8.9 Recovery of metals from industrial wastes

Presence of heavy metals and radio-nuclides in wastes is one of the common problems of many industries. The presence of such metals in wastes is not only harmful to the environment, but is also an economic loss. Removal and recovery of metals from wastes is, therefore, an important issue in pollutant control efforts. Biotechnology emerges as an interesting option for this purpose, especially when metals are present in relatively low concentration.

Sources of heavy metal pollution

Heavy metals comprise an ill-defined group of approximately 65 metallic elements of density greater than 5. They have diverse physical, chemical and biological properties, but generally, all of them exert toxic effects on microorganisms. Although elevated levels of toxic heavy metals occur in some natural locations, on an average, their presence is generally low. Most of the naturally occurring, high concentrations of heavy metals occur in an immobilised form in sediments and ores and are biologically unavailable. However, ore-mining and many other modern industries have disrupted natural bio-geochemical cycles and have caused increased deposition of heavy metals in the terrestrial and aquatic environment. Major sources of heavy metal pollution are combustion of fossil fuel; mineral mining and processing; nuclear and other industrial effluents and sludge; brewery and distillery wastes; biocides and preservatives, including organo-metallic compounds; wire manufacturing; light bulb manufacturing; plastic industry; power companies; leather industry; metal refining, cleaning, plating and processing; and dyeing industry. Common pollutants among the heavy metals are copper, zinc, cadmium, iron, mercury, lead, nickel, silver, gold and radioactive elements like uranium, thorium, etc. The acceptable value for heavy metal ions is usually less than 1.0 mg $litre^{-1}$. The concern for these pollutants comes from the following reasons:

1. Harmful effects on environment, particularly biotic community.
2. Public concern due to outbreaks of mercury, lead and cadmium poisoning.
3. Possibility of recovery of precious metals.
4. The presence of toxic metals negates the value of pulp mill fines, sludge, incinerator ash and municipal sludge as fertilisers, soil amendments and inert fills.

14.8.10 Biofuels

To ameliorate GHG emissions in the transport sector, there are no short to medium term alternatives but to combine further leaps forward in automobile fuel efficiency with biofuels. While first-generation corn-to-ethanol plants reduce GHG emissions by as much as 52% over petroleum-based fuels, ethanol made from cellulosic feedstocks, agricultural residues, or wood forest residues has the potential to reduce GHG emissions by as much as 129%, compared to gasoline. The CO_2 savings depend very much on the feedstock and conversion process.

14.9 Opportunities in biotechnology for next millennium

Biotechnology is a frontier area of science with a high promise for the welfare of humanity. New generation of bio-technology, developed as a result of intensive work in India, has opened up research of national relevance.

Enzyme industry is witnessing steady growth globally. It is estimated that the global industrial enzyme market has witnessed growth in excess of 100% in the last six years. The global enzyme market is divided into five main market sectors viz,: Detergents, textiles, starch, dairy and others.

The 'others' category is highly fragmented and includes diverse applications such as baking, brewing, fruit juice, animal feed, flavours, fats, oils, protein hydrolysis, paper, waste treatment, leather, diagnostics, pharmaceuticals/fine chemicals (biotransformations).

Many of the 'other' category enzymes are high value products. It is the 'others' category, which is witnessing the highest growth. It has achieved phenomenal growth of more than 500% in the last six years. In this category, baking and animal feed are considered to be the fastest-growing applications. Wastewater treatment has also a very good growth potential. Applications in diagnostics, pharmaceuticals and bio-transformations have very wide applications and are expected to capture very big enzyme market share in future. Detergents, textiles, starch and dairy application sectors, though growing steadily, are not recording growth as high as the 'others' sector. This is mainly because of the fact that the market of these enzymes is saturated and there is little growth of its end-user industries.

It is estimated that 90% of the industrial enzymes are produced in Europe, Japan and the U.S., but China and India are expected to become significant producers for enzyme markets in Asia. The major Indian markets are today in textile, detergent, leather, pharmaceuticals and dairy industry.

Section II

Pollution control in mechanical and electrical industries

15

Pollution control in mechanical industry

15.1 Introduction

Metal fabrication is the building of metal structures by cutting, bending and assembling processes. It is a value added process that involves the construction of machines and structures from various raw materials. Metal fabricators (companies that specialise in this process) are called fab shops.

Some of the raw materials used by metal fabricators are: plate metal, preformed and expanded metal, welding wire, fittings and castings. To begin, these raw materials need to be cut to the correct size. This is done with tools like a special band saw or with cutting torches. Metal can then be formed using dies. Tube bending machines and rolling machines are also used to bend and make round sections of metal. Parts are welded together and then after cooling the machine or structure is usually sand blasted and painted. This final product is thoroughly inspected before it is shipped off to the client.

Many metal fabricators specialise in specific processes based on their clients' needs and their own expertise. These could be punching, welding, forging, casting, brazing, shearing, drawing and spinning. Even higher levels of specialisation include electrical and hydraulics services.

15.2 Managing waste in the metal manufacturing industry

1. Reduce input and waste disposal costs.
2. Improve your environmental performance.
3. Be more competitive.

Focusing on waste can help you to address high raw material prices, rising costs of hazardous waste treatment and disposal and pressure to increase the sustainability of your operations.

15.2.1 Major waste generating activities

Metal manufacturers generate waste through activities such as: casting; machining and welding; parts assembly; surface finishing; quality control; and packaging. The major environmental wastes in this sector involve: the use and disposal of machine lubricants and coolants; aqueous and solvent cleaning systems and recycling of scrap. Hazardous wastes, such as solvents, pickling solutions and oils, can be expensive to manage as they require special waste treatment and disposal arrangements.

15.2.2 Benefits of reducing waste

Improving waste management can benefit your business and the environment by:

1. Reducing the cost of purchasing metals and other raw materials (e.g., through identifying reuse opportunities for waste products).
2. Minimising waste treatment and disposal costs (and possibly generating alternative income streams).
3. Increasing profitability (as wasted product equals lost profit).
4. Reducing environmental impacts associated with hazardous chemical use, waste disposal and consumption of raw materials.
5. Improving the reputation of your business and employee satisfaction through promoting an environmentally responsible image providing a safer workplace through process improvements and less waste handling.

15.2.3 Improving waste management practices

Implementing waste management improvements may require forward planning and some changes to the way your business operates. For example:

1. Proposed changes may need to be discussed with managers, workplace safety representatives, unions, insurers, investors, suppliers and customers to check that they will not impact negatively on productivity, security or safety and other standards.
2. Employee training and awareness may be required to successfully implement actions and support the introduction of new equipment or processes, such as recycling systems that alter the way wastes are handled.
3. Special licences or permits may be required by your business or contractors to store, treat, transport and dispose of hazardous wastes.
4. Results are more likely to be achieved and maintained if you have a written plan and targets agreed by all areas of management. Prioritise actions and consider beginning with the 'low-hanging fruit' for fast gains and to generate enthusiasm.
5. Monitoring waste generation and disposal (e.g., benchmarking production against raw material purchases) is important for environmental compliance, stock control and to measure (and reward) improvements.

The costs, savings and payback periods for the waste reduction options are shown in Table 15.1 are a rough guide only. They include estimates of up front costs such as capital, labour and installation, but do not include ongoing costs unless these are fundamental to the option itself (e.g., improved maintenance regimes). The suitability and benefits of each option depend on the nature and size of your business and the scale of application. You should also check that they comply with local environment, safety and other requirements. The waste hierarchy provides a framework for managing waste: avoid, reduce, reuse, recycle and dispose. Waste avoidance generally delivers the best financial and environmental outcomes.

Table 15.1: Waste reduction options.

Process change	*Equipment/plant upgrade*	
Option	*Saving*	*Waste hierarchy*
Purchase high-quality lubricants, coolants and metalworking fluids so you can use less and therefore generate less waste	Chemical use and hazardous waste disposal	Avoid
Extend the life of working fluids by using demineralised water for dilution. Implement a regular maintenance plan to optimise equipment operation and minimise loss of fluids	Chemical use and hazardous waste disposal	Avoid
Plan jobs where possible to minimise metal scrap by using the maximum percentage of sheets or blocks (e.g., provide a scrap size that is suitable for a subsequent smaller job)	Raw material use	Avoid
Dewater sludge before disposal (as often up to 80–90% water)	Hazardous waste disposal	Treat
Recover metals from plating and other sludges through specialist treatments or reuse in metal smelting feedstock	Hazardous waste disposal	Recycle
Minimise tanks wastes by ensuring parts are clean and drain times are optimised before they are moved to the next process	Wastewater treatment	Treat
Remove and store waste oil from cleaning and degreasing processes for recycling	Chemical use and hazardous waste disposal	Recycle
Use efficient spray painting systems (e.g., electrostatic or airless delivery methods) to reduce over spray, increase transfer efficiency and minimise solvent emissions	Paint and solvent use	Avoid
Talk to your waste management company about sending paint sludge from over spray and booth clean for energy recovery	Hazardous waste disposal	Energy recovery

(Cont'd...)

Process change	Equipment/plant upgrade	
Option	*Saving*	*Waste hierarchy*
Reduce packaging waste by buying in bulk or sourcing packaging that is recyclable or is taken back by suppliers	Packaging use	Recycle
Minimise cutting fluids content in metal scrap (e.g., by increasing swarf drainage times or placing a screen above the bottom of the swarf collection bin) to make both streams easier and safer to store and recycle (or filter and reuse fluids on-site). Compress softer metals (e.g., aluminium) into bricks in a bale press	Chemical use and hazardous waste disposal	Recycle
Cover tanks when not in use to reduce solvent and other losses	Chemical use	Avoid
Extend tank life and improve process efficiency by using air agitation and eductor-jet pumps and continuously filtering contaminants out of the tanks	Chemical use and wastewater	Avoid
Reduce carry-over in the plating process by counter current rinsing or installing spray rinses over plating baths	Wastewater	Avoid

15.3 Minimising industrial wastes from the fabricated metal products industries

The fabricated metal products sector includes farm machinery, packaging machinery and construction machinery. Other machines also has many highly specialised niches within the fabricated metals sector in areas such as bolts, nuts, screw and rivet, spring and crown and closure manufacturing. Additionally, there are a number of specialised small/machine fabrication shops that cater to automotive industry.

The fabricated metal products industries sector engages in the forming/shaping of metals and their surface preparation. While a large variety of alloys are processed in these facilities, they can be broadly categorised as ferrous and non-ferrous. Surface preparation includes a myriad of operations that range from electroplating to powder coating. The primary markets served by this sector include automotive, electronics, aerospace and consumer durables among others. In a number of plants, several such forming operations may be necessary to produce a final product. In addition to these operations, many processes categorised as machining are also frequently part of the manufacturing process. These include drilling, milling, turning and the like. These processes accomplish their objective through precise removal of metal and are termed metal removal processes.

In addition to basic forming/shaping processes, the surfaces of the metal parts are frequently manipulated in various ways to enhance functional properties. One common operation is painting. Other examples include electroplating, electroless plating, anodising and phosphating. These surface treatment operations frequently require the surfaces themselves to be clean prior to application- or adhesion can be impaired and functionality compromised. Solvents, alkali and acids are used in immersion, spray, or pickling processes to achieve cleanliness.

The combination of operations enumerated above and the use of energy in the form of compressed air, electrical motors, hot air, hot water/steam and the like result in gaseous, liquid and solid effluents. The objective of this chapter is to illustrate with a few examples some ways to minimise the generation of these wastes.

15.3.1 Common industrial wastes in the fabricated metal product industries

The most common wastes arising from the fabricated metal products sector are metal, paint, electroplating sludge, sludge from various processes, acids, alkali, used industrial fluids, volatile organic compounds, water and combustion products such as CO_2. The common processes and the wastes associated with them are shown in Table 15.2.

15.3.2 Minimising waste

A useful tool for achieving waste minimisation is to recognise that every situation can be reduced to four elements: people, process, management controls and externalities (Fig. 15.1).

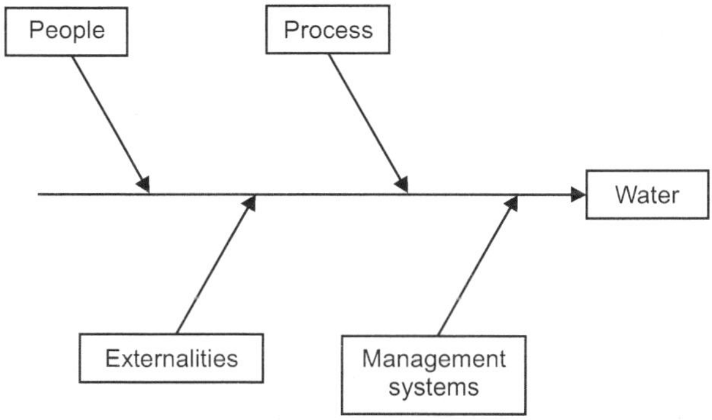

Figure 15.1: Waste minimisation tool.

Table 15.2: Process inputs and wastes.

Process	*Material input*	*Air emission*	*Process wastewater*	*Solid waste*
		Metal shaping		
Metal cutting/or forming	Cutting oils, degreasing and cleaning solvents, acids, alkali and heavy metals	Solvent wastes	Waste oils, spent acid, alkaline and solvent wastes	Metal chips, solvent still bottom wastes, metal bearing cutting fluid sludge
		Surface preparation		
Solvent degreasing, alkaline and acid cleaning	Solvents, alkali, acids	Solvents	Solvent, alkaline and acid waste	Ignitable wastes, solvent wastes, still bottoms
		Surface preparation		
Anodizing	Acids	Metal-ion bearing mists and acid mists	Acid wastes	Wastewater treatment sludge
Chemical conversion coating	Metals and acids	Metal-ion bearing mists and acid mists	Metal salts, acid and base wastes	Spent solutions, wastewater treatment sludge
Electroplating	Acid/alkaline solutions, heavy metal bearing solutions, cyanide bearing solutions	Metal-ion bearing mists and acid mists	Acid/alkaline cyanide and metal wastes	Metal and reactive wastes
Painting	Solvents and paints	Solvents	Solvent wastes	Still bottoms, sludge, paint solvents

15.3.3 Leveraging process improvements – an electroplating example

Electroplating, the process of depositing metal over a substrate through the use of electricity, is a commonly used technique to obtain both functional and decorative finishes. An example of a functional finish is chromium deposition for corrosion resistance. Achieving an antique brass finish on a steel part is an example of a decorative finish.

Quality coatings can only be achieved if the parts are cleaned thoroughly prior to being plated. Cleaning is accomplished using alkaline solutions formulated with surfactants to achieve removal of tenacious oil films and other types of soils. Following this, the surfaces are rinsed and then processed through acid solutions to remove oxide scales. Residual acid is removed by rinsing. Finally, the parts are plated, rinsed and subjected to further processing as desired. Electroplating can be done from both alkaline and acidic solutions.

The movement of parts through the sequence of steps results in carry-over of solutions from the chemical tanks to the rinse tanks. The carry-over causes contamination of the rinse tanks with acids and alkali. Rinsing ability is compromised as contaminants accumulate in rinse tanks. To counter the build-up of contaminants, the rinse water is bled either periodically or continuously and replaced with fresh water. As the rinse tanks contain hazardous metals and metal complexing agents such as cyanide, they require treatment prior to discharge. These treatments frequently are pH adjustment followed by heavy metal precipitation. Treatment for hexavalent chromium tends to be more complicated. In all these cases, the sludge formed is deemed hazardous and disposal is expensive.

15.4 Pollution control in metal fabricating industry

Pollution control in its broadest sense includes the maintenance of hygienic conditions within the plant itself, for the benefit of the employer as well as the employees. In many cases, the solution of this problem will result in considerable reduction or complete elimination of a pollution problem outside the environs of the plant. For this reason, considerable attention must be given to the solution of the in-plant problem. In general, this is primarily a problem of air pollution. Air contaminants in a metalworking plant may be classified on the basis of origin or method of formation into the following principal groups: (i) solid particulate matter – dusts, fumes and smoke, (ii) liquid particulate matter – mist and fog, (iii) non-particulate matter – vapours and gases.

In addition, these air contaminants may be classified by chemical nature – organic or inorganic, by optic properties – visible or invisible; by particle size

– submicroscopic, microscopic and macroscopic, by state – particulate or gaseous and by effect on humans – toxic or harmless. In metal fabricating plants, the range of the particles responsible for air contamination is considered to vary in mean size from ultramicroscopic (0.7 micron) for metallic fumes to microscopic (100 microns) for metal grinding particulate.

15.4.1 Machining of metals

Machining of metals invariably involves the use of coolants, cutting oils and lubricants, each of which is a potential pollutant of air or water or both. Lubricating oils can be separated within the plant by skimming off the top layer in a separator. Metal chips and fines can be pulled off from the bottom of the settling chamber.

Coolants such as soluble oils should be retained without excessive wastage, in order to minimise the pollution control requirements for the plant waste. However, in order to continue to reuse the coolant, it is necessary to prevent the breakdown due to the action of anaerobic bacteria. Aeration by use of air injection into a holding tank can prevent putrefaction so that the useful life of the coolant is extended. Coolant which clings to the cutting chips can be separated by centrifuging and can be returned to the coolant system.

15.4.2 Heat treatment of metals

Heat treatment is an operation or combination of operations that involves the heating and cooling of a solid metal or an alloy for the purpose of obtaining certain desired physical properties. These physical properties are attained through changes in the nature, form, size, or distribution of the structural constituents. These changes occur because of the effect of temperature on phase equilibrium and include grain growth, recrystallisation and diffusion or migration of atoms, or from changes in the composition of the material, such as carburising or nitriding. In the heat treatment process, the metal is subjected to a definite time-temperature cycle, which may be conveniently divided up into three parts: (i) heating, (ii) holding and (iii) cooling.

The chemical effect of the environment is of great importance during heating, as in carburising or nitriding, or the controlled environment (oxidising, reducing or neutral) inside the heat treating chamber of a furnace. Cooling may be done by immersion in a fluid (air, water, oil, molten salt or other media) after an elapsed period of time at the holding temperature in order to attain the desired properties. Heat treatment methods include hardening, annealing, stress relieving, tempering, normalising ferrous metals and refining grain of ferrous metals. Surface changes in steel by casehardening include such methods as carburising, nitriding, carbonitriding, cyaniding, flame hardening, induction hardening and siliconising.

Sources of air pollution in heat treatment areas may be attributed to:

1. Malfunctioning of the furnace combustion system.
2. Issuance of vapours and fumes because of unclean surfaces on the work.
3. Oil mist or vapour generation during cooling depending on the medium being used.
4. Molten salt pot fumes.
5. Toxic gases emitted from furnaces under special atmospheres.

During oil quenching, conditions are created which are similar to machining operations in that oil mist and vapours are released. However, these mists and vapours are released in far greater quantities than during machining, adding to the difficulty of abating this pollution. The primary source of toxic and hazardous emissions is the carbonitriding process known as 'gas cyaniding'. The cyanide salt decomposes into carbonates and if these carbonates are carried into the atmosphere uncontrolled they will combine with atmospheric moisture to form toxic compounds.

There is a widely held belief that heat treatment furnaces are major contributors to the emission of many toxic pollutants—cyanides, carbon monoxide, nitrogen, methane and metallic oxides. Effective prevention and control of emission resulting from heat treating operations may be accomplished by the following techniques :

1. Correction of operating procedures combined with selection of the proper fuel and burner, resulting in elimination of smoke.
2. Removal of organic material adhering to the work by degreasing prior to heat treatment.
3. Proper selection of the oils and control of oil bath temperature, which will materially reduce the mists and fumes generated during oil quenching.
4. Removal of fumes from molten salt pots by cloth arresters in baghouses. Baghouses may require constant duty operation because of the hygroscopic and corrosive properties of the salt fumes.
5. Use of flame curtains covering the ends of the heat treat furnace, preventing escape of the special atmosphere into the heat treating department.
6. Use of properly designed canopy hoods above the heat treat furnaces to capture effectively the convection updraft of hot air.

15.4.3 Metal joining processes

In metal joining processes, respiratory protection by means of adequate ventilation must be provided. Pollutants are created as a result of the use of the various base and coating materials in welding, cutting and soldering operations. Actual hazard or damage usually does not occur unless an individual inhales these materials in substantial amounts over an extended period of time. The generation of toxic materials will depend upon the type of welding, the filler, the base metals and the presence of contamination on the base metal or of volatile and reactive solvents in the air.

The following list shows the contaminants for the various welding processes:

1. Shielded metal arc welding.
2. Submerged arc welding.
3. Gas metal arc welding.
4. Resistance welding does not usually involve significant air pollutants, depending on the amount of metal release and its toxicity.
5. Base metals may be extremely hazardous if they have been coated with toxic meterials such as cadmium, lead, or mercury, or paints containing toxic meterials. Beryllium welding may produce an extremely hazardous atmosphere.
6. Some silver brazing alloys contain cadmium which may be volatilised if heated appreciably above its melting point. The yellow-brown fumes are extremely dangerous, producing severe and even fatal lung damage on short exposure. The amount of cadmium fumes produced depends largely on the temperature reached. Low concentrations have been reported when using a gas-air torch, higher with oxyacetylene.
7. Fluxes containing fluorides are frequently used in silver brazing and for brazing and welding aluminium, magnesium and their alloys.

15.4.4 Wastewater treatment in metal fabricating plants

Generally the wastewaters originating in metal fabricating plants are similar regardless of whether they originate from small shops or large production facilities. For the most part the wastewaters from such plants contain small amounts of metal particles, free and soluble oils and various cleaning compounds used in cleaning either the product or the shop itself. Some plants have wastewaters from air pollution control devices for painting operations. These residues are treated in a manner similar to free oils. Normally, the wastes are slightly acid or alkaline and they are usually opaque, milk-coloured and contain some free (nonsoluble) oil.

The simplest form of treating such wastes is by means of gravity skimming tanks. In many cases such skimming will be sufficient to permit discharge of the industrial waste to a sewer. Optimum gravity skimming tanks are generally designed along standards set forth by the American Petroleum Institute which relate tank dimensions to particle size, oil rise rates and detention time as well as to the nature of the type of oil to be skimmed. Such tanks operate best when supplied with a limited, uniform influent flow rate. As a result it is often wise to optimise the skimming tank design and then utilise large holding or equalising tanks upstream of the actual skimming tanks to permit pumps to deliver the wastewater to the gravity skimmers at a fixed design rate. The gravity skimming will remove free oils.

The metal particles will settle in the bottom of the tank and can be removed manually at infrequent intervals, or automatically by means of drag conveyors if such metal particles are deposited in significant quantities. If the metal fabricating plant wastewater contains considerable soluble oils, generally used as a machining coolant, further treatment will be required in addition to gravity skimming. Usually wastewaters from a metal fabricating plant will be contaminated with soluble oils, since even if soluble oils are not used in plant processes, they may well develop because of the mixing of free oils and metal cleaners or emulsifying agents when brought together in the wastewater collection system.

15.5 Pollutant associated with scrap recycling

The activities, pollutant sources and pollutants detailed in Table 15.3 are commonly found at scrap recycling and waste recycling facilities.

Table 15.3: Common activities, pollutant sources and associated pollutants at scrap recycling and waste recycling facilities.

Activity	*Pollutant source*	*Pollutant*
Scrap and waste recycling facilities (non-source separated, non-liquid recyclable materials)		
Stockpiling and storage of materials (including loading and unloading)	Leaking of various fluids from used automotive engines, radiators, brake fluid reservoirs, transmission housings, other vehicle parts and lead-acid from batteries. Deterioration/corrosion of materials	PCBs, oil and grease, lubricants, paint pigments or additives, heavy metals, ionizing radioactive isotopes, transmission and brake fluids, fuel, battery acid, lead acid, antifreeze, benzene, chemical residue, heating oil, petroleum products, solvents, ionizing radioactive isotopes, infectious/bacterial contamination, asbestos, metals, total Kjeldahl nitrogen (TKN), battery acid, oily wastes, chemical residue.

(Cont'd...)

Activity	*Pollutant source*	*Pollutant*
Material processing: Air pollution equipment (including incinerators, furnaces, wet scrubbers, filter houses and bag houses)	Normal equipment operations that include the collection and disposal of filter bag material and ash, process wastewater from scrubbers, accumulation of particulate matter around leaking joint connections, malfunctioning pumps and motors (e.g., leaking gaskets, seals or pipe connections, leaking oil-filled transformer casings)	Hydraulic fluids, oils, fuels, grease and other lubricants, accumulated particulate matter, chemical additives, and PCBs from oil-filled electrical equipment.
Material processing: Combustion engines	Spills and/or leaks from fuel tanks, spills/leaks from oil/ hydraulic fuel reservoirs, faulty/leaking hose connections, worn gaskets, leaking transmissions, crank-cases, and brake systems (if applicable), leaking battery casings and/or corroded terminals	Accumulated particulate matter, oil/lubricants, gas/ diesel fuel, fuel additives, antifreeze (ethylene glycol), battery acid, and products of incomplete combustion
Material processing: Material handling systems (forklifts, cranes and conveyors)	Spills and leaks from fuel tanks, hydraulic and oil reservoirs due to malfunction parts (e.g., worn gaskets and parts, leaking hose connections and faulty seals). Damaged or faulty electrical switches (mercury filled). Damaged or leaking battery casings, including exposed corroded battery terminals. Damaged or worn bearing housings.	Hydraulic fluids, oils, fuels and fuel additives, grease and other lubricants, accumulated particulate matter, chemical additives, mercury, lead, battery acid
Material processing: Stationary scrap processing facilities (balers, briquetters, shredders, shearers, compactors, engine block/ castiron breakers, wire chopper, turnings crusher)	Leaks from hydraulic reservoirs, hose and fitting connections, worn gaskets, spills or leaks from fuel tanks, particulates/residue from scrap processing, malfunctioning pumps and motors (e.g. leaking gaskets, seals or pipe	Heavy metals (e.g. zinc, copper, lead, cadmium, chromium) and hydraulic fluids, PCBs

(Cont'd...)

Activity	*Pollutant source*	*Pollutant*
	connections, leaking oil-filled transformer casings)	
Material processing: Hydraulic equipment and systems, balers/briquetter, shredders, shearers, compactors, engine block/cast iron breaker, wire chopper, turnings crusher	Particulate/residue from material processing, spills and/or leaks from fuel tanks, spills/leaks from oil/hydraulic fuel reservoirs, faulty/leaking hose connections/fittings, leaking gaskets	Hydraulic fluids/oils, lubricants, particulate matter from combustion engines, PCBs (oilfilled electrical equipment components), heavy metals (nonferrous, ferrous)
Material processing: Electrical control systems (transformers, electrical switch gear, motor starters)	Oil leakage from transformers, leakage from mercury float switches, faulty detection devices	PCBs, mercury (float switches), ionizing radioactive material (fire/smoke detection systems)
Material processing: Torch cutting	Residual/accumulated particulates	Heavy metal fragments, fines
Material handling systems	Spills and/or leaks from fuel tanks, spills/leaks from oil/hydraulic fuel reservoirs, faulty/leaking hose connections/fittings, leaking gaskets	Accumulated particulate matter (ferrous and nonferrous metals, plastics, rubber, other), oil/lubricants, PCBs (electrical equipment), mercury (electrical controls), lead/battery acids
Vehicle maintenance	Parts cleaning, waste disposal of rags, oil filters, air filters, batteries, hydraulic fluids, transmission fluids, brake fluids, coolants, lubricants, degreasers, spent solvents	Gas/diesel fuel, fuel heavy metals, brake fluids, additives, oil/lubricants, transmission fluids, chlorinated solvents, arsenic
Vehicle fueling	Spills and leaks during fuel transfer, spills due to 'topping off' tanks, runoff from fueling areas, washdown of fueling areas, leaking storage tanks, spills of oils, brake fluids, transmission fluids, engine coolants	Gas/diesel fuel, fuel additives, oil, lubricants, heavy metals
Vehicle and equipment cleaning and washing	Washing and steam cleaning	Solvent cleaners, oil/lubricants/additives, antifreeze (ethylene glycol)

(Cont'd...)

Activity	*Pollutant source*	*Pollutant*
Waste recycling facilities (liquid recyclable materials)		
Drum/individual container storage and handling	Leaks or spills due to faulty container/drum integrity (e.g. leaking seals or ports). Container materials incompatible with waste material. Improper stacking and storage of containers	Mineral spirits, industrial solvents, immersion cleaners, dry cleaner, solvents, paint solvents, spent antifreeze
Return and fill stations	Leaks, spills, or overflows from tanker truck transfer of wastes and hose drainage. Leaking pipes, valves, pumps, worn or deteriorated gaskets or seals	Mineral spirits, industrial solvents, immersion cleaners, dry cleaner, solvents, paint solvents, spent antifreeze
Storage tank operations	Overfill of storage tanks, leaking pipes, valves, worn or deteriorated pumps seals. Leaking underground storage tanks.	Mineral spirits, industrial solvents, immersion cleaners, dry cleaner, solvents, paint solvents, spent antifreeze
Material handling equipment	Leaking fuel lines, worn gaskets, leaking hydraulic lines and connections	Fuel, hydraulic fluid, oil and grease
Vehicle and equipment maintenance (if applicable)	Replacement of fluids such as transmission and brake fluids, antifreeze, oil and other lubricants, washdown of maintenance areas, dumping fluids down floor drains connected to storm sewer system, outside storage of fluids and oily rags and waste material	Oil and grease, fuel, accumulated particulate matter, antifreeze
Vehicle or equipment washing (if applicable)	Wash water or steam cleaning	Oil, detergents, chlorinated solvents, suspended solids and accumulated particulate matter
Recycling facilities		
Unknowing acceptance of nonrecyclable materials and/or small quantities of household hazardous wastes	Inbound recyclable materials	Dependant on material

(Cont'd...)

Activity	***Pollutant source***	***Pollutant***
Outdoor material storage	Deterioration of wastepaper and unprocessed aluminum beverage containers	Biochemical oxygen demand (BOD)
Processing and storage	Illicit connections or improper dumping to floor drains discharging to a storm sewer system Washing down tipping floor areas	Dependant on material
Vehicle maintenance	Replacement of fluids such as transmission and brake fluids, antifreeze, oil and other lubricants, washdown of maintenance areas, dumping fluids down floor drains connected to storm sewer system, outside storage of fluids and oily rags and waste material	Oil and grease, gas/diesel fuel, accumulated particulate matter, anti-freeze (ethylene glycol)

16

Pollution control in semiconductors and electronic industry

16.1 Introduction

A semiconductor is a tiny complex of electronic components and their connections that is produced in or on a small slice of material, which may be composed of silicon, germanium, gallium arsenide, gallium phosphate, gallium arsenic phosphide and semiconducting crystals. A large percentage of the ingots manufactured are silicon-based. Electrical current flows through a solid medium instead of a gas or vacuum as in other electrical devices.

The semiconductor industry is expanding to meet the demands of a technically oriented society that desires to process information faster and more efficiently. This industry is in the production of integrated circuits supplanting the production of electron tubes and discrete semiconductor devices, such as resistors, capacitors and transistors. The application of a semiconductor will determine the base material and the design, amount and complexity of the circuitry. The base material and circuitry engineering are factors in the amount of wastes that are produced during manufacturing. The PC board and semiconductor industries generate numerous waste streams (Table 16.1), including wastewater, spent process solutions, waste solvents, wastewater treatment sludges, solids and air emissions. The following section discusses the major waste streams generated by the PC board and semiconductor industries and the practices typically used to manage these wastes.

Wastes from PC board and semiconductor facilities can be managed onsite or offsite, depending on: (i) the characteristics of the waste, (ii) the resources of the facility, (iii) the availability and cost of offsite treatment and disposal facilities and (iv) regulatory requirements.

These industries typically treat rinse waters onsite and then discharge to the sewer and dispose of concentrated wastes—such as wastewater treatment sludges, process baths and waste solvents—offsite.

In the electronics industry, components are generally cleaned after soldering to remove contaminants, which originate from the fluxes used to promote the wetting necessary for the formation of good solder joints. The flux residue can interfere with future processes and reduce the aesthetics and reliability of a component. Traditionally, chlorinated, fluorinated and other halogenated solvents have been used to remove these residues.

Table 16.1: Typical wastes generated In the microelectronics industry.

Waste stream	*Generating process*	*Example of waste stream*
Acid solutions	Etching	Hydrofluoric acid etch of silicon wafers
		Sulphuric acid/peroxide removal of copper from PC boards
Solvents—nonhalogenated	Cleaning	Acetone cleaning of silicon wafers
	Photolithography	Xylene as solvent in photoresist
Metal-bearing solutions	Electroplating	Copper and solder plating of PC boards
	Stripping	Removal of solder and copper from PC boards
	Rinsing	Cleanout of metallisation equipment for semiconductors
		Removal of processing solution from plated parts
Solvents—halogenated	Drying	Freon drying of completed components
	Degreasing	1,1,1-Trichloroethane degreasing of components
Alkaline solutions	Cleaning	Alkaline detergents used for pre-plate cleaning of parts
	Stripping	Ammonium hydroxide cleaning of silicon wafers
		Sodium hydroxide stripping of photoresist
Sludges	Wastewater treatment	Calcium fluorides from neutralisation of hydrochloric acid
		Metal hydroxides from precipitation treatment of electroplating rinses
Scrap	Wafer fabrication	Silicon from slicing of semiconductor crystals
	Trimming	Fibreglass with solder and copper from PC boards
	Rejects from testing	Integrated circuits with heavy metals from doping and layering
Photoresists	Photolithography	Pattern definition during fabrication of semiconductors and PC boards
Fluxes	Soldering	Fluxes with copper and lead from PC board assembly/fabrication

Halogenated solvents are used because of their stability, ease of drying and effectiveness in removing oils. These same characteristics, which make these solvents effective in cleaning processes, have detrimental effects on the environment. Solvent evaporation has been investigated for its role in stratospheric ozone.

16.2 Environmental issues in semiconductors industry

Environmental issues in semiconductors and other electronics manufacturing projects primarily include the following:

1. Hazardous material used and waste management.
2. Air emissions.
3. Wastewater.
4. Energy use.
5. General process modifications.

16.2.1 Hazardous materials and wastes

Almost all processes in semiconductors and other electronics manufacturing generate hazardous or potentially hazardous wastes, such as spent deionised water (containing inorganic acid), spent solvents and developers (e.g., iso-paraffinic hydrocarbons), spent cleaning solutions, sludges from wastewater treatment, spent epoxy material (printed circuit board [PCB] and semiconductor manufacturing), spent cyanide solutions (electroplating) and soldering fluxes and metals residue (printed circuit board assembly [PCBA]).

In addition to the relevant measures on hazardous materials management indicated in the general Environmental Health and Safety (EHS) guidelines, specific pollution prevention techniques include process modifications and substitutes as follows:

1. Implementing process or equipment modifications, including:
   (a) Regenerate plating baths by activated carbon filtration to remove built-up organic contaminants, which reduces the volume of plating baths to be disposed off and reduces the need for new chemicals.
   (b) Adopt automated gas cabinet systems to control fugitive emissions of gases from cylinders, particularly during changes.
   (c) Use of lead solder replacements with substitutes, such as tin alloys and other lead-free solders.
2. Raw material substitution or elimination—for example, substitution of cyanide plating solutions (for gold plating in PCB industry) with acid sulphate copper, gold sulphite and electroless nickel; replacing Cr VI

with Cr III plating baths (PCB manufacturing, although the use of chromic plating baths is obsolete).

3. Hazardous substance and waste segregation, separation and preparation—for example, segregation of wastewater sludge by metal contaminants enhances waste recovery storage of plating chemicals to segregate incompatible substances, such as cyanides from acids and oxidising agents from combustibles.
4. Metal recovering and recycling, primarily in the semiconductor and PCBA sectors—for example, recovering copper and precious metals by electrolytic process; removing and recovering copper and tin from boards by electrolysis-chemical precipitation; recovery of arsenic and gallium from gallium arsenide (GaAs) processing wastes (through thermal separation of GaAs solid wastes and recovery from GaAs polishing wastes).
5. Reduce releases of perfluorooctane sulphonates (PFOS) in semiconductor manufacturing by phasing out non-critical uses of PFOS-based substances, such as some etching mixtures for which substitutes exist. For critical PFOS uses where no alternatives exist, such as shorter wavelength technologies used in the manufacture of semiconductors, controlled disposal of wastes should be carried out, particularly if incineration is involved.

Hazardous materials management is discussed in the General EHS Guidelines. Specific measures for this sector include:

1. Process chemicals storage areas should be regularly checked to identify leaks.
2. Underground piping should be in a double pipe, with a means of identifying leaks from the inner pipe.
3. Pipework carrying hazardous materials should be constructed of compatible materials and should be sufficiently supported, clearly labelled and installed with high-quality joints. Piping should also be designed with low point drains, high point vents and isolation valves every 30 meters maximum.
4. Waste spill containment trays should be used.

Solid and hazardous waste management is discussed in the General EHS Guidelines. In this sector include, all wastes with hazardous properties (such as spent deionised water, spent solvents, spent cleaning solutions, sludges from wastewater treatment, spent epoxy material and spent cyanide solutions, among others) should be clearly labelled and stored separately from general waste in dedicated and contained storage areas that are chemically resistant.

Safe storage and containment are essential due to the high reactivity and toxicity of industry waste and by-products, as also discussed in the following section on occupational health and safety.

16.2.2 Air emissions

The main emissions of concern generated by the semiconductors and electronics manufacturing industry include greenhouse gases, toxic, reactive and corrosive substances (for example, acid fumes, dopant, cleaning gases and volatile organic compounds [VOCs]), resulting from diffusion, cleaning and wet-etching processes.

There are three types of abatement systems for toxic and hazardous gases:

1. Point-of-use (POU) systems that are relatively small and typically dedicated to a single process tool. These systems can remove up to 99.99% of effluent gases. For example, a POU scrubber can remove arsine to less than 50 ppb. Six basic technology types are used for POU abatement of gaseous and particulate pollutants, including perfluorocarbon compounds (PFCs), as follows:
   (a) Wet scrubbing in semiconductor manufacturing, although it has a limited treatment range. Wet scrubbers are also used to treat acid gases and by-products of combustion/oxidation treatment.
   (b) Hot chemical beds in semiconductor manufacturing.
   (c) Combustion/oxidation through fuel burners or electrically heated chambers, often combined with wet scrubbers (semiconductor and PCBA manufacturing).
   (d) Plasma reactors in semiconductor manufacturing, although they have a limited treatment range and require additional downstream abatement units.
   (e) Cold adsorbers in PCBA industry.
   (f) Traps, filters, cyclones and precipitators in PCBA industry for removing solids and condensing vapours from exhaust stream.
2. House systems that are relatively much larger and placed outside a fab (semiconductors foundry) and can handle high flow rates of effluents from many different sources.
3. Emergency release scrubbers suitable for handling a large and sudden release of toxic gases are usually dedicated to the exhaust ventilation of gas cylinder storage areas. Emergency release scrubbers are aimed at preventing uncontrolled releases. However, most toxic gases can be controlled in special cabinets that are scrubbed or scrammed to atmosphere after careful monitoring of gas concentration to ensure that the gases are safely released with no impact on health and environment.

Perfluorocarbon compounds and other greenhouse gases

PFCs—including CF_4, C_2F_6 and C_3F_8—nitrogen trifluoride (NF_3) HFC-23 (CHF_3) and sulphur hexafluoride (SF_6) are used in semiconductor manufacturing, as cleaning gases in chemical vapour deposition (CVD) systems, in plasma etching and primarily, in thin film transistor-liquid crystal display (TFT-LCD) screens manufacturing. The main environmental issue associated with PFCs is their high global warming potential (GWP), which is linked to their long atmospheric life.

PFC emission reduction and control techniques include the following:

1. Process optimisation, especially in CVD cleaning processes.
2. Chemical substitution, for example, using c-C_4F_8 or NF_3 as a drop-in alternative chamber cleaning gas instead of C_2F_6 in a modified CVD chamber, minimising atmospheric emission.
3. Abatement, through dissociation of the molecules into non-PFC by-products, by combustion, catalytic decomposition, or plasma destruction systems (the latter is applicable to etch tools only, less than or equal to 200 mm). Thermal destruction technology can be applied to chamber cleaning and etching processes within a fab (POU applications) or fab-wide (end-of-pipe applications).
4. Capture and reuse of PFCs from exhaust streams, which is, however, a technically and economically challenging process.

Acid fumes

Potential emissions of acid fumes (mainly, hydrochloric acid and hydrofluoric acid) are related to the following processes in semiconductor manufacturing and PCBA industry:

1. Cleaning, etching and resist-stripping operations in semiconductors manufacturing.
2. Etching, during which hydrogen chloride vapours may be released.
3. Cleaning, surface preparations, cupric chloride etching and plating in PCB manufacturing.

Sulphuric acid aerosol emissions are also associated with the treatment of wafers with acid-etching mixtures. The most commonly used mixture contains sulphuric acid and hydrogen peroxide.

Acid fume emissions are reduced through the installation of horizontal (cross-flow) wet scrubbers or vertical (counterflow) wet scrubbers. Pollution prevention measures also include the following:

1. Use of a mist suppressant on bath solution surfaces and use of wetting agents (surfactants).

2. Reprocessing sulphuric acid used during wafer fabrication through heating and distillation to purify the acid stream, which is recovered and pumped back into wet stations.
3. Installation of plating bath covers and meshpad mist eliminators.

Volatile organic compounds

Volatile organic compounds (VOCs) are primarily used in semiconductor manufacturing and PCBA industry. VOCs may be released in most cleaning and photolithography processes, during resist-drying, developing and resist-stripping operations. Usually, VOC emissions are adsorbed onto activated carbon systems to facilitate recovery and/or treated by thermal oxidisers. Applicable pollution control techniques or add-on control devices used to control emissions consist of the following:

1. Regenerative thermal oxidisers, which are usually practical when process exhaust stream volumetric flow rates are above about 3000 scfm.
2. Zeolite rotor concentrators with recuperative thermal oxidisers, which are used to concentrate dilute VOC streams prior to sending them to a destruction or recovery device.
3. Fixed bed carbon adsorption with steam stripping for VOC recovery (either reuse or recycling off site).
4. Fluidised bed carbon adsorption with hot nitrogen desorption and VOC recovery (either reuse or recycling off site).
5. Fluidised bed polymer adsorption with hot nitrogen regeneration and VOC recovery where practical or with recuperative thermal oxidisers.

Nitrogen oxides

As in other industrial sectors, NO_x emissions in semiconductors manufacturing include by-products of combustion processes. These by-products are derived from heating system boilers, emergency stand by power generators and thermal oxidisers that reduce VOC emissions.

Dust

Drilling and routing processes during PCB manufacture generate significant amounts of dust, while semiconductor and PCBA industries are not significant emitters of dust. Some limited dust is generated by laser cutting, trimming, chemical mechanical polishing and backgrinding process in semiconductors manufacturing, as well as from the manufacturing of magnetic devices and passive components. Recommended control measures include:

1. Water sedimentation systems.
2. Abatement with bag filters or electrostatic precipitators.

16.2.3 Energy consumption

Because there are many thermal processes and wafers handling is highly mechanised, semiconductor manufacturing involves significant energy use, which demands optimal energy consumption. The use of specialised equipment, which combines improved performance efficiency and energy efficiency should be implemented, for example:

1. Air-handling equipment that controls humidity and temperature, allowing up to 25% of energy saved.
2. High-efficiency chillers.
3. Recovering heat from water condensers that use heat exchangers may allow a modern industrial facility to save up to 40% of its needs.

Advanced technologies in emissions abatement also provide new equipment with enhanced abatement efficiency and lower energy consumption.

16.2.4 Wastewater

Industrial process wastewater

Wastewater effluents may be impacted by organic and inorganic compounds, such as metals, acids and alkalis, cyanides and suspended solids. To minimise both water use and potential discharge impacts, rinse water should be recovered to return treated water to the process for reuse.

Process wastewater may include organic compounds, particularly non-chlorinated solvents (e.g., pyrrole-based, aminebased, fluoro/ether-based resists, isopropyl alcohol and tetramethylammonium hydroxide) from a number of semiconductor and PCBA manufacturing steps, including cleaning, resist drying, developing and resist stripping; metals from metallisation and CMP processes; acids and alkalis from spent cleaning solutions, process operations such as etching, cleaning and metallisation, among others; cyanides from metallisation processes and suspended solids from film residues and metallic particles (derived from photolithography, metallisation, backgrinding and dicing processes).

Process wastewater treatment

Since the semiconductors and electronics manufacturing operations use a diverse range of raw materials, chemicals and processes, wastewater treatment may require the use of unit operations specific to the manufacturing process in use and the specific contaminant. Techniques for treating industrial process wastewater in this sector include: (i) source segregation and pre-treatment of wastewater streams containing high concentrations of non-biodegradable compounds using phase separation such as solvent recovery, air stripping, chemical oxidation, adsorption processes, etc., (ii) reduction in heavy metals using chemical precipitation, coagulation and flocculation, electrochemical recovery, ion exchange, etc., (iii) chemical oxidation of cyanides and (iv) dewatering and disposal of residuals in designated hazardous waste landfills.

Additional engineering controls may be required for: (i) advanced metals removal using membrane filtration or other physical/chemical treatment technologies, (ii) removal of recalcitrant organics and halogenated organics using activated carbon or advanced chemical oxidation, (iii) reduction in effluent toxicity using appropriate technology (such as reverse osmosis, ion exchange, activated carbon, etc.), and (iv) containment and treatment of volatile organics stripped from various unit operations in the wastewater treatment system.

Other wastewater streams and water consumption

Guidance on the management of non-contaminated wastewater from utility operations, non-contaminated stormwater and sanitary sewage is provided in the General EHS Guidelines. Contaminated streams should be routed to the treatment system for industrial process wastewater. Recommendations to reduce water consumption, especially where it may be a limited natural resource, are provided in the General EHS Guidelines.

Printed circuit board (PCB) manufacturing

Several pollution prevention measures have been developed in the PCB manufacturing process. Examples of process modifications with environmental benefits include:

1. Board manufacture: Surface mount technology (SMT) rather than plated through-hole technology, injection molded substrate, additive plating.
2. Cleaning and surface preparation: Use of non-chelating cleaners, extend bath life, improve rinse efficiency, counter current cleaning and recycle/reuse cleaners and rinses.
3. Pattern printing and masking: Aqueous processable resist, screen printing to replace photolithography, ink-jet printing, dry photoresist, recycle/reuse photoresist strippers, segregate streams and recover metals.
4. Electroplating and electroless plating: Replace by mechanical board production, non-cyanide baths, extend bath life, recycle/reuse cleaners and rinses, improve rinse efficiency, countercurrent rinsing, segregate streams and recover metals.
5. Etching: Use differential plating, non-chelated etchants and non-chrome etchant, pattern versus panel plating, additive versus subtractive process and recycle/reuse etchants.

6. Metal recovery by regenerative electrowinning and ion exchange technologies results in a near zero effluent discharge for segregated metal bearing streams. Heavy metals are recovered to metal sheets, which eliminate 95% of sludge disposal. Metal-bearing sludges that are not treated for recovery of metals should be disposed in secure landfills.

16.3 Occupational health and safety

Occupational health and safety hazards in semiconductors and other electronic manufacturing projects primarily include the following:

1. Exposure to material released by substrates during handling or mechanical manipulation.
2. Exposure to hazardous process chemicals, including metallic powders.
3. Physical hazards and exposure to energy hazards (kinetic, electrical, pneumatic and hydraulic).
4. Exposure to ionising and nonionising radiation and lasers.

While silicon-based semiconductor substrates (silicon dioxide) are non toxic dust arising from their manufacture and use can be hazardous. However GaAs and Indium Phosphide (InP) substrates pose more serious health and physical impacts. The most common exposure pathway for GaAs and InP is inhalation of particulates. Because of the high toxicity of arsenic and indium, both of these compounds have low occupational exposure levels. InP is flammable and can react with water vapour and acids to form phosphine, a toxic and flammable gas. GaAs is a hazard when ground, cut, or polished.

Prevention and control of these hazards involve the adoption of engineering and administrative controls to safeguard workers.

17

Environmental aspects for e-wastes

17.1 Introduction

E-waste is any refuse created by discarded electronic devices and components as well as substances involved in their manufacture or use. The disposal of electronics is a growing problem because electronic equipment frequently contains hazardous substances. In a personal computer, for example, there may be lead in the Cathode Ray Tube (CRT) and soldering compound, mercury in switches and housing and cobalt in steel components, among other equally toxic substances.

Around the world, a number of initiatives have arisen to address the issue of e-waste, by promoting the reuse of electronic devices (e-cycling) and mandating that safer alternatives to hazardous substances be used in their manufacture whenever possible. In Europe, legislation has been drafted to deal with the problem, including the Waste from Electrical and Electronic Equipment (WEEE) Directive and the Basel Convention.

Consumer electronics have become an integral part of daily life and revolutionised the way we communicate, retrieve information and entertain ourselves. Between cell phones, computers, televisions, iPads and e-Readers, it is estimated that the average person in the United States uses 24 electronic products. Rapid technological advancements and growth in the electronics industry have led to a constant stream of new products and a resulting decrease in the life span of electronics. Globally, more than 70 million tonnes of e-waste were discarded in 2012 and 95 million tonnes are expected to be disposed in 2020. Europeans produce approximately 20 kilograms of e-waste/person/year, while U.S., residents produce about 7 kilograms of e-waste/person/year. This discrepancy may be attributed to the varying definitions of e-waste; in the U.S. electronic waste generally consists of information technology (IT) and telecommunications equipment, monitors and televisions, whereas in Europe it also includes large household appliances, cooling and freezing appliances and medical devices.

Therefore, responsible end of-life management of e-waste is imperative in order to recover valuable components and properly manage hazardous and toxic components. End-of-life management of e-waste includes reuse of functional electronics, refurbishment and repair of electronics, recovery of electronic components, recycling e-waste and disposal. Reuse, refurbishment or repair of electronic products is most desirable since this option increases the life span of the electronic product and higher resource efficiency. Recycling of electronics allows for precious and special metals to be recovered, reduces the environmental impact associated with electronic manufacturing from raw materials and ensures that hazardous and toxic substances are handled properly.

17.2 Environmental/resource factors

Circuit boards contain the highest value of precious metals in a computer, as well as most of the heavy metals. The components of a personal computer have the highest economic value, due to gold plated connectors, components, pins and transistors:

1. Motherboard (main circuit board).
2. Peripheral Component Interconnect (PCI) boards (connects to motherboard).
3. Random Access Memory (RAM) (long, rectangular small circuit boards).
4. Processor (large chip that plugs directly into the motherboard).

In addition to recovering precious metals, recycling electronics also reduces the environmental impact associated with primary production of electronic products. The primary production of precious and special metals, including energy intensive stages such as mining and smelting, has a significant impact on carbon dioxide emissions. Reuse and recovery of electronics reduces the environmental impact of these products, as well as the impact from primary production of metals and fractions found in electronics.

17.2.1 Public health factors

Discarded electronics contain a variety of toxic metals, including lead, cadmium, mercury, chromium and polyvinyl chlorides and thus the disposal of electronics poses a significant environmental and health risk when not properly handled. Although e-waste represents less than 2% of landfill mass, it contains 70% of the hazardous waste in heavy metals. The following hazardous components can be found in e-waste (Table 17.1). It is estimated that 50 to 80% of e-waste collected in developed nations is exported to developing countries such as China, India and Pakistan due to cheap labour and lenient environmental regulations. These developing nations lack the health and safety infrastructure to process and dispose of materials safely and consequently workers handle toxic metals without proper equipment. While there are operators in China who are licensed to process e-waste, the market is dominated by small-scale entities that are not authorised, nor properly equipped to treat e-scrap. Common techniques for processing e-waste in developing nations include manual dismantling of hazardous materials and open-air burning, which generates significant accounts of dioxins and furans if performed without proper emission control systems. Cyanide leaching is also a prevalent technique for processing e-waste in developing countries, posing a significant concern to worker well-being if the spent leaching solution is not properly disposed.

Table 17.1: Effects of e-waste constituent on health.

Source of e-wastes	*Constituent*	*Health effects*
Solder in printed circuit boards, glass panels and gaskets in computer monitors	Lead (PB)	• Damage to central and peripheral nervous systems, blood systems and kidney damage • Affects brain development of children
Chip resistors and semiconductors	Cadmium (CD)	• Toxic irreversible effects on human health. • Accumulates in kidney and liver. • Causes neural damage. • Teratoenic.
Relays and switches, printed circuit boards	Mercury (Hg)	• Chronic damage to the brain. • Respiratory and skin disorders due to bioaccumulation in fishes
Corrosion protection of untreated and galvanised steel plates, decorator or hardner for steel housings	Hexavalent chromium (Cr) VI	• Asthmatic bronchitis • DNA damage
Cabling and computer housing	Plastics including PVC	Burning produces dioxm. It causes • Reproductive and developmental problems • Immune system damage • Interefere with regulatory hormones
Plastic housing of electronic equipments and circuit boards	Brominated flameretardants (BFR)	• Disrupts endocrine system functions
Front panel of CRTs	Barium (Ba)	Short term exposure causes: • Muscle weakness • Damage to heart, liver and spleen.
Motherboard	Beryllium (Be)	• Carcinogenic (lung cancer) • Inhalation of fumes and dust. Causes chronic beryllium diseaseor beryllicosis • Skin diseases such as warts

Data security factors

Privacy protection concerns have also fuelled the processing of electronic waste. Confidential and personal data must be destroyed properly in order to ensure the safety of organisations and individuals information.

17.2.2 End-of-life options for e-waste

End-of life management options for electronic waste include:

1. Reuse of functional electronics.
2. Refurbishment and repair of electronics.
3. Reuse and recovery of electronic components.
4. End-processing for recovering metals.
5. Disposal.

Reuse, refurbishment or repair of electronic products is most desirable since this option increases the life span of the electronic product in order to achieve greater resource efficiency. However, in order to reuse electronics, the equipment must be functional and working. The minimum requirements for donation vary depending on the organisation receiving the electronics. Recycling of electronics allows for precious and special metals to be recovered, reduces the environmental impact associated with electronic manufacturing from raw materials and ensures that hazardous substances in electronics are handled correctly. It should be noted that reuse and recycling are not alternative options; reused products need to be recycled properly and efficiently at the end of their useful life.

E-waste processing can be broken into three major steps: (i) collection, (ii) sorting/dismantling/mechanical processing (including shredding, magnetic separation, etc.), and (iii) end-processing. Figure 17.1 shows e-waste processing steps.

Collection: Collection generally takes place at a regional or national level and is achieved through take-back programmes sponsored by retails and manufacturers of electronics, municipal drop-off collection centers and non-profit and for-profit collection programmes. There are many different entities which collect e-scrap for recycling, ranging from local municipal governments, to large waste management companies.

Sorting/dismantling and mechanical processing: Sorting, dismantling and pre-processing generally takes place at a regional level or national level and has the end goal of separating device streams into material streams, primarily metals, glass and plastics, for end-processing. The goal of this stage is to upgrade the valuable material content and remove and safely dispose of hazardous. It should be noted that the optimal level of pre-processing is dictated

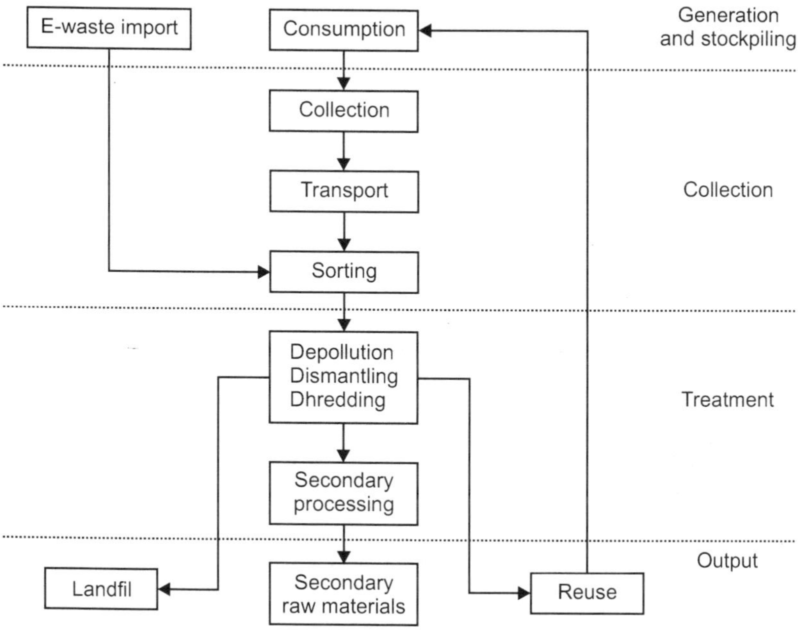

Figure 17.1: E-waste processing steps.

by the quality of feed requirements for end-processing. Excessive pre-processing not only adds cost, but also may lead to significant losses of precious metals. Therefore there is an optimal level of preprocessing that needs to be achieved.

Once components are separated, ferrous fractions are sent to steel plants for recovery of iron, aluminium fractions are sent to aluminium smelters and copper alloys are sent to an integrated smelter to recover precious metals, copper and other non-ferrous metals.

End processing: End-processing takes place at a global level and is dictated by the material stream. The goal of this step is to recover valuable components (i.e., precious metals) and remove impurities. Sampling and assaying is necessary in order to determine the composition and content of precious metals in the e-waste stream and to ensure that the optimum process is used to recover precious metals.

Pyrometallurgical is the primary method used to recover precious metals, however hydrometallurgical and biometallurgical methods have been gaining in popularity over the last two decades.

Collection, dismantling, pre-processing and recovery from the less complex parts of e-waste (e.g., ferrous, copper and aluminium) generally takes place at local or regional facilities. End processing of the more complex components of e-wastes (e.g., circuit boards, batteries, cell phones) commonly occurs in integrated copper smelters and takes place in a global context. These smelters use non-ferrous extractive metallurgy to separate complex fractions into their constituent metals. These plants are very costly to build and therefore it is not feasible or practical to build them in every country.

It is important to note that steps 1 and 2 (ref. section 17.2.2) have been developed with a focus on the electronic device stream (i.e., ICT equipment, C&F appliances and monitors and TV), whereas end processing technologies have been developed with a focus on the material stream. Sorting/dismantling/mechanical processing generally uses mechanical processes, whereas chemical processes are used in end processing. Methods for e-waste processing are shown in Table 17.2.

Table 17.2: Methods for e-waste processing.

Recycling stage	*Stream*	*Process*	*Level*
Collection	Device	Manual	Regional or National
Sorting/dismantling and mechanical processing	Device	Manual and mechanical	Regional or National
End-processing	Material	Chemical	Global

The processing sequence is dictated by the geographic location, type of device, grade of components and toxicity. For example, devices such as mobile phones and MP3 players do not always require shredding or dismantling processes and can be sent directly to an end-processor to recover the metals; whereas computers require manual dismantling and mechanical preprocessing in order to separate and sort the various fractions. It is very important to optimise all steps of the recycling stage. Generally speaking, the greater the number of steps in a recycling process, the greater the risk of losing precious metals. The processes employed during sorting/dismantling influence how e-waste is treated in end-processing steps. Attention needs to be paid to the interfaces between the many steps in the e-waste recycling change in order to ensure that the highest quality metals are recovered.

17.2.3 Financing the recycling of e-waste

Approaches to financing e-waste recycling depend on many factors including state legislation and policies, social preferences, commodity prices and available recycling facilities. Although many programmes offer 'free' recycling to the consumer, recycling is never truly free because of costs associated with collecting, transporting, processing, refurbishing and disposing electronic waste. Before discussing the current state of e-waste disposal in a global and national context, it is important to understand the two main financing models for e-waste collecting and recycling efforts; extended producer responsibility (otherwise known as manufacturers responsibility) and advance recycling fee.

17.2.4 Extended producer responsibility

Extended Producer Responsibility (EPR) assigns collection and recycling responsibility to the manufacturer. EPR is defined as 'an environmental protection strategy to reach an environmental objective of a decreased total environmental impact of a product, by making the manufacturer of the product responsible for the entire life-cycle of the product and especially for the take-back, recycling and final disposal'. The purpose of EPR is to promote social responsibility by encouraging manufacturers to take into account end-of-life management during the product design phase.

In addition to recycling e-waste, electronic manufactures can take the following actions to achieve a level of EPR:

1. Use recycled and environmentally friendly materials.
2. Design products that minimise resource use.
3. Reuse by-products and waste of manufacturing process.
4. Minimise packaging or use recyclable packaging.
5. Reduce toxic and hazardous substances used in the manufacturing process and product itself.
6. Recycle e-waste through certified electronic recyclers to ensure that e-waste is properly managed.

Currently the EPR approach is used in all European Union countries and twenty-three of the twenty-five states in the U.S. that have enacted e-waste legislation.

Advance recycling fee

The Advance Recycling Fee (ARF) is a fee paid by the customer at the point of purchase, depending on the size and type of the electronic. In California, the fee is then deposited into a state recycling fund, which is used to pay qualified e-waste collector and recyclers to cover the cost of managing e-waste. In South Korea, consumers are not required to pay a collection fee if they buy a replacement product; and the retailer collects the e-waste. ARF models are also in place in Switzerland, Belgium and select provinces in Canada.

17.3 Environmentally sound options for e-wastes management

E-waste is a popular, informal name for electronic products nearing the end of their useful life. E-wastes are considered dangerous, as certain components of some electronic products contain materials that are hazardous, depending on their condition and density. The hazardous content of these materials pose a threat to human health and environment.

Industrial revolution followed by the advances in information technology during the last century has radically changed people's lifestyle. Although this development has helped the human race, mismanagement has led to new problems of contamination and pollution. The technical prowess acquired during the last century has posed a new challenge in the management of wastes. For example, Personal Computers (PCs) contain certain components, which are highly toxic, such as chlorinated and brominated substances, toxic gases, toxic metals, biologically active materials, acids, plastics and plastic additives. The hazardous content of these materials pose an environmental and health threat. Thus proper management is necessary while disposing or recycling e-wastes.

17.4 Effects on environment and human health

Disposal of e-wastes is a particular problem faced in many regions across the globe. Computer wastes that are land filled produces contaminated leachates which eventually pollute the groundwater. Acids and sludge obtained from melting computer chips, if disposed on the ground causes acidification of soil. For example, Guiyu, Hong Kong a thriving area of illegal e-waste recycling is facing acute water shortages due to the contamination of water resources. This is due to disposal of recycling wastes such as acids, sludges, etc., in rivers. Now water is being transported from faraway towns to cater to the demands of the population. Incineration of e-wastes can emit toxic fumes and gases, thereby polluting the surrounding air. Improperly monitored landfills can cause environmental hazards. Mercury will leach when certain electronic devices, such as circuit breakers are destroyed. The same is true for Polychlorinated Biphenyls (PCBs) from condensers. When brominated flame retardant plastic or cadmium containing plastics are landfilled, both Polybrominated Diphenyl Ethers (PBDE) and cadmium may leach into the soil and groundwater. It has been found that significant amounts of lead ion are dissolved from broken lead containing glass, such as the cone glass of cathode ray tubes, gets mixed with acid waters and are a common occurrence in landfills. Not only does the leaching of mercury poses specific problems, the vapourisation of metallic mercury and dimethylene mercury, both part of Waste Electrical and Electronic Equipment (WEEE) is also of concern.

In addition, uncontrolled fires may arise at landfills and this could be a frequent occurrence in many countries. When exposed to fire, metals and other chemical substances, such as the extremely toxic dioxins and furans (TCDD-tetrachloro - dibenzo - dioxin. PCDDs - polychlorinated dibenzodioxins.

PBDDs-polybrominated dibenzo-dioxin and PCDFs - polychlorinated dibenzo furans) from halogenated flame retardant products and PCB containing condensers can be emitted. The most dangerous form of burning e-waste is the open-air burning of plastics in order to recover copper and other metals. The toxic fallout from open air burning affects both the local environment and broader global air currents, depositing highly toxic by-products in many places throughout the world.

If these electronic items are discarded with other household garbage, the toxics pose a threat to both health and vital components of the ecosystem. In view of the ill-effects of hazardous wastes to both environment and health, several countries exhorted the need for a global agreement to address the problems and challenges posed by hazardous waste. Also, in the late 1980s, a tightening of environmental regulations in industrialised countries led to a dramatic rise in the cost of hazardous waste disposal. Searching for cheaper ways to get rid of the wastes, 'toxic traders' began shipping hazardous waste to developing countries.

International outrage following these irresponsible activities led to the drafting and adoption of strategic plans and regulations at the Basel. The convention secretariat, in Geneva, Switzerland, facilitates and implementation of the convention and related agreements. It also provides assistance and guidelines on legal and technical issues, gathers statistical data and conducts training on the proper management of hazardous waste.

17.5 Basel convention

The fundamental aims of the basel convention are the control and reduction of transboundary movements of hazardous and other wastes including the prevention and minimisation of their generation, the environmentally sound management of such wastes and the active promotion of the transfer and use of technologies. A draft strategic plan has been proposed for the implementation of the basel convention.

The draft strategic plan takes into account existing regional plans, programmes or strategies, the decisions of the conference of the parties and its subsidiary bodies, ongoing project activities and process of international environmental governance and sustainable development. The draft requires action at all levels of society: training, information, communication, methodological tools, capacity building with financial support, transfer of know-how, knowledge and sound, proven cleaner technologies and processes to assist in the concrete implementation of the basel declaration. It also calls for the effective involvement and coordination by all concerned stakeholders as essential for achieving the aims of the basel declaration within the approach of common but differentiated responsibility.

A set of interrelated and mutually supportive strategies are proposed to support the concrete implementation of the activities as indicated below:

1. To involve experts in designing communication tools for creating awareness at the highest level to promote the aims of the basel declaration on environmentally sound management and the ratification and implementation of the basel convention, its amendments and protocol with the emphasis on the short-term activities.
2. To engage and stimulate a group of interested parties to assist the secretariat in exploring fund raising strategies including the preparation of projects and in making full use of expertise in non-governmental organisations and other institutions in joint projects.
3. To motivate selective partners among various stake holders to bring added value to making progress in the short-term.
4. To disseminate and make information easily accessible through the internet and other electronic and printed materials on the transfer of know-how, in particular through Basel Convention Regional Centers (BCRCs).
5. To undertake periodic review of activities in relation to the agreed indicators.
6. To collaborate with existing institutions and programmes to promote better use of cleaner technology and its transfer, methodology, economic instruments or policy to facilitate or support capacity-building for the environmentally sound management of hazardous and other wastes.

The Basel convention brought about a respite to the transboundary movement of hazardous waste. India and other countries have ratified the convention. However United States (U.S.) is not a party to the ban and is responsible for disposing hazardous waste, such as, e-waste to Asian countries even today. Developed countries such as U.S., should enforce stricter legislations in their own country for the prevention of this horrifying act.

In the European Union, where the annual quantity of electronic waste is likely to double in the next 12 years, the European parliament recently passed legislation that will require manufacturers to take back their electronic products when consumers discard them this is called extended producer responsibility. It also mandates a timetable for phasing out most toxic substances in electronic products.

17.6 Management of e-wastes

It is estimated that 75% of electronic items are stored due to uncertainty of how to manage it. These electronic junks lie unattended in houses, offices, warehouses, etc., and normally mixed with household wastes, which is finally disposed off at landfills. This necessitates implementable management measures. In industries management of e-waste should begin at the point of generation. This can be done by waste minimisation techniques and by sustainable product design. Waste minimisation in industries involves adopting:

1. Inventory management.
2. Production-process modification.
3. Volume reduction.
4. Recovery and reuse.

17.6.1 Inventory management

Proper control over the materials used in the manufacturing process is an important way to reduce waste generation. By reducing both the quantity of hazardous materials used in the process and the amount of excess raw materials in stock, the quantity of waste generated can be reduced. This can be done in two ways, i.e., establishing material-purchase review and control procedures and inventory tracking system.

Developing review procedures for all material purchased is the first step in establishing an inventory management programme. Procedures should require that all materials be approved prior to purchase. In the approval process all production materials are evaluated to examine if they contain hazardous constituents and whether alternative non-hazardous materials are available.

Another inventory management procedure for waste reduction is to ensure that only the needed quantity of a material is ordered. This will require the establishment of a strict inventory tracking system. Purchase procedures must be implemented which ensure that materials are ordered only on an as-needed basis and that only the amount needed for a specific period of time is ordered.

17.6.2 Production-process modification

Changes can be made in the production process, which will reduce waste generation. This reduction can be accomplished by changing the materials used to make the product or by the more efficient use of input materials in the production process or both. Potential waste minimisation techniques can be broken down into three categories: (i) improved operating and maintenance procedures, (ii) material change and (iii) process-equipment modification.

Improvements in the operation and maintenance of process equipment can result in significant waste reduction. This can be accomplished by reviewing current operational procedures or lack of procedures and examination of the production process for ways to improve its efficiency. Instituting standard operation procedures can optimise the use of raw materials in the production process and reduce the potential for materials to be lost through leaks and spills. A strict maintenance programme, which stresses corrective maintenance, can reduce waste generation caused by equipment failure. An employee-training programme is a key element of any waste reduction programme. Training should include correct operating and handling procedures, proper equipment use, recommended maintenance and inspection schedules, correct process control specifications and proper management of waste materials.

Hazardous materials used in either a product formulation or a production process may be replaced with a less hazardous or non-hazardous material. This is a very widely used technique and is applicable to most manufacturing processes. Implementation of this waste reduction technique may require only some minor process adjustments or it may require extensive new process equipment. For example, a circuit board manufacturer can replace solvent-based product with water-based flux and simultaneously replace solvent-vapour degreaser with detergent parts washer.

Installing more efficient process equipment or modifying existing equipment to take advantage of better production techniques can significantly reduce waste generation. New or updated equipment can use process materials more efficiently, producing less waste. Additionally such efficiency reduces the number of rejected or off-specification products, thereby reducing the amount of material which has to be reworked or disposed of. Modifying existing process equipment can be a very cost-effective method of reducing waste generation. In many cases the modification can just be relatively simple changes in the way the materials are handled within the process to ensure that they are not wasted. For example, in many electronic manufacturing operations, which involve coating a product, such as electroplating or painting, chemicals are used to strip off coating from rejected products so that they can be recoated. These chemicals, which can include acids, caustics, cyanides, etc., are often a hazardous waste and must be properly managed. By reducing the number of parts that have to be reworked, the quantity of waste can be significantly reduced.

17.6.3 Volume reduction

Volume reduction includes those techniques that remove the hazardous portion of a waste from a non-hazardous portion. These techniques are usually to reduce the volume and thus the cost of disposing of, a waste material. The techniques that can be used to reduce waste-stream volume can be divided into 2 general categories: source segregation and waste concentration. Segregation of wastes is in many cases a simple and economical technique for waste reduction. Wastes containing different types of metals can be treated separately so that the metal value in the sludge can be recovered. Concentration of a waste stream may increase the likelihood that the material can be recycled or reused. Methods include gravity and vacuum filtration, ultra filtration, reverse osmosis, freeze vapourisation, etc. For example, an electronic component manufacturer can use compaction equipments to reduce volume of waste cathode ray-tube.

17.6.4 Recovery and reuse

This technique could eliminate waste disposal costs, reduce raw material costs and provide income from a salable waste. Waste can be recovered on-site, or at an off-site recovery facility, or through inter industry exchange. A number of physical and chemical techniques are available to reclaim a waste material such as reverse osmosis, electrolysis, condensation, electrolytic recovery, filtration, centrifugation, etc. For example, a printed-circuit board manufacturer can use electrolytic recovery to reclaim metals from copper and tin-lead plating bath. However recycling of hazardous products has little environmental benefit- it simply moves the hazards into secondary products that eventually have to be disposed of. Unless the goal is to redesign the product to use non hazardous materials, such recycling is a false solution.

17.6.5 Sustainable product design

Minimisation of hazardous wastes should be at product design stage itself keeping in mind the following factors.

Rethink the product design: Efforts should be made to design a product with fewer amounts of hazardous materials. For example, the efforts to reduce material use are reflected in some new computer designs that are flatter, lighter and more integrated. Other companies propose centralised networks similar to the telephone system.

Use of renewable materials and energy: Bio-based plastics are plastics made with plant-based chemicals or plant-produced polymers rather than from petrochemicals. Bio-based toners, glues and inks are used more frequently. Solar computers also exist but they are currently very expensive.

Use of non-renewable materials that are safer: Because many of the materials used are non-renewable, designers could ensure the product is built for reuse, repair and/or upgradeability. Some computer manufacturers such as Dell and Gateway lease out their products thereby ensuring they get them back to further upgrade and lease out again.

17.7 Management options

Considering the severity of the problem, it is imperative that certain management options be adopted to handle the bulk e-wastes. Following are some of the management options suggested for the government, industries and the public.

17.7.1 Responsibilities of the government

Governments should set up regulatory agencies in each district, which are vested with the responsibility of co-ordinating and consolidating the regulatory functions of the various government authorities regarding hazardous substances.

Governments should be responsible for providing an adequate system of laws, controls and administrative procedures for hazardous waste management. Existing laws concerning e-waste disposal be reviewed and revamped. A comprehensive law that provides e-waste regulation and management and proper disposal of hazardous wastes is required.

Such a law should empower the agency to control, supervise and regulate the relevant activities of government departments.

Under this law, the agency concerned should:

1. Collect basic information on the materials from manufacturers, processors and importers and to maintain an inventory of these materials. The information should include toxicity and potential handful effects.
2. Identify potentially handful substances and require the industry to test them for adverse health and environmental effects.
3. Control risks from manufacture, processing, distribution, use and disposal of electronic wastes.
4. Encourage beneficial reuse of 'e-waste' and encouraging business activities that use waste. Set up programmes so as to promote recycling among citizens and businesses.
5. Educate e-waste generators on reuse/recycling options.

Governments must encourage research into the development and standard of hazardous waste management, environmental monitoring and the regulation of hazardous waste-disposal.

Governments should enforce strict regulations against dumping e-waste in the country by outsiders. Where the laws are flouted, stringent penalties must be imposed. In particular, custodial sentences should be preferred to paltry fines, which these outsiders/foreign nationals can pay.

Governments should enforce strict regulations and heavy fines levied on industries, which do not practice waste prevention and recovery in the production facilities.

Polluter pays principle and extended producer responsibility should be adopted.

Governments should encourage and support NGOs and other organisations to involve actively in solving the nation's e-waste problems.

Uncontrolled dumping is an unsatisfactory method for disposal of hazardous waste and should be phased out.

Governments should explore opportunities to partner with manufacturers and retailers to provide recycling services.

Responsibility and role of industries:

1. Generators of wastes should take responsibility to determine the output characteristics of wastes and if hazardous, should provide management options.
2. All personnel involved in handling e-waste in industries including those at the policy, management, control and operational levels, should be properly qualified and trained. Companies can adopt their own policies while handling e-wastes. Some of the policies are given below:
   (a) Use label materials to assist in recycling (particularly plastics).
   (b) Standardise components for easy disassembly.
   (c) Re-evaluate 'cheap products' use, make product cycle 'cheap' and so that it has no inherent value that would encourage a recycling infrastructure.
   (d) Create computer components and peripherals of biodegradable materials.
   (e) Utilise technology sharing particularly for manufacturing and de manufacturing. Encourage/promote/require green procurement for corporate buyers.
   (f) Look at green packaging options.
3. Companies can and should adopt waste minimisation techniques, which will make a significant reduction in the quantity of e-waste generated and thereby lessening the impact on the environment. It is a 'reverse production' system that designs infrastructure to recover and reuse every material contained within e-wastes - metals such as lead, copper, aluminium and gold and various plastics, glass and wire. Such a 'closed loop' manufacturing and recovery system offers a win-win situation for everyone - less of the Earth will be mined for raw materials and groundwater will be protected, researchers explain.
4. Manufacturers, distributors and retailers should undertake the responsibility of recycling/disposal of their own products.
5. Manufacturers of computer monitors, television sets and other electronic devices containing hazardous materials must be responsible for educating consumers and the general public regarding the potential threat to public health and the environment posed by their products. At minimum, all computer monitors, television sets and other electronic devices containing hazardous materials must be clearly labelled to identify environmental hazards and proper materials management.

17.7.2 Responsibilities of the citizen

Waste prevention is perhaps more preferred to any other waste management option including recycling. Donating electronics for reuse extends the lives of valuable products and keeps them out of the waste management system for a longer time. But care should be taken while donating such items, i.e., the items should be in working condition.

Reuse, in addition to being an environmentally preferable alternative, also benefits society. By donating used electronics, schools, non-profit organisations and lower-income families can afford to use equipment that they otherwise could not afford. E-wastes should never be disposed with garbage and other household wastes. This should be segregated at, the site and sold or donated to various organisations.

While buying electronic products opt for those that:

1. Are made with fewer toxic constituents.
2. Use recycled content.
3. Are energy efficient are designed for easy upgrading or disassembly utilise minimal packaging.
4. Offer leasing or take back options.
5. Have been certified by regulatory authorities. Customers should opt for upgrading their computers or other electronic items to the latest versions rather than buying new equipments.

NGOs should adopt a participatory approach in management of e-wastes.

17.8 Global e-waste management

Globally, more than 70 million tonnes of e-waste were discarded in 2012 and 95 million tonnes are expected to be disposed in 2020. It is estimated that 50% to 80% of e-waste from developed countries is exported to developing countries. While some governments are forbidding the export of e-waste to developing nations, exportation is on the rise due to economic incentives of informal recycling. Developed nations benefit from cheap labour costs in developing nations, while the imported e-waste creates jobs for developing nations and provides second hand products for reuse. Because the majority of e-waste is processed through informal recycling systems, there is limited data on the volumes of e-waste collected and treated through the formal sector.

The following section describes e-waste recycling policies in Europe and Japan, nations which take different approaches but both boast high recycling rates. Using the EPR concept, the European Union (EU) is reported to have achieved a recycling rate of about 35% (CRU). Japan uses a combination of EPR and ARF to achieve an estimated recycling rate of 75% (CRU).

17.8.1 European union e-waste recycling policies

The e-waste recycling rate for the European Union is significantly higher than the e-waste recycling rate of the U.S., at approximately 35% (CRU). The United Nations University estimates that e-waste generation in EU countries is rising by 8.3 to 9.1 million tonnes per year, while global e-waste generation is increasing about 40 million tonnes per year.

The goal of the WEEE directive is to increase the collection rate for discarded electronic and electrical products from 65% by 2012 and to 85% by 2019. The WEEE directive adopted regulations in five major categories: (i) electrical and electronic equipment (EEE) product design, (ii) e-waste collection, (iii) e-waste recovery, (iv) e-waste treatment and treatment financing and (v) EEE user awareness.

Producers of these goods are responsible for collection and recycling once the products reach the end of their useful life. Producers include primary manufacturers as well as companies that import and rebrand products. In many cases, distributors offer take-back schemes and producers come together to invest in central schemes, such as takeback facilities.

Under the WEEE directive, processors of e-scrap must comply with the following regulations:

1. Be an authorised treatment facility.
2. Have an environmental permit, a pollution prevention control permit or a waste management license.
3. Treat WEEE according to the guidance on best available treatment, recovery and recycling techniques.

17.8.2 E-waste recycling policy in Japan

In Japan, e-waste policies require manufacturers and importers to take-back electronics for end of-life management. Japan's 'Home Appliance Recycling Law' mandates that four types of household e-wastes be collected: televisions, refrigerators, washing machines and air conditioners.

Consumers are required to pay an end-of-life fee that covers a portion of the recycling and transportation costs. Consumers are required to bring e-waste to the establishment they purchased the product. Retailers then ship the products to designated collection sites and manufacturers are responsible for ensuring the e-scrap is recycled. Manufactures can also sell the e-scrap for reuse or pay to have another company recycle the waste. Japan's e-waste recycling rate is approximately 75% for products covered under the home appliances recycling law, due to the fact that greater financial responsibility is placed on the consumer (CRU).

17.8.3 Indian scenario

While the world is marvelling at the technological revolution, countries like India are facing an imminent danger. E-waste of developed countries, such as the U.S., dispose their wastes to India and other Asian countries. The magnitude of these problems is yet to be documented. However, groups like toxic links India are already working on collecting data that could be a step towards controlling this hazardous trade.

It is imperative that developing countries and India in particular wake up to the monopoly of the developed countries and set up appropriate management measures to prevent the hazards and mishaps due to mismanagement of e-wastes.

18

Pollution control in thermal power plants

18.1 Introduction

Thermal or heat pollution is common in the vicinity of power plants. The water used for cooling the working steam is led back to the source at an appreciably higher temperature, affecting the aquatic organisms. The use of nuclear power opens up the possibility of radioactive wastes leaking out, which might cause enormous damage to the biosphere around.

Thermal power plants have been found to affect environmental segments of the surrounding region very badly. Environmental deterioration is attributed to emission of large amount of SO_x, NO_x and SPM and RSPM which disperse over 25 Kms radius and cause respiratory and related ailments to human beings and animal kingdom. It also affects photosynthesis process, balance of minerals and micro and major nutrients in the plants, soil strata, structures and buildings get affected due to corrosive reactions.

Power generating units are mega project, which require not only huge capital investment but also various natural resources like, fossil fuels and water, thus create an immeasurable and everlasting impacts on the environment and generate tremendous stress in the local ecosystem in spite of stringent government norms to control and mitigate the damages to the environment by the power plants.

18.2 Air pollution from thermal power plants

A thermal power station involves large-scale combustion and the consequential flue gases are hazardous. Thermal power stations use generally coal, fuel oil or gas for combustion. Presently produced boilers use heavy oil firing to start with. Newer boilers are being designed for direct coal firing. Use of oil is on the decline, due to obvious reasons.

With the availability of more natural gas from indigenous sources attention has been on installation of gas-based power stations even though at present the population of gas-fired power stations are being increased only nominally. Still coal-fired ores are presently the mainstay.

The main objectionable constituents of flue gases from the boiler depend on quality of combustion and on the coal used. A typical coal analysis of Indian coal is typically made of carbon, ash, sulphur, volatile material, moisture.

18.2.1 Dust concentration

The content of ash in coal is very high. This leads to very high dust burden even with high efficiency of ash collection.

Dust concentration is reduced in the thermal power station by using high efficiency electrostatic precipitators. Equipped with modern EPMS control system they are able to have an efficiency of as high as 99.9%. The ESPs have to be designed and guaranteed for a value as prescribed by the concerned pollution board and also with an eye on future, as the allowable limits can be further brought down depending on the existing pollution levels in the area. When the ESP provided does not meet the required levels, additional fields have to be added to reduce the level. Spraying of sulphuric acid at high temperature has also been a method abroad to improve the efficiency of old ESPs. However this introduction needs to be carefully studied with respect to coal characteristics as well as stack outlet temperatures and stack outlet flue gas analysis for sulphur.

Electrostatic Precipitator Management System (EPMS) introduced for a systematic and sequential operation of the fields has also given good results in decreasing the dust level at ESP outlet.

18.2.2 Sulphur dioxide

The content of SO_2 in flue gases has been a very important factor abroad since the coal there has a sulphur content of 0.8 to 1.0%. However SO_2 is a very critical ingredient considering the possibility of formation of H_2SO_4 vapours and subsequently getting condensed in atmosphere, now popularly known as acid rain. Hence regulations are becoming very stringent and watch needs to be kept on this parameter. Desulphurisation plants are a solution to this problem but are highly capital intensive.

The flue gas outlet temperature at furnace also needs to be monitored as the corrosion can be very high with low ESP inlet temperatures.

Sulphur dioxide emissions can be reduced either by removing the sulphur from the fuel before combustion (oil desulphurisation or coal cleaning) or by removing the sulphur dioxide from the flue gases Flue-gas Desulphurisation (FGD).

Flue-gas desulphurisation processes (FGD)

The plants use aqueous scrubbers to remove the sulphur dioxide and a considerable amount of research was invested into dry methods of removing sulphur dioxide. These methods were of two types, the injection of an alkali into the combustion chamber and the use of solid adsorbents to take up the sulphur dioxide from the flue gas. These dry methods have the big advantage that the flue gases are not cooled so that extra heat energy is not required to reheat the flue gases. This reheating, which consumes a large amount of energy, is necessary for all wet processes in order to prevent acid corrosion of ductwork and to provide adequate plume buoyancy. However, in general, dry methods were not commercially successful and wet methods were more fully developed.

Wet processes

All wet processes are capable of removing 90% of the sulphur dioxide but if the removal is limited to about 60%, by not treating about one third of the flue gases then considerable energy savings can be made by the elimination of the necessity of reheating the flue gases.

Once-through systems: The first FGD units installed in the world at Battersea and Bankside were once-through processes which made use of the natural alkalinity of the river Thames. The only other process in this category suitable for use in the U.K. is that using sea-water as the scrubbing liquid. However, for a modern large power station the natural alkalinity of sea-water will not be sufficient and thus extra limestone will be required. The dispersion of effluents from the process in estuaries and the sea and the effects of these effluents on marine life are problems which have not yet been fully elucidated. A considerable amount of work has been carried out by the Central Electricity Generating Board (CEGB) on these and other aspects of sea-water washing. Much further work will be required before sea-water washing can be considered a fully viable process.

Throwaway processes: In these processes the absorbent liquor is recirculated in a closed loop. The reaction products are separated off and fresh absorbent is added. The only water usage is to cover losses by evaporation and that entrained with the product. In most cases a pre-cooler is used to cool the gases to avoid concentration changes in the absorber due to evaporation.

Lime/limestone processes: The vast majority of scrubbing systems installed, use lime or limestone as the alkali to neutralise the sulphur dioxide. The resulting sulphite/sulphate mixture is then disposed off in ponds or disused mines. This mixture is very difficult to dewater and forms a thixotropic solid commonly known as 'sludge'. Usually various fixation methods, including mixing with fly ash, cement, lime, soil, etc., are used to stabilise the sludge. However even with these treatments the resulting land will probably still be sterile and unsuitable for agricultural or amenity use. Because of environmental considerations these processes are considered to be unacceptable. In addition, with the advent of gypsum producing systems, there is little or no economic advantage in installing sludge producing systems.

Double-alkali processes: Problems of scaling in the scrubber have led to the search for other methods. One such method is to absorb the sulphur dioxide in some soluble medium and then treat the absorber effluent with lime or limestone in a reactor outside the scrubber loop. However a calcium sulphite product is again produced and so these processes would not be generally suitable for use.

Gypsum processes: Since the disposal of sulphite/sulphate sludges is difficult there has been an incentive to produce a more amenable product. Gypsum is such a product which can be readily disposed off to landfill or if produced in a suitable form can be sold for the manufacture of plaster boards or as a cement setting retardant.

Any of the methods outlined in the previous section can be made to produce gypsum by the addition of an oxidation step in which calcium sulphite is oxidised to gypsum by air injection. Oxidation is usually carried out at lower pH than that used for absorption. Intentional oxidation can also be carried out in the scrubber loop. Some of these processes produce gypsum of marketable quality and a survey has shown that there are adequate markets for gypsum for a strictly limited number of FGO plants only. Moreover, the gypsum product from any single station might have to be sent to several end usages.

Regenerative processes. In these processes the absorbent is chemically or thermally regenerated for reuse and a saleable product (liquefied SO_2, sulphur or sulphuric acid) is produced. Ideally, there should be no disposal problems but all the processes produce some by-products which means that there is still a need for absorbent make-up and for by-product disposal.

Wellman-lord process. The Wellman-Lord process is based on the absorption of sulphur dioxide in a concentrated solution of sodium sulphite and the regeneration of the resultant bisulphite (pyrosulphite) solution by steam stripping in an evaporator-crystalliser. The process produces a pure concentrated sulphur dioxide gas stream which can be treated to produce sulphur or sulphuric acid. Sodium sulphate must be disposed off as a by-product. The Wellman-Lord process is very popular in Japan.

Magnesium oxide process. In this process an aqueous slurry of magnesium hydroxide and magnesium sulphite is used to absorb sulphur dioxide. The resulting magnesium sulphite is then calcined to generate sulphur dioxide and to regenerate magnesium oxide. The concentration of sulphur dioxide produced is suitable for conversion to either sulphuric acid or sulphur. Since the magnesium salts are separated from the absorber as a solid and solid MgO is regenerated it is possible to separate the absorption and regeneration sites. In certain circumstances, economic savings can be made by using one regeneration complex to serve several sites.

Dry processes

Alkali injection into the combustion chamber: It is well known that not all the sulphur present in the coal ends up as sulphur dioxide in the flue gases after combustion. A significant proportion (5–15% with coals) reacts with alkaline elements in the coal and the sulphur is then fixed and removed with the ash. The proportion of sulphur removed in this way depends on the amount and alkalinity of the ash present in the coal.

In the dry injection method alkali (usually limestone or slaked lime) is injected into the hot boiler gases where it reacts with the sulphur oxides and is then removed after reaction. The equipment for preparation of the alkali, injection into the gas and for removing it together with the products is required.

In pulverised fuel boilers because of the high flame temperature and low residence times only very low removal efficiencies are achieved (~20%) even with a large excess of limestone. The process was thus never used on a large scale. In both fluidised bed boilers and lignite-fired boilers the flame temperature is much lower and the residence time is much longer so that removal efficiencies of upto 90% can be achieved.

Interest has recently been revived in direct injection because of research into new types of burners for the reduction of nitrogen oxides (NO_x) emissions. These burners give lower flame temperatures and, because of the staged introduction of air, part of the flame is reducing. This leads to the possibility of increasing limestone utilisation but even so it is unlikely that the maximum removal efficiency will be greater than 50% with a limestone stoichiometry of 2. Some of the latest results suggest that a stoichiometry as high as 4 might be required. This means that large amounts of limestone would be needed in a station and large quantities of product and calcined reagent would be required to be removed. This would create a difficult disposal problem.

Adsorption processes. Both the Shell copper oxide processes and the Bergbau-Forschung carbon adsorption processes can remove sulphur and nitrogen oxides. However neither of these processes has been fully developed to a large scale on coal-fired boilers.

Spray-drying processes. During the past few years, the use of spray driers for the adsorption of sulphur dioxide has become very popular. Already these processes are being installed on 3500 MW of power plant in the US. In these processes, a spray drier replaces the absorber and the sulphur dioxide absorption takes place in the droplets created by the atomiser of the spray drier. The absorbent used can be either lime or sodium carbonate or bicarbonate. A mixture of fully reacted and unreacted absorbent is produced (which is collected in an ESP or bag filter), some of which can be recycled to optimise the utilisation of the absorbent. The remainder is sent for disposal.

Several advantages are claimed for spray-dry absorption among which are the minimisation of reheat, no scaling or plugging, the production of a totally dry product and low investment. However, spray drying processes are most suitable for low to medium sulphur fuels. In addition large quantities of lime must be imported and large quantities of product and unused reagent are produced which are collected and intimately mixed with the fly-ash. This mixture must then be disposed off in a satisfactory manner. On a 2% sulphur coal only about 50–70% removal will generally be achieved with a lime stoichiometric ratio of 2. At present spray drying FGD is not a proven process on medium to high sulphur coals but a considerable amount of work is being carried out to investigate the use of this process on higher sulphur coals.

An interesting concept in spray-drying technology is the use of magnesium oxide as absorbent. The resulting product can then be regenerated as in the magnesium oxide process.

Effect of flue-gas desulphurisation on the environment and other industries: Two important aspects of FGD which have not been much examined are the effects on the environment caused by disposal of waste products from the processes and the effects that the vast import and export of materials from a large power station can have on other industries. These are problems which need very careful attention for each specific site.

In order to obtain system cost comparability between all the FGD systems it is necessary to assume a common commissioning date. In practice this is unlikely to be the outcome because all the FGD systems investigated would not be fitted to new power stations at the same time and the time-scales for any retrofit would be affected by different considerations from those affecting new plants.

Coal cleaning: A radical alternative to the direct processing of flue gases lies in the removal of sulphur from coal prior to burning. However, the primary aim of existing commercial coal preparation plants is to separate mineral matter from runoff of mine coal and to produce different grades of fuel with ash contents controlled according to market specifications. Any separation of sulphur which occurs is incidental. Power station fuel with an average of 16–20% ash represents the lowest quality end of the production spectrum at the other end of which is, for example, coal destined for domestic use with less than 5% ash.

The beneficiation processes currently in use exploit the differences in physical properties, mainly those of density and ease of wetting, between the coal constituents and can therefore extract only that part of the sulphur which exists in mineral form.

Over 60% of all treated coal is typically processed in a jig which depends upon the stratification of various density fractions in a water-borne bed which is kept in motion by alternate upward and downward currents of water generated by pulses of compressed air. The separation of the coal from the mineral matter is achieved principally during the upward current of water with clean coal transferred near to the surface and the dirt collecting at the bottom of the vessel. Such a separation is, of course, not complete and part of the fuel ends up in the intermediate layer of water from which it can be extracted as a product with medium to high ash content, the so-called middlings.

Consequently, the heavy pyrites are redistributed among the final products. The total sulphur content of the clean coal (expressed in terms of sulphur content related to the heating value) would be substantially lower (about half) than that of the raw feed coal, with most of the pyrite appearing in the reject and middlings. In order to make the process economic it is necessary to utilise the thermal value of the middlings while preventing the release of the associated sulphur into the atmosphere. This can be achieved by further crushing and reprocessing before combustion or by directing the middlings into boilers fitted with gas washing plant. In effect, the coal cleaning in this case would be used to boost the effectiveness of the expensive gas washing units, by putting a greater proportion of sulphur through them than would be the case if flue gases from only untreated coal were being processed.

In most of coals about 0.8% of the total sulphur content is chemically bound to the organic coal matter, with only the excess of sulphur over this value being present in the form of pyritic particles which are amenable to extraction by physical cleaning. As the average sulphur content is about 1.6% coal cleaning could not, even theoretically, remove more than 50% of the sulphur. Because of the inefficiency of any practical process industrial plants could more realistically achieve 20 to 30% removal. The higher limits of this range are associated only with the more highly sulphurous coals which have a correspondingly higher proportion of pyritic sulphur. In such cases where coal preparation would be combined with separate combustion of middlings the total sulphur retention could be raised to about 50%.

It must be emphasised that the ease of sulphur extraction, the achievable thermal yield and the resulting size composition of the product are all factors very specific for individual coals. The presently available experimental data cover only a small proportion of the production steam and a reliable technical and financial assessment of coal benefication potential would require much further investigation.

The applicability of pyrites removal is closely linked with the type of emission control regulations which might be enacted. If an overall reduction in current CEGB emissions in the 10 to 20% range were required, then it would probably be competitive with gas washing aimed at the same target.

However, the further development of improved coal preparation depends not only on its importance for emission control but also from its essential role in the production of substitute liquid fuels suitable for medium to large boilers. The emerging technology of production of stable coal-water slurries with solid loadings typically in the 65–75% range could offer a fuel cheaper than oil while retaining the advantages of handling liquids. The energy penalty associated with the increased water content is relatively small (only 2–3% reduction in the heating value of the original coal). The size of the potential market would be therefore to a great extent determined by the convenience of use of fuel slurries.

The design and the cost of the boiler with its auxiliary equipment is greatly affected by the level of the ash content (and its composition) which should ideally approach that encountered in heavy fuel oil (<0.5%). Present coal preparation cannot generally result in clean coals with less than 3–5% ash and further substantial reduction would be achieved only on a finely ground product which would also ensure preliminary mechanical liberation of pyritic particles. Provision of very low ash coal slurries would therefore most probably also achieve the limits of physical desulphurisation of coal.

18.2.3 Nitrous oxides

Nitrogen oxides are released into atmosphere when coal is burnt due to oxidation of nitrogenous compounds present in coal and some times by oxidation of atmospheric nitrogen during combustion process. Here one way of controlling NO_x would be through monitoring furnace temperature. Effect of NO_x would also be similar to SO_x. Combining with atmospheric moisture they tend to produce harmful acidic formations.

18.2.4 Carbon dioxide/carbon monoxide

During combustion in boiler furnace formation of carbon dioxide/carbon monoxide takes place and are released to atmosphere. They tend to raise levels of CO_2/CO in atmosphere. Massive green belts around power stations help in controlling effects of CO_2 in the atmosphere.

18.3 Air pollution prevention measures

There are two monitoring factors for air pollution. Firstly measurement of pollutant factors at the stack outlet or each unit flue gas outlets. However apart from the concerned industry, the other units in the surroundings also contribute to atmospheric pollution. Hence the second controlling area would be ambient air monitoring in the surroundings and to keep the overall pollutant levels within limits. The dust concentration level at stack outlet is at present limited to 115 mg/NM^3 at stack outlet. This measurement is taken generally at electrostatic precipitator outlet. This level can be achieved mainly by efficiency in operation of electrostatic precipitators. The measurement is done generally by manual samplers, two to four times every month depending on size of stations. These results need to be corrected with load and combustion conditions.

Efforts are also being made to establish other continuous measurement methods such as opacity monitors which reveal dust level by analysing optically the density of flue gases.

The ambient air dust concentration limits allowed depends on the location of power plant mainly considering urban and rural areas. Based on dispersion studies adequate monitoring stations where the pollutant levels are highest are identified.

The number of stations and frequency of testing are also fixed by pollution board regulations. Either permanent stations can be established in such locations or we can go for mobile stations. However the fixed station has the benefit of recording the values simultaneously at all locations representing similar operating condition. There are many types of instruments available for use at these stations.

Measurements of SO_2 and NO_x are also done both at flue gas outlet sampling point and ambient air measurement point and should be within the limits prescribed by pollution board.

18.4 Water pollution prevention and management

In a thermal power station water pollution goes on a more conventional system and norms are of general nature.

The sources of water effluent from a power station come from:

1. Boiler blow down.
2. CW system blow down/CW discharge.
3. Water treatment plant discharge.
4. Ash pond outlet.
5. Miscellaneous plant drains.

Boiler blow down will be of small quantity of the order of 25 to 30 cubic metres per hour for a 500 MW unit and does not contain any components beyond allowed limits.

In a closed cycle CW system a certain percentage of water has to be regularly blow down in order to maintain the quality of water in closed cycle. The effluent is totally normal as per analysis and does not need any treatment. This is often used in other processes in many plants.

However if the power station adopts an open cycle system the entire water quality of the order of 25–30000 cubic metres for every 200 MW will be led into the water body at a temperature upto 5–7°C higher than ambient. Depending on type of water body such as lake/river/sea, etc., and the type of marine life available therein the impact needs to be studied and proper action taken in this regard.

In water treatment plant all effluents are collected in a neutralising pit and after suitably treating to acceptable levels, same is led into drain. Thus it is not expected to contain any harmful ingredients if treated properly before discharge. Further the quantity is also very small to have any effect in the final effluent. In ash pond normal system is to pump ash in a slurry form into the pond and by providing suitable settling and filter system the entire ash gets settled and water goes out. Here obviously the emphasis on analysis of outlet water will be on total suspended solids which bears testimony to the effectiveness of filtering system. In addition to the same depending on the ash analysis the pH value is also likely to show alkalinity in nature. Hence a decision on treatment of effluents is necessary after air analysis of same.

Apart from all above there are many miscellaneous drains which individually though small finally add up to a fairly large quantity. The drains from coal handling area are likely to contain high level of suspended particulate matter for which if required we may have to provide filters.

The drains from fuel oil area should be ensured free from oil and grease. It is essential to provide oil traps for these drains before they join other drains.

The above explained points are various sources of plant drains. However ultimately the entire drain goes out of plant area after joining together in one or two outlets. Thus it is essential that weekly samples are taken regularly from these channels for testing and any abnormalities are immediately controlled from source which can be identified. Further regular observations are necessary for any visible abnormalities to take immediate corrective action.

In order to assess the impact of the total plant effluents to the surrounding water bodies sampling of surface water from the water body as well as ground water samples in the plant and ash pond surroundings are also necessary. This helps in identifying any adverse impacts of plant water effluents.

In general it can be said that water effluents from a thermal power station can be well controlled as not to cause any harmful effects on surroundings.

18.5 Coal handling

Coal handling system consisting of mills, crushers, conveyors, etc., is always a source of dust nuisance. Dust control in these areas can be done by spraying of water. This has to be controlled since wet coal is more difficult to handle and fire in the boiler. These buildings are also provided with dust extraction system to ensure a good working atmosphere.

Problems of dust also arise in coal yard where coal stacked creates dust problems due to winds. Hence a CHP plant and yard are located in such locations where wind blows away from the yard in opposite direction of main plant. Also water spray is used to control the dust. These also helps in controlling fires of instantaneous combustion in coal yard area.

18.6 Noise pollution

In rotating and reciprocating machines, noise is produced through vibration caused by imperfectly balanced parts; bearing noise, wind noise and other noises also exist. The amplitude of such noises varies with operating speed, usually increasing exponentially with speed. Noise frequencies cover a wide range since normally several harmonics of each fundamental are produced.

Electrical machines produce noise from magnetic as well as mechanical forces. Alternating current machines convert electrical to mechanical energy by cyclically changing magnetic forces which also cause vibration of the machine parts. These magnetic forces change in magnitude and direction as the machine rotates and air gaps and their magnetic reluctance change. The noise frequencies thus produced are related both to line frequency and its harmonics and to rotational speed. The entire pattern is quite complicated. In non-rotating machines (transformers, magnetic relays and switches), the noise frequencies are the line frequency and its harmonics and the frequencies of vibration of small parts which are driven into vibration when their resonant frequencies are near some driving frequency.

In many machines, more noise is produced by the material being handled than by the machine. In metal-cutting or grinding operations, much noise is produced by the cutting or abrading process and is radiated from both work-piece and machine.

Belt and screw conveyors are sometimes serious noise sources; they are large-area sources; their own parts vibrate and cause noise in operation and the material they handle produces noise when it is stirred, dropped, or scraped along its path of motion. Vibration from conveyors is conducted into supports and building structure as well. Feeding devices, as for automatic screw machines, often rattle loudly.

Jiggers, shakers, screens and other vibrating devices produce little audible noise in themselves (partly because their operating frequency is so low), but the material they handle produces much higher frequency noise. Ball mills, tumblers and the like produce noise from the many impacts of shaken or lifted-and-dropped pieces; their noise frequencies are often low and much mechanical vibration is around them. Industry uses many pneumatic tools. Some air motors are quite noisy, others less so. Exhausting air is a major noise maker and the manner in which it is handled has much to do with the noise produced. Exhausting or venting any gas (in fact, any process which involves high velocity and pressure changes) usually produces turbulence and noise. In liquids, turbulent flow is noisy because of cavitation.

Turbulence noise in gas is usually predominantly high frequency; cavitation noise in liquids is normally midrange to low frequency. Both types of noise can span several octaves in frequency range.

Gas and steam turbines produce high-frequency exhaust noise; steam turbines (for improved efficiency) usually exhaust their steam into a condenser; gas turbines sometimes feed their exhausts to mufflers. If such turbines are not enclosed, they can be extremely noisy; turbojet airplane engines are an example. Impact noises in industry are produced by many processes; materials handling, metal piercing, metal forming and metal fabrication are perhaps most important. Such noises vary widely because of machine design and location, energy involved in the operation and particularly because of the rate of exchange of energy.

18.6.1 Control of noise

Reduce speeds and pressures

Reducing the speed of rotating and moving parts in machines and mechanical systems results in smoother operation and lower noise output. Likewise, reducing pressure and flow velocities in air, gas and liquid circulation systems lessens turbulence, resulting in decreased noise radiation.

Reduce frictional resistance

Reducing friction between rotating, sliding, or moving parts in mechanical systems frequently results in smoother operation and lower noise output. Similarly, reducing flow resistance in fluid distribution systems results in less noise radiation.

Reduce noise leakage

In many cases, machine cabinets can be made into rather effective soundproof enclosures through simple design changes and the application of some sound-absorbing treatment.

Provide mufflers/silencers

There is no real distinction between mufflers and silencers. They are often used interchangeably. They are in effect acoustical filters and are used when fluid flow noise is to be reduced. The devices can be classified into two fundamental groups: absorptive mufflers and reactive mufflers. An absorptive muffler is one whose noise reduction is determined mainly by the presence of fibrous or porous materials, which absorb the sound.

Barriers and panels

Placing barriers, screens, or deflectors in the noise path can be an effective way of reducing noise transmission, provided that the barriers are large enough in size and depending upon whether the noise is high frequency or low-frequency.

18.7 Heat pollution

The major effect comes from cooling water discharge from a power station. In a station adopting open system of cooling, water at temperatures 5°C to 7°C higher than ambient temperature are led to an external source such as river, pond, sea, etc. In such cases study of ecological effects of this water discharge have to be studied before adopting the system. Sometimes cooling towers are adopted even in open cycle to limit the temperature of discharged water. However this problem is not encountered in a closed system.

The heat from equipment is controlled by effective insulation. Insulation helps in not only controlling heat generated but also in saving heat for the process. The insulation is designed and guaranteed for a surface temperature of 55°C–65°C. This is also a safety necessity. In spite of such effective insulation the areas of working have to be properly ventilated to contain the temperatures in working areas within working limits.

The major heat input to atmosphere from a thermal power station comes mainly from the flue gases going out of chimney. Their minimum temperature is limited by the condensation temperature of sulphuric acid and nitric acid formations from vapours.

Any temperature lower than this leads to corrosion of equipment on flue gas path and chimney walls. As a protection measure, the top one third of a chimney is lined with acid resistant bricks. The heat led into atmosphere has a tendency to generally raise the temperature which can be controlled only by development of large green belts.

18.8 Operational control and equipment maintenance

It is also necessary to highlight the importance of proper plant operation and maintenance. The quality of stack emission depends very much on the operating parameters of the plant. A proper balanced combustion is a must for less impure flue gases. By improper combustion in boiler the unburnt carbon particles in the exhaust flue gases increase. Similarly oil support for combustion also gives a bad smoke with higher sulphur oxides content. Similarly maintenance of electrostatic precipitators with all fields available for dust collection is an important factor.

Maintenance of proper operating parameters in ESP as well as maintenance of collecting and rapping mechanisms is an important factor. Similarly maintenance of equipment in other areas also leads to limiting the harmful discharges from the plant.

Fuel handling area with its spillage is another potential polluting area. In this area maintenance and clearing oil traps regularly is very important to prevent oil leakage into plant effluents along with drain.

18.8.1 Retrofits

With the emission limitations becoming more and more stringent day by day it becomes necessary to carry out modifications in old power stations which had been designed for generally higher emission values. It is very common to introduce more efficient electrostatic precipitators in old plants. Apart from above, methods to increase collection efficiently of ESPs is also a practice. Improvement in instrumentation as a part of measurement and increase in operation efficiency are other steps.

This aspect in old thermal power stations has to be reviewed thoroughly and new modifications have to be brought forth.

18.9 Power generation from thermal power plants

Modern steam power plants, whether using nuclear or fossil fuel, are variations of the Rankine steam power cycle, which consists of a boiler, a turbine, a generator and a condenser. The steam from the boiler when expanded through a turbine, turns the turbine shaft, which is coupled to an electric power generator. The spent steam and the turbine exhaust is passed through a condenser (Fig. 18.1). Detailed flowsheet of steam and electric power generation plant based on fossil fuel and sources of wastes generation is shown in Fig. 18.2.

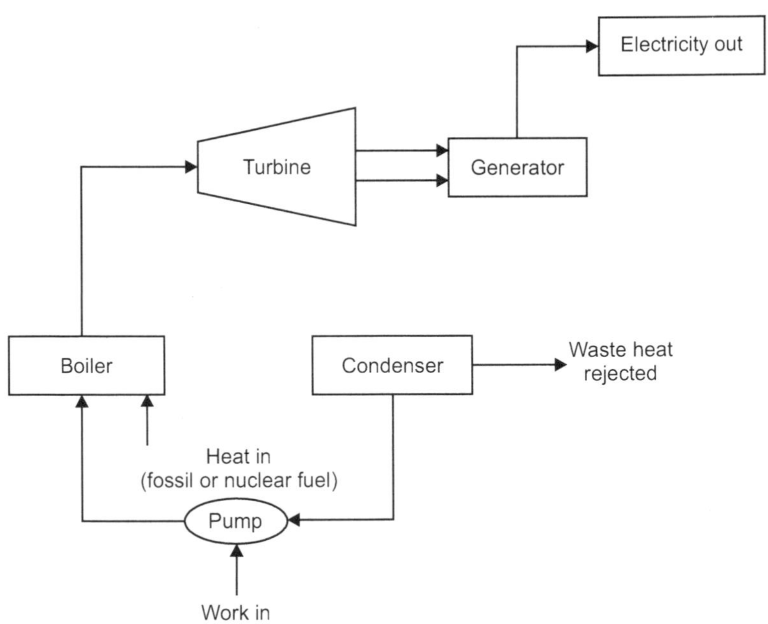

Figure 18.1: Rankine steam power cycle.

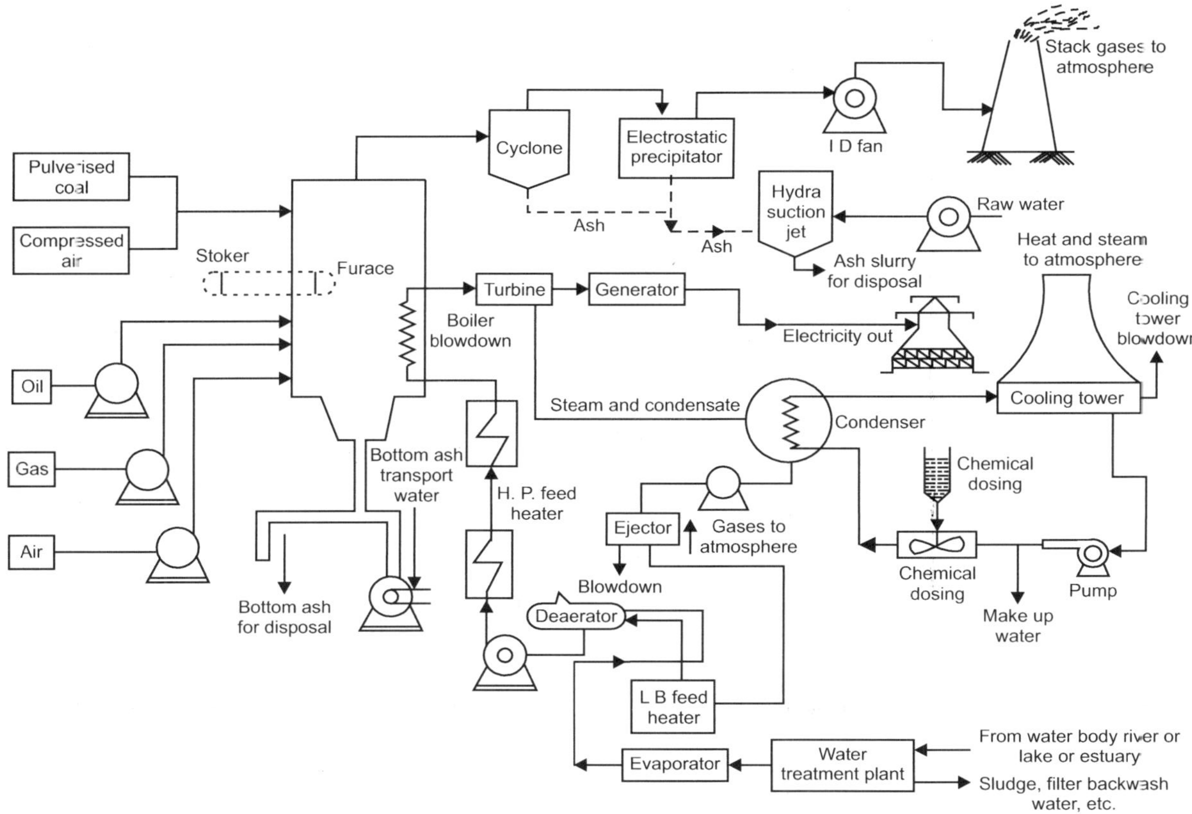

Figure 18.2: Flowsheet of steam and electric power generation based on fossil fuel.

It is evident that the major sources of wastewater generation in thermal power plants are:

1. Cooling water from condensers.
2. Overflow from ash ponds.

The other sources are:

1. Boiler blow down.
2. Cooling tower blow down.
3. Wastewater from regeneration of ion-exchange columns.
4. Rainfall-runoff from coal pile storage.
5. Runoff from coal handling and preparation yards.
6. Wastewater from the water treatment plants, like filter back washing.
7. Oil pollution from oil storage and handling yards.

The water use and the wastewater generation in the thermal power plants are predominantly influenced by the type of cooling system adopted, as consume a these systems tremendous volume of water when compared with cooling towers.

18.10 Impact of thermal power plants on water and land

18.10.1 Impact on water

The water requirement for a coal-based power plant is about 0.005 m^3/kwh to 0.18 m^3/kwh. At STPS, the water requirement has been marginally reduced from about 0.18 m^3/kWh to 0.15 m^3/kwh after the installation of a treatment facility for the ash pond decant.

Still the water requirement of 0.15 m^3/kwh = 150 litres per unit of electricity is very high compared to the domestic requirement of water of a big city.

Ash pond decant contains harmful heavy metals like B, As, Hg which have a tendency to leach out over a period of time. Due to this the ground water gets polluted and becomes unsuitable for domestic use.

The exposure of employees to high noise levels is very high in the coal based thermal power plant. Moreover, the increased transportation activities due to the operation of the power plant leads to an increase in noise levels in the adjacent localities.

18.10.2 Impact on land

The land requirement per mega watt of installed capacity for coal, gas and hydroelectric power plants is 0.1–4.7 ha., 0.26 ha. and 6.6 ha., respectively. In case of coal based power plants the land requirement is generally near the area to the coal mines. While in the case of gas-based it is any suitable land where the pipeline can be taken economically. Land requirement of hydroelectric power plants is generally hilly terrain and valleys.

18.10.3 Biological and thermal impact

The effect on biological environment can be divided into two parts, viz., the effect on flora and the effect on fauna. Effect on flora is due to two main reasons, land acquisition and due to flue gas emissions. Land acquisition leads to loss of habitat of many species.

The wastewater being at higher temperature (by 4–5°C) when discharged can harm the local aquatic biota. The primary effects of thermal pollution are direct thermal shocks, changes in dissolved oxygen and the redistribution of organisms in the local community. Because water can absorb thermal energy with only small changes in temperature, most aquatic organisms have developed enzyme systems that operate in only narrow ranges of temperature. These stenothermic organisms can be killed by sudden temperature changes that are beyond the tolerance limits of their metabolic systems. Periodic heat treatments used to keep the cooling system clear of fouling organisms that clog the intake pipes can cause fish mortality.

18.10.4 Socioeconomic impact

The effect of power plants on the socioeconomic environment is based on three parameters, viz. Resettlement and Rehabilitation (R&R), effect on local civic amenities and work related hazards to employees of the power plants. The development of civic amenities due to the setting up of any power project is directly proportional to the size of the project. The same has been observed to be the highest for the coal based plants followed by the natural gas based plant and lastly the hydroelectric plant. The coal based plant has the highest number of accidents due to hazardous working conditions.

18.11 Fly ash from thermal power plants

Fly ash is one of the coal combustion products, composed of the fine particles that are driven out of the boiler with the flue gases. Ash that falls in the bottom of the boiler is called bottom ash. In modern coal-fired power plants, fly ash is generally captured by electrostatic precipitators or other particle filtration equipment before the flue gases reach the chimneys. Together with bottom ash removed from the bottom of the boiler, it is known as coal ash. Depending upon the source and make-up of the coal being burned, the components of fly ash vary considerably, but all fly ash includes substantial amounts of silicon dioxide (SiO_2) (both amorphous and crystalline), aluminium oxide (Al_2O_3)

and calcium oxide (CaO), the main mineral compounds in coal-bearing rock strata. Constituents depend upon the specific coal bed make-up but may include one or more of the following elements or substances found in trace concentrations (up to hundreds ppm): arsenic, beryllium, boron, cadmium, chromium, hexavalent chromium, cobalt, lead, manganese, mercury, molybdenum, selenium, strontium, thallium and vanadium, along with very small concentrations of dioxins and PAH compounds.

In India, unlike in most of the developed countries, ash content in the coal used for power generation is 30–40%. High ash coal means more wear and tear of the plant and machinery, low thermal efficiency of the boiler, slogging, choking and scaling of the furnace and most serious of them all, generation of a large amount of fly ash. India ranks fourth in the world in the production of coal ash as by-product waste after USSR, U.S., and China, in that order. Fly ash is defined in cement and concrete terminology as the 'finely divided residue resulting from the combustion of ground or powdered coal, which is transported from the fire box through the boiler by flue gases'. Fly ash is fine glass powder, the particles of which are generally spherical in shape and range in size from 0.5 to 100 μm. Fly ash is classified into two types according to the type of coal used. Anthracite and bituminous coal produces fly ash classified as class F. Class C fly ash is produced by burning lignite or sub-bituminous coal. Class C fly ash has self-cementing properties.

18.11.1 Effluent and disposal

Disposal and management of fly ash is a major problem in coal-fired thermal power plants. Fly ash emissions from a variety of coal combustion units show a wide range of composition. All elements below atomic number 92 are present in coal ash. A 500 MW thermal power plant releases 200 mt SO_2, 70 T NO_2 and 500 T fly ash approximately every day. Particulate Matter (PM) considered as a source of air pollution constitutes fly ash. The fine particles of fly ash reach the pulmonary region of the lungs and remain there for long periods of time; they behave like cumulative poisons. The submicron particles enter deeper into the lungs and are deposited on the alveolar walls where the metals could be transferred to the blood plasma across the cell membrane. The residual particles being silica (40–73%) cause silicosis. All the heavy metals (Ni, Cd, Sb, As, Cr, Pb, etc.), generally found in fly ash are toxic in nature. Table 18.1 provides a list of diseases caused due to the presence of these toxic metals.

Fly ash can be disposed-off in a dry or wet state. Studies show that wet disposal of this waste does not protect the environment from migration of metal into the soil. Heavy metals cannot be degraded biologically into harmless products like other organic waste. Studies also show that coal ash satisfies the criteria for landfill disposal, according to the environmental agency of Japan.

Table 18.1: Diseases due to the presence of heavy metals in fly ash.

Metal	*Content (ppm)*	*Diseases*
Nickel (Ni)	77.6	Respiratory problem, lung cancer
Cadmium (Cd)	3.4	Anaemia, hepatic disorder
Antimony (Sb)	4.5	Gastroenteritis
Arsenic (As)	43.4	Skin cancer, dermatitis
Chromium (Cr)	136	Cancer
Lead (Pb)	56	Anaemia

According to the hazardous waste management and handling rule of 1989, fly ash is considered as non-hazardous. With the present practice of fly-ash disposal in ash ponds (generally in the form of slurry), the total land required for ash disposal would be about 82,200 ha by the year 2020 at an estimated 0.6 ha per MW. Fly ash can be treated as a by-product rather than waste.

18.11.2 Utilisation of fly ash

The ash generated from volcanoes was used extensively in the construction of Roman structures. The colosseum (constructed in AD 100) is a classic example of durability achieved using volcanic ash. Fly ash is generated in artificial volcanoes (coal-fired). Volcanic ash acts just like fly ash obtained from coal-fired thermal power plants.

Broadly, fly ash utilisation programmes can be viewed from two angles, i.e., mitigating environmental effects and addressing disposal problems (low value – high volume utilisation).

Following are some of the potential areas of use of fly ash.

Fly ash bricks

Fly ash bricks have a number of advantages over the conventional burnt clay bricks. Unglased tiles for use on footpaths can also be made from it. Awareness among the public is required and the government has to provide special incentives for this purpose.

Fly ash in manufacture of cement

Fly ash is suitable for use as pozzolana. In the presence of moisture, it reacts chemically with calcium hydroxide at room temperature to form compounds possessing cementitious properties. Fly ash has a high amount of silica and alumina in reactive form. These reactive elements complement the hydration chemistry of cement. On hydration, cement produces C–S–H gel and free lime, i.e., $Ca(OH)_2$. The C–S–H gel binds the aggregates together and strengthens the concrete. Water, sulphates and CO_2 present in the environment attack the free lime causing deterioration of the concrete. A cement technologists observed that the reactive elements present in fly ash convert the problematic free lime into durable concrete. The difference between fly ash and Portland cement becomes apparent under a microscope. Fly ash particles are almost totally spherical in shape, allowing them to flow and blend freely in mixtures. This property make fly ash a desirable admixture for concrete.

Fly ash in distemper

Distemper manufactured with fly ash as a replacement for white cement has been used in several buildings all over the world.

Fly ash-based ceramics

The National Metallurgical Laboratory, Jamshedpur has developed a process to produce ceramics from fly ash having superior resistance to abrasion.

Fly ash as fertiliser

Fly ash provides the uptake of vital nutrients/minerals (Ca, Mg, Fe, Zn, Mo, S and Se) by crops and vegetation and can be considered as a potential growth improver. It serves as a good fertiliser.

Fly ash in road construction

The use of fly ash in large quantities making the road base and surfacing can result in low value – high volume utilisation.

To sum up it has been recognised worldwide that the utilisation of an enormous amount of fossil fuels has created various adverse effects on the environment, including acid rain and global warming. An increase in the average global temperature of approximately 0.56 K has been measured over the past century (global warming). Gases with three or more atoms that have higher heat capacities than those of O_2 and N_2 cause the greenhouse effect. Carbon dioxide (CO_2) is a main greenhouse gas associated with global climate change. The disposal, management and proper utilisation of waste products has become a concern for the scientists and environmentalists.

Proper management of solid-waste fly ash from thermal power plants is necessary to safeguard our environment. Because of high cost involved in road transportation for the dumping of fly ash, it is advisable to explore all its possible applications. Concentrated efforts are needed to utilise fly ash in the manufacture of building bricks, cement and ceramics and mitigate the unemployment problem as well.

18.12 Thermal waste treatment

Thermal waste treatment faces many challenges of high ash related compounds from waste fuels. Vagaries in its composition that are very heterogeneous lead to problems faced during incineration apart of course high energy consumption. However, new technologies and techniques have become more efficient addressing some of these problems too. Despite this, it is advisable to use this treatment only for waste streams which cannot be effectively treated through other alternate techniques.

Thermal treatment and especially incineration of waste has been utilised for many years. Initially, the main intention was to reduce the volume to enable more space at the deposit sites. In many cases this was performed as open dump fires and unfortunately this method is still applied in some countries today. Identified toxic emissions with the flue gases and leaching of the solid residues led to a more controlled combustion and deposit of the ashes.

Compared to conventional fuels (e.g., coal and wood), waste fuels are generally more non-homogeneous. Properties of the waste vary, for example, by local sources, season and business activity. Common for waste fuels mixes are the high alkali (especially sodium), sulphur and chlorine content compared to most other fuels. These elements are reactive and could cause deposits and corrosion on the heat exchanger surfaces in the boiler, which leads to higher demand and cost of maintenance, reduced efficiency and material losses or even unplanned production stops. Waste fuels may also contain other elements which could cause problems. Among those are heavy metals such as zinc, lead and mercury. They can vaporise at rather low temperatures and may react with available chlorine to further increase deposits and corrosion or pass through the gas cleaning stages and thus be emitted with the flue gases. Other materials, such as copper and iron, act as a catalyst for dioxin formation when the flue gases are cooled in the convection path and are especially active in the range of 250–400°C. Therefore, to burn waste successfully, the boiler must have wide fuel flexibility and be equipped with sufficient gas cleaning.

18.12.1 Waste combustion technologies

The most common technology in the world today is the grate fired boiler. In the grate boiler, the concept is to have fuel bed on top of a grate while primary air is entering from below. Most of the primary air will be consume when passing through the fuel bed due to the combustion process and additional air is supplied above the bed in several stages to ensure a complete combustion while minimising the formation of nitrogen oxides. The fuel is transported on the sloped grate by moving elements and the speed is adjusted in order for the fuel to reach burn-out just before the end of the grate.

Another technology, that is not as widespread in waste combustion but has a number of advantages such as, for example fuel load flexibility and lower investment cost, is the fluidised bed boiler. In a fluidised bed boiler, there is a bed of inert material (often sand) at the bottom of the combustion chamber.

The bed rests upon an air distributor, which allows the air to pass into the bed from below but prevents the bed particles from pouring down. The nozzles are designed to distribute the primary air evenly over the bed cross section. As a result of the air acting on the sand particles, the bed becomes fluidised which gives a good mixing between the fuel and the air and distributes the heat more evenly. Drawbacks with this technology are the high running cost due to fresh sand and higher degree of pre-treatment (crushing) needed of the incoming waste.

18.12.2 Ashes

The ash content of an average waste fuel is around 20% but can be anything between 10% and 50%. It consists of the inorganic elements of the fuel such as: silica, calcium, sulphur, potassium, sodium, aluminium, magnesium, titanium and iron. Additional impurities like glass, ceramics, soil and gravel may also be present. Most of the ash-forming matter is inert during combustion, but some could form compounds with low melting and evaporation temperatures, for example, alkali silicates, chlorides and sulphates.

18.12.3 Modern waste-to-energy (WTE) boilers

It is possible to improve the traditional (WTE) boiler. One such improvement is to produce both heat and electricity in a combined heat and power plant. This increases the energy efficiency of the plant and also increases the revenues. If district heating is available, it is possible to increase the thermal efficiency even further by utilising a flue gas condenser. This unit condenses the water vapour in the flue gases and produces warm water, but at the cost of a slightly higher electricity demand and a cleaning system for the condensed water. On the other hand, the condensation of flue gases also reduces emissions of particulates further and removes all acid gases. Another opportunity for improving the boiler is to reduce the existing temperature of the flue gases, which means that more heat is transferred in both the economiser and the air pre-heater. It is important that the temperature is kept above the dew point of HCl and sulphuric acids to prevent corrosion.

There exist also a number of alternatives for increasing the electrical efficiency. Among those are: reducing the temperature (and thus the pressure) in the condenser after the steam turbine, increasing the pressure and temperature of the steam, introducing more advanced combustion control and including a reheating cycle. The reduced temperature in the condenser gives a higher production of electricity and a lower loss of energy but is often determined by the natural conditions and availability of cooling water. Increasing the temperature and pressure of the steam can also increase the production of electricity but is limited by the corrosion of the superheater tubes. Advanced combustion control (i.e., process control with recirculation of the flue gases) provides better utilisation of the fuel because it ensures a complete burnout and reduces the needed amount of excess oxygen. Finally, the reheating cycle is a way of increasing the production of electricity by reheating. The steam after the high pressure stages in the turbine.

One of the largest plants is the AEB Amsterdam plant in the Netherlands, comprising a total of six lines with a capacity of almost 1,400,000 tpa. At the plant, two of the lines operate at 420°C and 130 bar. An electrical efficiency of 30.6% is achieved using a number of measures which includes: reheating the steam after the high-pressure stage using saturated boiler steam, reducing the oxygen level from 8%–11% to 6% by recirculating the flue gases, lowering the outlet temperature of the flue gases to 180°C from the 200°C–220°C previously used, pre-heating the condensate in the scrubbers (including the flue gas condenser) and using more corrosive-resistant materials. These measures aside, the plant is situated close to the harbour and uses cooling water at 25°C, which lowers the condenser pressure to 0.03 bar. The Zabalgarbi plant in Spain is another example of a plant with high electrical efficiency. The steam pressure used here is 100 bar but the temperature is only 330°C. These are not ideal conditions for a steam turbine, as heavy condensation would occur already in the high-pressure section.

The plant however utilises a nearby gas turbine combined cycle plant to provide extra heat (i.e., by using the flue gases after the gas turbine). The final temperature reaches 540°C with the pressure being kept constant at 100 bar. With this adaption an efficiency of 42% is reached (this includes a reheating of the steam after the high-pressure stage).

18.12.4 Other thermal treatment solutions

In order to overcome the shortcomings of the combustion process regarding electrical efficiency or material recycling other methods has been suggested. Malkow, presents several thermal treatment processes that were about to be commercialised. It contained both gasification and pyrolysis processes and it was concluded that they possess the potential of increasing the energy efficiency and decreasing both the emissions and amount of hazardous residues. Arena also lists a number of potential advantages of applying thermal treatment to waste, which include the possibility of recovering metals from ash fractions, the destruction of organic contaminates and a reduced environmental burden. A key hurdle associated with incineration is the high risk of corrosion which limits the steam properties and thereby the electrical efficiency.

Higher steam temperatures could be used if gasification together with an intermediate gas-cleaning stage is carried out before combustion. Gasification also provides opportunities for producing a syngas that could be used in a gas turbine with an even higher electrical efficiency or for other purposes, such as manufacturing chemicals.

Pyrolysis could be applied, either with the focus of producing pyrolysis oil (for use as a fuel or, after further processing, as automotive fuel) or as a pre-treatment prior to gasification/incineration. Aside from the energy perspective, the lack of oxygen in the process prevents the metals from oxidising which is beneficial for its subsequent value. It can also be used to reduce already oxidised metals but this requires high temperatures. The benefit of having a relatively low temperature is that different inorganic compounds can be removed (e.g., by leaching) without the application of strong acids, as is the case when there is substantial agglomeration. Neither of these processes has been successfully competitive across the world but specific requirements in some countries regarding low levels of dioxins in the ashes and high landfill costs have, for example, led to a commercial usage of the gasification processes. In those cases, the gasification is performed at high temperature and the ashes form a slag that after a rapid cooling could be used as a construction material. Several initiatives exist and the largest commercial technology is the fluidised waste gasification in Lahti, Finland by Valmet. It consists of a two-stage unit. First is a gasifier, which is followed, after a high temperature alkali filter, by a gas boiler. This allows for substantially higher steam temperatures (above 550°C) compared to common waste combustion (around 400°C) and thus the electrical efficiency increases.

Looking at the situation today, a higher degree of material recovery is wanted which means that even the remaining ash fraction should be minimised. Stricter regulations, decreasing demand for material for landfill construction and landfill coverage and current processes that are time and/or space limited are enforcing a development in this area. An improved pre-treatment to remove different inert material such as glass, sand, gravel, soil and metals could be initialised to reduce the ash related issues and to minimise the ash production.

To sum up, thermal waste is an increasing problem around the globe, as the amount tends to increase with both population and economics. The composition looks different among different countries and regions but in general there is a high amount of ash related compounds compared to other fuels. This creates problems both during the incineration process (mainly deposits and corrosion which decrease the electrical efficiency and increase the cost of maintenance) and afterwards (high levels on unwanted substances leads to a need of controlled landfill sites). Other thermal treatment processes has been suggested that can overcome some of these issues by operating in different temperature levels and with less available oxygen. It is however important to ensure a safe process with low emissions of toxic substances where especially dioxins and furans have been problematic in the early days of waste incineration. Thermal treatment of the waste should in the end only be used for the waste streams that are not possible to effectively presort and recycle in order to reduce the environmental burden associated with virgin feedstock.

18.13 Environmental impact assessment

An Environmental Impact Assessment (EIA) is an assessment of the possible impacts that a proposed project may have on the environment, consisting of the environmental, social and economic aspects. The purpose of the assessment is to ensure that decision makers consider the environmental impacts when deciding whether or not to proceed with a project.

The International Association for Impact Assessment (IAIA) defines an environmental impact assessment as 'the process of identifying, predicting, evaluating and mitigating the biophysical, social and other relevant effects of development proposals prior to major decisions being taken and commitments made. EIAs are unique in that they do not require adherence to a predetermined environmental outcome, but rather they require decision makers to account for environmental values in their decisions and to justify those decisions in light of detailed environmental studies and public comments on the potential environmental impacts.

18.14 Environment impact assessment of thermal power plant for sustainable development

Thermal power plants are the major source of generation of electricity for any developing country. Around 60% of electricity generation in our country is met by thermal power plants. Fuel is blow into the combustible chamber of the boiler where it is burnt at high temperature where Heat energy converts water into steam. High energy steam is passed through the turbine and the steam creates force on the turbine causing the shaft to rotate at high speed. A generator is coupled at one end of the turbine shaft which generates power.

18.14.1 Environmental issues related with coal based power sector during construction phase

Environmental issues related with coal based power sector are shown in Table 18.2.

Environmental impact during operational stage

Among thermal based power generation, coal based power plants are highest in: air pollution, waste generation, water consumption, emission of mercury, greenhouse emission.

Table 18.2: Environmental issues related with coal based power sector.

Activities/issues	*Impacts*
Change in land use pattern/site clearing	• Erosion • Loss of biodiversity • Loss or change of soil quality and quantity • Huge diversion and acquisition of land in case of power plant with captive mine
Civil works such as earth moving and building of structures	• Dust pollution • Noise pollution

Impact of thermal power plant on water source

Water intensive:

1. New thermal power plant of 500 MW installed capacity requires – around 14 million m^3 of water per annum.
2. High impact on river and ground water.
3. Water demand for the once-through system is 30 to 50 times that of a closed cycle system.

Use of open cycle system has been prohibited due to more efficiency of closed cycle system than open cycle system and because of economical reasons.

As already discussed thermal power plants are the largest emitter of: (i) mercury (ii) particulates matter, (iii) gaseous emission, (iv) sulphur dioxide, (v) oxides of nitrogen, (vi) carbon monoxide, (vii) carbon dioxide and (viii) hydrocarbon.

Remediation measures in thermal power plant

Air pollution control – Point source:

1. For boiler stacks – ESP/Bag house.
2. Coal crusher – Bag filter.
3. Coal mill – Bag filter.

SO_x and NO_x control technology are given in Table 18.3.

Table 18.3: SO_x and NO_x control technology.

SO_x control technology	***NO_x control technology***
Use of alternative fuel	NH_3 injection
Use of low sulphur containing fuel	Use of low NO_x burner
Lime dosing	Flue gas recirculation
High stack	

18.15 New technological practices

18.15.1 Super critical technology

1. Larger unit size (more than 500 MW) Higher thermal efficiency (of 5% and above).
2. Low gaseous and soot emissions.

18.15.2 Fluidised bed combustion (CFBC/PFBC)

1. Can burn wide range of coals and other fuels such as pet coke, lignite, etc.
2. Higher thermal efficiency (>40%).

18.15.3 Water conservation techniques

Big power plant - closed loop instead of open loop, small power plant - air cooling instead of water cooling, recycle and reuse of process and effluent water, zero discharge in case of small plant, demineralised backwash water shall be treated with RO and reuse in process, collection of rain water which can be used for many purposes

18.15.4 Practices to reduce transportation impact

Encourage bulk transportation by train, only pollution certified vehicles should be engaged in transportation, appropriate infrastructure for vehicles such as concrete or pitched road, separate approach road for transportation of raw materials such as coal, diesel, caustic, etc.

18.15.5 Practices for soil management

Provision for topsoil storage and reuse, separate stacking of topsoil with adequate collection rain all around, topsoil storage heap should be covered with grasses and bushes to avoid erosion, removing vegetative cover only from the specific site on which construction has to take place.

Section III

Pollution control in metallurgical and mining industry

19

Pollution control in iron and steel industry

19.1 Introduction

The iron and steel industry is one of the oldest and largest manufacturing operations in world. There are three basic types of steel manufacturers: integrated mills, mini-mills and specialty-steel mills.

Integrated steel mills use iron ore and coal as the raw materials for manufacturing various steel products. Mini-mills produce steel from scrap, typically using electric arc furnaces. Speciality-steel mills are similar in operation to mini-mills; they manufacture stainless, tool and high-alloy steels. Few people outside the steel industry are aware of the enormous volumes of water and air required to produce iron and steel from raw materials. About 3.5 T of air must be supplied to the Blast furnace to reduce the ore to metal and this amount of air produces about 5 T of Blast furnace gas. So, it is apparent that the tonnage of air and gas far exceeds the solid charge to the furnace and the products tapped from the furnace.

Generally iron and steel plants spend roughly 15% of their capital expenditures on environmental projects. Despite this significant progress, further improvements in pollution prevention technologies are needed for iron and steel plants to reduce costs, improve profitability and facilitate compliance with changing federal regulations. The industry also faces substantially tighter regulations under the clean water act. EPA has proposed revisions to Effluent Limitations Guidelines (ELGs) for the iron and steel point source category that have been in place since the early 1980s.

Many steel plants are already required to reduce effluents below the previously established ELG limits because of the need to meet state and federal water quality standards, particularly in the great lakes area as a result of the great lakes water quality initiative. Meeting these new and changing regulations requires effective R&D efforts throughout many different aspects of the steel making process. To target future research, it is useful to examine the various processes and the pollutants they produce. Although the coking process has been widely considered to pose the biggest environmental challenge, environmental regulations affect all stages of the manufacturing and forming process.

The following sections discuss the potential pollutant output and development needs of each of the major processes: coke making, iron making, steel making, refining, casting, forming and finishing.

19.2 Emissions from coke making

Coke making represents perhaps the greatest environmental concern to the steel industry. Although much progress has been made, coke making emissions continue to be a target of federal environmental regulations. Continual efforts to reduce emissions are necessary to ensure the stability of U.S., coke making; some by-products present an additional concern. Figure 19.1 illustrates the coke making process with its major inputs and outputs.

Strict environmental regulations in the United States, including the clean air act amendments of 1990, threaten the economic viability of domestic coke production. While improvements to contain emissions have been incorporated into recent oven designs, many older coke making facilities cannot be retrofitted economically to meet current and proposed environmental standards. Tighter environmental regulations threaten to accelerate plant closures, in turn reducing domestic production capacity by approximately 30% by the year 2025. Other developed and developing nations are expected to follow the lead of the United States with tighter environmental regulations that will affect worldwide coke production. At present, no new by-product coke making technology is guaranteed to be environmentally compliant. In fact, new MACT standards provide an incentive for non-recovery coke ovens for new facilities. In the future, U.S. dependence on foreign sources is expected to increase.

19.2.1 Air emissions from coke making

Typical coke making air emissions include ammonia, benzene-soluble organics, benzene, particulates, SO_x and VOCs. Coke plant operators have initiated significant efforts to control these emissions and have spent millions of dollars on environmental control systems, improved operating and maintenance practices and increased personnel training. These efforts have substantially reduced coke making air emissions over the last three decades.

The industry's efforts have reduced emissions by more than 95% for benzene and VOCs from by-product sources, particulate matter emissions from charging and pushing and SO_x emissions from the combustion of 'cleaned' coke oven gas. To reduce air emissions, the industry installed many types of new and improved environmental hardware. Training and, more importantly, the conscientious efforts of skilled operating personnel have also substantially reduced emissions of benzene-soluble organics from coke plant doors, lids, offtakes and charging emissions sources.

19.2.2 Coke making effluents

The typical volume of process wastewater generated at a well-controlled coke plant is approximately 150 gallons/T of coke produced. About 25 to 35 gallons/T

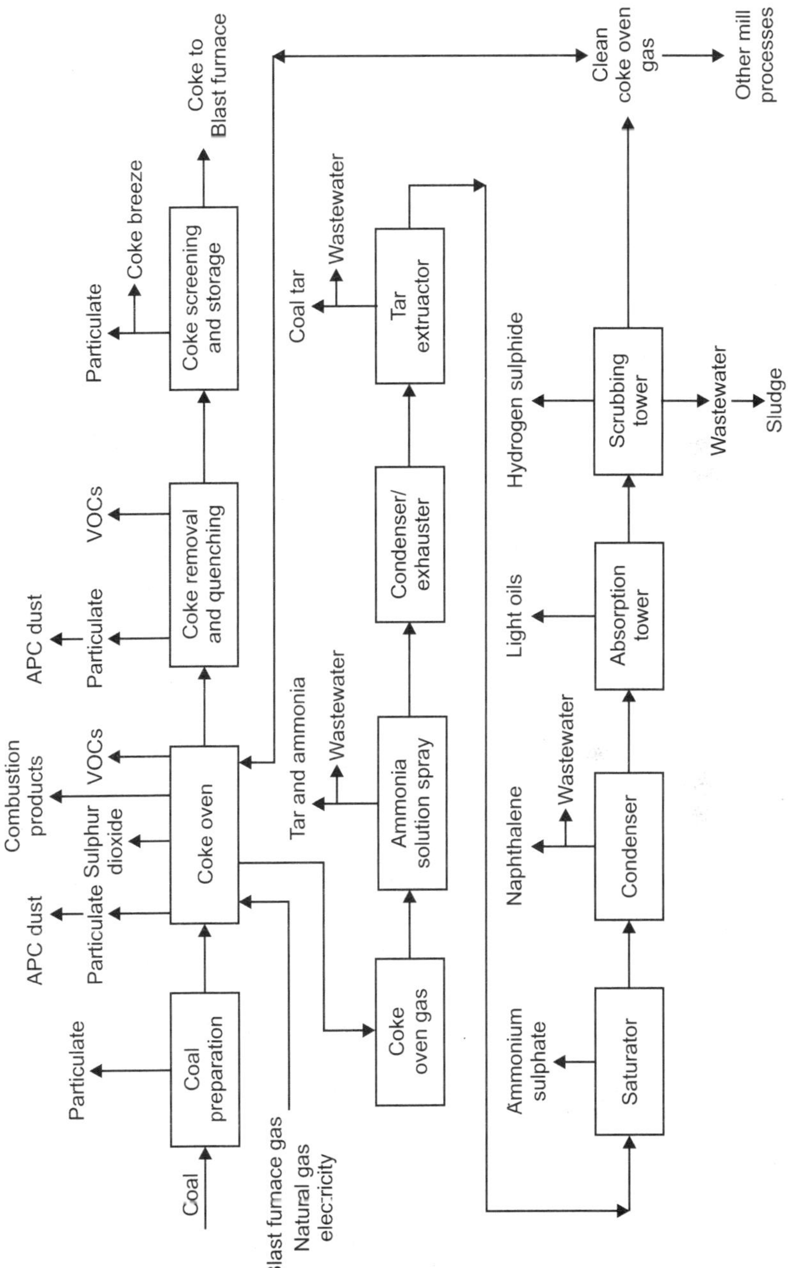

Figure 19.1: Cokemaking flow diagram.

are generated as waste ammonia liquor from moisture contained in the charged coal. The balance is from the steam used in distilling ammonia from the waste liquor, light oil recovery and other processes.

Coke making wastewaters contain significant levels of oil and grease, ammonia, nitrogen, cyanides, thiocyanates, phenols, benzenes, toluene, xylene, other aromatic volatile components and polynuclear aromatic compounds. Wastewaters also contain trace amounts of the toxic metals antimony, arsenic and selenium. The amount of each pollutant generated depends on the by-product process, specific facility equipment, practices and the range of constituents in the coals used.

19.2.3 Coke making by-products

In the by-product coking process, volatile components are collected as unpurified 'foul' gas containing water vapour, tar, light oils, coal dust, heavy hydrocarbons and complex carbon compounds. Condensable materials, such as tar, light oils, ammonia and naphthalene are removed, recovered and processed as gas and coal chemical by-products. Finally, sulphur is removed, leaving clean, desulphurised coke oven gas. The cleaned, desulphurised gas is then used as fuel for coke ovens or other plant combustion processes, or it is sold to nearby facilities.

19.2.4 Hazardous coke making wastes

As shown in Fig. 19.1, seven listed hazardous wastes are associated with coke making under the Resource Conservation and Recovery Act (RCRA). These include residues from coal tar recovery, tar storage tanks, light oil processing units, wastewater sump residues and naphthalene collection and recovery.

Process residues from coal tar recovery are generated when uncondensed gas enters a primary cooler. Condensates from the primary cooler flow into the tar collecting sump and are discharged in the flushing liquor decanter. Tar collection sump residue or sludge accumulates at the bottom of the collecting sump and must be recycled periodically as an individual stream, through the flushing liquor decanter, or back to the coke oven. Tar storage tank residues are produced when residuals settle out of the crude coal tar collected as a coking by-product. The residues are periodically removed from the storage tanks and are recycled to the oven or landfilled.

Residues from light oil processing units collect in a light-oil scrubber and light-oil stripping still. Resin is a related waste that accumulates from cleaning wash oil used in the light-oil recovery process. Wash-oil muck, residue from a wash-oil purifier or decanter, is periodically removed and recycled to the coke oven, reclaimed off-site, or used as Blast furnace or boiler fuel.

Wastewater sump residues accumulate in the bottom of a sump used to provide sufficient quiescent residence time for oil and water to separate during light oil recovery. These settled solids are removed periodically and either recycled to the oven or landfilled off-site. Residues from naphthalene collection and recovery accumulate at the bottom of a skimmer sump where naphthalene is mechanically skimmed off the surface.

Residues also accumulate in the hot and cold sumps used as collection or surge vessels and on cooling tower surfaces. Residues from wastes are recycled to the decanter or sometimes the oven.

19.2.5 Coke making environmental trends and drivers

Meeting the constantly changing environmental regulations is important to ensure the stability of U.S., coke making. The federal government has already promulgated coke plant air emission regulations for doors, offtakes, lids, coal charging and collection mains that call for significantly more stringent emission requirements. In addition, standards are also being proposed for pushing, quenching and battery combustion stack emissions. Moreover, tighter effluent requirements are being proposed for coke making wastewaters.

These regulatory requirements challenge operating management and employees as they deal with ageing batteries. Even the installation of new batteries as greenfield sites or increased capacity at conventional existing coke oven by-product plants will present environmental challenges because of new EPA regulations.

19.2.6 New and emerging coke making technologies

R&D needs of coke making are given below:

1. New ways to reduce emissions from leaking doors, lids and offtakes.
2. Improved quench cars.
3. Improved capture devices or quenching methods.
4. Improved refractories for walls or better ways to seal existing leaks through walls.
5. Refractories or oven designs to accommodate operation as negative-pressure batteries.
6. New, more environmentally friendly coke making technologies.
7. Improved or higher valued chemicals from coke oven gas and by-products.

19.3 Iron and steel making processes

Three process steps—coking, iron-making and steel-making—are fundamental to the iron and steel industry.

19.3.1 Coking

Coking is the process of converting coal to pure carbon or coke. The coking process yielding coke as well as by-products, such as coal tar, ammonium sulphate and light oils. A typical coke plant consists of two major areas; the coke oven batteries and the by-product recovery plant.

19.3.2 Iron-making

In iron-making, coke is combined with iron ore and limestone and supplied to the top of a Blast furnace (Fig. 19.2). Air is heated to about 1800°F and blown counter currently from the bottom of the furnace through circumferential openings called tuyeres. In the furnace, coke reacts with iron ore, releasing iron and generating CO and CO_2 gas.

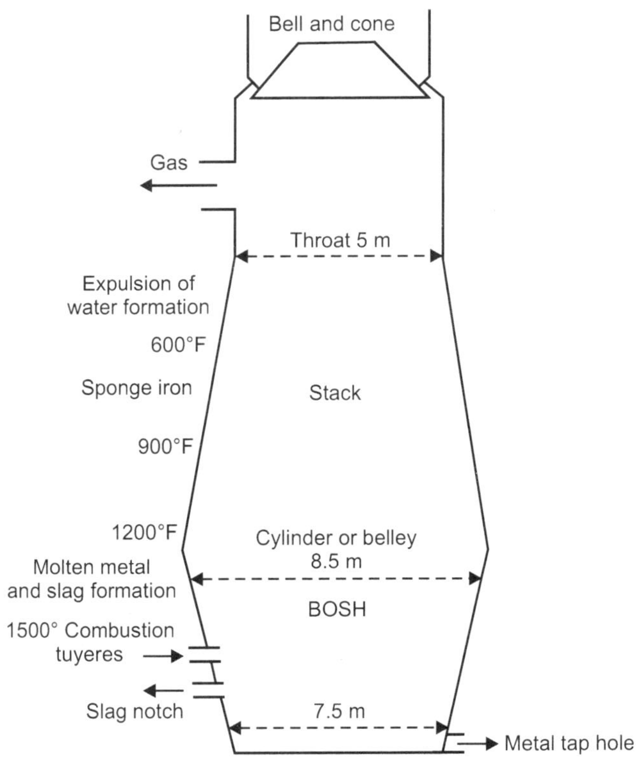

Figure 19.2: Blast furnace.

Iron sinks to the furnace hearth. Impurities in the charge combine with the limestone, forming slag that floats to the top of the iron. The combustible exhaust gases from the furnace have considerable heating value and are used to pre-heat the incoming air and to generate steam. Blast furnaces produce iron, commonly known as pig iron, with a carbon content of about 4%.

19.3.3 Steel making

Steel is an alloy of iron, containing less than 1% carbon. Steel is made by three different methods: the Basic Oxygen Furnace (BOF), the Open Hearth Furnace (OHF) and the Electric Arc Furnace (EAF) methods. These furnaces reduce impurities, especially the carbon content by oxidation and add various alloying elements, depending on the desired grade steel. Because BOF and EAF are currently the most widely-used techniques, only these have been described briefly here. In the BOF method, (Fig. 19.3), hot molten pig iron from the Blast furnace, scrap steel, mill scale and slag conditioning materials (such as limestone, burned lime, dolomite and fluorspar) are charged to the furnace. Oxygen is injected into the furnace through a lance positioned a few

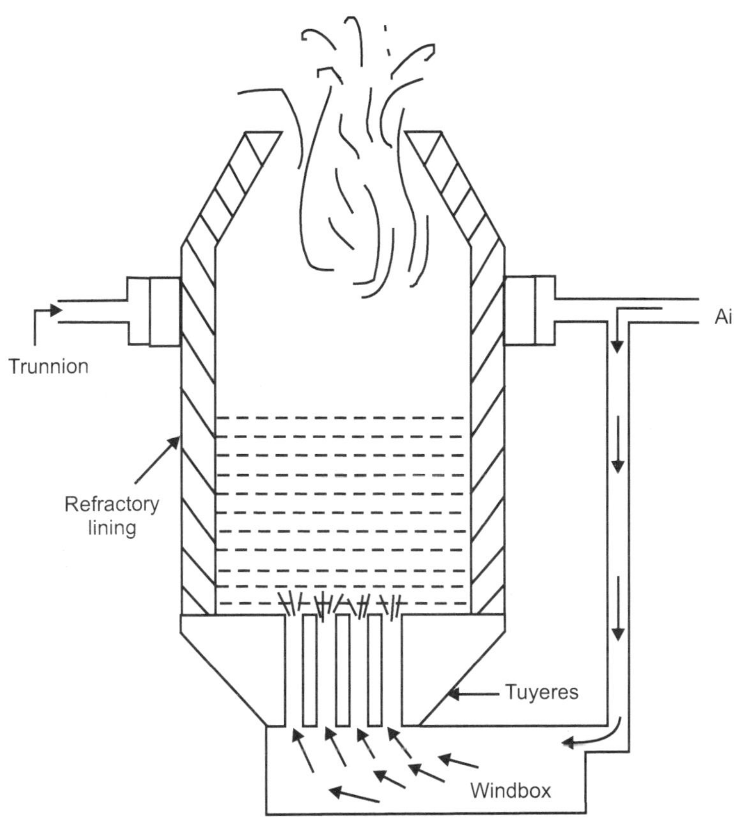

Figure 19.3: Bassemer converter.

inches over the charge. Oxygen reacts to oxidise the charge, which releases large amounts of heat that melts the charge. During the melting process, the fluxing materials, called slag, collect and float on top of the molten metal. The slag is decanted through a spout while the molten steel is either transferred to a teeming station or to a continuous casting station.

EAFs are used for manufacturing common grades of low-carbon steel as well as special alloy and tool steels. A chief advantage of the EAF method is that it is not dependent on hot molten pig iron from Blast furnaces and can typically operate, if necessary, with scrap steel as the sole iron source in the initial charge to the furnace. EAFs also need the same or similar flux conditioning materials as are required by BOFs. Slag and molten steel separation is similar to that in a BOF.

19.3.4 Open hearth furnaces

The open hearth furnace produces steel (below 1% carbon) from a charge of steel scrap, iron (about 4% carbon) and slagging materials, such as lime and fluorspar. These furnaces produce about 100 to 300 T per heat of 8–12 hour duration (Fig. 19.4). As in the Blast furnace, open hearth furnaces use large quantities of water for cooling, generally ranging from about 750 gpm for small furnaces to as much as 1500 gpm for larger units.

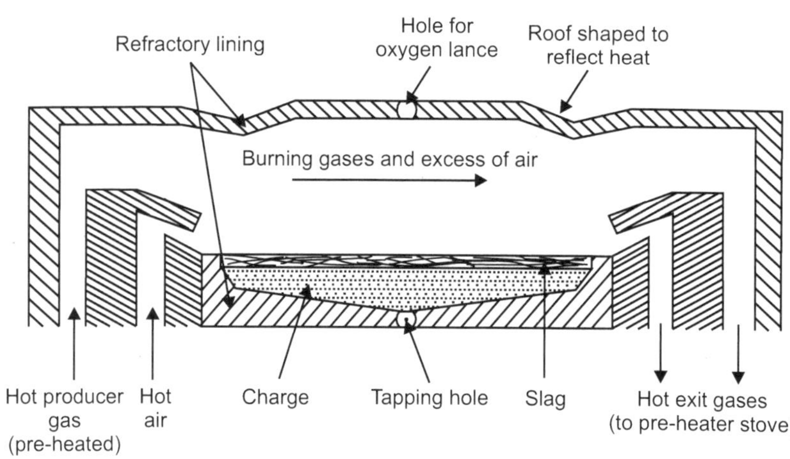

Figure 19.4: The open hearth furnace.

19.4 Dust particles and wastewater from blast furnace

In the Blast furnace, the gases from the open hearth furnaces contain a high concentration of dust, which may be removed by washing, precipitation equipment or both. This dust is removed from the wash water by clarifiers and may be combined with the recovered dust from the Blast furnaces for sintering if the zinc content is sufficiently low. The dust loading varies appreciably during the heat from a range of about 0.2 to 0.6 grain/standard cubic foot of gas (scf) during charging to about 2 to 3 grains/scf several hours after start of oxygen lancing.

The dust particles from open hearth furnaces are much finer than those from the Blast furnace and chemical coagulation is needed to separate them from the wastewater. Oxygen lanced furnaces discharge a higher dust load than the older, conventionally-fired furnaces. Whereas the wash water from the Blast furnace is alkaline, that from open hearth scrubbers is quite acidic, so special materials of construction are required in closing open hearth washer systems. However, because the pH of the circulated water is in the range of 2.5–3.0, the system can be completely closed without fear of deposits, the only water loss being that associated with the collected sludge of filter cake.

It may be necessary to select the scrap charge for the open hearth furnace carefully if the dust is recovered for sintering, since a high zinc content can be damaging to the refractory lining of the Blast furnace when the sinter is recharged to the furnace. The zinc, which comes from galvanised sheet or white metal in the scrap, may pass through the refractory and sublime on the metal shell, gradually expanding and loosening the refractory.

A third source of wastewater from the open hearth is the blow-down from the waste heat boilers, which recover heat from the exit gas and produce by-product steam. Water from the continuous blow down is generally of such small volume that it does not significantly affect the composition of the collected wastewater from the open hearth area, where this includes the furnace cooling water. However, mass blow down from the boilers may show up as periodic high pH or high conductivity in the combined wastewater. If there is a separate water treatment plant to supply make-up for the waste heat boilers, wastes from regeneration of the water treatment system may also constitute one of the pollution loads from the Blast furnace area.

19.4.1 Basic oxygen furnaces

The basic oxygen furnaces (BOF) shop is a facility for conversion of iron into steel at much higher rates than those attainable in the open hearth. In the conversion of iron to steel in the basic oxygen furnace, the molten metal charge is reacted with oxygen, introduced through a water-cooled lance, which burns off the impurities in a period of about 20 to 25 min. Total heat requires about 50 min compared with the 8–12 hours required for steel-making in the open hearth furnace. Most BOF units produce in the range of 100–300 T of steel per heat. Because of the high heat release, gases leaving the furnace hood during lancing are very hot. The hood may be cooled with circulating water, or heat may be recovered from this gas through a boiler mounted directly above the hood. Because the heat is released at an uneven rate and because little heat is released during the charging and tapping period, the flow of steam from the generating unit is irregular and must be modulated by discharge through an accumulator. This then releases steam at a relatively steady rate into the plant steam system at a pressure below that of the waste heat boiler.

Because of the very high temperature service, the lance cooling water is specially treated to prevent corrosion or scale formation; this circuit may be completely closed and heat extracted from the lance cooling water through a secondary cooling water circuit using a surface type heat exchanger.

The gas in such furnaces is cleaned in equipment similar to that used in the open hearth, although there are variations in methods of wash-in and removing accumulated dust from the washers and precipitators.

The dust loading is usually appreciably higher in the BOF discharge (about 50 to 60 lb/T of steel and in the range of 0.5–5.0 grains/scf in the gas entering the spark box) than in either the Blast furnace gas (about 30 lb/T) or the open hearth (about 20 lb/T). There is appreciable variation in temperature and pH of the wash water during the heat, as different ingredients are added to the charge during the oxygen blow.

Other important iron and steel operations

Other common and important operations in an iron and steel plant are sintering, vacuum degassing, casting, rolling and finishing.

Sintering: Sintering is the cementing of Blast furnace fines, mill scale and flue dust that cannot be directly charged to a Blast furnace. In this process, the fines are mixed with fine coal in a travelling gate, which makes them usable as a direct charge to the Blast furnace.

In this process, molten steel is subjected to a vacuum to remove gaseous impurities, such as hydrogen, nitrogen and oxygen.

Casting: At the teeming station, molten steel is poured into molds to form ingots. After the ingots are cooled and the molds are stripped, the ingots are transferred to a primary hot rolling mill where the ingots are reheated in soaking pit furnaces and rolled into billets, blooms, or slabs. The continuous casting operation eliminates the hot rolling operation by casting the molten metal directly into thin slabs or other desired shapes.

Hot-forming: In the secondary hot rolling mill, slabs or blooms are reduced to billets, shapes or strips.

Cold-rolling: In cold rolling, unheated hot-rolled products are passed through rolls to produce steel with smooth dense surfaces and enhanced mechanical properties. Cold-rolled steel products are not as thick as those produced by hot rolling.

Surface preparation: The rolled-steel products are pickled in acid to prepare the surfaces for subsequent application of corrosion-protection coatings. Molten salt baths are used to descale stainless-steel products.

19.5 Emissions from iron making

This section focuses on the sintering and Blast furnace iron making processes. Although sintering plants offer significant benefits, environmental concerns have stymied greater use of traditional plants. NO_x, VOC and CO_2 control are the major areas requiring further R&D. Figures 19.5 and 19.6 illustrate the sintering and Blast furnace processes, respectively, with their major inputs and outputs.

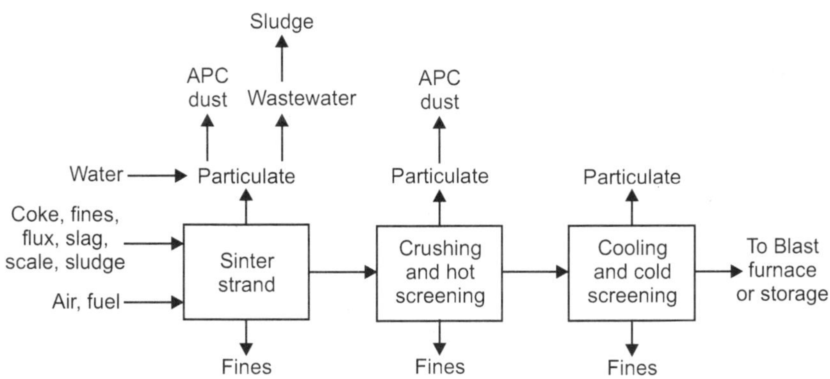

Figure 19.5: Sintering flow diagram.

19.5.1 Iron making emissions

Sinter plant

Sinter plant emissions are generated from raw material handling, wind-box exhaust, sinter discharge (associated sinter crushers and hot screens) and the cooler and cold screen. Wind-box exhaust is the primary source of particulate emissions - mainly iron oxides, sulphur oxides, carbonaceous compounds and chlorides. Fluorides, ammonia and various metal compounds may also be present. At the discharge end, emissions are mainly iron and calcium oxides.

Sinter strand wind-box emissions are commonly controlled by cyclone cleaners followed by a dry or wet electrostatic precipitator, high-pressure drop wet scrubber, or baghouse. Crusher and hot screen emissions, the next largest emission source, are controlled by a hood and baghouse or scrubber.

Other material-handling operations also generate emissions. Baghouses are used to capture particulate matter generated during conveyor transport and loading or unloading of sinter plant feedstocks and product. This dust is either

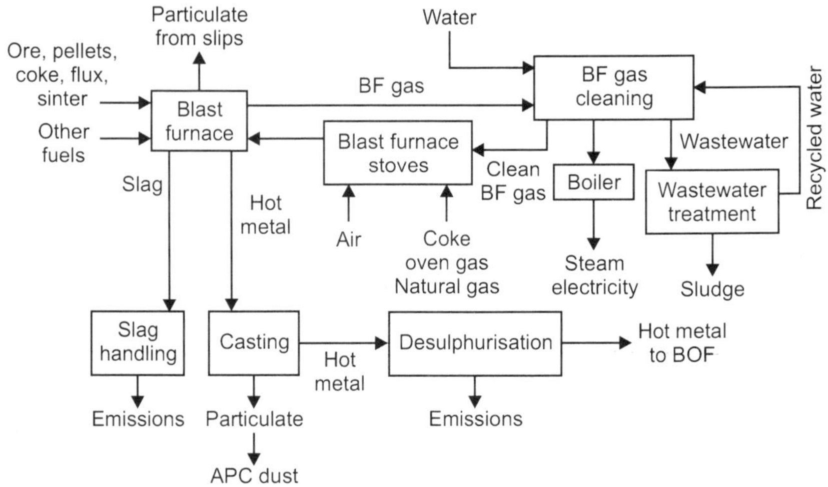

Figure 19.6: Blast furnace flow diagram.

recycled as feedstock to the sinter plant or landfilled as solid waste. Although the chemical composition of the dust is highly dependent upon the sintering practice, typical components include Fe, C, S, Fe_2O_3, SiO_2, Al_2O_3, CaO and MgO. Sinter plants have also been identified as potential sources of small concentrations of dioxin emissions. However, the sources and formation of dioxins is not well understood and more study is needed.

Blast furnace

Particulate emissions are primarily generated during the casting of molten iron and slag from the Blast furnace. During casting, molten iron and slag flow out of a taphole at the base of the furnace into runners that lead to transport ladles. When the molten iron and slag contact air, particulate emissions are generated. Emissions also are generated by drilling and plugging the taphole and by using an oxygen lance to open a clogged taphole.

During the casting operation, iron oxides, magnesium oxide are generated as particulate matter. Casting emissions are captured in a high canopy or local hood and exhausted to a cleaning device, generally a baghouse. Flame suppression techniques can minimise the amount of fume generated.

Dust captured from casting emissions is much less than that collected by a Blast furnace gas cleaning system. If hot metal desulphurisation is used to remove or alter the sulphur compounds in the hot metal, emissions are controlled by hoods and the exhaust gases are discharged through a baghouse to control particulate matter. This operation usually occurs at a facility located between the Blast furnace and BOF facilities.

19.5.2 Iron making effluents

Sintering

Wastewaters are generated from the use of wet air pollution control scrubbers to clean the wind-box and discharge ends of sinter machines. Typical flows for wet air pollution control devices are 1000 gallons/T, with discharge (blow down) rates of 50 to 100 gallons/T for the better controlled water treatment plants. Wet-scrubber wastewater treatment includes removal of heavy solids by sedimentation, recycle of clarifier or thickener water overflows and metals precipitation treatment for blow downs. Some sinter plants are operated with once-through water in the wet-scrubber wastewater treatment plants. The principal constituents can include suspended solids, oil and grease, ammonia-nitrogen, cyanide, phenolic compounds and heavy metals, such as lead, zinc, arsenic, cadmium, copper, chromium and selenium.

Blast furnace

Nearly all of the wastewater generated from Blast furnace operations is direct contact water used in the gas coolers and high-energy wet scrubbers that clean the Blast furnace gas. Typical water requirements are 6000 gallons/T of iron, which is recycled. These waters are typically treated by clarifiers or thickeners to remove suspended solids and the overflows are recycled to the gas scrubbers. The final blow down from the recirculated system is then treated to remove additional heavy metals and other non-solid pollutants, such as cyanide and, typically less than 100 gallons/T, ammonia.

19.5.3 Iron making by-products

The primary by-products generated during the production of molten iron include Blast furnace gas, slag, air pollution control dust (flue dust) and Blast furnace filter cake (wastewater treatment plant sludge). Between 2 and 3 T of Blast furnace gas, containing up to 50 kg of dust, are generated for each ton of iron. Blast furnace gas contains up to 40% CO and 6% H. The main solid components of the flue dust include Fe, C, SiO_2, Al_2O_3, CaO and MgO. Blast furnace flue gas is emitted at the top of the furnace as particulate-laden combustion gases composed primarily of CO. About 60% of the particulate matter is removed from the gas stream by cyclonic vortex separation (i.e., dust catcher) of the heavy particles (flue dust). Fine particulate matter is subsequently removed in a two-stage cleaning process with a wet scrubber and a high-energy venturi impact scrubber or electrostatic precipitator. This two-stage cleaning process produces a gas containing less than 0.05 grams/m^3 of particulate matter. The cleaned gas is used as a fuel to pre-heat the Blast air for the furnace or to generate steam in an adjacent boiler house.

Dust and sludge

Blast furnace dust is often recycled through the sinter plant. At steel making operations without sintering plants, Blast furnace dust is sometimes mixed with other by-product residues, briquetted and recycled into the Blast furnace. In other steel making operations, the dust and sludge are landfilled, stockpiled or sold.

Slag

The Blast furnace also yields a slag that contains oxides of silicon, aluminium, calcium and magnesium, along with other trace elements. Blast furnace slag amounts to 20 to 40% of molten iron production by weight.

19.6 Iron making environmental trends and drivers

Although the technology exists for efficient particulate matter collection systems, opportunities exist for improving capture and collection systems, wastewater treatment and reducing water use.

Sintering poses a unique environmental challenge. Sintering is the traditional and proven method for recycling iron-bearing secondary materials produced by iron making and steel making facilities. It can also produce a material that replaces iron pellets and adds stability to Blast furnace operation. Japanese Blast furnaces and at least one U.S., plant, for example, operate on virtually 100% sinter feed. As more emphasis is placed on recycling and recovering waste or secondary iron-bearing materials, plants with sintering operations will have an advantage over those that do not.

Most sinter plants are equipped with high-energy wet scrubbers, fabric filters, or wet electrostatic precipitators for collection of wind-box emissions. However, some of these devices have not always performed well across the range of feed materials that operators would like to introduce into sintering. Therefore, opportunities exist for developing new environmental control technologies. New emphasis on fine particles and NO_x emissions will place further pressures on the ability of traditional air pollution control devices. Concern over potential dioxin emissions and restrictions on oily sinter plant feed materials are also relevant issues.

19.6.1 Iron making research and development needs and opportunities

R&D needs

1. Improved capture and collection systems and improved wastewater treatment for sintering.
2. Understanding of NO_x and toxic air pollutants emissions, formation and control.

3. Alternative cost-competitive agglomeration processes.
4. Coke substitutes.
5. Greater understanding of formation and mitigation of NO_x emissions in stoves.
6. Methods of reducing total sulphur emissions.
7. Improved methods of cast house emission controls.
8. Improved methods of slag quenching, including collection and treatment of excess slag quenching water.
9. Methods for de-oiling sinter plant feed materials.

Fundamental research is needed to provide more information on the formation and mitigation of CO from sintering, reduction of NO_x and sulphur emissions and formation and control of organic and inorganic toxic air pollutants.

When conducting R&D on new and emerging iron making technologies, information on environmental aspects and energy consumption must be an integral part of that work in order to assess alternative technologies on the basis of environmental issues, operating costs, capital costs and productivity. Data on particulate and gaseous air pollutant generation, wastewater generation and treatability, solid and hazardous wastes and energy consumption, including CO_2 emissions, should all be considered.

Specific environmental control R&D opportunities in iron making include:

1. NO_x and VOC control.
2. Energy use and CO_2 emission reduction.

NO_x and VOC control

NO_x and VOC controls have grown in importance as ozone control strategies have been developed for ozone non-attainment areas and ozone transport regions. Because NO_x is a precursor for fine particles, new fine particle standards are likely to place even greater emphasis on sinter plants as a source of potential NO_x.

Although a lower flame temperature makes NO_x formation less problematic for Blast furnace gas than for other fuels, increased concern over NO_x emissions presents challenges for combustion practices and devices that reduce NO_x emissions in boilers. A comprehensive study of NO_x control at sinter plants is needed to develop ways of reducing these emissions. The emission, formation and control of NO_x and organic and inorganic toxic air pollutants, including dioxins, need to be understood. As a result of the interest in ozone strategies, sinter plants may have to limit VOC components in feed materials or increase the efficiency of air pollution control systems. A study of VOC components, including dioxins, is also needed.

Energy use and CO_2 emission reduction

Energy consumption and CO_2 emissions in existing and emerging agglomeration and iron making technologies pose opportunities for reductions. Government concern over greenhouse gas emissions, particularly CO_2 and the related impetus for reducing energy consumption, suggest the need for developing ways to reduce energy consumption and CO_2 generation in existing and emerging agglomeration and iron making technologies.

19.7 Emissions from basic oxygen furnace (BOF)

Environmental standards for air and water emissions, including those for BOF's, are becoming more stringent. Consequently, current environmental controls may not be adequate in the future. In particular, better air-cleaning and in-shop work environments are needed. Figure 19.7 illustrates BOF steel making with its major inputs and outputs.

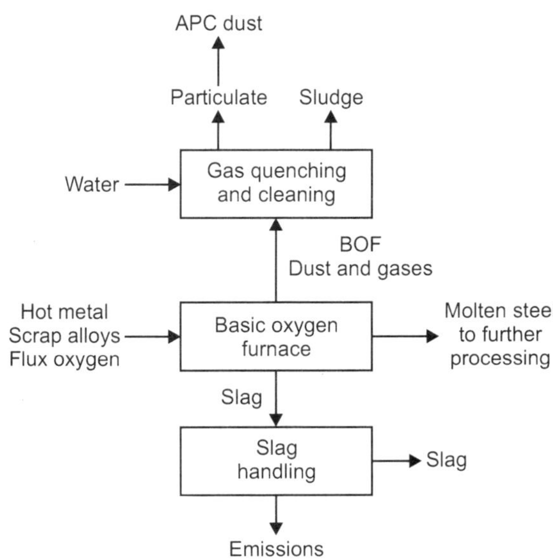

Figure 19.7: BOF steelmaking flow diagram showing various emissions.

19.7.1 BOF emissions

Sources

Emissions are generated during each of the five major BOF steel making and refining operations: charging, melting, refining, tapping and slag handling. The most significant emissions from BOF steel making occur during the oxygen blow period. The principal compounds generated are iron oxides and lime.

Control systems

The particulate-laden combustion gases and fume (a very fine iron oxide) created during oxygen blow periods are removed from the furnace by evacuation through a large collection main. These hot gases are typically treated by one of three air pollution control methods:

1. Semi-wet: Water is added to condition furnace off-gas temperature and humidity prior to processing the gas in electrostatic precipitators or baghouses.
2. Wet-open combustion: Excess air is drawn into the hood of the off-gas exhaust system, allowing CO to combust prior to high-energy wet scrubbing for air pollution particulate control.
3. Wet-suppressed combustion: Excess air is excluded from the off-gas collection system prior to high energy wet scrubbing for air pollution particulate control, thus suppressing the combustion of CO until after scrubbing.

Charging and tapping emissions are controlled by a variety of evacuation systems and operating practices. Charging hoods, tapside enclosures and full furnace enclosures are used to capture these emissions. The particulate emissions are then exhausted to a primary gas cleaner or a second gas cleaner.

As in the case of the Blast furnace, most trace hazardous air pollutants generated in the BOF are heavy metals, including cadmium, chromium, lead, manganese and nickel.

Zinc compounds are also present in BOF fume in varying amounts depending on the zinc content of the scrap used in the process. Other factors affecting HAP emissions from the BOF include the degree of oxidation of the molten steel and the amount of time required to process the melt.

19.7.2 BOF effluents

The gases and fumes released during BOF steel making are quenched with water to reduce their temperature prior to being treated in air pollution control systems. The three major types of off-gas control systems generate wastewater streams containing total suspended solids and metals (primarily lead and zinc, but also trace amounts of arsenic, cadmium, copper, chromium and selenium).

Suppressed-combustion and open-combustion systems use, respectively, about 1000 and 1100 gallons of water/T of steel, but typically less than 100 gallons/T is discharged. Standard treatment consists of sedimentation in clarifiers or thickeners. Blow down treatment consists of metals precipitation.

BOF steel making by-products include BOF slag, air pollution control dust and water treatment plant sludge.

Slag

BOF slag is composed of calcium silicates and ferrites combined with fused oxides of iron, aluminium, manganese, calcium and magnesium. After molten BOF slag is removed from the furnace, it is cooled and processed to recover the high metallic portions (iron and manganese) for use in the sinter plant or as a flux in Blast furnaces. Some slag may also be recycled to steel making under certain conditions.

The remaining nonferrous fraction is then crushed and sized for reuse either within the steel works or as a by-product.

BOF slag differs from Blast furnace slag in composition (hydration of up to 10% can occur because of calcium oxide and other oxides are present in BOF slag); therefore, its uses are more limited than Blast furnace slag. The lower metallic aggregate is typically used in construction applications such as railroad ballast or unconfined highway base or shoulders, where its expansion will not create a problem. It is also used as an additive in cement kilns.

Dust and sludge

The BOF dust and sludge collected in the air pollution control system represent two of the three largest-volume wastes typically land-disposed by iron and steel plants. Together with slag and water treatment plant sludge, these wastes represent more than 93% of all wastes stored, recycled, or disposed.

Particulate matter is separated from combustion gases generated during BOF steel making by one of the three means described in Section 19.5.1. Cleaned gases from suppressed combustion systems are ignited before release to the atmosphere to prevent CO emissions in the uncombusted gas. These gases can also be recovered for heating value.

19.7.3 BOF steel making environmental trends and drivers

Air emissions

Regulatory pressures to reduce releases of toxic and hazardous substances in all media, including air, water and solid wastes, will continue. The issue of global warming may lead to increased requirements to reduce emissions of greenhouse gases from fossil-fuel combustion. This increased pressure on CO_2 emissions affects the steel industry because the BOF steel making process inherently generates CO_2.

During BOF steel making, carbon in the hot metal is oxidised and emitted from the process as CO_2. This carbon oxidation provides the heat that converts the iron into steel and is the basis for BOF steel making. Until another method of steel making is developed, BOF steel making will continue to emit CO_2.

Fugitive emissions

Fugitive emissions from a BOF come from the following:

1. Scrap handling.
2. Ore and flux handling.
3. Hot metal transfer operations.
4. Hot metal charging operations.
5. Tapping.
6. Teeming operations.

Each state has regulations governing fugitive emissions; it is anticipated that there will be continued regulatory pressure to minimise these emissions.

Water discharges

Stringent EPA and state water quality standards will likely reduce the allowable discharge limits for the discharges from BOF steel making facilities in the near future. Such a reduction will probably be directed at the discharge of metals such as lead and zinc. Reducing metal-laden wastewater discharges may require additional water pollution control facilities and water recycling for some BOF steel making operations.

The primary wastes collected from a BOF are slag dust and fumes collected either by wet scrubbers or dry electrostatic precipitators or baghouses. Baghouse dust is also collected from operations such as hot metal reladling, hot metal desulphurisation and slag skimming.

By current EPA definitions and test protocols, these wastes are not hazardous and can be disposed of in nonhazardous landfills. However, with increasing government emphasis on pollution prevention and waste minimisation programmes and with the increasing scarcity of landfill space, steelmakers will have to reduce the volume of these wastes, increase recycling, or develop additional processing to increase their use as products.

19.7.4 New and emerging BOF steel making technologies

No new and emerging air or water pollution control technologies for BOF steel making operations are anticipated. In waste management, developing technologies will increase the amount of BOF slag, BOF fume and other steel making revert materials that are recycled or used as other products.

Some companies are processing BOF fumes with other steel plant wastes for recycle to the BOF. Others are recycling BOF fumes with low zinc content to the Blast furnace via the sintering plant. Over the years, a number of processes have been studied to remove the zinc from the BOF fume so more of it can be recycled to the Blast furnace via the sinter plant and some of these processes may be viable in the future.

BOF slag

Research is needed to find either alternate uses of BOF slag (with the calcium oxide content and hydration effects) or an economical way to remove the calcium oxide. Research is also needed to remove phosphorous from BOF slag because phosphorus content is a limiting factor to increased use of this material. Removing phosphorus from the Blast furnace pig iron prior to steel making would help increase the amount of this slag that can be recycled to the iron and steel making process. Some companies recycle a portion of their BOF slag to the Blast furnace in order to supply calcium oxide flux and reclaim the iron oxide content of the slag.

Zinc content

Research into the mechanisms of zinc ferrite formation during the steel making process to prevent its formation is needed, making it much easier to remove the zinc from fume. BOF fumes contain varying amounts of zinc, lead and other metals. The primary source of these metals is the scrap used in the steel making process. If the scrap contains a large amount of zinc, as does much of galvanised steel scrap, the zinc content of the fume can be high enough to prevent its recycle to the Blast furnace.

The presence of zinc in the feed materials of a Blast furnace can cause serious damage to Blast furnace refractories, which will shorten the life of a Blast furnace.

Continued research is needed to develop economical ways to remove zinc from scrap since zinc is a commercial product. This would help decrease the amount of zinc in BOF fumes and thus promote increased recycling of these fumes. Research to determine the distribution of zinc throughout the recycle system could lead to a better understanding of the chemical mechanisms at work, which would improve understanding of this recycling technique.

Some companies recycle BOF fume directly to steel making after mixing and agglomerating it with other iron oxide materials, such as mill scale. When recycling zinc-bearing fumes directly to steel making, the zinc content of the fume, slag, or steel will increase substantially unless a bleed stream is provided to keep the zinc build-up in the system at manageable levels. In facilities using this recycling practice, zinc build-up has not occurred.

Research will likely be conducted to determine the effect of zinc on the physical, chemical and metallurgical properties of steel. When zinc-bearing scrap is used in steel making, or when zinc-bearing fumes are recycled to steel making, the zinc concentration in the steel produced also increases. Analysis has shown that some steel samples from these systems contain several hundred parts per million of zinc.

BOF steel making

1. Methods of removing phosphorus from the Blast furnace pig iron prior to steel making.
2. Economical ways to remove zinc from scrap.
3. Understanding of zinc ferrite formation mechanism.
4. Determination of zinc distribution throughout the recycle system.
5. Determination of zinc's effect on the properties of steel.

19.7.5 Steel making - electric arc furnace

Over the past 30 years, electric arc furnace fume systems have evolved considerably from simple systems aimed at improving the ambient work environment around the furnace to sophisticated systems aimed at controlling not only particulate emissions, but also toxic gases. Environmental regulations have stimulated progress in primary and secondary fume control technology from side draft hoods, furnace roof hoods and rudimentary fourth hole extraction systems to today's sophisticated direct evacuation systems and fugitive emission control systems. Modern fume systems are now designed to minimise the formation of gaseous pollutants and ensure that others are destroyed before exiting the system. Figure 19.8 illustrates electric arc furnace steel making with its major inputs and outputs.

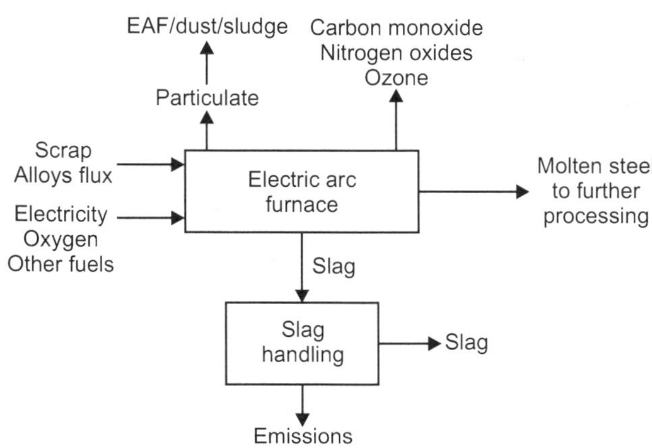

Figure 19.8: Electric arc furnace steelmaking flow diagram.

EAF emissions

Sources: Emissions are generated during each of the five major EAF processes: charging, melting, refining, tapping and slag handling. During the EAF process, oxide and other metal forms are volatilised in the presence of intense heat and turbulence inside the furnace. Also, carbon from the addition of coal, iron and steel scrap and graphite electrodes react with injected air or oxygen. Consequently, the primary constituents in EAF emissions are particulate matter and gases (specifically CO_2, CO and NO_x). Note that use of EAF for steel production provides the single most effective means of reducing CO_2 emissions due to the significantly lower energy requirements of melting scrap compared to smelting ore.

Dioxins and furans have become a major concern over the past few years. Dioxins and furans are combustion by-products and the prevention of these emissions depends strongly on control of the combustion process. Some concern has been expressed over possible dioxin formation in EAFs.

Iron oxide is the predominant constituent of the particulate matter emitted during melting. In addition, small amounts of NO_x and ozone are generated during melting. During refining, the primary particulate compound emitted is calcium oxide from the slag. These particulate emissions from melting and refining account for about 90% of total EAF emissions. The remaining 10% of emissions are generated during charging and tapping.

Emissions from charging scrap are difficult to quantify because they depend on the grade of scrap utilised. Scrap emissions usually contain iron and other metallic oxides from alloys in the scrap metal. Iron oxides and oxides from the fluxes are the primary constituents of slag handling emissions. During tapping, iron oxide is the major particulate compound emitted.

EAF effluents

Most electric arc furnaces are operated with dry gas cleaning systems, which have no process wastewater discharges. A small number of wet and semi-wet systems also exist. The water flows and pollutants of concern for EAFs with wet and semi-wet air cleaning systems are similar to those of wet BOFs, but the metal content (primarily lead, zinc, arsenic, cadmium, copper, chromium and selenium) is higher because of the 100% scrap charge and higher use of post-consumer scrap.

EAF wastewater treatment operations are similar to those for the wet BOFs, including sedimentation in clarifiers or thickeners and water recycling.

EAF by-products

Slag: The major non-hazardous by-product generated during EAF steel making is slag. The primary components in EAF slag are CaO, SiO_2, FeO, MgO and Al_2O_3. EAF slag is managed similar to BOF slag. Cooled, solidified slag is crushed and screened to recover metallics for recycle or reuse and the lower metallic aggregate is used in construction applications.

Hazardous EAF wastes

Dust and sludge: EAF dust consists of particulate matter and gases produced during the EAF process and subsequently conveyed into a gas cleaning system. The particulate matter that is removed from emissions in a dry system is called EAF dust. Particulate matter removed from emissions in a wet system is called EAF sludge.

EAF steel making environmental trends and drivers

In BOF steel making, new ambient air quality standards for fine particulate matter (PM-2.5) may also lead to tighter requirements for emissions associated with EAF steel making and related fugitive emissions permits and compliance assurance monitoring rules may require new continuous process and emission monitoring facilities and practices. New ambient air standards for ozone will also lead to increase regulatory attention of NO_x, which contribute to the ozone formation in the atmosphere.

Increased national and international emphasis on climate change and the relationship between CO_2 emissions and global warming potential may also lead to increase regulatory pressure to reduce CO_2 associated with EAF production. Because new source performance standard's are technology-based and technology continues to improve, tighter particulate emission standards, monitoring requirements and CO or NO_x emission limitations are all possible considerations in the near future.

Dusts and sludges from EAF air pollution control facilities have been a designated hazardous waste and standards exist for the acceptable treatment and disposal of these wastes and their treatment residuals. Nevertheless, the high costs of treatment and disposal of waste continue to present challenges for more cost-effective means of treatment and recovery.

20

Pollution control in non-ferrous industry

20.1 Introduction

There are two groups of metals; ferrous and non-ferrous. Ferrous metals contain iron, for example carbon steel, stainless steel (both alloys: mixtures of metals) and wrought iron. Non-ferrous metals don't contain iron, for example aluminium, brass, copper and titanium. There are also non-ferrous metals as alloys as alloys, e.g., brass is an alloy of copper and zinc. The non-ferrous industry is dominated by aluminium.

Non-ferrous metals are specified for structural applications requiring reduced weight, higher strength, nonmagnetic properties, higher melting points or resistance to chemical and atmospheric corrosion. They are also specified for electrical and electronic applications.

This chapter discusses pollution prevention aspects of non-ferrous industry such as aluminium, copper, lead and zinc.

20.2 Aluminium industry

The production of aluminium begins with the mining and beneficiation of bauxite. At the mine (usually of the surface type), bauxite ore is removed to a crusher. The crushed ore is then screened and stockpiled, ready for delivery to an alumina plant. In some cases, ore is upgraded by beneficiation (washing, size classification and separation of liquids and solids) to remove unwanted materials such as clay and silica.

At the alumina plant, the bauxite ore is further crushed or ground to the correct particle size for efficient extraction of the alumina through digestion by hot sodium hydroxide liquor. After removal of 'red mud' (the insoluble part of the bauxite) and fine solids from the process liquor, aluminium trihydrate crystals are precipitated and calcined in rotary kilns or fluidised bed calciners to produce alumina (Al_2O_3).

Primary aluminium is produced by the electrolytic reduction of the alumina. The alumina is dissolved in a molten bath of fluoride compounds (the electrolyte) and an electric current is passed through the bath, causing the alumina to dissociate to form liquid aluminium and oxygen. The oxygen reacts with carbon in the electrode to produce carbon dioxide and carbon monoxide. Molten aluminium collects in the bottom of the individual cells or pots and is removed under vacuum into tapping crucibles. There are two prominent technologies for aluminium smelting: prebake and soderberg. Raw materials for secondary aluminium production are scrap, chips and dross. Pre-treatment of scrap by shredding, sieving, magnetic separation, drying and so on is designed to remove undesirable substances that affect both aluminium quality and air emissions. The prevailing process for secondary aluminium production is smelting in rotary kilns under a salt cover. Salt slag can be processed and reutilised. Other processes (smelting in induction furnaces and hearth furnaces) need no or substantially less salt and are associated with lower energy demand, but they are only suitable for high-grade scrap.

Depending on the desired application, additional refining may be necessary. For demagging (removal of magnesium from the melt), hazardous substances such as chlorine and hexachloroethane are often used, which may produce dioxins and dibenzofurans. Other, less hazardous methods, such as adding chlorine salts, are available. Because it is difficult to remove alloying elements such as copper and zinc from an aluminium melt, separate collection and separate reutilisation of different grades of aluminium scrap are necessary. It should be noted that secondary aluminium production uses substantially less energy than primary production—less than 10–20 gigajoules per metric T (GJ/T) of aluminium produced, compared with 164 GJ/T for primary production (mine to aluminium metal).

20.3 Waste characteristics

At the bauxite production facilities, dust is emitted to the atmosphere from dryers and materials-handling equipment, through vehicular movement and from blasting. Although the dust is not hazardous, it can be a nuisance if containment systems are not in place, especially on the dryers and handling equipment. Other air emissions could include nitrogen oxides (NO_x), sulphur dioxide (SO_2) and other products of combustion from the bauxite dryers.

Ore washing and beneficiation may yield process wastewaters containing suspended solids. Runoff from precipitation may also contain suspended solids. At the alumina plant, air emissions can include bauxite dust from handling and processing; limestone dust from limestone handling, burnt lime dust from conveyors and bins, alumina dust from materials handling, red mud dust and sodium salts from red mud stacks (impoundments), caustic aerosols from cooling towers and products of combustion such as sulphur dioxide and nitrogen oxides from boilers, calciners, mobile equipment and kilns. The calciners may also emit alumina dust and the kilns, burnt lime dust.

Although alumina plants do not normally discharge effluents, heavy rainfalls can result in surface runoff that exceeds what the plant can use in the process. The excess may require treatment. The main solid waste from the alumina plant is red mud (as much as 2 T of mud per T of alumina produced), which contains oxides of alumina, silicon, iron, titanium, sodium, calcium and other elements. The pH is 10–12. Disposal is to an impoundment.

Hazardous wastes from the alumina plant include spent sulphuric acid from descaling in tanks and pipes. Salt cake may be produced from liquor purification if this is practiced.

In the aluminium smelter, air emissions include alumina dust from handling facilities; coke dust from coke handling; gaseous and particulate fluorides; sulphur and carbon dioxides and various dusts from the electrolytic reduction cells; gaseous and particulate fluorides; sulphur dioxide; tar vapour and carbon particulates from the baking furnace; coke dust, tars and polynuclear aromatic hydrocarbons (PAHs) from the green carbon and anode-forming plant; carbon dust from the rodding room; and fluxing emissions and carbon oxides from smelting, anode production, casting and finishing. The electrolytic reduction cells (pot line) are the major source of the air emissions, with the gaseous and particulate fluorides being of prime concern. The anode effect associated with electrolysis also results in emissions of carbon tetrafluoride (CF_4) and carbon hexafluoride (C_2F_6), which are greenhouse gases, of concern because of their potential for global warming. Emissions numbers that have been reported for uncontrolled gases from smelters are 20–80 kilograms per ton of product (kg/T) for particulates, 6–12 kg/T for hydrogen fluoride and 6–10 kg/T for fluoride particulates. Corresponding concentrations are 200–800 milligrams per cubic meter (mg/m^3); 60–120 mg/m^3 and 60–100 mg/m^3. These values are for a prebaked-technology plant built in 1983.

An aluminium smelter produces 40–60 kg of mixed solid wastes per ton of product, with spent cathodes (spent pot and cell linings) being the major fraction. The linings consist of 50% refractory material and 50% carbon. Over the useful life of the linings, the carbon becomes impregnated with aluminium and silicon oxides (averaging 16% of the carbon lining), fluorides (34% of the lining) and cyanide compounds (about 400 parts per million). Contaminant levels in the refractories portion of linings that have failed are generally low. Other by-products for disposal include skim, dross, fluxing slags and road sweepings.

Atmospheric emissions from secondary aluminium melting include hydrogen chloride and fluorine compounds. Demagging may lead to emissions of chlorine, hexachloroethane, chlorinated benzenes and dioxins and furans. Chlorinated compounds may also result from the melting of aluminium scrap that is coated with plastic.

Salt slag processing emits hydrogen and methane. Solid wastes from the production of secondary aluminium include particulates, pot lining refractory material and salt slag. Particulate emissions, possibly containing heavy metals, are also associated with secondary aluminium production.

20.3.1 Pollution prevention and control

Pollution prevention is always preferred to the use of end-of-pipe pollution control facilities. Therefore every attempt should be made to incorporate cleaner production processes and facilities to limit, at source, the quantity of pollutants generated.

In the bauxite mine, where beneficiation and ore washing are practiced, a tailings slurry of 7–9% solids is produced for disposal. The preferred technology is to concentrate these tailings and dispose of them in the mined-out area. A concentration of 25–30% can be achieved through gravity settling in a tailings pond. The tailings can be further concentrated, using a thickener, to 30–50%, yielding a substantially volume reduced slurry.

The alumina plant discharges red mud in a slurry of 25–30% solids and this also presents an opportunity to reduce disposal volumes. Today's technology, in the form of high-efficiency deep thickeners and large-diameter conventional thickeners, can produce a mud of 50–60% solids concentration. The lime used in the process forms insoluble solids that leave the plant along with the red mud. These lime-based solids can be minimised by recycling the lime used as a filtering aid to digestion to displace the fresh lime that is normally added at this point. Finally, effluent volume from the alumina plant can be minimised or eliminated by good design and operating practices: reducing the water added to the process, segregating condensates and recycling to the process and using rainwater in the process.

Using the prebake technology rather than the Soderberg technology for aluminium smelting is a significant pollution prevention measure. In the smelter, computer controls and point feeding of aluminium oxide to the centerline of the cell help reduce emissions, including emissions of organic fluorides such as CF_4, which can be held at less than 0.1 kg/T aluminium. Energy consumption is typically 14 megawatt hours per ton (MWh/T) of aluminium, with prebake technology. (Soderberg technology uses 17.5 MWh/T). Gas collection efficiencies for the prebake process is better than for the Soderberg process: 98% vs. 90%. Dry scrubber systems using aluminium oxide as the adsorbent for the cell gas permits the recycling of fluorides. The use of low-sulphur tars for baking anodes helps control SO_2 emissions. Spent pot linings are removed after they fail, typically because of cracking or heaving of the lining. The age of the pot linings can vary from 3 to 10 years. By improving the life of the lining through better construction and operating techniques, discharge of pollutants can be reduced. Note that part of the pot lining carbon can be recycled when the pots are relined.

Emissions of organic compounds from secondary aluminium production can be reduced by thoroughly removing coatings, paint, oils, greases and the like from raw feed materials before they enter the melt process.

20.3.2 Treatment technologies

At bauxite facilities, the major sources of dust emissions are the dryers and emissions are controlled with electrostatic precipitators (ESPs) or baghouse dust collectors. Removal efficiencies of 99% are achievable. Dust from conveyors and material transfer points is controlled by hoods and enclosures. Dust from truck movement can be minimised by treating road surfaces and by ensuring that vehicles do not drop material as they travel. Dusting from stockpiled material can be minimised by the use of water sprays or by enclosure in a building.

At the alumina plant, pollution control for the various production and service areas is implemented as follows:

1. Bauxite and limestone handling and storage: Dust emissions are controlled by baghouses.
2. Lime kilns: Dust emissions are controlled by baghouse systems. Kiln fuels can be selected to reduce SO_2 emissions; however, this is not normally a problem, since most of the sulphur dioxide that is formed is absorbed in the kiln.
3. Calciners: Alumina dust losses are controlled by ESPs; SO_2 and NO_x emissions are reduced to acceptable levels by contact with the alumina.
4. Red mud disposal: The mud impoundment area must be lined with impervious clay prior to use to prevent leakage. Water spraying of the mud stack may be required to prevent fine dust from being blown off the stack. Longer term treatment of the mud may include reclamation of the mud, neutralisation, covering with topsoil and planting with vegetation.

In the smelter, primary emissions from the reduction cells are controlled by collection and treatment using dry sorbent injection; fabric filters or electrostatic precipitators are used for controlling particulate matter. Primary emissions comprise 97.5% of total cell emissions; the balance consists of secondary emissions that escape into the potroom and leave the building through roof ventilators.

Wet scrubbing of the primary emissions can also be used, but large volumes of toxic waste liquors will need to be treated or disposed off. Secondary emissions result from the periodic replacement of anodes and other operations; the fumes escape when the cell hood panels have been temporarily removed. While wet scrubbing can be used to control the release of secondary fumes, the high-volume, low-concentration gases offer low scrubbing efficiencies, have high capital and operating costs and produce large volumes of liquid effluents for treatment. Wet scrubbing is seldom used for secondary fume control in the prebake process.

When anodes are baked on site, the dry scrubbing system using aluminium oxide as the adsorbent is used. It has the advantage of being free of waste products and all enriched alumina and absorbed material are recycled directly to the reduction cells. Dry scrubbing may be combined with incineration for controlling emissions of tar and Volatile Organic Compounds (VOCs) and to recover energy. Wet scrubbing can also be used but is not recommended, since a liquid effluent, high in fluorides and hydrocarbons, will require treatment and disposal. Dry scrubber systems applied to the pot fumes and to the anode baking furnace result in the capture of 97% of all fluorides from the process. The aluminium smelter solid wastes, in the form of spent pot lining, are disposed of in engineered landfills that feature clay or synthetic lining of disposal pits, provision of soil layers for covering and sealing and control and treatment of any leachate. Treatment processes are available to reduce hazards associated with spent pot lining prior to disposal of the lining in a landfill. Other solid wastes such as bath skimmings are sold for recycling, while spalled refractories and other chemically stable materials are disposed of in landfill sites.

Modern smelters using good industrial practices are able to achieve the following in terms of pollutant loads (all values are expressed on an annualised basis): hydrogen fluoride, 0.2–0.4 kg/T, total fluoride, 0.3–0.6 kg/T, particulates, 1 kg/T, sulphur dioxide, 1 kg/T and nitrogen oxides, 0.5 kg/T. CF_4 emissions should be less than 0.1 kg/T.

For secondary aluminium production, the principal treatment technology downstream of the melting furnace is dry sorbent injection using lime, followed by fabric filters. Waste gases from salt slag processing should be filtered as well. Waste gases from aluminium scrap pre-treatment that contain organic compounds of concern may be treated by postcombustion.

Air emissions

The air emissions levels presented in Table 20.1 should be achieved.

Table 20.1: Air emissions from aluminium smelting.

(Milligrams per normal cubic meter)

Parameter	*Maximum value*
Particulate matter	30
Hydrogen fluoride	1
Total fluoride	2
VOCs	20

Liquid effluents

If there is a process effluent from the aluminium smelter, the effluent emissions levels presented in Table 20.2 should be achieved.

Table 20.2: Liquid effluents from aluminium smelting.
(Milligrams per litre, except pH and temperature)

Parameter	*Maximum value*
pH	6–9
TSS	50
Fluoride	20
Aluminium	0.2
COD	150
Hydrocarbons	5
Temperature increase	≤3°Ca

Ambient noise

Noise abatement measures should achieve either the levels given below or a maximum increase in background levels of 3 decibels (measured on the A scale) [dB(A)]. Measurements are to be taken at noise receptors located outside the project property boundary.

	Maximum allowable log equivalent (hourly measurements), in dB(A)	
Receptor	*Day (07:00–22:00)*	*Night (22:00–07:00)*
Residential, Institutional, Educational	55	45
Industrial, Commercial	70	70

20.4 Monitoring of emissions and effluents

Air emissions should be monitored regularly for particulate matter and fluorides. Hydrocarbon emissions should be monitored annually on the anode plant and baking furnaces. Liquid effluents should be monitored weekly for pH, total suspended solids, fluoride and aluminium and at least monthly for other parameters. Monitoring data should be analysed and reviewed at regular intervals and compared with the operating standards so that any necessary corrective actions can be taken. Records of monitoring results should be kept in an acceptable format. The results should be reported to the responsible authorities and relevant parties, as required.

20.4.1 Key issues related to emissions

The key production and control practices that will lead to compliance with emissions requirements are summarised here.

Bauxite production

1. Concentrate bauxite tailings prior to disposal.
2. Control dust emissions at the bauxite mine and in the alumina plant by using ESPs and baghouses.

Bauxite handling dust control

Bauxite handling facilities are usually very large systems that are used intermittently when bauxite ships arrive at a loading facility. The need for bauxite dust control should therefore be carefully evaluated on a case-by-case basis. Since some bauxites can be highly dusty, while others will generate almost no dust at all, the sources of bauxite should be established and control equipment designed accordingly. In all cases, care must be taken to ensure that the collection device is compatible with the wide range of materials that exhibit quite different properties when exposed to moisture contents that may only differ by several percentage points.

Bauxite dust, when relatively dry, can be and is controlled by the use of conventional dust collection equipment such as baghouses and mechanical collectors. The dry material, although somewhat abrasive, is relatively easy to capture and the sizing of dust collection equipment is routine. The degree of agitation that the material receives and the distance of freefall dictates the dust loadings that will occur when a given volume of air is removed from a transfer facility. Inlet concentrations to control equipment can range from 5 to 15 g/m^3. The great difference in particle size has made it necessary and wise to size dust collection equipment on the conservative side.

A major problem associate with equipment collection from bauxite handling facilities is the impact of moisture on the material to be collected. If wet dust is drawn into a duct and bag collector, the material will cake and eventually plug the dust blind the bags and stick to belts. When this occurs, a major effort is required to clean the entire system and make the collection workable at reasonable pressure drops. Large baghouse installations handling bauxite dust must be designed to overcome pressure drops. Large baghouse installations handling bauxite dust must be designed to overcome problems associated with high relative humidity situations that occur when slightly moist bauxite is handled.

Wet capture techniques generally have not be used successfully to control bauxite dust. The high energy consumption, equipment sizing and most of all, disposing of the wet material in an effective manner have steered engineers away from using the wet-type collectors on bauxite handling facilities.

Alumina calciner emission control

The most effective control systems for these emissions are electrostatic precipitators. In some kiln systems, the hot end (product discharge end) is controlled by cyclones, multiclones, or electrostatic precipitators. The 'cool end' of the kiln is usually equipped with an electrostatic precipitator. Fluid-bed calciners are complex systems which rely on product movement through a series of heating chambers and cyclones to maintain appropriate temperature levels during the drying process. All fluid-bed calciners are equipped with electrostatic precipitators as the final dust removal device.

Alumina plant

Thicken and concentrate red mud in the alumina plant, using high-efficiency thickeners and then dispose of it in engineered and managed stacks.

Primary aluminium smelting

1. Give preference to the prebake process for smelting.
2. Use computers to control the bath and limit anode effects.
3. Incinerate baking furnace gases for energy recovery.
4. Use dry scrubber systems with aluminium oxide absorbent for control of emissions from reduction cells and from anode bake ovens.
5. Maximise the reuse of spent pot linings.
6. Dispose of non-reusable spent pot linings in engineered landfills.

20.5 Utilisation of red mud

Red mud is disposed as dry or semi dry material in red mud pond or abandoned bauxite mines and as slurry having a high solid concentration of 30–60% and with a high ionic strength. The environmental concerns relate to two aspects: very large quantity of the red mud generated and its causticity.

Problems associated with the disposal of red mud waste include:

1. Its high pH (10.5–12.5).
2. Alkali seepage into underground water.
3. Instability of storage.
4. Alkaline air borne dust impact on plant life.
5. Vast areas of land consumed.

Up to 2 T of liquid with a significant alkalinity of 5–20 g/L caustic (as Na_2CO_3) accompany every ton of red mud solids.

20.5.1 Storage and disposal of red mud

Red mud waste is usually managed by discharge into engineered or natural impoundment reservoirs, with subsequent dewatering by gravity-driven consolidation and sometimes followed by capping for closure. Red mud disposal methods include traditional Closed Cycle Disposal (CCD) methods and Modified Closed Cycle Disposal (MCCD). A new class of Dry Stacking (DS) technology has also emerged which requires much less land. Due to various problems associated with disposal of red mud, it may cause economical as well as ecological problem in near future.

Red mud disposal

Safe treatment and storage of high volume industrial waste streams pose unique waste management challenges. Seawater discharge, lagooning, dry stacking and dry disposal are the methods currently in use for the disposal of bauxite residue.

In seawater discharge, after washing and thickening process of red mud, the slurry is disposed directly via a pipeline into the deep sea. This process reduces environmental impact of land disposal but may release toxic metals to the marine environment and increase the turbidity of the sea due to the fine mud and the formation of colloidal magnesium and aluminium compounds. Nevertheless, French and Japanese practices have favoured disposal at sea as the best option on economic and environmental grounds. In Japan, the alumina plants are restricted to available land area for disposal of residues and so have discharged the residue into the deep sea. The plants of Gardanne Alumina in France and Aluminium De Greece in Viotia, Greece still use marine dumping but are now pursuing other alternatives.

Lagooning is the conventional disposal method in which the residue slurry is directly pumped into land- based ponds. This consists of the construction of clay- lined dams into which bauxite residue slurry is simply pumped and allowed to dry naturally. This minimises the liquor leakage to the underlying water. The red mud ponds are lined with soil and bentonite. This process requires lowest capital cost, suppresses dust generation but requires substantial storage land and increases environmental hazards such as contact of humans and wild life with caustic liquor and contamination of ground water. Most of the alumina refineries till 1975 were using lagooning method for red mud disposal but some of them such as Pinjarra, Kwinana and Wagerup refineries in Australia have shifted to Dry stacking method. Queensland Alumina Ltd (Australia) after treatment of its red mud with seawater and CVG Bauxilium (Venezuela) still use wet disposal method by disposing their red mud in lagoons. In dry stacking method, the residue slurry is thickened to 48–55% solids and discharged in thin layers, dewatered and air dried before discharge of next layer on it. After the consolidation of paste to about 65%, it can be safely stacked. This reduces the area of disposal but may increase dust generation and requires funds for its long-term closure.

This method have been successfully applied at the MOTIM plant in Hungary. The original wet disposal method at NALCO, India has been replaced by Thickened Tailings Disposal (TTD) system. Dry disposal is a method in which the residue is filtered to a dry cake (>65% solids) and the material is washed on the filter with water or steam to recover soda and minimise the alkalinity of residue. Without further treatment, the dry residue is carried by truck or conveyor to the disposal site. This reduces the storage area but requires installation and operation of filtration plant. Solids contents of greater than 75% have been achieved with Bokela hyperbaric filtration technology at the Stade plant in Germany. Even with the excellent washing performance offered by hyperbaric steam filtration, significant alkalinity remains associated with the solids because of the complex nature of red mud. Hence these hazards associated with alkalinity may be further reduced by employing suitable methods of neutralising the red mud slurry.

20.5.2 Red mud neutralisation

Neutralisation of red mud will help to reduce the environmental impact caused due to its storage and also lessen significantly the ongoing management of the deposits after closure. It will also open opportunities for reuse of the residue which to date have been prevented because of the high pH. The cost of neutralisation will, to some degree at least, be offset by a reduction in the need for long-term management of the residue deposits. Instead of accruing funds to deal with a future liability, the funds can be invested in process improvements, which reduce or remove the liability. As per the Guidelines of Australian and New Zealand Environment and Conservation Council (ANZEX) and Agriculture and Resource Management Council of Australia and New Zealand (ARMCANZ), the liquor being strongly alkaline with a high pH, requires neutralisation to a pH below 9 with an optimum value of 8.5–8.9 before becoming environmentally benign. Neutralisation of red mud to pH around 8.0 is optimal because the chemically adsorbed Na is released, alkaline buffer minerals are neutralised and toxic metals are insoluble at this pH.

Efforts are being carried out to study the amelioration of red mud by possibly incorporating a pH-reduction processing step during disposal of red mud and include studies on processes based on acid neutralisation, CO_2 treatment, seawater neutralisation, bioleaching and sintering.

Acid neutralisation

Various aqueous acidic solutions have been considered for neutralisation of alkalinity, including acidic industrial wastewater. The use of carbonic acid has also been considered. A number of studies have been done to assess the feasibility of treating bauxite residue with acid as for instance on Kwinana red mud slurry. Large volumes of reagent are required to fully neutralise the residue at a relatively high cost, even if spent (waste) acid could be used. The use of acid also introduces large volumes of impurities to the process water stream (sulphate in the case of sulphuric acid, chloride in the case of hydrochloric acid). It is therefore likely that the return of any water from the residue deposits to the production process will be unacceptable without further treatment to remove these added impurities.

Treating red mud with acidic Spent Pickling Solutions (SPSs), derived from the steel making process, provides a coagulant – a mixture of aluminium and iron salts- for wastewater treatment.

CO_2 treatment

Gas phase CO_2 or CO_2-containing flue gas has been bubbled through aqueous slurries to form carbonic acid in the aqueous phase. Mechanisms of neutralisation of red mud by carbon dioxide gas have been studied. The carbonic acid reacts with basic components of the red mud, lowering its pH. At the short contact times which industrial process rates demand, only a fraction of the alkaline material in red mud is neutralised using gaseous CO_2. Hence although the pH of the aqueous phase drops rapidly upon exposure to CO_2 gas, it soon rises again to unacceptable levels as additional alkaline material leaches from the mud. The pH of water exposed to gaseous CO_2 is not likely to drop below 5.5 (approximately) and hence the rate of neutralisation of the solids in the aqueous slurry is typically not fast enough to satisfy industrial needs. Hence researchers have investigated the use of high-pressure liquid carbon dioxide rather than vapour phase carbon dioxide for the pH reduction of red mud. A laboratory study on neutralisation of red mud using CO_2 in multiple cycles has been investigated.

Seawater neutralisation

When seawater is added to caustic red mud, the pH of the mixture is reduced causing hydroxide, carbonate or hydroxycarbonate minerals to be precipitated. Average seawater contains 965 gm of water and 35 gm of salts (i.e., 3.5% salinity). The concentration of various salt ions in seawater is 55% Chlorine (Cl^-), 30.6% sodium (Na^+), 7.7% sulphate (SO_4^{-2}), 3.65% magnesium (Mg^{2+}), 1.17% calcium (Ca^{2+}), 1.13% potassium (K^+) and 0.7% others. Seawater neutralisation does not eliminate hydroxide from the system but converts the readily soluble, strongly caustic wastes into less soluble, weakly alkaline solids. The carbonate and bicarbonate alkalinity of the waste is removed primarily by reaction with calcium to form aragonite and calcite. The neutralising effect of the calcium and the magnesium ions is initially large but decreases rapidly as pH 8.5 is approached and calcium and magnesium carbonates precipitate. Neutralisation is considered to be complete when the liquid that can be separated from the treated red mud has a pH less than 9.0 and a total alkalinity less than 200 mg/L (as calcium carbonate equivalent alkalinity) and decant of seawater neutralised red mud can be safely discharged to the marine environment.

Bioleaching

Bioremediation of bauxite residue in Western Australia by Alcoa of Australia has been carried out by adding some organic substrate to the red mud for growth of micro-organisms which generate different organic acids and CO_2 (in some cases) which in turn neutralise the red mud. Similar work has also been carried out by using microbes.

Sintering

Sintering of residue can be carried out to fix all leachable soda, but the cost would be very high due to the elevated energy consumption required for high temperature sintering of red mud. But the mechanism can be made use of in making bricks and blocks from red mud.

20.5.3 Application of red mud

A considerable research has been done on the utilisation of red mud as a raw material for production of a range of products. It can be used as a constructional/building material in bricks, blocks, light weight aggregates, in cement industry as cements and special cements and in concrete industry. Bauxite residues can be used for soil remediation, as geopolymers and as a clay material. It can also be used as an additive to cements, mortars and concretes, construction of dykes and as ceramic/refractory product. In iron and steel industry it can be used after recovery of iron and titania. In environmental field, it can be utilised in pollution control by acting as adsorbent for cleaning of industrial gases, as synthetic coagulants in wastewater treatment and as a catalyst especially for coal hydrogenation. Red mud can as well be used in paints and pigments.

Building materials

Among the uses standing out, are those reported on the utilisation of red mud for building materials production such as cement, bricks, roofing tiles and glass-ceramics. The bulk production of building materials could eliminate the disposal problem. Red mud is considered as a raw material for production of these materials.

Preparation of construction materials from bauxite residues: A successful pilot project of a road embankment construction using Greek bauxite residue has been carried out by laboratory of road engineering of the Aristotle University of Thessalniki, Greece. The performance of the embankment with regards to its deformability was studied by means of the elastic behaviour theory. This is an attractive option with a high potential for large volume reuse of red mud use. Bauxite residues have other options for its reuse in preparation of construction materials as stated below:

Geopolymers

Geopolymer is a term covering a class of synthetic aluminosilicate materials with potential use in a number of areas, essentially as a replacement for Portland cement and for advanced high-tech composites and ceramic applications. The geopolymerisation process involves a chemical reaction between red mud and alkali metal silicate solution under highly alkaline conditions. The product of this reaction is an amorphous to semi-crystalline polymeric structure, which binds the individual particles of red mud transforming the initial granular material to a compact and strong one. The potential use of red mud for synthesis of inorganic polymeric materials through a geopolymerisation process was studied to use it in the construction sector as artificial structural elements such as massive bricks. Red mud was reacted with fly ash, sodium silicate via geopolymerisation reaction to get red mud geopolymers which are a viable cementitious material that can be used in roadway constructions.

Giannopoulou and others studied the geopolymerisation of the red mud and the slag generated in the ferronickel production, in order to develop inorganic polymeric materials with advanced mechanical and physical properties. The inorganic polymeric materials produced by the geopolymerisation of the red mud developed compressive strength up to 21 MPa and presented water absorption lower than 3%. They stated that red mud may be viewed as alternatives in the industrial sectors of construction and building materials.

Clay material

Investigations of the use of red mud and fly ash for the production of heavy clay products have been extensively undertaken at the Central Building Research Institute, Roorkee, India. Ekrem studied the potential use of red mud for the preparation of stabilisation material. The test results show that compacted clay samples containing red mud and cement–red mud additives have a high compressive strength, decreased hydraulic conductivity and swelling percentage as compared to natural clay samples. Consequently, it was concluded that red mud and cement – red mud materials can be successfully used for the stabilisation of clay liners in geotechnical applications. Study on the exploitation of red mud as a clay additive for the ceramic industry or as a compound for self-binding mortars in the fabrication of stoneware was carried out at National Institute of Technology, Rourkela, Orissa, India. A study carried out by Pontikes and others was aimed at using bauxite residue in heavy clay industry in which the plasticity of clay mixtures with bauxite residue and polymer addition was evaluated. They found that addition of 30 wt% bauxite residue substituting the clay mixture increases the maximum cohesion of the mixture. To make its use as a traditional ceramic, behaviour of bauxite residue was studied in different firing atmospheres (Air, N_2, Ar/4% H_2), for different maximum temperature (950–1050°C) and different soaking time (30–300 min).

Cements

Red mud from HINDALCO, Renukoot, India was investigated by Ghosh and others for its application in cements and they found that cements made from lime + red mud + bauxite + gypsum exhibit strengths comparable or superior to ordinary Portland cement (OPC). It was stated that as red mud is very rich in iron, red mud can be used as cheap pigment for coloured concrete. Also a uniform and durable coloured concrete could be obtained using white cement interground with 11% of burnt red mud. The red colouration could be enhanced by calcination in the range of 600 to 800°C. They found that such operation transforms the aluminium hydroxides (goethite and boehmite) and clays minerals into pozzolanic admixtures that are able to consume the calcium hydroxide produced by cement hydration. Thus, it is possible to develop a new admixture for concrete: a pozzolanic pigment. Tsakiridis and others in Greece studied the addition of red mud residue by 1% in the raw mix for the production of Portland cement and found that it did not affect either the sintering or the hydration process and concluded that the red mud can be utilised as a raw material in cement production, at no cost to the producer, contributing in reduction of the process cost.

Preparation of building materials from bauxite residues: Vast usage of red mud can be made in preparation of building materials such as ceramics, glass ceramic products, fired and non-fired bricks and concretes.

Ceramics

Red mud is made into useful ceramics articles by mixing 51–90% by weight of red mud with 49–10% by weight of at least one mineral and/or silicate containing material, shaping the mixture and firing it at a temperature of 950°–1250°C. The investigators have successfully converted red mud into glass ceramic products which involves addition of a small quantity of glass former along with traces of nucleating agents to a specific mixture of red mud, fly ash, followed by melting at around 1200°C and vitrification by cooling. The feasibility of recycling red mud and fly ash by producing glasses and glass-ceramics has also been investigated by Yanga and others. Glass has been obtained by melting red mud from Shandong Province in China with different additives. Suitable thermal treatments were employed to convert the obtained glass into nano-crystal glass-ceramics. X-ray diffraction (XRD) patterns showed that the main crystalline phase in both the glass-ceramics is wollastonite

($CaSiO_3$). These crystals are homogeneously dispersed within the parent glass, with an average crystal size of less than 100 nm. The size of nano-crystals varies when different thermal processes were used. These glass-ceramics have potential for a wide range of construction application.

Fired building materials

United States Patent 3886244 claims a process for manufacturing fired bricks wherein 50–90 wt% of red mud can be used along clay and a water fixing agent. The raw bricks are dried with heated gases at a temperature below 70°C and subsequently fired at a temperature between 900°–1100°C. Efforts have been made at Central Building Research Institute (CBRI), India to produce burnt clay bricks by partially replacing the clay with red mud (from the Indian Aluminium Company), lime and fly-ash.

Non-fired building materials

Efforts have also been made at CBRI to incorporate a small percentage of lime in red mud and compress the mix at optimum moisture content in the form of bricks with the purpose of examining their strength and stability to the erosive action of water. A maximum wet compressive strength of 3.75 MPa with 5% lime and 4.22 MPa with 8% lime has been obtained after 28 days of casting and humid curing of these bricks in the month of August. Studies were carried out at Jamaica Bauxite Institute and the University of Toronto using red mud to make bricks for inexpensive housing. The red mud was pressed into bricks using a standard brick press, immersed in sodium silicate followed by drying in the sun.

Non-fired bricks by mixing red mud, Portland cement and river sand were also made by the researchers at the institute. Liu and others studied the recovery of iron from Bayer red mud with direct reduction roasting process followed by magnetic separation and then building materials were prepared from aluminosilicate residues.

Then brick specimens were prepared with aluminosilicate residues and hydrated lime and the mean compressive strength of specimens was 24.10 MPa. It was indicated that main mineral phase nepheline ($NaAlSiO_4$) in aluminosilicate residues transformed into gehlenite ($Ca_2Al_2SiO_7$) in brick specimens as demonstrated by X-ray diffraction (XRD) technology. Combining the recovery of iron with the reuse of aluminosilicate residues, it can realise zero-discharge of red mud from Bayer process. Unsintered bricks have been developed from red mud disposed from Chinese sintering alumina process cured at ambient conditions. The optimal proportions of red mud brick are suggested as the following: 25–40% red mud, 18–28% fly ash, 30–35% sand, 8–10% lime, 1–3% gypsum and about 1% Portland cement.

Concrete industry

Red mud from Birac Alumina Industry, Serbia was tested as a pigment for use in the building material industry for standard concrete mixtures. Red mud was added as a pigment in various proportions (dried, not ground, ground, calcinated) to concrete mixes of standard test blocks (ground limestone, cement and water). The idea to use red mud as pigment was based on extremely fine particles of red mud (upon sieving: 0.147 mm up to 4 wt%, 0.058 mm up to 25 wt% and the majority smaller than 10 microns) and a characteristic red colour. Compressive strengths from 14.83 to 27.77 MPa of the blocks that contained red mud between 1 and 32% were considered satisfactory. The reported tests have shown that neutralised, dried, calcined and ground red mud is usable as pigment in the building materials industry. Red oxide pigment containing about 70 % iron oxide was prepared from NALCO red mud by after hot water leaching filtration, drying and sieving.

Application in pollution control

The interesting applications of red mud are however in the environmental field, after adequate neutralisation, for the remediation of contaminated sites and treatment of contaminated liquid waste.

Wastewater treatment: Red mud presents a promising application in water treatment for removal of toxic heavy metal and metalloid ions, inorganic anions such as nitrate, fluoride and phosphate, as well as organics including dyes, phenolic compounds and bacteria. The researchers have used acid and acid-thermal treated raw red mud to develop effective adsorbents to remove phosphate from aqueous solution. Study on the use of red mud for removal of dyes from textile effluents has also been conducted. Efforts have been made to use red mud for the removal of chlorophenols from wastewater. Neutralised red mud in batch adsorption technique was used for the removal of phenol from aqueous phase. Tor and others have also used granular red mud for removal of fluoride from water. Removal of boron from aqueous solution has also been studied by using neutralised red mud. Red mud has been converted into an inexpensive and efficient adsorbent to remove cadmium, zinc, lead and chromium from aqueous solutions. Brunori and others studied the possibility of reusing treated red mud (through the technology patented by Virotec International, consisting of a seawater treatment for pH neutralisation) in the Eurallumina SpA bauxite refinery, located in Sardinia (Italy) for treating contaminated waters and soils. Researchers have investigated the effectiveness of using thermally activated seawater neutralised red mud for the removal of arsenate, vanadate and molybdate in individual and mixed solutions. They found that thermally activated seawater neutralised red mud removes at least twice the concentration of anionic species than thermally activated red mud alone, due to the formation of 40–60% hydrotalcite during the neutralisation process in seawater neutralised red mud. Hydrotalcite structure in the seawater neutralised red mud has been determined to consist of magnesium and aluminium with a ratio between 3.5:1 and 4:1. Removal of arsenate from aqueous solutions has also been studied by other researchers. Fuhrman and others studied arsenic removal from water using 4 sorbents namely seawater-neutralised red mud (Bauxsol), acid treated Bauxsol (ATB), activated Bauxsol (AB), Bauxsol coated sand (BCS) and activated Bauxsol coated sand (ABCS). The affinity of the developed sorbents towards arsenic in a decreasing order is AB > ATB >ABCS > BCS > Bauxsol and sorptive capacity of all tested sorbents compares well with conventional sorbents such as activated alumina and ferric oxides. The removal of arsenate using seawater neutralised red mud is sensitive to several parameters such as pH, ionic strength, adsorbent dosage, initial arsenate concentration and the source water composition. Arsenate adsorption is favoured by slightly alkaline pH values with maximum adsorption recorded at pH 8.5. Hofstede and others have made use of bauxite refining residue to reduce the mobility of heavy metals in municipal waste compost. A US Patent Application 20090234174 shows that a neutralised and activated red mud is suitable for heavy metals remediation in soil and water. Entrapped metals are not easily exchangeable and removable. However, more investigation would be needed to further understand the metal trapping mechanisms of red mud. Seymer and Kirkpatrick of Kaiser Aluminium and Chemical Corporation and Tulane University have successfully developed and tested bauxite residue as liquid waste absorbent. They have researched soil synthesis as well as the use of red mud to reduce or eliminate sewage pathogens.

They have shown that 0.5 mg/L red mud was sufficient for near complete removal of metals such as silver, arsenic, barium, cadmium, mercury and lead but not selenium at an initial water pH of 8.0 and at contact/reaction times as low as one minute. Cadmium and selenium were present at a concentration of 0.5 mg/L while other metals at 2.0 mg/L in the wastewater. Selenium removal is very pH dependant with an optimum pH around 6.0.

A laboratory investigation to evaluate the capacity of red mud to inhibit acid mine drainage has been carried out. The investigators have studied the effectiveness of covers and liners made of red mud and/or cement kiln dust for limiting acid mine drainage. It has been proposed to use red mud that is very alkaline to neutralise acidic tailings. Previous experiments showed that red mud has a good neutralising capacity for a short time, but the long-term neutralisation potential is uncertain. So brine was added to red mud to verify if it can improve long-term alkalinity retention of red mud. McConchie investigated that the sea water-neutralised red mud can strip all trace metals in cyanide spills and neutralise the pH.

Absorption and purification of acid waste gases with bauxite residues: Red mud can be used to neutralise acid forming gases produced during coal combustion. Studies have been carried out on absorption of SO_2 on red mud (Sumitomo scrubbing process). Also studies on CO_2 sequestration by red mud are being carried out to neutralise red mud as explained earlier which would help in absorption of CO_2 and purification of flue gases from thermal power plant.

Red mud as a coagulant, adsorbent and catalyst

Red mud can also be employed as catalysts for hydrogenation, hydrodechlorination and hydrocarbon oxidation. It has also been studied as a support in catalytic wet oxidation of organic substances present in industrial wastewaters.

Use of red mud as a catalyst can be a good alternative to the existing commercial catalysts. Its properties such as iron content in form of ferric oxide (Fe_2O_3), high surface area, sintering resistance, resistance to poisoning and low cost makes it an attractive potential catalyst for many reactions. U.S. patent 4017425 describes a method developed for the red mud to be used as adsorbent, catalyst, ion-exchanging substance and clarifying substance particularly with respect to the catalytic cracking, decolourisation of hydrocarbon, clarification of waste gas and adsorption processes. The method comprises digesting red mud with acid, before adjusting the pH of the acid digested mixture comprising the sludge product to above 4, removing the residue acid employed from the gelating product with washing and heat treating the product to provide an active red mud. Cakici and others studied the utilisation of red mud as catalyst in conversion of waste oil and waste plastics to fuel in comparison with a commercial hydrocracking catalyst (silica-alumina) and a commercial hydrotreating catalyst (Ni–Mo/alumina). Garg and others have made a comparison of the catalytic activity of pyrite, red mud and flue dust and based on selective analysis showed that red mud was the most desirable disposable catalyst in the conversion of coal and oil production. Novel applications of red mud as a coagulant and adsorbent for water and gas treatment as well as catalyst for some industrial processes have been reviewed by Shaobin and others.

Recovery of metals

The analysis of red mud shows that iron is the major constituent of red mud and hence much work has been carried out till now for its recovery. Some red mud also contains titania in substantial amount which if successfully recovered has the most potential value. Iron can be obtained as value-added product and alumina and soda can be recycled in the process.

Red mud generated from Guangxi Province (China) was treated with Chinese coal and coal-sort for direct reduction of iron. Bauxite of this province was treated by Bayer process on alumina first – iron second basis due to its composition (~27 wt % Al_2O_3 and ~43 wt% Fe_2O_3) and factors related to reduction performance were reported including quality/property of coal. An extensive study on the possibility of magnetic separation of red mud from Fria deposit (Guinea) reported that ~ 85 % of the iron present in red mud was recovered at 0.06 Tesla magnetic intensity. Best result was obtained by treating –125 μm + 90 μm size fraction. Red mud was mixed with dolomite and coke to make pellets and sintered (1100°C) followed by smelting (1500°C) to produce pig iron. The slag was further treated with sulphuric acid followed by solvent extraction of iron, then silica and alumina. Pigment grade titania was also recovered from the slag. Laboratory-scale research has been focused by on the recovery of titanium from red mud in which the leaching process is based on the extraction of this element with diluted sulphuric acid from red mud under atmospheric conditions and without using any preliminary treatment. Leaching followed by solvent extraction was tried in Japan using sulphuric acid and some solvents like diisopropyl ether, DP-10R or PC-88A (Daihachi Chemical Industry Co., Ltd.) to recover iron and titania respectively. At the end, iron, titania and alumina were separated. Red mud of Alcoa Alumina, Kaiser Alumina and Reynolds Metals (all in USA) were reacted with different reductants (sawdust, bagasse, etc.), at a temperature of about 350°C to reduce different forms of iron to magnetite followed by magnetic separation to produce high iron containing and soda free product/material for further usage. Studies were done on Jamaican red mud to recover all possible metal values: at first alumina was recovered by soda-ash sintering process followed by reduction (partial or complete) of iron to magnetic/metal phase followed by magnetic separation to separate iron and titania from the non-magnetic portion. Different parametric conditions are also highlighted in the paper. A patent was also filed in the same field claiming all possible metal values recovery by reacting red mud with acid followed by selective precipitation of salts at different pH. Iron mineral transformation during thermal treatment of red mud has been studied. Liquid-liquid extraction (LLE) of iron and titanium by bis-(2-ethyl-hexyl) phosphoric acid has been studied. Studies have been carried out to investigate the optimum condition for sulphuric acid leaching of iron from red mud and a diffusion model has also been developed to support the study. Dissolution kinetics of iron and aluminium from red mud in sulphuric acid solution for different parameters such as calcination temperature, acid concentration, agitation rate, particle size and time have been studied by researchers. Red mud of Shandong province of China has also been tried for reduction roasting in presence of proper additives (reduction enhancer such as $CaCO_3$, $MgCO_3$, etc.), and an encouraging metallisation ratio of 96.97% has been reported with scope for better usage for non-magnetic fraction in building material industry because of removal of iron and increase in percentage of aluminosilicate compounds which helps to set the building material more strongly in presence of lime. Krause and Rohm have patented a process wherein iron oxide in red mud has been reduced to magnetite by suitable hydrocarbon and was then recovered. A process Mud to Money claims to recover virtually all of the residual alumina from bauxite residue at attractive economics. The inventors claim the environmental benefits of this technology include a reduction in residue generation per ton alumina by some 8% and a reduction of bauxite consumption per ton of alumina by some 4%. Work on microbiological leaching of aluminium from red mud with selective fungi has been carried out. United States Patent 3876749 claims a process wherein the red mud is mixed with a reducing agent, separating it into molten steel and molten slag, reacting the slag with CaO, leaching out the useful aluminates for recirculation to the Bayer process and utilising the remaining calcium silicates in cement manufacture.

For soda recovery, a patent relates to a process for the treatment of red mud and in particular relates to a process capable of both ameliorating the pH of red mud and allowing soda recovery from red mud by passing carbon dioxide through a stream of red mud. United States Patent 4045537 discloses a process for recovering the caustic and alumina values from red mud utilising the so-called lime-soda-sinter process wherein a carbonaceous material such as coke is included in the sintering operation and leaching is carried out without any intermediate iron separation step. WO/1997/029992 relates to a method for recovering soda and/or alumina values from red mud from DSP (desilicated product) formed in a Bayer process, the method comprising mechanically activating the DSP to induce a mechano-chemical reaction. Any reagent which is thermodynamically capable of reacting with DSP to solubilise soda and/or alumina values may be used. Suitable reagents include oxides and hydroxides such as CaO, NaOH and $Ca(OH)_2$.

In addition to compounds of main elements, red mud also comprises of small quantities of rare earth elements such as Yttrium (Y), Scandium (Sc) and Lanthanides (Ln). SO_2 dissolved in water can be introduced in red mud slurry to selectively dissolve the rare earth elements while leaving iron substantially undissolved in the red mud. An innovative method for the recovery of rare earth elements from the red mud and separation of (Sc) was developed on a laboratory and pilot scale by Aluminium of Greece (Pechiney group) in Greece.

Soil remediation with bauxite residues

Soil amendment is a technique used to create fertile topsoil by increasing the soil's ability to retain moisture and nutrients and filter some contaminants, such as heavy metals, before they infiltrate the groundwater. Soil amendment involves adding an agent to the soil to improve its structure, porosity, water holding capacity and nutrient recycling capacity. Potential amendment agents in an urban environment include compost, organic rich soils, loam soils, natural clay, crushed limestone and gypsum.

Soil amendment agents 'are generally distinguished from fertilisers' by having a lower nutrient content and a greater ability to retain and recycle both moisture and nutrients.

The Department of Agriculture, Western Australia has been working with Alcoa World Alumina Australia Ltd for more than ten years investigating the potential to use bauxite refining residues as soil amendments for the poor, acidic, sandy soils of the Swan Coastal Plain in south west Australia. Extensive laboratory, field and catchment-scale trials have shown the ability of soil amendment with fine bauxite refining residue (now trademarked in this context as Alkaloam™) to reduce the leaching of nutrients to sensitive regional waterways by up to 75%, whilst increasing pasture productivity by up to 25% (up to 200% in well-controlled experimental situations). The potential applications of bauxite residue in soil/sediment remediation and soil/sediment stabilisation have been investigated.

Bauxite residue was mixed with a variety of soil types such as acid soils, saline soils, organic rich material and silicate soil. Appropriate pH conditions were achieved to promote vegetation growth. Preliminary studies have also been carried out at Louisiana State University, to investigate the use of red mud to enhance coastal wetlands.

Other uses

Along with successfully developing and testing bauxite residue as liquid waste absorbent, Seymer and Kirkpatrick of Kaiser Aluminium and Chemical Corporation at their Gramercy Louisiana Plant along with red mud as liquid waste absorbent have also studied red mud as landfill cover material and as levee construction material. A novel process for making radiation- shielding materials utilising red mud has been developed by adopting ceramic- chemical processing route using phosphate bonding.

Efforts were made to utilise red mud for developing plasma spray coatings (ceramic and cermet) on metal substrates, stainless steel, mild steel, Cu and Al. As red mud consists of metal oxides of iron, titanium, silicon, aluminium it was felt that red mud can possibly be spray coated. Building Material and Technology Promotion Council of India (BMPTC) has produced composite from red mud, polymer and natural fibres, called Red Mud Jute Fibre Polymer Composite (RFPC), to replace wood in the wood based panel products in the building industry.

Rehabilitation of red mud pond

Red mud ponds and abandoned bauxite mine pits can be rehabilitated through vegetation. The ecological rehabilitation of red mud residue deposits is complicated by many factors, including its hazardous nature, extremely high pH and salinity, poor water-holding capacity and extremely low microbial activity. Hence, caustic properties of red mud are to be modified using suitable modifiers for the growth of proper flora and fauna on it. Vegetation cover will not only prevent deterioration of soil erosion but also act as method of suppressing dust generation due to the dried red mud. In this process, bulk utilisation of red mud would be possible.

Work carried out for rehabilitation of red mud pond for an alumina refinery situated at Belgaum (Karnataka, India) show that a combination of 55% red mud, 25% FYM (farmyard manure), 15% gypsum and 5% vegetative dry dust, inoculated with both bacteria and mycorrhizae, resulted in good responses from three tree species—kikar (*Acacia nilotica*), karanj (*Pongamia pinnata*) and vilayati babul (*Prosopis juliflora*)—while other two species—drek (*Melia azedarach*) and Israeli babul (*Acacia tortilis*)—did not survive in any of the treatment combinations. Among the grass species, para grass (*Brachiaria mutica*), signal grass (*B. decumbens*) and shrubby stylo grass (*Stylosanthes scabra*) performed well in the same treatment combination as the trees, along with sesban (*Sesbania sesban*), a legume species.

The effectiveness of various industrial wastes and low cost chemicals such as gypsum, sewage sludge, ferrous sulphate, ammonium sulphate, ammonium nitrate and calcium phosphate as ameliorants for red mud to develop and maintain a low cost, self sustaining vegetation cover has been studied by many researchers. It was found that the addition of 5% or more gypsum reduced the pH, electrical conductivity and sodium and aluminium content of the soil, as well as providing a continuous supply of calcium ions, thus reducing the exchangeable sodium percentage and was effective in treating the soil to permit revegetation by *Agropyron elongatum* (tall wheat grass) and *Cynodon dactylon* (Bermuda grass). The survival of the plant species *Cynodon dactylon* (bermudagrass), Atriplex nummalari (oldman saltbush) and Atriplex canescens (fourwing saltbush) in red mud indicated that it was more vigorous with gypsum amendments. Recently researchers after reviewing neutralisation and utilisation methods have modified dried red mud with different amenders and utilised it for growth of ornamental plants.

20.6 Copper industry

The main environmental issues for the primary copper production are the potential emission to air of dust and metals/metal compounds and of sulphur dioxide from roasting and smelting sulphide concentrates or using sulphur-containing fuels or other materials. The pyrometallurgical processes are potential sources of dust and metals from furnaces, reactors and the transfer of molten metal. The main environmental issues for the secondary copper production are also related to the off-gases from the various furnaces and transfers that contain dust, metals and volatile organic components. There is also the potential for the formation of dioxins due to the presence of small amounts of chlorine in the secondary raw materials.

In summary the main issues for the production processes of copper comprise various components such as SO_2, dust, metal compounds, organic compounds, wastewater (metal compounds), residues such as furnace linings, sludge, filter dust and slag. Dioxin formation during treatment of secondary copper materials is also an issue.

20.7 Emissions to air

Furnace design, the use of suitable pre-treatment methods and process control are identified as important features of best available techniques for pollution prevention and process control. The use of furnace sealing and enclosed transfers and storage is important in preventing fugitive emissions.

20.7.1 Emissions to environment

Emissions to the environment depend on the collection or abatement systems that are used. The hierarchy of gas collection from all of the process stages is: process optimisation and minimisation of emissions, sealed reactors and furnaces and targeted fume collection.

The potential sources of emissions to air are summarised in Table 20.3, which also gives a review of prevention and treatment methods.

Table 20.3: Summary of sources and treatment/abatement options.

Process stage	*Component in off-gas*	*Treatment method*
Materials handling and storage	Dust and metals	Correct storage, handling and transfer. Dust collection and fabric filter if necessary
Grinding, drying	Dust and metals	Process operation. Gas collection and fabric filter
Sintering/roasting smelting converting	VOCs, dioxins	After burner, adsorbent or activated carbon addition
fire refining	Dust and metal compounds	Gas collection, gas cleaning in EP or fabric filter, heat recovery
	Carbon monoxide	After burner if necessary

(Cont'd...)

Process stage	*Component in off-gas*	*Treatment method*
	Sulphur dioxide	Sulphuric acid plant (for sulphidic ores) or scrubber
	Dust and metals	Gas collection, cooling and fabric filter
Slag treatment	Sulphur dioxide	Scrubber
	Carbon monoxide	After burner
	Dust and metals	Gas collection and fabric filter
Thermal refining	Sulphur dioxide	Scrubber if necessary
Metal powder production	Dust and metals	Gas collection and fabric filter
Melting and casting	Dust and metals	Gas collection and fabric filter.
	VOCs, dioxins (organic feed)	After burner (Carbon injection)

Note: Dust arrestment using a fabric filter may require the removal of hot particles to prevent fires. Hot electrostatic precipitators would be used in a gas cleaning system prior to a sulphuric acid plant or for wet gases.

Sulphur capture is an important requirement when sulphidic ores or concentrates are roasted or smelted. The sulphur dioxide produced by the process is collected and can be recovered as sulphur, gypsum (if no cross-media effects) or sulphur dioxide or can be converted to sulphuric acid.

The process choice depends on the existence of local markets for sulphur dioxide. The production of sulphuric acid in a double contact sulphuric acid plant with a minimum of four passes, or in a single contact plant with gypsum production from the tail gas and using a modern catalyst, are considered to be best available technique.

Plant configuration will depend on the concentration of sulphur dioxide produced in the roasting or smelting stage.

Process gases from primary and secondary copper production are captured and than cleaned in electrostatic precipitators and fabric filters and so the emissions of dust and metal compounds are reduced. Gas cleaning using wet scrubbers and wet electrostatic precipitators is particularly effective for process gases that undergo sulphur recovery in a sulphuric acid plant. After burners and carbon absorption are used to remove dioxins and VOCs. Uncaptured gases or fugitive emissions, however, are not treated.

Dust emissions also occur from storage, handling and the pre-treatment of raw materials where fugitive dust emissions also play an important role. This is true for both primary and secondary production, as their significance can be much greater than captured and abated emissions. Careful plant design and process operations are needed to capture and treat process gases where fugitive emissions are significant.

For primary copper the reverberatory furnace is not considered to be best available technique. The other major influences are the blending of the raw materials, process control, management and the collection of fume.

20.7.2 Process residues

Process residues are produced at various stages of the process and are highly dependent on the constituents of the raw materials:

1. Filter dusts (using fabric filters or EPs or using scrubbers or wet EPs) can be recycled within the same plant or used for the recovery of other metals; dust and sweepings from raw materials handling - feed for the main process.
2. Slag from smelting, converting, refining and melting furnaces - construction material after slag treatment or recycle to smelter, return to process after treatment, metal recovery, recovery of salt and other material.
3. Electrolyte bleed, anode remnants and anode slime from electro-refining - return to converter or recovery of precious metals.
4. Catalyst, acid sludge, weak acid from sulphuric acid plan – regeneration, safe disposal, leaching or disposal, respectivelly.
5. Refractory from furnace linings - use as slagging agent or disposal.

In copper industry like as other non-ferrous metals production, furnace design, the use of suitable pre-treatment methods and process control are identified as important features of best available technique.

The use of raw material blending to optimise the process prevents inappropriate material being used and maximises process efficiency. Sampling and analysis of feed materials and the segregation of some materials are important factors in this technique.

Good design, maintenance and monitoring are important for all process and abatement stages. Sampling and monitoring of emissions to the environment should be carried out according to national or international standard methods. Important parameters that can be used for the control of process or abatement should be monitored. Continuous monitoring of key parameters should be carried out if practical.

20.8 Lead and zinc industry

Lead and zinc can be produced pyrometallurgically or hydrometallurgically, depending on the type of ore used as a charge. In the pyrometallurgical process, ore concentrate containing lead, zinc, or both is fed, in some cases after sintering, into a primary smelter. Lead concentrations can be 50–70% and the sulphur content of sulphidic ores is in the range of 15–20%.

Zinc concentration is in the range of 40–60%, with sulphur content in sulphidic ores in the range of 26–34%. Ores with a mixture of lead and zinc concentrate usually have lower respective metal concentrations. During sintering, a blast of hot air or oxygen is used to oxidise the sulphur present in the feed to sulphur dioxide (SO_2). Blast furnaces are used in conventional processes for reduction and refining of lead compounds to produce lead.

20.8.1 Primary lead processing

The conventional pyrometallurgical primary lead production process consists of four steps: sintering, smelting, drossing and refining. A feedstock made up mainly of lead concentrate is fed into a sintering machine. Other raw materials may be added, including iron, silica, limestone flux, coke, soda, ash, pyrite, zinc, caustic and particulates gathered from pollution control devices. The sintering feed, along with coke, is fed into a blast furnace for reducing, where the carbon also acts as a fuel and smelts the lead-containing materials. The molten lead flows to the bottom of the furnace, where four layers form: 'speiss' (the lightest material, basically arsenic and antimony), 'matte' (copper sulphide and other metal sulphides), blast furnace slag (primarily silicates) and lead bullion (98% by weight). All layers are then drained off. The speiss and matte are sold to copper smelters for recovery of copper and precious metals. The blast furnace slag, which contains zinc, iron, silica and lime, is stored in piles and is partially recycled. Sulphur oxide emissions are generated in blast furnaces from small quantities of residual lead sulphide and lead sulphates in the sinter feed. Rough lead bullion from the blast furnace usually requires preliminary treatment in kettles before undergoing refining operations. During drossing, the bullion is agitated in a drossing kettle and cooled to just above its freezing point, 370°–425°C (700°–800°F). A dross composed of lead oxide, along with copper, antimony and other elements, floats to the top and solidifies above the molten lead. The dross is removed and is fed into a dross furnace for recovery of the nonlead mineral values.

The lead bullion is refined using pyrometallurgical methods to remove any remaining nonlead materials (e.g., gold, silver, bismuth, zinc and metal oxides such as oxides of antimony, arsenic, tin and copper). The lead is refined in a cast-iron kettle in five stages. First, antimony, tin and arsenic are removed. Next, gold and silver are removed by adding zinc. The lead is then refined by vacuum removal of zinc. Refining continues with the addition of calcium and magnesium, which combine with bismuth to form an insoluble compound that is skimmed from the kettle. In the final step, caustic soda, nitrates, or both may be added to remove any remaining traces of metal impurities. The refined lead will have a purity of 99.90–99.99%. It may be mixed with other metals to form alloys, or it may be directly cast into shapes.

20.8.2 Secondary lead processing

The secondary production of lead begins with the recovery of old scrap from worn-out, damaged, or obsolete products and with new scrap. The chief source of old scrap is lead-acid batteries; other sources include cable coverings, pipe, sheet and other lead-bearing metals. Solder, a tin-based alloy, may be recovered from the processing of circuit boards for use as lead charge.

Prior to smelting, batteries are usually broken up and sorted into their constituent products. Fractions of cleaned plastic (such as polypropylene) case are recycled into battery cases or other products. The dilute sulphuric acid is either neutralised for disposal or recycled to the local acid market. One of the three main smelting processes is then used to reduce the lead fractions and produce lead bullion. Most domestic battery scrap is processed in blast furnaces, rotary furnaces, or reverberatory furnaces. A reverberatory furnace is more suitable for processing fine particles and may be operated in conjunction with a blast furnace. Blast furnaces produce hard lead from charges containing siliceous slag from previous runs (about 4.5% of the charge), scrap iron (about 4.5%), limestone (about 3%) and coke (about 5.5%). The remaining 82.5% of the charge is made up of oxides, pot furnace refining drosses and reverberatory slag. The proportions of rerun slags, limestone and coke vary but can run as high as 8% for slags, 10% for limestone and 8% for coke.

The processing capacity of the blast furnace ranges from 20 to 80 metric tonnes per day (tpd). Newer secondary recovery plants use lead paste desulphurisation to reduce sulphur dioxide emissions and generation of waste sludge during smelting. Battery paste containing lead sulphate and lead oxide is desulphurised with soda ash, yielding market-grade sodium sulphate as a by-product. The desulphurised paste is processed in a reverberatory furnace and the lead carbonate product may then be treated in a short rotary furnace. The battery grids and posts are processed separately in a rotary smelter.

20.9 Waste characteristics in zinc manufacturing

In the most common hydrometallurgical process for zinc manufacturing, the ore is leached with sulphuric acid to extract the lead/zinc. These processes can operate at atmospheric pressure or as pressure leach circuits. Lead/zinc is recovered from solution by electrowinning, a process similar to electrolytic refining. The process most commonly used for low-grade deposits is heap leaching. Imperial smelting is also used for zinc ores.

20.9.1 Air pollutants and particulate matter

The principal air pollutants emitted from the processes are particulate matter and sulphur dioxide (SO_2). Fugitive emissions occur at furnace openings and from launders, casting molds and ladles carrying molten materials, which release sulphur dioxide and volatile substances into the working environment. Additional fugitive particulate emissions occur from materials handling and transport of ores and concentrates. Some vapours are produced in hydrometallurgy and in various refining processes. The principal constituents of the particulate matter are lead/zinc and iron oxides, but oxides of metals such as arsenic, antimony, cadmium, copper and mercury are also present, along with metallic sulphates. Dust from raw materials handling contains metals, mainly in sulphidic form, although chlorides, fluorides and metals in other chemical forms may be present. Off-gases contain fine dust particles and volatile impurities such as arsenic, fluorine and mercury. Air emissions for processes with few controls may be of the order of 30 kilograms lead or zinc per metric tonne (kg/T) of lead or zinc produced. The presence of metals in vapour form is dependent on temperature. Leaching processes will generate acid vapours, while refining processes result in Products of Incomplete Combustion (PICs). Emissions of arsine, chlorine and hydrogen chloride vapours and acid mists are associated with electrorefining.

Wastewaters are generated by wet air scrubbers and cooling water. Scrubber effluents may contain lead/zinc, arsenic and other metals. In the electrolytic refining process, by-products such as gold and silver are collected as slimes and are subsequently recovered. Sources of wastewater include spent electrolytic baths, slimes recovery, spent acid from hydrometallurgy processes, cooling water, air scrubbers, washdowns and stormwater. Pollutants include dissolved and suspended solids, metals and oil and grease.

The larger proportion of the solid waste is discarded slag from the smelter. Discard slag may contain 0.5–0.7% lead/zinc and is frequently used as fill or for sandblasting. Slags with higher lead/zinc) content—say, 15% zinc—can be sent for metals recovery. Leaching processes produce residues, while effluent treatment results in sludges that require appropriate disposal. The smelting process typically produces less than 3 T of solid waste per ton of lead/zinc produced.

Pollution prevention and control

The most effective pollution prevention option is to choose a process that entails lower energy usage and lower emissions. Modern flash-smelting processes save energy, compared with the conventional sintering and blast furnace process. Process gas streams containing over 5% sulphur dioxide are usually used to manufacture sulphuric acid. The smelting furnace will generate gas streams with SO_2 concentrations ranging from 0.5% to 10%, depending on the method used. It is important, therefore, to select a process that uses oxygen-enriched air or pure oxygen. The aim is to save energy and raise the

SO_2 content of the process gas stream by reducing the total volume of the stream, thus permitting efficient fixation of sulphur dioxide. Processes should be operated to maximise the concentration of the sulphur dioxide. An added benefit is the reduction (or elimination) of nitrogen oxides (NO_x).

1. Use doghouse enclosures where appropriate; use hoods to collect fugitive emissions.
2. Mix strong acidic gases with weak ones to facilitate production of sulphuric acid from sulphur oxides, thereby avoiding the release of weak acidic gases.
3. Maximise the recovery of sulphur by operating the furnaces to increase the SO_2 content of the flue gas and by providing efficient sulphur conversion. Use a double-contact, double-absorption process.
4. Desulphurise paste with caustic soda or soda ash to reduce SO_2 emissions.
5. Use energy-efficient measures such as waste heat recovery from process gases to reduce fuel usage and associated emissions.
6. Recover acid, plastics and other materials when handling battery scrap in secondary lead production.
7. Recycle condensates, rainwater and excess process water for washing, for dust control, for gas scrubbing and for other process applications where water quality is not of particular concern.
8. Give preference to natural gas over heavy fuel oil for use as fuel and to coke with lower sulphur content.
9. Use low-NO_x burners.
10. Use suspension or fluidised bed roasters, where appropriate, to achieve high SO_2 concentrations when roasting zinc sulphides.
11. Recover and reuse iron-bearing residues from zinc production for use in the steel or construction industries.
12. Give preference to fabric filters over wet scrubbers or wet electrostatic precipitators (ESPs) for dust control.

Good housekeeping practices are key to minimising losses and preventing fugitive emissions. Losses and emissions are minimised by enclosed buildings, covered conveyors and transfer points and dust collection equipment. Yards should be paved and runoff water routed to settling ponds.

Implementation of cleaner production processes and pollution prevention measures can yield both economic and environmental benefits.

The target pollutant load for lead and zinc smelting operations for particulate matter is 0.5 kg/T of concentrated ore processed. ESPs are used to recover dust. Pollutant load factors for lead in air emissions are 0.08 kg/T from roasting, 0.08 kg/T from smelting and 0.13 kg/T from refining. A double-contact, double-absorption plant should emit no more than 2 kg of sulphur dioxide per ton of sulphuric acid produced, based on a conversion efficiency of 99.7%. Sulphur dioxide should be recovered to produce sulphuric acid, thus yielding a marketable product and reducing SO_2 emissions. Fugitive emissions are controlled by using enclosed conveyors.

Treatment technologies

ESPs and baghouses are used for product recovery and for the control of particulate emissions. Dust that is captured but not recycled will need to be disposed of in a secure landfill or in another acceptable manner.

Arsenic trioxide or pentoxide is in vapour form because of the high gas temperatures and must be condensed by gas cooling so that it can be removed in fabric filters.

Collection and treatment of vent gases by alkali scrubbing may be required when sulphur dioxide is not being recovered in an acid plant.

Effluent treatment of process bleed streams, filter backwash waters, boiler blow down and other streams is required to reduce suspended and dissolved solids and heavy metals and to adjust pH. Residues that result from treatment are recycled to other industries such as the construction industry, sent to settling ponds (provided that groundwater and surface water contamination is not a concern), or disposed of in a secure landfill.

Slag should be either landfilled or granulated and sold for use in building materials.

Air emissions: The air emissions levels presented in Table 20.4 should be achieved. The environmental assessment should address the buildup of heavy metals from particulate fallout in the vicinity of the plant over its projected life.

Table 20.4: Emissions from lead/zinc smelting.

(Milligrams per normal cubic meter)

Parameter	*Maximum value*
Sulphur dioxide	400
Arsenic	0.1
Cadmium	0.05
Copper	0.5
Lead	0.5
Mercury	0.05
Zinc	1.0
Particulates	20

Liquid effluents: The effluent emissions levels presented in Table 20.5 should be achieved.

Table 20.5: Effluents from lead/zinc smelting.
(Milligrams per litre, except for pH and temperature)

Parameter	*Maximum value*
pH	6–9
TSS	20
Arsenic	0.1
Cadmium	0.1
Copper	0.5
Iron	3.5
Lead	0.1
Mercury	0.01
Zinc	2.0
Total metals[a]	5
Temperature increase	≤ 3°C[b]

[a] Includes arsenic, beryllium, cadmium chromium, gold, lead, mercury, nickel, selenium, silver, thallium and vanadium.
[b] The effluent should result in a temperature increase of no more than 3° C at the edge of the zone where initial mixing and dilution take place. Where the zone is not defined, use 100 meters from the point of discharge.

Monitoring of emissions and effluents

Air emissions: Air emissions should be monitored continuously for sulphur dioxide and particulate matter. Other air emissions parameters should be monitored monthly. Fugitive emissions should be monitored annually.

Liquid effluents: Liquid effluents should be monitored daily for pH and total suspended solids and at least weekly for all other parameters.

Solid wastes: All solid waste, tailings and leachates should be monitored for toxic metals. Contamination of groundwater and surface waters should be avoided. Monitoring data should be analysed and reviewed at regular intervals and compared with the operating standards so that any necessary corrective actions can be taken. Records of monitoring results should be kept in an acceptable format. The results should be reported to the responsible authorities and relevant parties, as required.

20.9.2 Key issues related to emissions and effluents

The key production and control practices that will lead to compliance with emissions requirements can be summarised as follows:

1. Give preference to the flash-smelting process where appropriate.
2. Choose oxygen enrichment processes that allow higher SO_2 concentrations in smelter gases to assist in sulphur recovery; use the double-contact, double-absorption process.

3. Improve energy efficiency to reduce fuel usage and associated emissions; use low-NO_x burners; give preference to natural gas as fuel.
4. Reduce air emissions of toxic metals to acceptable levels.
5. Maximise the recovery of dust and minimise fugitive emissions; use hoods and doghouse enclosures.
6. Reduce effluent discharge by maximising wastewater recycling.
7. Avoid contamination of groundwater and surface waters by leaching of toxic metals from tailings, process residues, slag and other wastes.

21

Pollution control in foundry industry

21.1 Introduction

Foundries melt ferrous and non-ferrous metals and alloys and reshape them into products at or near their finished shape through the pouring and solidification of the molten metal or alloy into a mold. The foundry industry is a differentiated and diverse industry. It consists of a wide range of installations, from small to very large; each with a combination of technologies and unit operations selected to suit the input, size of series and types of product produced in the specific installation. The organisation within the sector is based on the type of metal input, with the main distinction being made between ferrous and non-ferrous foundries.

The main markets served by the foundry industry are the automotive, general engineering and construction sectors.

The high dependence on the automotive sector has a major influence on activities in the foundry sector and concerns various aspects, such as economy, location, quality standards, environmental standards, new developments, etc. One example of this dependency is that the automotive industry's shift towards lighter vehicles, is reflected in the foundry industry by an increased demand (and thus market) for aluminium and magnesium casting, thus enabling the growth of these sectors.

21.2 Type of foundries

Besides the metal type (i.e., ferrous/non-ferrous) the foundry layout is largely dependent on the size of castings and the series size. A small series foundry is termed a 'jobbing foundry' and a large series one is termed a 'series foundry'. Foundries may be also be classified according to the type of metal manufactured, i.e., either a ferrous or a non-ferrous foundry. There is a large difference between a zinc foundry producing a large series of frames for toy cars and a cast iron foundry producing rotor housings for wind turbines. Foundries apply different degrees of automation according to their series size and the repeatability of the work.

Concerning applied techniques, the main distinction is made by the type of melting furnace used (e.g., cupola, electrical, rotary) and the type of mold (e.g., sand molding, die-casting).

21.3 Environmental issues

The foundry industry is a major player in the recycling of metals. Steel, cast iron and aluminium scrap can all be remelted into new products. The possible negative environmental effects of foundries result from the presence of a thermal process and the use of mineral additives. The environmental effects of a foundry process therefore mainly relate to the exhaust and off-gases and to the reuse or disposal of mineral residues.

21.3.1 Air emissions

Noxious emissions from the melting and treatment of metals are generally related to the use of additives and fuels or to impurities in the feed. The use of cokes as fuels or the heating of crucibles with gas or oil-fired burners can cause emissions of combustion products. Also the application of additives in metal treatment processes generates reaction products. The presence of impurities (e.g., oil and paint) in scrap used for re-melting can potentially cause the production of the products of incomplete combustion or recombination and dust. Also any dust generated may contain metal and metal oxides. The evaporation of elements with a high vapour pressure occurs during melting and small particles of metal escape from the bath. Metallic particles are also generated during grinding and finishing operations. In the making of molds and cores, various additives are used to bind the sand. In the binding of the sand and pouring of the metal, reaction and decomposition products are generated. These include both inorganic and organic products. The generation of decomposition products further continues during the casting cooling and de-molding operations.

Dust and particles' releases are a general issue in all stages of the foundry process and for all processes used. Dust is generated in the production and processing of sand molds and cores, as well as in the finishing of the castings (both from lost molds and permanent molds).

In the foundry process, emissions to air are not limited to one (or several) fixed point(s). The process involves various emission sources (e.g., from hot castings, sand, hot metal). A key issue in emission reduction is not only to treat the exhaust and off-gas flow, but also to capture it.

21.3.2 Residues

Sand molding involves the use of large sand volumes, with sand-to-liquid-metal weight ratios generally ranging from 1:1 up to 20:1. At the end of the molding process the used sand can be regenerated, reused or disposed off. Additional mineral residues such as slag and dross are generated in the melting stage when removing impurities from the melt. These should also be considered for either reuse or disposal.

21.3.3 Energy

Since foundries deal with a thermal process, energy efficiency and management of the generated heat are important environmental aspects. However, due to the high amount of transport and handling of the heat carrier (i.e., the metal) and due to its slow cooling, the recovery of heat is not always straightforward.

21.3.4 Water

In most foundries, water management involves an internal circulation of water, but a major part of the water still evaporates. The water is generally used in the cooling systems of electric furnaces (induction or arc) and cupola furnaces. In general, the outgoing waste water stream is therefore very small. For (high) pressure die-casting, a waste water stream is formed, which needs treatment to remove organic (phenol, oil) compounds before disposal.

21.3.5 Applied processes and techniques in foundries

The process can be divided into the following major activities:

1. Melting and metal treatment: The melting shop.
2. Preparation of molds: The molding shop.
3. Casting of the molten metal into the mold, cooling for solidification and removing the casting from the mold the casting shop.
4. Finishing of the raw casting the finishing shop.

Starting from foundry scrap (selected scrap according to a certain chemical composition) or ingots, the foundry produces finished castings. Usually these are components which will require further treatment or assembly to yield a final product.

On the molding side, a basic distinction is made between permanent and lost molds. Foundries casting in permanent molds, buy these metal molds (dies) externally, but typically operate an in-house mold repair and maintenance shop. Foundries casting in lost molds, often buy wooden, metal or plastic patterns (for their mold design) and operate an in-house pattern maintenance and repair shop. Molds, cores and lost models are generally produced as part of the foundry process.

Traditionally in the foundry sector the main distinction made is between ferrous and non-ferrous foundries. This is mainly because the applied processes in both sectors differ. Non-ferrous foundries often apply die-casting techniques. These allow a better surface finish, which is important for many of the aluminium and brass applications. Due to the high cooling rate, castings with a high mechanical strength are produced. However, this technique does not allow the production of massive or large pieces, which require sand casting techniques.

Iron casting

Cast iron is an iron-carbon alloy, containing usually between 2.4 and 4% carbon. The minimum carbon content is 1.8%. Silicon, manganese, sulphur and phosphorus are also present in various amounts. Special grades of iron are produced which contain various levels of nickel, chrome and other metals. Due to its high carbon content, cast iron has a low melting point and a good casting ability as compared to steel. Its ductility is low and does not allow rolling or forging. Variations in properties can be achieved by varying the ratio of carbon to silicon, by alloying and by heat treatment.

Steel casting

Steel is a material of which the (mass) content of iron is bigger than that of any other element, with a carbon content generally lower than 2% and which also usually contains other elements. A limited number of chromium steel types may contain over 2% carbon, but 2% is the usual cut-off limit used to distinguish steel from cast iron. One particularly useful aspect of steel is that it can be hot worked. Low alloy cast steel contains elements such as Mn, Cr, Ni and Mo in amounts less than 5%. High alloy steel includes more than 5% of alloying elements, e.g., 12% Cr and 8% Ni. Special steel grades are produced with enhanced properties, such as higher strength, higher magnetic permeability, better resistance to corrosion, fatigue or wear and improved behaviour during welding or at high or low temperatures.

Aluminium casting

About two-thirds of all aluminium castings are used in the automotive industry, e.g., in cars, buses, lorries, trains and aircraft. The need to reduce vehicle fuel consumption and weight has increased the interest in aluminium.

Many different types of melting furnaces are used in aluminium foundries the choice depending on individual requirements. Directly and indirectly heated furnaces, using fuel and electricity, are applied.

Magnesium casting

Magnesium alloy castings are used for aerospace, automotive and electronic applications. The main advantage for using them is their light weight; typically magnesium alloys have a density of 1.8 g/mL compared with 2.7 g/mL for aluminium alloys.

Copper casting

Copper is cast in the form of various groups of alloys, each having copper as the main element.

Zinc casting

Zinc casting almost exclusively uses the pressure die-casting technique. The zinc alloy is almost exclusively melted in a casting machine, of the hot chamber type. In rare cases and only when high production capacity is needed, centralised melting may be applied.

Lead casting

Lead is a low melting (melting point 327°C) heavy metal. Lead is relatively soft, corrosion-resistant and has good self-lubricating properties.

21.4 Cupola furnaces

The cupola is a refractory lined shaft furnace where the metal charge is heated by the combustion of coke, which takes place in the lower part of the shaft (the 'hearth'). Combustion air, supplied by fans, is injected in the hearth through a number of exhaust nozzles (tuyères). A ring and control valve allows a controlled and evenly distributed flow of combustion air through the tuyères. The metal (pig iron, steel scrap, scrap iron, foundry returns), coke, alloying elements (e.g., FeSi, SiC), slag forming (SiO_2) and fluxing agents (e.g., $CaCO_3$) are added to the shaft through a charging door at the upper part of the shaft. The combustion gases move upward from the hearth and exchange heat with the charge, before leaving the furnace through the cupola stack.

When the pre-heated charge reaches the combustion zone, the metallic parts melt due to the high temperatures and the charged coke starts to burn in the presence of oxygen. The molten metal droplets run through the coke bed and gather in the zone called the well, which is below the combustion zone. All the impurities are trapped in the slag, which is mostly formed by SiO_2, CaO, Al_2O_3 and FeO. Fluxing agents lower the melting point and the viscosity of the slag. Due to its lower density, the slag floats on the molten metal in the well. Once the liquid metal in the well has reached a certain level, a tap-hole is opened. The metal flows discontinuously through the tap hole, via a refractory lined channel or launder into a separate collection vessel or ladle.

Alternatively, the molten metal can be continuously directed to a holding furnace. The slag is tapped separately by means of a dam and a slag spout placed at a higher level. It is collected discontinuously in pots, or continuously granulated in a water stream, or in a special installation for dry granulation.

In its basic configuration, the cupola is called a Cold Blast Cupola (CBC). This is a cupola which uses the blast at atmospheric pressure and at normal environmental temperature. Basic structure of a cupola is shown in Fig. 21.1.

In order to reduce the CO-emission and increase the energy efficiency, the conversion of cold blast furnaces to hot blast furnaces may be considered.

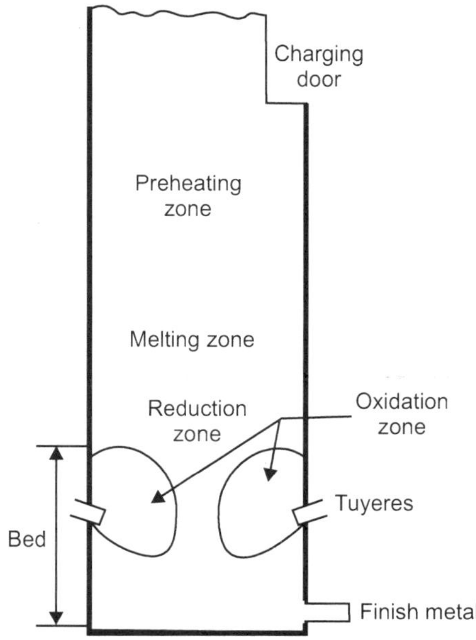

Important zones of a cupola

Figure 21.1: Basic structure of a cupola.

However, for medium-sized foundries handling less than 2000 T/month of good castings, the hot blast cupola will be difficult to consider because of the large investment it requires. The use of a hot blast cupola is also restricted to continuous melting, otherwise the running costs are too high.

The cold blast cupola prevails for some types of production, e.g., enamelled iron castings, kitchenware, heating appliances, counterweights.

Hot blast cupola furnace: In order to optimise the efficiency of the cupola furnace, the combustion air may be pre-heated. This principle is used in the Hot Blast Cupola (HBC), i.e., a cupola which uses a pre-heated blast.

21.4.1 Nature of atmospheric emissions

Cupolas can be charged with a wide range of materials, many of which may contain loose particles such as rust, sand and non-ferrous materials. The metallurgical-coke can break and produce small pieces, as can the added fluxing materials. Breakage and mechanical abrasion during charge preparation, as well as during charging itself, generate particles, some of which are immediately emitted. During melting, abrasion of the charge against the refractory lining will also generate dust. A third source of particulate matter is coke ash, generated in the melting zone, which is not trapped by the slag phase.

Particulate matter of various sources, if light enough, can be entrained in the combustion gases of the cupola. Under certain conditions metallurgical-fume may be generated from the melting zone, leading to a visible plume from the cupola stack. The smoke particles consist of submicron agglomerates of spherical soot particles and metallic oxides, such as ZnO, PbO, etc., if the metals are present in the charged steel or iron scrap, such as in galvanised or painted scrap. The smoke emission will increase with the proportion of coke and contaminants in the charge, the blast temperature and the oxygen injection rate. Carbonaceous smoke is airborne matter formed by the incomplete combustion of organic matter in the cupola. Scrap contaminants such as oil and grease, wood, textiles and, rubber will form oily vapours in the stack gases. Vapours and partially burnt organic matter may carry unpleasant smells.

Again, scrap cleanliness and its nature significantly effect the nature of the emissions. The burning of coke creates odorous gas emissions containing CO_2, CO and SO_2. Decreasing the proportion of the coke charge (by increasing the thermal efficiency) or (partial or complete) substitution of the coke by natural gas can help reduce the levels of these substances.

21.5 Electric arc furnace (EAF)

Electric arc furnaces may be categorised as direct arc or indirect arc. Both types of units are suited for the melting of high melting point alloys such as steels. They may be lined with acid or basic refractories.

21.5.1 Direct arc furnaces

Direct arc furnaces are very popular for the melting of alloy steels and range in size from a few kilograms, for laboratory units, to in excess of 100 T per batch. Typically units found in foundries are in the range of 1 to 10 T. The furnace generally consists of a cylindrical steel shell which is lined with acid or basic refractories. The roof which can normally swing away to facilitate charging, generally contains three carbon electrodes operating on a high tension three-phase power supply.

These electrodes protrude vertically through the roof and an electric current passes directly through them and into the metal bath. The distance between the electrodes and the metal bath is automatically controlled and determines the power input into the bath. These furnaces generally have a door at the back for alloying, oxygen lancing and slag removal purposes and a pouring spout at the front. The entire unit is capable of being tilted for discharge of the melt through the pouring spout.

Some of the advantages of direct arc furnaces include high melt rates, high pouring temperatures and excellent control of melt chemistry.

21.5.2 Indirect arc furnaces

Generally consist of a horizontal barrel shape steel shell lined with refractories. Melting is effected by the arcing between two horizontally opposed carbon electrodes. Heating is via radiation from the arc to the charge. The barrel shaped shell is designed to rotate and reverse through approximately 180°C in order to avoid excessive heating of the refractories above the melt level and to increase the melting efficiency of the unit. Indirect arc furnaces are suitable for melting a wide range of alloys but are particularly popular for the production of copper base alloys. The units operate on a single-phase power supply and hence the size is usually limited to relatively small units.

21.5.3 Nature of atmospheric emissions in EAF

The emissions from EAFs originate from the charging, melting and refining operations and during tapping of the furnace.

During charging, dust and dirt will be emitted from the open furnace body. When charging a hot furnace (for instance when melting with a molten heel in particular), any combustibles such as grease, paint or oil ignite and give rise to smoke plumes of burnt and partially burnt organic material and dust particles. The mechanical abrasion of the furnace lining also generates additional dust.

During melting, heating of the scrap generates metal oxide fumes that significantly increase during the decarburisation treatment. The injection of oxygen gas into the molten metal develops significant quantities of iron oxide fumes, which leave the furnace as red clouds. The addition of slag forming materials increases the furnace emission, but only in small quantities and only for a short time.

Minor emissions occur during the transfer of the molten metal into a ladle or holding furnace. Emissions from the melting operation itself are referred to as primary emissions. Secondary emissions are the fumes and dust originating from the charging and tapping. In considering the nature and quantities of the emissions, the different sources of dust and fume emissions, as described above, demonstrate that large differences exist between foundries, depending on the cleanliness of the charged material, the applied charging procedure, the composition of the charge, the refining treatments and on the additions to the melt. Since no coal or fuel is burned in the furnace the emissions solely depend on these parameters.

21.6 Induction furnace (IF)

The principle of induction melting is that a high voltage electrical source from a primary coil induces a low voltage, high current in the metal, or secondary coil. Induction heating is simply a method of transferring heat energy. Induction furnaces are ideal for melting and alloying a wide variety of metals with minimum melt losses, however, little refining of the metal is possible. There are two main types of induction furnace and coreless and channel.

21.6.1 Coreless induction furnaces

The heart of the coreless induction furnace is the coil, which consists of a hollow section of heavy duty, high conductivity copper tubing which is wound into a helical coil. Coil shape is contained within a steel shell and magnetic shielding is used to prevent heating of the supporting shell. To protect it from overheating, the coil is water-cooled, the water is recirculated and cooled in a cooling tower. The shell is supported on trunnions on which the furnace tils to facilitate pouring.

The crucible is formed by ramming a granular refractory between the coil and a hollow internal former which is melted away with the first heat leaving a sintered lining.

The power cubmicle converts the voltage and frequency of main supply, that is required for electrical melting. Frequencies used in induction melting vary from 50 cycles per second (mains frequency) to 10,000 cycles per second (high frequency). The higher the operating frequency, the greater the maximum amount of power that can be applied to a furnace of given capacity and the lower the amount of turbulence induced.

When the charge material is molten, the interaction of the magnetic field and the electrical currents flowing in the induction coil produce a stirring action within the molten metal. This stirring action forces the molten metal to rise upwards in the centre causing the characteristic meniscus on the surface of the metal. The degree of stirring action is influenced by the power and frequency applied as well as the size and shape of the coil and the density and viscosity of the molten metal. The stirring action within the bath is important as it helps with mixing of alloys and melting of turnings as well as homogenising of temperature throughout the furnace. Excessive stirring can increase gas pick up, lining wear and oxidation of alloys.

The coreless induction furnace has largely replaced the crucible furnace, especially for melting of high melting point alloys. The coreless induction furnace is commonly used to melt all grades of steels and irons as well as many non-ferrous alloys. The furnace is ideal for remelting and alloying because of the high degree of control over temperature while the induction current provides good circulation of the melt.

21.6.2 Channel induction furnaces

The channel induction furnace consists of a refractory lined steel shell which contains the molten metal. Attached to the steel shell and connected by a throat is an induction unit which forms the melting component of the furnace. The induction unit consists of an iron core in the form of a ring around which a primary induction coil is wound. This assembly forms a simple transformer in which the molten metal loops comprises the secondary component. The heat generated within the loop causes the metal to circulate into the main well of the furnace. The circulation of the molten metal effects a useful stirring action in the melt. Channel induction furnaces are commonly used for melting low melting point alloys and or as a holding and superheating unit for higher melting point alloys such as cast iron.

Channel induction furnaces can be used as holders for metal melted off peak in coreless induction units thereby reducing total melting costs by avoiding peak demand charges.

21.6.3 Nature of emissions in induction furnace (IF)

Since no coal or fuel is burned in the induction furnace and no refining procedures are executed, the emissions solely depend on the cleanliness and the composition of the charged material. Two major categories of emissions can be distinguished. The first and major, category relates to the charge cleanliness, e.g., rust, dirt, foundry sand, paint, oil, galvanised or soldered metal, all of which are elements which give rise to the emission of dust and fumes (organic or metallic). The second category relates to chemical reactions at high temperatures, (e.g., while holding or adjusting the metal composition), which can give rise to metallurgical fume due to oxidation.

Additionally the refractory lining (acid SiO_2 based, neutral Al_2O_3 based, or basic MgO based) may add a small amount of dust particles to the emission.

Waste gases from cupola

Coke fired cupola gas is composed primarily of N_2, CO_2, H_2O and CO, with smaller amounts of SO_2. On conventional cupolas, where the off-gas is collected above the charging door, a distinction has to be made between the condition of the flue-gases below and above the charging door, since ambient air is entrained through the open charge door. This input significantly changes the total airflow.

If the cupola gases are hot enough and if there is enough CO present, the gases may burn spontaneously together with drawn-in air ($CO + O_2 — 2CO_2$) and temperatures may rise to 900°C. Little or no CO will then be left in the exhaust. If no combustion takes place, the air intake will result in a cooling effect, of between 100 and 300°C and the CO/CO_2 balance will remain unchanged. The temperature of the gases just below the charging door is primarily dependent on the charge height; the input of ambient air is determined by the fan capacity or the natural draught available.

Cupola slag

Slags contain oxides that float upon the melt and which arise from impurities in the feed material, wearing of the furnace refractory and from the ash of coke and from the melting loss of the metal charge. Slags are bonded through the addition of binders, such as vermiculite.

Cupola slag consists of 30% refractory material, 10% sand (from internal scrap), 40% CaO (flux), 10% cokes ashes and 10% burn-off material.

An important aspect of cupola slag is its high SiO_2-content. After quenching, the slag has a vitrified structure. This generates an inert non-leaching material.

Waste refractory

The cupola furnace has the specific feature that the lining material (quartz-clay mixture) of the furnace at the melting zone only lasts for one melting campaign. The main part of the refractory is converted into slag. The amount to be taken out and disposed of as waste is much smaller than the applied amount.

Wastewater

The amount of process water used in foundries is small and mainly depends on the selection of techniques for dedusting. Waste water is generated mainly in dust removal and waste gas treatment systems, which are applied in the melting shop, molding material preparation and reclamation and in the cleaning shop. Waste water is also generated in core production, if wet scrubbers are used (depending of the core-making technique and bonding agents used). In a minority of plants, water is used for wet sand regeneration.

Furthermore, water is generally used for cupola shell cooling and for the cooling baths for (die) castings.

Possible sources of waste water are:

1. Scrap storage area and site-drainage.
2. Sand preparation.
3. Wet scrubbers used for dedusting in various areas of the foundry.
4. Cooling of pressure die casting machines and tools.
5. Vibratory finishing (slide grinding).
6. Cooling baths used for heat treatment.

The production-specific waste water quantity averages 0.5 m^3/t good castings.

Wastewater from wet scrubbers used in cupola melting

Wet scrubbers are in use in melting shops to clean the cupola off-gas. Wet scrubbers use water to remove dust particles (average loading: 10–15 g/Nm^3)

and gases such as sulphur dioxide from cupola off-gas. In the settling tank, a large part of the solid particles are captured and thereby almost completely removed from the scrubbing water.

The partly acid gases washed out of the cupola off-gas, such as sulphur dioxide, accumulate in the scrubbing water, causing its salt concentration to increase (e.g., the formation of e.g., sodium sulphate) and its pH to decrease. This salt accumulation is supported by evaporation losses, so it is necessary to draw the water off from time to time. Waste water from wet dedusting systems in melting shops mainly contains:

1. Solids such as oxides of silicon, iron and aluminium, calcium carbonates and cyanides.
2. Heavy metals, although with the exception of zinc, these are present only in very low concentrations.
3. Organic pollutants, which can enter waste water via soiled scrap.

21.6.4 Cokeless cupola furnace

In the cokeless cupola, the metal charge is heated by the combustion of natural gas. Instead of the traditional coke bed, a bed of refractory spheres on a water-cooled grid supports the metal charge. The molten metal droplets run through this bed and are collected in the well at the bottom of the furnace. The life of the spheres exposed to the superheating temperatures of the metal is limited. A cokeless cupola is therefore run at a reduced temperature (1400°C instead of 1500°C) and superheating of the liquid metal is done in an attached gas-fired or induction furnace (duplex operation).

An important operational factor is that the cokeless cupola has to be run continuously. There is a high risk of clogging and blocking on the cooled grate bars. In case there is a need to stop the metal flow (e.g., due to problems in the molding shop), the power of the burners can be reduced to 35–40%, in order to compensate for wall heat-losses. The need for continuity needs to be balanced with the usage of refractory in the melting zone. Therefore, melting campaigns of one week are typically used.

The oxidising atmosphere and relatively low flame temperature cause increased oxidation losses. This limits the possibility to feed in steel. A maximum amount of 35% of steel is used in the production of nodular iron, though 20% can be considered general practice. The quality of the feed needs to be well controlled since the cokeless cupola is more susceptible to bridging than the coke-fired cupola.

In the production of nodular iron, an important advantage of the cokeless cupola is that there is no resulphurisation, so the melt may be used immediately after recarburisation. Besides a higher thermal efficiency, this furnace has other environmental advantages. The combustion of natural gas instead of cokes has the following consequences for the flue-gas.

Odour abatement

Odour emissions are mainly associated with processes involving sand binders. The actual odorous products of pyrolysis may vary according to the type of system used, but generally the phenolic breakdown products, i.e., cresols and xylenols, are the most common source of odour complaint, due to their very low odour detection thresholds. The dispersion of odours during casting, cooling and shake-out involves mixing with large volumes of air, which makes collection and treatment difficult. The use of inorganic binders, such as sodium silicate, can substantially reduce emissions. No totally effective proven method of eliminating foundry generated smells is known to be currently available. A general approach is to ensure good ventilation and a rate of air change that ensures that emissions are quickly and efficiently dispersed to the atmosphere. Process-integrated measures include the substitution of binders or binder solvents. End-of-pipe techniques aim to abate pyrolysis products and amines. These include adsorption techniques, post combustion, wet scrubbers and biofilters.

Reuse of dusts from the green sand circuit in mold making

Dust is collected through the exhaust filtration from the shake-out installation and from the dosing and handling stations for dry green sand. The collected dust contains active binder compounds and may be recycled into the green sand circuit.

Dust and solid residues: treatment and reuse

A good segregation of materials is essential to facilitate opportunities for recovery, recycling and reuse. Waste should always be recycled or recovered unless a satisfactory justification has been accepted by the regulator that recovery is 'technically and economically impossible'.

In general, the waste streams comprise:

1. Raw material dusts collected in bag or cartridge filters.
2. Slag from desulphurisation.
3. Slag/dross from melting.
4. Melting dust and fume collected in a filter plant.
5. Casting dust and fume collected in a filter plant.
6. Fettling dusts collected in an abatement plant.
7. Chips and turnings from fettling.
8. Used abrasives from shot blasting.

9. Scrubber liquors and sludges and output from the effluent treatment plant.
10. Refractory waste from launders and ladles.
11. Sand.
12. Chemical and oil containers.
13. General inert industrial waste.

Waste should be recovered wherever practicable. Other than sand, the most significant wastes are:

1. Slag from the melting and metal treatment processes.
2. Dust collected from abatement plants.
3. Collected sludge.
4. Refractory waste.

Minimisation of slag forming

The production of slag can be minimised using in-process measures, such as:

1. Using clean scrap.
2. Using a lower metal temperature.
3. Preventing overshoots (temporary high temperatures).
4. Preventing long standing times of molten metal in the melting furnace.
5. Making adequate use of fluxes.
6. Making adequate use/choice of refractory lining.
7. Applying water cooling of the furnace wall to avoid usage of refractory lining.

21.6.5 Noise reduction

The foundry process contains various point sources of noise. These include:

1. Scrap handling.
2. Furnace charging.
3. Burners.
4. HPDC automates.
5. Shake-out.
6. Grit-blasting.
7. Core (and mold) shooting.
8. Finishing.
9. All motors and hydraulic systems.
10. Transport (tipping, loading, etc.).

The overall reduction of noise levels involves developing a noise reduction plan. Here each of the sources needs to be checked and evaluated.

Alternative techniques with lower noise levels may be applied and/or point sources may be enclosed.

General measures to control noise

General measures to control noise include:

1. Using noise reducing flaps on all outside doors and closing all doors (and keeping them shut as much as possible), especially during the night.
2. Actively blowing air into the foundry hall. This causes a small increase in the indoor pressure and keeps the noise inside.
3. Enclosing fans, insulating of ventilation pipes and using of dampers.
4. Minimising the number of transport activities during the night.

21.6.6 Heat efficiency of cupola

Heat efficiency = (Heat content of molten metal/total heat input) × 100.

The heat efficiency is calculated by using the above formula after calculating heat balance by using heat input and heat output obtained for each item per ton of molten metal.

Energy efficiency calculation

1. Parameters measured:
   (a) Bed coke.
   (b) Charged coke.
   (c) Metallics charged.
   (d) Melting time (blower operating hours).
2. Melting rate (tph).
3. Charged coke consumption (%).

21.7 Improving the heat efficiency of a cupola

Hot blast cupola: In hot blast cupola the temperature of exhaust gas of a cupola is as high as 800°C, making it possible to pre-heat blast air up to as high as 400°C by heat exchange. In addition, both the sensible and latent heat of exhaust gas can be recycled for pre-heating blast air by combustion of CO gas included in exhaust gas. When blast air is pre-heated to 300°C or higher, the sensible heat of blast air is added to heat input, activating combustion of coke, leading to the rise in combustion temperature (Fig. 21.2), thus improving heat efficiency and economy owing to the reduction in the coke ratio and increase in melting speed. Moreover, in the upper part of the combustion zone, CO_2 gas due to coke is deoxidised by high temperature, realising a highly reductive atmosphere, thus decreasing the oxidation loss of metal.

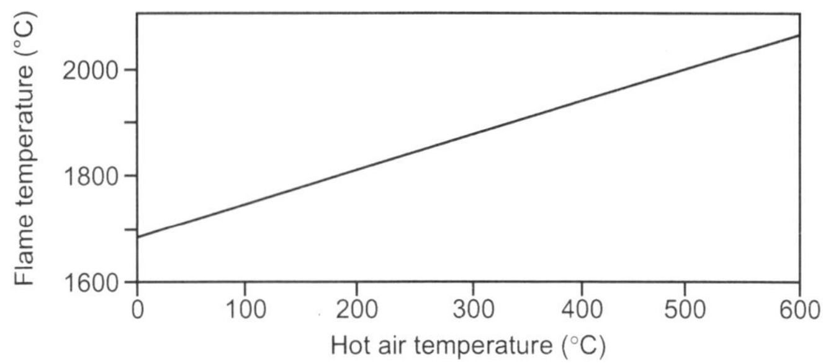

Figure 21.2: Hot air temperature and coke flame temperature.

In cold blast operation, combustion is hindered around tuyeres by cold air, while in hot blast operation, the maximum temperature zone comes down just above tuyeres (Fig. 21.3).

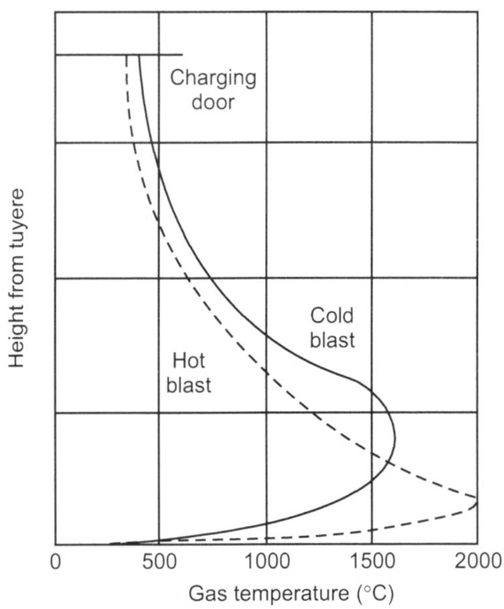

Figure 21.3: Furnace temperature distribution in cold blast and hot blast operation.

21.7.1 Cupola furnace: waste heat utilisation

The need to cool cupola off-gases before they enter the bag filter leads to the possibility of attaching to a secondary user and applying waste heat utilisation. The secondary user may be, e.g., (i) a steam boiler, (ii) a thermal oil circuit, (iii) a heating circuit and (iv) a hot water circuit.

21.8 Recommendations for future work

Techniques for VOC-abatement: VOC-emissions are generated by the thermal decomposition of the binders upon pouring, in coating the cores and molds and in the melting of dirty scrap and are one of the key environmental issues for the sector. VOC emissions in foundries typically involve large-volume, low-level flows, which represent a significant load to the environment and may result in an odour emission. Information is needed on techniques for the efficient capture and treatment of these flows, together with associated emission and economic data. The prevention of VOC emissions through the use of alternative binder and coating materials, also deserves further study. Data should be gathered and reported from the implementation of these techniques in operational practice.

Melting of non-ferrous metals: The discussion on non-ferrous metal melting is based on the melting of clean ingots and clean internal return material. On this basis, low emission levels are presumed. There is a need for more complete information on both guided and fugitive emissions (e.g., of metals, organics) from non-ferrous metal melting in foundries, based on operational practice and expressed both as an emission level (mass/exhaust volume) and as a mass flow (mass/output of metal or castings).

Energy efficiency: Examples of techniques for heat recovery for induction and cupola furnaces are given in this chapter. There is a need for additional examples of heat recovery systems, with performance data. Information should be gathered concerning energy consumption and energy losses for all types of furnaces, related to specific techniques to increase the energy efficiency.

22

Pollution control in mining industry

22.1 Introduction

Mining is the extraction of valuable minerals or other geological materials from the earth from an orebody, lode, vein, seam, reef or placer deposits which forms the mineralised package of economic interest to the miner. Mining operations usually create a negative environmental impact, both during the mining activity and after the mine has closed. Hence, most of the world's nations have passed regulations to decrease the impact. Worker safety has long been a concern as well and modern practices have significantly improved safety in mines.

22.2 Phases of a mining project

There are different phases of a mining project, beginning with mineral ore exploration and ending with the post-closure period. Each phase of mining is associated with different sets of environmental impacts.

The majority of mining projects involve the extraction of ore deposits such as copper, nickel, cobalt, gold, silver, lead, zinc, molybdenum and platinum.

22.2.1 Exploration

A mining project can only commence with knowledge of the extent and value of the mineral ore deposit. Information about the location and value of the mineral ore deposit is obtained during the exploration phase. This phase includes surveys, field studies and drilling test boreholes and other exploratory excavations. The exploratory phase may involve clearing of wide areas of vegetation (typically in lines), to allow the entry of heavy vehicles mounted with drilling rigs. Many countries require a separate Environmental Impact Assessment (EIA) for the exploratory phase of a mining project because the impacts of this phase can be profound and because further phases of mining may not ensue if exploration fails to find sufficient quantities of high-grade mineral ore deposits.

22.2.2 Development

If the mineral ore exploration phase proves that there is a large enough mineral ore deposit, of sufficient grade, then the project proponent may begin to plan for the development of the mine. This phase of the mining project has several distinct components.

22.2.3 Construction of access roads

The construction of access roads, either to provide heavy equipment and supplies to the mine site or to ship out processed metals and ores, can have substantial environmental impacts, especially if access roads cut through ecologically sensitive areas or are near previously isolated communities. If a proposed mining project involves the construction of any access roads, then the Environmental Impact Assessment (EIA) for the project must include a comprehensive assessment of the environmental and social impacts of these roads.

Site preparation and clearing

If a mine site is located in a remote, undeveloped area, the project proponent may need to begin by clearing land for the construction of staging areas that would house project personnel and equipment. Even before any land is mined, activities associated with site preparation and clearing can have significant environmental impacts, especially if they are within or adjacent to ecologically sensitive areas. The EIA must assess, separately, the impacts associated with site preparation and clearing.

22.2.4 Active mining

Once a mining company has constructed access roads and prepared staging areas that would house project personnel and equipment, mining may commence. All types of active mining share a common aspect: the extraction and concentration (or beneficiation) of a metal from the earth. Proposed mining projects differ considerably in the proposed method for extracting and concentrating the metallic ore.

In almost every case, metallic ores are buried under a layer of ordinary soil or rock (called 'overburden' or 'waste rock') that must be moved or excavated to allow access to the ore deposit. The first way in which proposed mining projects differ is the proposed method of moving or excavating the overburden. What follows are brief descriptions of the most common methods.

Open-pit mining

Open-pit mining is a type of strip mining in which the ore deposit extends very deep in the ground, necessitating the removal of layer upon layer of overburden and ore. In many cases, logging of trees and clear-cutting or burning of vegetation above the ore deposit may precede removal of the overburden. The use of heavy machinery, usually bulldozers and dump trucks, is the most common means of removing overburden. Open-pit mining often involves the removal of natively vegetated areas and is therefore among the most environmentally destructive types of mining, especially within tropical forests. Because open-pit mining is employed for ore deposits at a substantial depth underground, it usually involves the creation of a pit that extends below the groundwater table. In this case, groundwater must be pumped out of the pit to allow mining to take place. A pit lake usually forms at some point in time after mining stops and the groundwater pumps are turned off.

Placer mining

Placer mining is used when the metal of interest is associated with sediment in a stream bed or flood plain. Bulldozers, dredges, or hydraulic jets of water (a process called 'hydraulic mining') are used to extract the ore. Placer mining is usually aimed at removing gold from stream sediments and flood plains. Because placer mining often occurs within a streambed, it is an environmentally destructive type of mining, releasing large quantities of sediment that can impact surface water for several miles downstream of the placer mine.

Underground mining

In underground mining, a minimal amount of overburden is removed to gain access to the ore deposit. Access to this ore deposit is gained by tunnels or shafts. Tunnels or shafts lead to a more horizontal network of underground tunnels that directly access the ore. In an underground mining method called 'stoping' or 'block caving,' sections or blocks of rock are removed in vertical strips that leave a connected underground cavity that is usually filled with cemented aggregate and waste rock.

Although underground mining is a less environmentally destructive means of gaining access to an ore deposit, it is often more costly and entails greater safety risks than strip mining, including open-pit mining. While most large scale mining projects involve open-pit mining, many large underground mines are in operation around the world.

Reworking of inactive or abandoned mines and tailings

Some mining projects involve the reworking of waste piles (often tailings) from inactive or abandoned mines or older waste piles at active mines. Typically, this is proposed when more efficient methods of metal beneficiation have made it economical to re-extract metals from old mining waste. The material from the piles may be sent to processing facilities on-site or off-site. Mining projects that only involve the reworking of abandoned mine waste piles avoid the environmental impacts of open-pit mining and placer mining, but still entail environmental impacts associated with purification (beneficiation) of metals from the waste piles.

22.2.5 Disposal of overburden and waste rock

In almost every project, metallic ores are buried under a layer of ordinary soil or rock (called 'overburden' or 'waste rock') that must be moved or excavated to allow access to the metallic ore deposit. For most mining projects, the quantity of overburden generated by mining is enormous. The ratio of the quantity of overburden to the quantity of mineral ore (called the 'strip ratio') is usually greater than one and can be much higher. For example, if a proposed mining project involves the extraction of 100 million metric tons of mineral ore, then the proposed mining project could generate more than one billion metric tons of overburden and waste rock.

These high-volume wastes, sometimes containing significant levels of toxic substances, are usually deposited on-site, either in piles on the surface or as backfill in open pits, or within underground mines. Therefore, the EIA for a proposed mining project must carefully assess the management options and associated impacts of overburden disposal.

22.2.6 Ore extraction

After a mining company has removed overburden, extraction of the mineral ore begins using specialised heavy equipment and machinery, such as loaders, haulers and dump trucks, which transport the ore to processing facilities using haul roads. This activity creates a unique set of environmental impacts, such as emissions of fugitive dust from haul roads, which an EIA for a proposed mining project should assess separately.

22.2.7 Beneficiation

Although metallic ores contain elevated levels of metals, they generate large quantities of waste. For example, the copper content of a good grade copper ore may be only one quarter of one per cent. The gold content of a good grade gold ore may be only a few one-hundredths of a per cent. Therefore, the next step in mining is grinding (or milling) the ore and separating the relatively small quantities of metal from the nonmetallic material of the ore in a process called 'beneficiation.' Milling is one of the most costly parts of beneficiation and results in very fine particles that allow better extraction of the metal. However, milling also allows a more complete release of contaminants when these particles become tailings. Tailings are what remains following milling of the ore to fine particles and extraction of the valuable metal(s).

Beneficiation includes physical and/or chemical separation techniques such as gravity concentration, magnetic separation, electrostatic separation, flotation, solvent extraction, electrowinning, leaching, precipitation and amalgamation (often involving the use of mercury). Wastes from these processes include waste rock dumps, tailings, heap leach materials (for gold and silver operations) and dump leach materials (for copper leach operations).

Leaching involving the use of cyanide is a kind of beneficiation process, usually used with gold, silver and copper ores, that merits separate attention because of the serious environmental and public safety impacts. With leaching, finely ground ore is deposited in a large pile (called a 'leach pile') on top of an impermeable pad and a solution containing cyanide is sprayed on top of the pile. The cyanide solution dissolves the desired metals and the 'pregnant' solution containing the metal is collected from the bottom of the pile using a system of pipes.

22.2.8 Tailings disposal

As previously discussed, even high-grade mineral ores consist almost entirely of non-metallic materials and often contain undesired toxic metals (such as cadmium, lead and arsenic). The beneficiation process generates high-volume waste called 'tailings,' the residue of an ore that remains after it has been milled and the desired metals have been extracted [e.g., with cyanide (gold) or sulphuric acid (copper)].

If a mining project involves the extraction of a few hundred million metric tons of mineral ore, then the mine project will generate a similar quantity of tailings. How a mining company disposes of this high-volume toxic waste material is one of the central questions that will determine whether a proposed mining project is environmentally acceptable. The key long-term goal of tailings disposal and management is to prevent the mobilisation and release into the environment of toxic constituents of the tailings.

The first option (a tailings pond) is by far the most commonly used option, but the second option (dry tailings disposal) is, in most circumstances, the environmentally-preferable option. The third option (submarine tailings disposal) is sometimes proposed with mine sites located near deep sea environments, or in rare instances in freshwater lakes. Submarine tailings disposal has a poor environmental record in the few instances where it has been practiced.

Before the adoption of environmental laws and standards, many mining companies simply dumped tailings in the nearest convenient location, including nearby rivers and streams. Some of the worst environmental consequences of mining have been associated with the open dumping of tailings, a practice now nearly universally rejected.

22.2.9 Site reclamation and closure

When active mining ceases, mine facilities and the site are reclaimed and closed. The goal of mine site reclamation and closure should always be to return the site to a condition that most resembles the pre-mining condition. Mines that are notorious for their immense impact on the environment often made impacts only during the closure phase, when active mining operations ceased. These impacts can persist for decades and even centuries. Therefore, the EIA for every proposed mining project must include a detailed discussion of the mine reclamation and closure plan offered by the mining proponent.

Mine reclamation and closure plans must describe in sufficient detail how the mining company will restore the site to a condition that most resembles pre-mining environmental quality; how it will prevent – in perpetuity – the release of toxic contaminants from various mine facilities (such as abandoned open pits and tailings impoundments) and how funds will be set aside to insure that the costs of reclamation and closure will be paid for.

22.3 Environmental and social impacts of mining

This section describes the most important environmental impacts of mining projects.

22.3.1 Impacts on water resources

Perhaps the most significant impact of a mining project is its effects on water quality and availability of water resources within the project area. Key questions are whether surface and groundwater supplies will remain fit for human consumption and whether the quality of surface waters in the project area will remain adequate to support native aquatic life and terrestrial wildlife.

Acid mine drainage and contaminant leaching

The potential for acid mine drainage is a key question. The answer will determine whether a proposed mining project is environmentally acceptable. When mined materials (such as the walls of open pits and underground mines, tailings, waste rock and heap and dump leach materials) are excavated and exposed to oxygen and water, acid can form if iron sulphide minerals (especially pyrite or 'fools gold') are abundant and there is an insufficient amount of neutralising material to counteract the acid formation. The acid will, in turn, leach or dissolve metals and other contaminants from mined materials and form a solution that is acidic, high in sulphate and metal-rich (including elevated concentrations of cadmium, copper, lead, zinc, arsenic, etc.)

Leaching of toxic constituents, such as arsenic, selenium and metals, can occur even if acidic conditions are not present. Elevated levels of cyanide and nitrogen compounds (ammonia, nitrate, nitrite) can also be found in waters at mine sites, from heap leaching and blasting.

Acid drainage and contaminant leaching is the most important source of water quality impacts related to metallic ore mining. As Earthworks explains:

'Acid mine drainage is considered one of mining's most serious threats to water resources. A mine with acid mine drainage has the potential for long-term devastating impacts on rivers, streams and aquatic life.

Acid mine drainage is a concern at many metal mines, because metals such as gold, copper, silver and molybdenum, are often found in rock with sulphide minerals. When the sulphides in the rock are excavated and exposed to water and air during mining, they form sulphuric acid. This acidic water can dissolve other harmful metals in the surrounding rock. If uncontrolled, the acid mine drainage may runoff into streams or rivers or leach into groundwater. Acid mine drainage may be released from any part of the mine where sulphides are exposed to air and water, including waste rock piles, tailings, open pits, underground tunnels and leach pads.

Harm to fish and other aquatic life: If mine waste is acid-generating, the impacts to fish, animals and plants can be severe. Many streams impacted by acid mine drainage have a pH value of 4 or lower – similar to battery acid. Plants, animals and fish are unlikely to survive in streams such as this.

Toxic metals: Acid mine drainage also dissolves toxic metals, such as copper, aluminium, cadmium, arsenic, lead and mercury, from the surrounding rock. These metals, particularly the iron, may coat the stream bottom with an orange-red coloured slime called yellow boy. Even in very small amounts, metals can be toxic to humans and wildlife. Carried in water, the metals can travel far, contaminating streams and groundwater for great distances. The impacts to aquatic life may range from immediate fish kills to sublethal, impacts affecting growth, behaviour or the ability to reproduce.

Metals are particularly problematic because they do not break down in the environment. They settle to the bottom and persist in the stream for long periods of time, providing a long-term source of contamination to the aquatic insects that live there and the fish that feed on them.

Perpetual pollution: Acid mine drainage is particularly harmful because it can continue indefinitely causing damage long after mining has ended. Due to the severity of water quality impacts from acid mine drainage, many hard rock mines across the west require water treatment in perpetuity. Even with existing technology, acid mine drainage is virtually impossible to stop once the reactions begin. To permit an acid generating mine means that future generations will take responsibility for a mine that must be managed for possibly hundreds of years.

Erosion of soils and mine wastes into surface waters

For most mining projects, the potential of soil and sediment eroding into and degrading surface water quality is a serious problem.

According to various case studies: 'Because of the large area of land disturbed by mining operations and the large quantities of earthen materials exposed at sites, erosion can be a major concern at hard rock mining sites.' Consequently, erosion control must be considered from the beginning of operations through completion of reclamation. Erosion may cause significant loading of sediments (and any entrained chemical pollutants) to nearby water bodies, especially during severe storm events and high snow melt periods.

Sediment-laden surface runoff typically originates as sheet flow and collects in rills, natural channels or gullies, or artificial conveyances. The ultimate deposition of the sediment may occur in surface waters or it may be deposited within the flood plains of a stream valley.

Historically, erosion and sedimentation processes have caused the build-up of thick layers of mineral fines and sediment within regional flood plains and the alteration of aquatic habitat and the loss of storage capacity within surface waters. The main factors influencing erosion includes the volume and velocity of runoff from precipitation events, the rate of precipitation infiltration downward through the soil, the amount of vegetative cover, the slope length or the distance from the point of origin of overland flow to the point where deposition begins and operational erosion control structures.

Major sources of erosion/sediment loading at mining sites can include open pit areas, heap and dump leaches, waste rock and overburden piles, tailings piles and dams, haul roads and access roads, ore stockpiles, vehicle and equipment maintenance areas, exploration areas and reclamation areas. A further concern is that exposed materials from mining operations (mine workings, wastes, contaminated soils, etc.), may contribute sediments with chemical pollutants, principally heavy metals. The variability in natural site conditions (e.g., geology, vegetation, topography, climate and proximity to and characteristics of surface waters), combined with significant differences in the quantities and characteristics of exposed materials at mines, preclude any generalisation of the quantities and characteristics of sediment loading.

The types of impacts associated with erosion and sedimentation are numerous, typically producing both short-term and long term impacts. In surface waters, elevated concentrations of particulate matter in the water column can produce both chronic and acute toxic effects in fish.

Sediments deposited in layers in flood plains or terrestrial ecosystems can produce many impacts associated with surface waters, ground water and terrestrial ecosystems. Minerals associated with deposited sediments may depress the pH of surface runoff thereby mobilising heavy metals that can infiltrate into the surrounding subsoil or can be carried away to nearby surface waters. The associated impacts could include substantial pH depression or metals loading to surface waters and/or persistent contamination of ground water sources. Contaminated sediments may also lower the pH of soils to the extent that vegetation and suitable habitat are lost. 'Beyond the potential for pollutant impacts on human and aquatic life, there are potential physical impacts associated with the increased runoff velocities and volumes from new land disturbance activities. Increased velocities and volumes can lead to downstream flooding, scouring of stream channels and structural damage to bridge footings and culvert entries. In areas where air emissions have deposited acidic particles and the native vegetation has been destroyed, runoff has the potential to increase the rate of erosion and lead to removal of soil from the affected area. This is particularly true where the landscape is characterised by steep and rocky slopes. Once the soils have been removed, it is difficult for the slope to be revegetated either naturally or with human assistance.

Environmental aspects summarises the problem as follows:

Potentially adverse effects of inadequate minesite water management and design include: Unacceptably high levels of suspended solids (non-filterable residue) and dissolved solids (filterable residue) in surface runoff [and] bed and bank erosion in waterways. It is self-evident that a sediment and erosion control plan is a fundamental component of a minesite water management plan.

Impacts of tailing impoundments, waste rock, heap leach and dump leach facilities

The impacts of wet tailings impoundments, waste rock, heap leach and dump leach facilities on water quality can be severe. These impacts include contamination of groundwater beneath these facilities and surface waters. Toxic substances can leach from these facilities, percolate through the ground and contaminate groundwater, especially if the bottom of these facilities are not fitted with an impermeable liner.

Tailings (a by-product of metallic ore processing) is a high-volume waste that can contain harmful quantities of toxic substances, including arsenic, lead, cadmium, chromium, nickel and cyanide (if cyanide leaching is used). Although it is rarely the environmentally preferable option, most mining companies dispose of tailings by mixing them with water (to form a slurry) and disposing of the slurry behind a tall dam in a large wet tailings impoundment. Because the ore is usually extracted as a slurry, the resulting waste contains large amounts of water and generally forms ponds at the top of the tailings dams that can be a threat to wildlife. Cyanide tailings in precious metals mines are particularly dangerous.

Ultimately, tailing ponds will either dry, in arid climates or may release contaminated water, in wet climates. In both cases, specific management techniques are required to close these waste repositories and reduce environmental threats.

During periods of heavy rain, more water may enter a tailings impoundment than it has the capacity to contain, necessitating the release of tailings impoundment effluent. Since this effluent can contain toxic substances, the release of this effluent can seriously degrade water quality of surrounding rivers and streams, especially if the effluent is not treated prior to discharge.

Dozens of dam breaks at wet tailings impoundments have created some of the worst environmental consequences of all industrial accidents. When wet tailings impoundments fail, they release large quantities of toxic waters that can kill aquatic life and poison drinking water supplies for many miles downstream of the impoundment.

Impacts of mine dewatering

When an open pit intersects the water table, groundwater flows into the open pit. For mining to proceed, mining companies must pump and discharge this water to another location. Pumping and discharging mine water causes a unique set of environmental impacts that are well described in a study commissioned by Smith: 'Mine water is produced when the water table is higher than the underground mine workings or the depth of an open pit surface mine.' When this occurs, the water must be pumped out of the mine. Alternatively, water may be pumped from wells surrounding the mine to create a cone of depression in the ground water table, thereby reducing infiltration. When the mine is operational, mine water must be continually removed from the mine to facilitate the removal of the ore. However, once mining operations end, the removal and management of mine water often end, resulting in possible accumulation in rock fractures, shafts, tunnels and open pits and uncontrolled releases to the environment.

Ground water drawdown and associated impacts to surface waters and nearby wetlands can be a serious concern in some areas. Impacts from ground water drawdown may include reduction or elimination of surface water flows; degradation of surface water quality and beneficial uses; degradation of habitat (not only riparian zones, springs and other wetland habitats, but also upland habitats such as greasewood as ground water levels decline below the deep root zone), reduced or eliminated production in domestic supply wells, water quality/quantity problems associated with discharge of the pumped ground water back into surface waters downstream from the dewatered area.

The impacts could last for many decades. While dewatering is occurring, discharge of the pumped water, after appropriate treatment, can often be used to mitigate adverse effects on surface waters. However, when dewatering ceases, the cones of depression may take many decades to recharge and may continue to reduce surface flows. Mitigation measures that rely on the use of pumped water to create wetlands may only last as long as dewatering occurs.

22.3.2 Impacts of mining projects on air quality

Airborne emissions occur during each stage of the mine cycle, but especially during exploration, development, construction and operational activities. Mining operations mobilise large amounts of material and waste piles containing small size particles are easily dispersed by the wind.

The largest sources of air pollution in mining operations are:

1. Particulate matter transported by the wind as a result of excavations, blasting, transportation of materials, wind erosion (more frequent in open-pit mining), fugitive dust from tailings facilities, stockpiles, waste dumps and haul roads. Exhaust emissions from mobile sources (cars, trucks, heavy equipment) raise these particulate levels.
2. Gas emissions from the combustion of fuels in stationary and mobile sources, explosions and mineral processing.

Once pollutants enter the atmosphere, they undergo physical and chemical changes before reaching a receptor (Fig. 22.1). These pollutants can cause serious effects to people's health and to the environment.

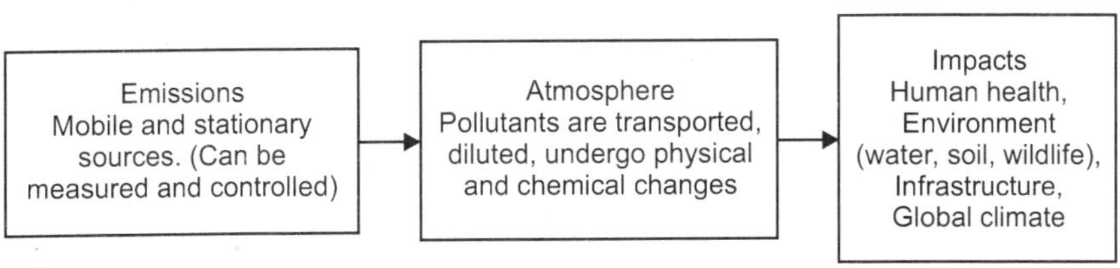

Figure 22.1: Pollutants-their effects on environment and health.

Large-scale mining has the potential to contribute significantly to air pollution, especially in the operation phase. All activities during ore extraction, processing, handling and transport depend on equipment, generators, processes and materials that generate hazardous air pollutants such as particulate matter, heavy metals, carbon monoxide, sulphur dioxide and nitrogen oxides.

Mobile sources

Mobile sources of air pollutants include heavy vehicles used in excavation operations, cars that transport personnel at the mining site and trucks that transport mining materials. The level of polluting emissions from these sources depends on the fuel and conditions of the equipment. Even though individual emissions can be relatively small, collectively these emissions can be of real concern. In addition, mobile sources are a major source of particulate matter, carbon monoxide and volatile organic compounds that contribute significantly to the formation of ground-level ozone.

Stationary sources

The main gaseous emissions are from combustion of fuels in power generation installations and drying, roasting and smelting operations. Many producers of precious metals smelt metal on-site, prior to shipping to off-site refineries. Typically, gold and silver is produced in melting/fluxing furnaces that may produce elevated levels of airborne mercury, arsenic, sulphur dioxide and other metals.

Fugitive emissions

The U.S. Environmental Protection Agency defines 'fugitive emissions' as 'those emissions which could not reasonably pass through a stack, chimney, vent or other functionally-equivalent opening. Common sources of fugitive emissions include: storage and handling of materials; mine processing; fugitive dust, blasting, construction activities and roadways associated with mining activities; leach pads and tailing piles and ponds and waste rock piles. Sources and characteristics of fugitive emissions dust in mining operations vary in each case, as do their impacts. Impacts are difficult to predict and calculate but should be considered since they could be a significant source of hazardous air pollutants.

Incidental releases of mercury

Mercury is commonly present in gold ore. Although concentrations vary substantially, even within a specific ore deposit, mercury is found in gold ore and associated waste materials. If the content of mercury in a gold ore is 10 mg/kg and one million tons of ore is processed at a particular mine (not unusual concentrations), 10 T of mercury are potentially released to the environment. This is a major source of mercury and should be controlled. In some gold mining projects, gold-containing ore is crushed and then, if necessary, heated and oxidised in roasters or autoclaves to remove sulphur and carbonaceous material that affects gold recovery. Mercury that is present in the ore is vapourised, particularly in roasters, which are some of the largest sources of mercury emitted to the atmosphere. Following roasting or autoclaving, the ore is mixed with water and reacted with a cyanide leach solution, where gold and mercury are dissolved and solids removed via filtration. The purified solution is sent to an electrowinning process, where the gold is recovered. In this process, mercury must also be recovered and collected. If not collected by air pollution control devices, this mercury could be released to the atmosphere and impact the environment and public health.

Volatilisation of mercury from active heaps and tailings facilities has recently been identified as another substantial source of mercury emitted to the atmosphere. This process should be assessed and controlled. Overall, mercury present in gold ore may be released to the land (in disposed air pollution control wastes and spent ore tailings), to the air (not removed by air pollution control devices, or from tailings or heaps) or in the gold product (i.e., as an impurity).

Noise and vibration

Noise pollution associated with mining may include noise from vehicle engines, loading and unloading of rock into steel dumpers, chutes, power generation and other sources. Cumulative impacts of shoveling, ripping, drilling, blasting, transport, crushing, grinding and stock-piling can significantly affect wildlife and nearby residents. Vibrations are associated with many types of equipment used in mining operations, but blasting is considered the major source. Vibration has affected the stability of infrastructures, buildings and homes of people living near large-scale open-pit mining operations. According to a study commissioned by the European Union in 2010: Shocks and vibrations as a result of blasting in connection with mining can lead to noise, dust and collapse of structures in surrounding inhabited areas. The animal life, on which the local population may depend, might also be disturbed.

22.3.3 Impacts of mining projects on wildlife

Wildlife is a broad term that refers to all plants and any animals (or other organisms) that are not domesticated. Mining affects the environment and associated biota through the removal of vegetation and topsoil, the displacement of fauna, the release of pollutants and the generation of noise.

Habitat loss

Wildlife species live in communities that depend on each other. Survival of these species can depend on soil conditions, local climate, altitude and other features of the local habitat. Mining causes direct and indirect damage to wildlife. The impacts stem primarily from disturbing, removing and redistributing the land surface. Some impacts are short-term and confined to the mine site; others may have far-reaching, long-term effects.

The most direct effect on wildlife is destruction or displacement of species in areas of excavation and piling of mine wastes. Mobile wildlife species, like game animals, birds and predators, leave these areas. More sedentary animals, like invertebrates, many reptiles, burrowing rodents and small mammals, may be more severely affected.

If streams, lakes, ponds or marshes are filled or drained, fish, aquatic invertebrates and amphibians are severely impacted. Food supplies for predators are reduced by the disappearance of these land and water species. Many wildlife species are highly dependent on vegetation growing in natural drainages.

This vegetation provides essential food, nesting sites and cover for escape from predators. Any activity that destroys vegetation near ponds, reservoirs, marshes and wetlands reduces the quality and quantity of habitat essential for waterfowl, shore birds and many terrestrial species.

The habitat requirements of many animal species do not permit them to adjust to changes created by land disturbance. These changes reduce living space. The degree to which animals tolerate human competition for space varies. Some species tolerate very little disturbance. In instances where a particularly critical habitat is restricted, such as a lake, pond or primary breeding area, a species could be eliminated.

Surface mining can degrade aquatic habitats with impacts felt many miles from a mining site. For example, sediment contamination of rivers and streams is common with surface mining.

Habitat fragmentation

Habitat fragmentation occurs when large areas of land are broken up into smaller and smaller patches, making dispersal by native species from one patch to another difficult or impossible and cutting off migratory routes. Isolation may lead to local decline of species, or genetic effects such as inbreeding. Species that require large patches of forest simply disappear.

22.3.4 Impacts of mining projects on soil quality

Mining can contaminate soils over a large area. Agricultural activities near a mining project may be particularly affected. According to a study commissioned by the European Union: Mining operations routinely modify the surrounding landscape by exposing previously undisturbed earthen materials. Erosion of exposed soils, extracted mineral ores, tailings and fine material in waste rock piles can result in substantial sediment loading to surface waters and drainage ways. In addition, spills and leaks of hazardous materials and the deposition of contaminated windblown dust can lead to soil contamination.

Soil contamination: Human health and environmental risks from soils generally fall into two categories: (i) contaminated soil resulting from wind blown dust and (ii) soils contaminated from chemical spills and residues.

Fugitive dust can pose significant environmental problems at some mines. The inherent toxicity of the dust depends upon the proximity of environmental receptors and type of ore being mined.

High levels of arsenic, lead and radionucleides in windblown dust usually pose the greatest risk. Soils contaminated from chemical spills and residues at mine sites may pose a direct contact risk when these materials are misused as fill materials, ornamental landscaping or soil supplements.

22.3.5 Impacts of mining projects on social values

The social impacts of large-scale mining projects are controversial and complex. Mineral development can create wealth, but it can also cause considerable disruption. Mining projects may create jobs, roads, schools and increase the demands of goods and services in remote and impoverished areas, but the benefits and costs may be unevenly shared. If communities feel they are being unfairly treated or inadequately compensated, mining projects can lead to social tension and violent conflict.

EIAs can underestimate or even ignore the impacts of mining projects on local people. Communities feel particularly vulnerable when linkages with authorities and other sectors of the economy are weak or when environmental impacts of mining (soil, air and water pollution) affect the subsistence and livelihood of local people.

Power differentials can leave a sense of helplessness when communities confront the potential for change induced by large and powerful companies. The EIA process should enforce mechanisms that enable local communities to play effective roles in decision making. Mineral activities must ensure that the basic rights of the individual and communities affected are upheld and not infringed upon. These must include the right to control and use land; the right to clean water, a safe environment and livelihood; the right to be free from intimidation and violence; and the right to be fairly compensated for loss.

Human displacement and resettlement

According to the International Institute for Environment and Development: The displacement of settled communities is a significant cause of resentment and conflict associated with large-scale mineral development. Entire communities may be uprooted and forced to shift elsewhere, often into purpose-built settlements not necessarily of their own choosing. Besides losing their homes, communities may also lose their land and thus their livelihoods. Community institutions and power relations may also be disrupted. Displaced communities are often settled in areas without adequate resources or are left near the mine, where they may bear the brunt of pollution and contamination. Forced resettlement can be particularly disastrous for indigenous communities who have strong cultural and spiritual ties to the lands of their ancestors and who may find it difficult to survive when these are broken.

Impacts of migration

According to the International Institute for Environment and Development: One of the most significant impacts of mining activity is the migration of people into a mine area, particularly in remote parts of developing countries where the mine represents the single most important economic activity.

Sudden increases in population can also lead to pressures on land, water and other resources as well as bringing problems of sanitation and waste disposal.

Migration effects may extend far beyond the immediate vicinity of the mine. Improved infrastructure can also bring an influx of settlers.

Lost access to clean water

Impacts on water quality and quantity are among the most contentious aspects of mining projects. Companies insist that the use of modern technologies will ensure environmentally friendly mining practices. However, evidence of the negative environmental impacts of past mining activity causes local and downstream populations to worry that new mining activities will adversely affect their water supply.

There are major stakes in these conflicts, affecting everything from local livelihood sustainability to the solvency of national governments. Fears for water quantity and quality have triggered numerous and sometimes violent conflicts between miners and communities.

Impacts on livelihoods

When mining activities are not adequately managed, the result is degraded soils, water, biodiversity and forest resources, which are critical to the subsistence of local people. When contamination is not controlled, the cost of the contamination is transferred to other economic activities, such as agriculture and fishing. The situation is made worse when mining activities take place in areas inhabited by populations historically marginalised, discriminated against, or excluded.

Impacts on public health

EIAs of mining projects often underestimate the potential health risks of mining projects. Hazardous substances and wastes in water, air and soil can have serious, negative impacts on public health. The World Health Organisation (WHO) defines health as a 'state of complete physical, mental and social well-being and not merely the absence of disease or infirmity.

The term 'hazardous substances' is broad and includes all substances that can be harmful to people and/or the environment. Because of the quantity, concentration, or physical, chemical or infectious characteristics, hazardous substances may: (i) cause or contribute to an increase of mortality or an increase in serious irreversible or incapacitating illness and (ii) pose a substantial present or potential hazard to human health or the environment when improperly treated, stored, transported, disposed of or otherwise managed.

Frequent public health problems related to mining activities include:

1. Water: Surface and ground water contamination with metals and elements; microbiological contamination from sewage and wastes in campsites and mine worker residential areas.
2. Air: Exposure to high concentrations of sulphur dioxide, particulate matter, heavy metals, including lead, mercury and cadmium.
3. Soil: Deposition of toxic elements from air emissions.

Mining activities can suddenly affect quality of life and the physical, mental and social well-being of local communities. Improvised mining towns and camps often threaten food availability and safety, increasing the risk of malnourishment.

Indirect effects of mining on public health can include increased incidence of tuberculosis, asthma, chronic bronchitis and gastrointestinal diseases.

Impacts to cultural and aesthetic resources

Mining activities can cause direct and indirect impacts to cultural resources. Direct impacts can result from construction and other mining activities. Indirect impacts can result from soil erosion and increased accessibility to current or proposed mining sites. Mining projects can affect sacred landscapes, historical infrastructures and natural landmarks. Potential impacts include:

1. Complete destruction of the resource through surface disturbance or excavation.
2. Degradation or destruction, due to topographic or hydrological pattern changes, or from soil movement (removal, erosion, sedimentation).
3. Unauthorised removal of artifacts or vandalism as a result of increased access to previously inaccessible areas.
4. Visual impacts due to clearing of vegetation, large excavations, dust and the presence of large-scale equipment and vehicles.

22.3.6 Climate change considerations

Every EIA for a project that has the potential to change the global carbon budget should include an assessment of a project's carbon impact. Large-scale mining projects have the potential to alter global carbon in at least the following ways:

Lost CO_2 uptake: Lost CO_2 uptake by forests and vegetation that is cleared. Many large-scale mining projects are proposed in heavily forested areas of tropical regions that are critical for absorbing atmospheric carbon dioxide (CO_2) and maintaining a healthy balance between CO_2 emissions and CO_2 uptake. Some mining projects propose long-term or even permanent destruction of tropical forests. EIAs for mining projects must include a careful accounting of how any proposed disturbance of tropical forests will alter the carbon budget. The EIA should also include an analysis of the potential for the host country to lose funding from international consortiums that have and will be established to conserve tropical forests.

CO_2 emitted by machines: CO_2 emitted by machines (e.g., diesel powered heavy vehicles) involved in extracting and transporting ore. The EIA should include a quantitative estimate of CO_2 emissions from machines and vehicles that will be needed during the life of the mining project. These estimates can be based on the rate of fuel consumption (typically diesel fuel) multiplied by a conversion factor that relates units (typically litres or gallons) of fuel that is consumed and units (typically metric tons) of CO_2 that is emitted.

CO_2 emitted by the processing of ore into metal: CO_2 emitted by the processing of ore into metal (for example, by pyro-metallurgical versus hydro-metallurgical techniques). An example is found in an assessment by CSIRO minerals of Australia which used the life cycle assessment methodology to estimate the life cycle emissions of greenhouse gases from copper and nickel production, including mining. This assessment found that life cycle greenhouse gas emissions from copper and nickel production range from 3.3 kilograms (kg) of CO_2 per kg of metal for copper produced by smelting to 16.1 kg of CO_2 per kg of metal for nickel produced by pressure acid leaching followed by solvent extraction and electrowinning. The bottom line is that metal mining generates more than 1 kg of greenhouse gas for every 1 kg of metal that is produced and this does not take into account lost carbon uptake of cleared forests.

22.4 Pollution in coal mining

Coal is an important fossil fuel for generation of electricity and for other industrial purposes. Its importance in electricity generation has become more prominent after increase in international price of crude oil. Therefore, coal mining is now essential part of civilisation. Traditionally, coal mining and coal fired power plants are considered to be most polluted industry. We discuss below the environmental pollution issues involved with coal mining activities and ways to mitigate them.

22.4.1 Environmental problems in mining

Environment problems related to coal mines are discussed below:

Air pollution

Air pollution in coal mines is mainly due to the fugitive emission of particulate matter and gases including methane (CH_4), sulphur dioxide (SO_2) and oxides of nitrogen (NO_x). The mining operations like drilling, blasting, movement of the heavy earth moving machinery on haul roads, collection, transportation and handling of coal, screening, sizing and segregation units are the major sources of such emissions. Underground mine fire is also a major source of air pollution in some of the coal fields. High levels of suspended particulate matter increase respiratory diseases such as chronic bronchitis and asthma cases while gaseous emissions contribute towards global warming besides causing health hazards to the exposed population.

Methane emission from coal mining depends on the mining methods, depth of coal mining, coal quality and entrapped gas content in coal seams.

Water pollution

The major source of water pollution in the coal mines is the carry over of the suspended solids in the drainage system of the mine sump water and storm water drainage. In some of the coal mines, acidic water is also found in the underground aquifers. In addition, wastewater from coal preparation plant and mine water are other sources of water pollution.

Land degradation

The opencast coal mines are developed at the surface, because of that these mines are also called surface coal mines. The overburden, i.e., the rock or soil overlaid the coal seam, are removed before extraction of coal. This overburden is dumped on surface, preferably on mined-out or decoaled area. Therefore, this type of mining requires quite large area on surface. Many a times, large forest areas are transferred for coal mining purpose. The land degradation is the result of creation and expansion of opencast coal mines. The aspect of land degradation in underground coal mines is due to subsidence over the underground cavity resulted from underground caving.

Noise pollution

Main sources of noise pollution are blasting, movement of heavy earth moving machines, drilling and coal handling plants, etc.

Solid wastes

Major source of solid waste in a coal mine is the overburden. Segregation of the stones in the coal handling plants and the coal breeze also contribute to the solid waste generation. Overburden to coal ratio in the open cast mining is about 2 m^3/T of coal or sometime more. Therefore, the quantum of overburden generated and its proper management is the main concern area in dealing with the environmental issue of opencast coal mines.

Deforestation

As explained, the requirement of land for a big opencast coal projects are quite large. Many of the forest area, many a times, are converted to mining field. Therefore, large forest areas are deforested to make a way for large opencast coal mines. The unscientific mining practices undertaken result in large degradation of land in the form of subsidence, underground goaf filled with water, mine fires, destruction of vegetation, generation of wind blow dust, etc. To mitigate above environmental problems several control measures, generally, are adopted. Some of the control measures are discussed below:

Subsidence: Subsidence of surface takes place due to extraction of coal by underground mining. Subsidence is exhibited by cracks on surface and lowering of land in the worked out areas compared to surroundings. The surface is rehabilitated by dozing and sealing of cracks followed by plantation of trees. The subsided areas with medium-sized depressions are ideal for developing water pools and sustain green vegetation and also to meet the water needs of local people.

Abandoned mines: The mined-out areas are to be backfilled and then rehabilitated for development of vegetation. In the quarried areas water reservoir is developed for water harvesting. The big voids created by open-pit mining cause land degradation. These voids can be gainfully utilised to serve as water reservoirs. This water provides moisture for vegetation in the surroundings areas. The water is used for domestic supply after necessary treatment. Irrigation to nearby agricultural land also may be thought off.

External overburden dump: The external dump area presents an unaesthetic appearance unless rehabilitated. Vegetative rehabilitation of these dumps prevents erosion and also improves aesthetics.

Mine fire: The measures for controlling the mine fires, include dozing, levelling and blanketing with soil to prevent the entry of oxygen and to stabilise the land for vegetal growth.

Water and air pollution control: Mine water is pumped to a lagoon, which acts as a sedimentation pond. The overflow water, which is fairly clean, is drained out to natural drain or used for dust suppression activities. Similarly, washery effluent is recirculated through thickener and slime ponds. For reducing air pollution, water spraying and sprinkling is done on the haul / transport roads to suppress the dust generation.

22.5 Environment impact assessment of coal mining

Environmental impact assessment is a decision making tool to predict the effect of a proposed activity/project on the environment, to compare various alternatives for a project and to identify best combination of economic, environmental and social costs and benefits. Coal mining is the process of extracting the coal from the deep underground mines in the earth crust. Coal mining is one of the most illegal activities going on in the country. At the same time it totally neglects the role of EIA in its pre-operational, during mining and post operational plans. Here various case studies of coal mining are done and the impact of coal mining to the environment is looked into with some measures to mitigate this problem with proper use of technology and generation of awareness. The impact of coal mining on land, water, health of workers, air and social impact on surrounding villages and cities is evaluated and environment management plan of coal mining is developed. Therefore, the coal has to be used sustainably as coal reserves are depleting rapidly.

Coal is a non-renewable fossil fuel formed from the remains of plants that lived and died about 100 to 400 million years ago. Depth, thickness and configuration of minerals resource decide the selection of mining technology (open cast or underground). Minerals exploitation progresses through four stages: Prospecting means search for deposits. Exploration: Once the deposit is assured, this is done to assess the size, shape, location and economic value of the deposit. Development work of preparing access to the deposit so that the coal can be extracted exploitation means extraction of coal. Depending on the content of volatiles, coal is classified into lignite, sub-bituminous coal, bituminous coal and anthracite. Open cast mining and underground mining (room pillar, long wall, drift, shaft, slope mining) are some types of coal mining. Proper environment impact assessment of coal mining is very important and here we take some issues which should be considered and some mitigation measures are discussed.

22.5.1 Environment and social issues related to coal mining

Coal mining has significant impact on land and land use, some of the land related impact are: Loss of biodiversity, economic loss or loss of livelihood due to displacement and encroachment of agricultural land and Impact on water resource (in term of water availability and quality). Open cast mine has significant impact on land as compared to underground mine.

1. Land: Around 4 ha of land is damaged for every million tons of coal mined by the surface mining. For instance, a capacity of 10 million tons opencast coal mine in 20 years has a potential to destroy around 800 ha of land.
2. Air: Mainly in the form of fugitive dust. Most mining operations generate dust such as drilling, blasting, vehicles movement on haul roads, collection, transportation and handling of coal, screening, sizing and segregation and storage. For example, a coal stack of 50,000 T can generate 250 T of fugitive dust even if assuming loss of only 0.5% as fugitive dust.
3. Water: Breaching of groundwater affects the local water availability. High risk to alter the water quality of areas, e.g., low pH, increase in total solids, TDS and heavy metal concentration. Acid mine drainage produced by the leaching of sulphide minerals present in the coal leaves a direct

impact on drinking water quality, aquatic life and corrosion of equipment and structures. Trace of arsenic and hexavalent chromium around 0.05 ppm has been found.

4. Solid wastes: Overburden of organic material and soil that overlie a mineral deposit. Overburden generation is denoted by stripping ratio which is the ratio of overburden that needs to be removed to the amount of ore removed. Low stripping ratio translates into low quantities of waste. Mining operation results in excavation of large quantities of top soil. It is precious as it holds nutrients and is essential for successful rehabilitation and afforestation. Top soil management and its reuse are important. Poor storage can lead to run-off.
5. Noise and vibration: Cumulative effect of mining activities produces considerable noise like blasting, drilling, crushing and movement of vehicles. Blasting results in ground vibrations and if there are human habitation nearby, it can destroy property and houses.
6. Occupational health: The workers in mine have high risk of the occupational diseases: Pneumoconiosis: Due to inhalation of coal dust. Can cause severe lung problems and lung cancer, Dust allergy and asthma, Noise hazard such as temporary or permanent hearing loss, headache and high blood pressure.
7. Mine fires: An issue for not only the workers but also the people living in adjoining area is a major problem.
8. Socio-economic factor: Coal mining affects communities at multiple levels.

During mine's opening: Displacement, loss of livelihood.

During mine's operation: Pollution impact (air, water, noise, vibration), health impact.

Mine closure: Sudden economic halt and land contamination

22.5.2 Mitigation and environment management plan

Environmental Management Plan (EMP) is a framework for the implementation and execution of mitigation measures and alternatives. Ideally EMP should covers all phases of project development, i.e., preconstruction, operation of mine and decommissioning of the mine. It is a documented plan containing details of impacts, recommended mitigation and monitoring measures, etc., and legal document based on which the performance is monitored.

1. Top soil management: The best practices for topsoil management is to scrap the topsoil prior to drilling and blasting, scraped topsoil should be used immediately for plantation/agriculture, If it is not possible to use

the topsoil immediately, then it should be stacked at a designated area, Storage must be done in a pyramidal form, with garland drains all around.

2. Overburden management: If an external overburden dump is unavoidable, then it should be stabilised with biological reclamation. Excavation from a new pit should begin after an existing pit has been exhausted. This would ensure that the overburden and interburden generated is used for backfilling the exhausted pit, instead of being dumped elsewhere. Till a pit is exhausted, the overburden should be compacted and stacked in specified locations in lowlying, non-mineralised zones within the lease area. Vegetation should be planted over the dump slopes as early as possible. The height and slope of the overburden dumps should be maintained to prevent slope failure. Sedimentation tanks should be constructed to treat run-off from external overburden dumps. For external overburden dumps, the bench height should not exceed 10 meters and the final dump height should not be more than 60 meters.
3. Subsidence management: Planned subsidence by considering surface structures and human lives, as in the case of long-wall mining. Preparation of a subsidence management plan and its approval by the regulatory agency. The plan should ensure the following: Simultaneous stowing of the decoaled area. Compensation to and rehabilitation of the affected people. Fencing of the subsidence zone during active mining operation to prevent unintended entry into the affected area. Reclaiming the subsided area by afforestation. Regular monitoring and inspection of subsidence area to detect any subsidence and taking the necessary steps. Preparing a subsidence monitoring programme that covers the impact of subsidence on surface and groundwater (quality and quantity) and its management.
4. Management of water pollution: Use of mine seepage water (after treatment) for different purposes, construction of gabbion wall, garland drain, siltation pond, wherever appropriate and coal stockpiles, overburden and topsoil should be selectively placed in a stable area which is less prone to erosion.
5. Mitigation measures for air pollution: Fixed and mobile water spraying on the haulage roads. Asphalting of the road junction meeting public roads. Covering of the trucks carrying ore with tarpaulin. Wheel wash system. 9 inch free board for all the trucks. Green belt around mining area as physical barrier.
6. Mitigation of noise and vibration: Use of closed and advanced blasting technology like shock tube technology conducting blast only during day time, use of hydraulic drills and provision of sound insulated chambers for workers deployed on machines producing higher levels of noise like

bulldozers, drills, etc. Enclosing crusher units in covered buildings to minimise sound propagation. Providing silencers or enclosures for noise generating machines such DG sets, compressors, etc. Creating a green belt around potential noise prone area and provision of protective devices like earmuffs/earplugs to workers, who are exposed to high level of noise and reducing the exposure time of workers by practicing worker rotation.

7. Mine closure plan: A good mine closure plan should include the following: A detailed final closure plan to create productive and sustainable land use. The plan must be accepted by mine owners, regulating agencies and local communities, A plan to protect the health and safety of the surrounding habitat and A plan to eliminate/contain all possible sources of pollution post-mining.
8. Mitigation of occupational health and safety issues.

Table 22.1 shows mitigation of occupational health and safety issues.

Table 22.1: Mitigation of occupational health and safety issues.

Activity	*Mitigation measures*
Drilling and blasting	Driller shall be equipped with a closed cabin to reduce exposure to noise and dust. In addition, the operators and other workers should be provided with masks, helmets, gloves and earplugs.
Safety zone	Provisions should be made for a buffer zone between the local habitation and the mine lease in the form of a green belt of suitable depth. Restricted entry, use of sirens and cordoning of the blasting area are some of the good practices to avoid accidents.
Workers health surveillance	Health survey programmes for workers and local community. Regular training and awareness of employees to be conducted to meet health and safety objectives
Mine inundation	Mine inundation may lead to a serious disaster if a river is flowing close to the mine pit. Hence a buffer space of a suitable width should be maintained, which should be followed by construction of embankments after considering high flood level. In underground mines, if the shaft is located in a low lying area or is vulnerable to flood risk, garland drains should be provided around the shaft.
Mine fires	Try to extinguish the new fires and try to isolate existing fires. Try to take a scientific approach to mining practices and give due consideration to proposed fire prevention plans. Carry out monitoring using new scientific techniques. Risk of explosions in underground mines can be managed by good mine planning, methane gas monitoring, good ventilation and controlling dust levels.

Industrial Pollution and Its Control

Volume II

Industrial Pollution and Its Control

Volume II

S. C. Bhatia
BE (Chemical), BITS Pilani & MBA

Edited by

Pramod Kumar
Assistant Professor & Head of Department – Biotechnology
Hindustan College of Science and Technology, (Mathura – UP)

and

Sarvesh Devraj
B.Tech (Mechanical), UPTU
M.Tech (Renewable Energy Engineering & Management), TERI University,
(Research Associate – TERI, New Delhi)

Published by Woodhead Publishing India Pvt. Ltd.
Woodhead Publishing India Pvt. Ltd.,
303, Vardaan House, 7/28, Ansari Road,
Daryaganj, New Delhi - 110002, India
www.woodheadpublishingindia.com

First published 2017, Woodhead Publishing India Pvt. Ltd.

Woodhead Publishing India Pvt. Ltd. ISBN: 978-93-85059-24-7
Woodhead Publishing India Pvt. Ltd. e-ISBN: 978-93-85059-82-7

Typeset by Asian Enterprises, Vasant Kunj, New Delhi
Printed and bound India by Replika Press Pvt. Ltd.

Contents

Volume I

Section IV

Pollution control in cement, ceramic and glass industry

23

Pollution control in cement industry

23.1 Introduction

The world has become increasingly aware of the need to preserve and conserve resources. Control of pollution is not only a legislative requirement, but has also become a tool for competitiveness. Those companies that choose to implement sound pollution prevention methods are likely to enjoy better business than those that do not, mainly because the former will generally be regarded as socially responsible. Therefore the technological advancement that is being witnessed everyday is an opportunity for industry to minimise waste and become greener and more lucrative. There is great potential for cement industries to minimise emissions.

The cement industry contributes significantly to the imbalances of the environment; in particular air quality. The key environmental emissions are nitrogen oxides (NO_x), sulphur dioxide (SO_2) and grey dust. Industrial plant smokestacks from cement and construction companies are some of the biggest contributors to poor air quality, especially in urban developments. The principal aim in pollution control in the cement industry is to minimise the increase in ambient particulate levels by reducing the mass load emitted from the stacks, from fugitive emissions and from other sources.

23.2 Manufacture of cement

Cement is produced from raw materials such as limestone, chalk, shale, clay and sand. These raw materials are quarried, crushed, finely ground and blended to the correct chemical composition. After the mining, grinding and homogenisation of raw materials, the process of calcination is followed by burning the resulting calcium oxide together with silica, alumina and ferrous oxide at high temperatures to form clinker; the clinker is then ground or milled together with other constituents (as gypsum, slag, etc.), to produce cement. The main stages in cement production are discussed below.

23.2.1 Quarrying (raw material acquisition)

Most of the raw materials used are extracted from the earth through mining and quarrying and can be divided into the following groups: lime, silica, alumina and iron. Limestone (calcium carbonate – $CaCO_3$) is the predominant raw material therefore most plants are situated near a limestone quarry or receive this material from a source via inexpensive transportation. The plant must minimise the transportation cost since one third of the limestone is converted to carbon dioxide (CO_2) during the pyro-processing and is subsequently lost. Quarry operations consist of drilling, blasting, excavating, handling, loading, hauling, crushing, screening, stockpiling and storing.

Raw materials preparation

Raw milling involves mixing the extracted raw materials to obtain the correct chemical configuration and grinding them to achieve the proper particle-size to ensure optimal fuel efficiency in the cement kiln and strength in the final concrete product. Three types of processes may be used: the dry process, the wet process, or the semidry process. If the dry process is used, the raw materials are dried using impact dryers, drum dryers, paddle-equipped rapid dryers, or air separators, before grinding, or in the grinding process itself. In the wet process, water is added during grinding. In the semidry process the materials are formed into pellets with the addition of water in a pelletising device.

Clinker burning

In pyro-processing, the raw mix is heated to produce cement clinkers. Clinkers are hard, grey, spherical nodules with diameters ranging from 0.32–5.0 cm created from the chemical reactions (sintering) between the raw materials. The pyro-processing system involves three steps: drying or pre-heating, calcining (a heating process in which calcium oxide is formed) and burning (sintering). The pyro-processing takes place in the burning/kiln department. The raw mix is supplied to the system as a slurry (wet process), a powder (dry process) or as moist pellets (semidry process). All systems use a rotary kiln and contain the burning stage and all or part of the calcining stage. For the wet and dry processes, all pyro-processing operations take place in the rotary kiln, while drying and pre-heating and some of the calcination is performed outside the kiln on moving grates supplied with hot kiln gases.

Cement grinding

This stage is also known as finish milling. Here the clinker is ground with other materials (which impart special characteristics to the finished product) into a fine powder. Gypsum and/or natural anhydrite are added to regulate the setting time of the cement. Other chemicals, such as those which regulate fluidity or air entrainment, may also be added. Material that has not been completely ground is sent through the system again.

Cement packaging and dispatch

The finished product is transferred using bucket elevators and conveyors to storage silos. Most of the cement is transported to customers in bulk by railway, trucks and in bags (normally 50 kg bags). Cement is mostly used in mortar and concrete in the construction industry.

23.3 Environmental impacts of cement manufacture

Cement manufacturing is a 'high volume process' and correspondingly requires adequate quantities of resources, that is, raw materials, thermal fuels and electrical power. The main environmental (air quality) impacts of the manufacture of cement in general are related to the categories discussed below.

23.3.1 Gaseous atmospheric emissions of CO_2, NO_x, SO_2, volatile organic compounds (VOCs) and others

Carbon dioxide is released during the production of clinker, a component of cement, in which calcium carbonate ($CaCO_3$) is heated in a rotary kiln to induce a series of complex chemical reactions. Specifically, CO_2 is released as a by-product during calcination, which occurs in the upper, cooler end of the kiln or a precalciner, at temperatures of 600–900°C and results in the conversion of carbonates to oxides. The simplified stoichiometric relationship is as follows:

$$CaCO_3 + \text{Heat} \rightarrow CaO + CO_2$$

Additional air pollutants emitted include such materials as sulphur oxides and nitrogen oxides generated from the kiln and drying processes. Sulphur dioxide is generated from the sulphur compounds in the ores and the combusted fuel and varies in amount produced from plant to plant. The efficiency of particulate control devices is inconclusive as the result of variables such as feed sulphur content, temperature, moisture and feed chemical composition, in addition to alkali and sulphur content of the raw materials and fuel. The combustion of fuel in rotary cement kilns generates nitrogen oxides from the nitrogen in the fuel and incoming combustion air. The amount emitted depends on several factors including fuel type, nitrogen content and combustion temperature. Both sulphur dioxide and some of the nitrogen oxide react with the alkaline cement and are removed from the gas stream. Volatile organic carbon compounds (VOCs) are a class of chemicals that are emitted directly to the air as a result of evaporation or another type of volatilisation. Sources include stored gasoline, stored solvents and other industrial chemicals and certain industrial processes. Incomplete combustion of fuels of many types is also an important source of VOC discharge to the ambient air.

The principal harmful effects of VOCs are toxicity, possible contribution to smog via photochemical reactions in the atmosphere and possible contribution to the 'greenhouse effect' and consequent global warming. Examples are Polychlorinated Dibenzodioxins (PCDDs) and Polychlorinated

Dibenzofurans (PCDFs) which comprise a family of 210 chemically related organic compounds containing from one to eight chlorine atoms. PCDDs and PCDFs are commonly and colloquially referred to as PCDD/Fs.

23.3.2 Dust

Dust emissions originate mainly from the raw mills, the kiln system, the clinker cooler and the cement mills. A general feature of these process steps is that hot exhaust gas or exhaust air is passing through pulverised material resulting in an intimately dispersed mixture of gas and particulates. The nature of the particulates generated is linked to the source material itself, that is, raw materials (partly calcined), clinker or cement (Karstensen). Dust emissions have been linked to respiratory problems such as Tuberculosis.

23.3.3 Fugitive emission (FE)

By definition, fugitive emission are diffuse and typically come from many small sources as opposed to well-defined specific stack emission. Fugitive Emission are also to diffuse nature in the sense that these are low level emissions from a large area, as in the case of wind erosion from storage piles or dust from inplant vehicular traffic. Many of the fugitive emissions are intermittent as well, e.g., unloading railcars, slag dumping processes, mechanical turbulence caused due to traffic movement, etc.

Because of the intermittent and diffuse nature of fugitive emission, their measurement is difficult. Briefly describing the monitoring techniques of fugitive emission sampling can be: (i) quasi-stack sampling where a collection hood near the source is sampled, (ii) roof monitor sampling in which a large enclosed area as a building is sampled and (iii) upwind-downwind sampling where fugitive emissions are back calculated from particulate measurement around the source perimeter.

Accuracy of such methods is high enough for such conducive analysis. Because of suitability to Indian industrial process conditions, techniques of upwind-downwind sampling can be adopted for FE monitoring from different activity sources of a cement plant under study.

23.3.4 Stack emissions

Stack emission can be monitored for particulate matter and other relevant pollutants, namely limestone and coal crushers, coal mills, kilns, boilers, cement mills and packing plant. However, for analysis in the present context only kiln emissions are taken. This is because emission standards promulgated in India at this stage, are mainly designed for controlling emissions from kilns alone as being the single major source in cement mills.

Sampling and analytical procedures followed for particulate emission sampling should be in accordance with EPA technique using indigenous sampling train. The sampling ports in the test ducts should be made according to standard norms. Isokinetic sampling can be attained through dry gas meter and rotameter assembly to cross-check the duct velocity at each traverse sampling point.

Dust samples can be collected on cellulose or allundum thimbles at each traverse point. The thimble dust thus collected, was determined gravimetrically taking adequate precaution for moisture elimination. The sampling period varies between 5 to 10 minutes at each traverse point of all the kiln ducts.

23.3.5 Bad odour

Foul smell is sometimes a direct result of the gases emitted during cement manufacturing. Moreover, since cement manufacture has life threatening impacts to plants and animals, the manufacturing process then directly and indirectly gives rise to offensive smells as the dead plants and animals decay.

23.4 Air pollution control technologies and equipment

Controlling particulate emissions from sources other than the kiln usually entails capturing the dust using a hood or other partial enclosure and transporting it through a series of ducts to the collectors. The type of dust collector used is based on factors such as particle size, dust loading, flow rate, moisture content and gas temperature. The best disposal method for collected dust is to send it through the kiln creating the clinker. However, if the alkali content of the raw materials is too high, the dust must be discarded, or must be pre-treated before introduction into the kiln. The highest allowable alkali content is 0.6% (as sodium oxide).

23.4.1 Flexible pulse jet filters

Raw gas enters the filter compartments via inlet ducts equipped with guide vanes that distribute the gas uniformly across the filter bags. This arrangement creates a downward gravimetric gas flow along the filter bags, precipitating the dust into the hopper below. In contrast to filters where raw gas enters through dampers located in the hoppers, the design of the flexible pulse jet filter gas distribution system prevents the creation of high can velocities (or vertical, upward gas flow). High gas velocities prevent fine particulate from settling into the hoppers during on-line cleaning cycles. The raw gas is filtered by the fabric from the outside and the clean gas exits at the top of the bag. The fan is located on the clean gas outlet side of the filter.

23.4.2 Electrostatic precipitators

The cleaning process in the ESP consists of three successive distinct and inter-related stages, viz. charging, collecting and removal of the particles. The particles on reaching the collecting electrodes, neutralise the precipitate and thereafter are removed by means of rapping or vibration in the case of dry precipitators. The gas composition affects the ESP chiefly through difference in ion mobility. Increase in gas temperature also increases the ion mobility and thus decreases the dielectric strength of the gas. Another important factor for electrostatic precipitation is the resistivity of the dust. If the resistivity is too low the electrical charges on the particle will leave them very rapidly thereby causing re-entrainment. If the resistivity is too high, the field strength and the voltage gradient on the precipitated layer will also be too high and this will result in a breakdown of the dust layer on the collected particles. This process is known as 'Back Corona' or back ionisation, which has to be taken care in the operation of ESP in case of high resistivity dust.

Electrostatic precipitators use electrostatic forces to separate the dust from the exhaust gas. By means of discharge electrodes, the dust particles are negatively charged and can be separated on corresponding collecting electrodes. The particles are then discharged from the collecting electrodes to dust hoppers by electrode rapping.

23.4.3 Wet scrubbers

In a wet scrubber, the polluted gas stream is brought into contact with the scrubbing liquid, by spraying it with the liquid, by forcing it through a pool of liquid, or by some other contact method, so as to remove the pollutants. Scrubbers can be designed to collect particulate matter and/or gaseous pollutants. Wet scrubbers remove dust particles by capturing them in liquid droplets. Wet scrubbers remove pollutant gases by dissolving or absorbing them into the liquid.

23.4.4 Ordinary bag house method

This is a filtration method and is one of the oldest and most efficient methods of particulate control. The most commonly-used filtration device is known as a bag-house and consists of fabric bags through which the air stream is directed as shown in Fig. 23.1. Particles become trapped in the fibre mesh on the fabric bags, as well as the filter cake which is subsequently formed.

The dust is collected on the outside of the bags and a high filtration velocity is admissible mainly due to the effective cleaning mechanism utilised through the pulse given to the bags being fast, strong and short, which gives a good cleaning effort. Very good outlet emission has been achieved even in India by using the modern-day bag filters in cement plants.

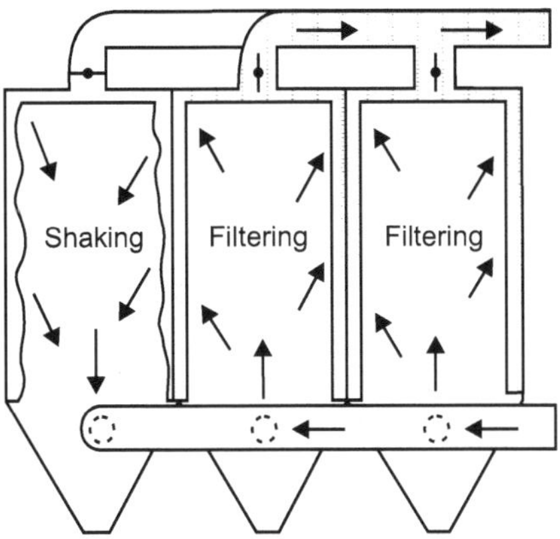

Figure 23.1: Bag-house filtration method.

23.5 Computer modelling for quantification of carbon dioxide (CO_2) emissions

Database systems can be taken advantage of, together with visual basic applications to quantify or estimate and display CO_2 emissions based on composition of the $CaCO_3$ and amounts of lost cement kiln dust. Estimating emissions generally involves two emission factors: an emission factor for clinker production and an emission factor for Cement Kiln Dust (CKD) production.

23.5.1 Clinker emission factor

The clinker emission factor is the product of the fraction of lime in the clinker multiplied by the ratio of the mass of CO_2 released per unit of lime. This factor can be estimated as illustrated below:

$$EF_{clinker} = \text{Fraction CaO} \times 0.785$$

The multiplication factor (0.785) is the molecular weight ratio of CO_2 to CaO in the raw material mineral calcite ($CaCO_3$), from which most or all of the CaO in clinker is derived. The result gives the amount of carbon dioxide (in tonnes) emitted per tonne of clinker. Databases and visual basic can be used to process and store data and to provide a graphical user interface for monitoring the emissions as production runs. Any deviations above company and legal limits can be easily observed and corrected.

23.5.2 Cement kiln dust (CKD)

CO_2 is also emitted during the calcination of Cement Kiln Dust (CKD) in the kiln. CKD is a by-product of the kiln process and a portion of the CKD is placed back in the kiln and incorporated into the clinker. The remaining portion is lost – placed in a landfill or used for other purposes. The lost CKD represents additional CO_2 emissions not accounted for in the clinker emissions estimate. The recommended method to estimate the additional CO_2 emissions from the lost CKD is to multiply an emission factor by the amount of lost CKD. The CO_2 from the lost CKD is generally equivalent to about 2–6% of the total CO_2 emitted from clinker production.

23.6 Methodology

Thermocouples and Opacimeters are the main apparatus that were used to measure/quantify the emissions. Thermocouples were used to measure exit gas temperature whilst opacimeters were used to quantify particulates.

23.6.1 Opacimeter working principle

Two types of opacimeters are used – online and an offline. The basic working principle of an opacimeter is that exhaust gas from the chimney is directed into the opacimeter chamber. Opacity is an optical property that refers to the ability to stop light from being transmitted. The basic principle of the opacimeter is that light is emitted from a light source and a sensor some distance away registers the intensity of the light. If a sample with opacity more than 0% is placed in between the light source and the sensor the measured light intensity will decrease. Through calibration the measured intensity can be correlated to the opacity of the sample. When there is a perfect transparent matter, e.g., air, in between the opacity is 0% and in the opposite case, where no light is transmitted, the opacity is 100% .

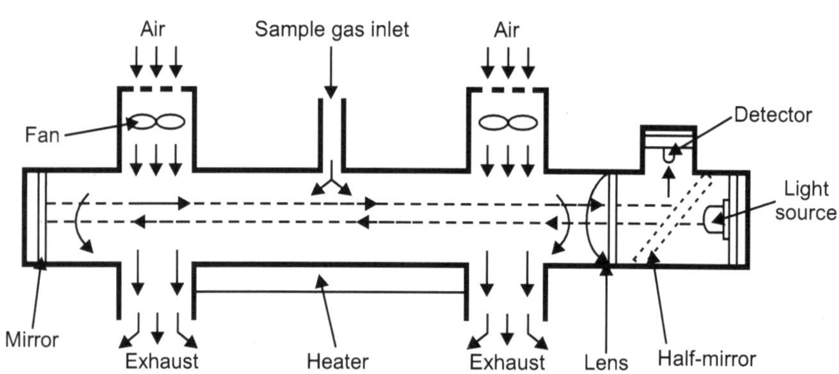

Figure 23.2: Opacimeter schematic.

23.6.2 Thermocouple working principle

A thermocouple is a device made by two different wires joined at one end, called the junction end, or measuring end. The two wires are called thermo-elements or legs of the thermocouple. The other end of the thermocouple is called the tail end or reference end. This arrangement is shown in Fig. 23.3 below. Because of the temperature difference between junction end and tail end a voltage difference can be measured between the two thermo-elements at the tail end. Thus, the thermocouple is a temperature voltage transducer.

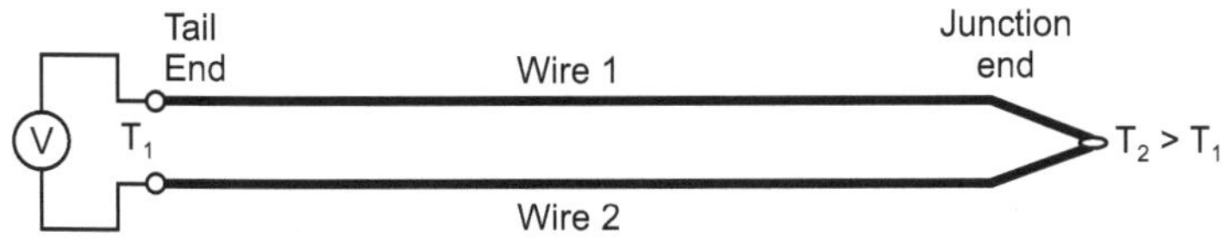

Figure 23.3: Diagram of thermocouple.

23.6.3 Gas conditioning towers

Exhaust gases leaving the kiln is at very high temperature which destabilises the ecosystem by killing creatures which cannot survive such high temperature. Conditioning towers in the cement industry are used for cooling the exhaust gases from the kiln before it is fed into the precipitators. The exhaust hot gases are fed from the top of gas cooling tower and are cooled by injected water; the gases are drawn through the tower by a fan. Conditioning towers in the cement industry treat exhaust gases with temperatures between 300 and 400°C, cooling them down to 150°C. After the sampled points have been analysed and are found to be above the required concentration level, the source of emisions is investigated and an appropriate technical solution is recommend. The random sampling can be scheduled and the process becomes continuous throughout the life span of the plant.

23.6.4 Stage by stage pollution control techniques for cement industry

The various techniques that can be used to control pollution in the cement manufacturing process can be easily identified by considering the major process stages as outlined below.

23.6.5 Raw materials acquisition and handling

During raw material acquisition the primary air pollutant emitted is particulate matter. Particulate matter is emitted from the quarrying, handling, loading, unloading and transport of raw materials.

The following methods are used to control particulate emissions generated from the quarry and handling of raw materials:

1. Fabric filters (pulse-jet or reverse-air/shaker).
2. Equipment enclosures.

3. Water sprays – to suppress dust.
4. Mechanical collectors.
5. Chemical dust suppressants.
6. Paving (where possible).

Dust that is collected by these means is restored to the process. This therefore means that the reduce, reuse, recycle and recover (4Rs) techniques can be employed to minimise and manage waste during raw materials acquisition.

23.6.6 Clinkerisation (pyro-processing)

The main pyro-processing system emissions are nitrogen, carbon dioxide, water, oxygen, nitrogen oxides, sulphur oxides, carbon monoxide and hydrocarbons. Cement kiln dust is also produced. The cement kiln itself has been designated as Best Available Control Technology (BACT) for the control of SO_2. The highly alkaline conditions of the kiln system enable it to capture up to 95% of the possible SO_2 emissions. However, if sulphide (pyrites) is present in the kiln feed, this absorption rate can decline to as low as 50%. Therefore, sulphur emissions can be decreased through careful selection of raw materials. Options are available to move from coal fuel to oil or gas fuel as these will result in considerably lower sulphur oxides and carbon monoxide emissions. Possible areas of exploitation for the control of NO_x are as follows:

1. Stable kiln operation (reduces long term NO_x emissions).
2. Staged combustion for precalciner kilns.
3. Recirculation of the flue gas (oxygen deficient air in the rotary kiln).
4. Alternative/low-nitrogen fuels.

23.6.7 Cement loading and despatch

In the shipping department particulate matter is emitted from the silos and the handling and loading operations. Active and passive fabric filters can be used to collect this dust. To ensure dust-free loading onto the transport vessel, a flexible loading spout consisting of concentric tubes is used. The outermost tube seals the delivery spout to the transport vehicle. The product is then delivered through the inner tube and displaced air drawn up the outer tube to a filter. At distribution terminals, fabric filters are again used and the collected dust is returned to the product.

23.7 Recommendations

1. Rapid flue gas quenching or other measures to minimise post-furnace particulate residence time in the critical temperature zone.
2. Use of formation inhibitors.

3. End-of-pipe flue gas cleaning techniques for removal or catalytic decomposition of emissions.
4. Measuring equipment and instruments such as opacimeters should be upgraded to enhance accuracy of measurements.

23.8 Further opportunities for improvement of emission in cement industry

Carbon monoxide emission (up to 0.1 ppm) is due to incomplete combustion during clinkerisation. There is need to look at alternative fuel composition and raw materials residence time in the kiln. The levels of SO_x emissions for cement industry should be below 5% shows an indication of a good furnace/kiln design. However there is still need to completely eliminate SO_x emissions into the natural atmosphere. Catalytic reduction of SO_x emissions at end of pipe can also be taken advantage of to further reduce such emissions.

24

Pollution control in ceramic industry

24.1 Introduction

A ceramic is an inorganic, nonmetallic solid material comprising metal, nonmetal or metalloid atoms primarily held in ionic and covalent bonds. Ceramics is a diverse industry and contains several categories of products, including sanitary ware, refractories, cement, advanced ceramics and ceramic tiles. Ceramic products like crockery, sanitary ware, tiles, etc., play a very important role in our daily life. This is because, apart from their decorative look, ceramic products are primarily hygiene products. This is also one of the chief reasons for their wide usage in bathrooms and kitchens in modern households to medical centers, laboratories, milk booths, schools, public conveniences, etc.

24.2 Ceramic manufacturing process

Figure 24.1 shows the general process flow diagram of ceramic manufacturing process.

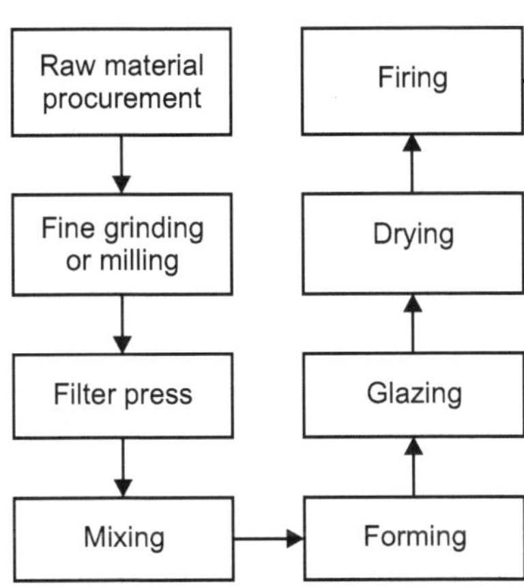

Figure 24.1: Process flow diagram of ceramic manufacturing process.

Raw materials procurement and weighing: The raw materials used in the manufacture of ceramics range from relatively impure clay materials mined from natural deposits to ultrahigh purity powders prepared by chemical synthesis. Naturally occurring raw materials used to manufacture ceramics include silica, sand, quartz, flint, silicates and aluminosilicates. The first step in the process is to weigh the raw materials required to manufacture a ceramic tile including all types of frit, feldspar and various clays. All the raw materials are accurately weighed, so that the quality of the product can be stabilised.

Fine grinding and milling: The basic beneficiation processes include crushing, grinding and sizing or classification. Primary crushing is used to reduce the size of coarse materials, such as clays, down to approximately 1 to 5 centimeters. The most common types of crushers used are jaw crushers, cone crushers, gyratory crushers and roll crushers. Secondary crushing or grinding reduces particle size down to approximately 1 millimeter in diameter. Fine grinding or milling reduces the particle size down to as low as 1.0 micrometer in diameter. Ball mills are the most commonly used piece of equipment for milling.

Blending and kneading: Technical know-how is the most important in this process second to the firing process and the final product depends on this blending process. The yield, quality and workability are also affected by this process. The major points in the kneading process is how to knead the material uniformly with water and how to mix various materials.

Forming: Metal molds and plaster molds patterns are used in most cases of ceramics forming. Since the ceramics will shrink when fired, it is necessary to take it into account when determining the shape of these molds. Thus, special production know-how is required to produce a mold with a complicated shape.

Drying: The drying process is an important process affecting the product yield and requires a longer time than many of other processes. Inappropriate drying may cause the products to be cut or broken; this makes it essential to find out appropriate conditions of temperature, humidity and time. The waste heat of the kiln is generally used as the heat source for drying. If this is insufficient, the hot air generator is also used to make up for the insufficiency.

Firing: Firing for the pottery and refractories is actually sintering which causes the crystals to be combined with one another. Firing is terminated before the material gets molten. In contrast, temperature is raised until the fired product is molten completely; this is called glass. A kiln is used to manufacture the former product, while a furnace is used to produce the latter.

24.3 Environmental issues

Depending on the specific production processes, plants manufacturing ceramic products cause emissions to be released into air, water and land (waste). Additionally, the environment can be affected by noise and unpleasant smells. The type and quantity of air pollution, wastes and waste water depend on different parameters.

These parameters are, e.g., the raw materials used, the auxiliary agents employed, the fuels used and the production methods:

1. Emissions to air: Particulate matter/dust, soot, gaseous emissions (carbon oxides, nitrogen oxides, sulphur oxides, inorganic fluorine and chlorine compounds, organic compounds and heavy metals) can arise from the manufacture of ceramic products.
2. Emissions to water: Process waste water mainly contains mineral components (insoluble particulate matter) and also further inorganic materials, small quantities of numerous organic materials as well as some heavy metals.
3. Process losses/waste: Process losses originating from the manufacture of ceramic products, mainly consist of different kinds of sludge, broken ware, used plaster molds, used sorption agents, solid residues (dust, ashes) and packaging waste.
4. Energy consumption/CO_2 emissions: All sectors of the ceramic industry are energy intensive, as a key part of the process involves drying followed by firing to temperatures of between 800 and 2000°C. Today natural gas, liquefied petroleum gas (propane and butane) and fuel oil are mainly used for firing, while heavy fuel oil, Liquefied Natural Gas (LNG), biogas/biomass, electricity and solid fuels (e.g., coal, petroleum coke) can also play a role as energy sources for burners.

24.4 Emissions – general considerations

24.4.1 Emissions to air

Dust (particulate matter)

The processing of clays and other ceramic raw materials inevitably leads to dust formation – especially in the case of dry materials. Drying, (including spray drying), comminution (grinding, milling) screening, mixing, conveying can all result in a release of fine dust, necessitating dust extraction equipment. Some dust also forms during decorating and firing of the ware and during machining or finishing operations on fired ware. Dust is one of the main pollutants, in quantitave terms, from the ceramic processes.

Gaseous emissions

The gaseous compounds released during drying, calcining and firing are derived from the raw materials as well as fuels contribute gaseous pollutants. Fluorine compounds represent one of the main typical pollutants from ceramic processes. Gaseous emissions relevant to the ceramic industry are discussed below.

Sulphur dioxide and other sulphur compounds

The concentration of SO_x (mainly SO_2) in flue-gases is closely related to the sulphur content of the raw material and of the fuel. Ceramic materials may contain sulphur in the form of pyrite (FeS_2), gypsum and other sulphates and organic sulphur compounds. Gaseous fuels are virtually sulphur-free, but solid fuels and fuel oils contribute sulphur oxides on combustion. It should be noted that basic compounds from raw materials (e.g., CaO formed by dissociation of $CaCO_3$ during firing) can reduce sulphur emissions by reacting with sulphur oxides. The reaction products are retained within the body of the ware.

Oxides of nitrogen and other nitrogen compounds

NO_x is mainly produced by thermal 'fixation' of nitrogen and oxygen from the combustion air. This reaction is favoured by high temperatures (>1200°C) and by excess oxygen. Fixation can occur within hot flames, even when the actual kiln temperature is below 1200°C. Nitrogen compounds present in fuels (mainly solid or liquid types) or in organic additives, form NO_x during combustion at much lower temperatures.

Carbon monoxide (and carbon dioxide): Carbon monoxide, CO, arises from the combustion of organic matter in the ceramic body – especially under low oxygen conditions. It may also be formed by the reaction of 'fixed carbon' in the body with carbon dioxide (CO_2) released by the thermal dissociation of alkali and earth alkali carbonates, for instance calcium or magnesium carbonate during firing:

$$CaCO_3 \rightarrow CaO + CO_2$$

$$C + CO_2 \rightarrow 2CO$$

(Carbon dioxide forms as above and is the main oxide of carbon formed during the combustion of fossil fuels).

Volatile organic compounds (VOCs): Ceramic raw materials may themselves contain organic matter and a wide range of organic materials are added in the form of binding agents, pore-forming agents, drying aids, adhesives, body fuels, burnout additives, etc.

During the early heating process, carbonisation of organic compounds may occur with the release of a complex range of VOCs.

Metals and their compounds: The heavy metal content of most ceramic raw materials is very low and causes no emission problems. Exceptions occur in the case of ceramic pigments and glaze materials, but present practice is to use coloured compounds ('stain' containing pigments), which are stable at high temperatures and inert in silicate systems, where the metallic oxides have a stable crystal structure like spinel or zircon. These types of glaze are also subject to extremely short firing cycles, minimising the risk of volatility.

Heavy fuel oils and solid fuels may contain low levels of nickel and vanadium, but tests have shown these elements to be largely absorbed by a reaction with the product during firing.

Chlorine and its compounds: Most clays contain trace levels of chloride, often derived from an original marine formation, but also additives or water containing chloride are possible sources of hydrochloric acid (HCl) emissions (e.g., the Cl content of water added during the preparation of the raw materials may be in the range of 50–100 mg/L or even higher). Due to the decomposition of mineral salts containing chloride at temperatures in excess of 850°C and of organic compounds containing chloride at temperatures of between 450 and 550°C, HCl occurs during the firing process in the flue-gases of a kiln.

Fluorine and its compounds: Nearly all natural raw materials contain fractional amounts of fluoride (which readily substitutes for OH groups in clays and hydrous minerals). Hydrofluoric acid (HF) comes mainly from the decomposition of these fluorosilicates present in clay material.

The emission of HF takes place in accordance with two different mechanisms:

1. By direct decomposition of the fluoride minerals which depends heavily on the type of clay (e.g., smectite at temperatures from 550°C, illite at temperatures from 750°C, apatite at temperatures from 600 to 700°C).
2. By decomposition of CaF_2 at temperatures in excess of 900°C, the reaction is favoured by the presence of water vapour.

Emissions to water

Water is a very important raw material in the ceramic manufacturing industries, but the amount used varies greatly between sectors and processes. The water added directly to ceramic body mixes does not lead to a waste water problem, as it is subsequently evaporated into the air during the drying and firing stages. Process waste water is generated mainly when clay materials are flushed out and suspended in flowing water during the manufacturing process and equipment cleaning, but emissions to water also occur during the operation of wet off-gas scrubbers.

Process waste water occurs in small quantities in the manufacture of bricks and roof tiles, vitrified clay pipes and refractory products, if surface treatment such as glazing and engobing, or wet grinding is carried out. Additional amounts of water arise in the cleaning process of the mixing, engobing and glazing units and of the molds. In expanded clay aggregates production, usually no process waste water is generated, only the cooling water for the kiln system passes through an oil separator and is returned to the cooling circuit. In the wall and floor tiles, table- and ornamentalware, sanitaryware, technical ceramics and inorganic bonded abrasives sectors, process waste water occurs mainly as cleaning water in preparation units, in casting units, in the glazing and decoration process or as grinding water in subsequent treatment.

Effluents which arise as a result of compound processing and equipment cleaning usually contain the same raw materials and auxiliary materials as those used in the process concerned. As a rule, these compounds are water insoluble.

The waste water in the process mostly shows turbidity and colouring owing to the very fine suspended particles of glaze and clay minerals. From a chemical point of view, these are characterised by the presence of:

1. Suspended solids: clays, frits and insoluble silicates in general.
2. Dissolved anions: sulphates.
3. Suspended and dissolved heavy metals, e.g., lead and zinc.
4. Boron in small quantities.
5. Traces of organic matter (screen printing vehicles and glues used in glazing operations).

Process losses/waste

Process losses/waste originating from the manufacture of ceramic products, mainly consist of the following materials:

1. Different kinds of sludge originate from the process waste water treatment facilities of water arising in the processes involved in cleaning body preparation, glaze preparation and application equipment and also from wet grinding. The quantity and composition of sludge varies considerably as, besides arising in the different production processes, a great variety of raw materials tends to be used (different glazes, frits, clays, etc.), even in the same facility, which produce significant fluctuations in sludge composition.
2. Broken material/ware arises from shaping, drying, firing, subsequent treatment and broken refractory material.
3. Dust arises from off-gas cleaning units.
4. Used plaster molds arise from shaping processes.
5. Used sorption agents (granular limestone, limestone dust) arise from flue-gas cleaning systems.
6. Packaging waste (plastic, wood, metal, paper, etc.), arises in the packaging step.
7. Solid residues, e.g., ashes arise from firing with solid fuels.

Parts of the accumulated process losses mentioned above, can be recycled and reused within the plant due to product specifications or process requirements. Materials, which cannot be recycled internally, leave the plant to be used in other industries or to be supplied to external waste recycling or waste disposal facilities.

Emissions of noise

Noise occurs in the several steps during the manufacturing processes of ceramic products, mainly by operating noise generating machinery and by carrying out noisy working procedures.

24.5 Consumption – general considerations

24.5.1 Energy consumption in ceramic manufacturing

Two types of energy are used in the ceramic industry, electrical energy and chemical energy. The electrical energy is used in two different ways, mechanical energy when used in the motor and fan of the machine and thermal energy when used to heat the kilns and furnaces. The chemical energy of petroleum fuel is all converted into thermal energy through combustion reaction. Energy used in the ceramic industry is predominantly occupied by petroleum energy. The drying and firing processes in ceramic production use much energy, furthermore, electric machineries (e.g., motor, fan) use electric energy. Of these processes, the firing process is the greatest energy consumer.

Firing is an essential process in the ceramic production process and heat (combusted exhaust gas, heat when cooling, namely, waste heat) produced in this process is used as a heat source to dry the formed product or as the secondary air for combustion, thus leading to energy conservation.

24.5.2 Drying process

The drying process in the ceramic industry is the greatest energy consumer second to the firing process. Drying means loss of moisture from the surface of the substance by evaporation and the drying speed depends on the temperature and humidity. When the substance is dried and moisture is lost, particles are put close to each other, resulting in shrinkage. The constant rate drying period is the period when balance is kept between moisture shifting from inside the substance to the surface and moisture evaporation from the surface. The first falling drying rate period is the period when moisture shifts from inside the substance to the surface with reduced moisture evaporating from the surface. The second falling drying rate period is the period when evaporation takes place inside and vapour diffuses to the surface, without moisture shifting from inside the substance. Cracks due to drying is caused during the period when the green body shrinks that is, when the temperature gradient is steep under drying conditions between the constant rate drying period and first falling drying rate period or when the temperature is excessively low. It is essential to have a correct understanding on these conditions before starting the drying process.

The rate of drying is controlled by:

1. Temperature: The higher the temperature, the faster the drying.
2. Relative humidity: The lower the humidity, the faster the drying.
3. Airflow: The higher the airflow, the faster the drying.
4. Product design features: Surface area and thickness both affect drying.

These parameters largely determine the size of the dryer and technology used. The high latent heat of water means that large amounts of energy are sometimes used inefficiently in drying. Energy efficiency involves process optimisation, monitoring and maintenance.

24.5.3 Water consumption

Water is used in virtually all ceramic processes and good quality water is essential for the preparation of clays and glaze slips; clay bodies for extrusion and 'muds' for molding; preparation of spray dried powders; wet grinding/ milling; washing operations. Water is used for different purposes: basically, as a raw material for the body and its preparation, as a raw material for glazes, as a washing fluid and as a cooling medium. Good quality water is required for glazes, washing and cooling purposes. Low quality water – like untreated process waste water from washing operations in the glaze section – can be suitable for the body. Process waste water reuse techniques take these different qualities into account.

24.5.4 Raw materials consumption

A vast range of raw materials is consumed by the ceramic industry. These include the main body forming materials, involving high tonnages and various additives, binders and decorative surface-applied materials which are used on a lesser scale.

24.6 Process waste water

24.6.1 Water used as a raw material

Water is a very important raw material in the ceramic manufacturing industries, but the amount used varies greatly between sectors and processes. The water added directly to ceramic body mixes does not lead to a waste water problem, as it is subsequently evaporated into the air during the drying and firing stages. Process waste water is generated mainly when clay materials are flushed out and suspended in running water during the manufacturing process steps.

24.6.2 Water used as a heat exchange vehicle

Water performs this function in cooling hydraulic systems and compressors, etc. Water used for this purpose must be clean and show low hardness to prevent scaling in the heat exchangers. The water used can circulate in closed circuits after simple cooling and/or cleaning operations and thus, water consumption corresponds to the amount of evaporated water. Because water in closed circuits for cooling operations is, in most cases, chemically conditioned to avoid corrosion or fermentation of organic matter, treatment with production waste water is usually not possible.

24.6.3 Water used as a scrubbing agent

Water is also used in off-gas scrubbers (wet flue-gas cleaning systems and wet dust separators). In these systems, recycled process waste water can be used, which has been treated by a simple physical procedure (settling with or without prior chemical treatment) and can be circulated or treated again.

24.6.4 Water used as a cleaning agent

Water is used to clean the facilities, especially raw material preparation units, molds and other casting units, glazing lines, engobing and other decoration units. Cleaning is the operation in which most water is used and which requires suitable management to achieve savings and avoid process waste water. Water consumption can be reduced if the water is treated and reused several times in cleaning.

24.6.5 Objectives and solutions for the reduction of process wastewater (emissions and consumption)

The objectives for treating the water arising in ceramic manufacturing processes involve lowering water consumption and achieving minimum process waste water emissions. To reach these objectives, process optimisation measures and process waste water treatment systems are employed.

Process optimisation

Minimising water consumption is fundamental and to achieve this reduction, the following process optimisation measures can be implemented:

1. Acting on the water circuit, installing automatic valves that prevent leaks of water when it is no longer needed.
2. Installation of a high pressure system in the plant for cleaning purposes (or high pressure cleaning equipment).
3. Switching from wet off-gas cleaning systems to alternative, non-water consuming systems (cleaning by dry off-gas cleaning systems).

4. Installation of '*in situ*' waste glaze collection systems.
5. Installation of slip conveying piping systems.
6. Separate collection of process waste water streams from different process steps.
7. Reuse of process waste water in the same process step, in particular repeated reuse of the cleaning water after suitable treatment.

Systems of process waste water treatment

The main process waste water treatment systems are as follows:

Homogenisation: Homogenisation tanks are used to obtain a consistent composition in the water to be treated and suppress, as far as possible, problems relating to variations in the constituents. Using such tanks yields improvement in all subsequent treatments, as the resulting homogeneity facilitates control of product additions and consistency in the operating facilities.

Aeration: This is a physical process that is frequently used in water treatment for different purposes, such as oxidation of the materials to facilitate subsequent flocculation, oxygenation of the organic compounds present in the process waste water, elimination of odours, etc. Aerating equipment may involve surface stirrers or turbines.

Sedimentation (settling): This is the partial separation of solid particles from a liquid by gravity. There are various types of settling tanks; these may be rectangular, round or lamellar.

Filtration: Filtration involves the separation of suspended solids from a liquid, by putting the suspension through a porous medium that retains the solids and allows the liquid to flow through. The types used in the ceramic industry are in-depth filters, filter presses and rotating vacuum filters.

Activated carbon absorption: This treatment is based on carbons ability to strongly absorb organic molecules present in water. This is a very suitable system for removing nonbiodegradable organic substances.

Chemical precipitation: This is a process for eliminating different dissolved elements by precipitation as insoluble compounds, using reagents such as lime.

Coagulation and flocculation: The purpose of this treatment is to break up colloidal suspensions and produce particle agglomeration, e.g., by use of alums or polyelectrolyte and/or a combination of lime and metal salts.

Ion exchange and reverse osmosis: These processes serve to remove boron from the cleaning water coming from the glaze and application sections. Reverse osmosis is also applied for reducing the amount of process waste water for discharge.

By applying these measures or a combination of them, significant reductions of process waste water emissions and lower water consumption can be achieved.

Cross-media effects

Sedimentation/filter residues have to be disposed of, if no reuse is possible (especially if flocculation, precipitation or absorption agents are used).

Operational date and applicability

Process wastewater treatment systems can be applied in all sectors of the ceramic industry. Figure 24.2 shows process wastewater purification plant.

1. If the process waste water is to be reused in the body preparation process, no treatment will in principle, be required, but a homogenising tank will be needed to ensure keeping the most consistent possible characteristics.
2. If water is to be reused in cleaning the facilities, water quality will need to be higher, so that sedimentation is required, following by aerating, with or without subsequent chemical treatment to remove odours.

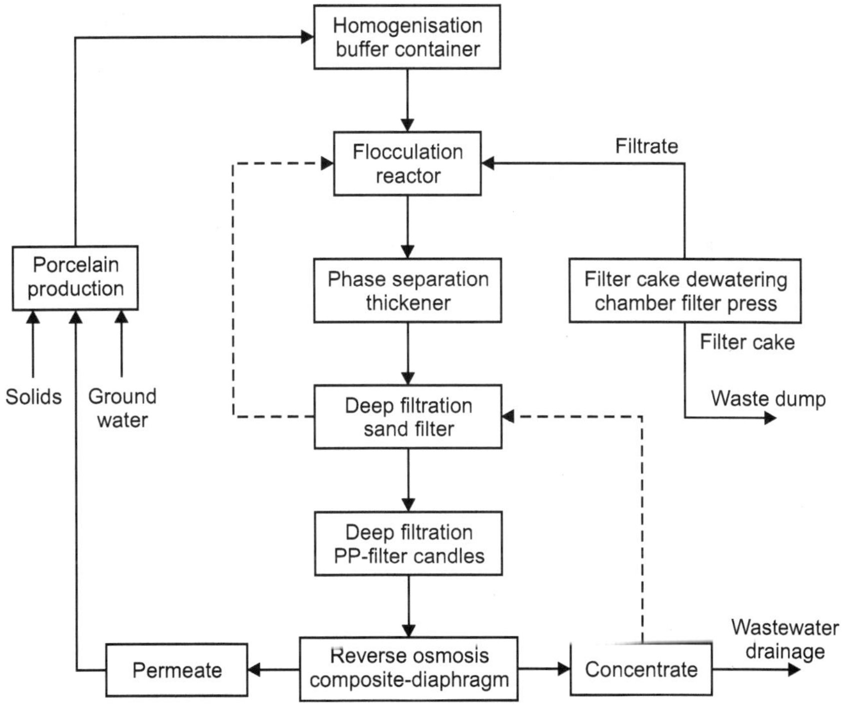

Figure 24.2: Flow diagram of a process wastewater purification plant.

3. Excess process waste water which has to be disposed of externally normally requires a combination of homogenisation, flocculation, sedimentation and filtration. Also subsequent reverse osmosis is applied for reducing the amount of discharged process wastewater.

As an example, the above figure shows a process wastewater purification plant of a porcelain tableware manufacturer which consists of a combination of the process waste water treatment systems mentioned above.

Economics

If the process waste water has to be disposed of externally, the compliance with the limits specified in the existing laws requires an expensive combination of several treatment steps including sedimentation, chemical precipitation, flocculation and a final process, based on either ion exchange or reverse osmosis, as mentioned above.

If the cleaned water from the process waste water treatment systems can be reused, this leads to reduced fresh water costs. With a combination of process wastewater recycling/reuse measures and process optimisation measures for minimisation of water consumption, costs associated with the disposal of waste materials from the above treatments can be saved.

24.7 Process losses/waste

24.7.1 Sludge arising in the manufacture of ceramic products

Sludge recycling systems

Recycling involves sludge reuse in the ceramic bodies as is the case with aqueous suspensions and slips containing ceramic materials, for example for the manufacturing of wall and floor tiles.

Sludge recycling systems can be easily implemented in facilities with raw materials preparation by wet milling, as the sludge can be directly used without needing any subsequent treatment, or just simple physical or physicochemical treatments, with the additional advantage of being able to use the water that it contains as milling water. If a dry body preparation process is involved, although the sludge addition is no problem, management is more complicated as the sludge needs to be dried first. If sludge has to be dried for reuse, energy consumption will rise. Attention should also be paid to possible negative consequences from the fact that pollutants coming from a given phase/section of the process/plant, such as glaze preparation and application, are introduced in a different phase/section, such as ceramic body preparation. The controls of gaseous emissions and process waste water represent a suitable solution to prevent such negative consequences.

Sludge reuse in other products

Sludge may be reused in sectors of the ceramic industry which are different from the ceramic sector generating the sludge, because this may involve technically interesting or economically advantageous solutions. For instance, sludge from the manufacture of houschold ceramics or sanitaryware can be used as a raw material/additive input in clay block making and in the manufacture of expanded clay aggregates.

This leads, on the one hand, to the avoidance of waste and, on the other hand, to raw material savings.

If sludge has to be dried for reuse, energy consumption will rise. Attention should also be paid to possible negative consequences from the fact that pollutants coming from a given phase/section/sector of the process/plant/industry, such as glaze preparation and application in the manufacture of household ceramics, are introduced in a different phase/section/sector, such as body preparation in a brickworks. The controls of gaseous emissions and process waste water represent a suitable solution to prevent such negative consequences.

24.7.2 Solid process losses/solid waste

Dust collected while loading, unloading, conveying, mechanically handling and processing raw materials can normally be reused as raw material. For instance, in the storage of raw materials, any filter dust can be fed back directly to the production process or to the silo, when a local air cleaning by a silo top-filter is used. But the direct feedback may not be possible when central dedusting units are operated, because of the mixing of different raw materials.

Also, process losses, which are generated before firing, can normally be reused as raw materials by adding them to the raw material mixture. Cuttings, used plaster molds and also dust are the main process losses arising from the shaping process. The cut pieces from the press are often pushed from the pressing table, collected under the press and directly fed back to the material storage of the press. Furthermore, dust or cut pieces are often fed back as batch components in a casting slip or in stoneware masses. Used plaster molds can be reused in the cement industry or, after crushing and milling, partly in the fertiliser industry.

Dust originating from a flue-gas cleaning system can only be reused under certain circumstances, because this dust can contain high concentrations of sulphur and fluorine. When added to thc raw material mixture, this dust can result in increased emissions of HF and SO_x during the firing process. Also dust from a flue-gas cleaning system can contain adsorbent particles like lime – for instance, if an adsorption system containing lime is applied – which influences the product properties. Therefore, a separate conduction of different off-gas streams facilitates the optimal utilisation of dusty process losses. In most cases for the manufacture of expanded clay aggregates, the dust can be recycled. Gypsum generated during flue-gas cleaning, especially in the manufacture of expanded clay aggregates or in the brick and roof tiles industry, cannot be reused in the process, but is used in the cement industry as a hardening regulator. Materials, which cannot be recycled internally, leave the plant to be used in other industries or to be supplied to external waste recycling or waste disposal facilities.

Fired products which are unsuitable for sale (broken ware) and broken firing auxiliaries as well as broken refractory material from kilns, can in some cases after being ground, be reused as raw materials. Broken ware which cannot be reused in the process, can sometimes be used as a raw material input in other industries, for example broken inorganic bonded abrasive wheels are used as refractory material in steel manufacturing furnaces or as a filler in road construction. Broken bricks can be used, for example, as substrates for green roofs, tennis courts or as aggregates for special concretes.

Compared to waste disposal in landfills, the reuse of solid process losses entails:

1. A reduction of raw materials used.
2. A reduction of pollutants released into the environment.
3. Solid process losses reused leads to avoid – and therefore save – the disposal costs and also some raw material costs.

General considerations on plaster molds, firing auxiliaries and broken ware – avoidance/replacement/reduction

The application of modern techniques in the shaping process provides considerable potential to prevent waste. Processes such as slip casting in plaster molds can be replaced by pressure slip casting units with polymer molds. With this method, the use of plaster molds is avoided, additionally a raw material saving of up to 20% is achievable and the occurrence of white sludge is reduced. The use of isostatic presses with polymer molds also avoids the need for plaster molds.

In some sectors, for instance in the manufacture of roof tiles, the use of closed metal molds instead of open plaster molds is possible. However, the replacement of plaster molds is only possible if a new casting unit is built or the old one is totally renewed, which is expensive and, therefore especially for low volume business, difficult to carry out.

In some cases quality problems can appear by using plaster molds because sharp edges tend to occur. Furthermore, the metal molds have a significantly lower exhaustion rate of the water coming out of the clay during molding compared to plaster molds.

The amount of used plaster molds can also be reduced by an increase of the lifetime of the plaster molds. The operation of automatic plaster mixers and vacuum plaster mixers permits the production of harder plaster molds, which means that the lifetime of the molds is two to three times longer. However, for instance, in the wet pressing process for clay roof tiles, it is usually not possible to increase the hardness of the plaster. An optimum between the porosity and surface hardness has to be found to conform with the requirements of water removal and wear resistance.

In the firing process, there are also measures for a direct prevention of waste by the use of modern techniques. If, instead of capsule or rack tunnel kiln firing, fast firing systems like roller hearth kilns are used, the amount of (used) firing auxiliaries can be reduced, but it has to be taken into consideration in this context, that the lifetime of firing auxiliaries depends on the temperature (which is typically higher in roller hearth kilns) and on handling (which is more frequent in fast firing, because firing time is shorter). Additionally, fast firing creates more rejected ceramic products – for instance tiles – when the shape is complex, because it is primarily suited for flat products and therefore, it is not applicable in all cases.

24.8 Emerging techniques for ceramic manufacturing

24.8.1 Radiant tube burners

Reduction of water vapour levels in the kiln gases usually results in lower emission rates for fluoride, because the basic mechanism for the release of fluorine from clay minerals is pyrohydrolysis. This reaction occurs at temperatures of 800°C and upwards.

In laboratory tests, it was found that reducing the water content in the kiln atmosphere, did result in reduced emissions of HF as well as of SO_x. In practice, a reduction of the water content in the kiln atmosphere would be technically difficult to achieve, given that water is produced during the combustion of the fossil fuels used to heat the kiln. This production of water can be avoided by heating the kiln indirectly with gas-fired radiant-tube burners.

For application in the ceramic industry, radiant-tube burners are made of silicon carbide with a high thermal conductivity and a high heat resistance to strongly fluctuating temperatures. With these radiant-tube burners, the kilns are heated indirectly (except the pre-heating zone of the kiln) and the flame of the burner is placed in a heat resistant tube, where combustion takes place. Heat transfer from the burner to the process is mainly carried out by thermal radiation, the corresponding high values for the heat flux are in a range of between 70–120 kW/m^2. The radiant-tube burners are, in principle, applicable in the wall and floor tiles, table- and ornamentalware, sanitaryware, technical ceramics and inorganic bonded abrasives industries, but the production scale in the sectors bricks and roof tiles, vitrified clay pipes, refractory products and expanded clay aggregates is too large. Researchers have found that they can be applied in roller hearth kilns and shuttle kilns, although the technology is not yet proven for tunnel kilns.

24.8.2 Microwave assisted firing and microwave dryers

The firing or sintering of ceramic ware is a critical step in the manufacturing process. Heavy loads on large kiln cars make the heat transfer from the outside to the middle of the setting and into the centre of individual bricks difficult. Temperature gradients can lead to thermal stresses and damaged products – surface temperatures are frequently higher than those in the centre of the unit or setting.

Considerable research has been conducted on applying microwave energy in the firing of ceramic ware. This process heats the product directly – including the centre of the units. In order to avoid undue heat loss to the kiln structure, microwave energy has been used in combination with conventional heating such as gas or electrical energy.

Technical problems, including safety aspects still need to be solved before the process can be applied economically on full-scale manufacture kilns and also relatively high electrical energy costs have to be taken into account. However, experimental work has indicated a number of significant benefits which should be developed in the future:

1. Minimisation of thermal stresses throughout the firing cycle.
2. Product throughputs considerably increased, i.e., much shorter firing cycles.
3. Energy consumption for firing considerably reduced – but there may be less excess heat available for drying purposes.
4. Reduction of solid process losses/solid waste.
5. Quality improvements, including significant gain in mechanical properties.
6. Enhanced binder removal (from refractory products).
7. Reduced emissions based upon less energy consumption and higher production output.
8. Lower fluoride emissions – which are closely related to the time the product spends above 800°C.

24.8.3 New type of drying system for refractory products

The drying of large size refractory building components such as ladle lids or the roofs of electric arc furnaces can be very time consuming and energy intensive. By placing heat resistant stainless steel foils (with heating temperatures of up to 1100°C) or carbon fibres (at lower drying temperatures of less than 250°C) as the heating element into the refractory mix which needs to be poured, the component can be dried from the inside out. The foils or carbon fibres adapt to the structure and cause no interference.

As drying takes place from the inside out, the water moves in the same direction as the temperature front. This reduces drying times considerably.

Controlling the setting temperature improves the quality of the building components substantially, resulting in more even drying with fewer dry cracks and improved structural stability. Improvements were achieved in the following parameters:

1. Cold crushing strength : + 50%.
2. Bending strength : + 50%
3. Abrasion resistance : + 70%
4. Extension of service life : + 50%.

This drying technique results in considerable energy savings, which can be further increased – particularly in the case of very large building components by insulating the refractory product during drying.

24.9 Advanced process wastewater management with integrated glaze recovery

An innovative model waste water treatment system has been built at a ceramics factory, resulting in considerable reductions in pollutants and savings in the manufacture of sanitaryware, tableware and wall and floor tiles.

The new process waste water system has five modules:

1. Microfiltration: The process waste water from the largest glazing process (white glazing, 80% of total glaze consumption) is passed through a microfiltration plant for glaze recovery. The recovered glaze is returned to glaze preparation.
2. Treatment of a tributary process waste water stream: The process waste water from paste preparation in wall and floor tile manufacture, which is low in volume but highly contaminated, is passed to a separate treatment plant. The resulting sewage sludge, which contains high quality components which help increase the sinter ability of the ceramic body, is sent to paste preparation while the effluent goes to the municipal waste water treatment plant for removal of biodegradable constituents.
3. Inclusion of existing sedimentation tanks: The process waste water from glaze preparation is passed through a system of sedimentation tanks, which have been upgraded by additional measures and are monitored automatically. The resulting sewage sludge is also returned to paste

preparation in wall and floor tile manufacture while the process waste water is passed to the new, substantially smaller process waste water treatment plant.

4. Construction of a new central process waste water treatment plant: The previous process waste water treatment plant handled up to 3500 m^3 of process waste water per day. The developed shed structures in the wall and floor tile works area made it impossible to separate rainwater and production wastewater. The new centralised plant, which represents current state of the art, collects the process waste water from the factory and pumps it to the new system over ground, so that no mixing with rainwater occurs. An adjustment to the process waste water volumes of 700–800 m^3/day, the reduced pollutant loads in process waste water streams and improved technology have made it possible to design it as a fully automated plant whose operation requires only 20% of the original personnel needs and generates considerable savings in energy and flocculants. Part of the sewage sludge from this plant can also be reused in tile production. As its pollutant loads are lower and as it also has lower water content (25% maximum) because of a new chamber filter press, the remainder can be stored at the factory's own disposal site without any risk. Tests showed that the material is also suitable to be used as a component of the mineral sealing layer of landfill final cover systems.
5. Separation of rainwater: As process waste water is separated from the underground sewerage system, costly modernisation of old sewers and shafts is not required. Reduction and automation regarding the use of flocculants promises considerable material and cost savings.

The redesigning of the system has the following environmental advantages:

1. Recovery of white glaze and reuse within the process.
2. Saving of landfill capacity.
3. Use of biocides and activated charcoal filters is no longer necessary.
4. Highly contaminated process waste water is treated separately.
5. In plant recycling of sewage sludge.
6. Biologically polluted process waste water fractions are fed to the municipal waste water treatment plant.
7. No risk from polluted process waste water in the old sewerage system.
8. Energy savings due to adaptation of the plants.
9. Savings in flocculants for sludge separation.
10. Reduction of personnel costs to 20% of the original number of staff required through automation.

11. Reduction of energy costs.
12. Reduction in chemical additives.
13. Recovery of reusable glaze.
14. Reduction of landfill costs.

24.9.1 Potential disadvantages

Although the system presented here can be characterised as a model system, the risks that could result when circumstances differ from those described, may lead to the following economic and qualitative disadvantages:

1. Energy costs for the filtration technology may exceed the savings achieved from recovery of raw materials (depending on the original rate of raw material loss).
2. Incorrect design of the microfiltration plant with the consequence of excessive investment costs or insufficient capacity.
3. Long phase of research into the use of recovered glaze (colour corrections may be necessary).
4. Mixing of tributary streams is possible to a limited extent only, otherwise the saving effects from high quality recovery are no longer achieved.
5. High technical competence is necessary regarding the overall system.
6. Redundancy measures have to be taken into account in the planning phase.

24.10 General considerations concerning noise

This section deals with possibilities for the reduction of noise occurring in the several steps during the manufacturing processes of ceramic products. Some noise generating installations/techniques are discussed below.

A reduction of sound emissions can often be achieved by directly applying measures at the source of the noise. The main sources of noise are, for example, pneumatic filter cleaning systems, compressors, motors of the preparation units as well as the handling units. Noise protection can be achieved by enclosure of the noisy unit or by building up noise protection walls. Double walls or sheathing in a double shelled construction are very efficient because, additionally, the air between the first and the second wall guarantees a higher noise protection level.

Vibrations and noise from several facilities (for example presses, crushing and mixing facilities) cannot be reduced efficiently by the above-mentioned measures, so the transfer of vibrations and noise has to be avoided by vibration insulation. Metal suspensions, rubber-metal connections and components made of felt, rubber, cork as well as a vibration insulation of the whole base with a layer of bitumen or an individual engine bed are efficient measures to reduce vibration and noise. Further measures to reduce noise emissions at the units are the use of silencers at the source of noise and the replacement of fast turning fans by larger fans with a slower rotation. If the above-mentioned noise protection measures cannot be applied and if a transfer of noisy units inside the building is not possible, secondary noise protection measures – often at the building itself – have to be carried out. This can be achieved by thicker walls and sound insulation of the windows (multi-glass windows, which also help by saving heating costs), which have to kept closed during noisy operations. In this context, it has to be mentioned however, that a necessity for closed windows often leads to the installation of (expensive and energy consuming) ventilation systems because of workplace conditions. Also, a transfer of windows, gates and noisy outdoor installations to a direction away from neighbouring residential areas is possible.

The operation manner of the employees also has an effect on noise emissions. Gates have to be closed if steady through traffic is not necessary and cautious driving with trucks and forklifts at the site also reduces the noise emissions. Beyond this, time-limiting of noise intensive work, e.g., the tipping of broken ware or forklift traffic in the open air in the evening or at night, has to be avoided. Also the regular maintenance of the units by greasing, as well as the timely replacement of silencers, leads to a reduction of noise.

Because the driving force of a noise protection measure is normally a legal requirement (protection of neighbourhood or workplace), from the economic point of view it is mainly a matter of expense, especially if construction measures are involved.

25

Pollution control in glass industry

25.1 Introduction

The glass industry is extremely diverse, both in the products made and the manufacturing techniques employed. Products range from intricate hand-made lead crystal goblets to huge volumes of float glass produced for the construction and automotive industries. Manufacturing techniques vary from small electrically heated furnaces in the ceramic fibre sector to cross-fired regenerative furnaces in the flat glass sector, producing up to 700 T/day.

The glass industry is essentially a commodity industry, although many ways of adding value to high volume products have been developed to ensure the industry remains competitive. Over 80% of the industry output is sold to other industries and the glass industry as a whole is very dependent on the building industry and the food and beverage industry. However, some of the smaller volume sectors produce high value technical or consumer products.

25.2 Manufacture of glass

Glass was formed naturally from common elements in the earth's crust long before anyone ever thought of experimenting with its composition, molding its shape, or putting it to the myriad of uses that it enjoys in the world today. Glass technology has evolved for six thousand years and some of today's principles date back to early times.

This includes what is today known about the structure of glass, its composition, properties, method of manufacture and uses.

Most glass articles are manufactured by a process in which raw materials are converted at high temperatures to a homogeneous melt that is then formed into the articles. The flow diagram 25.1 summarises the details of conventional glass manufacturing. The vapour deposition of SiO_2 from a flame fed with $SiCl_4$ and oxygen is the basis for manufacturing high-purity glass used for blanks that are redrawn into optical-waveguide fibres. Fused silica items that cannot be formed from viscous melts of SiO_2 or quartz are prepared by vapour deposition. Raw materials are selected according to purity, supply, pollution potential, ease of melting and cost. Sand is the most common ingredient. Limestone is the source of calcium and magnesium. Powdered anthracite coal is a common reducing agent and common colourants for glass include iron, chromium, cerium, cobalt and nickel.

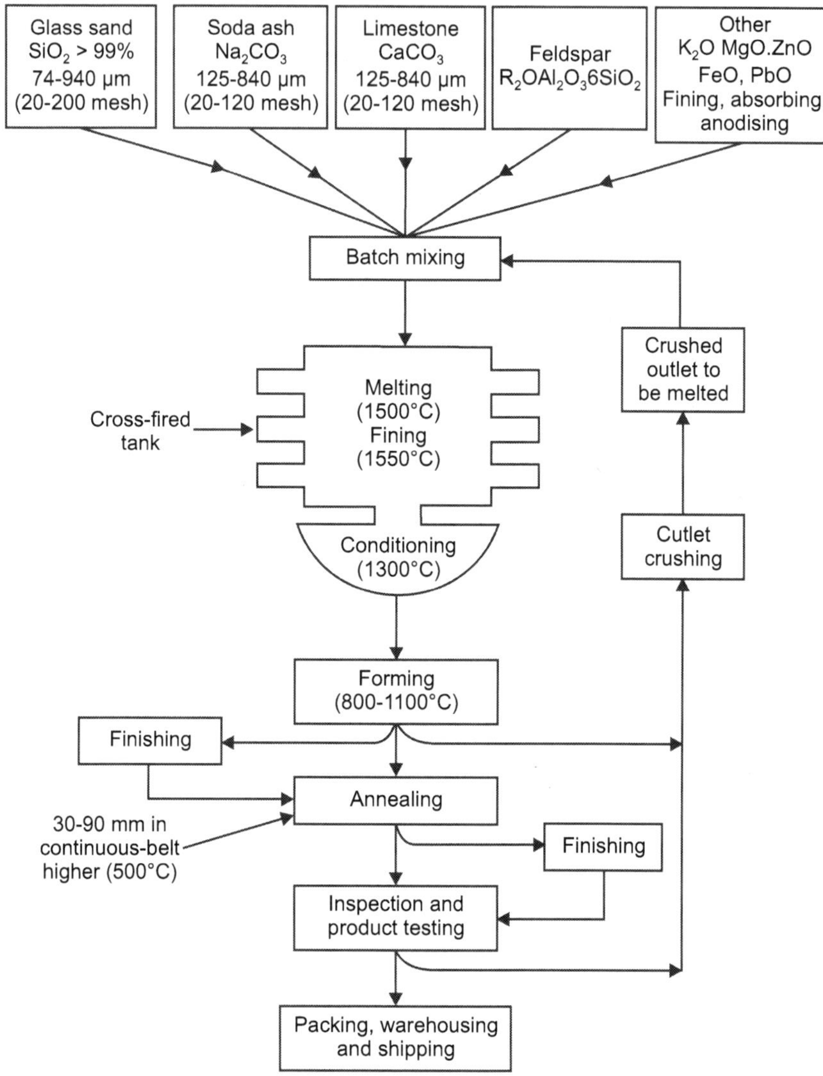

Figure 25.1: Glass manufacture. Temperatures are for common soda-lime glass. Other glasses may require appreciately different temperatures.

Melting and fining depends on the batch materials interacting with each other at the proper time and in the proper order. Thus, extreme care must be taken to obtain materials of optimum grain size, to weigh them carefully and mix them intimately. The efficiency of the melting operation and the uniformity and quality of the glass product are very often determined in the mix house. Batch handling systems vary widely through the industry, from manual to fully automatic. Melting units range from small pot furnaces for manual production to large and continuous tanks for rapid machine forming. The two common feeding designs used today are the screw feeder and the reciprocating pusher. Control devices have advanced from manual control to sophisticated computer-assisted operation. Radiation pyrometers in conjunction with thermocouples monitor furnace temperatures at several locations. Natural gas, oil and electricity are the primary sources of energy; propane is used as a backup reserve in emergencies.

Molten glass is either molded, drawn, rolled, or quenched, depending on desired shape and use. Bottles, dishes, optical lenses, television picture tubes, etc., are formed by blowing, pressing, casting, and/or spinning the glass against a mold to cool it and to set its final shape. Window glass, tubing, rods and fibre are formed by freely drawing the glass in air (or in a bath of molten tin as in the float process) until it sets up and can be cut to length. Art glass is usually hand-formed by freely blowing and shaping it while soft. Glass that is intended to be crushed into powder, called frit, is quenched between water-cooled rollers or ladled or poured directly into water and then dried (dri-gauged). Simultaneous forming of two glasses to produce a single article is possible if the two have similar viscosities in the appropriate temperature range.

Glass articles formed at high temperatures must be cooled in order to reduce the strain and associated stress caused by temperature gradients in the glass to a low level to prevent damage during finishing and subsequent use. The continuous solidification of glass on cooling and the methods of reducing strain are unique to glasses and do not apply to crystalline materials that undergo phase transitions. Lehring processes include annealing, tempering, densifying or compacting and post heat treatments. The term annealing generally refers to removal of stress and terms such as fine annealing or trimming of optical glass imply structural changes associated with lehring.

In secondary forming, a piece of preformed glass is reheated and reworked into the finished product. Secondary operations include mechanical finishing, chemical finishing and cleaning.

Glass manufacturing is among the most energy-intensive industries.

The bulk of energy consumed in the glass manufacturing industry comes from natural gas combustion used to heat furnaces to melt raw materials to form glass. These furnaces are mainly natural gas-fired, but there are a small number of electrically-powered furnaces. Many glass furnaces also use electric boosting (supplementary electric heating systems) to increase throughput and quality. After the melting and refining process is complete, the glass is formed and finished to create the final product. Specific manufacturing processes depend on the intended product and can include annealing (slow cooling), tempering, coating and polishing, which require additional energy.

Different types of glass include:

1. Container glass for packaging foods, beverages and cosmetics.
2. Flat glass for residential and commercial construction, automotive applications, tabletops and mirrors.
3. Fibreglass for insulation and roofing.
4. Specialty glass that is pressed and blown glass for tableware, cookware, lighting, laboratory equipment and other uses.
5. Purchased glass serves as an input to produce glass containers, flat glass, glassware and high-value products such as some optical lenses and fibre optic cable.

There is substantial potential for energy efficiency improvements in glass manufacturing. Estimates range from 20–25%, mainly focused on the extremely energy-intensive melting and refining process.

Energy used in other key glass-making processes, such as forming, is highly variable depending on the product, so those processes don't have the same potential for efficiency gains.

25.2.1 Glass products and their characteristics

Glass products and their characteristics, as well as the production and processing routes involved in their production show large variations. However, batch preparation, melting and refining, conditioning and forming and finishing (annealing, tempering, polishing or coating) steps are found in virtually all glass plants and are the most important from an energy point of view.

Melting furnaces – employing combustion-heating (with air-fuel or oxy-fuel burners), direct electrical heating, or a combination of the two (electric boosting) – are the major energy users. In general, the energy necessary for melting glass may account for over 75% of the total energy (in terms of final energy) requirements of glass manufacture, with an average of about 65% of the total energy input when considering all the sectors of the glass industry. Typically, melting furnaces operate with an overall efficiency of 50–60%, where structural and flue gas losses represent 20–25% and 25–35% of losses, respectively. Melting furnaces are also the most important area for efficiency improvement, followed by refining and conditioning.

25.3 Environmental issues in glass industry

The main issue is the control of dust from the handling of fine materials. The main raw materials for melting are glass forming materials (e.g., silica sand, cullet), intermediate/modifying materials (e.g., soda ash, limestone, feldspar) and colouring/decolouring agents (e.g., iron chromite, iron oxide).

Melting, the combination of the individual raw materials at high temperature to form a molten glass, is the central phase in the production of glass. The melting process is a complex combination of chemical reactions and physical processes and melting can be divided into several stages: heating; primary melting; fining and homogenisation and conditioning.

The main melting techniques are summarised below. Glass making is a very energy intensive activity and the choice of energy source, heating technique and heat recovery method are central to the design of the furnace. The same choices are also some of the most important factors affecting the environmental performance and energy efficiency of the melting operation. The three main energy sources for glass making are natural gas, fuel oil and electricity.

Regenerative furnaces utilise regenerative heat recovery systems. Burners are usually positioned in or below combustion air/waste gas ports. The heat in the waste gases is used to pre-heat air prior to combustion, by passing the waste gases through a chamber containing refractory material, which absorbs the heat. The furnace only fires on one side at a time. After about twenty minutes, the firing is reversed and the combustion air is passed through the chamber previously heated by the waste gases. Pre-heat temperatures up to 1400°C may be attained leading to very high thermal efficiencies. In the cross-fired regenerative furnace, combustion ports and burners are positioned along the sides of the furnace and the regenerator chambers are located either side of the furnace. In the end-fired regenerative furnace the principles of operation are the same, however, the two regenerative chambers are situated at one end of the furnace.

Recuperative furnaces utilise heat exchangers (termed recuperators) for heat recovery, with continuous pre-heat of combustion air by the waste gases. Air pre-heat temperatures are limited to around 800°C for metallic recuperators. The specific melting capacity (per unit of melter area) of recuperative furnaces is around 30% lower than for a regenerative furnace. The burners are located along each side of the furnace, transverse to the flow of glass and fire continuously from both sides. This type of furnace is primarily used where high flexibility of operation is required with minimum initial capital outlay, particularly where the scale of operation is too small to make the use of regenerators economically viable. It is more appropriate to small capacity installations although higher capacity furnaces (up to 400 T per day) are not uncommon.

Oxy-fuel firing involves the replacement of the combustion air with oxygen (>90% purity). The elimination of the majority of the nitrogen from the combustion atmosphere reduces the volume of the waste gases by about two thirds. Therefore, furnace energy savings are possible because it is not necessary to heat the atmospheric nitrogen to the temperature of the flames.

The formation of thermal NO_x is also greatly reduced. In general, oxy-fuel furnaces have the same basic design as unit melters, with multiple lateral burners and a single waste gas exhaust port. However, furnaces designed for oxygen combustion do not utilise heat recovery systems to pre-heat the oxygen supply to the burners.

Electric furnaces consist of a refractory lined box supported by a steel frame, with electrodes inserted either from the side, the top or more usually the bottom of the furnace. Energy for melting is provided by resistive heating as the current passes through the molten glass. The technique is commonly applied in small furnaces particularly for special glass. There is an upper size limit to the economic viability of electric furnaces, which depends on the cost of electricity compared with fossil fuels. The replacement of fossil fuels in the furnace eliminates the formation of combustion products.

Combined fossil fuel and electric melting can take two forms: predominantly fossil fuel firing with electric boost; or predominantly electrical heating with a fossil fuel support. Electric boosting is a method of adding extra heat to a glass furnace by passing an electric current through electrodes in the bottom of the tank. A less common technique is the use of gas or oil as a support fuel for a principally electrically heated furnace.

Discontinuous batch melters are used where smaller amounts of glass are required, particularly if the glass formulation changes regularly. In these instances pot furnaces or day tanks are used to melt specific batches of raw material. Many glass processes of this type would not fall under the control of Integrated Pollution Prevention Control (IPPC) because they are likely to be less than 20 T per day melting capacity. Basically a pot furnace consists of a lower section to pre-heat the combustion air and an upper section which holds the pots and serves as the melting chamber. Day tanks are further developed from pot furnaces to have larger capacities, in the region of 10 T per day. Structurally they more closely resemble the quadrangle of a conventional furnace, but are still refilled with batch each day.

Special melter designs have been produced to improve efficiency and environmental performance. The best known of this type of furnace are the $LoNO_x$ melter and the flex melter.

Container glass is a diverse sector and almost all of the melting techniques described above are found. The forming process is carried out in two stages, the initial forming of the blank either by pressing with a plunger or by blowing with compressed air and the final molding operation by blowing to obtain the finished hollow shape. These two processes are thus respectively termed 'press and blow' and 'blow and blow'. Container production is almost exclusively by Individual Section (IS) machines.

Flat glass is produced almost exclusively with cross-fired regenerative furnaces. The basic principle of the float process is to pour the molten glass onto a bath of molten tin and to form a ribbon with the upper and lower surfaces becoming parallel under the influence of gravity and surface tension. From the exit of the float bath the glass ribbon is passed through the annealing lehr, gradually cooling the glass to reduce residual stresses. On-line coatings can be applied to improve the performance of the product (e.g., low emissivity glazing).

Continuous filament glass fibre is produced using recuperative or oxy-fuel fired furnaces. The glass flows from the furnace to the forehearths where it passes through bushings at the base. The glass is drawn through the bushing tips to form continuous filaments. The filaments are drawn together and pass over a roller or belt, which applies an aqueous coating to each filament. The coated filaments are gathered together into bundles (strands) for further processing.

Domestic glass is a diverse sector involving a wide range of products and processes. Ranging from intricate handmade lead crystal, to high volume, mechanised methods used for mass produced tableware. Almost all of the melting techniques, described above are used in the sector, from pot furnaces to large regenerative furnaces. The forming processes are automatic processing, hand made or semi-automatic processing and following production the basic items can be subjected to cold finishing operations (e.g., lead crystal is often cut and polished).

Special glass is also a very diverse sector, covering a wide range of products that can differ considerably in composition, method of manufacture and use. The most common techniques are recuperative furnaces, oxy-gas furnaces, regenerative furnaces, electric melters and day tanks. The wide product range means that many forming techniques are used within the sector. Some of the most important are: press and blow production; rolling; pressing, ribbon process; tube extrusion; the drawing process; and dissolution (water glass).

Glass wool furnaces are usually either electric melters, gas fired recuperative furnaces, or oxyfuel furnaces. Molten glass flows along a forehearth and through single orifice bushings into rotary centrifugal spinners. Fibreising is by centrifugal action with attenuation by hot flame gases. An aqueous phenolic resin solution is sprayed onto the fibres. The resin coated fibre is drawn under suction onto a moving conveyor and then passes through an oven to dry and cure the product.

Stone wool is usually produced with the coke fired hot blast cupola. The molten material gathers in the bottom of the furnace and flows out along a short trough onto the spinning machine. Air is used to attenuate the fibres and to direct them onto the collection belts. An aqueous phenolic resin solution is sprayed onto the fibres by a series of spray nozzles. The remainder of the process is essentially as for glass wool.

Ceramic fibre is produced exclusively using electric furnaces. The melt is fibreised either by high speed wheels or a high pressure air jet and the fibres are drawn on to a collection belt. The product can be baled at this point or processed into blanket to be baled as product or needle felted. Further downstream processing may also be carried out.

Frit production utilises both continuous furnaces and batch furnaces. It is common for small batches to be produced for a wide range of formulations. Frit furnaces are generally natural gas or oil fired and many frit plants use oxy-fuel firing. Continuous furnaces can be cross-fired or endfired with a single burner. Discontinuous batch furnaces are box shaped or cylindrical refractory lined vessels, mounted to allow a degree of rotation. The melt can be quenched directly in a water bath, or can be cooled between water cooled rollers to produce a flake product.

25.4 Emission levels from glass industry

The major environmental challenges for the glass industry are emissions to air and energy consumption. Glass making is a high temperature, energy intensive activity, resulting in the emission of products of combustion and the high temperature oxidation of atmospheric nitrogen, i.e., sulphur dioxide, carbon dioxide and oxides of nitrogen. Furnace emissions also contain dust and lower levels of metals. The main emissions arising from melting activities within the glass industry are summarised in the Table 25.1.

Table 25.1: Emissions from melting activities.

Emission	*Source*
Particulate matter	Condensation of volatile batch components.
	Carry over of fine material in the batch.
	Product of combustion of some fossil fuels.
Oxides of nitrogen	Thermal NO_x due to high melter temperatures.
	Decomposition of nitrogen compounds in the batch materials.
	Oxidation of nitrogen contained in fuels.
Oxides of sulphur	Sulphur in fuel.
	Decomposition of sulphur compounds in the batch materials.
	Oxidation of hydrogen sulphide in hot blast cupola operations.
Chlorides/HCl	Present as an impurity in some raw materials, particularly man made sodium carbonate.
	NaCl used as a raw material in some special glasses.

(Cont'd...)

Emission	Source
Fluorides/HF	Present as a minor impurity in some raw materials. Added as a raw material in the production of enamel frit to provide certain properties in the finished product. Added as a raw material in the continuous filament glass fibre industry and in some glass batches to improve melting, or to produce certain properties in the glass, e.g., opalescence. Where fluorides are added to the batch, typically as fluorspar, uncontrolled releases can be very high.
Heavy metals (e.g., V, Ni, Cr, Se, Pb, Co, Sb, As, Cd)	Present as minor impurities in some raw materials, post consumer cullet and fuels. Used in fluxes and colouring agents in the frit industry (predominantly lead and cadmium). Used in some special glass formulations (e.g., lead crystal and some coloured glasses). Selenium is used as a colorant (bronze glass), or as a decolourising agent in some clear glasses.
Carbon dioxide	Combustion product. Emitted after decomposition of carbonates in the batch materials (e.g., soda ash, limestone).
Carbon monoxide	Product of incomplete combustion, particularly in hot blast cupolas.
Hydrogen sulphide	Formed from raw material or fuel sulphur in hot blast cupolas due to the reducing conditions found in parts of the furnace.

25.4.1 Summary of emissions to atmosphere arising from melting activities

The emissions from downstream activities can vary greatly between the different sectors. Although there are similarities in the melting techniques used in many sectors the downstream activities tend to be specific. Emissions to air can arise from: coating application and/or drying; secondary processing (e.g., cutting, polishing, etc.), and from some product forming operations (e.g., mineral wool and ceramic fibre).

In general, emissions to the water environment are relatively low and there are few major issues that are specific to the glass industry.

A characteristic of most of the sectors is that the great majority of internally generated glass waste is recycled back to the furnace. The main exceptions to this are the continuous filament sector, the ceramic fibre sector and producers of very quality sensitive products in the special glass and domestic glass sectors. The mineral wool and frits sectors show a wide variation in the amount of waste recycled to the furnace ranging from nothing to 100% for some stone wool plants.

25.5 Emission control techniques

Many of the sectors within the glass industry utilise large continuous furnaces with lifetimes up to twelve years. These furnaces represent a large capital commitment and the continuous operation of the furnace and the periodic rebuild provide a natural cycle of investment in the process. Major changes of melting technology are most economically implemented if coincided with furnace rebuilds and this can also be true for complex secondary abatement measures. However, many improvements to the operation of the furnace, including the installation of secondary techniques, are possible during the operating campaign.

This summary briefly outlines the main techniques for controlling each substance emitted from melting activities and from some of the downstream operations. It concentrates predominantly on emissions to air as these are generally the most significant emissions from glass processes.

25.5.1 Particulate matter

Techniques for controlling particulate emissions include secondary measures, generally electrostatic precipitators and bag filters and primary measures. The Electrostatic Precipitator (EP) consists of a series of high voltage discharge electrodes and corresponding collector electrodes. Particles are charged and subsequently separated from the gas stream under the influence of the electric field. EPs are very effective in collecting dust in the range 0.1 μm to 10 μm and overall collection efficiency can be 95–99%. Actual performance varies depending mainly on waste gas characteristics and EP design. In principle, this technique is applicable to all new and existing installations in all sectors (except stone wool cupolas due to the risk of explosion).

In most applications a modern well designed two or three stage EP could be expected to achieve 20 mg/m^3. Where high efficiency designs are used or where favourable conditions exist lower emission levels are often possible. Costs vary greatly, depending heavily on required performance and waste gas volume.

Bag filter systems use a fabric membrane which is permeable to gas but which will retain the dust. Dust is deposited on and within the fabric and as the surface layer builds up it becomes the dominating filter medium. The direction of gas flow can be either from the inside of the bag to the outside, or from the outside to the inside. Fabric filters are highly efficient and a collection efficiency of 95–99% would be expected. Particulate emissions between 0.1 mg/m^3 and 5 mg/m^3 can be achieved and levels consistently below 10 mg/m^3, could be expected in most applications. The ability to achieve such low levels can be important if dusts contain significant levels of metals and low metal emissions must be achieved. In principle, bag filters are applicable to all new and existing installations in all sectors. However, due to their potential to blind in certain circumstances, they are not the preferred choice in all applications. In most cases there are technical solutions to these difficulties, but there may be an associated cost.

Primary control techniques are based mainly on raw material changes and furnace/firing modifications. In most applications primary techniques cannot achieve emission levels comparable with bag filters and EPs.

25.5.2 Oxides of nitrogen (NO_x)

The most appropriate techniques for controlling emissions of NO_x are generally: primary measures, oxy-fuel melting, chemical reduction by fuel, selective catalytic reduction and selective non-catalytic reduction.

Primary measures can be split into two main types 'conventional' combustion modifications and then special furnace designs or optimised combustion design packages. Oxy-firing is also a primary technique but is dealt with separately due to its specific nature. Conventional combustion modifications are usually based on: reduced air/fuel ratio, reduced pre-heat temperature, staged combustion and low NO_x burners; or a combination of these techniques.

Capital costs are generally quite low and operating costs are often reduced due to reduced fuel usage and improved combustion. A great deal of progress has been made in this area, but the emission reductions achievable clearly depend on the starting point. NO_x reductions of 40–60% are not uncommon and emission levels of less than 650–1100 mg/Nm^3 have been achieved in some applications.

Special furnace designs have been developed which reduce NO_x emissions, e.g., the LoNO_x melter. These designs have been very successful but there are certain process constraints that limit their applicability. The FENIX process is a combustion optimisation package based on primary measures, that is tailored for a specific furnace.

Oxy-fuel firing involves the replacement of the combustion air with oxygen. The elimination of the majority of the nitrogen from the combustion atmosphere reduces the volume of the waste gases by about two thirds. Therefore, energy savings are possible because it is not necessary to heat the atmospheric nitrogen to the temperature of the flames. The formation of thermal NO_x is greatly reduced, because the only nitrogen present in the combustion atmosphere is the residual nitrogen in the oxygen/fuel, nitrogen from nitrate breakdown and that from any parasitic air.

The principle of oxy-fuel firing is well established and in principle can be considered as applicable to the industry as a whole. However, the technique is still considered by some sectors (particularly flat glass and domestic glass), as a developing technology with potentially high financial risk. Considerable development work is being undertaken and the technique is becoming more widely accepted as the number of plants increases. The issues surrounding this technique are very complex. The economic competitiveness of the technique depends largely on the scale of energy savings (and the relative costs of alternative abatement techniques) compared with the cost of oxygen. Both the technical and economic viability of the technique are heavily dependent on site specific issues.

Chemical reduction by fuel describes those techniques where fuel is added to the waste gas stream to chemically reduce NO_x to N_2 through a series of reactions. The fuel does not burn but pyrolyses to form radicals which react with the components of the waste gas. The two main techniques that have been developed for use in the glass industry are the 3R process and the reburning process. Both of these techniques are currently restricted to regenerative furnaces. The 3R process has been fully developed for application within the industry and the Reburning process has been run at full plant scale and has shown promising results. The 3R process can achieve emission levels of less than 500 mg/Nm^3 corresponding to an increase in fuel usage of 6–10%. The Reburning process is hoped to achieve comparable emission levels following development. The increased energy usage for both techniques can be greatly reduced by the use of energy recovery systems and by combining the techniques with primary measures.

Selective Catalytic Reduction (SCR) involves reacting NO_x with ammonia in a catalytic bed generally at around 400°C. Most applications in the glass industry will require a 3-stage system including dust abatement and acid gas scrubbing. Systems are normally designed to achieve 75–95% reductions and in general emission levels of less than 500 mg/Nm^3 are achievable. The cost of SCR depends mainly on the waste gas volume and the desired NO_x reduction. However, there are a number of issues that can limit the applicability of the technique in certain cases. For example, the technique has not been proven for heavy fuel oil fired glass furnaces, glass wool or continuous filament glass fibre.

Selective non-catalytic reduction (SNCR) operates on the same basis as SCR but the reactions take place at higher temperature (800–1100°C) without the need for a catalyst. SNCR does not require dust abatement or acid gas scrubbing. Reduction efficiencies of 30–70% are generally achievable, the critical factor is the availability of sufficient ammonia at the correct temperature window. In principle, the technique is applicable to all glass processes including new and existing plants. The main limitation to the applicability of SNCR is whether the reagent can be introduced at a point in the waste gas system where the correct temperature can be maintained for an adequate reaction time. This is of particular relevance in existing plants and for regenerative furnaces.

25.5.3 Oxides of sulphur (SO_x)

The main techniques for controlling SO_x emissions are fuel selection, batch formulation and acid gas scrubbing. In oil fired processes the main source of SO_x is the oxidation of sulphur in the fuel. The amount of SO_x derived from the batch materials varies depending on the glass type but in general wherever oil is burned the SO_x emissions from the fuel outweigh those from the batch materials. The most obvious way to reduce SO_x emissions is to reduce the sulphur content of the fuel. Fuel oil is available in various sulphur grades (<1%, <2%, <3% and >3%) and natural gas is essentially sulphur free. The conversion to a lower sulphur content fuel does not generally result in any increased costs except the higher fuel price. The conversion to gas firing requires different burners and a range of other modifications. However, where natural gas is burned SO_x emissions will usually be lower and where oil is burned a sulphur level of 1% or less is considered best available technology (BAT). Burning higher sulphur content fuels may also represent BAT if abatement is used to achieve equivalent emission levels.

In conventional glass making, sulphates are the main source of SO_x emissions from batch materials. Sulphates are the most widely used fining agents and are also important oxidising agents. In most modern glass furnaces the levels of batch sulphates have been reduced to the minimum practicable levels, which vary depending on the glass type.

In stone wool production important sources of SO_2 emission (in addition to coke) are the use of blast furnace slag and cement bound briquettes in the batch. The availability of low sulphur coke and slag are restricted by very limited supply within economical transport distances. Slag can generally be eliminated from most batches, with the exception of the production of limited amounts of white fibre for specific applications. The use of cement bound waste briquettes involves a balance between waste minimisation and SO_x emission reduction, which will often depend on specific priorities and must be considered in association with the use of acid gas scrubbing.

The operating principles of dry and semi-dry scrubbing are the same. The reactive material (the absorbent) is introduced to and dispersed in the waste gas stream. This material reacts with the SO_x species to form a solid, which must be removed from the waste gas stream by an electrostatic precipitator or bag filter system. The absorbents chosen to remove SO_x are also effective in removing other acidic gases. In the dry process the absorbent is a dry powder (usually $Ca(OH)_2$, $NaHCO_3$, or $Na_2(CO)_3$). In the semi-dry process the absorbent (usually Na_2CO_3, CaO or $Ca(OH)_2$) is added as a suspension or solution and water evaporation cools the gas stream. The reductions achieved with the techniques depend on a number of factors including waste gas temperature, the amount and type of absorbent added (or more precisely the molar ratio between reactant and pollutants) and the dispersion of the absorbent.

The complete recycling of filter dust, including the sulphated waste, is often considered to be a reasonable environmental and economic option, where it is technically possible. The overall reduction of SO_x emissions is limited (by mass balance considerations) to the reduction at source obtained by substituting sulphate in the raw materials by filter dust. (Clearly this is in addition to other appropriate primary measures to reduce overall sulphur input.) Therefore, in order to reduce acid gas emissions, it may be necessary to consider an external disposal route for a portion of the collected material. The determination of what represents the best protection of the environment as a whole can often be site-specific and may involve balancing the potentially conflicting priorities of waste minimisation and SO_x emission reduction.

With closed loop filter dust recycling, the SO_x emission levels observed today, are generally in the range of 200–800 mg/Nm^3 for natural gas firing and 800–1600 mg/Nm^3 with 1% S fuel oil. The majority of installed SO_x scrubbing systems operate with dry lime scrubbing at a temperature of around 400°C, which is the waste gas temperature obtained from an efficient regenerative type furnace. At these temperatures, an SO_x reduction of around 50% can be achieved. An improved SO_x reduction rate can be reached at temperatures around 200°C and with a humid atmosphere but this involves further considerations.

25.5.4 Fluorides (HF) and chlorides (HCl)

HF and HCl emissions generally arise from the volatilisation of fluorides and chlorides in the batch materials, either present as impurities or added intentionally to provide specific product or processing characteristics to the glass. The main techniques for the reduction of these emissions are batch modification or scrubbing. Where the halides are present as impurities emissions can generally be controlled by raw material selection, although scrubbing is often used either where raw material selection is not sufficient or where scrubbing is used to control other substances. Where halides are used to impart specific characteristics there are two main approaches scrubbing or batch reformulation to achieve the same characteristics by other means. Particular successes with reformulation have been achieved in continuous filament glass fibre.

25.6 Emissions from non-melting activities

Emissions from downstream processing are sector specific. With the exception of the mineral wool sector emissions are generally much lower than from the melting activities. Abatement techniques are generally oxidation.

In mineral wool processes there is the potential for substantial emissions from the application and curing of organic resin based binder systems.

25.6.1 Emissions of water

In general, emissions to the water environment are relatively low and there are few major issues that are specific to the glass industry. Water is used mainly for cleaning and cooling and can be readily recycled or treated using standard techniques. Specific issues of organic contamination can arise from mineral wool and continuous filament glass fibre processes. Issues of heavy metals (particularly lead) can arise from special glass, frits and domestic glass processes. The Table 25.2 identifies the main potential techniques for controlling emissions of water.

Table 25.2: Main potential techniques for controlling emissions of water.

Physical/chemical treatment	
• Screening	• Neutralisation
• Skimming	• Aeration
• Settlement	• Precipitation
• Centrifuge	• Coagulation and flocculation
• Filtration	
Biological treatment	
• Activated sludge	
• Biofiltration	

25.6.2 Solid wastes

A characteristic of the glass industry is that most of the activities produce relatively low levels of solid waste. Most of the processes do not have significant inherent by-product streams. The main process residues are unused raw materials, waste glass that has not been converted into the product and waste product. Other solid wastes include waste refractory and dust collected in abatement equipment or flues. Non-fibrous waste is generally readily recycled to the process and techniques are under development for recycling other wastes. The extent of waste recycling is increasing as financial incentives develop, particularly increased disposal costs.

25.6.3 Energy

Glass making is a very energy intensive process and the choices of energy source, heating technique and heat recovery method are central to the design of the furnace and to the economic performance of the process. The same choices are also some of the most important factors affecting the environmental performance and energy efficiency of the melting operation. In general, the energy necessary for melting glass accounts for over 75% of the total energy requirements of glass manufacture. The cost of energy for melting is one of the largest operational costs for glass installations and there is a significant incentive for operators to reduce energy use.

Section V

Pollution control in food and agriculture industries

26

Pollution control in food processing industry: A review

26.1 Introduction

The food processing industry has special concerns about the health and safety of the consumer.

Key resources used by the food processing industry include the water, raw materials and energy. Traditionally, the food processing industry has been a large water user. Water is used as an ingredient, an initial and intermediate cleaning source, an efficient transportation conveyor of raw materials and the principal agent used in sanitising plant machinery and areas. Although water use will always be a part of the food processing industry, it has become the principal target for pollution prevention, source reduction practices.

The key environmental issues for the food industry include the following:

Wastewater: Primary issues of concern are Biochemical Oxygen Demand (BOD); Total Suspended Solids (TSS); excessive nutrient loading, namely nitrogen and phosphorus compounds; pathogenic organisms, which are a result of animal processing; and residual chlorine and pesticide levels.

Solid waste: Primary issues of concern include both organic and packaging waste. Organic waste, that is, the rinds, seeds, skin and bones from raw materials, results from processing operations. Inorganic wastes typically include excessive packaging items that are, plastic, glass and metal. Organic wastes are finding ever-increasing markets for resale and companies are slowly switching to more biodegradable and recyclable products for packaging. Excessive packaging has been reduced and recyclable products such as aluminium, glass and High-Density Polyethylene (HDPE) are being used where applicable. The food processing factories should follow the major technological innovations in the industry, including those in clean technologies and processes. Clean technologies include:

1. Advanced wastewater treatment practices: Use of wastewater technologies beyond conventional secondary treatment.
2. Improved packaging: Use of less excessive and more environmentally friendly packaging products.
3. Improved sensors and process control: Use of advanced techniques to control specific portions of the manufacturing process to reduce wastes and increase productivity.

4. Food irradiation: Use of radiation to kill pathogenic micro-organisms.
5. Water and wastewater reduction (closed loop/zero emission systems): Reduction or total elimination of effluent from the manufacturing process

26.2 Pollution from food processing

Food processing can be divided into four major sectors including fruit and vegetables; meat, poultry and seafood; beverage and bottling and dairy operations. All of these sectors consume huge amount of water for processing food. A considerable part of these waters are potential wastewaters to be treated for safe disposal to the environment.

Table 26.1 shows typical rates of water use for various food processing sectors. An abundant and inexpensive source of water is a requirement for success in the food processing industry. This coincides with the same need for water resources in agricultural farmland activities.

Table 26.1: Typical rates for water use for various industries.

Industry	*Range of flow gal/T product*
Fruits and vegetables	
Green beans	12,000–17,000
Peaches and pears	3600–4800
Other fruits and vegetables	960–8400
Food and beverage	
Beer	2400–3840
Bread	480–960
Meat packing	3600–4800
Milk products	2400–4800
Whiskey	14,400–19,200

26.2.1 Fruit and vegetable food processing sector

The primary steps in processing fruits and vegetables include:

1. General cleaning and dirt removal.
2. Removal of leaves, skin and seeds.
3. Blanching.
4. Washing and cooling.
5. Packaging.
6. Cleanup.

Wastewater and solid wastes are the primary area of pollution control within the fruit and vegetable food processing industry. Their wastewater is high in suspended solids and organic sugars and starches and may contain residual pesticides. Solid wastes include organic materials from mechanical preparation processes that is, rinds, seeds and skins from raw materials. For the most part, solid waste that is not resold as animal feed is handled by conventional biological treatment or composting. The total amount of material generated is a function of the amount of raw material moved through a facility, for example, for a given weight of apples processed comes a set amount of peel and seed waste.

Attempts to decrease solid waste streams have not been an area of great development for pollution prevention opportunities and clean technologies. Pre-treatment opportunities intended to reduce the amount of raw materials lost to the waste stream have been an area of clean technology development. For the most part, the majority of clean technology advances and research have been in reducing the volume of wastewater generated in food processing operations.

Most fruit and vegetable processors use traditional biological means to treat their wastewater. Advancements in the degradation chemistries of pesticides have aided in reducing their quantities and toxicity in process wastewater.

Washing fresh produce (also known as surface treatment) can reduce the overall potential for microbial food safety hazards. This is an important step since most microbial contamination is on the surface of fruits and vegetables. If pathogens are not removed, inactivated, or otherwise controlled, they can spread to surrounding produce, potentially contaminating a significant proportion of the produce.

Sanitizers or anti-microbials in wash water and other processing water may be useful in reducing pathogens on the surface of produce and/or reducing pathogen build-up in water.

The effectiveness of a sanitiser depends on its chemical and physical nature, treatment conditions (such as water temperature, pH and contact time), resistance of pathogens and the nature of the fruit or vegetable surface. Chlorine is a commonly used anti-microbial.

Chlorine dioxide, trisodium phosphate, organic acids and ozone have also been studied for use as anti-microbials in produce wash water. All chemical substances that contact food must be used in accordance with FDA and EPA regulations.

26.2.2 Meat, poultry and seafood sector

The primary steps in processing livestock include:

1. Rendering and bleeding.
2. Scalding and/or skin removal.

3. Internal organ evisceration.
4. Washing, chilling and cooling.
5. Packaging.
6. Cleanup.

Meat, poultry and seafood facilities offer a more difficult waste stream to treat. The killing and rendering processes create blood by-products and waste streams, which are extremely high in BOD. These facilities are very prone to disease spread by pathogenic organisms carried and transmitted by livestock, poultry and seafood. This segment of the food processing industry is by far the most regulated and monitored.

Waste streams vary per facility, but they can be generalised into the following: process wastewaters; carcasses and skeleton waste, rejected or unsatisfactory animals, fats, oils and greases (FOG), animal feces, blood and eviscerated organs. The primary avenue for removal of solid waste has been its use in animal feed, cosmetics and fertilisers. These solid wastes are high in protein and nitrogen content. They are excellent sources for recycled fish feed and pet food. Skeleton remains from meat processing are converted into bonemeal, which is an excellent source of phosphorus for fertilisers. FOG waste (typically from industrial fisheries) is used as a base raw material in the cosmetics industry.

26.2.3 Beverage and fermentation sector

The primary steps in processing beverages are:

1. Raw material handling and processing.
2. Mixing, fermentation, and/or cooking.
3. Cooling.
4. Bottling and packaging.
5. Cleanup.

Wastewater and solid waste are the primary waste streams for the beverage and fermentation sector. Solid wastes result from spent grains and materials used in the fermentation process. Wastewater volume of 'soft drink processes' is lower than in other food processing sectors, but fermentation processes are higher in BOD and overall wastewater volume compared to other food processing sectors. Ozone technology has proven very useful in the beverage market since the earliest 20th century. In bottled water plants, ozone can be used to disinfect product water without leaving any residual taste or odour. At beverage plants, ozone can reduce or eliminate the need for chemical or high temperature disinfections during clean-in-place (CIP) cycles, reducing downtime and chemical costs.

26.2.4 Dairy sector

A majority of the waste milk in dairy wastewaters comes from start-up and shutdown operations performed in the High-Temperature Short Time (HTST) pasteurisation process. This waste is pure milk raw material mixed with water. Another waste stream of the dairy sector is from equipment and tank-cleaning wastewaters. These waste streams contain waste milk and sanitary cleaners and are one of the principal waste constituents of dairy wastewater. Over time, milk waste degrades to form corrosive lactic and formic acids. Approximately 90% of a dairy's wastewater load is milk.

26.2.5 Can cooker products

Water plays a role in most of the problems associated with metal food containers after processing. Whether steam, hot water or cold water, each can serve as the vehicle to transport undesirable substances. It is important to understand, when designing an effective, comprehensive water treatment programme, how these mechanisms chemically interact.

26.2.6 Wastewater from food processing industries

Food processing wastewater can be characterised as nontoxic, because it contains few hazardous and persistent compounds. With the exception of some toxic cleaning products, wastewater from food processing facilities is organic and can be treated by conventional biological technologies. Part of the problem with the food processing industry's use and discharge of large amounts of water is that it is located in rural areas in which the water treatment systems (i.e., potable and wastewater systems) are designed to serve small populations. As a result, one medium-sized plant can have a major effect on local water supply and surface water quality. Large food processing plants will typically use more than 1,000,000 gallons of potable water per day.

Another contaminant of food processing wastewaters, particularly from meat-, poultry- and seafood processing facilities, is pathogenic organisms. Wastewaters with high pathogenic levels must be disinfected prior to discharge. Typically, chlorine (free or combined) is used to disinfect these wastewaters. Ozone, ultraviolet (UV) radiation and other nontraditional disinfection processes are gaining acceptance due to stricter regulations on the amount of residual chlorine levels in discharged wastewaters.

The pH of a wastewater is of paramount importance to a receiving stream and POTW. Biological micro organisms, used in wastewater treatment, are sensitive to extreme fluctuations in pH. Companies that are found to be the responsible polluter are fined and/or ordered to shut down operations until their pH level meets acceptable values. Wastewater discharge values that range from 5 to 9 on the pH logarithmic scale are usually acceptable.

Low pH values are more damaging to a receiving stream and POTW biological treatment process. The food processing industry utilises water to meet its individual day-to-day needs. Fifty per cent of the water used in the fruit and vegetable sector is for washing and rinsing. Water is the primary ingredient in products for the beverage and fermentation sector and dairies utilise water as the standard cleaning agent for process machinery.

26.2.7 Defining load using BOD_5 and COD

Chemical Oxygen Demand (COD) and biochemical oxygen demand (BOD_5) are common measurements used to determine water quality. They measure the strength of the waste stream by measuring the oxygen required to stabilise the wastes. The five-day biochemical oxygen demand (BOD_5) value is used as a gauge to measure the level of treatment needed to discharge a wastewater safely to a receiving water treatment center. The BOD for all food processing wastewater is relatively high compared to other industries. A high BOD level indicates that a wastewater contains elevated amounts of organic material, dissolved and/or suspended solids, minerals, nitrogen and phosphorus.

COD and BOD_5 are important to the food processing industry because they can be used to indicate lost product and wasteful practices. High BOD_5 and COD levels indicate increased amounts of product lost to the waste stream. Measurements at various process locations can help locate sources of waste.

26.2.8 Relating COD to BOD_5

At any point in a particular food processing operation, the relationship between BOD_5 and COD is fairly consistent. However, the ratio's of these two measures varies widely with the type of product (Table 26.2).

Table 26.2: Typical values of BOD_5 and COD for different food plant wastewater.

Type of processor	*BOD_5 (mg/L)*	*COD (mg/L)*	*BOD_5/COD*
Bakery products	3200	7000	0.46
Dairy processing	2700	4700	0.57
Jams and jellies	2400	4000	0.60
Meat packing	1433	2746	0.52
Meat specialties	530	900	0.59
Poultry processor	1306	1581	0.83

26.3 Environmental protection

26.3.1 Source reduction

The most effective method of environmental protection and reducing your disposal costs is to decrease the volume of waste material and by-products generated in the production process. If less waste is generated, then less material needs to be disposed of. Source reduction should be the most logical starting point for reducing disposal costs since your company is in business to produce a saleable product, not waste materials or by-products.

Examples of source reduction include:

1. Use brooms and scrapers to clean floors and equipment while they are dry before washing them down with water.
2. Use high-pressure spray washes during cleanup to conserve water.
3. Dedicate mixing lines to certain products to reduce changeover cleanups.
4. Minimise spills and leaks on the production line to prevent raw materials from becoming wastes.

26.3.2 Management alternatives

If source reduction is not a viable solution, management alternatives exist, including:

1. Using the food by-product as an animal feed.
2. Composting or land spreading the food by-product.

Animal feed

Feeding food by-products directly to livestock allows for former wastes to be useful again. In addition, the quantity of liquid and solid waste is reduced when by-products are fed to livestock rather than being disposed of in landfills or wastewater treatment plants.

Composting and land spreading

When it is impractical to feed by-products to livestock, both composting and land spreading the food waste are viable alternatives. Both methods degrade food by-products into a useful soil additive called 'humus.' Composting degrades by-products above ground in a concentrated area, while land spreading degrades by-products beneath the soil in a cultivated field.

Composting: With proper management, food by-products can be kept out of the landfill and instead be composted and added to the soil at appropriate rates. Composting has the following benefits:

1. Low transportation costs: The by-products can be composted on site and the resulting humus can have a volume and weight reduction of up to 40%.
2. Low capital investment: Composting is a batch process that can be done by using either a mound or a windrow system. In both systems all the by-products are managed to accelerate biological breakdown.

3. Good for seasonal processors: For a company (such as a cannery) that only processes food for several months a year, composting may be a suitable alternative to animal feeding or land filling. Livestock producers may be unwilling to switch to a livestock feed that is only available for a short period.
4. Long shelf life: Humus can be stored without spoiling and applied to enrich the soil as needed.

Land spreading: If your company has sufficient land, it is possible to incorporate food by-products directly into the soil on site . A farmer can be paid to take the by-products to a suitable field. Again, with proper management, food is kept out of the landfill and is used to enhance the soil. Land spreading has the following benefits:

1. A separate compost facility is not necessary.
2. The finished product does not have to be stored.
3. The finished product does not have to be transported. It is left in the soil as a plant nutrient.

26.4 Clean technology developments

Because wastewater generation is the industry's biggest area of concern, the following clean technologies focus on source reduction, recycling, reuse and treatment of wastewater.

Clean technologies are defined as 'manufacturing processes or product technologies that reduce pollution or waste, energy use, or material use in comparison to the technologies that they replace.' The food processing industry has special concerns about the health and safety of the consumer. It should be noted that some of the technologies outlined in the report target both human health and environmental pollution issues.

Common source reduction methods employed at most plants include improving good housekeeping practices, making process modifications, substituting more environmentally friendly raw materials and segregating waste streams. Some simple cost-effective means of achieving source reduction include installing automatic shut-off valves, using low-flow or air injected faucets/spray cleaners, switching from chemical caustic peeling processes to mechanical peeling and converting from water to mechanical conveyance of raw materials through a production line.

26.4.1 Advanced wastewater treatment practices

Advanced wastewater treatment is defined as any treatment beyond secondary (or biological) treatment. These treatment practices are employed to target specific discharge constituents that are of concern. Typically, pathogens, suspended solids, dissolved solids, nitrogen and phosphorus are removed in advanced wastewater treatment. The following is a listing of some technologies being used in advanced wastewater treatment.

1. Membrane applications.
2. Disinfection.
3. Charge separation.
4. Other separation practices.

Membrane applications

Membrane applications focus on separating water from contaminants, using semi permeable membranes and applied pressure differentials. In generic terms, they work like window screens that let air but not insects and other larger objects pass through. The smaller the screen holes, the smaller the objects need to be to pass through. Pressure is applied to reverse the natural equilibrium between the clean water and wastewater. The basic principle of natural equilibrium is that the clean water tends to migrate to the wastewater side to equalise the concentrations across the membrane. Mechanical pressure is used to force water molecules from the wastewater side to the clean water side and, thus, a 'high-tech' filtration of the wastewater occurs. In the past, the energy needed to apply the pressure and the fragility of the membrane surface made use of these alternatives economically unjustifiable. There are varying degrees of membrane filtration. Microfiltration, Ultrafiltration (UF) and Reverse Osmosis (RO) are the current membrane systems used commercially.

The filtering capabilities of each (i.e., ability to filter based on contaminant particle size) decrease respectively. Microfiltration is only recommended for removing particles from 0.05 to 2 microns in size, UF is used for particles and suspended solids from 0.005–0.1 microns and RO is used for particles, suspended solids and dissolved solids in the Angstrom range (e.g., molecular weight above 200).

Problems with membrane applications include befouling of the membrane and fragility of the membrane surface. Toxic synthetic compounds can oxidise the surface of the membrane, thus, destroying it. New innovations in membrane technology have advanced the 'cleanability' and reuse of membranes. The use of stainless steel and ceramic materials for membranes has greatly improved their use in advanced wastewater treatment.

Sanitary conditions have always been a concern for food products created in the manufacturing process. In recent years, they have also become a requirement of wastewater effluent. As for water treatment practices, disinfection through chlorination has been the quickest means of disinfecting wastewater. Disinfection has come under criticism due to chlorination by-products and toxicity concerns that residual chlorine pose to aquatic life. The two principal means of disinfecting wastewater without using chlorination are ozone disinfection or UV disinfection. Ozonation works on the same principle as chlorination but leaves no residual in the treated wastewater and does not produce the magnitude of disinfection by-products that chlorination produces. UV disinfection is even more environmentally friendly than ozone but requires more space and cleaner wastewater to be effective. Both technologies require high capital and operating costs.

Charge separation

Charge separation involves separating uncharged water molecules and charged contaminants, such as nitrogen compounds and phosphates (i.e., NH_4^+, NO_2^-, NO_3^- and PO_4^{-3}). Electro-coagulation is starting to be an economical way of removing charged particles from wastewater, utilising charge separation. Ion exchange is widely used to filter wastewater through cationic and anionic resins to remove the wastewater's charged ions of concern. Ion exchange replaces the waste particles with a donor ion from the resin. The resins eventually reach a capacity at which all the ions have been replaced or exchanged. The resin manufacturer typically recycles spent resin. Problems with using ion exchange are that it requires monitoring for breakthrough contamination and pH fluctuations can greatly affect the removal rates of specific ions (e.g., a pH greater than 9.3 makes ammonium removal inefficient). Also, resins remove ions selectively, meaning the greater the charge differential from neutrality, the greater the exchange attraction between the resin and the charged contaminant (e.g., Ca^{+2} will be removed before NH_4^+).

Other separation practices

Other separation practices include using centrifugal and gravity mechanisms to separate and remove contaminants from a wastewater. Air flotation systems use diffused pumped air to lift suspended solids and FOG wastes to the surface of a wastewater for removal. Skimmers and mechanical devices are then employed to separate waste from the surface. Problems with using either of these methods include capital costs to modify current treatment processes and increased operational energy costs.

With the exception of centrifugal and gravity separation, all these advanced treatments require a wastewater influent that is low in turbidity.

Benefits

Studies have shown that membrane applications can be less energy intensive than evaporation and distillation operations and take up less space. The technology gives better control of the process effluent. Unlike chemical precipitation, membrane technology does not produce a sludge disposal problem, but it does produce a concentrated brine solution. The main benefit of disinfecting wastewater is that it improves and protects water quality of and aquatic life in the receiving water. Similar to membrane applications, ion exchange does not produce a chemical sludge and, like disinfection, it protects the water quality of receiving water and decreases the nutrient-loading problems that cause eutrophication in receiving waters.

Electro-coagulation is beginning to receive attention as a treatment option and is expected to increase in use in the food processing industry.

Centrifugal and gravity separation processes are placed before any of the preceding advanced operations. This ensures that a cleaner, less turbid wastewater reaches these advanced operations. As stated earlier, the recovered FOG is a resalable by-product. Use of any of these advanced processes improves the final wastewater effluent quality and also increases the likelihood of recycling renovated process water.

26.4.2 Water and wastewater reduction (closed loop/zero emission systems)

An increasingly viable option for companies is the 'zero-discharge' system. Many food processing facilities are looking to pre-treatment options that can help reduce the amount of lost product. Once a part of the food product is lost to a waste stream, it represents a decrease in product utilisation and an increase in treatment costs. A large capital expenditure and a customised treatment solution are required to handle a zero-discharge option. Furthermore, the uniqueness of the various food processing operations makes it impossible to find 'off-the-shelf' treatment designs to fit a user's needs. A more plausible approach is that of achieving zero emissions. As noted earlier, the 'zero emissions' strategy relies on a network of companies utilising each other's waste streams. The strategy is a more economically efficient system than a 'closed loop' because the waste products do not have to be fully treated. Although facilities are moving toward decreased effluent quantities, material mass balances still dictate that process residuals such as sludge will require management and possibly off-site disposal. Both zero discharge and zero emission systems achieve better effluent water quality and have fewer negative impacts on the environment.

26.5 Future trends

26.5.1 Regulations and standards

International standards developed by the Geneva based International Organisation of Standardisation, called ISO 14000, represent the latest attempts to provide a global environmental management system. ISO 14000 was intended to help organisations manage and evaluate the environmental aspects of their operations without being prescriptive. The International Organisation of Standardisation intends to provide companies with a framework to comply with both domestic and foreign environmental regulations. ISO 14000 contains sections calling for implementation of pollution prevention programmes and many U.S., companies are evaluating the pros and cons of becoming fully certified in ISO 14001. Furthermore, EPA is talking about easing reporting requirements for U.S., companies that earn ISO 14001 certification.

26.5.2 Industry trends

There are several ongoing trends and research and development activities apparent within the food processing community in the areas of pollution prevention and clean technology implementation.

Solid waste reduction

Companies will continue to look at ways to reduce solid waste generation, use less or reusable packaging and use biodegradable packing products. Excessive packaging has been reduced and recyclable products such as aluminium, glass and HDPE are expected to continue being used to a wider degree in packaging situations.

Mechanical versus chemical processing

Companies will show increased consideration for using mechanical methods for food processing (e.g., the fruit and vegetable sector). Mechanical processing can be used to perform many of the same functions as chemical processing. The costs and benefits of using mechanical versus chemical processing will be further quantified to aid in decision making.

26.5.3 Pre-treatment options, water conservation and wastewater reduction

Pre-treatment opportunities and water conservation will continue to be principal targets for pollution prevention source reduction practices in the food processing industry. Pre-treatment options look to minimise the loss of raw materials to the food processing waste streams. Water used in conveying materials, facility cleanup, or other non-ingredient uses will be reduced, which in turn will reduce the wastewater volume from food processing facilities. Wastewater treatment will continue to be the pollution prevention treatment focus for food processing companies. The industry will continue to implement advanced innovative techniques to lessen the environmental impact of food processing discharge wastewaters.

27

Pollution control in wheat, pulse and rice mills

27.1 Introduction

Pulses occupy important place in the world food and nutrition economy. Pulses are an important constituent in the diet and are primary sources of protein. They also provide substantial quantity of minerals and vitamins to the diet.

Wheat is commonly used all over the country. Numerous type of qualities of wheat products and by-products of wheat like wheat flour, suji, maida and dalia, etc., are produced by the wheat flour and processing mills. The wheat flour is used in domestic purposes for making chapatti. Wheat in combination of maida is used for making bread, biscuit, etc.

Rice is also a staple food in India. It is processed from paddy both at household level and industrial level. It is either used in raw or parboiled form. There are many varieties of paddy cultivated in different parts of the country.

Primary milling of rice, wheat and pulses is the most important activity in food grains. These grains are grown in almost all parts of world. Today, due to industrialisation and global competitive market trend, it has emerged as a major industrial activity in small medium scale sector to cater to the needs of increasing population.

27.2 Environmental pollution aspects in pulse/wheat/rice mills

The raw material received by the mills commonly contain much fine dust in addition to long fibre shaped dust particles. In the conveying equipments, the rubbing friction of the raw-material creates fine dust particles. Dust is generated at the transfer points of the enclosed conveying equipment, carried through bucket elevators and emitted at the discharge of the conveyed materials.

The main pollutant of concern in pulse, wheat & rice milling facilities is particulate matter emission generated from material handling, cleaning, milling or packing operations. These operations are typically located inside the mill buildings. Dust released from these operations spread inside the shop floor environment of the factory. Dust extraction systems are generally used to collect the dust and to improve the shop floor environment. Dust emitted inside the mill may settle on the floor and wall surfaces, or may be emitted to the environment through doors and windows.

The environmental aspects associated with the pulse, wheat and rice mills from the view point of air pollution, water pollution and noise, etc., are discussed below.

27.3 Air pollution

27.3.1 Pulse making mills

During the process of the pulse making, dust is generated/emitted from various points, which spreads into the surrounding areas and pollute the shop floor environment.

Sources of dust emissions

1. Cleaning section: In the cleaning section, fine dust is generated at various points as detailed below:
   (a) While unloading the pulses.
   (b) At different stages of lifting and discharging of pulses through bucket elevator and pipes.
   (c) During rotary screening.
   (d) During scratching of pulses for breaking of tips in roller type m/c.
   (e) At discharge points of the rotary screens.
   (f) De-stoner machines.
   (g) Blown away dust/bhussi.
2. Milling section: In this section, the dust generated is lesser compared to the cleaning section, as most of the dust has already been cleaned in the cleaning section. The dust emanating from this section is mainly due to breaking of chana and its movement. The points of generation are:
   (a) The bucket elevator discharge points.
   (b) Chakki discharge (milling/grinding).
   (c) Chillka collection/storage room.
   (d) The reel machines.
   (e) The grinding chakki/machines.
   (f) The rotary screen classifiers.

The dust and powder emitted from the cleaning and milling sections need to be separately extracted and captured in separate air pollution control systems. The dust collected from the cleaning section primarily contains impurities like soil, silica, stones, etc., which needs to be disposed, whereas, the powder collected from the milling section is a by-product which can be sold as animal feed.

27.3.2 Pulse grinding mills

In the pulse grinding mills, since the raw material is clean dal, there is hardly any dust emission at the sections like bucket elevator discharge point or classifying screen. After grinding, the besan is pneumatically transported and separated in the centrifugal/cyclone type equipment and the fine besan escaping the centrifuge is collected in filter bags. No fine besan is allowed to escape from leakages, etc., as it would be loss of product. There are no fugitive emission sources observed in these mills.

Sources of dust emissions

In the pulse grinding mills, the source of emissions could be leakages in the system if any. No stack is provided discharging emissions to environment.

27.4 Wheat flour mill

27.4.1 Sources of dust emissions

Cleaning section

1. While unloading the wheat.
2. At different stages of lifting and discharging of wheat through bucket elevator and screw conveyor.
3. At reel machine.
4. During vibrating screening.
5. At de-stoner machines.
6. At scourer machine.
7. At aspirator machine.
8. At magnetic separator.
9. At toggle cylinder.

Milling section

1. At bucket elevator discharge points.
2. At grinding mill discharge.
3. At purifier machine.
4. At plansifter machine.

27.4.2 Wheat processing (roasting) mill

In the wheat roasting mill, the dust emission occur primarily from the cleaning section and fumes/combustion gases are emitted from the roaster.

The points of dust/fume emissions are as given below:

Sources of dust emissions

1. Raw wheat unloading point.
2. Wheat conveying bucket elevator discharge points.
3. Rotary screen.
4. Roaster feeding end and discharge end.
5. Points of free fall of wheat through pipes.

Sources of fumes/gas emissions

1. Oily fumes from the material discharge end of the roaster.
2. Combustion gas emission through leakages in the roaster walls and roofs.

The dust/fume generated from above mentioned sources need to be extracted/sucked from various point as close to the source of generation as possible and it should be directed through efficient control equipment.

27.5 Rice mill

27.5.1 Sources of dust emissions

Cleaning section

1. While unloading the paddy.
2. At different stages of lifting and discharging of paddy rice through bucket elevator.
3. During pre-cleaning.
4. During paddy cleaning in paddy cleaner/vibrating screen.
5. At destoner machine.
6. At magnetic separator.

Milling section

1. At different stages of lifting and discharging of paddy/rice through bucket elevator.
2. During dehusking in rubber roll.
3. At Aspirator used for husk removal.
4. During polishing of rice.
5. During grading of rice in rice grader.

Boiler house: Flue gas through boiler stack.

27.6 Water pollution

27.6.1 Pulse making mills

In the pulse making mills, there is no wastewater generation from process, since whatever little quantity of water if used for conditioning the pulses gets consumed/evaporated completely. However, the storm (rain) water from the rooftops may carry the organic residual material comprising of pulses/shells, etc., lying on the roof terrace during sun-drying. This storm water is generally let out to the public sewer directly.

27.6.2 Rice mills

In rice mill, the wastewater is usually generated in the parboiled rice mills from the soaking operation. In addition, in few older mills where mechanical cleaning facilities are not available, the cleaning is done by washing of paddy prior to further processing and this leads to generation of wastewater. In the parboiled type of rice mills, wastewater is also generated from the boiler section in the form of boiler blow down, backwash of DM plant/softener and from discharge of wet scrubber type air pollution control equipments installed for the controlling of boiler stack emission. In the dry rice grinding mills, there is no use of water in the process and therefore there is no generation of wastewater.

27.7 Noise pollution

Noise is generated from various material handling and process equipments, which are highlighted in Table 27.1.

Table 27.1: Typical noise pollution sources in pulse/wheat/rice mills.

Types of mill	*Section*	*Points of noise emission*
Pulse	Cleaning	Vibrating screen Reel machine Destoner Bucket elevator
	Milling	Bucket elevator Chakki
	Cleaning	Separator Scourer Vibratory screen Reel machine Destoner

(Cont'd...)

Types of mill	*Section*	*Points of noise emission*
Wheat	Milling	Packing machine Roller mill Purifier Planshifter Blower Bran finisher Milling fan ID Fan
	Cleaning	Destoner Vibratory screen Bucker elevator Dryer
Rice	Milling	De-stoner Polisher Colour sorter Length grader Blower Dehusker Paddy separator Thickness grader Whitener Packing section Paddy cleaner
	Boiler house	ID fans, soot blowing

Apart from the above sections, noise is also generated from the diesel generating sets.

27.8 Emission control practices in U.S.

27.8.1 Techniques used to reduce emissions

1. Controls are often used at unloading and loading areas.
2. Control systems include cyclones, fabric filters, baffles in unloading pits, choke unloading, dust covers, or belt wipers, oil suppression and use of dead-boxes or specially designed spouts for grain loading.
3. Deepening the trough of the open-belt conveyors.
4. Slowing the conveyor speed.
5. Increasing the size of leg belt buckets.
6. Slowing leg velocity.
7. In addition, for modern elevators, process vents are tied to ventilation systems and exhausted to air pollution control systems.

27.8.2 General types of measures to reduce emissions

Process modifications for emission prevention/reduction

1. Enclosing the receiving area to the degree practicable.
2. Specifying dust-tight cleaning and processing equipment.
3. Providing lip-type shaft seals at bearings on conveyor and other equipment housings.
4. Providing flanged inlets and outlets on all spouting, transitions and miscellaneous hoppers.
5. Full enclosure and sealing of all areas in contact with products handled.
6. By reducing grain free fall distances and grain velocities.
7. Extraction of the entrained air and dust liberated at the point of grain free falls to a dust collector.

27.9 Emission control practices in Germany

1. All equipment handling materials and machinery should be enclosed and fitted with appropriate aspiration connections.
2. Mass separators or filter separators may be used.
3. Dust from aspiration and cleaning systems to be collected and bagged.
4. Dust waste and granular cleaning waste is treated and added to mill after products (bran).
5. The live pests from cleaning waste to be destroyed immediately.

27.9.1 Noise control requirement in Germany

1. Precautions must be taken to protect employees and local residents.
2. Structural measures, such as the lining of ceilings and walls with sound-proofing materials, must be taken.
3. Vibration isolation materials must be used for machine foundations.

27.10 Suitable treatment technology

27.10.1 Primary treatment

Screen

The wastewater generated after washing of wheat carry floating waste agricultural material (like stem, husk, etc.). These materials clog the drain and obstruct during pumping of wastewater from one treatment unit into other. It may lead to frequent breakdown of pump. Therefore, to avoid these difficulties, it is recommend to install a two stages screening system consisting of coarse and fine screens.

To facilitate manual cleaning of the screens, a screen chamber should be constructed based on the peak flow. The location of screen chamber should be below the level of the main drain. The suspended/floating solids retained on the screen would be manually removed as and when required and would be disposed off as solid waste in a safe manner. The screened wastewater should either flow by gravity or pumped to the settling tank/primary clarifier. In case of pumping, a pump sump should be constructed to collect and transfer wastewater coming from screen to the settling tank/primary clarifier.

Oil and grease trap

The oil and grease concentration in the wastewater is in the range of 5~10 mg/L. It is meeting the normal discharge limit and it won't affect the biological treatment. Therefore, there may not be a need to install oil and grease trap.

Chemical treatment

The pH of the wastewater generated from washing operation of wheat flour mill is in the range of 6.4~7.5. There is no possibility of the presence of heavy metals. Therefore, there is no need for going in for chemical treatment.

Settling tank/primary clarifier

The suspended solid concentration varies between 215~325 mg/L. The desirable suspended solid concentration should be less than 600 or 200 or 100 mg/L for discharge into public sewer or land for irrigation or inland surface water respectively. Therefore, a settling tank/primary clarifier is required to be installed for the settlement of suspended solids for discharging either on land or surface water.

27.10.2 Secondary biological system

The BOD concentration of the wastewater varies in the range of 450~600 mg/L. The desirable BOD concentration should be 350 or 100 or 30 mg/L for discharge into Public Sewer or land for irrigation or inland surface water respectively. Therefore desirable BOD efficiency varies between 42 to 95%. This could be achieved by means of different modes of activated Sludge Process (ASP) such as high rate, conventional, extended aeration process, etc. A sedimentation tank/secondary clarifier should be installed close to the aeration tank for solid liquid separation and sludge recirculation, so as to maintain the desired range of MLSS concentration

27.10.3 Tertiary treatment system

There is no need to go for tertiary treatment of the wastewater. The proposed limit of discharge could be achieved through primary and secondary treatments, discussed above.

28

Pollution control in vegetable and palm oil industry

28.1 Introduction

The production process of vegetable oil involves the removal of oil from plant components, typically seeds. This can be done via mechanical extraction using an oil mill or chemical extraction using a solvent. The extracted oil can then be purified and, if required, refined or chemically altered.

Mechanical extraction: Oils can be removed via mechanical extraction, termed 'crushing' or 'pressing.' This method is typically used to produce the more traditional oils (e.g., olive, coconut, etc.), and it is preferred by most 'health-food' customers in the United States and in Europe.

Solvent extraction: The processing of vegetable oil in commercial applications is commonly done by chemical extraction, using solvent extracts, which produces higher yields and is quicker and less expensive. The most common solvent is petroleum-derived hexane. This technique is used for most of the 'newer' industrial oils such as soyabean and corn oils. Supercritical carbon dioxide can be used as a non-toxic alternative to other solvents.

Hydrogenation: Oils may be partially hydrogenated to produce various ingredient oils. Hardening of vegetable oil is done by raising a blend of vegetable oil and a catalyst in near-vacuum to very high temperatures and introducing hydrogen.

This causes the carbon atoms of the oil to break double-bonds with other carbons, each carbon forming a new single-bond with a hydrogen atom. Adding these hydrogen atoms to the oil makes it more solid, raises the smoke point and makes the oil more stable.

Deodorisation: In the processing of edible oils, the oil is heated under vacuum to near the smoke point and water is introduced at the bottom of the oil. The water immediately is converted to steam, which bubbles through the oil, carrying with it any chemicals which are water-soluble. The steam sparging removes impurities that can impart unwanted flavours and odours to the oil. Deodorisation is key to the manufacture of vegetable oils. Nearly all soyabean, corn and canola oils found on supermarket shelves go through a deodorisation stage that removes trace amounts of odours and flavours and lightens the colour of the oil.

28.2 Environmental issues related to vegetable oils

Environmental issues associated with the operational phase of vegetable oil production and processing primarily include the following:

1. Solid waste and by-products.
2. Water consumption and management.
3. Energy consumption and management.
4. Atmospheric emissions.
5. Greenhouse gas emissions.
6. Hazardous materials.

28.2.1 Solid waste and by-products

Vegetable oil processing activities generate significant quantities of organic solid waste, residues and by-products, such as Empty Fruit Bunches (EFBs) and waste palm kernels from palm oil processing or olive oil cake and pulp from olive processing. The amount of waste generated depends on the quality of the raw materials and on process efficiency. Wastes, residues and by-products may be used for producing commercially viable by-products or for energy generation. Other solid wastes from the vegetable oil manufacturing process include soap stock and spent acids from chemical refining of crude oil; spent bleaching earth containing gums, metals and pigments; deodoriser distillate from the steam distillation of refined edible oils; mucilage from degumming; and spent catalysts and filtering aid from the hardening process.

Recommended techniques for minimising the volume of solid waste and by-products for disposal include the following:

1. Reduce product losses through better production/storage control (e.g., monitor and adjust air humidity to prevent product losses caused by the formation of molds on edible materials).
2. Collect residues from the raw material preparation phase for conditioning (drying) and reprocessing (grinding) to yield by-products (e.g., animal feed).
3. Return waste and residues to fields to assist in soil nutrient management; for example, EFBs from oil palm plantations with tree trimmings are a valuable soil amendment and/or can be composted with vegetable oil wastewater effluent.
4. Use waste and residues for energy generation in the project plant's boiler(s). Note, however, that relatively high atmospheric emissions [such as particulate emissions (PM)] are possible when burning crop residues and potential fire risks (e.g., from combustible dust) may arise from handling, storing and processing crop residues; as such, expert advice on fuel characteristics and boiler design should be solicited when planning to use biofuels in this manner.

5. Investigate the following options for the responsible disposal of spent bleaching earth:
   (a) Use as fertiliser, if not contaminated with heavy metals such as nickel, pesticide residues, or other contaminants.
   (b) Recover non-food-grade oils from spent bleaching earth that could be used in other applications (feedstock for conversion to biodiesel or in bio-lubricants).
   (c) Avoid direct recycling on agricultural land. Add spent earth to other organic waste and compost to avoid contact with air and risk of spontaneous combustion of spent bleaching earth.
   (d) Consider use as a feedstock for brick, block and cement manufacturing.
6. Investigate the following options for the use of distillates [e.g., free fatty acids and volatile organic compounds (VOCs)], depending on the level of contaminants (pesticides and/or residues):
   (a) Use free fatty acid as animal feed if uncontaminated.
   (b) Apply as a feedstock for chemical industry processes (e.g., antioxidants).
   (c) Use as fuel for energy production.
7. The nickel catalyst from hydrogenation should be either:
   (a) Recycled and recovered for reuse as a nickel catalyst or as nickel metal, salt, or other application.
   (b) Stored and disposed of according to the hazardous waste management guidelines.
8. Manage filtering aid mixed with nickel in accordance with the recommendations for nickel catalyst.
9. Use uncontaminated sludge and effluent from on-site wastewater treatment as fertiliser in agricultural applications or as a supplemental boiler fuel. Dispose of contaminated sludge from wastewater treatment at a sanitary landfill or by incineration. Incineration should only be conducted in permitted facilities operating under internationally recognised standards for pollution prevention and control.

28.2.2 Water consumption and management

Vegetable oil facilities require significant amounts of water for crude oil production (cooling water), chemical neutralisation processes and subsequent washing and deodorisation. Sector-specific recommendations to reduce water consumption, optimise water use efficiency and reduce subsequent wastewater volumes include the following:

1. When economically viable, consider the use of physical refining instead of chemical refining to reduce water consumption.

2. Replace water-based conveyor systems by mechanical systems (augers or conveyors).
3. Apply cleaning-in-place (CIP) procedures to help reduce chemical, water and energy consumption in cleaning operations.
4. Recover and reuse condensate from heating processes.
5. Upgrade equipment water sprays (e.g., to include jets or nozzles).
6. Use dry cleanup techniques before rinsing floors.
7. Manually clean vessels before rinsing to remove solids for recovery or disposal.
8. Use high-pressure, low-volume washing systems and auto shut-off valves.
9. Vegetable oil processing wastewater generated during oil washing and neutralisation may have a high content of organic material and, subsequently, a high biochemical oxygen demand (BOD) and chemical oxygen demand (COD). Wastewater may also have a high content of suspended solids, organic nitrogen and oil and fat and may contain pesticide residues from the treatment of the raw materials. Recommended measures to reduce contaminant loading includes the following:
   (a) Install spill collection trays to collect solids at appropriate places in the production line.
   (b) Use emulsion breaking techniques, [e.g., dissolved air flotation (DAF)], to segregate high BOD and COD oils from wastewater.
10. Use grids to cover drains in the production area to prevent solid wastes and concentrated liquids from entering the wastewater stream.
11. Select disinfection chemicals to match the cleaning operation being applied on the process equipment to the type of problem. Caustics (e.g., lye) are typically used for polymerised fat and acids are used for lime deposit acids.
12. Apply cleaning chemicals using the correct dose and application method.
13. Properly treat and discharge cleaning solutions (e.g., through a soap-splitting process) to separate oil and fatty acids from the water phase and then pass through a fat trap.
14. When appropriate and feasible, reduce phosphoric acid in degumming operations through the use of improved neutralisation processes or alternative methods, such as enzymatic degumming (this reduces the phosphorus load in the wastewater and also brings about a slight reduction in sludge quantities).

Process wastewater treatment

Techniques for treating industrial process wastewater in this sector include: grease traps, skimmers or oil water separators for the removal of floatable solids; flow and load equalisation; sedimentation for suspended solids reduction using clarifiers, biological treatment—typically anaerobic, followed by aerobic treatment—for the reduction of soluble organic matter BOD, biological nutrient removal for reduction in nitrogen and phosphorus; chlorination of effluent when disinfection is required and dewatering and disposal of residuals. In some instances, composting or land application of wastewater treatment residuals of acceptable quality may be possible. Additional engineering controls may be required to contain and neutralise nuisance odours.

28.2.3 Energy consumption and management

Vegetable oil processing facilities use energy to heat water and produce steam both for process applications (especially for soap splitting and deodorisation) and cleaning processes. Other common energy consumption systems include refrigeration and compressed air. In addition to the energy conservation sector-specific recommendations include the following:

1. Improve uniformity of feed to stabilise and reduce energy requirements.
2. Increase efficiency of air removal in sterilisation vessels to improve heat transfer.
3. Identify and implement opportunities for process heat exchange, e.g., optimised oil-oil heat exchangers in continuous deodorisation.
4. Reduce stripping steam consumption by improving process efficiency; e.g., improve stripping tray design. Where possible, consider technologies such as dry ice condensing systems that may lower energy consumption.
5. Consider co-generation (combined heat and power (CHP)) to improve energy efficiency.
6. Consider more advanced approaches—such as the use of enzymes—for processes such as degumming and oil recovery.
7. Where feasible, use anaerobic digestion for wastewater treatment and capture methane for heat and/or power production.

28.2.4 Atmospheric emissions

Process emissions

Particulate matter (dust) and VOCs are the principal emissions from vegetable oil production and processing. Dust results from the processing, including cleaning, screening and crushing, of raw materials, whereas VOC emissions are caused by the use of oil-extraction solvents, normally hexane. Several sources within vegetable oil processing plants generate solvent emissions, including the solvent-recovery unit, the meal dryer and cooler, leaks in piping and vents and product storage. Additional emissions will result from the refining process if a fractionation method is used. Small quantities of solvent may be present in the crude vegetable oil if the oil has been extracted by a solvent and will volatilise during the oil refining process, particularly during deodorisation. Odour emissions are produced by multiple sources (e.g., cookers, soap splitting and vacuum generation).

Recommended management techniques to prevent and control VOCs include the following:

1. Process improvements, for example:
   (a) Optimise recovery of solvents by distilling the oil from the extractor.
   (b) Back-vent to the solvent delivery tanks during bulk storage tank filling.
   (c) Improve exhaust air collection systems.
   (d) Implement leak prevention systems.
2. Adoption of abatement technologies:
   (a) Recover solvent vapours where feasible, primarily through the use of counter current flow desolventiser – toaster in vegetable oil extraction.
   (b) Use a condenser, a reboiler and a gravity separator to treat condensates with high solvent content, to reduce both solvent emissions and the risk of explosions in the sewer.
   (c) Treat hexane-laden air from the condenser/reboiler process with a mineral oil scrubber.
   (d) Consider cryogenic condensation in the solvent fractionation process. Best practice approaches use a closed loop process in which 99.9% of the solvent input is reused.

The recommended management techniques to prevent and control dust and odours include the following:

1. Ensure proper maintenance of cleaning, screening and crushing equipment—including in any ventilation and air handling systems—to reduce emissions of fugitive dust and avoid the use of compressed air or steam for cleaning.
2. Install cyclones and/or fabric filters or electrostatic precipitators on selected vents—including meal dryers, coolers and grinders—to remove odour emissions.
3. Reduce odour emissions (e.g., from soap splitting, cookers in the extraction process, vacuum systems and pressurised systems) with a caustic, alkaline, or ozone scrubber system, or incinerate the gas in a boiler plant or in separate incinerator systems.

Combustion products

Vegetable oil processing plants are large energy consumers, making use of auxiliary boilers for the generation of steam energy. Emissions related to the operation of these steam energy sources typically consist of combustion by-products, such as NO_x, SO_x, PM, VOCs and greenhouse gases (CO_2). Recommended management strategies include adoption of a combined strategy, which involves a reduction in energy demand, use of cleaner fuels and the application of emissions controls, where required.

28.2.5 Greenhouse gas emissions (GHG)

Vegetable oil processing produces GHG emissions through the use of fossil energy.

1. The high nutrient loading of wastewater can be a source of methane (CH_4) when treated or disposed of anaerobically. It can also be a source of nitrous oxide (N_2O) emissions associated with the degradation of nitrogen components in the wastewater (e.g., urea, nitrate and protein). Recommended measures to prevent and control non-fossil-fuel-related GHG emissions include:
   (a) Avoid open anaerobic conditions for wastewater treatment by ensuring a regular programme of operational maintenance in the wastewater treatment system.
   (b) Consider biological methods of wastewater treatment, such as anaerobic digestion and methane capture; use of waste effluent for irrigation; co-composting of by-products, where appropriate (e.g., oil palm empty fruit bunches with palm oil mill effluent nutrient waste or olive mill waste residue with wastewater) and detoxification by nitrogen fixation.

28.2.6 Hazardous materials

Vegetable oil processing involves the transport, storage and use of bulk quantities of acids, alkalis, solvents and hydrogen during extraction and refining. Their transport, storage and handling provide opportunities for spills or other types of releases with potentially negative impacts on soil and water resources. Their flammability and other potentially hazardous characteristics also present a risk of fire and explosions.

Chemical hazards

Operators in vegetable oil facilities may be exposed to hazardous substances via, inhalation of hexane or other solvents used for extraction; inhalation of toxic chemicals (e.g., sodium methylate can cause burns on the skin and lung tissue if inhaled); eye or skin exposure to acids or bases; inhalation of dust from the transportation of raw materials (e.g., seeds and beans to the crushing plant); inhalation of dust from meal treatment and shipment; inhalation of dust from bleaching earth, filter aid and nickel catalyst and inhalation of aflatoxins present in raw materials.

Additional industry-related recommendations include the following:

1. In oil extraction areas, ensure that there is adequate air circulation to reduce the concentration of solvents.
2. Provide ventilation, especially at workstations devoted to raw-material handling, milling, handling of bleaching earth and use of solvents.
3. Maintain air concentrations of VOCs below 10% of lower explosive limits. For hexane, the lower explosive limit is 1.1% volume per volume (v/v) and the upper explosive limit is 7.5% v/v.
4. Ensure proper distillation of oil after extraction for effective solvent removal.
5. Prevent leaks and spills of oils in the extraction plant.
6. Control the flash-point temperature of the incoming extracted oils and use temperature control for all facilities receiving solvent-extracted oils.
7. When feasible, use hot water, rather than solvents, to facilitate cleaning.

Physical hazards

Physical hazards in vegetable oil production and processing facilities are similar to those present in other industry sectors and include the potential for falls caused by slippery floors and stairs; injuries caused by unprotected machinery or moving parts; hazards associated with potential collisions with internal transport, such as trucks; and accidental contact with conveyor systems, such as those used in crushing plants and in the removal of spent earth.

Electrical hazards: Electrical systems are a source of danger for workers that can lead to injuries or fatalities.

Risk of fire and explosion

Risks of fire and explosion occur at different stages of vegetable oil production and processing and can lead to loss of property, as well as possible injury or fatalities among project workers. Sector-specific risks are related to the combustibility of vegetable oil and the high volumes of combustible dust present both in grain and oil-seeds handling and in storage facilities. The control and removal of this dust and the control or removal of potential ignition sources are key to eliminating the explosion hazard. The storage of grains and seeds represents a combustion risk, owing to the potential for self-heating and ignition. Silo safety for these products, as well as for oil storage, is critical. Vegetable oil facilities also present the risk of explosions resulting from the volatilisation of solvent dissolved in the oil (e.g., hexane), along with the risk of fire from spent bleaching earth with a high iodine-value oil, high ambient temperature and high circulation-draft of air.

28.3 Environmental issues related to palm oil mills

Palm oil is one of the two most important vegetable oils in the world's oil and fats market following soyabeans. Oil palm (*Elaeis guineensis*) is the most productive oil producing plant in the world.

Crude palm oil contains fatty acid ester of glycerol commonly referred to as triglycerides, therefore, contributing to the worlds need of edible oil and fats. It is composed of approximately 50% saturated fats (primarily palmitic acid) and 40% unsaturated fats (principally linolenic and oleic acid); a unique composition if compared with other major fats. The distinctive colour of the oil is due to the fat soluble carotenoids (pigment) which are also responsible for its vitamins E (tocopherols and tocotrienols) content. There are several stages of processing the extraction of palm oil from fresh fruit bunches. These include sterilisation, bunch stripping, digestion, oil extraction and finally clarification and purifications; each process with its own various unit operations. Palm oil production process is shown in Fig. 28.1. These extraction and purification processes generate different kinds of wastes.

28.3.1 Oil palm waste products

The oil palm mills generate many by-products and wastes beside the liquid wastes that have been mentioned, that may have a significant impact on the environment if they are not properly dealt with. The most common among these by-products is the empty fruit bunch. The empty bunch is a solid waste product of the oil palm milling process and has a high moisture content of approximately 55–65% and high silica content, from 25% of the total palm fruit bunch. The treated empty bunches are mechanically crushed (de-watered and de-oiled) in the process, but are rich in major nutrients and contained reasonable amounts of trace elements. They have a value when returned to the field to be applied as mulch for the enrichment of soil. However, it was noted that over application of the effluent must be avoided as it may result in anaerobic conditions in the soil by formation of an impervious coat of organic matter on the soil surface.

Air emission from the oil palm mills are from the boilers and incinerators and are mainly gases with particulates such as tar and soot droplets of 20–100 microns and a dust load of about 3000 to 4000 mg/nm. Incomplete combustion of the boiler and incinerator produce dark smoke resulting from burning a mixture of solid waste fuels such as shell, fibre and some times empty bunches. These boiler fly ashes are also a waste in them and also pose problems of disposal. In the bid to achieve a zero discharge of the palm oil mill, boiler fly ash have been used to reduce the BOD, TSS, colour and other contaminants from POME before discharge. Boiler fly ash has also been used in the removal of heavy metals from other industrial effluents.

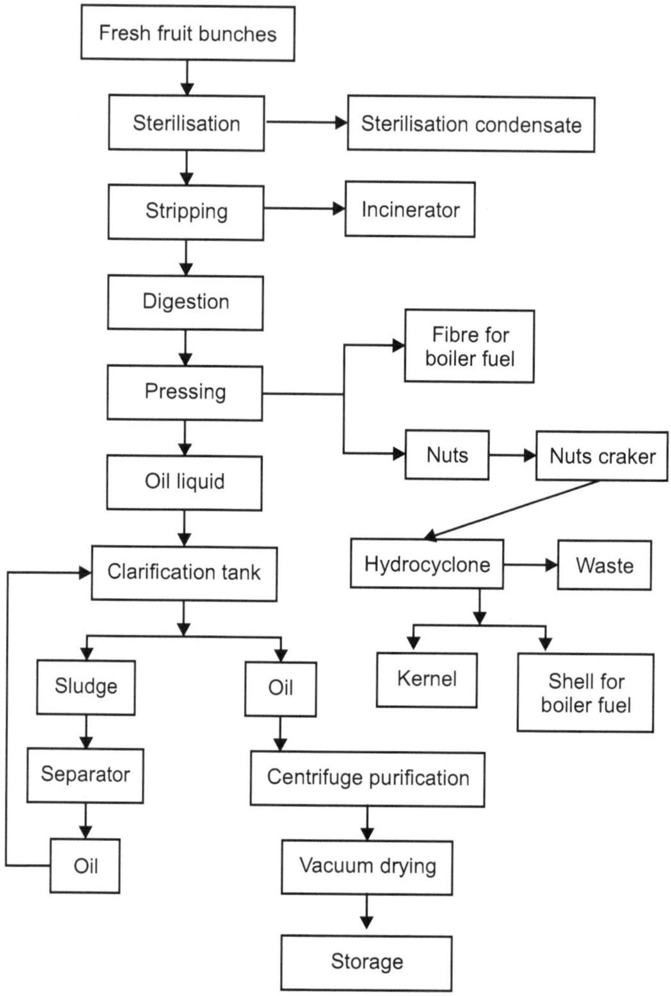

Figure 28.1: Flow diagram of the palm oil mill process.

28.3.2 Palm oil mill effluent (POME)

Effluent water is defined as water discharged from industry, which contains soluble materials that are injurious to the environment. Such soluble materials may be gases such as CH_4, SO_2, NH_3, halogens or soluble liquids or solids which contain ions of either organic or inorganic origin and with their concentration above the threshold value. Since these compounds are harmful to the environment, it becomes necessary that effluents water should be treated or purified before discharged into the environment. Thus, the major objective of industrial effluents treatment is to reduce the amount of these potentially toxic compounds to their acceptable threshold limit value (TLV), according to some standards of the Federal Environmental Protection Agency (FEPA); World Health Organisation (WHO); Department of Petroleum Resources (DPR), etc.

Characteristics of industrial effluents

Generally, the characteristics of industrial effluents are given as follows:

1. Soluble organics resulting in dissolved oxygen depletion in streams and estuaries and/or causing taste and odour.
2. Organic suspended solids resulting in dissolved oxygen depletion.
3. Inert suspended solids causing turbidity and resulting in bottom sediments.
4. Toxic substances and heavy metals.
5. Oil and floating materials.
6. Dissolved salts particularly phosphates, chlorides and nitrates.

Specifically, Palm Oil Mill Effluent (POME), is a general phrase referring to the effluent from the final stages of palm oil production in the mill. It includes various liquids, dirties, residual oil and suspended solids. POME in its untreated form is a very high strength waste, depending on the operation of the process, that is; informal, semi-formal and formal processes, the biological oxygen demand of these wastes ranges from 25000 to 35000 mg/L. It contains about 94% water. POME actually is the sum total of liquid waste which cannot be easily or immediately reprocessed for extraction of useful products and is run down the mill internal drain system to the so called effluent (or sludge) pit.

Sources of wastes

POME is generated from various points during processing in an oil mill. These include:

1. Clarification sludge.
2. Sterilisation condensates.
3. Fruit washing water.
4. Hydro cyclone drain-off.
5. Various boiler blows down, tank and decanters drain.

The composition of the effluent from these various sources, are mainly water, oil, solids (suspended and dissolved) and sand.

28.3.3 Management of palm oil mill effluent

Effluent management involves the typical handling of liquid waste. The mechanical technique often involves sedimentation, filtration and decolorisation of effluent. Mechanical technique is normally at the first stage of purification process to remove suspended solid particles. This is called primary treatment. The commonly used devices include sieve, sedimentation bed and filter. Physicochemical technique involves coagulation of finely dispersed and suspended solid particles, adsorption of the dissolved impurities such as heavy metals, selective crystallisation, reverse osmosis and ion-exchange processes. Reverse osmosis is most often used at the final stage of effluent treatment. Secondary treatment is biological process following primary treatment. The forms of secondary biological process include activated sludge, tricking filters, contact stabilisation, etc. There are widely known methods of effluent treatment in palm oil mill industries. These include the following:

Tank digestion and facultative ponds

In this system, raw effluent after oil trapping is pumped to a closed tank which has a retention time of about twenty days. The liquid is mixed by means of horizontal stirrers. The methane gas (CH_4) generated is flared off into the atmosphere, but the flaring of the CH_4 is unacceptable and calls for improvement on this method. Digested liquid is discharged into a holding pond before it is disposed on land.

Tank digestion and mechanical aeration

This group consists of cooling/acidification ponds, an anaerobic digestions tank and an aeration pond. Raw effluent after oil trapping is pumped to the acidification pond through a cooling tower and retained for one to two days. It is then mixed with an equal volume of liquid from the anaerobic digester before it is fed back to the digester and the achievement recorded indicates that the effluent water has been treated. The hydraulic retention time of the digester is about twenty days. The digested liquid is discharged to an aeration pond with two floating aerators. The liquid is aerated for twenty days before it is discharged.

Decanter and facultative ponds

In a few mills, decanters are used to separate the fruits juice after pressing into liquid and solid phase, the liquid which is mainly oil is fed to the conventional clarification process. The water resulting from the clarification station is recycled. The solid is either disposed off on land or is dried in a rotary drier to about 10% moisture and then used as fuel. Thus, the effluent which consists of only the steriliser condensate and waste from the hydro cyclone is greatly reduced in volume and is treated in a series of ponds.

Anaerobic and facultative ponds

This system consists of a series of ponds connected in series for different purposes. The effluent after oil trapping is retained in an acidification buffering pond for about two or three pond with a hydraulic retention time of thirty to eighty days depending on the mills. This digested liquid is further treated in a series of facultative ponds before it is discharged. In some cases, part of the digested liquid is recycled to the acidification and buffering pond. The total hydraulic retention time of the system ranges from 75 to 120 days.

Antra system

The treatment consists of a combination of mechanical chemical process and ponds. The raw effluent after oil trapping is separated into water and solid phases using a three- phase decanter. The oil is returned to the main line while the solid is dried in a rotary drier after the filter press. The water containing dissolved and suspended solids is treated with coagulants and flocculants to remove as much solids as possible before it is fed to an anaerobic digester which has a hydraulic retention time of about ten days. The digested liquid is further treated in an aeration tower and then oxidised.

Process treatment of the effluent water

Figure 28.2 shows the flow chart for the treatment of the effluent water showing the different stages. This is divided into the acidic, methanogenic and aerobic phases.

Acidic phase: This is the first phase of the anaerobic digestion process. It is a very rapid process whereby acid bacteria converts the organic components of the waste into volatile fatty acids (VFA) which in turn acts as substrate for the next phase of the anaerobic process. The pH of the system is depressed during VFA formation. The phase is not susceptible to environmental influence in that changes in environmental conditions like temperature do not affect its

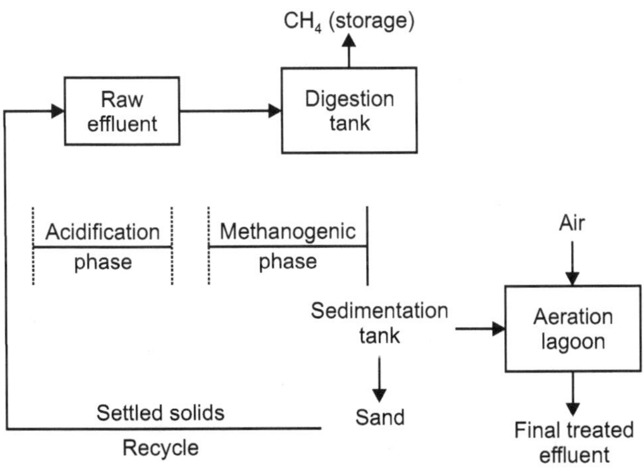

Figure 28.2: Flow chart for process treatment of the effluent water.

required performance, so open ponds are suitable and cost effective. At the start, anaerobic liquid is run down to the pond and then mixed with clarification waste and then pumped over cooling tower in the ratio of 1:1 to 1.5:1. The mixture is left overnight to react. Recycling of anaerobic liquid helps to supply seed bacteria for continuous acidification, cools hot effluent and improves the pH.

Methanogenic phase: Methanogenic phase is susceptible to the environment in that changes in environmental conditions like temperature would affect its required performance. Thus, it is best carried out in a tank with so many advantages which are found using a tank digester and this includes:

1. The tank can be dislodged easily and efficiently – uncontrolled build up of organic sludge's would lead to loss of retention volume and eventual system failure.
2. Rainwater dilution is reduced as volume/surface area ratio is high. Rain water is acidic and harmful on anaerobic process.
3. There can be no subsurface inflows or outflows into the system hence the process can be monitored very easily. In lagoons, the same uniformity is not possible.
4. High gas production rates over a low surface area results in the break-up of scum formation which, if allowed to build-up will lead to loss of retention time. Disadvantages can come in the corrosion of tank, which can be controlled by better protective paint work.

The digestion tank is completely closed and the biogas ratio (methane/CO_2) produced is higher in the two phase mode of operation than in the single phase indicating higher digesters rate and this gas is trapped and stored in a floating roof storage tank and may be used as energy source. The liquid discharged from the methanogenic phase is transferred to a sedimentation tank and offer over night settlement. The settled solids are recycled to the acidification ponds and the supernatant is discharged to the aerobic lagoon. To give room for maintenance, three tanks with a combined full load retention time of 20 days can be used.

The efficiency of an anaerobic system cannot be calculated by considering only the input and output BOD. The build-up of organic matter in the system has to be considered. The digester tanks have to be dislodged periodically to minimise sand accumulations, such sludge have a high nitrogen content which are good replacement for fertiliser and withdrawal of 0.1 T of sludge/T of feed is allowed.

Aerobic phase: For a 20 T fresh fruit Bunch per hour (FFB/hr) oil mill, an aerobic lagoon with 20 days retention at 0.1 kg BOD: kg Mixed Liquid Suspended Solid (MLSS), a minimum of two 11 KW mechanical aerators can be used.

An extended aeration process is advantageous in the following ways:

1. Operation is simple and the problem of solid generation and handling are reduced.
2. Nitrogen destruction efficiencies are high.
3. Power requirement is not critical.
4. Construction costs are low.
5. Land usage is reasonable.

After sedimentation the discharge from the digester is dumped in the aeration lagoon at the start. Twin aerators, operate continuously to provide mixing and oxygen transfers. The lagoon discharge is passed through a sedimentation tank and the settled suspended sludge is at present recycled to the acidification pond but can be used as fertiliser because of its high nitrogen content.

It is essential that the lagoon content are well mixed as failure will result to a build up of facultative condition in unstirred parts of the system. The process streams are shown in Fig. 28.3.

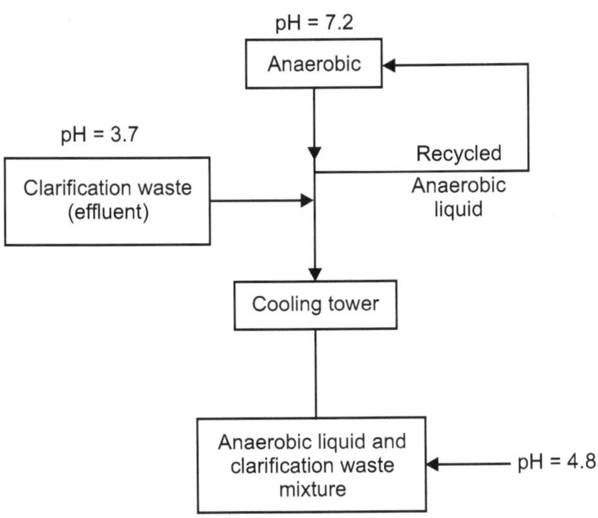

Figure 28.3: Flow diagram of the process showing stream.

Process flow of the biological treatment plant

The liquid discharged from the tank is transferred to a sedimentation tank. The supernatant liquid from this tank overflows to the aeration lagoon. The concentrated digested liquid is recirculated into the digestion tank, to maintain a constant level of suspended solids in the digestion tank. When solids level exceeds the desired concentration, some of it is taken off into the sludge storage tank and sent to the decanter for dewatering. The sludge cake produced is mixed with fibre until a moisture content of about 60% is obtained. This mixture is then placed in the composting tank. This process biological treatment plant is shown in Fig. 28.4.

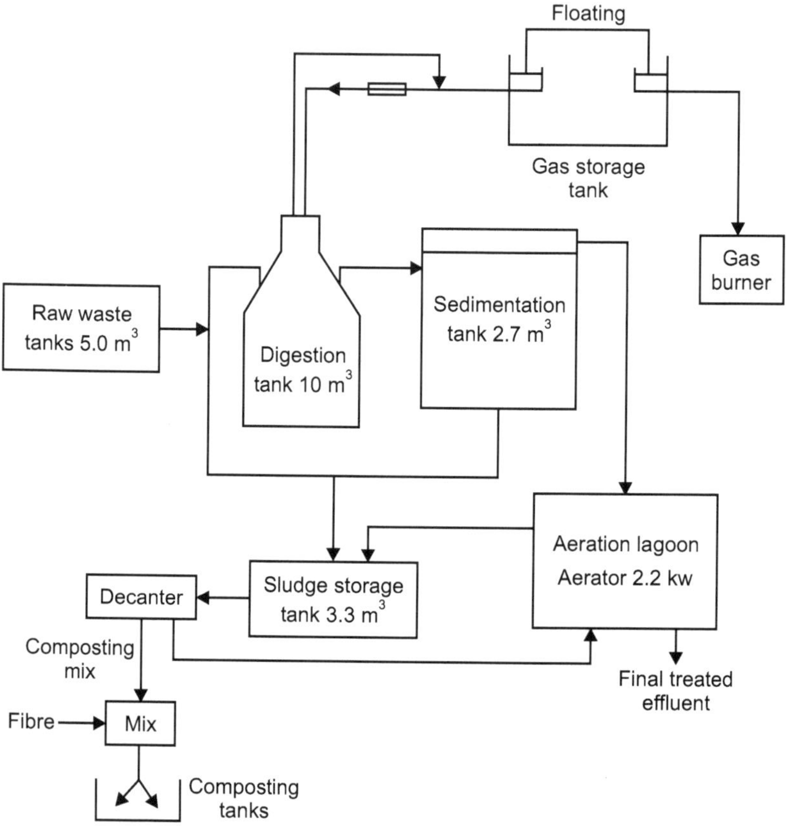

Figure 28.4: Process for biological treatment plant.

Smith, suggests that biological treatment methods is one of the most common ways of treating the effluent water. Also, the extent of conversion of the organic matter is measured by the ratio of BOD to COD. The ratio BOD: COD > 0.6 is ideal for biological waste treatment. It was also found out that the aerobic treatment method is more effective and has a high process rate and enhances maximum destruction of carcinogenic products. Biological treatment of the effluent is carried out in filter bed, biological pond, bio-filter and aeration are commonly used. Bio-filters are concrete wall reservoirs with perforated bottom, filled with packing of various sizes and inhabited by micro-organisms. The microbes form a thin layer on packing surface.

This effluent water is evenly distributed on the filter bed-layer to ensure contact with the microbes. In all these effluent types, it is evident that anaerobic digestion is the most attractive biological process as it requires minimum power input. Unfortunately, it operates within a relatively narrow range of physical conditions. However, the process is more stable with wider range if the acidic and methanogenic phases of digestion are separated.

After anaerobic digestion, the waste is normally still too strong for discharge to waterway and extended aeration and sedimentation provides a supernatant of low BOD. The settled solids from the anaerobic and aerobic digestion process are higher in nitrogen content than raw waste and are eminently suitable for use as fertiliser.

What actually happens is that the different palm oil mill effluent management methods are selected to optimise the BOD, the COD, total solids (TS), suspended solids (SS), oil and grease (O&G), ammonical nitrogen, pH, temperature and cost. This effluent as stated earlier comes from three major streams, i.e., steriliser condensate which contains oil, hydro cyclone waste which comes from kernel operation and finally separator sludge.

All these amount to about 2.5 T/T of oil product and are highly polluting. Realising the seriousness of the problem, there became the need to regulate the effluent.

Sludge treatment: The sludge fraction discharge from the tanks usually contains oil >15%. The palm oil mill effluent ex-sludge is made up of mainly two other components in addition to the oil, which are water 93–94% and solids 3–4%, but the composition may very widely. If the oil is not recovered some what loses will be incurred in the form of effluent waste. The sludge fraction is therefore subjected to further treatment to recover the oil before discarding to the effluent pit.

The sludge fraction from the settling tank is routed to a holding tank (sludge tank), from where it is pumped to the sand cyclone. The sand cyclone separates the sand from the stream which will cause severe wear on down stream equipment. The stream is then routed to a rotating brush strainer where fibres and other solids are isolated by the sweeping operation of the rotating brush strainer. This is to avoid choking and blocking of the centrifuge (called sludge separator) which is responsible for this work.

These sludge separators are equipped with paring discs, sprawt wear resistant tungsten carbide nozzles and conical separating plates. When in operation, the centrifugal force exerted by splitting of the feed into a number of streams by the conical separating plates, causes the separation achieved eventually. The oil passes inward and the water and dirt passes outwards, part of water and dirt escapes over a gravity ring the size of which determines the position of the interface between the aqueous and oily zones.

Oil fraction is usually recycled back to the settling tank while the water and dirt is discarded to the effluent. Efficiency of oil recovery in a sludge centrifuge is dependent on oil content of recycled stream over total oil content of feed. More especially, efficiency of sludge separator is gauged by the residual oil content of the centrifuge waste which includes dirties, H_2O, heavy sludge, sand, etc. The variables affecting centrifuge efficiency are feed rate, nozzle diameter, feed temperature, dilution (viscosity) of feed oil and particles.

29

Pollution control in dairy industry

29.1 Introduction

Dairy products include milk and any of the foods made from milk, including butter, cheese, ice cream, yogurt and condensed and dried milk. Milk has been used by humans since the beginning of recorded time to provide both fresh and storable nutritious foods. In few countries almost half the milk produced is consumed as fresh pasteurised whole, low-fat or skim milk.

29.2 Processing

29.2.1 Pasteurisation

Pasteurisation is most important in all dairy processing. It is the biological safeguard which ensures that all potential pathogens are destroyed. Extensive studies have determined that heating milk to 63°C (145°F) for 30 minutes or 72°C (161°F) for 15 seconds kills the most resistant harmful bacteria. In actual practice these temperatures and times are exceeded, thereby not only ensuring safety but also extending shelf life.

Most milk today is pasteurised by the continuous High Temperature Short Time (HTST) method (72°C or 161°F for 15 seconds or above). The HTST method is conducted in a series of stainless steel plates and tubes, with the hot pasteurised milk on one side of the plate being cooled by the incoming raw milk on the other side. This 'regeneration' can be more than 90% efficient and greatly reduces the cost of heating and cooling.

There are many fail-safe controls on an approved pasteuriser system to ensure that all milk is completely heated for the full time and temperature requirement. If the monitoring instruments detect that something is wrong, an automatic flow diversion valve will prevent the milk from moving on to the next processing stage.

Higher temperatures and sometimes longer holding times are required for the pasteurisation of milk or cream with a high fat or sugar content.

Pasteurised milk is not sterile and is expected to contain small numbers of harmless bacteria. Therefore, the milk must be immediately cooled to below 4.4°C (40°F) and protected from any outside contamination.

The shelf life for high-quality pasteurised milk is about 14 days when properly refrigerated.

Extended shelf life can be achieved through ultra pasteurisation. In this case, milk is heated to 138°C (280°F) for two seconds and aseptically placed in sterile conventional milk containers. Ultra pasteurised milk and cream must be refrigerated and will last at least 45 days. This process does minimal damage to the flavour and extends the shelf life of slow-selling products such as cream, eggnog and lactose-reduced milks. Ultrahigh Temperature (UHT) pasteurisation is the same heating process as ultra pasteurisation (138°C or 280°F for two seconds), but the milk then goes into a more substantial container—either a sterile five-layer laminated 'box' or a metal can. This milk can be stored without refrigeration and has a shelf life of six months to a year. Products handled in this manner do not taste as fresh, but they are useful as an emergency supply or when refrigeration is not available.

29.3 Environmental aspects

Pre-treatment of effluents: Pre-treatment of effluents consists of screening, flow equalisation, neutralisation and air flotation (to remove fats and solids); it is normally followed by biological treatment. If space is available, land treatment or pond systems are potential treatment methods. Other possible biological treatment systems include trickling filters, rotating biological contactors and activated sludge treatment.

Air emissions: Odour controls (such as absorbents/biofilters on exhaust systems) should be implemented where necessary to achieve acceptable odour quality for nearby residents. Fabric filters should be used to control dust from milk powder production to below 50 milligrams per normal cubic meter (mg/Nm^3). Odour control by ventilation and scrubbing may be required where cheese is stored or melted. Dust control at milk powder plants is provided by fabric filters.

Ambient noise: Noise abatement measures should achieve either the levels given below or a maximum increase in background levels of 3 decibels (measured on the A scale) [dB(A)]. Measurements are to be taken at noise receptors located outside the project property boundary.

	Maximum allowable log equivalent (hourly measurements), in dB(A)	
Receptor	*Day (07:00–22:00)*	*Night (22:00–07:00)*
Residential, Institutional, Educational	55	45
Industrial, Commercial	70	70

29.4 Wastewater treatment in dairy industry

The dairy industry involves processing raw milk into products such as consumer milk, butter, cheese, yogurt, condensed milk, dried milk (milk powder) and ice cream, using processes such as chilling, pasteurisation and homogenisation. The typical by-products of milk are buttermilk, whey and their derivatives. The effluents are generated from milk processing through milk spillage, drippings, washing of cans, tankers bottles, utensil and equipment's and floors. The dairy industry generate on an average 2.5–3.0 litres of wastewater per litre of milk processed. Generally this wastewater contains large quantities of fat, casein, lactose and inorganic salts, besides detergents, sanitizers, etc., used for washing. These all contribute largely towards their high biological oxygen demand (BOD), chemical oxygen demand (COD) and oil and grease much higher than the permissible limits. Among the biological treatments trickling filter and activated sludge process involve more economy high power requirement, more chemical consumption and large area requirement. Use of a dairy wastewater for irrigation after primary treatment in an aerated lagoon may also be good for the disposal of dairy wastes.

The food industry have one of the highest consumptions of water and is one of the biggest producers of effluents per unit of production, in addition they generate a large volume of sludge during biological treatment. The dairy industry is one of those sector, in which the cleaning silos, tanks, homogenisers, pipe sand, heat exchangers other equipment, engenders a large amount of effluents with a high organic load. This organic load is basically constituted of milk (raw material and dairy products), reflecting an effluent with high levels of chemical oxygen demand (COD), biochemical oxygen demand (BOD), oils and grease, nitrogen and phosphorus. The automatic cleaning system – CIP (cleaning in place) – discard rinse waters with pH varying between 1.0 and 13.0, further complicating the question of treatment. BOD is directly related to milk wastes (90% to 94% of the effluent BOD) and in some cases losses can reach 2% of the volume processed by the industry.

In order to reduce the effects of industrial sector pollutants, the end-of-pipe treatment techniques have been improved at the same time prevention measures are being implemented in order to minimise the production of residues. End-of-pipe control captures wastewater after its generation, enabling its discharge into environment. These are peripheral solutions that focus primarily on the chemical, biological and physical treatment of terminal streams. However, they address the symptoms and not the true causes of the environmental problems and therefore they are not cost effective or sustainable. The essential feature of the pollution prevention programme is the concept of reduction at sources, based on the idea that the generation of pollutant can be reduced or eliminated by increasing efficiency in the use of raw materials, energy, water and some other resources. Cleaner production intends to integrate the production aims in order to reduce the quantity and toxicity of residues and discharges. Pollution prevention or source generation reduction refers to any processor technology that seeks the reduction or elimination of the volume, concentration or toxicity of generating source residues.

The concept of cleaner production involves the reduction of negative environmental impacts throughout the products life cycle, from extraction of raw material to its final use.

The dairy is a multiproduct factory and its wastewater treatment process is based on five steps: (a) screening, (b) sand trap/oil and grease separation in a tank, (c) flow equalisation in a tank, (d) an activated sludge process, (e) tertiary treatment in three facultative lagoons.

However, the process is almost overloaded and requires a more complete diagnosis. On the other hand, minimisation of the pollution index indicator must be evaluated, not only in terms of final treatment, but also as an opportunity to reduce production costs, by optimising them and increasing process efficiency and profit.

29.5 Sources of wastes

The liquid waste from a large dairy originate from the following sections of plants- receiving station, bottling plant, cheese plant, butter plant , casein plant, condensed milk plant, dried milk plant and ice cream plant. Waste also comes from water softening plant and from bottle and can washing plants. At the receiving station the milk is received from the farms and after inspection the same is emptied into large containers for transport to bottling or other processing's. The empty cans are rinsed, washed sterilised and are returned to the farmers. At the bottling point, the raw milk delivered by the receiving station is stored. The processing includes cooling, clarification, filtration, pasteurisation and bottling. In the above two sections, the liquid wastes originate out of rinse and washings of bottles, cans and equipment's and thus contain milk drippings and chemicals used for cleaning containers and equipment's.

The skimmed milk may now be sent for bottling for human consumptions or for further processing in the dairy for other products like non-fat milk powders. Milk powders are produced by evaporation followed by drying by either roller process or spray process. The dry milk plant wastes consist chiefly of wash waters used to clean containers and equipment's. The soured or spoiled milk and sometimes the skimmed milks are processed to produce caseins used for preparation of some plastics; the process involves the coagulation and precipitation of the caseins by the addition of some minerals acids. The waste from the section includes whey, washings and the chemicals used for precipitation. Very large dairies also produce condensed milk and ice creams. In addition to the wastes from all the above milk processing's units, some amount uncontaminated cooling water comes as wastes; these are very often recirculated. The dairy wastes are very often discharge intermittently; the nature and composition of waste also depends on the types of products produced and the size of the plants.

29.6 Waste characteristics

Dairy effluents contain dissolved sugars proteins and fats and possibly residues of additives. The key parameters are biochemical oxygen demand (BOD), with an average ranging from 0.8 to 2.5 kilograms per metric ton (kg/T) of milk in the untreated effluent, chemical oxygen demand (COD), which is normally about 1.5 times the BOD level, total suspended solids, at 100–1000 milligrams per litre (mg/L), total dissolved solids: phosphorus (10–100 mg/L) and nitrogen (about 6% of the BOD level). Cream, butter, cheese and whey production are major sources of BOD in wastewater. The waste load equivalents of specific milk constituents are: 1 kg of milk fat = 3 kg COD, 1 kg of lactose = 1.13 kg COD and 1 kg protein = 1.36 kg COD.

The wastewater may contain pathogens from contaminated materials or production processes. A dairy often generates odours and, in some cases, dust, which need to be controlled. Most of the solid wastes can be processed into other products and by-products.

29.6.1 Effects of wastes on the receiving streams/sewers

Composition of the wastewater of typical dairy industries is shown in Table 29.1.

Table 29.1: Composition of the wastewater of typical dairy industries.

Item	*Value (influent)*
pH	7.2
Alkalinity	600 mg/L as caco3
Total dissolved solids	1060 mg/L
Suspended solids	760 mg/L
BOD	1240 mg/L
COD	84 mg/L
Total nitrogen	84 mg/L
Phosphorus	11.7 mg/L
Oil and grease	290 mg/L
Chloride	105 mg/L

This is also slightly alkaline when fresh. Dairy effluents decompose rapidly and deplete the dissolved oxygen level of the receiving streams immediately resulting in anaerobic conditions and release of strong foul odour due to nuisance conditions. The casein precipitation from dairy waste decomposes further into a highly odorous black sludge. At certain dilutions the dairy waste is found to be toxic for fish and other aquatic living being and becomes breeding place for flies and mosquitoes. Dairy effluent contains soluble organics and suspended solids, they degrade to promote release of gases, odour, imparts colour or turbidity and promotes eutrophication.

29.7 Treatment of the dairy wastes

Dairy wastewater have low COD and BOD ratio and it can be treated efficiently by biological processes. These wastes contain sufficient nutrients for bacterial growth and this can be prevented by:

1. The prevention of spills, leakages and dropping of milks from cans.
2. The requirement of water can be minimised during washes.
3. By segregating the uncontaminated cooling water and recycling the same.
4. Utilisation of butter milk and whey for the production of dairy by-products.

Both high rates tricking filters and activated sludge plants can be operated very effectively for complete treatment of dairy waste. But these conventional methods involve much skilled persons and special type of equipment's. On the other hand the low cost treatment method like oxidation ditches is also used commonly. Use of dairy waste for irrigation after primary treatment in an aeration lagoon may also be good answer for disposal of dairy waste.

29.8 Dairy effluent

pH: It is a term used to express the intensity of the acid or alkaline condition of a solution. It is a way of expressing the hydrogen-ion concentration or the hydrogen-ion activity. Pure water is said to be neutral, with a pH close to 7.0 at 25°C (77°F). Solutions with a pH less than 7 are said to be acidic and solutions with a pH greater than 7 are said to be basic or alkaline.

Chemical oxygen demand (COD): The COD test is widely used as a means of measuring the organic strength of effluents. This test allows measurement of waste of a waste in terms of the total quantity of oxygen required for oxidation to CO_2 and H_2O. During the determination of COD, organic matter is converted to carbon dioxide and water regardless of the biological assimilability. The dichromate reflux method is preferred over procedures using other oxidants (e.g., potassium permanganate) because of its superior oxidising ability, applicability to a wide variety of samples and ease of manipulation.

Biochemical oxygen demand (BOD): It is defined as the amount of oxygen required by bacteria while stabilising decomposable organic matter under aerobic conditions. The BOD test is widely used to determine the pollution strength of domestic and industrial wastes in terms of oxygen that they will require if discharged into natural water courses in which aerobic condition exist. It is not a precise quantitative test, although it is widely used as an indication of the organic quality of water. It is most commonly expressed in milligrams of oxygen consumed per litre of sample during 5 days of incubation at 20°C. The BOD is used for measuring the oxygen consumed by living organisms (mainly bacteria) while utilising the organic matter present in wastewater.

Total dissolved solids (TDS): It is a measure of the combined content of all organic and inorganic substances present in a liquid in molecular, ionised or micro-granular (colloidal solution) suspended form. Generally the operational definition is that the solids must be small enough to survive filtration through a sieve the size mm of two micrometer. The principal application of TDS is in the study of water quality for streams, rivers and lakes, although TDS is not generally considered a primary pollutant. It is used as an indication of aesthetic characteristics of drinking water and as an aggregate indicator of the presence of a broad array of chemical contaminants.

Suspended solids (SS): It refers to small solid particles which remain in suspension form in water as a colloid. It is used as one of the indicator of water quality. It is sometimes abbreviated SS, but is not to be confused with settleable solids, which contribute to the blocking of sewer pipes.

Oil and grease: Dissolved or emulsified oil and grease is extracted from water by intimate contact with an extracting solvent. Some extractable, especially unsaturated fats and fatty acids oxidise readily; hence special precautions regarding temperature and solvent vapour displacement are included to minimise this effect. Organic solvents shaken with some samples may form an emulsion that is very difficult to break. This method includes a means for handling such emulsions.

Sulphate: In inorganic chemistry, a sulphate is a salt of sulphuric acid. The sulphate ion is a polyatomic anion with the empirical formula SO^{2-4} and a molecular mass of 96.06 daltons (96.06 g/mol); it consists of a central sulphur atom surrounded by four equivalent oxygen atoms in a tetrahedral arrangement. Many examples of ionic sulphates are known and many of these are highly soluble in water. Exceptions include calcium sulphate, strontium sulphate, lead (II) sulphate and barium sulphate, which are poorly soluble.

Chloride: The chloride ion is formed when the element chlorine picks up one electron to form an anion (negatively-charged ion) Cl^-. The salts of hydrochloric acid (HCL) contain chloride ions and can also be called chlorides. The word chloride can also refer to a chemical compound in which one or more chlorine atoms are covalently bonded in the molecule. This means that chlorides can be either inorganic or organic compounds. The simplest example of an inorganic covalently-bonded chloride is hydrogen chloride, (HCL). A simple example of an organic covalently-bonded (an organochloride) chloride is chloromethane (CH_3Cl), often called methyl chloride.

29.9 Prevention of wastes

Cleaning by means of the clean in place (CIP) system and the reuse and the recycling of water are examples of processes which reduce the volumetric coefficient. It minimised water consumption, in most of the processes; however, there were some exceptions, such as a few trucks that were not adapted with a 'spray bowl' for washing by the CIP system. In addition, the spray dryer was operated manually rather than automatically, which consumed large amounts of water. The action proposed in two ways:

1. A reduction in water consumption.
2. Minimisation of the organic load.

Water reuse and recycling is a reality in the dairy industry. Many processes, such as centrifugal separation with cooling water in a closed circuit recycled water. The filling machines (for milk packaging) can be cooled with recovered water and the evaporated water (from the milk evaporator for the production of powdered milk) can be used for cleaning trucks and outside floors. The retentate from the reverse osmosis system (used for desalination of boiler feed water) can be mixed into the water supply reservoir. The effluent considered water consumption while the water evaporated in the boiler and cooling towers are not computed in the material balance. The sum of these preventive actions account for a 10% decrease in total effluents generated. The industry can take action to reduce of effluents loads, which is reflected by the low BOD, nitrogen, phosphorus and oil and grease coefficients. The actions taken should be for separation of discharged milk by automatic ejection of sludge in the centrifugal separators; segregation of whey from butter for use in animal feed and recirculation of the first rinse water from the evaporators at the beginning of the process, reducing the organic load of fluids and dry products In spite of the great concern of the industry to minimise waste, there are still opportunities for the reduction of previous coefficients. The recovery of solids in the first rinse could be a pollution prevention action. There are examples of milk solids recovered by the use of membrane separation processes (reverse osmosis) used for production ice-cream and milk desserts. Three direct results were obtained: minimisation of impact of the effluent generated; the production of casein and reuse of the permeated stream, which is of a high quality enough to be used for drinking water. A central system for treatment of these rinse waters could be installed, recovering the milk solids, mainly from the reception and the fluid products sector. This is the direct result of the production of pasteurised butter and cream, which generates effluents with high values for this parameter. In this case, the simple separation of the first rinse water and its use for animal feed would be beneficial in the reduction of organic load. Thus, the membrane separation process is a promising alternative to the recovery of nutrients found in the effluents.

In the light of above mentioned facts the industry should have comprehensive treatment system consisting of primary or secondary and/or tertiary treatment as is warranted with reference to influent quality and operate and maintained the same continuously so as to achieve the quality of the treated effluent.

Quality of raw milk: The collected raw milk contains bacteria. Bacteria release enzymes which is harmful for human being. Acid forming bacteria consumes sugar in the milk and forms acid and gases. In short they decompose the milk within very few time, so total quality of milk depends upon the number of initial bacterial count.

29.9.1 Method for counting bacterial growth

Method for counting bacterial growth can be done by chemical and electrical methods.

Chemical method

1. Standard plate count method.
2. Litmus paper test.
3. MBRT test (methyl blue reduction time).

Electrical method

1. Milk barometer.
2. Online system.

29.10 Land treatment and irrigation

29.10.1 Land treatment

Dairy wastewater, along with a wide variety of other food processing wastewaters, has been successfully applied to land in the past. Interest in the land application of wastes is also increasing as a direct result of the general move of regulatory authorities to restrict waste disposal into rivers, lakes and the ocean, but also because of the high costs of incineration and landfilling. Nutrients such as N and P that are contained in biodegradable processing wastewaters make these wastes attractive as organic fertilisers, especially since research has shown that inorganic fertilisers might not be enough to stem soil degradation and erosion in certain parts of the world.

Land application of these effluents may, however, be limited by the presence of toxic substances, high salt concentrations or extreme pH values. It might be, according to Wendorff, the most economical option for dairy industries located in rural areas.

29.10.2 Irrigation

The distribution of dairy wastewaters by irrigation can be achieved through spray nozzles over flat terrain or through a ridge and furrow system. The nature of the soil, topography of the land and the waste characteristics influence the specific choice of irrigation method. Wastewater would typically percolate through the soil, during which time organic substances are degraded by the heterotrophic microbial population naturally present in the soil. An application period followed by a rest period (in a 1:4 ratio) is generally recommended.

Specific wastewater characteristics can have an adverse effect on a spray irrigation system that should also be considered. Suspended solids, for instance, may clog spray nozzles and render the soil surface impermeable, while wastewater with an extreme pH or high salinity might be detrimental to crop cover. Highly saline wastewater might further cause soil dispersion and a subsequent decrease in drainage and aeration, as a result of ion exchange with sodium replacing magnesium and calcium in the soil.

Generally, it was found that hydraulic conductivity, microbial content and *N*-cycling processes all increased substantially in long-term irrigated soils. Since increases in infiltration as well as biochemical processing were noted in all the irrigated soils, most of the changes in soil properties were considered to be beneficial.

29.11 Sludge disposal

Different types of sludge arise from the treatment of dairy wastewaters. These include: (i) sludge produced during primary sedimentation of raw effluents (the amounts of which are usually low), (ii) sludge produced during the precipitation of suspended solids after chemical treatment of raw wastewaters, (iii) stabilised sludge resulting from the biological treatment processes, which can be either aerobic or anaerobic and (iv) sludge generated during tertiary treatment of wastewater for final suspended solid or nutrient removal after biological treatment.

Primary sedimentation of dairy wastewater for BOD reduction is not usually an efficient process, so in most cases the settleable solids reach the next stage in the treatment process directly. An important advantage of anaerobic processes is that the sludge generated is considerably less than the amount produced by aerobic processes and it is easier to dewater. Final wastewater polishing after biological treatment usually involves chemical treatment of the wastewater with calcium, iron or aluminium salts to remove dissolved nutrients such as nitrogen and phosphorus. The removal of dissolved phosphorus can have a considerable impact on the amount of sludge produced during this stage of treatment.

The application of dairy sludge as fertiliser has certain advantages when compared to municipal sludge. It is a valuable source of nitrogen and phosphorous, although some addition of potassium might be required to provide a good balance of nutrients. Sludge from different factories will also contain different levels of nutrients depending on the specific products manufactured. Dairy sludge seldom contains the same pathogenic bacterial load as domestic sludge and also has considerably lower heavy metal concentrations. The recognition of dairy sludge as a fertiliser does, however, depend on local regulations. Some countries have limited the amount of sludge that can be applied as fertiliser to prevent nitrates from leaching into groundwater sources.

It is generally agreed that disposal of sludge by land spraying or as fertiliser is the least expensive method. If the transport and disposal of liquid sludge cannot be done within reasonable costs, other treatment options such as sludge thickening, dewatering, drying or incineration must be considered. Gravity thickeners are most commonly used for sludge thickening, while the types of dewatering machines most commonly applied are rotary drum vacuum filters, filter presses, belt presses and decanter centrifuges.

29.12 Pollution prevention

Reduction of wastewater pollution levels may be achieved by more efficiently controlling water and product wastage in dairy processing plants. Comparisons of daily water consumption records vs. the amount of milk processed will give an early indication of hidden water losses that could result from defective subfloor and underground piping. An important principle is to prevent wastage of product rather than flush it away afterwards. Spilled solid material such as curd from the cheese production area and spilled dry product from the milk powder production areas should be collected and treated as solid waste rather than flushing them down the drain.

Small changes could also be made to dairy manufacturing processes to reduce wastewater pollution loads. In the cheese production area, milk spillage can be restricted by not filling open cheese vats all the way to the rim. Whey could also be collected sparingly and used in commercial applications instead of discharging it as waste.

Manual scraping of all accessible areas after a butter production run and before cleaning starts would greatly reduce the amount of residual cream and butter that would enter the wastewater stream. In the milk powder production area, the condensate formed could be reused as cooling water (after circulation through the cooling tower) or as feed water to the boiler. Returned product could be emptied into containers and used as animal feed. Milk and product spillage can further be restricted by regular maintenance of fittings, valves and seals and by equipping fillers with drip and spill savers. Pollution levels could also be limited by allowing pipes, tanks and transport tankers adequate time to drain before being rinsed with water.

The management of dairy wastes becomes an ever-increasing concern, treatment strategies will need to be based on state and local regulations. Because the dairy industry is a major water user and wastewater generator, it is a potential candidate for wastewater reuse. Purified wastewater can be utilised in boilers and cooling systems as well as for washing plants and so on. Even if the purified wastewater is initially not reused, the dairy industry will still benefit directly from in-house wastewater treatment, since levies charged for wastewater reception will be significantly reduced.

Before selecting any treatment method, a complete process evaluation should be undertaken along with economic analysis. This should include the wastewater composition, concentrations, volumes generated and treatment susceptibility, as well as the environmental impact of the solution to be adopted. All options are expensive, but an economic analysis may indicate that slightly higher maintenance costs may be less than increased operating costs. What is appropriate for one site may be unsuitable for another.

The most useful processes are those that can be operated with a minimum of supervision and are inexpensive to construct or even mobile enough to be moved from site to site. The changing quantity and quality of dairy wastewater must also be included in the design and operational procedures. Since no single process for treatment of dairy wastewater is by itself capable of complying with the minimum effluent discharge requirements, it is necessary to choose a combined process especially designed to treat a specific dairy wastewater.

30

Pollution control in bakery industry

30.1 Introduction

Food industry plays a significant role in country's economic development. Processed agricultural products account for about 30% of the processing industry's, out of which 90% is produced by the food industry.

The food industry is a complex and global collective of diverse businesses that supply most of the food consumed by the world population.

Agriculture is the process of producing food, feeding products, fibre and other desired products by the cultivation of certain plants and the raising of domesticated animals (livestock).

Food processing includes the methods and techniques used to transform raw ingredients into food for human consumption. Food processing takes clean, harvested or slaughtered and butchered components and use them to produce marketable food products. There are several different ways in which food can be produced. Modern food production is defined by sophisticated technologies. These include many areas. Agricultural machinery, originally led by the tractor, has practically eliminated human labour in many areas of production. Biotechnology is driving much change, in areas as diverse as agrochemicals, plant breeding and food processing.

30.2 Bakery industry

The bakery industry comprises mainly of bread, biscuits, cakes and pastries manufacturing units. Bread is the product of baking a mixture of flour, water, salt, yeast and other ingredients. The basic process involves mixing of ingredients until the flour is converted into a stiff paste or dough, followed by baking the dough into a loaf. Mixing has two functions: (i) to evenly distribute the various ingredients and (ii) allow the development of a protein (gluten) network to give the best bread possible. Each dough has an optimum mixing time, depending on the flour and mixing method used. Process flow diagram of bakery industry is shown in Fig. 30.1.

Rising (fermentation): Once the bread is mixed, it is then left to rise (ferment). As fermentation takes place, the dough slowly changes from a rough dense mass lacking extensibility and with poor gas holding properties into a smooth, extensible dough with good gas holding properties. The yeast cells grow, the gluten protein pieces stick together to form networks and alcohol

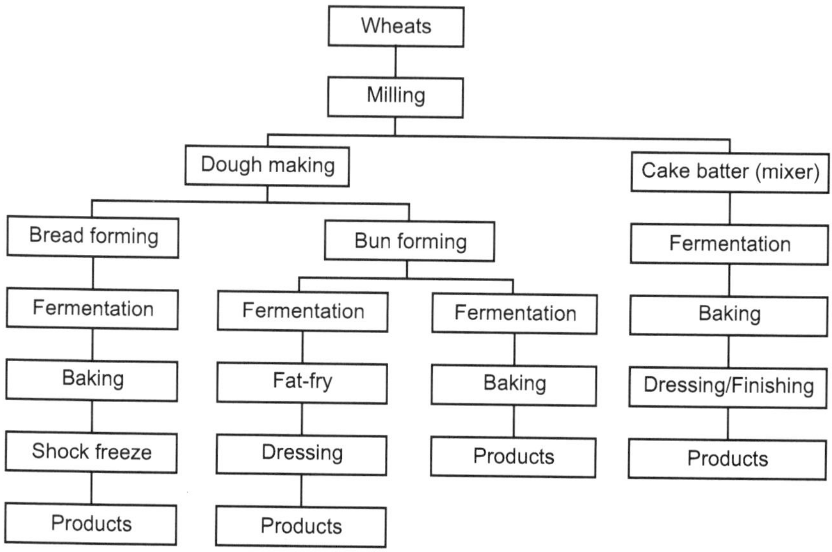

Figure 30.1: Process flow diagram of bakery industry.

and carbon dioxide are formed from the breakdown of carbohydrates (starch, sugars) that are found naturally in the flour. The yeast uses sugars in much the same way as we do, i.e., it breaks sugar down into carbon dioxide and water. Enzymes present in yeast and flour also help to speed up this reaction. The energy which is released is used by the yeast for growth and activity. In a bread dough where the oxygen supply is limited, the yeast can only partially breakdown the sugar. Alcohol and carbon dioxide are produced in this process known as alcoholic fermentation.

The carbon dioxide produced in these reactions causes the dough to rise and the alcohol produced mostly evaporates from the dough during the baking process. During fermentation, each yeast cell forms a centre around which carbon dioxide bubbles are formed. Thousands of tiny bubbles, each surrounded by a thin film of gluten, form cells inside the dough piece.

Kneading: Any large gas holes that may have formed during rising are released by kneading. A more even distribution of both gas bubbles and temperature also results. The dough is then allowed to rise again and is kneaded, if required by the particular production process being used.

Second rising: During the final rising, the dough again fills with more bubbles of gas and once this has proceeded far enough, the doughs are transferred to the oven for baking.

Baking: The baking process transforms an unpalatable dough into a light, readily digestible, porous flavourful product. During baking the yeast dies at

46°C and so does not use the extra sugars produced between 46–75°C for food. These sugars are then available to sweeten the bread crumb and produce the attractive brown crust colour. As baking continues, the internal loaf temperature increases to reach approximately 98°C. The loaf is not completely baked until this internal temperature is reached. Weight is lost by evaporation of moisture and alcohol from the crust and interior of the loaf. Steam is produced because the loaf surface reaches 100°C. As the moisture is driven off, the crust heats up and eventually reaches the same temperature as the oven. Sugars and other products, some formed by breakdown of some of the proteins present, blend to form the attractive colour of the crust. These are known as 'browning' reactions and occur at a very fast rate above 160°C. They are the principal causes of the crust colour formation.

Cooling: In bakeries, bread is cooled quickly when it leaves the oven. The crust temperature is over 200°C and the internal temperature of the crumb is about 98°C. The loaf is full of saturated steam which also must be given time to evaporate. The whole loaf is cooled to about 35°C before slicing and wrapping can occur without damaging the loaf. A moist substance like bread loses heat through evaporation of water from its surface. The rate of evaporation is affected by air temperature and the movement of cool air around the loaf. In a bakery there are special cooling areas to ensure efficient cooling takes place before the bread is sliced and wrapped.

30.3 Environmental aspects of bakery industry

Even though bakery industry is a low polluting industry but for the ovens, boilers, hot water generators, DG sets they all come under the air pollution act due to emission of flue gases through stack. Acts prescribes limits on the emission and stack height and in addition to it covers noise pollution with it. Various countries have different values. Bigger plants with more than two or three production line of bread or biscuit ovens with six or seven zones can emit considerable amount of flue gases which consists of carbon monoxides, sulphur oxides, nitrogen oxides which are dangerous for health and if not monitored can result in polluting neighbourhood.

30.3.1 Air pollution

Air pollution can be minimised with following steps:

1. Use flue gas analyser for stack monitoring in regular intervals.
2. Do regular burner maintenance.
3. Substitute if possible diesel oil to LPG or natural gas.
4. Use high pressure monoblock burners.
5. Get low sulphur diesel oil.

Dust

Flour production workers are usually harmed by dust pollution. Lengthy exposure time at a high exposure level can cause serious skin and respiration diseases. The control approaches include prevention of the leakage of flour power, provision of labour protection instruments and post-treatment. Filters and scrubbers are commonly used.

Refrigerant

In the chilling, freezing storage or transport of bakery products, a large amount of refrigerant is used. Chlorofluorocarbons (CFCs) and Hydrochlorofluorocarbons (HCFCs) are the common refrigerants and can damage the ozone layer. They can be retained in the air for approximately 100 years. Owing to the significantly negative environmental effects, replacement chemicals such as Hydrofluorocarbons (HFC) have been developed and used. Another measure is the prevention of the refrigerant leakage.

30.3.2 VOC emissions

The primary emission source at a bakery is the oven, which produces VOCs and carbon dioxide from the release of compounds formed in the yeast fermentation process. The main VOC emitted from bakery operations is ethanol. In yeast leavened products, the yeast metabolises sugar and produces carbon dioxide gas bubbles in the dough which causes the product to rise. The ethanol produced by yeast metabolism is generally a liquid at temperatures below 77°C (170°F) and is not emitted in large amounts until the dough is exposed to high temperatures in the oven around 240°C (464°F). Therefore, the bakery oven is the main source of ethanol. Ethanol and other VOCs combine in the atmosphere to form smog and therefore ethanol is regulated as an ozone precursor under EPA's criteria pollutant programme.

The primary source of VOC emissions at a bakery is the oven. Screening measurements taken at mixers, fermentation vessels, comfort hoods, proof boxes, oven exhausts, cooling area exhausts and packaging areas suggests that greater than 90% of VOC emissions are from the oven.

30.3.3 Noise pollution

Noise pollution can be controlled by following steps.

Safe limit prescribed in Act is 70 db in night time and 75 db in day time as CPCB standards:

1. Monitor noise through decibel meter.
2. DG set room to be to be lined with acoustic lining.
3. Maintain equipment with maintenance schedules.

4. Use equipments with enclosure.
5. Use screw compressors instead of reciprocating compressor.
6. Silencers in stack for DG sets.
7. Put anti-vibration pad where ever possible for machinery.

Ear plugs can help to effectively reduce the suffering. Other noise control measures include the reduction of source noise, use of noise enclosures, reduction of reverberation and reduction of exposure time.

30.3.4 Wastewater

The ratio of water consumed to products is about 10 in common food industry, much higher than that of 5 in the chemical industry and 2 in the paper and textiles industry. Normally, half of the water is used in the process, while the remainder is used for washing purposes (e.g., of equipment, floor and containers).

Different products can lead to different amounts of wastewater produced. The wastewater from cake plants has higher strength than that from bread plants. The pH is in acidic to neutral ranges, while the 5-day biochemical oxygen demand (BOD_5) is from a few hundred to a few thousand mg/L, which is much higher than that from the domestic wastewater. The suspended solids (SS) from cake plants is very high. Grease from the bakery industry is generally high, which results from the production operations. The waste strength and flow rate are very much dependent on the operations, the size of the plants and the number of workers. Generally speaking, in the plants with products of bread, bun and roll, which are termed as dry baking, production equipment (e.g., mixing vats and baking pans) are cleaned dry and floors are swept before washing down.

The wastewater from cleanup has low strength and mainly contains flour and grease. On the other hand, cake production generates higher strength waste, which contains grease, sugar, flour, filling ingredients and detergents.

Due to the nature of the operation, the wastewater strength changes at different operational times.

Bakery wastewater lacks nutrients; the low nutrient value gives BOD_5:N:P of 284:1:2. This indicates that to obtain better biological treatment results, extra nutrients must be added to the system. The existence of oil and grease also retards the mass transfer of oxygen. The toxicity of excess detergent used in cleaning operations can decrease the biological treatment efficiency. Therefore, the pre-treatment of wastewater is always needed.

30.3.5 Solid wastes

Solid wastes generated from bakery industries are principally waste dough and out-of-specified products and package waste. Solid waste is the loss of raw materials, which may be recovered by cooking waste dough to produce bread crumbs and by passing cooked product onto pig farmers for fodder.

30.4 Treatment of bakery wastes

Generally, bakery industry waste is nontoxic. It can be divided into liquid waste, solid waste and gaseous waste. In the liquid phase, there are high contents of organic pollutants including chemical oxygen demand (COD), BOD_5, as well as fats, oils and greases (FOG) and SS. Wastewater is normally treated by physical, chemical and biological processes.

30.4.1 Pre-treatment systems

Pre-treatment or primary treatment is a series of physical and chemical operations, which precondition the wastewater as well as remove some of the wastes. The treatment is normally arranged in the following order: screening, flow equalisation and neutralisation, optional FOG separation, optional acidification, coagulation-sedimentation and dissolved air flotation.

In the bakery industry, pre-treatment is always required because the waste contains high SS and floatable FOG. Pre-treatment can reduce the pollutant loading in the subsequent biological and/or chemical treatment processes; it can also protect process equipment. In addition, pre-treatment is economically preferable in the total process view as compared to biological and chemical treatment.

It is usually economical to use a flow equalisation tank to meet the peak discharge demand. However, too long a retention time may result in an anaerobic environment.

Screening

Screening is used to remove coarse particles in the influent. Coarse screen, fine screen. Smaller opening can have a better removal efficiency; however, operational problems such as clogging and higher head lost are always observed.

FOG separation

As wastewater may contain high amount of FOG. The FOG can be separated and recovered for possible reuse, as well as reduce difficulties in the subsequent biological treatment.

Acidification

Acidification is optional, depending on the characteristics of the waste. Owing to the presence of FOG, acid (e.g., concentrated H_2SO_4) is added into the acidification tank; hydrolysis of organics can occur, which enhances the biotreatability. Grove designed a treatment system using nitric acid to break the grease emulsions followed by an activated sludge process. A BOD_5 reduction of 99% and an effluent BOD_5 of less than 12 mg/L can be obtained at a loading of 40 lb BOD_5/1000 ft^3 and detention time of 87 hour. The nitric acid also furnished nitrogen for proper nutrient balance for the biodegradation.

Coagulation-flocculation

Coagulation is used to destabilise the stable fine SS, while flocculation is used to grow the destabilised SS, so that the SS become heavier and larger enough to settle down. The coagulation-flocculation process can be used to remove fine SS from bakery wastewater. It normally acts as a preconditioning process for sedimentation and/or dissolved air flotation.

The wastewater is preconditioned by coagulants such as alum. The pH and coagulant dosage are important in the treatment results.

Sedimentation

Sedimentation, also called clarification, has a working mechanism based on the density difference between SS and the water, allowing SS with larger particle sizes to more easily settle down. Rectangular tanks, circular tanks, combination flocculator-clarifiers and stacked multilevel clarifiers can be used.

Dissolved air flotation (DAF)

Dissolved Air Flotation (DAF) is usually implemented by pumping compressed air bubbles to remove fine SS and FOG in the bakery wastewater. The wastewater is first stored in an air pressured, closed tank. Through the pressure-reduction valves, it enters the flotation tank. Due to the sudden reduction in pressure, air bubbles form and rise to the surface in the tank. The SS and FOG adhere to the fine air bubbles and are carried upwards. Dosages of coagulant and control of pH are important in the removal of BOD_5, COD, FOG and SS. Other influential factors include the solids content and air/solids ratio. With the DAF treatment, 48.6% of COD and 69.8% of SS can be removed in 10 minutes at a pressure of 4 kg/cm^2 and pH 6.0. Mulligan used DAF as a pre-treatment approach for bakery waste. At operating pressures of 40–60 psi, grease reductions of 90–97% were achieved. The BOD_5 and SS removal efficiencies were 33–62% and 59–90%, respectively.

Biological treatment

The biological treatment removes the dissolved and particulate biodegradable components in the wastewater. It is a core part of the secondary biological treatment system. Micro-organisms are used to decompose the organic wastes.

With regard to different growth types, biological systems can be classified as suspended growth or attached growth systems. Biological treatment can also be classified by oxygen utilisation: aerobic, anaerobic and facultative. In an aerobic system, the organic matter is decomposed to carbon dioxide, water and a series of simple compounds. If the system is anaerobic, the final products are carbon dioxide and methane.

Compared to anaerobic treatment, the aerobic biological process has better quality effluent, easier operation, shorter solid retention time, but higher cost for aeration and more excess sludge. When treating high-load influent (COD > 4000 mg/L), the aerobic biological treatment becomes less economic than the anaerobic system. To maintain good system performance, the anaerobic biological system requires more complex operations. In most cases, the anaerobic system is used as a pre-treatment process.

Suspended growth systems (e.g., activated sludge process) and attached growth systems (e.g., trickling filter) are two of the main biological wastewater treatment processes. The activated sludge process is most commonly used in treatment of wastewater. The trickling filter is easy to control and has less excess sludge. It has higher resistance loading and low energy cost. However, high operational cost is its major disadvantage. In addition, it is more sensitive to temperature and has odour problems. Comprehensive considerations must be taken into account when selecting a suitable system.

30.5 Aerobic treatment

30.5.1 Activated sludge process

In the activated sludge process, suspended growth micro-organisms are employed. A typical activated sludge process consists of a pre-treatment process (mainly screening and clarification), aeration tank (bioreactor), final sedimentation and excess sludge treatment (anaerobic treatment and dewatering process). The final sedimentation separates micro-organisms from the water solution. In order to enhance the performance result, most of the sludge from the sedimentation is recycled back to the aeration tank(s), while the remaining is sent to anaerobic sludge treatment.

The performance of activated sludge processes is affected by influent characteristics, bioreactor configuration and operational parameters. The influent characteristics are wastewater flow rate, organic concentration (BOD_5 and COD), nutrient compositions (nitrogen and phosphorus), FOG, alkalinity, heavy metals, toxins, pH and temperature. Configurations of the bioreactor include PFR, CSTR, SBR, membrane bioreactor (MBR) and so on. Operational parameters in the treatment are biomass concentration [mixed liquor volatile suspended solids concentration (MLVSS) and volatile suspended solids (VSS)], organic load, food to micro-organisms (F/M), dissolved oxygen (DO), sludge retention time (SRT), hydraulic retention time (HRT), sludge return ratio and surface hydraulic flow load. Among them, SRT and DO are the most important control parameters and can significantly affect the treatment results. A suitable SRT can be achieved by judicious sludge wasting from the final clarifier. The DO in the aeration tank should be maintained at a level slightly above 2 mg/L.

Owing to the high organic content, it is not recommended that bakery wastewater be directly treated by aerobic treatment processes.

30.5.2 Trickling filter process

Aerobic attached-growth processes include trickling filters (biotower) and Rotating Biological Contactors (RBC). In these processes, micro-organisms are attached onto solid media and form a layer of biofilm. The organic pollutants are first adsorbed to the biofilm surface, oxidation reactions then occur, which break the complex organics into a group of simple compounds, such as water, carbon dioxide and nitrate. In addition, the energy released from the oxidation together with the organics in the waste is used for maintenance of micro-organisms as well as synthesis of new micro-organisms.

The trickling filter can be used to treat bakery wastewater. Solid media such as crushed rock and stone, wood and chemical-resistant plastic media are randomly packed in the reactor. In trickling filter surface area and porosity are two important parameters of filter media. A large surface area can cause accumulation of a large amount of biomass and result in high treatment efficiency; large porosity would lead to higher oxygen transfer rate and less blockage. A common problem in trickling filter systems is the excess growth of micro-organisms, which can cause serious blockage in the medium and reduce the porosity.

30.6 Anaerobic biological treatment

In addition to accommodating organic waste treatment, anaerobic treatment can produce methane, which can be used for production of electricity. The disadvantages, however, include complexity in operation, sensitivity to temperature and toxicity, time-consuming in startup and susceptibility to process upset.

Anaerobic processes are suitable for a variety of bakery wastewater. For example, an anaerobic contactor was successfully used to treat wastewater from a production facility of snack cake items. Alternatively, a subsequent aerobic treatment can be used to further reduce the waste strength and the effluent can then be discharged to a watercourse.

30.6.1 Solid waste management

Bakery solid waste includes stale bakery products, dropped raw materials (e.g., dough) and packages. The most simple and common way is to directly transport these to landfill or incineration. Landfill can cause the waste to decompose, which eventually leads to production of methane (a greenhouse gas) and groundwater pollution (organic compounds and heavy metals). Incineration of bakery waste can also release nitrogen oxide gases. Reclamation of the bakery waste can play an important role in its management. The waste consists primarily of stale bread, bread rolls and cookies—all of which contain high energy and can be fed directly to animals, such as swine and cattle. Another application is to use the waste for production of valuable products.

30.6.2 Cleaner production in the bakery industry

The production of bakery products involves many operation units that may cause a variety of wastes. Most bakery industries are of small or medium size and are often located in densely populated areas, which makes environmental problems more critical. Nevertheless, the conventional 'end-of-pipe' treatment philosophy has its restrictions in dealing with these problems. It only addresses the result of inefficient and wasteful production processes and should be considered only as a final option. Manufacturing will always cause direct or indirect pollution of the environment. It is hard to realise 'zero discharge' and waste treatment is always expensive. Cleaner Production (CP) has two key components: maximisation of waste reduction and minimisation of raw material usage and energy consumption.

Cleaner production results from one or a combination of conserving raw materials, water and energy; eliminating toxic and dangerous raw materials; and reducing the quantity and toxicity of all emissions and wastes at source during the production process. It aims to reduce the environmental, health and safety impacts of products over their entire life-cycles, from raw materials extraction, through manufacturing and use, to the 'ultimate' disposal of the product. It implies the incorporation of environmental concerns into designing and delivering services.

In the CP process, raw materials, water and energy should be conserved, their emission or wastage should be reduced and application of toxic raw materials must be avoided. It is also important to reduce the negative impacts during the whole production life-cycle, from the design of the production to the final waste disposal.

31

Pollution control in poultry industry

31.1 Introduction

This chapter analyses the environmental impacts arising from intensive poultry production, evaluating such impacts across the food chain. The chapter also discusses technical options to mitigate environmental impacts, such as improvements to farm management, animal waste management and nutrition management, along with options to reduce the impacts of intensive feed production.

The use of large facilities associated with higher concentrations of poultry, has given rise to environmental concerns that are not only limited to the local production settings, but extend to environmental problems at regional and global scales. The obvious and often limited, impacts observed at production-site level, thus, tend to obscure much larger impacts on the regional and global environment.

31.2 Production and processing units in poultry

This section provides an overview of environmental concerns at the local level, arising from two point sources: the animal production site and the abattoir. At this level, impacts are usually directly observed by farmers, neighbours and policy-makers.

31.2.1 Animal production units

Local disturbances (e.g., odour, flies and rodents) and landscape degradation are typical local negative amenities in the surroundings of poultry farms. Pollution of soil and water with nutrients, pathogens and heavy metals is generally caused by poor manure-management and occurs where manure is stored. Water and soil pollution related to poultry litter is, however, generally not an issue at the production site, as poultry manure is only directly discharged into the environment in exceptional conditions. Indeed, the high nutrient content and low water content of poultry litter make it a valuable input to agriculture. Manure is either recycled on cropland belonging to the animal farm or marketed. In the usual setup, an intermediary or a processor collects manure from poultry farms. Manure is either resold rough or processed into compost or pellets. Manure products are used as fertiliser, or as animal feed especially for fish and cattle.

Poultry facilities are a source of odour and attract flies, rodents and other pests that create local nuisances and carry disease. Odour emissions from poultry farms adversely affect the life of people living in the vicinity. Odour associated with poultry operations comes from fresh and decomposing waste products such as manure, carcasses, feathers and bedding/litter. On-farm odour is mainly emitted from poultry buildings and manure and storage facilities. Odour from animal feeding operations is not caused by a single compound, but is rather the result of a large number of contributing compounds including ammonia (NH_3), Volatile Organic Compounds (VOCs) and hydrogen sulphide (H_2S). Of the several manure-based compounds which produce odour, the most commonly reported is ammonia. Ammonia gas has a sharp and pungent odour and can act as an irritant when present in elevated concentrations.

Odour is a local issue, which is hardly quantifiable; the impact greatly depends on the subjective perception of populations neighbouring the farm. It is, therefore, difficult to evaluate the maximum distance over which odorous gas travels; however, odour problems are generally concentrated within 500 metres of the farm. Although generally not causing any public-health concern, odours can represent a strong local problem that is frequently reported by farms' neighbours as the most disturbing environmental impact.

The emission of odours mostly depends on the frequency of animal-house cleaning, on the temperature and humidity of the manure, on the type of manure storage and on air movements. For these reasons it is generally higher in waterfowl farms than in chicken farms. Flies are an additional concern for residents living near poultry facilities. In addition to the nuisance they cause, flies and mosquitoes can transmit diseases, such as cholera, dysentery, typhoid, malaria, filaria and dengue fever. Although less often reported than flies and mosquitoes, rats and similar pests are also a local nuisance associated with poultry production. As with flies and mosquitoes, they can be a vector for disease transmission. Their presence is mainly related to animal-feed management and especially to storage and losses from feeding systems.

Pesticides used to control pests (e.g., parasites and disease vectors) and predators have been reported to cause pollution when they enter groundwater and surface water. Active molecules or their degradation products enter ecosystems in solution, in emulsion or bound to soil particles and may, in some instances, impair the uses of surface waters and groundwater.

The trend to larger production units and their regional concentration, certainly has the potential to adversely affect surrounding land use and the appearance of the landscape.

Massive industrial poultry installations can create an adverse aesthetic impact. Impact on land use in highly concentrated areas is manifested through conflict with development needs and in some areas with rural tourism.

Improper disposal of poultry carcasses can contribute to water-quality problems especially in areas prone to flooding or where there is a shallow water table. Methods for the disposal of poultry carcasses include burial, incineration, composting and rendering. In the case of recent Highly Pathogenic Avian Influenza (HPAI) outbreaks, the disposal of large numbers of infected birds has presented new and complex problems associated with environmental contamination. Large volumes of carcasses can generate excessive amounts of leachate and other pollutants, increasing the potential for environmental contamination.

31.3 Slaughterhouse

The most significant environmental issue resulting from slaughterhouse operations is the discharge of wastewater into the environment. Like many other food-processing activities, the necessity for hygiene and quality control in meat processing results in high water usage and consequently high levels of wastewater generation. Poultry processing activities require large amounts of high-quality water for process cleaning and cooling. Typical water usage in poultry slaughterhouses ranges between 6 and 30 cubic metres/T of product. Large quantities of water are consumed in poultry slaughterhouses for evisceration, cleaning and washing operations.

Process wastewater generated during these activities typically has high biochemical and chemical oxygen demand (BOD and COD_3) due to the presence of organic materials such as blood, fat, flesh and excreta. In addition, process wastewater may contain high levels of nitrogen, phosphorus and residues of chemicals such as chlorine used for washing and disinfection, as well as various pathogens including *Salmonella* and *Campylobacter*.

Poultry by-products and waste may contain up to 100 different species of micro-organisms, including pathogens, in contaminated feathers, feet and intestinal contents. Typical values for wastewater produced from poultry processing are 6.8 kg BOD/T Live Weight Killed (LWK) and 3.5 kg suspended solids per tonne of LWK.

Poultry slaughterhouses release large amounts of waste into the environment, polluting land and surface waters as well as posing a serious human-health risk. The discharge of biodegradable organic compounds may cause a strong reduction of the amount of dissolved oxygen in surface waters, which in turn may lead to reduced levels of activity or even death of aquatic life.

Macronutrients (nitrogen, phosphorus) may cause eutrophication of the affected water bodies. Excessive algal growth and subsequent dying off and mineralisation of these algae may lead to the death of aquatic life because of oxygen depletion.

Slaughterhouses are usually located in urban or peri-urban locations, where transport costs to markets are minimised and where there is abundant labour supply. This situation increases the risk of environmental impacts: first, because slaughterhouses often lack the land required to set up waste-management facilities; second, because the pollutants that are emitted add to those emitted by other human activities; and third, because neighbouring communities are directly affected by surface-water and groundwater contamination.

31.4 Watershed-level pollution associated with waste management

Intensification of production and the geographical concentration of production units often results in environmental concerns. The decoupling of crop and livestock production through the migration of livestock production away from crop activities into areas with little or no agricultural land leads to high levels of environmental impact – mainly related to manure mismanagement and nutrient overloads.

31.4.1 Poultry manure

Poultry manure contains considerable amounts of nutrients such as nitrogen, phosphorus and other excreted substances such as hormones, antibiotics, pathogens and heavy metals which are introduced through feed. Leaching and runoff of these substances has the potential to result in contamination of surface water and groundwater resources.

Nutrients

Animals reared in intensive production systems consume a considerable amount of protein and other nitrogen-containing substances in their diets. The conversion of dietary nitrogen to animal products is relatively inefficient, 50 to 80% of the nitrogen is excreted. Nitrogen is excreted in both organic and inorganic compounds. Nitrogen emissions from manure take four main forms: ammonia (NH_3), dinitrogen (N_2), nitrous oxide (N_2O) and nitrate (NO_3^-).

Phosphorus is an essential element for animal growth. Unlike nitrogen, phosphorus is relatively stable once attached to soil particles and does not leach through the soil into groundwater. It does not pose any environmental risks except as a nutrient; it limits biological activity in water resources and builds up in soil when applied in excess. Phosphorus emissions from manure occur in one main form: phosphate.

Heavy metals

Manure contains appreciable quantities of potentially toxic metals such as arsenic, copper and zinc. In excess, these elements can become toxic to plants, can adversely affect organisms that feed on these plants and can enter water systems through surface run-off and leaching. Trace elements are introduced into poultry diets either involuntarily through contaminated feedstuffs or voluntarily, as feed additives used to supply animals' requirements or – in much greater proportions – as veterinary medicines or growth promoters.

Pathogens

Manure also contains pathogens which may potentially affect soil and water resources, particularly if poorly managed. Parasites such as *Cryptosporidium* and *Giardia* spp., can easily spread from manure to water supplies and can remain viable in the environment for long periods of time.

31.5 Poultry wastes

Poultry processing is a speciality segment of the meat industry. Although poultry plants bear a resemblance in general production sequences to meat processing plants, they have specific distinctions and problems, including the characteristics and treatment of wastewaters.

31.5.1 Processing plants

The poultry processing plant should be highly mechanised having high production facilities. Conveyor speeds of 35–50 turkeys per minute or 100 broilers per minute are common. The sequence of processing operations is: live bird receiving, killing, bleeding, plucking and eviscerating, chilling and packing in ice or freezing for shipment. Inventories of live birds are kept low by scheduling incoming truck shipments in a manner to reduce holding time and inventory level. It is good practice to receive and process poultry on the same day, without hold-over of live birds. Storage of dressed ice-packed or frozen poultry is minimal also, with a significant reduction of holding inventory.

The egg processing plant cannot be typified because any or all egg processing operations may be found in any location. The fresh shell egg is a complete food and package in itself. Processing consists of candling for blood spots and other inside defects, examination of the shell for defects, washing, grading and cartoning. Hand methods are in general use; however, where quantity and quality permit, bulk candlers, sizers and automatic car toners are being installed. Egg breaking operations consist of separating the yolks from the whites, which are then either frozen separately or blended to a customer's specifications and frozen. The breaking of eggs is, in part, a salvage operation of the shell egg business where cracked shells or otherwise substandard conditions are encountered. Further processing of eggs or yolks and whites through spray drying is common. These dried egg products are used by bakeries, home cake or bake mixes and similar applications.

Blood wastes

The major source of pollution in poultry processing is generally the waste blood, which contains feathers, dirt and some manure. The majority of poultry processors attempt to recover free-draining blood. For good blood recovery, an area must be restricted to the collection of the blood, allowing for 2 minutes drainage time from the kill station. Even with the best of recovery practices, some blood is introduced into the plant sewer system. In addition, cleanup operations in the kill and bleed area send congealed blood and other matter to the plant sewers.

Approximately 6% of the poultry body weight is blood, 4% drainable in the bleeding area, leaving about 2% by weight in the bird. Tests have shown that chickens contain sufficient blood to equal a 5-day BOD of about 17.5 lb per 1000 chickens processed. In a good poultry plant operation, where blood wastes are carefully salvaged in the killing area, approximately 3–5 lb BOD per 1000 chickens may be sent to the plant wastes collectively from all operations.

Defeathering wastes

In the customary processing progression, the next wastes result from defeathering operations. These wastewaters consist mainly of the continuous scalder overflow, continuous water spray flow from the defeathering machines and final cleanup operations from dumping of scalder water and wash down of floors and equipment. Where the sudden release of water from a tank or vat might cause sewer overflow, it is good practice to provide a restricted drain opening.

Prior to entry into the defeathering machine, the bird is immersed in a scalder to relax feather follicles for easier feather removal. The scalder, operating on a continuous feed of approximately 1 quart of water per bird, therefore provides a steady flow to the plant sewer system. This waste is high in pollutional strength and contains feathers, some blood and dirt.

Defeathering machines are equipped with beaters (rubber fingers) that force the feathers from the skin. Simultaneously with the beating, warm water is applied to lubricate the rubber fingers, to prevent epidermis removal and to carry or 'flow away' the feathers from the machine area. The continuous flow of water used in the defeathering machines is high in volume but relatively low in pollutional strength.

Solid wastes

Following defeathering, the eviscerating and cutting operations produce solid wastes consisting of feet, heads, inedible viscera, crops and windpipes, lungs, grit, sand, gravel, trimmings, grease and blood. Lungs are generally removed by suction to a storage tank and then released to a flow-away system. In a hand removal operation, the lungs are discarded at the viscera trough with the remainder of the wastes.

31.5.2 Environmental issues

Gerber and others summarise some of the major potential impacts of intensive livestock production on land and water resources:

1. Eutrophication of surface waters, caused by the input of organic substances and nutrients either through wastewater from production, runoff or leakages from storage and handling facilities – affecting aquatic ecosystems and drinking water quality.
2. Leaching of nitrate and possible pathogen transfers to groundwater – affecting the quality of drinking water.
3. Accumulation of nutrients and other elements in soil due to continuous application of excess quantities of manure.
4. Impacts of pollution on nutrient-sensitive ecosystems resulting in biodiversity losses.

In most cases, structural changes in the production system have a rather negative impact on manure management practices. In particular, growth in the scale of production and geographical concentration in the vicinity of urban areas, cause dramatic land – livestock imbalances, hampering manure recycling options. Indeed, in such conditions, transport costs associated with carrying manure back to the field are prohibitive.

31.5.3 Waste disposal

The location of the poultry processing plant dictates the extent and method of waste treatment. In the search for a new plant site, the economics of waste disposal are an important factor. In locations where a municipal treatment plant is available, the plant pre-treatment provisions depend on the municipal plant's available extra capacity and efficiency.

Pre-treatment

Regardless of the treatment system, pre-treatment at the plant is necessary. The first pre-treatment stage is the use of screens to remove the larger materials from the defeathering and eviscerating flow-away systems.

The reuse of water of defeathering and eviscerating flow away water for the removal of feathers, but flow-away water from either system may not be reused for eviscerating flow-away. Therefore, these two wastes are pre-treated separately. The feather flow-away wastes, which are high in water content, are passed over a screen of 20–50 mesh. This is generally a rotary or a vibrating screen from which the feathers are discharged directly to a container or into a conveyor. Water from this screening is collected in a sump; some is recirculated by pump and reintroduced into the flow-away flumes of the defeathering operations. The offal flow-away wastes from the eviscerating area are discharged into a collection tank and are pumped over a rotary or vibrating screen of 20–50 mesh. The screened water is discharged into the feather water collection sump. Screened offal material is discharged to a bin or container for further disposal.

Primary material disposal: Feathers and offal, in plants not processing these materials further; are commonly removed by a truck trailer, in which the feathers and eviscerating wastes are separately compartmented. Where permitted, these materials may then be buried or incinerated. More often, poultry plants not further processing these primary waste materials sell the separated materials to rendering companies for further processing to feather meal, cracklings for animal feeds and other by-products. Innovations and improvements in rendering have resulted in a market for primary waste materials in many locations.

Grease removal: In addition to the screening of feathers and offal, the treatment of wastes should include grease removal. Wastes from all poultry plants contain grease that may be removed by skimming in a primary settling basin, by a grease trap or by air flotation.

Egg processing plants have no special or unusual wastes because whole eggs, whites or yolks introduced to sewers would be a loss of revenue. The only waste produced is egg shells, which are handled as common trash. Shells are accumulated in drums or directly in trucks and hauled to dumps.

Complete treatment of poultry process wastes

Methods available to the poultry processor for the complete treatment of the wastes are trickling filter, activated sludge, extended aeration, lagooning, land irrigation, sand filtration and chemical treatment.

Settling tanks: In all treatment systems, the floatable and sinkable materials must be separated in sedimentation tanks. The sludge requires additional handling by drying, lagooning, digestion or burial. For each plant design, settling rates and sludge volumes are determined by laboratory tests.

Trickling filters: Trickling filter systems in general, because of the high construction cost and operational requirements, are not often used where the processing plant is treating its own sewage. Trickling filter plant design for poultry wastes should not materially differ from standard practices for other biological wastes.

Activated sludge: The conditions outlined in regard to the trickling filter plant apply also to the activated sludge process. The construction costs of an activated sludge plant to obtain the same results as a trickling filter process are generally higher. The extended aeration process is a modification of activated sludge and utilises a 24 hours aeration or longer in the presence of activated sludge, followed by settling of the mixture. Part of the liquids discharged from the settled sludge tank are returned to the aerator. Lagooning may be required for additional purification of effluent before discharge into the natural waterway. The stream pollution regulations govern this action.

Lagoons: Lagoons or ponds are the most common type of waste treatment for poultry wastes. The lagoons generally require large areas of land and should be isolated from populated regions. Five-day BOD loadings vary from 20 to 150 lb per acre per day, depending upon pollution control regulations.

Although lagooning is common, there is much still to be learned about it. Many states are running tests on mechanical agitation, heating of the aerobic ponds and other experimental practices; and some industries, in connection with state authorities, are testing other types of modified lagooning designs to provide design data for future installations.

31.6 Environmental impacts related to feed production

31.6.1 Intensive agriculture

Intensification of feed production affects land and water resources through pollution caused by the intensive use of mineral fertiliser, pesticides and herbicides to maintain high crop yields. Intensive feed production also contributes to air pollution. The application of nitrogen fertiliser to cropland is a major source of air pollution through the volatilisation of ammonia.

31.6.2 Intrusion into natural habitats

Increases in feed production, have to some extent been related to the expansion of cropland dedicated to feed. Feed production to satisfy the feed demand of intensive systems indirectly affects the global land base through changes in land use.

31.6.3 Erosion of biodiversity

Feed production is also driving biodiversity erosion through the conversion of natural habitats and the over exploitation of non-renewable resources for feed production. Intensive feed production contributes to biodiversity loss through land use and land-use change and modification of natural ecosystems and habitats.

31.6.4 Over exploitation of natural resources

The production of fishmeal for the poultry sector is an important factor contributing to the over exploitation of fisheries. The expansion of the aquaculture sector and its demand for fishmeal as a feed ingredient has led to a reduction in the use of fishmeal by the poultry sector.

31.6.5 Climate change

The relatively high energy input in intensive livestock systems has given rise to concerns regarding greenhouse gas emissions and climate change. The energy consumption of industrially produced poultry is relevant because of the production of carbon dioxide (CO_2) along the production chain. Carbon dioxide emissions are produced by the burning of fossil fuels during animal production and slaughter and transport of processed and refrigerated products, but importantly also through land use and land-use change and the use of inputs for the production of feed.

On-farm energy consumption

On-farm energy consumption includes direct and indirect energy input – direct energy refers to fossil energy used for the production process (e.g., energy input for poultry housing systems) and indirect energy to that used as an integral part of the production process (e.g., feed processing).

Due to a lack of information on energy use for processing, this estimation of on-farm fossil fuel consumption is limited to quantifying energy use associated with poultry housing.

The energy used for heating, ventilation and air conditioning systems typically accounts for the largest quantity of energy used in intensive poultry operations. Animal housing facilities are therefore potential sources of carbon dioxide emissions from intensive poultry farms. Other sources of carbon dioxide emissions include energy used for feed preparation, on-farm transport and burning of waste.

Generally, on layer farms, artificial heating of housing is not commonly applied, due to the low temperature needs of birds and the high stocking density. The activities that require energy are ventilation, feed distribution, lighting and egg collection, sorting and preservation.

On broiler farms, the main energy consumption is related to local heating, feed distribution and housing ventilation. Quantification of the energy consumption of intensive poultry farms is a complex undertaking because systems are not homogeneous. The amount of energy consumed varies with the technologies applied, the production characteristics of the farms and climatic conditions.

Carbon dioxide emissions from slaughtering

Poultry processing facilities use energy to heat water and produce steam for process applications and cleaning and for the operation of mechanical and electrical equipment, refrigeration and air compressors. In poultry abattoirs, fossil fuel is mainly used for process heat, while electricity is used for the operation of machinery and for refrigeration, ventilation, lighting and the production of compressed air.

31.6.6 Greenhouse gases emissions from feed production

Emissions of greenhouse gases such as carbon dioxide and nitrous oxide are influenced in an indirect way by intensification of feed production, which requires energy input for the production of mineral fertiliser and the subsequent use of this fertiliser in the feed production process.

Carbon dioxide (CO_2)

This greenhouse gas is produced by the burning of fossil fuels during the manufacture of fertiliser.

Nitrous oxide (N_2O)

Poultry production is indirectly associated with the greenhouse gas nitrous oxide because of the sector's high concentrate-feed requirements and the related emissions from arable land due to the use of nitrogen fertiliser.

Overall, intensive poultry production (indirectly and directly) contributes an estimated 3% of the total anthropogenic greenhouse gas and is responsible for about 2% of the total greenhouse gas emissions from the livestock sector. This estimate however does not include emissions from land use and land-use change associated with feed production or emissions related to transport of feed.

31.6.7 Odour emissions

Odour emissions can be controlled by:

1. Minimising the surface of manure in contact with air – frequent collection of litter (once a week in dry seasons and twice a week in rainy seasons), closed storage (bags or closed sheds).
2. Cooling animal manure, achieved as a positive side effect of cooling the animal houses – cooling systems can be equipped with biofilters and air scrubbers that trap odours from the ventilation airflow.
3. Lowering litter's water content – achieved by the incorporation of hydrophilic products such as hashes, rice husk, peanut husk, dust or sawdust.
4. Applying deodorant products to feed or directly to animal houses.
5. Building wind protection structures.

Flies and mosquitoes

The proliferation of flies and mosquitoes can be controlled by:

1. Minimising the surface of manure in contact with air – frequent collection of litter (once a week in dry seasons and twice in rainy seasons, i.e., at shorter intervals than the length of the larvae development cycle), closed storage (bags or closed sheds).
2. Lowering litter's water content – achieved by the incorporation of hydrophilic products such as hashes, rice husks, peanut husks, dust, sawdust or available dry crop residues.
3. Applying insecticides (this practice may however have significant public health-related side effects).
4. Building wind protection structures.
5. Positioning nets around the farm.

Rat proliferation can be controlled by:

1. Minimising feed losses during storage and feeding.
2. Raising cats or keeping snakes in cages close to the poultry barn to scare rats.
3. Use of poison or traps.

Visual impact and landscaping can be improved by:

1. Use of screening trees around the farm facility to reduce the visual impact of farm infrastructure and of noise, dust, light and odour.
2. Use of the natural topography and terrain of the site and the existing vegetative cover to maximise visual screening.
3. Use of construction materials that minimise visual impact.

31.7 Animal waste management

Soil and water pollution is controlled through the implementation of good fertilisation practices. In brief: environmental risks are reduced when manure is applied in amounts and at times that correspond to crop or fishpond uptake. Water pollution is often an acute problem in waterfowl production, especially when the flock is concentrated on relatively small ponds. There is currently a lack of information with regard to the effects of waterfowl production on surface water and groundwater resources.

Water- and food-borne disease propagation can be prevented by:

1. Storing manure in closed buildings or bags – a storage system allows producers to hold manure until a convenient and optimum time for use; storing poultry manure in closed buildings reduces the emissions of gaseous compounds to the air and the risk of environmental contamination as compared to the risk associated with leaving manure exposed.

2. Storing the manure for one to two months before its application on land or fish ponds.
3. Composting manure – potentially reduces or even eliminates certain pathogens and fly larvae and improves the handling characteristics of manure and other residues by reducing their volume, weight and moisture content (most manure and other organic residues usually contain high nitrogen content and are, therefore, subject to nitrogen loss during composting).
4. Drying (with machine or by spreading out) – minimises the moisture content of manure, inhibits chemical reactions and thus reduces emissions (the best way to prevent ammonia emissions from poultry litter and manure is to reduce microbial decomposition, which can be accomplished by drying the freshly produced manure as soon as possible and keeping it dry).
5. Timing and rate of manure application – this is a critical management factor; manure must be applied at the correct time of year to prevent losses to surface water, groundwater and the atmosphere and to optimise the utilisation of manure nutrients by growing plants; proper timing is a function of several variables, including weather, soil conditions and stage of crop growth.
6. Dead-bird management and disposal, which must comply with legally accepted practices including rendering, composing, incineration and burial; a contingency plan should be in place for disposal of large numbers of dead birds in the event of disease outbreaks; in addition, consideration should be given to impacts on the physical environment – e.g., burial pits should be at least 3 metres above the maximum groundwater table.

32

Pollution control in meat industry

32.1 Introduction

Meat is animal flesh that is eaten as food. Meat is mainly composed of water, protein and fat and is usually eaten together with other food. It is edible raw, but is normally eaten after it has been cooked and seasoned or processed in a variety of ways. Unprocessed meat will spoil or rot within hours or days as a result of infection with and decomposition by bacteria and fungi.

Most often, meat refers to skeletal muscle and associated fat and other tissues, but it may also describe other edible tissues such as offal.

The meat processing and rendering industry includes the slaughter of animals and fowl, processing of the carcasses into cured, canned and other meat products and the rendering of inedible and discarded remains into useful by-products such as lards and oils.

Processed meat products are those in which the properties of fresh meat have been modified by the use of one or more procedures, such as grinding, adding of seasoning, alteration of colour or heat treatment.

It includes curing by smoking and any treatment or process resulting in a substantial change in the natural state of the meat but does not include boning, cutting, cleaning or trimming.

There is a concept of inevitable association between processing and preservation of meat. This is because some processing method result in products which have long shelf life, for example salted and smoke meat is processed for flavour and it is also preserved by the presence of sale, smoke and heat treatment. On the contrary raw, seasoned beef sausage is processed but it is not preserved. A wide range of processes are used. Table 32.1 provides information on water usage in the industry.

Table 32.1: Typical water usage in the meat industry.
(Cubic meters per metric tonne of product)

Process	*Water use*
Slaughterhouse	
Pigs	1.5–10
Cattle	2.5–40
Poultry	6–30
Meat processing	2–60

32.2 Fermented meat

Fermented meat is a type of dried meat that has been prepared and dehydrated according to specific techniques that alter some of the chemical make-up of normally perishable foods. Fermentation typically involves the introduction of bacteria or yeast that convert certain meat nutrients into mixtures of carbon dioxide and alcohol. This process of making fermented meat can both increase the flavour intensity and prevent the end product from spoiling for relatively long periods of time. Successful fermentation of meat can only be accomplished under specific environmental conditions in order for the end product to be completely safe to eat. Various culinary traditions often have their own unique types of fermented meat such as salami or chorizo sausage.

Meat must be fermented in an environment without oxygen, which is frequently called an anaerobic condition. Removing the oxygen is essential for the meat to correctly dehydrate without the possibility of harmful bacterial invasion. The fermentation catalyst bacteria is usually introduced in controlled amounts. This type of bacteria is usually one of several varieties derived from lactic acid. Cooks with experience in fermenting frequently use a strain of lactic acid bacteria called pediococus cerevisiae. This important step is normally accomplished by sealing the cuts of meat to be dried in an airtight container.

Various flavours of salami are normally made through this basic fermentation process. This kind of dehydrated sausage is often made with a combination of catalyst bacteria and other additives called nitrates. The inclusion of nitrates can often lessen the chances of contamination by allowing the beneficial lactic acid bacteria to grow and spread through the meat at a faster rate. Once the technique of fermenting meat has been perfected, the process is normally not limited to meats such as beef or pork. Some types of seafood can also be successfully fermented, although health professionals sometimes advise that this practice should be approached carefully. Fermented fish can often have high rates of bacterial contamination that can lead to serious food-borne illnesses such as botulism.

Some of the benefits of fermented meat include economic practicality and environmental necessity. People who live in remote geographic areas with harsh winters often ferment large stores of meat due to limited access to other food sources during the worst weather conditions.

Fermented meats are generally able to stand up to extreme temperatures at a better rate than other foods.

32.3 Waste characteristics

The meat industry has the potential for generating large quantities of solid wastes and wastewater with a Biochemical Oxygen Demand (BOD) of 600 milligrams per litre (mg/L). BOD can be as high as 8000 mg/L or 10–20 kilograms per metric tonne (kg/L) of slaughtered animal; and suspended solids levels can be 800 mg/L and higher. In some cases, offensive odours may occur. The amounts of wastewater generated and the pollutant load depend on the kind of meat being processed. For example, the processing of gut has a significant impact on the quantity and quality (as measured by levels of BOD and of Chemical Oxygen Demand (COD), of wastewater generated. The wastewater from a slaughterhouse can contain blood, manure, hair, fat, feathers and bones. The wastewater may be at a high temperature and may contain organic material and nitrogen, as well as such pathogens as salmonella and shigella bacteria, parasite eggs and amoebic cysts. Pesticide residues may be present from treatment of animals or their feed. Chloride levels from curing and pickling may be very high—up to 77,000 mg/L. Smoking operations can release toxic organics into air. Rendering is an evaporative process that produces a condensate stream with a foul odour.

All slaughtering wastes (generally, 35% of the animal weight) can be used as by-products or for rendering. The only significant solid waste going for disposal is the manure from animal transport and handling areas.

32.3.1 Pollution prevention and control

Separation of product from wastes at each stage is essential for maximising product recovery and reducing waste loads. The materials being handled are all putrescible; hence, cleanliness is essential. Water management should achieve the necessary cleanliness without waste. The amounts and strength of wastes can be reduced by good practices such as dry removal of solid wastes and installation of screens on wastewater collection channels.

In-plant measures that can be used to reduce the odour nuisance and the generation of solid and liquid wastes from the production processes include the following:

1. Recover and process blood into useful by-products. Allow enough time for blood draining (at least seven minutes).
2. Process paunches and intestines and utilise fat and slime.
3. Minimise water consumed in production by, for example, using taps with automatic shut-off, using high water pressure and improving the process layout.
4. Eliminate wet transport (pumping) of wastes (for example, intestines and feathers) to minimise water consumption.
5. Reduce the liquid waste load by preventing any solid wastes or concentrated liquids from entering the wastewater stream.

6. Cover collection channels in the production area with grids to reduce the amount of solids entering the wastewater.
7. Separate cooling water from process water and wastewaters and recirculate cooling water.
8. Implement dry precleaning of equipment and production areas prior to wet cleaning.
9. Equip the outlets of wastewater channels with screens and fat traps to recover and reduce the concentration of coarse material and fat in the combined wastewater stream.
10. Optimise the use of detergents and disinfectants in washing water.
11. Remove manure (from the stockyard and from intestine processing) in solid form.
12. Dispose of hair and bones to the rendering plants.
13. Reduce air emissions from ham processing through some degree of air recirculation, after filtering.
14. Isolate and ventilate all sources of odorous emissions. Oxidants such as nitrates can be added to wastes to reduce odour.

In rendering plants, odour is the most important air pollution issue. To reduce odour:

1. Minimise the stock of raw material and store it in a cold, closed, well-ventilated place.
2. Pasteurise the raw material before processing it in order to halt biological processes that generate odour.
3. Install all equipment in closed spaces and operate under partial or total vacuum.
4. Keep all working and storage areas clean.

32.3.2 Treatment technologies

Wastewaters from meat processing are suitable for biological treatment and (except for the very odorous rendering wastewater) could be discharged to a municipal sewer system after flow equalisation, if the capacity exists. Sewer authorities usually require pre-treatment of the wastewater before it is discharged into the sewer. Screens and fat traps are the minimum means of pre-treatment in any system. Flotation, in some cases aided by chemical addition, may also be carried out to remove suspended solids and emulsified fats, which can be returned to the rendering plant. The choice of an appropriate biological treatment system will be influenced by a number of factors, including wastewater load and the need to minimise odours. Rendering wastewater typically has a very high organic and nitrogen load.

Extended aeration is an effective form of treatment, but care must be taken to minimise odours. Disinfection of the final effluent may be required if high levels of bacteria are detected. Ponding is a simple solution but requires considerable space. Chemical methods, usually based on chlorine compounds, are an alternative. Biofilters, carbon filters and scrubbers are used to control odours and air emissions from several processes, including ham processing and rendering. Recycling exhaust gases from smoking may be feasible in cases where operations are not carried out manually and smoke inhalation by workers is not of concern.

32.3.3 Air emissions

Odour controls should be implemented, where necessary, to minimise odour impacts on nearby residents. Particulate matter emissions of smokehouses should be kept below 150 milligrams per normal cubic meter (mg/Nm^3), with a carbon content of less than 50 mg/Nm^3.

32.3.4 Liquid effluents

The liquid effluent levels presented in Table 32.2 should be achieved.

Table 32.2: Effluents from meat processing and rendering industry.
(Milligrams per litre, except for pH and bacteria)

Parameter	*Maximum value*
pH	6–9
BOD	50
COD	250
TSS	50
Oil and grease	10
Nitrogen (total)	10
Total phosphorus	5
Coliform bacteria	400 MPN/100 mL

Note: Effluent requirements are for direct discharge to surface waters. MPN, most probable number.

32.4 Processing facilities and wastes generated

As a direct result of its operation, a slaughterhouse generates waste comprised of the animal parts that have no perceived value to the slaughterhouse operator. It also generates wastewater as a result of washing carcasses, processing offal and from cleaning equipment and the fabric of the building. Additionally, wastes and wastewaters are also generated from the stockyards, any rendering process, cooling facilities for refrigeration, compressors and pumps, vehicle wash facilities, wash rooms, canteen and possibly laundry facilities.

32.4.1 Waste characteristics and quantities generated

In general the characteristics of the solid wastes generated reflect the type of animal being killed, but the composition within a particular type of operation is similar regardless of the size of the plant. Once on the slaughter line, the quantity of waste generated depends on the number of animals slaughtered and the type of animal.

Wastewater flow

Water is used in the slaughterhouse for carcass washing after hide removal from cattle, calves and sheep and after hair removal from hogs. It is also used to clean the inside of the carcass after evisceration and for cleaning and sanitising equipment and facilities both during and after the killing operation.

Wastewater characteristics

Effluents from slaughterhouses and packing houses are usually heavily loaded with solids, floatable matter (fat), blood, manure and a variety of organic compounds originating from proteins. As already stated the composition of effluents depends very much on the type of production and facilities. The main sources of water contamination are from lairage, slaughtering, hide or hair removal, paunch handling, carcass washing, rendering, trimming and cleanup operations.

These contain a variety of readily biodegradable organic compounds, primarily fats and proteins, present in both particulate and dissolved forms. The wastewater has a high strength, in terms of biochemical oxygen demand (BOD), chemical oxygen demand (COD), suspended solids (SS), nitrogen and phosphorus, compared to domestic wastewaters. The actual concentration will depend on in-plant control of water use, by-products recovery, waste separation source and plant management. In general, blood and intestinal contents arising from the killing floor and the gut room, together with manure from stockyard and holding pens, are separated, as best as possible, from the aqueous stream and treated as solid wastes. This can never be 100% successful, however and these components are the major contributors to the organic load in the wastewater, together with solubilised fat and meat trimmings.

32.4.2 Wastewater minimisation

The overall waste load arising from a slaughterhouse is determined principally by the type and number of animals slaughtered. The partitioning of this load between the solid and aqueous phases will depend very much upon the operational practices adopted, however and there are measures that can be taken to minimise wastewater generation and the aqueous pollution load.

One of the major contributors to organic load is blood, which has a COD of about 400000 mg/L and washing down of dispersed blood can be a major cause of high effluent strength. Minimisation can be achieved by having efficient blood collection troughs allowing collection from the carcass over several minutes. Likewise the trough should be designed to allow separate drainage to a collection tank of the blood and the first flush of washwater. Only residual blood should enter a second drain for collection of the main portion of the washwater.

An efficient blood recovery system could reduce the aqueous pollution load by as much as 40% compared to a plant of similar size that allows the blood to flow to waste.

The second area where high organic loads into the wastewater system can arise is in the gut room. Most cattle and sheep abattoirs clean the paunch (rumen), manyplies (omasum) and reed (abomasum) for tripe production. A common method of preparation is to flush out the gut manure from the punctured organs over a mechanical screen and allow water to transport the gut manure to the effluent treatment system.

Within the slaughtering area and cutting rooms, measures should be adopted to minimise meat scraps and fatty tissue entering the floor drains. Once in the drains these break down due to turbulence, pumping or other mechanical actions (e.g., on screens), leading to an increase in effluent COD. These measures include using fine mesh covers to drains, encouraging operators to use collection receptacles for trimmings and using well-designed equipment with catch trays. Importantly, a 'dry' cleaning of the area to remove solid material, for example using cyclonic vacuum cleaners, should take place before any washdown.

Other methods can also be employed to minimise water usage. These will not in themselves reduce the organic load entering the wastewater treatment system, but will reduce the volume requiring treatment and possibly influence the choice of treatment system to be employed.

Water use minimisation methods include:

1. The use of directional spray nozzles in carcass washing, which can reduce water consumption by as much as 20%.
2. Use of steam condensation systems in place of scald tanks for hair and nail removal.
3. Fitting washdown hoses with trigger grips.
4. Appropriate choice of cleaning agents.
5. Reuse of clear water (e.g., chiller water) for the primary washdown of holding pens.

32.5 Wastewater treatment processes

The degree of wastewater treatment required will depend on the proposed type of discharge. Wastewaters received into the sewer system are likely to need less treatment than those having direct discharge into a watercourse. In European countries, direct discharges have to comply with the Urban Wastewater Treatment Directive and other water quality directives. In the United States the EPA has proposed Effluent Limitations Guidelines (ELGs) and standards for the Meat and Poultry Products industries with direct discharge. These proposed ELGs will apply to existing and new Meat and Poultry Products (MPP) facilities and are based on the well-tested concepts of 'Best Practicable Control technology currently available' (BPT), the 'Best Conventional Pollutant control technology' (BCT), the 'Best Available Technology economically achievable' (BAT) and the 'best available demonstrated control technology for New Source Performance Standards' (NSPS). In summary, the technologies proposed to meet these requirements use, in the main, a system based on a treatment series comprising flow equalisation, dissolved air flotation and secondary biological treatment for all slaughterhouses; and require nitrification for small installations and additional denitrification for complex slaughterhouses.

There is some potential, however, for segregation of wastewaters allowing specific individual pre-treatments to be undertaken or, in some cases, bypass of less contaminated streams. Depending on local conditions and regulations, water from boiler houses and refrigerating systems may be segregated and discharged directly or used for outside cleaning operations.

32.5.1 Primary treatment

In primary treatment grease removal is a common first stage in slaughterhouse wastewater treatment, with grease traps in some situations being an integral part of the drainage system from the processing areas. Where the option is taken to have a single point of removal, this can be accomplished in one of two ways: by using a baffled tank or by Dissolved Air Flotation (DAF).

The flotation process is dependent upon the release of sufficient air from the pressurised fluid when the pressure is reduced to atmospheric. The nature of the release is also important, in that the bubbles must be of reasonably constant dimensions (not greater than 130 microns) and in sufficient numbers to provide blanket coverage of the retaining vessel. In practice, the bubble size and uniform coverage give the appearance of white water. The efficiency of the process depends upon bubble size, the concentration of fats and grease to be separated, their specific gravity, the quantity of the pressurised gas and the geometry of the reaction vessel.

Dissolved air flotation has become a well-established unit operation in the treatment of abattoir wastes, primarily as it is effective at removing fats from the aqueous stream within a short retention time (20–30 minutes), thus preventing the development of acidity.

Chemical treatment can improve the pollution removal efficiency of a DAF unit and typically ferric chloride is used to precipitate proteins and polymers used to aid coagulation.

32.5.2 Secondary treatment

Secondary treatment aims to reduce the BOD of the wastewater by removing the organic matter that remains after primary treatment. This is primarily in a soluble form. Secondary treatment can utilise physical and chemical unit processes, but for the treatment of meat wastes biological treatment is usually favoured.

32.5.3 Biological secondary treatment

Using biological treatment, more than 90% efficiency can be achieved in pollutant removal from slaughterhouse wastes. Commonly used systems include lagoons (aerobic and anaerobic), conventional activated sludge, extended aeration, oxidation ditches, sequencing batch reactors and anaerobic digestion. A series of anaerobic biological processes followed by aerobic biological processes is often useful for sequential reduction of the BOD load in the most economic manner, although either process can be used separately. As noted above, slaughterhouse wastewaters vary in strength considerably depending on a number of factors. For a given type of animal, however, this variation is primarily due to the quantity of water used within the abattoir, as the pollution load (as expressed as BOD) is relatively constant on the basis of live weight slaughtered. Hence, the more economical an abattoir is in its use of water, the stronger the effluent will be and *vice versa*.

Anaerobic treatment

Anaerobic digestion is a popular method for treating meat industry wastes. Anaerobic processes operate in the absence of oxygen and the final products are mixed gases of methane and carbon dioxide and a stabilised sludge. Anaerobic digestion of organic materials to methane and carbon dioxide is a complicated biological and chemical process that involves three stages: hydrolysis, acetogenesis and finally methanogenesis. During the first stage, complex compounds are hydrolysed to smaller chain intermediates. In the second stage acetogenic bacteria convert these intermediates to organic acids and then ultimately to methane and carbon dioxide via the methanogenesis phase.

In the United States, anaerobic systems using simple lagoons are by far the most common method of treating abattoir wastewater. These are not particularly suitable for use in the heavily populated regions of western Europe due to the land area required and also because of the difficulties of controlling odours in the urban areas where abattoirs are usually located.

Aerobic treatment

Aerobic processes can roughly be divided into two basic types: those that maintain the biomass in suspension (activated sludge and its variants) and those that retain the biomass on a support medium (biological filters and its variants). There is no doubt that either basic type is suitable for the treatment of slaughterhouse wastewater and their use is well documented in research done by Broils and Broughton, where aerobic processes are compared with anaerobic ones. In selecting an aerobic process a number of factors need to be taken into account.

The most common aerobic biological processes used for the treatment of meat industry wastes are biological filtration, activated sludge plants, waste stabilisation ponds and aerated lagoons.

Waste stabilisation ponds: A waste stabilisation pond (WSP) is the simplest method of aerobic biological treatment and can be regarded as bringing about the natural purification processes occurring in a river in a more restricted time and space. They are often used in countries where plenty of land is available and weather conditions are favourable.

Biological filters: Biological filters can also be used for treating meat industry wastes. In this process the aerobic micro-organisms grow as a slime or film that is supported on the surface of the filter medium. The wastewater is applied to the surface and trickles down while air percolates upwards through the medium and supplies the oxygen required for purification. The treated water along with any microbial film that breaks away from the support medium collects in an under-drain and passes to a secondary sedimentation tank where the biological solids are separated. Trickling filters require primary treatment for removal of settleable solids and oil and grease to reduce the organic load and prevent the system blocking.

Activated sludge: The activated sludge process has been successfully used for the treatment of wastewaters from the meat industry for many decades. It generally has a lower capital cost than standard-rate percolating filters and occupies substantially less space than lagoon or pond systems. In the activated sludge process the wastewaters are mixed with a suspension of aerobic micro-organisms (activated sludge) and aerated. After aeration, the mixed liquor passes to a settlement tank where the activated sludge settles and is returned to the plant inlet to treat the incoming waste. The supernatant liquid in the settlement tank is discharged as plant effluent. Air can be supplied to the plant by a variety of means, including blowing air into the mixed liquor through diffusers; mechanical surface aeration, and floor-mounted sparge pipes.

All the methods are satisfactory provided that they are properly designed to meet the required concentration of dissolved oxygen in the mixed liquor (greater than 0.5 mg/L) and to maintain the sludge in suspension; for nitrification to occur it may be necessary to maintain dissolved oxygen concentrations above 2.0 mg/L.

32.6 Solid wastes

If good operational practice is followed in the slaughterhouse, the solids and organic loading entering the aqueous phase can be minimised. The separated solids still require treatment prior to disposal, however and traditional rendering of some of these fractions is uneconomic because of the high water and low fat content. These fractions are the gut manures, the manure and bedding material from holding pens, material from the wastewater screens and traps on surface drains, sedimentation or DAF sludge and possibly hair where no market exists for this material. Other high-protein and fat-containing residues such as trimmings, non-edible offal and skeletal material can be rendered to extract tallow and then dried to produce meat and bone meal.

32.6.1 Composting

Composting can be defined as the biological degradation of organic materials under aerobic conditions into relatively stable products, brought about by the action of a variety of micro-organisms such as fungi, bacteria and protozoa. The process of composting may be divided into two main stages: stabilisation and maturation.

During stabilisation, three phases may be distinguished: first a phase of rising temperature, secondly the thermophilic phase where high temperature is maintained and thirdly, the mesophilic phase where the temperature gradually decreases to ambient.

During the first phase, a vigorous multiplication of mesophilic bacteria is observed and a transformation of easily oxidised carbon compounds, rich in energy, to compounds of lower molecular structure takes place. Excess energy results in a rapid rise in temperature and leads to the thermophilic phase when less easily degraded organic compounds are attacked. When the energy source is depleted the temperature decreases gradually to ambient. Actinomycetes and fungi become active in the mesophilic phase, during which biological degradation of the remaining organic compounds is slowly continued. At this stage the composting material is stabilised but not yet matured. During maturation, mineralisation of organic matter continues at a relatively slow rate until a carbon:nitrogen (C:N) ratio of 10:1 is reached and the volatile matter content falls below 50%. Only then is the compost completely matured.

Composting of dewatered sedimentation tank solids from a slaughterhouse in mechanically turned open windrows was described by Supapong. The material was kept in windrows for 40 days and the temperature fluctuated between 149 and 158°F (65–70°C) for up to 3 weeks.

The resulting product was a friable, odourless and microbiologically satisfactory material whose bulk was only one-fifth of the original volume. It contained 0.5 and 3% by weight of phosphorus and nitrogen, respectively and was an acceptable soil conditioner.

33

Pollution control in seafood food industry

33.1 Introduction

Seafood is any form of sea life regarded as food by humans. Seafood prominently includes fish and shellfish. Shellfish include various species of molluscs, crustaceans and echinoderms. Historically, sea mammals such as whales and dolphins have been consumed as food, though that happens to a lesser extent in modern times. Edible sea plants, such as some seaweeds and microalgae, are widely eaten as seafood around the world, especially in Asia. In North America, although not generally in the United Kingdom, the term 'seafood' is extended to fresh water organisms eaten by humans, so all edible aquatic life may be referred to as seafood.

The world seafood industry plays a significant role in the economic and social wellbeing of nations, as well as in the feeding of a significant part of the world's population. Fishing and fish farming has emerged as one of the major food processing occupations of mankind. In ancient times, economically and socially backward people were employed in this profession. The advent of modern mechanised fishing vessels has brought vast changes in the attitude of the public fishing and seafood processing. From low income and socially backward communities the profession has shifted to the hands of industrialists and technologists. Today fishing and processing activities provide employment to millions of people around the world.

Fresh fish cannot be kept for a long time without refrigerating, so we have to process the fish for preservation. There are several different ways of processing fish, mainly according to the kinds of fish.

Sun-dried fish: Sun drying is a popular way of processing small fish, such as Utaka, Usipa, Mathemba and small size of Mcheni. To dry the fish, you have to put the fish on the reads-made drying rack and turn the fish once or twice a day to dry evenly. It takes 3–4 days to finish the process and the fish should be taken away from the rack at night or in the rain.

Boil-dried fish: This type of processing is only used to process Usipa. The fish is first boiled in water and then put on the rack to be dried under the sun. This process takes less time than the sun drying (for 2–3 days) and the taste is also distinctive.

Roast-dried fish: This type of processing is especially for the flat fish like Utaka, Mayani and Chendamwamba. First the fish is dried for a few hours and then roasted on a steel pan to cook both sides. The fish are put back to the rack to dry again for rest of the day. This process can be completed for 1–2 days and also produces a different taste.

Smoked fish: Smoking is the only way to process the bigger fishes such as Chambo, ampango, Mlamba and bigger size of Mcheni. Small fishes are also smoked because of its fairly distinctive taste and of the shorter time for processing.

Fish are first dried for about half a day and then put on the wired oven overnight. Big fish like Kampango, Chambo or Mlamba are opened and gutted before drying. Smoking is especially favoured in rainy season, for the process is not affected from the rain.

Fish smoking and forest management: Compared to other types of fish processing, smoking requires fairly large amount of firewood. To keep the fire for smoking fish for one night, you may need almost the same amount of firewood that you use for cooking for one week.

33.2 Fermented fish

Fermentation is one method of fish curing in which the development of a distinctive flavour in the final product is the principal objective. Therefore, this product is mainly used as a condiment in the preparation of traditional sauces. Fermentation alone as a curing process does not preserve fish because it results in the breakdown of fish muscle. For this reason, fermentation is often combined with salting and/or drying in order to reduce water activity and retard or eliminate the growth of proteolytical and putrefying bacteria. In Southeast Asia the fermentation process often lasts for several months and the final product is usually a paste, sauce or liquid. In Africa, however, fish fermentation lasts from a few hours to about two weeks. Under such conditions, fermentation is usually partial and the muscle structure is not broken down completely. Consequently, the fish retains its original form of whole or cut pieces after fermentation and can be eaten as food fish or used as a condiment.

Fish fermentation is generally accompanied by the development of a mild or strong odour which becomes associated with the final product. Fermented fishery products in Africa may either be soft with a high moisture content, semi-dry or very dry. Some products are also heavily salted and dried whilst others are dried without any salting. Some types of fermented products have a rancid taste. The fermented fish includes salted and/or sun-dried fishery products in which some degree of desirable enzymatic or microbiological change has taken place in the fish muscle. Fermented fish is, therefore, any fishery product which has undergone degradative changes through enzymatic or microbiological activity either in the presence or absence of salt.

33.3 Environmental issues related to seafood processing

Environmental issues in fish processing industries primarily include the following: Water consumption and wastewater generation, solid waste generation and by-products production, emission to aır and energy consumption.

33.3.1 Water consumption

Most seafood processors have a high baseline water use for cleaning plant and equipment. Therefore, water use per unit product decreases rapidly as production volume increases. Major sources of water consumption include: fish storage and transport; cleaning, freezing and thawing; preparation of brines; equipment sprays; offal transport; cooling water; steam generation; and equipment and floor cleaning. Water consumption in fish processing operations has traditionally been high to achieve effective sanitation. Several factors affect water use, including: the type of product processed, the scale of the operation, the process used and the level of water minimisation practices in place. General cleaning contributes significantly to total water demand so smaller-scale sites tend to have significantly higher water use per unit of production.

33.3.2 Energy Consumption

Seafood processing industries consumes large quantities of electrical energy. Most of the power is used for magnetic induction equipment, such as electric motors (compressors for freezers, cold stores, ice-making machines, water pumps, etc.), and lighting that requires magnetic ballasts, air-conditioning. For fish and fish meal processing energy is required for cooling, cooking, sterilizing, drying, evaporation, can cleaning, fork-lifting. For wastewater treatment energy is applied for pumping and aerating. Energy consumption depends on various factors like age and scale of plant, the level of automation and the range of products. Processes which involve heating, such as canning and fish meal production need more energy than other processes.

33.3.3 Effluent discharge

Sources of effluent from fish processing include the handling and storage of raw fish prior to processing, fluming of fish and product around the plant, defrosting, gutting, scaling, portioning and filleting of fish and the washing of fish products.

Effluent streams generated from seafood processing contain high loads of organic matter due to the presence of oils, proteins and suspended solids. They can also contain high levels of phosphates and nitrates. In canning operations, effluent is also discharged from the draining of cans after pre-cooking, from the spillage of sauces, brines and oil in the can filling process and from the condensate generated during pre-cooking.

Effluent quality highly depends on the type of fish being processed and type of processing undertaken. Pollution loads generated from the processing of oily fish species are much higher than from white fish species, due to the high oil content and the fact that these species are usually not gutted or cleaned on the fishing vessel.

If the effluent streams described above are discharged without treatment into water bodies, the pollutants they contain can cause eutrophication and oxygen depletion.

In addition, fish processing industries have been known to pollute nearby beaches and shores by releasing wastewater containing oils. Since oil floats on water, it can end up on the surrounding coastline.

33.3.4 Emission to air

Point-source emission

These emissions are exhausted into a vent or stack and emitted through a single point source to the atmosphere. The major air pollution sources in a typical seafood industry are from combustion sources like boiler and generators for electric power. Boiler is used for steam supply during pre-cooking and sterilisation process. The examples of fuels used in the boilers are electricity, fuel oil, coal and LPG.

Odour

Odour is often the most significant form of air pollution in fish processing. Major sources include storage sites for processing waste, cooking by-products during fish meal production, fish drying processes and odour emitted during filling and emptying of bulk tanks and silos. Fish quality may deteriorate under the anaerobic conditions found in onboard storage on fishing processing facilities. This deterioration causes the formation of odorous compounds such as ammonia, mercaptans and hydrogen sulphide gas.

33.3.5 Noise pollution

Noise is not a significant problem in seafood industries. Noise may be generated during cutting, pre-cooking, filing and weighing the can, can seaming and sterilisation. The noise measurements at any sources in the process do not exceed the standard of maximum sound level (140 dB), but they have more value than equivalent continuous sound level 24 hours. Sources of noise pollution are given in Table 33.1

Table 33.1: Sources of noise pollution

Source	*Causes of noise*	*Emission, dB*
Pre-cooking	Steam blow out from vent	80–85
Filling and weighing the can	Collision of cans	92–93
Can seaming	Seamer operating to seam the can	86–87
Sterilisation	Venting process, steam is continuously blowing out through the hole	90–95

33.3.6 Solid waste generation

Solid waste is mainly in the form of organic wastes generated in the production processes. It consists of fish shells and heads from the seafood processing. The majority of captured species are ground fish and it is estimated that only 25–50% of the raw material is utilised for primary products. The remaining 50–75% of the raw material is considered processing waste and is utilised for low-valued products or disposed. Seafood processing activities generate potentially large quantities of organic waste and by-products from inedible fish parts and endoskeleton shell parts from the crustacean peeling process. The waste generation depends on the species and the process. If coal is used as a fuel in the boiler, ash will be generated depends on the ash content of the coal. Generally, 80% of the ash becomes fly ash and the remaining 20% will be bottom ash. Source and quantities of solid waste are given in Table 33.2.

Table 33.2: Source and quantities of solid waste.

Process	*Waste generation (kg/T)*
Eviscerating	240
Cleaning	400
Seaming	50
Total	
Organic	640
Inorganic	50

33.4 Waste treatment practices

Wastewater from seafood processing operations can be very high in biochemical oxygen demand (BOD), fat, oil and grease (FOG) and nitrogen content. White fish filleting processes typically produce 12.5–37.5 kg of BOD for every tonne of product.

BOD is derived mainly from the butchering process and general cleaning and nitrogen originates predominantly from blood in the wastewater stream.

It is difficult to generalise the magnitude of the problem created by these wastewater streams, as the impact depends on the strength of the effluent, the rate of discharge and the assimilatory capacity of the receiving water body.

In general, the wastewater of seafood processing waste water can be characterised by its physico-chemical parameters, organics, nitrogen and phosphorus contents. Important pollutant parameters of the wastewater are five-day biochemical oxygen demand (BOD_5), chemical oxygen demand (COD), total suspended solids (TSS), fats, oil and grease and water usage.

33.5 Wastewater treatment methods

Seafood processing facilities typically employ a combination of primary and secondary wastewater treatments, depending on the degree to which organic materials are collected separately or mixed into the effluent stream. Wastewater from seafood processing facilities typically has a very high organic and nitrogen load which can be effectively treated in aerobic or anaerobic systems, including lagoons. However, care should be taken to reduce odours including those from extended biological and/or chemical phosphorus removal. Effluent standard for wastewater of different countries are given Table 33.3.

Table 33.3: Effluent standard for wastewater of different countries.

Parameters	*Units*	*Germany**	*Thailand*	*Sri Lanka*	*Bangladesh*
pH	pH	5.5–9	5.5–9	6–8.5	6–9
BOD_5	mg/L	25	20	100	50
COD	mg/L	110	120	250	250
Nitrogen	mg/L	25 (NH_4 – N) and 25 (TKN)	100	50 (NH_4 – N)	10 (TN)
Total phosphorous	mg/L	2	–	–	5
Oil, fats and grease	mg/L	–	5	20	10
Total suspended solids	mg/L	–	50	150	50
Total dissolved solids	mg/L	–	3000–5000	–	05
Temperature	°C	–	<40	<45	–

* Specially for seafood industry

33.5.1 Primary treatment

This wastewater contains considerable amounts of insoluble suspended matter, which can be removed from the waste stream by chemical and physical means. For optimum waste removal, primary treatment is recommended prior to a biological treatment process or land application. For seafood-processing wastewater, the primary treatment processes are screening, sedimentation, flow equalisation and dissolved air flotation. These unit operations will generally remove up to 85% of the total suspended solids and 65% of the BOD_5 and COD per cent in the wastewater.

Screening

The removal of relatively large solids (0.7 mm or larger) can be achieved by screening. This is one of the most popular treatment systems used by food-processing plants, because it can reduce the amount of solids being discharged quickly.

Sedimentation

Sedimentation separates solids from water using gravity settling of the heavier solid particles. In the simplest form of sedimentation, particles that are heavier than water settle to the bottom of a tank or basin. This operation is conducted not only as part of the primary treatment, but also in the secondary treatment for separation of solids generated in biological treatments, such as activated sludge or trickling filters.

Flow equalisation

A flow equalisation step follows the screening and sedimentation processes and precedes the dissolved air flotation (DAF) unit. Flow equalisation is important in reducing hydraulic loading in the waste stream.

Separation of oil and grease

Seafood processing wastewaters contain variable amounts of oil and grease, which depend on the process used, the species processed and the operational procedure. Gravitational separation may be used to remove oil and grease, provided that the oil particles are large enough to float towards the surface and are not emulsified otherwise, the emulsion must be first broken by pH adjustment.

Flotation

Flotation is one of the most effective removal systems for suspensions that contain oil and grease. The most common procedure is that of dissolved air flotation (DAF), which is a waste treatment process in which oil, grease and other suspended matter are removed from a waste stream. This treatment process has been in use for many years and has been most successful in removing oil from waste streams.

33.5.2 Biological treatment

To complete the treatment of the seafood-processing wastewaters, the waste stream must be further processed by biological treatment. Biological treatment involves the use of micro-organisms to remove dissolved nutrients from a discharge. Organic and nitrogenous compounds in the discharge can serve as nutrients for rapid microbial growth under aerobic, anaerobic or facultative conditions. The three conditions differ in the way they use oxygen. Aerobic micro-organisms require oxygen for their metabolism, whereas anaerobic micro-organisms grow in absence of oxygen; the facultative micro-organism can proliferate either in absence or presence of oxygen although using different metabolic processes.

Biological treatment systems can convert approximately one-third of the colloidal and dissolved organic matter into stable end products and convert the remaining two-thirds into microbial cells that can be removed through gravity separation.

The biological treatment processes used for wastewater treatment are broadly classified as aerobic and anaerobic treatments. Aerobic and facultative micro-organisms predominate in aerobic treatments, while only anaerobic micro-organisms are used for the anaerobic treatments.

Aerobic process

In seafood processing wastewaters, the need for adding nutrients (the most common being nitrogen and phosphorus) seldom occurs, but an adequate provision of oxygen is essential for successful operation. The most common aerobic processes are activated sludge systems, lagoons, trickling filters and rotating disc contactors.

Activated sludge systems: In an activated sludge treatment system, an acclimatised, mixed, biological growth of micro-organisms (sludge) interacts with organic materials in the wastewater in the presence of excess dissolved oxygen and nutrients (nitrogen and phosphorus). The micro-organisms convert the soluble organic compounds to carbon dioxide and cellular materials. Oxygen is obtained from applied air, which also maintains adequate mixing. The effluent is settled to separate biological solids and a portion of the sludge is recycled; the excess is wasted for further treatment such as dewatering.

Most of the activated sludge systems utilised in the seafood processing industry are of the extended aeration types: that is, they combine long aeration times with low applied organic loadings. The detention times are 1 to 2 days. The suspended solids concentrations are maintained at moderate levels to facilitate treatment of the low-strength wastes, which usually have a BOD_5 of less than 800 mg/L.

It is usually necessary to provide primary treatment and flow equalisation prior to the activated sludge process, to ensure optimum operation.

Aerated lagoons: Aerated lagoons are used where sufficient land is not available for seasonal retention or land application and economics do not justify an activated sludge system. Efficient biological treatment can be achieved by the use of the aerated lagoon system. The major difference with respect to activated sludge systems is that the aerated lagoons are basins, normally excavated in earth and operated without solids recycling into the system.

Two types of aerated lagoons are commonly used in seafood-processing wastewater treatment: completely mixed lagoons and facultative lagoons. In the completely mixed lagoon, the concentrations of solids and dissolved oxygen are uniformly maintained and neither the incoming solids nor the biomass of micro-organisms settle, whereas in the facultative lagoons, the power input is reduced, causing accumulation of solids in the bottom that undergo anaerobic decomposition, while the upper portions are maintained in an aerobic state.

In the aerated lagoon a stabilisation/polishing ponds system is commonly used to improve the effluent treated in the aerated lagoon. This system depends on the action of aerobic bacteria on the soluble organics contained in the waste stream. The organic carbon is converted to carbon dioxide and bacterial cells.

Trickling filters

The trickling filter is one of the most common attached cell (biofilm) processes. Unlike the activated sludge and aerated lagoons processes, which have biomass in suspension, most of the biomass in trickling filters are attached to certain support media over which they grow.

Rotating biological contactors (RBC)

Increasingly stringent requirements for the removal of organic and inorganic substances from waste-water have necessitated the development of innovative, cost-effective waste-water treatment alternatives in recent years. The aerobic rotating biological contactor (RBC) is one of the biological processes for the treatment of organic waste-water. It is another type of attached growth process that combines advantages of biological fixed-film (short hydraulic retention time, high biomass concentration, low energy cost, easy operation and insensitivity to toxic substance shock loads) and partial stir. Therefore, the aerobic RBC reactor is widely employed to treat both domestic and industrial wastewater.

33.5.3 Anaerobic treatment

Anaerobic biological treatment has been applied to high BOD or COD waste solutions in a variety of ways. Treatment proceeds with degradation of the organic matter, in suspension or in a solution of continuous flow of gaseous products, mainly methane and carbon dioxide, which constitute most of the reaction products and biomass. Anaerobic treatment is the result of several reactions: the organic load present in the wastewater is first converted to soluble organic material, which in turn is consumed by acid-producing bacteria to produce volatile fatty acids, plus carbon dioxide and hydrogen. The methane-producing bacteria consume these products to produce methane and carbon dioxide.

33.5.4 Physico-chemical treatments

Coagulation/flocculation

Coagulation or flocculation tanks are used to improve the treatability of wastewater and to remove grease and scum from wastewater. In coagulation operations, a chemical substance is added to an organic colloidal suspension to destabilise it by reducing forces that keep them apart, that is, to reduce the surface charges responsible for particle repulsions. This reduction in charges is essential for flocculation, which has the purpose of clustering fine matter to facilitate its removal. Particles of larger size are then settled and clarified effluent is obtained.

Several steps are involved in the coagulation process. First, coagulant is added to the effluent and mixing proceeds rapidly and with high intensity. The purpose is to obtain intimate mixing of the coagulant with the wastewater, thereby increasing the effectiveness of destabilisation of particles and initiating coagulation. A second stage follows in which flocculation occurs for a period of up to 30 minutes. In the latter case, the suspension is stirred slowly to increase the possibility of contact between coagulating particles and to facilitate the development of large flocs. These flocs are then transferred to a clarification basin in which they settle and are removed from the bottom while the clarified effluent overflows.

Electrocoagulation

It reduces organic levels in various food- and fish-processing waste streams. During testing, an electric charge was passed through a spent solution in order to destabilise and coagulate contaminants for easy separation.

Disinfection

Disinfection of seafood processing wastewater is a process by which disease-causing organisms are destroyed or rendered inactive. Most disinfection systems work in one of the following four ways: (i) damage to the cell wall, (ii) alteration of cell permeability, (iii) alteration of the colloidal nature of protoplasm and (iv) inhibition of enzyme activity.

Chlorination: Chlorination is a process commonly used in both industrial and domestic wastewaters for various reasons. In fisheries effluents, however, its primary purpose is to destroy bacteria or algae or to inhibit their growth.

A problem that may occur during chlorination of fisheries effluents is the formation of chloramines. These wastewaters may contain appreciable amounts of ammonia and volatile amines, which react with chlorine to give chloramines, resulting in an increased demand for chlorine to achieve a desired degree of disinfection.

Ozonation: Ozone (O_3) is a strong oxidising agent that has been used for disinfection due to its bactericidal properties and its potential for removal of viruses. It is produced by discharging air or oxygen across a narrow gap with application of a high voltage.

Ultraviolet (UV) radiation: Disinfection can also be accomplished by using ultraviolet (UV) radiation as a disinfection agent. UV radiation disinfects by penetrating the cell wall of pathogens with UV light and completely destroying the cell and/or rendering it unable to reproduce.

However, a UV radiation system might have only limited value to seafood processing wastewater without adequate TSS removal, because the effectiveness decreases when solids in the discharge block the light.

33.6 Land disposal of wastewater

Generally, several methods are used for land application, including irrigation, surface ponding, groundwater recharge by injection wells and subsurface percolation. Although each of these methods may be used in particular circumstances for specific seafood processing waste streams, the irrigation method is most frequently used. Irrigation processes may be further divided into four subcategories according to the rates of application and ultimate disposal of liquid. These are overland flow, normal irrigation, high-rate irrigation and infiltration percolation.

The treatability of seafood processing wastewater by land application has been shown to be excellent for both infiltration and overland flow systems. With respect to organic carbon removal, both systems have achieved pollutant removal efficiencies of approximately 98 and 84%, respectively. The advantage of higher efficiency obtained with the infiltration system is offset somewhat by the more expensive and complicated distribution system involved. Moreover, the overland flow system is less likely to pollute potable water supplies.

33.6.1 Potential problems in land application with seafood processing wastewater

Two potential problems may be encountered with land application of seafood processing wastewaters the presence of disease-producing bacteria and unfavourable sodium absorption ratios of the soil. A key to minimising the risk of spreading disease-producing bacteria can be accomplished by using low-pressure wastewater distribution systems to reduce the aerosol drift of the water spray. With respect to unfavourable sodium absorption ratios associated with the soil type, the seafood processor should be aware that clay-containing soils will cause the most serious sodium absorption problem.

33.7 Use of fish and shellfish wastes as fertilisers and feedstuffs

The practicality and economic attractiveness of recovering seafood wastes as fertilisers and feedstuffs is directly dependent on the waste composition, including protein, fat, mineral and moisture content. The characteristics of seafoods and seafood wastes are summarised below.

Product yields in fish and shellfish processing range from nearly 100% for whole rendered fish to a low of 15% for some crabs. The portions not incorporated as final products or by-products become waste.

This waste is removed from seafood processing plants as either wastewater or solid waste. Wastewater (fresh or sea water) solutions contain dissolved materials (proteins and breakdown products) and suspended solids consisting of bone, shell, or flesh and foreign material. Solid wastes consist of flesh, shell, bone, cartilage and viscera. Many of these waste materials have sufficient nutrients to be valuable as fertilisers and feedstuffs.

The recovery of seafood wastes as fertilisers and feedstuffs is generally directed at the solid waste fraction, since most wastewater solutions (washwaters, etc.), are too dilute for recovery. Increasingly, however, wastewater streams are being treated to meet effluent requirements for discharges. Typically, screening, settling or dissolved air flotation methods are utilised and this results in the production of additional quantities of solid waste.

Section VI

Polluction control in chemical process industries

34

Pollution control in chemical process industries: A review

34.1 Introduction

In any country, economic development and environment protection should go hand in hand. There cannot be economic development (i.e., the improvement of the standard of living of the common man) without increasing the production of goods and services. Every such activity of production has its own associated problem in the balance of ecology and environment in general. Chemical industries, in particular, are being looked at with awe and suspicion in this respect. It should be the responsibility of the chemical manufacturers to use only such technologies which contribute the least to the upsetting or downgradation of the ecology and environment. Efforts are to be made in improving the technologies to better the performance.

The environmental impact of an existing process can be reduced by using clean technologies, which will, in turn, mean higher costs. Manufacturers wanting to determine the environmental implication of their activities need to look beyond the factory fences and consider the impact of obtaining the raw materials and subsequent use of their products. This is due to the awareness created in the minds of people after major disasters like dioxin release in Italy, the Isocyanate tragedy of Bhopal, etc., and the legal and third party liability suits against the manufacturers. The strategies in developing clean technologies (different than the clean-up technologies stated above) should take into account the life cycle of materials, likely changes envisaged in customer practices, environmental regulation/legislation, etc.

The environmental friendly technologies or clean technologies should aim at satisfying the following basic scientific findings:

1. Handle substances (chemicals as raw material, intermediate or finished products) which are least harmful to living organisms (human, animal or plant life), i.e., which are not toxic, corrosive or flammable/explosive.
2. Reactions/processes which do not involve much of transfer of energy (neither too exothermic nor endothermic) with the transfer level of energy being not far away from environmental status, i.e., not at a very high temperature or high pressure process.
3. Suitable moderating/control mechanism at two levels, with back up system to bring the situation back to normal in the least time interval.

To further complicate the pollution control aspects of the chemical industry, the industry is constantly changing, using different raw materials and new processes and producing new products.

In many cases, the process and its wastes are unique and pollution abatement systems must be individually engineered. However, many processes are free of wastes and many plants report no waterborne wastes.

34.2 Air pollution and water pollution

34.2.1 Air pollution

The air pollution control measures applied in chemical manufacturing are 'for the most part' the everyday application of chemical engineering unit operations using mechanical collectors, bag filters, electrostatic precipitators, wet collectors, catalytic oxidation or reduction units and direct flame incinerators or adsorbers, either alone or in combination.

The selection of the control device for the particular process is a function of the degree of control necessary, the nature of the effluent, the capital and operating costs of the control equipment, the effect of the control system on the process itself, reliability and the ultimate disposal of the collected waste material. In addition, process or equipment modifications may be specifically incorporated to eliminate or minimise the emission of air pollutants. This is the preferred procedure and is particularly applicable to new plants or processes. Thus, the control device alone may not fully indicate the attention given to air pollution control.

34.2.2 Water pollution

The large number of commercial chemicals and the diversity of their effect on water make it impractical control to generalise measures for the entire chemical industry. The wastes from a chemical plant may be inorganic, organic, insoluble, soluble, inert, toxic, or any combination thereof. Both organic and inorganic wastes have water quality effects essentially different from sanitary sewage and cannot be described properly in sanitary sewage terms such as the fictitious 'population equivalents'. Simple yardsticks such as oxygen demand, acidity, or dissolved solids also are not always applicable.

The abatement technique applied by the chemical industry for its pollution problems bears the stamp of its own technology. Most waste treatment facilities are unique and individually conceived and constructed. Minimising or eliminating wastes is an important consideration in any new process or plant. Water pollution control should be considered at every stage of chemical plant development, from research through plant design or modification and operation.

To protect the beneficial uses of the receiving waters adequately, it is necessary to minimise the waste, characterise the effects of the waste on the receiving waters and have knowledge of the assimilative capacity of the receiving waters.

Waste characterisation

An intimate knowledge of the characteristics of waste stream is necessary to protect the receiving waters and to treat the wastes adequately. Needed information includes, but is not limited to, biological oxygen demand (BOD), toxicity, suspended and settleable matter, insoluble oil films, taste, odour, pH, temperature and volume of the waste stream. Some of these characteristics may vary with concentration, temperature, or acclimatisation of bacteria.

Ideally, each waste stream should be studied individually. Trace constituents may have synergistic effects, which adversely affect biota either in the treatment plant or in receiving waters. Some ions in the waste may react and precipitate in the receiving waters. The analytical and experimental methods used to determine the various parameters have been standardised and their use is recommended.

Waste effluents from organic chemical plants are usually a heterogeneous mixture, whose composition will not be reliably known until the plant has been in operation for some time. In considering a new process, evaluation of the projected waste streams is necessary. Sources of data may be an existing plant, estimates by analysis of pilot plant effluents, testing of synthetic mixtures and estimates based on anticipated composition of wastes. Much research is needed in this area, not only to fill existing gaps, but to develop and refine techniques which will provide reliable and useful results. Much of the existing information cannot be reconciled because of variations in purpose and method. It should be emphasised that literature searches are no substitute for experimental determination for a specific waste stream.

Process changes/modifications

Wastes and their treatment are non-profit operating cost items and prevention of wastes has the potential of reducing costs, reducing or eliminating the capital and operating costs of abatement facilities and possibly eliminating refractory or relatively toxic wastes. A number of techniques are available to minimise or eliminate wastes from a given process.

Segregation of process streams is often indicated. For example, cooling water may be discharged separately, rather than be contaminated by a small volume of waste. The concentrated waste may have raw material values or may have heating values and can be burned. A concentrated waste stream can generally be treated more economically than a dilute stream.

Direct contact condensers may produce dilute waste streams. Indirect condensation or cooling can often be justified on the basis of eliminating possible water pollution. A change in raw materials is often beneficial. The use of a higher-purity or different raw material may eliminate a troublesome waste stream. If in-process neutralisation is required, different waste characteristics result from the use of lime, caustic soda, or ammonia, or the use of sulphuric, hydrochloric, or nitric acid. An economic evaluation of total costs, including waste control, may indicate if a change in reagents is warranted.

Receiving waters

In order to prevent deleterious effects on the water receiving wastes, the nature of the receiving body must be known. Important considerations are: beneficial uses of the water resource, its quality, its flow rate and the variability of its flow. For existing plants, it is wise to keep in touch with regulatory agencies and stay up-to-date with their requirements.

The treatment method selected for a given plant will depend on the properties of the waste(s), its volume and the flow rate and the character and important downstream uses of the receiving waters.

Treatment methods

Many chemical wastes are compatible with sanitary sewage and can be discharged into sewage systems for subsequent treatment in the municipal treatment plant. Generally, such wastes are governed by local sewer ordinances which regulate the discharge of industrial wastes and usually specify the manner by which the waste may be admitted. Pre-treatment may be required to remove toxic substances, flammable compounds, heavy metals or to adjust pH prior to sewering. Various treatment methods such as physical, chemical and biological are already discussed in industrial water pollution.

34.3 Environmental friendly technologies

Green chemistry is the utilisation of a set of principles and practice renders control, regulation, clean-up and that reduces or eliminates the use or generation of remediation unnecessary and the resultant environmental hazardous substances in the design, manufacture and benefit can be expressed in turns of economic impact application of chemical products.

Green chemistry is the utilisation of a set of principles that reduces or eliminates the use or generation of hazardous substances in the design, manufacture and application of chemical products.

Green chemistry also involves reducing other associated environmental impacts, in particular reducing the amount of energy used in chemical processes.

Ideally, the application of green chemistry principles and practice renders control, regulation, clean-up and remediation unnecessary and the resultant environmental benefit can be expressed in turns of economic impact. It is science based non regulatory and economically driven approach to achieving the goals of environmental protection. The major difference between a green chemistry approach to environmental issues and more traditional approaches is that green chemistry utilises the creativity of the scientists and engineers to develop novel and benign approaches to processes from the start rather than relying on regulatory restrictions after the process has been discovered to be toxic or polluting.

Green chemistry is a tool for chemists, chemical engineers and others who design materials to help move society toward the goal of sustainability. Design at the molecular level allows decisions to be made that impact how materials will be processed, used and managed at the end of their life.

The 12 principles of green chemistry: The 12 principles focus on reducing the volumes of chemicals used and pollution prevention.

Waste minimisation and prevention: It is better to prevent waste than to treat or clean up waste after it has been created. Generally speaking, waste minimisation involves the reduction of waste toxicity by reducing the volume or quantity of highly toxic chemical constituents through substitution, recycling, recovery and reuse efforts. It is better to prevent waste than clean it up after-the-fact.

34.3.1 Waste reduction possibilities

Inputs

1. Eco friendly solvents, high purity reagents, recyclable auxiliaries, less hazardous materials.

Production

1. Change time, temperature and pressure, reactor types, mixing, heat transfer.
2. New route, appropriate cleaning.

Discharges

Reduce water volume, improved scrubbers, wastewater clean up, mineralisation of organics.

By-products

Maximise use, research and development, marketing, site integration.

Atom economy: Synthetic methods should be designed to maximise the incorporation of all materials used in the process into the final product.

Less hazardous chemical syntheses: Wherever practicable, synthetic methods should be designed to use and generate substances that possess little or no toxicity to human health and the environment. Synthetic methodologies must be designed such that the chemicals used and by-product, if generated, are not or less harmful to human health and the environment.

Designing safer chemicals: Chemical products should be designed to affect their desired function while minimising toxicity. The design of chemicals with minimal toxicity reduce the potential risk to human health and the environment; decrease the costs of production and site remediation; and increasing team commitment to workplace health and safety. For instance, prefer public transport instead of own vehicles, thereby minimising the CO_2 emission, use recyclable paper in order to minimise the burden on natural resource and also to lessen the amount of toxic products coming out after bleach during paper production. Develops efficient methods of converting solar energy into chemical energy and electrical energy to avoid the necessity of generating power from nuclear plants, which produces a lot of nuclear waste, gaseous emission and chemical pollutants.

Safer solvents and auxiliaries: The use of auxiliary substances (e.g., solvents, separation agents, etc.) should be made unnecessary wherever possible and innocuous when used. The idea of 'green' solvents expresses the goal to minimise the environmental impact resulting from the use of solvents in chemical production. Many organic solvents such as benzene, chloroform, toluene, carbon tetrachloride, etc., are Volatile Organic Compounds (VOCs) and it means that their high volatility, very useful for industrial applications, contributes both to increase the risks of fire and explosion and to facilitate the release in the atmosphere in which these solvents can act as air pollutants causing ozone depletion, photochemical smog and global warming.

In the interest of green chemistry, commonly used five main solvent systems which are currently considered as 'green' are: (i) solventless systems, (ii) water, (iii) ionic liquids, (iv) fluorous solvents and (v) supercritical fluids. Examples of applied green chemistry are supercritical water oxidation, on water reactions and dry media reactions.

Design for energy efficiency: Energy requirements of chemical processes should be recognised for their environmental and economic impacts and should be minimised. If possible, synthetic methods should be conducted at ambient temperature and pressure. Rising consumption and heavy future demand on energy that is primarily generated from petroleum and depleting resources has raised serious concerns in the international community. The solution would not lie in digging in more deeper to use up all the available resources, instead it lies in designing energy efficient processes and generating alternative sources of energy production. In line with this, chemist can design reactions that could take place at moderate temperature using catalyst or other methods, thereby reducing more demand on energy.

Use of renewable feedstocks: A raw material or feedstock should be renewable rather than depleting whenever technically and economically practicable. Renewable feedstocks are often made from agricultural products or from wastes of other processes; depleting feedstock are made from fossil fuels or are mined. Significant developments in using renewable feedstocks to make fuels and chemicals are blooming and efforts are being put in to produce organic chemicals; related products to be obtained from natural resources other than petroleum and depleting resources. This is not a very novel field to mankind, because since long ethanol has been developed from variety of sources like sugarcane (molasses), beet roots, grapes etc. So, ethanol producers are looked at and examine which are developing alternative downstream products aside from fuel ethanol. Diesel is the commonest form of transportation fuel with a demand higher than for petrol. Jatropha curcus, the chosen feedstock, has been under intensive development over the past few years and the volume of planting is slowly beginning to approach levels whereby it can start to fulfill its promise as a key biodiesel feedstock.

Another example is alternative, bio-based plastics PLA (polylactic acid) is one plastic that is being made from renewable feedstocks such as corn and potato waste. Methylene chloride, a common solvent in organic synthesis, is a suspected carcinogen made from fossil fuel based raw materials. Methyltetrahydrofuran has been identified as an alternative with similar solvent properties that can be made from renewable, bio-based resources. Benzene used in the commercial sythesis of adipic acid which is required in the manufacture of nylon, plasticisers and lubricants, has been replaced to some extent by the renewable and nontoxic glucose and the reaction is carried out in water.

Reduce derivatives: Unnecessary derivatisation (use of blocking groups, protection/deprotection and temporary modification of physical/chemical processes) should be minimised or avoided if possible, because such steps require additional reagents and can generate waste. In many traditional chemical reaction based routes, elegant multistep syntheses were designed that employed increasingly clever 'protecting groups' that would temporarily block the reactivity of a specific functional group until a deprotecting group was introduced to remove it. Unnecessary derivatisation (use of blocking groups, protection/deprotection, temporary modification of physical/chemical processes) should be minimised or avoided from an environmental impact perspective if possible, because such steps can generate more waste. Once such process designed in the industry is in Polaroid films, where researchers sought to release hydroquinone at elevated pH, which being highly basic trends to cleave covalent protecting groups also. Hence, a non-covalent protecting group in the form of co-crystal was developed which minimised lot of waste thus making the process green.

Catalysis: Catalytic reagents (as selective as possible) are superior to stoichiometric reagents. Catalytic chemistry is one of the most important aspects of eco-friendly chemistry which promotes most of the green chemistry goals in terms of atom efficiency, lower energy use, higher yields, attaining high levels of selectivity at minimal waste through biocatalysis most of which is fast biodegradable and non polluting, decreased use of separating and processing agents and the activation of inert material, thereby reducing reliance on toxic materials. It is for this central role it plays that catalysis is referred to as a foundational pillar of green chemistry. Catalytic reagents also eliminate need of stoichiometric amounts of it in the reaction.

New approaches in the design and development of second generation titanium oxide photo catalysts which can operate effectively under visible light and/or solar beam irradiation, have drawn attention to the vital need for totally new environmentally friendly, clean chemical technologies and processes, the most important challenge facing chemical scientists in the field of green chemistry.

Design for degradation: Chemical products should be designed so that at the end of their function they break down into innocuous degradation products and do not persist in the environment. Not only do we want materials and products to come from renewable resources, but we would also like them to not persist in the environment. There is no question that many products we use in our daily lives are far too persistent. Plastics do not degrade in our landfills and pharmaceutical drugs such as antibiotics build up in our water streams. This principle seeks to design products in such a way that they perform their intended function. Simple green products are designed to provide cleaning function with chemicals that are nontoxic and biodegradable. The surfactants, chemicals that remove soil and suspend it in water, in simple green are designed to maximise effective cleaning while minimising health and environmental impacts. The cleaners are a milder alternative to traditional cleaning products, with less impact on the user and the environment. The use of a water solvent provides a safer alternative to the volatile hydrocarbon solvents typically used to remove grease and oil.

Real-time analysis for pollution prevention: Analytical methodologies need to be further developed to allow for real-time, in-process monitoring and control prior to the formation of hazardous substances. There are two aspects of this principle time and materials. Real time analysis for a chemist is the process of 'checking the progress of chemical reactions as it happens.' If better, more responsive monitors can be designed, then the use of 'just-in-time' reagents and techniques can be employed to improve process yields, product consistency, use of raw materials that will minimise the environmental toll.

Also there is a need to improve analytical techniques to consume less material. New chromatographic methods (HPLC) that use less solvent or do not require complex mixtures of solvents need to be developed.

Inherently safer chemistry for accident prevention: Substances and the form of a substance used in a chemical process should be chosen to minimise the potential for chemical accidents, including releases, explosions and fires.

34.4 Emission control technologies and techniques for the chemical processing industry

As is the case with many industrial manufacturing operations, emissions are best destroyed through the use of oxidation technologies. How destruction occurs and to what level and efficiency, is often determined by the process conditions. Very few industries use custom oxidiser solutions like the chemical processing industry. This section covers some of the technologies and techniques being applied on the wide variety of chemical processing applications.

Today, most manufacturers worldwide in the Chemical Processing Industry (CPI), are required to be compliant with regulations like the United States' National Emission Standard for Hazardous Air Pollutants (NESHAP) for chemical manufacturing. Similar regulations are being adapted worldwide as environmental initiatives gain popularity. The primary offgas emissions from this industry include Volatile Organic Compounds (VOCs) as well as Hazardous Air Pollutants (HAPs). VOCs are defined as any organic chemical compound that contains the element carbon; excluding carbon monoxide (CO), carbon dioxide (CO_2), carbonic acid, carbonates, metallic carbides and methane. HAPs are a VOC with additional harmful properties. When left untreated, these emissions degrade in the presence of sunlight and contribute to low-lying ozone or smog. VOCs and HAPs are also known to cause respiratory ailments, heart conditions, birth defects, nervous system damage and cancer in humans.

Starting in the early 1970's, the idea of oxidation for emission abatement was developed due to the awareness of the harmful effects of VOCs and HAPs. New pollution control regulations were soon to follow. Proper oxidation converts hydrocarbons to CO_2 and water (H_2O) vapour. There are three conditions that must be met in order for oxidation to occur. They are often referred to as the three 'T's', temperature, time and turbulence. Based on the VOCs that need to be destroyed, there is an elevated temperature that must be reached before the compounds can be oxidised. The time relates to how long a compound needs to be at temperature in order for it to be oxidised. Lastly, turbulence is a fixed condition built into the equipment design that ensures a proper mixture of VOCs and oxygen for combustion.

Specific compounds and desired destruction rate efficiency determine temperature and residence time.

Process emissions from batch and continuous chemical manufacturing applications can vary greatly in volume and composition. Some of the sources of VOCs and HAPs in the chemical processing industry include:

1. Phenol production.
2. Ethylene oxide - sterilisation.
3. Formaldehyde production.
4. Organic chemical manufacturing.
5. PA (Phthalic Anhydride) production.
6. PTA (Pure Terepthalic Acid) production.
7. PET (Polyethylene Terephthalate) production.
8. Specialty chemicals.

With so many applications and sources of off-gasses, there is a wide variety of abatement technologies being applied in the CPI. Depending on the process conditions, chemical processors have used technologies such as flares, vapour combustors, catalytic and thermal oxidisers. The catalytic and Regenerative Thermal Oxidisers (RTO) have been the most widely applied emission abatement technologies.

34.4.1 Catalytic oxidation

Catalytic oxidation is a popular choice in the CPI and has been used extensively for many years. Employing catalyst into the oxidation process allows the compounds to break apart at lower temperatures and reduces the supplemental fuel requirements of the system. In operation, VOC and HAP laden process gas is either pushed or pulled into a metal heat exchanger inside the oxidiser via a system fan. The contaminated air stream is progressively heated while travelling through the heat exchanger towards the combustion chamber. At the burner, the process gas is raised to the catalyst operating temperature. As the heated gas passes through the catalyst, an exothermic (heat releasing) reaction takes place and the pollutants are converted to carbon dioxide, water vapour and heat.

There are two major categories of oxidation catalyst:

1. Precious/noble metal: platinum, palladium, rhodium, iridium.
2. Base metal: manganese dioxide, copper oxide, chromium oxide.

The different catalyst substrates include:

1. Pellets, beads, spheres.
2. Monolith - ceramic! stainless steel.

Once oxidised across the catalyst, the hot purified air passes through the opposite side of the heat exchanger and releases thermal energy to pre-heat the incoming air. The heat exchanger reduces the auxiliary fuel requirement, thus saving operating cost and allowing self-sustaining operation with no auxiliary fuel usage at Lower Explosive Limits (LEL) of 8–12%. Finally, the cooler, purified air is exhausted to the atmosphere through the system stack. Typically the reactor and internal heat exchanger are fully welded, leak -tested and constructed of stainless steel with high-density insulation.

Within the chemical industry, there are many situations where processors prefer catalytic systems to treat their process emissions. For example, carbon monoxide, aromatic compounds and alkenes are better expelled by passing over heated catalyst. A specialty ceramic material manufacturer in the United States operates several customer-built furnaces.

These high temperature kilns are a major source of air pollutants and therefore require treatment downstream of the process. After a thorough evaluation of their unique process conditions, they chose a catalytic oxidiser as the best solution for their emission control needs.

In this application, not only was the process challenging but the system also needed to be installed on a mezzanine with weight restrictions. Rather than design a heavy RTO to fit into the tight space, a catalytic oxidiser was designed and built to remove the VOCs and HAPs from their exhaust. The process exhaust was already quite warm; it was within the temperature range normally associated with catalytic oxidation temperatures (600°F/315.556°C or higher). Since the process exhaust temperature was already hot enough to sustain high levels of oxidation with the catalyst, the oxidiser was supplied without a heater or heat exchanger. This allowed for a very simple method of controlling the emissions from the process with the reduced capital and installation costs. Nearing the end of its life cycle, an old RTO at the same facility also needed to be replaced. This ceramic manufacturer was so pleased with their new catalytic system that they decided to replace the aging RTO with another catalytic oxidiser. They wanted a technology that would not only keep them in compliance, but do it without the pressure pulse issues that had been negatively impacting the kilns from their old RTO supplied by Smith Engineering.

A second catalytic system with multiple heat exchangers to utilise the heat generated during combustion was designed and manufactured. The added benefit of this control system is that it recirculates oxidiser Products of Combustion (POC), namely heat back into their process. A secondary heat exchanger downstream of the oxidiser cools the treated stream so that it can supply the process with tempered air. To enable re-entry, oxygen concentrations leaving the catalytic oxidiser are maintained at very low levels with customer designed electronic control systems, supply fans and ductwork. Maintaining low oxygen levels in this supply gas reduces the amount of nitrogen required by the process and therefore lowers operating cost for the production line. In addition, a primary heat exchanger within the oxidiser was designed to minimise fuel consumption of the abatement device itself, essentially using heat from the exothermic reaction to pre-heat incoming pollutants.

The 20,000 standard cubic feet per minute (SCFM) (31,542 Nm^3/hr) catalytic oxidiser was installed indoors near the process equipment. To keep it under negative pressure at all times, the oxidiser was manufactured in an induced-draft-fan configuration. Pulling gases through the system with the main fan ensures that no oxygen-depleted gases are emitted into the building where personnel were present.

Theoretically, catalytic oxidisers have a long life span and they can operate for many years if routine maintenance is continued. However, on most catalytic oxidisers, catalyst deactivation can begin within 3–8 years. Deactivation occurs due to constant and sometimes specific chemical exposure.

Mercury, lead and cadmium are major contributors of deactivation. Other factors contributing to the decrease in catalytic activity are accumulation of inorganic scale, phosphorous (lube oils), dirt and sintering that causes larger crystals to agglomerate from catalyst particles and catalyst substrate. Depending on the application, catalyst could be replaced more than once over the unit's lifespan. Catalyst testing and catalyst rejuvenation services are often required to prolong oxidiser efficiency and ensure minimal downtime.

34.4.2 Regenerative thermal oxidiser (RTO)

The RTO is a widely used and also popular choice for emission abatement in the CPI. Its popularity derives from the oxidiser's ability to reuse up to 97% of its thermal energy and its high destruction rate efficiency. A typical airflow range on an RTO is between 3,000 SCFM (4731.3 Nm^3/hr) – 70,000 SCFM (110,397 Nm^3/hr) on a single unit. On a multiple unit system it can start at 70,000 SCFM all the way up to 500,000 SCFM (788,550 Nm^3/hr).

Solvent laden process air is either pushed or pulled into the inlet manifold of the oxidiser via a system fan. Flow control or poppet valves then direct this gas into energy recovery chambers where it is pre-heated. The process gas and contaminants are progressively heated in the ceramic media beds as they move toward the combustion chamber. There are three different types of heat recovery media that can be found in an RTO design:

1. Random packing.

2. Extruded honeycomb monolith.
3. Multi-layered media.

Once oxidised in the combustion chamber, the hot purified air releases thermal energy as it passes through the media bed in the outlet flow direction. The outlet bed is heated and the gas is cooled so that the outlet gas temperature is only slightly higher than the process inlet temperature. Poppet valves alternate the airflow direction into the media beds to maximise energy recovery within the oxidiser. The high energy recovery within these oxidisers reduces the auxiliary fuel requirement and saves operating costs. The regenerative thermal oxidiser technology consistently achieves destruction efficiencies over 99%. Self-sustaining operation, with no auxiliary fuel usage, routinely occurs under low emission loading conditions (3% LEL or greater).

Over the past decade, RTOs have been increasingly applied in the CPI as technological advances have increased its applicability. Newer designs handle a wider range of emission constituents, concentrations and loadings. This allows the RTO, when it best fits the needs of the application, to be used on more chemical process exhausts than before. RTOs can now be used on CPI processes where the catalytic oxidiser or flare were once the only option. It differs from the catalytic, recuperative and direct-fired oxidiser technologies in that destruction is achieved with little to no auxiliary fuel required once the RTO is brought to temperature.

Advances in the RTO design include a Hot Gas Bypass (HGB), which allows the device to process highly concentrated streams while preventing potential damage to the heat recovery media. Another development to the RTO is Supplemental Fuel Injection (SFI). This redundant fuel delivery system lowers operating cost by reducing the use of burner combustion blowers. During system operation, the oxidiser system burners are disabled and natural gas is injected at the RTO inlet as necessary to maintain chamber temperature. This provides a secondary benefit and ultralow NO_x emissions with flameless operation. Scalability is another aspect of the RTO that make these systems popular with CPI processors. This allows companies the opportunity for future plant expansion and process line additions without needing to install a new oxidiser. In addition, RTOs generally need fewer replacement parts and with an annual preventative maintenance plan, can have a lifespan well over 15 years.

34.4.3 Managing PTA - A case study

A common source of VOCs and HAPs in the CPI is PTA. This is a white powder-substance used in the production of polyester products and plastics. Pure Terephthalic Acid (PTA) is created in a pressurised reactor where acetic acid and xylene are chemically combined in the presence of catalyst. CO_2, CO, water vapour, methyl acetate, unreacted xylene, acetaldehyde and small concentrations of methyl bromide are gas-phase by-products of the PTA reaction. These gaseous emissions are sent to a pressurised absorber and scrubbed with water to reduce emissions. The air that remains after having gone through this process gets routed to an abatement device.

One PTA plant in the middle east incorporated RTOs into their chemical process and by doing so, saved nearly one million dollars. This plant was looking to expand their production from 420 MT to 700 MT/year. They had an existing catalytic oxidiser with two reactor modules designed for 67,000 SCFM (106,000 Nm^3/hr) installed onsite, but they decided installing two RTOs was the best option for them. These RTOs have the capacity to handle 127,000 SCFM (200,000 Nm^3/hr) of total airflow. The RTOs share common components such as a main system stack and electrical controls, but are capable of operating independently during routine maintenance or slow production.

These newly installed RTOs can collectively emit 5000 fewer tons per year of CO_2 emissions and 5 T each year of NO_x emissions than the catalytic oxidiser. This difference corresponds to approximately 10 million metric British thermal unit (MMBTU)/hr of reduced auxiliary fuel consumption. The plant essentially doubled the capacity of the abatement device, lowered operating costs and reduced their carbon footprint all while achieving greater than 98% destruction of the targeted pollutants. In addition, there are no ongoing catalyst replacement costs, saving a significant amount of money on maintenance expenditures.

34.4.4 Emission control technology

There are some important design considerations to evaluate before choosing a specific technology:

1. Properly identify the air stream.
2. Temperature and residence time.
3. Needed/preferred materials of construction.
4. Induced/forced draft design.
5. Dew point considerations.

Process emissions from batch and continuous chemical manufacturing applications can vary greatly in volume and composition. In addition, green business practices don't always coincide with energy reduction strategies, which is often the case with emission abatement products. Therefore, it is important to consult with a professional to ensure the correct abatement technology for your process conditions, budget constraints, energy demands and compliance objectives. Very few processes are identical and not one specific technology choice can be applied on all applications.

35

Pollution control in fertiliser industry

35.1 Introduction

Fertilisers can be classified as nitrogenous fertilisers, phosphatic fertilisers and complex fertilisers. Figure 35.1 shows flow diagram for effluent treatment of a complex fertiliser plant. Plants may be producing only nitrogenous fertilisers, like urea, ammonium sulphate, ammonium nitrate and ammonium chloride or only phosphatic fertilisers like super phosphates.

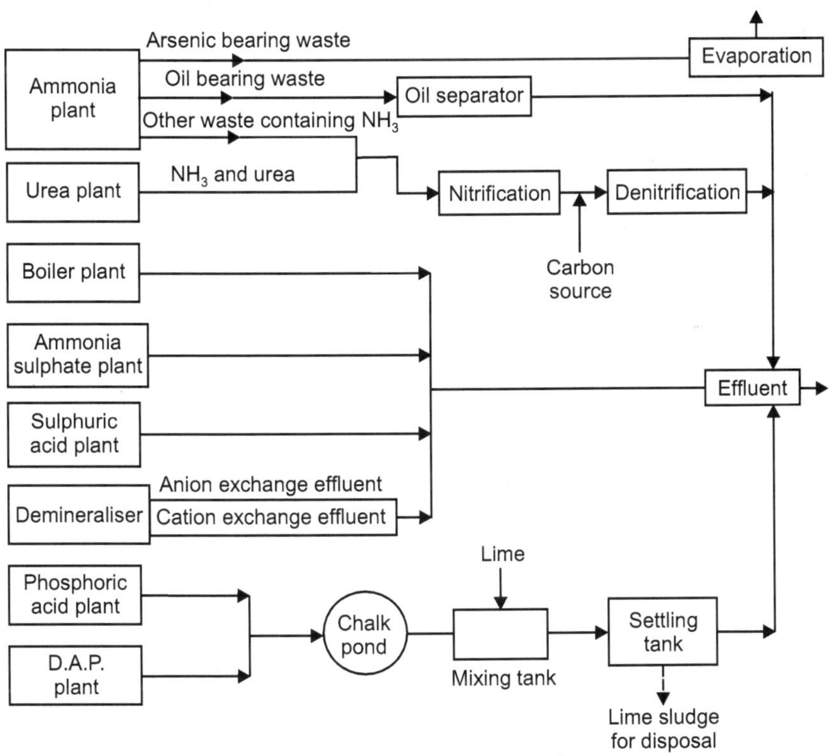

Figure 35.1: Flow diagram for effluent treatment of a complex fertiliser plant.

There are plants where complex fertilisers containing both nitrogen and phosphates, like ammonium phosphate and ammonium sulphate are produced. Also, some fertiliser units are only involved in mining and undertake no other activity. Organic fertilisation is accomplished through the addition of organic waste to the soil. Organic waste can include livestock manures, crop residues, compost and sewage sludge. The benefits of organic materials to the soil are twofold. First, the materials replenish the soil with nutrients. As organic material decomposes, the minerals within it become available to plants growing within the soil. The decomposition is accomplished by micro-organisms in the soil.

Inorganic fertilisation consists of adding chemical fertilisers to the soil. The inorganic fertilisers are commercially manufactured and are formulated to contain a vast array of nutrient compositions and concentrations.

Some fertilisers contain one main nutrient source, while others contain multiple sources. The main nutrients in mixed inorganic fertilisers are nitrogen, phosphorus and potassium; however, they often contain micro-nutrients as well. Fertiliser production is one of the most energy intensive processes. Energy is consumed in the form of natural gas, associated gas, naphtha, fuel oil, low sulphur heavy stock and coal. The choice of the feedstock is dependent on the availability of feedstock and the plant location. Production of ammonia has greatest impact on energy use in fertiliser production. It accounts for 80% of the energy consumption for nitrogenous fertiliser. The feedstock mix used for ammonia production has changed over the past. The shift towards the increased use of natural/associated gas and naphtha is beneficial, in that these feedstocks are more efficient and less polluting than heavy fuels like fuel oil and coal. Furthermore, capacity utilisation in gas based plants is generally higher than in other plants. Therefore, gas and naphtha present the preferred feedstocks for nitrogenous fertiliser production.

Energy intensity in fertiliser plants has decreased over time. This decrease is due to advances in process technology and catalysts, better stream sizes of urea plants and increased capacity utilisation. Capacity utilisation is important as losses and waste heat are of about the same magnitude no matter how much is actually produced in a plant at a specific point of time.

The production of phosphatic fertiliser requires much less energy than nitrogenous fertiliser. Depending on the fertiliser product, energy consumption varied from negative input for sulphuric acid to around 1.64 GJ/T of fertiliser for phosphoric acid. For sulphuric acid the energy input is negative since more steam (in energy equivalents) is generated in waste heat boilers than is needed as an input. The age of the technology, the scale of the plant and management practices have a large impact on energy efficiency of the overall process. Energy savings potential are highest for naphtha and fuel oil/LSHS based plants especially for plants built before 1980.

Optimal use and management of nitrogen (N) and other fertilisers is important to maximise return to the energy consumed in the manufacture, transportation and application for corn production. Efficiency gains can be achieved by avoiding losses during and after field application, applying recommended rates and substituting manure and legume N where energy has already been captured. Choice of N application rate is important for maximising economic return and minimising environmental loss. It is also important for maximising net energy return through crop capture of sunlight and grain/stover production.

35.2 Emission from fertiliser industry

The fertiliser industry implements wet scrubbers, condensate strippers and recovery and recycle methods to control ammonia emissions. Coke manufacturers use scrubbers and capture devices to collect and treat ammonia emissions. Livestock management facilities use wet scrubbers and good maintenance procedures to reduce emissions from animal manure. Facilities using ammonia as a refrigerant use capture devices and good management practices to safeguard against potential hazards due to leaks and spills of ammonia. Unlike other sources, emission of ammonia from fossil fuel combustion sources results from the implementation of post-combustion controls using ammonia or urea to decrease nitrogen oxide emissions. In this process unreacted ammonia exits the source in the effluent, representing a condition known as ammonia slip. Using good operating practices, the level of ammonia slip can be reduced.

The various control technologies available to control ammonia emissions include both add-on control devices and pollution prevention techniques. The wet scrubber, specifically the packed tower scrubber, has been successfully used to control ammonia emissions, demonstrating control efficiencies up to 99%. Condensers are also used to remove ammonia by converting the gas to a liquid. Many times the condensate is returned to the process for reuse. Ammonia recycle, in which the ammonia is retrieved from the exhaust gas stream and returned to the process, is a common pollution prevention method. The recycling method is often used in the manufacture of fertilisers, where ammonia is a feedstock and the implementation of recycling benefits the process in addition to controlling emissions. In the case of fossil fuel combustion where ammonia is emitted as a result of the controls implemented to reduce nitrogen oxide emissions, careful application of the nitrogen oxide control method will limit the amount of ammonia that will slip through the process. With regards to ammonia leakage, good management practices can reduce the amount of ammonia that escapes. For those leaks that cannot be prevented, capture devices such as hoods may be used to collect the ammonia gas, which is then conveyed to a control device for treatment. Both Particulate Matter (PM) and gaseous air emissions are generated as a result of the application of chemical fertilisers. Emissions may occur during application, shortly after application and for extended periods following application.

35.2.1 Emission mechanisms

Emissions from the application of fertiliser generally are attributed to four different mechanisms: (i) soil reactions with the applied fertiliser generating increased gaseous emissions including NO_x, N_2O, NH_3 and SO_2, (ii) volatilisation of the fertiliser immediately behind the vehicle generating gaseous emissions of NH_3 and the fertiliser itself, (iii) soil disturbance generating PM emissions where soil particles and other materials in the soil become airborne and (iv) volatilisation of the fertiliser immediately above the solid fertiliser trailer generating gaseous emissions of NH_3 and other fertilisers. Emissions attributed to the first mechanism are often called latent emissions, while those attributed to the other three mechanisms are called immediate emissions.

Emissions that result from the reactions between the soil and the applied fertiliser are believed to be higher than emissions that result from the other three mechanisms. Consequently, most of the data available on emissions from fertiliser application are estimates of emissions from soil-fertiliser reactions.

35.2.2 Particulate matter

Particulate matter emissions of solid fertiliser compounds are primarily generated along with windblown dust during broadcast application. Constituents of gaseous, fluid, or solid fertilisers and manure (or their reaction products) attached to soil particles may also become airborne much later as a result of soil disturbances caused by wind (i.e., wind erosion) or mechanical operations (e.g., tilling).

35.2.3 Gaseous air emissions

Gaseous air emissions from fertiliser application can occur either immediately, as a result of the volatilisation of the fertiliser itself, or after a period of time, as a result of the biological/chemical transformation of the fertiliser and subsequent release of gases to the atmosphere. The transformation products are generally oxidised forms of either nitrogen, sulphur, or phosphorus. Data on emissions related to the application of micronutrients, which are trace elements such as boron, chlorine, copper, iron manganese, sodium, molydenum and zinc that are essential for plant growth, are insufficient to permit an analysis, so they are not discussed.

Because emissions from fertiliser application are generated via the four primary mechanisms and because emission rates associated with each of these four mechanisms are affected by a variety of physical, chemical and biological processes, characterising emission rates for a particular application scenario or time period is complicated. The subsections below present an overview of five classes of factors that are described as affecting emissions. These five broad classes are biological and chemical reactions in the soil, other biological activities, soil conditions, climate and nutrient management (the form, placement and timing of fertiliser application).

Because of the complexity of the emission mechanisms and the interaction of many of the factors, data are insufficient to estimate the magnitude of the effects of most of the factors. For most fertilisers, it is believed that emissions increase significantly as moisture contents are raised via rainfall or irrigation. Also, emissions are directly related to ambient temperatures. Hourly emission rates exhibit diurnal patterns that follow temperature patterns and emissions are higher during summer months. The effects of both temperature and moisture are interrelated with other biological and chemical factors discussed below.

Biological and chemical reactions affecting air emissions from fertiliser

Naturally occurring biological and chemical reactions in the soil that affect air emissions from fertiliser application are primarily related to either the nitrogen cycle (Fig. 35.2) or the sulphur cycle (Fig. 35.3), depending on the type of fertiliser applied. These reactions generate four gases (N_2O, NO, NO_2 and SO_2) that can have an adverse effect on air quality when their concentrations are higher than can be maintained in the soil by the natural equilibrium between the soil and air. Both the nitrogen and sulphur cycles are part of a complex overall equilibrium between inorganic and organic solids, air, water and micro-

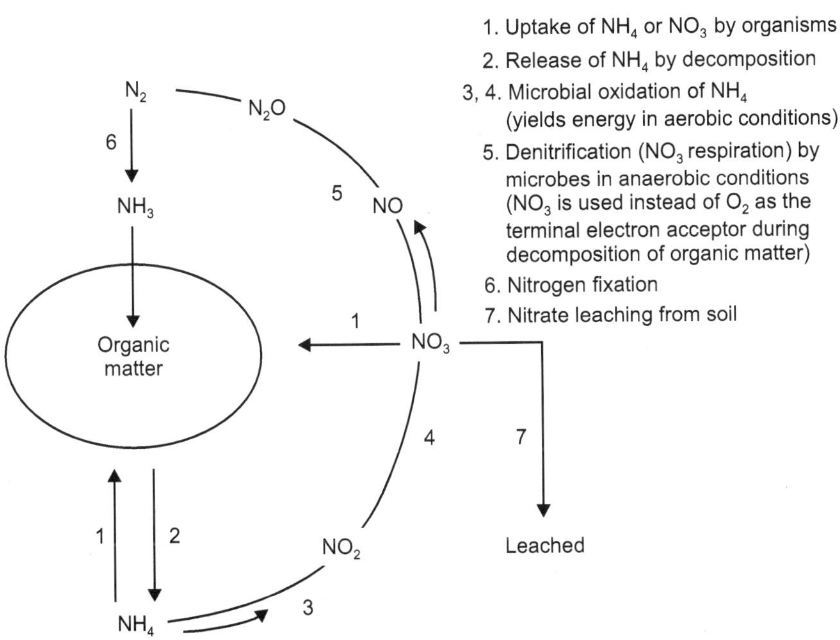

Figure 35.2: Nitrogen cycle.

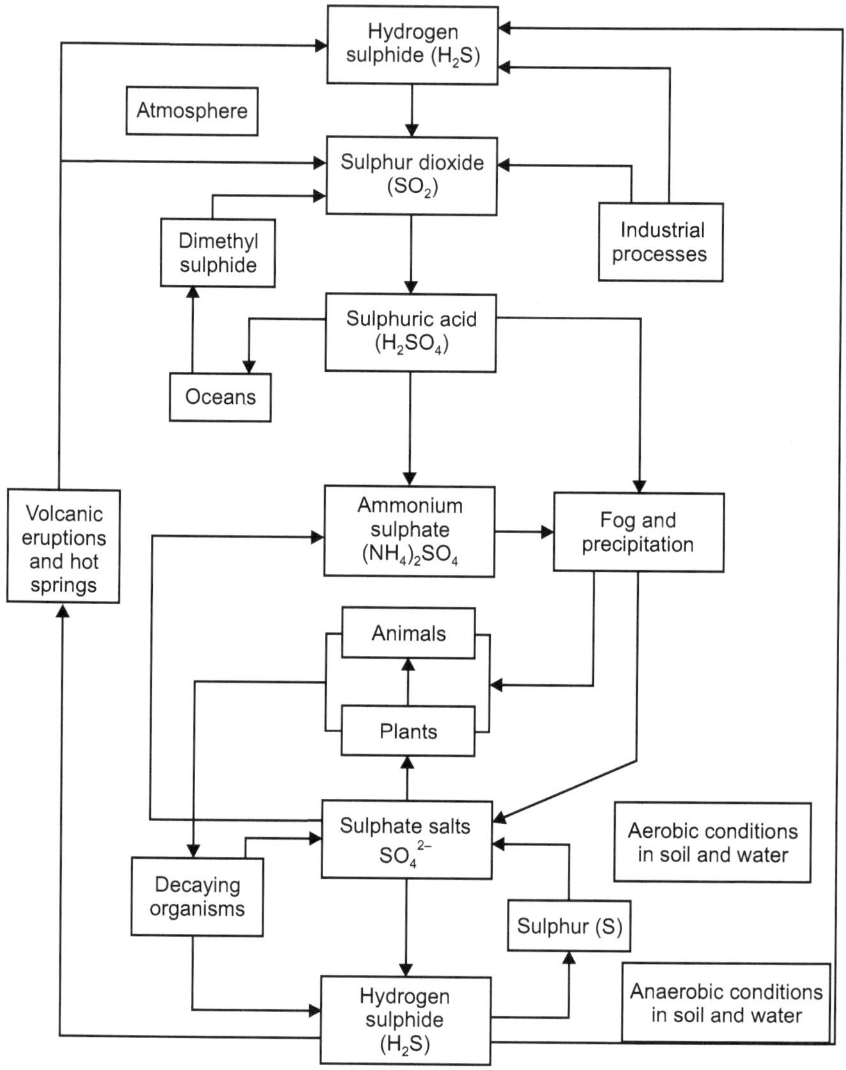

Figure 35.3: Sulphur cycle.

organisms. When one or more of these reactions is affected, the entire equilibrium is also affected.

Biological reactions: For several elements, notably carbon, nitrogen and sulphur, microbial reactions almost totally determine the soil reaction rates. Biochemical and microbial reactions are primarily catalytic processes affected by soil mineral composition, climate, gas exchange with the atmosphere and energy from photosynthesis. Three gases (NH_3, N_2O and hydrogen sulphide [H_2S]) that are precursors to the gases noted above are generated from three separate biological processes: nitrogen fixation, denitrification and the hydrogen sulphide reaction. A brief summary of these processes and the factors that affect them follow.

Nitrogen fixation: Nitrogen fixation is a process that reduces elemental nitrogen (N_2) from the atmosphere to NH_3 through a series of reactions catalysed by soil microflora. Factors that affect nitrogen fixation include the presence and type of organotropic bacteria, the presence or absence of air (or oxygen) in the soil matrix as related to the bacteria, the photosynthetic capability of bacteria and algae and the absence of hydrogen gas.

Denitrification/nitrification: Denitrification is a process that reduces nitrates to nitrogen in one or more reaction steps. One reaction produces N_2O. The reverse process, which is called nitrification, starts with either NH_3 or N_2 and oxidises it to nitrates through a series of reactions. Different microflora and molecular oxygen (O_2) are required for nitrification. Factors that affect nitrification and denitrification include the microflora, level of oxygen in the soil, the moisture content of the soil, the temperature and the available food energy sources for the microbes. Available information indicates that both nitrification and denitrification contribute to soil nitrogen compound emissions.

Hydrogen sulphide: Under anaerobic conditions, sulphates are reduced to H_2S. Factors that increase the generation of hydrogen sulphide include flooding, presence of sulphur reducing bacteria and the absence of oxygen.

Chemical reactions: The chemical reactions of fertilisers with soil are usually a series of reactions that occur under conditions closely related to those affecting the presence of microflora. This section summarises the factors that affect emissions as they are related to the fertiliser. Three specific processes are discussed: NH_3 volatilisation, reduction of nitrates and reduction of sulphates.

Ammonia volatilisation: Ammonium is normally stored in soil as a complex with carbonate ions or sulphate ions and is readily absorbed by plant roots. Ammonia volatilises more readily when the soil lacks these anions. Ammonia volatilisation also increases with flooding, high soil pH, the presence of high levels of calcium and high or elevated temperatures. Flooding mobilises the NH_3 and carries it to the surface where it is readily volatilised into the atmosphere. Soils with high pH (basic soils) react with ammonium ions to generate water and NH_3 gas. Calcium forms insoluble precipitates with sulphates and carbonates, thus reducing the anions available for complexing with ammonium ions. Ammonia emissions also increase with temperature. Under drying conditions, especially with increasing wind speed, soils with high moisture content enhance NH_3 volatilisation, especially with urea-containing materials.

Reduction of nitrates: Generally, nitrate is a soluble anion found in the soil solution and is readily absorbed by plant roots. However, these nitrate compounds can undergo reduction reactions to produce less soluble oxides of nitrogen and increase emissions of NO_x. The magnitude and rate of nitrate reduction in soils is increased with increasing quantities of decomposable organic matter, soil moisture content (decreasing soil aeration), soil pH, soil temperature and soil nitrate content.

Reduction of sulphates: Sulphates are loosely bound to the soil as salts and are readily absorbed by plant roots. However, chemical reduction of these sulphates to SO_2 or H_2S act to increase sulphur-related emissions. Factors that increase sulphur-related emissions by increasing the rate of these reduction reactions include flooding, the presence of key minerals and other anions, the concentration of sulphate ions, the type of clay and clay content in the soil and the type and quantity of soil organic matter. In general, the presence of more tightly bound anions within the soil increases sulphur-related emissions because of the reduced concentration of available cations.

Other biological activities

Because most emissions from fertiliser application are related to the ecological and chemical reactions related to the sulphur and nitrogen cycles, any biological factor that influences these biological and chemical reactions can influence the quantity and rate of gaseous emissions. For example, earthworms and other soil organisms can provide channels through the soil that enhance water and nutrient transport, which in turn effect nitrification and denitrification reactions. Other biological factors that affect emissions can be related to soil micro-organisms, surface plants and animal activity at the site.

Micro-organisms compete effectively with plants for available nitrogen and other nutrients. Without the application of certain nutrients, especially nitrogen, plant growth can be severely reduced because of microbial competition for nitrogen. In addition, any factor that reduces plant yield potential (pests, diseases, water and nutrient stress and many others) will reduce recovery of applied nitrogen and may potentially increase gaseous emission of nitrogen. When the supply of nitrates is high, the presence of growing plants can enhance denitrification because the population of denitrifier micro-organisms is greater than in root-free soil.

The presence of animals in grassland ecosystems enhances gaseous losses of nitrogen through volatilisation and denitrification of nitrogen in urine. These losses can be greater than those observed for urea with similar nitrogen content applied to the pasture.

Soil conditions

Physical and chemical conditions of soil, including pH, texture, moisture content and temperature, will affect air emissions from fertiliser application. Fine, well-aggregated soils are generally well-suited for optimum plant growth and nutrient use and thus reduce the potential for gaseous emissions. Poorly aggregated soils with genetic or management-related hardpans (compacted soil layers) reduce root penetration and water movement and may enhance gaseous emissions. Variations in soil properties between or within fields used to quantify gaseous emissions is one reason for wide variation in many of the test results [up to 50% Relative Standard Deviation (RSD)].

Moisture content of the soil is an important factor in emissions generation. As soil moisture content approaches saturation, the rate of denitrification greatly increases. Fluctuating soil moisture content, by frequent irrigation or rainfall, also enhances gaseous nitrogen emissions. When soil moisture is above the maximum moisture content (the point at which the voids between soil grains are filled with water), air emissions may be reduced because ammonium and nitrate in the soil solution are diluted and also may be transported into the ground and/or surface water systems.

Because they affect biological and chemical reaction rates, soil chemical conditions also affect gaseous emissions. Important chemical properties include the soil solution pH, the Cation Exchange Capacity (CEC) of the soil and the concentration of nutrients in the soil that potentially could be released to the atmosphere through numerous biological and chemical reactions. The CEC is defined as the capacity of the soil to adsorb or hold cations (Ca^{+2}, Mg^{+2}, Al^{+3}, K^+, Na^+, NH_4^+). Soils with a high CEC will adsorb more NH_4^+ and thus, exhibit lower NH_3 volatilisation potential than soils with a low CEC. Basic soils generally increase the release of nitrogen as NH_3 and nitrogen oxides (NO_x) (including NO and NO_2) and N_2O and convert all other nutrients to less soluble forms. Acidic soils (<5.0–5.5 pH) reduce NO_x emissions, but also may reduce plant growth due to aluminum toxicity.

Climate

Climatic conditions that affect emission rates through their influence on biological and chemical reaction rates include moisture, temperature and wind speed. Climatic conditions can impact all four of the emission mechanisms discussed above (section 35.2.1). Conditions that reduce oxygen content of the soil (increasing soil moisture, temperature changes, etc.), generally increase the emission of gas to the atmosphere above normal background levels. Even in well-aerated soils, denitrification still occurs in anaerobic microsites within soil aggregates. During short periods of saturation of the surface soil following rainfall or irrigation, the rate of denitrification greatly increases until drainage occurs and an aerated condition returns. Nitrogen and other soluble nutrients can subsequently be removed with the drainage water. Volatilisation losses of nitrogen generally are enhanced when wet soils are subject to drying conditions. Increasing wind velocity enhances volatilisation and under flooded conditions also increases denitrification. Ammonium volatilisation increases as soil temperature rises, emissions generally increase throughout the day relative to an increase in soil temperature. Also, daily peak emissions will increase throughout the summer season as compared to the other three seasons. Denitrification also increases with rising temperature.

Nutrient management

In addition to influencing the quantity of nutrient absorbed or used by the plant, the nutrient source and the rate, method and time of application can influence the magnitude and rate of gaseous emissions of the nutrient. Nutrient management can impact the emission mechanisms. It is important to recognise that any source of nitrate or ammonium nitrogen in the soil can participate in biological or chemical reactions that result in the formation of nitrogen gases. For example, nitrogen mineralised from manure or legume residues can be emitted to the atmosphere by the same reactions involved with gaseous emissions from nitrogen.

Compared to other nitrogen sources, NH_3 volatilisation is usually greater with urea or urea containing fertilisers (e.g., urea, ammonium nitrate) and manures. Ammonia loss with anhydrous NH_3 is usually not significant because this source must be injected 10 to 25 cm (4 to 10 in.) below the soil surface.

Generally, only small quantities of NH_3 are volatilised from ammonium-containing fertilisers (diammonium phosphate, monoammonium phosphate, ammonium sulphate and ammonium nitrate). However, NH_3 volatilisation can be significant with surface broadcast applications of diammonium phosphate and ammonium sulphate on calcareous or high pH soils. When these two fertilisers are applied to high pH soils, formation of calcium sulphate or calcium phosphate reaction products occurs, which increases the ammonium concentration in the soil solution and ultimately, the ammonium volatilisation potential. Generally, increasing the application rate increases the potential for gaseous emission and leaching of nitrogen. Therefore, identifying the correct nitrogen rate for optimum production will maximise the quantity of applied nitrogen recovered by the plant and minimise the potential environmental impact of nitrogen use. Again, this phenomenon holds for fertiliser, manure and legume nitrogen sources.

Compared to surface broadcast-applied nitrogen, any nitrogen containing fertiliser or manure that is applied to the subsurface will reduce the quantity of nitrogen emitted to the atmosphere. However, gaseous emissions related to volatilisation and denitrification still occur regardless of placement. In high pH soils and/or in zero tillage and reduced tillage systems (where crop residue covers the soil surface) on soils of any pH, subsurface placement of nitrogen fertiliser will enhance nitrogen recovery by the crop and reduce the potential for gaseous emissions. Surface broadcast nitrogen is usually incorporated into the soil with tillage shortly after application. Incorporation of nitrogen fertilisers will generally reduce potential gaseous emissions (especially with urea-containing sources) compared to not incorporating the nitrogen fertiliser or manure; however, with or without incorporation, nitrogen emissions are generally higher with surface broadcast nitrogen than with subsurface applications because broadcasting maximises the quantity of soil in contact with the nitrogen.

35.3 Emission control technology

The processes that generate emissions from fertiliser application are so complex and depend on a number of soil and climatic properties via complex relationships that have not been characterised quantitatively, the emission reduction potential of alternative management practices cannot be quantified at this time. However, the best form of emission control identified to date is through appropriate 'nutrient management.' Here, nutrient management is defined as the form, placement and timing of the fertiliser application relative to the crops' need for fertiliser.

Appropriate nutrient management requires not only appropriate quantities of fertiliser but also timing of the application. Maximising the quantity of nitrogen recovered by the plant requires that the nitrogen be applied as close to the time of maximum nitrogen demand as is possible. Therefore, split applications (part of the nitrogen is applied before planting and part is applied during an early crop growth stage) will maximise crop recovery and minimise gaseous emissions of the applied nitrogen. Since gaseous emissions can increase with increasing temperature, nitrogen application at cooler times during the year or during the day will reduce the potential for gaseous nitrogen loss.

Because a substantial quantity of emissions from fertiliser applications is related to the denitrification process, management techniques that reduce denitrification potential also will increase nitrogen utilisation and decrease emissions. Additives to fertiliser nitrogen sources that reduce or inhibit nitrification or urea hydrolysis (N-Scrvc, DCD and others) may reduce the potential for gaseous nitrogen emissions. Use of Encapsulated Calcium Carbide (ECC) has been shown to be effective in the inhibition of nitrification and the reduction of N_2O and N_2 emissions from irrigated corn and wheat fields as well as flooded rice fields. It was not effective for dry land wheat fields.

35.4 Wastewater from fertiliser plants

A variety of waste streams are discharged from fertiliser plants in the form of:

1. Processing chemicals like sulphuric acid.
2. Process intermediates like ammonia, phosphoric acid, etc.
3. Final products like urea, ammonium sulphate, ammonium phosphate, etc.

In addition, oil bearing wastes from compressor houses of ammonia and urea plants and some portion of cooling water and wash water from the scrubbing towers for the purification of gases, also come as waste.

Wash water from the scrubbing towers may contain toxic substances like arsenic, monoethanolamine, potassium carbonate, etc., in a nitrogenous fertiliser plant, while that in a phosphatic fertiliser plant may contain a mixture of carbonic acid, hydrofluoric acid and fluosilicic acid. Both alkaline and acidic wastes are also expected from the boiler feed water treatment plant, the wastes being generated during the regeneration of anion and cation exchanger units.

Previously, waste cooling water contained toxic elements like chromates, zinc, etc., which were used for corrosion control. The development of non-chromate technology using quaternary ammonium compounds has eliminated these toxic substances. Additional pollutants like phenol and cyanide will be introduced in the list of pollutants in a fertiliser plant where ammonia is derived from the waste ammoniacal liquor of coke ovens. Average characteristics of wastewater from a typical fertiliser plant, producing both nitrogenous and phosphatic fertiliser, is given in the Table 35.1.

Table 35.1: Composition of wastes of a complex fertiliser plant.

Item	*Values*
pH	7.5 to 9.5
Total solids	5400 mg/L
Ammonia nitrogen	700 mg/L
Urea nitrogen	600 mg/L
Phosphate	75 mg/L
Fluoride	15 mg/L
Arsenic	1.5 mg/L

35.4.1 Effects of wastes on receiving streams

All the components of waste from the fertiliser plants induce adverse effects in streams. Acids and alkalies can destroy normal aquatic life. Arsenic, fluorides and ammonium salts are found to be toxic to fish. Amines are not only toxic to fish but also exert a high oxygen and chlorine demand. Presence of different types of salts renders the stream unfit for use as a source of drinking water in the downstream side. Nitrogen and other nutrient content of the waste encourages growth of aquatic plants in the stream.

35.4.2 Treatment of fertiliser wastewater

Major pollutants in fertiliser wastewater for which treatment is necessary include oil, arsenic, ammonia, urea, phosphate and fluoride. Oil is removed in a gravity separator. Arsenic containing waste is segregated and after its concentration, the solid waste is disposed of in a safe place. Phosphate and fluoride bearing wastes are also segregated and chemically coagulated by lime; clarified effluent, which still contains some amount of phosphate and fluoride, is diluted by mixing with other wastes.

Several alternatives are there for the treatment of ammonia bearing wastes, including:

1. Steam stripping.
2. Air stripping in towers.
3. Lagooning after pH adjustment.
4. Biological nitrification and denitrification.

For all practical purposes, 'steam stripping' for ammonia removal from fertiliser wastes has been found to be uneconomical. Removal of ammonia gas from the solution in an air stripping tower, packed with red wood stakes, is found to be a very efficient method.

Very encouraging results are obtained from some laboratory and pilot plant studies conducted in U.K. for the removal of ammonia by simply lagooning the waste. It was found that considerable reduction in the ammonia content can be accomplished just by retaining the ammoniacal waste in an earthen tank, about 1 m deep, for a day or two, after pre-treatment of the waste by lime to increase the pH to 11.0. However, no reduction in urea content was observed within this period in wastes containing urea; thus waste containing both urea and ammonia required to be retained in the lagoon for a longer period, to allow urea to decompose to ammonia first.

Biological nitrification involves oxidation of ammonia to nitrate, via nitrite under aerobic conditions; this is followed by denitrification of the nitrified effluent under anaerobic condition, in which gaseous N_2 and N_2O are the end products and are released into the atmosphere. The denitrification requires addition of some quantity of carbonaceous matter in the reactor. In all the ammonia removal methods described above, urea remains untouched. If urea removal is required, wastes containing urea must be retained for a sufficiently long time in an earthen lagoon to allow it to decompose first to ammonia. Pollution control aspects related to Single Super Phosphate (SSP) fertiliser, straight nitrogeneous fertiliser and complex fertilisers are discussed below.

35.5 Pollution control in single super phosphate (SSP) plant

Single Super Phosphate (SSP) is the most popular and the cheapest phosphatic fertiliser. There is still a wide gap between the total installed capacity and the actual production. Fluorine emitted as a result of rock phosphate–acid reaction is a major pollutant in an SSP plant.

The importance of single super phosphate as an important source of phosphatic fertiliser in various countries have already been accepted. This importance is further increased as it provides some important micro-nutrients like sulphur and calcium in addition to P_2O_5. It is all the more crucial that SSP production, which is going to have increasing share of growing phosphatic demand in various countries should be allowed in the plants which have taken adequate precautions to protect the environment around them by treating emissions from the plant to the extent that technology can economically support.

Single super phosphate is mainly manufactured by a continuous process. Ground rock phosphate and diluted sulphuric acid are mixed in a Broadfield mixer and passed through a closed slow moving conveyor (Den). Most SSP manufacturers are using the above-mentioned process. During the acidulation process, some acidic fumes containing fluorine are emitted from the Den and the mixer. They are to be scrubbed to remove toxic gases. Hydrogen-silico-fluoride formed in the scrubber can be used to make sodium-silico-fluoride, a useful by-product.

To remove toxic gases, in older plants, these are channelised through zig-zag type scrubbing tower in series and a diluted solution of hydrofluoro silicic acid is sprayed through different nozzles mounted on top of the towers. Suction in mixer and Den is maintained by a 6000 cfm capacity centrifugal fan (Fig. 35.4). Water is recirculated by a pump from a small storage sump of hydrofluoro silicic acid. Most SSP plants earlier used such technology to removes toxic gases, but due to certain earlier used problems, this scrubbing tower method failed. The problems were mainly:

1. Choking of nozzles.
2. Scrubbing efficiency was quite low as fluorine recovery was low.
3. Choking in tower and duct due to silica.
4. Temperature of scrubbing liquid getting increased.

To overcome the above difficulties and increase scrubbing efficiency, the double-stage venturi scrubber method is used. In this, acidic fumes are scrubbed in a very advanced two-stage venturi scrubber, connected in series along with a cyclonic separator. In this case, water is used as an absorbent for the acidic fumes and fluorides. The fresh water for absorption purpose is added in the final stage of scrubbing. A schematic diagram of this is shown in Fig. 35.5.

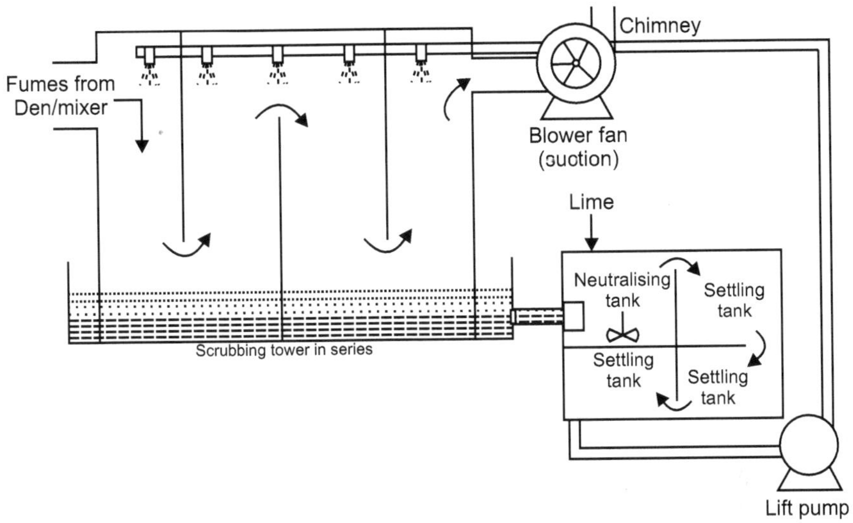

Figure 35.4: Scrubbing of fluoride fumes in series of scrubber towers.

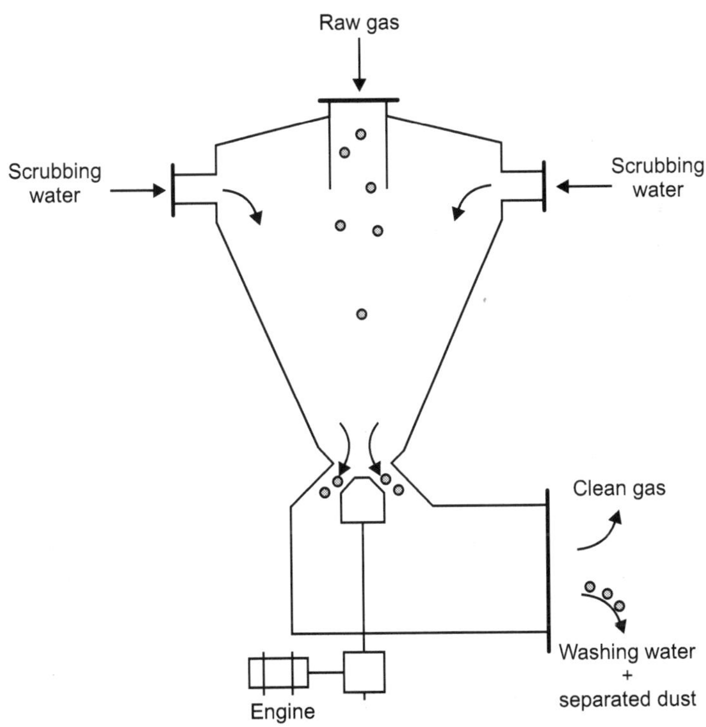

Figure 35.5: Venturi scrubber.

The reaction between rock phosphate and sulphuric acid takes place as follows:

$$Ca_9 (PO_4)_6.CaF_2 + 7H_2SO_4 \rightarrow 3CaH_4(PO_4)_2 + 7CaSO_4 + 2HF$$

The reaction indicates that 23.61 kg of HF can be emitted per tonne of super phosphate manufacturing. Scrubbing with mildly alkaline medium is known to perform at an efficiency level of 99.9% in arresting fluoride in plants.

The controlled emission can be less than 0.04 kg of fluorine per tonne of rock phosphate processed.

All the scrubbing system equipment are made of rubber-lined mild steel.

35.5.1 Method of treatment, utilisation and disposal of effluent

The main source of fluorides and phosphates in an SSP plant are scrubber liquors from various unit operations, involving scrubbing of the off gases. In the effluent, fluorides are present as fluorosilicic acid, with small amount of phosphoric acid and hydrofluoric acid.

For the removal of fluorides and phosphates, two stages of treatment–with chalk, followed by lime or double lime treatment are adopted. In the first stage, the effluent is treated with chalk or finely crushed calcium carbonate at a pH of about 3.0. The requirement of calcium carbonate is 3 to 3.5 g for 1 g of fluorine and 0.6 to 0.7 g for 1 g of phosphate.

The following reactions takes place:

$$SiO_2 + 6HF \rightarrow H_2SiF_6 + 2H_2O$$

$$H_2SiF_6 + 3CaCO_3 \rightarrow 3CaF_2 + SiO_2 + CO_2 + H_2O$$

$$2H_3PO_4 + CaCO_3 \rightarrow Ca(H_2PO_4)_2 + H_2O + CO_2$$

In the above reactions, almost all the fluorides are precipitated as calcium fluoride; silica is also precipitated. However, most of the phosphates remain in the solution as mono-calcium-phosphate. In the second stage, the product of the first stage is further treated with lime at a pH of about 7.0 to 8.5.

In this reaction, calcium hydroxide requirement is 2.1 to 2.8 g for 1 g of residual fluorine and 1.0 to 1.1 g of residual phosphate. It is believed that in the second stage, the following reactions take place:

$$H_2SiF_6 + 3Ca (OH)_2 \rightarrow 3CaF_2 + SiO_2 + 4H_2O$$

$$3Ca (H_2PO_4)_2 + 7Ca (OH)_2 \rightarrow Ca_5 OH (PO_4)_3 + 12H_2O$$

In the second stage, residual fluorides and phosphates from the first stage are converted into insoluble calcium fluoride and calcium hydroxy apatite, which at a pH of around 8.5 are precipitated. The overall fluoride and phosphate removal efficiency is above 99% in the above two-stage treatment.

In double lime treatment, lime is used in place of chalk or powdered calcium carbonate as indicated earlier in the two-stage treatment. The effluent after the second stage reaction is transferred to a settler for the removal of fluoride and phosphate precipitates and the overflow water is discharged or pumped out in the scrubbing process. The settled sludges are removed periodically and dumped or used as fill for low-lying areas. Various process units required for the treatment system are:

1. Neutralisation tank.
2. Settling chamber.
3. pH adjustment tank.
4. Settling chamber.
5. Lime mixing/storage tank.
6. Effluent lifting pump.

35.6 Straight nitrogenous fertilisers

Urea and salts of ammonia are referred to as straight fertilisers and if combined with other nutrients such as phosphates and potash, they are called complex/ mixed fertilisers.

Nitrogenous fertiliser plants use a large quantity of water mainly for process cooling, steam generation and process use, resulting in wastewater generation at various points in the manufacturing process. In an ammonia plant, if partial oxidation process is used, then the carbon of hydrocarbon feedstock is not completely combusted. As such, unburnt carbon as soot is generated. In most plants, inplant provisions are incorporated for recycle of 80% of the carbon generated. The rest 20% of carbon produced is required to be disposed of. Further, cyanide is formed due to a side reaction during gasification. Both carbon and cyanide come out in the wash water. Inplant system is usually provided for removal of major quantity of cyanide in the stripper, where the stripping is done by steam and acidic gases. However, the stripper bottom effluent containing some residual cyanide is continuously discharged. In case of fuel oil feedstock, the effluent discharged contains suspended carbon, sulphide, formate, ammonia and metals like vanadium, nickel, iron, etc., along with cyanide. Where the fuel oil feedstock contains high sulphur (e.g., HSHS containing 3–4% sulphur), a sulphur recovery plant is incorporated.

Where coal is used as feedstock, considerable quantity of fly ash is discharged in the wash water of the gasification section. This wash water contains a small quantity of cyanide also.

In the steam reformation process, steam in excess of stoichiomctric requirement is supplied in the primary reformer. When the gas is cooled, the excess steam condenses and forms process condensate, which contains ammonia, methanol and some organics. In a modern plant, built-in system is provided for removal of major quantity of ammonia by stripping. This makes the condensate fit for use as boiler feed water, with or without polishing. However, when the contaminants in the condensate are high, the condensate is unfit for use as boiler feed water and is discharged as effluent.

In the carbon dioxide recovery section, though the absorption towers are operated in closed circuit, some absorbent chemicals find their way out in the cooler condensate, sludge (formed in the process) and in leakage and spillage from the system. The type of pollutant depends on the absorbent solution used.

In the urea plant, the urea solution with associated ammonia and carbon dioxide is concentrated in vacuum concentrators. During evaporation, the vapours condense to form contaminated condensate. This condensate contains urea, ammonia and carbon dioxide. This is the main source of pollution in a urea plant. In modern urea plants, a built-in facility—thermal hydrolyser stripper—is provided whereby a major quantity of contaminants are removed and recycled before the discharge of the condensate.

Another effluent stream is generated from the carbon dioxide compression section, which mainly contains oil. The ammonia and carbamate solution plunger pump leakage also form an effluent stream, which contains small quantity of ammonia, urea and oil.

35.6.1 Disposal of the final treated effluent and monitoring

Disposal of the final treated effluent to the receiving water system is an important aspect in the pollution control system. The detailed procedure to be adopted for disposal system is given as under:

1. It is desired that all the effluents after treatment shall be routed to a properly lined guard pond for equalisation and final control.
2. The guard pond should have two compartments, each of at least four hours capacity. All the effluent streams shall be connected to these compartments by a parallel connection system. One compartment of the guard pond shall be used for the routine disposal of effluent, while the other compartment will remain empty and be utilised when effluent do not conform the limits.
3. In the guard pond, an automatic monitoring system for flow and the relevant pollutants shall be provided with high-level alarm system. The parameters necessary for automatic monitoring are pH, ammoniacal nitrogen, nitrate nitrogen and hexavalent chromium. Monitoring of nitrate nitrogen is applicable to the industries where nitric acid is produced and used for production of fertilisers and the monitoring of hexavalent chromium is applicable to the industries where chromate-based inhibitors are used for cooling water conditioning.

4. When pollutants in the final effluent exceed the stipulated figures as indicated by the alarm system and the effluent stream responsible cannot be identified, all the effluent streams shall flow the empty compartment of the guard pond. The effluent thus stored should be treated before discharge.
5. The area around the guard pond shall be developed with proper road connection and lighting system so that it can be approached easily at any time.
6. Till the continuous monitoring system as indicated above are not installed, the industry shall collect grab samples at a four-hour interval and analyse these for pH and ammoniacal nitrogen.
7. The other parameters as relevant to the industry concerned such as total kjeldahl nitrogen, hexavalent chromium, total chromium cyanide, nitrate nitrogen, vanadium, arsenic, suspended solids and oil and grease should be analysed in the grab sample collected once a day at fixed hours.
8. For effective appraisal of the performance of treatment units, the industry shall monitor the concerned parameters at least once a shift before and after treatment.

35.7 Complex fertiliser industries

Various grades of complex fertilisers containing nitrogen and phosphorus are manufactured with and without addition of potash (NP and NPK). The ingredients required for complex fertilisers are ammonia, phosphoric acid, nitric acid, etc., for NP grade fertilisers, which are either manufactured by the industry or procured from outside.

Various types of complex fertilisers are:

1. Diammonium phosphate (DAP): DAP is manufactured by reacting ammonia with phosphoric acid. After preliminary neutralisation of ammonia with phosphoric acid, the slurry is sent to the granulator where it is further ammoniated. Any unreacted ammonia is scrubbed with phosphoric acid and returned to the pre-neutraliser. The granulated material is then dried, screened and cooled (Fig. 35.6).
2. Nitrophosphate (NP): Nitrophosphate (NP) is manufactured by digesting rock phosphate with nitric acid. The calcium nitrate formed is cooled, crystallised and separated by filtration/centrifugation. The mother liquor is neutralised with ammonia in the neutraliser. Calcium nitrate crystals are reacted with ammonia and carbon dioxide, where ammonium nitrate and calcium carbonate (chalk) are formed. Chalk is separated by filtration. Mother liquor of chalk is also taken to the neutraliser. Neutralised mother liquor is evaporated and then prilled in the prilling tower.

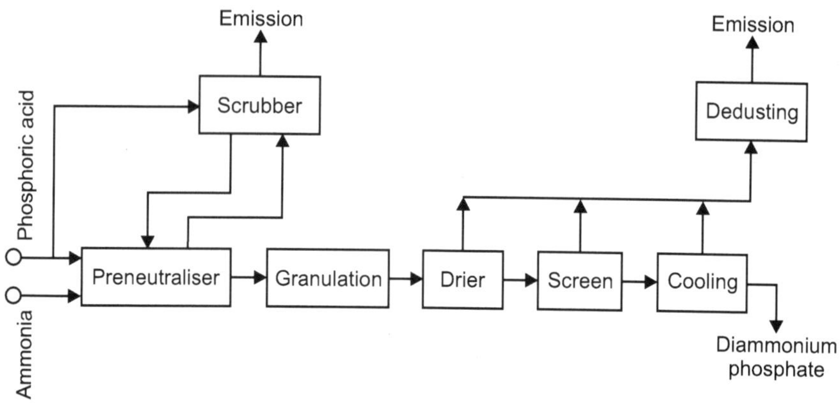

Figure 35.6: Production of diammonium phosphate (DAP) and emissions generated.

3. Urea ammonium phosphate (UAP): In the manufacturing of urea ammonium phosphate, ammonia and phosphoric acid are neutralised. The ammonium phosphate slurry formed is taken to the granulator where urea and some filler are added. Granulated material is dried, screened and cooled.
4. Ammonium phosphate sulphate (APS): Ammonium phosphate sulphate (APS) is produced by neutralisation of sulphuric acid and phosphoric acid with ammonia and then granulating the slurry. Alternatively, it may be produced by addition of ammonium sulphate to phosphoric acid and then ammoniating the mixture. The mixture is dried, screened and cooled.
5. NPK fertilisers (NPK): NPK complex fertiliser is produced by reacting ammonia with phosphoric acid in the required proportion in a pre-neutraliser. The slurry formed is taken to the granulator. Ammonia, potash, filler and also urea depending upon the grade of the product are added in the granulator and the product is dried, screened and cooled (Fig. 35.7).

35.7.1 Wastewater generation

In general, dilute phosphoric acid is used as the scrubbing medium for scrubbing of fumes and also dust. The scrubbed liquor is reused in the process. In some cases, where water is used as scrubbing medium, the purge liquor from the scrubber is either used in the process or discharged. In addition, other effluents generated are cooling water blow down and spillages/leakages. The main pollutants generated from complex fertiliser units are ammonia, fluoride, phosphate and suspended solids. Where urea or nitric acid is used in the process, the effluent generated will contain urea or nitrate as the case may be. Where phosphoric acid is manufactured, fluoride and phosphate bearing

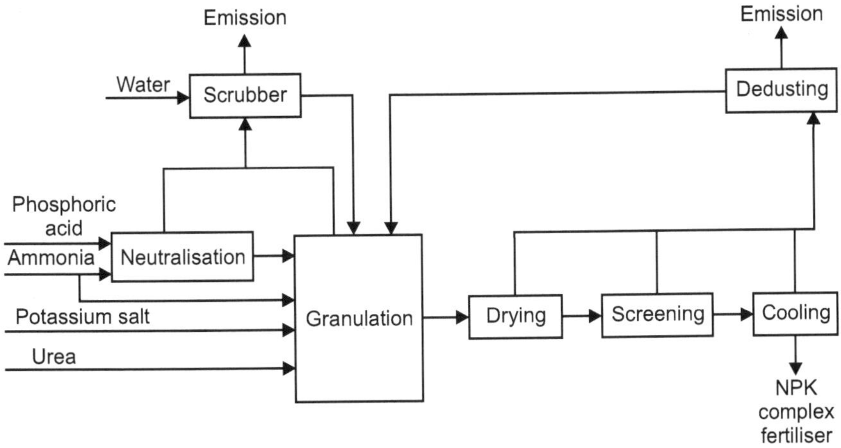

Figure 35.7: Production of NPK fertilisers and emissions generated.

effluents are generated. Treatment for ammonia, fluoride, phosphate and nitrate is required in case of complex fertiliser, depending upon fertiliser produced.

35.7.2 Disposal of treated effluent

Disposal of the final treated effluent to the receiving water system is an important aspect in the pollution control system. The detailed procedure to be adopted for disposal system is:

1. It is desired that all the effluents after treatment shall be routed to a properly lined guard pond for equalisation and final control.
2. The guard pond should have two compartments, each of at least four-hour capacity. All the effluent streams shall be connected to these compartments by a parallel connection system. One compartment of the guard pond shall be used for the routine disposal of effluent, while the other will remain empty, to be utilised when effluent do not conform the limits. Effluent of the guard pond shall overflow a weir system for measurement of flow.
3. In the guard pond, automatic monitoring system for flow and the relevant pollutants shall be provided with high-level alarm system. The parameters necessary for automatic monitoring are pH, ammoniacal nitrogen, nitrate nitrogen and hexavalent chromium. Monitoring of nitrate nitrogen is applicable to the industries where nitric acid is produced and used for production of fertilisers and the monitoring of hexavalent chromium is applicable to the industries where chromate-based inhibitors are used for cooling water conditioning. The fluoride should also be continuously monitored where phosphoric acid is manufactured.

4. When the pollutants in the final effluent exceed the stipulated figures as indicated by the alarm system, such effluent stream shall be diverted to the empty compartment of the guard pond and simultaneously necessary corrective measures shall be adopted in the affected plants (process plant or effluent treatment system). In case, the stream responsible for high pollutant level cannot be identified, all the effluent streams shall flow into the empty compartment of the guard pond where it should be treated before discharge.
5. The area around the guard pond shall be developed with proper road connection and lighting system so that the location can be approached easily at any time.
6. Till the time, the continuous monitoring systems as indicated above are not installed, the industry shall collect grab samples every four hours and analyse them for pH and ammoniacal nitrogen.
7. The other parameters as relevant to the industry concerned such as TKN, hexavalent chromium, total chromium, cyanide, nitrate nitrogen, vanadium, arsenic, fluoride, phosphate, suspended solids and oil and grease should be analysed in the grab sample collected once a day at a fixed hour.
8. For effective appraisal of performance of treatment units, the industry shall monitor the concerned parameters at least a shift before and after treatment.

36

Pollution control in petroleum refinery

36.1 Introduction

The exploration and production sector of the petroleum business explores, develops and extracts crude oil and gas resources from natural formations beneath the earth. Exploration involves mapping geological features and drilling test wells in areas where there is a high probability of finding reserves. Proven oil fields are then developed through a well drilling programme, which leads to a producing field. The extraction techniques used depend on the geological characteristics of the area, the type and formation of the oil and/or gas deposits and the design of the well field.

Crude oil refining operations involve extracting useful petroleum products (e.g., gasoline). Crude oil contains fractions of naphthas, jet fuel, gasoline, diesel fuel, gas oils, lubrication oils and asphalt. Refineries extract these fractions thermally or catalytically.

36.2 Refinery operations

Crude oils are complex mixtures of hydrocarbons of varying molecular weight and structure. These hydrocarbons range from simple highly volatile substances to complex waxes and asphaltic compounds. The final petroleum products are obtained from crude oil through a series of operations, viz., topping, thermal cracking, catalytic cracking, catalytic reforming, etc., These basic operations, however, vary from plant-to-plant, depending upon the properties of the crude oil and desired grade of the final product. The basic refinery operations as mentioned above are shown diagramatically in Fig. 36.1.

In general, crude oil is first subjected to fractional distillation in the process known as 'topping'. The products obtained are called raw products and include raw gasoline, raw naphtha, raw kerosene, gas oil, fuel oil, etc.

Now these intermediate refinery products are again treated to yield various finished market products as per requirements. The operations practised include 'catalytic cracking' or 'thermal cracking' and further purification processes like 'acid treatment', 'sweatening treatment', 'hydrobe sulphurisation', etc.

The decomposition of heavy or high boiling petroleum distillates like gas oil and fuel oil to lighter products like gasoline is called cracking. These reactions take place practically at the atmospheric pressure and at a high temperature. In catalytic cracking, however, an acid type solid catalyst (such

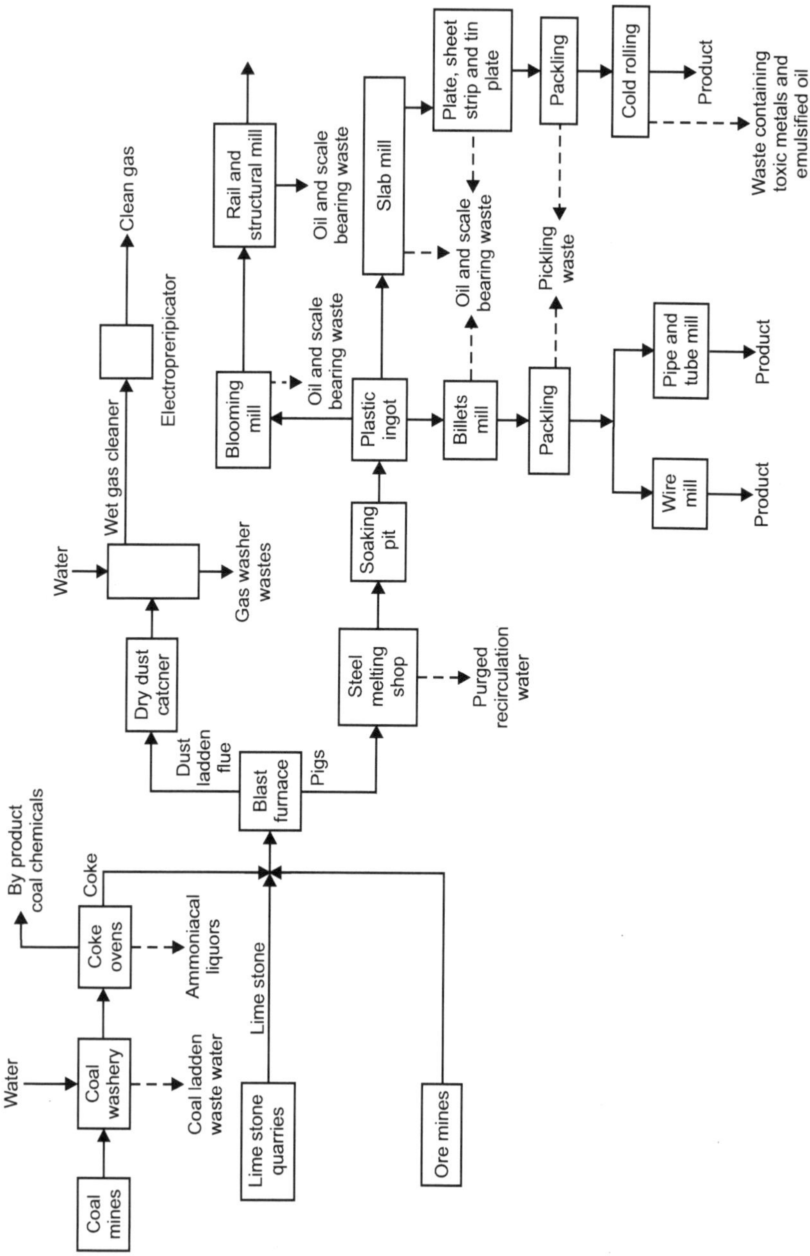

Figure 36.1: Basic refinery operations.

as synthetic silica alumina) is introduced into the reactor. A residue of heavy black material, known as coke, is obtained out of the process of cracking. In the catalytic cracking process, however, the asphaltic or tar-like products get adsorbed on to the surface of the catalysts in the form of coke.

The wax distillates of crude oil may be used as raw material for the manufacture of light lubricating oils. The process of their manufacture consists of chilling the distillates and filtering the wax from the oil.

The products thus obtained, either as final product or as feedstock for further processing, may contain certain undesirable constituents, which must be either removed or converted into less harmful compounds by further refining processes. Such refining processes include: (i) removal of hydrogen sulphide gas and mercaptans by washing with strong caustic soda solution, (ii) adsorption of hydrogen sulphide gas in an alkaline absorber liquid (which is immiscible with the product) followed by the recovery of the hydrogen sulphide gas, in the process known as regenerative process, (iii) conversion of mercaptans to less harmful desulphide, employing solutions like lead oxide in caustic soda together with sulphur, or sodium hypochlorite or copper chloride, in a process known as 'sweatening' and (iv) conversion of alkylsulphides and thiophenes and practically all other sulphur compounds into hydrogen sulphide in a catalytic reaction in the process known as 'hydrodesulphurisation'—hydrogen sulphide thus formed is recovered by the regenerative process.

Crude oil is sometimes pre-treated with caustic soda along with water to remove salts. Water may be removed by passing the product through a bed of calcium chloride or a similar drying agent.

The brine, which comes out of the desalting operation and the spent drying agent, used for the removal of the water, goes into the sewer. But wherever possible, the removed impurities are recovered and sold or are used for manufacture of other products; process chemicals are reused after the recovery of the impurities. As such, these impurities do not join the wastewater stream directly, but may be introduced to the flow through condensate and wash waters that come into contact with products at some stage of the processing.

36.2.1 Sources and types of wastes

Various wastes are generated during petroleum exploration, production and refining. Large quantities of production wastes – low-toxicity solid waste and wastewater – are produced during exploration and production.

The waste stream from refineries is smaller but is considerably more toxic, containing listed and characteristic hazardous wastes. In addition, poor housekeeping practices during exploration, production and refining can produce excessive waste.

Waste from exploration and production

Production wastes can be grouped broadly into two classes: wastes related to drilling and well completion, including drilling mud, drill cuttings and chemical additives and wastes related to oil production, primarily produced water. The volume of produced water far exceeds the volume of drilling wastes.

Treatment and disposal: Treatment and disposal of oil drilling wastes take place either on or off the drilling site. Wastes that are regulated by CPCB must be tested for hazardous waste characteristics before disposal. Wastes can be treated and disposed of on the site in reserve pits, by land spreading, or through annular disposal.

Liquid wastes can be treated in closed systems or treated and discharged to surface water. Offsite treatment and disposal methods include disposal in centralised pits, reinjection and application on commercial landfarms. Drilling muds can be reconditioned. The majority of produced water is disposed of underground through injection wells.

36.3 Strategies for preventing pollution

The petroleum industry has no control over the composition of the crude oil they extract. Most waste generated during production is the result of this basic lack of control over the composition of the crude oil. Lower-quality crude oil requires more processing and generates more waste than high-quality crude.

Although source reduction is the preferred strategy in overall waste management, the petroleum industry has limited opportunities in this regard. Therefore, the key elements of pollution prevention in petroleum production and exploration are recycling and careful handling of waste.

Pollution prevention techniques used during production include:

1. Collecting used solvents, hydraulic fluids and motor oils that are generated during drilling operations and shipping them to a petroleum recycler.
2. Using smaller reserve pits for disposal of drilling muds and produced water.
3. Segregating drilling muds for disposal as non-hazardous waste before they enter a hydrocarbon zone.
4. Recycling drilling muds in a closed-loop system.
5. Solidifying drilling wastes in reserve pits.

Although implementing pollution prevention practices during production is not mandated by regulation, several factors provide some motivation to do so. These factors include the high cost of commercial disposal, the potential for future liability for environmental damages and the difficulty of obtaining disposal permits.

36.3.1 Solid wastes from refineries

Solid wastes from the refinery can be categorise in order of increasing difficulty in disposal.

1. Inert dry solid: Trash, silt, spent cracking catalyst
2. Combustible dry solid: Trash, waste paper, scrap lumber.
3. Sludge containing water and solid: Water softner, sludge and sanitary sludge.
4. Sludge containing oil: Spent clays.
5. Sludges containing oil water: Tank bottoms, oil water separator and solids.

Solid wastes

Crude oil supply. Crude oil generally contains some basic substance and water, which is composed of mixture of water, iron rust, iron sulphide, clay, etc., produced with crude oil or accumulated during transit of crude oil.

Solid from surface water: Normally oil refineries are provided with common sewer system for oily water and surface drainage. Hence, the refineries have disposal problem with silt carried into drainage by surface water. Also the storm water overloads the capacity of oil water separators.

Solid from reaction in wastewater treatment unit: Sludges are formed in the wastewater treatment units.

Catalytic cracking catalyst and solid catalyst: This is an inert granular solid containing small amount of hydrocarbons and carbon. Other solid catalysts are used in hydrogenation and other unit operations.

Solids from coking operation: Coke fines are generated when the furnace is water washed and are carried to the water treatment unit.

Sludges: Sludges are generated at different places. They usually contain waxy material, oils, water, metals, etc. Sludge from the bottom of the tanks are different in characteristics for different crudes and oils. These may have a very high contents of waxes, water and oil.

Disposal of solid wastes

The solid wastes in the refinery are naturally obtained in the form of sludges containing solids, water or oil which pose problems in its treatment and disposal. The treatment disposal method depends on the type of sludge and its constituents.

Disposal of water treatment sludge: This type of sludge is not allowed to flow in oily water drain. It is usually dewatered by gravity in sludge thickeners or ponds. Ultimate disposal is on land.

The possibilities of its reuse are as follows:

1. Sludge may be acid treated to reduce volume requiring final disposal and to recover the flocculants.
2. Sludge may be utilised to neutralise certain wastewaters and in some cases can be dissolved in the water supply to combat corrosion. They are sometimes used as wetting agent in wastewater flocculation.
3. Sludges may be used in cement manufacture.

Sludges containing oil and solids: The most oily sludges that are produced in refinery treatment process are:

1. Spent contact clays.
2. Percolation clays from treatment of lubricating oil.

Dewatering and deoiling of sludge can be carried out by gravity settling, centrifugation and filteration. The concentrate so obtained is disposed off by land filling, incineration or other suitable methods.

Biological decomposition of solid wastes

This is used when sufficient land area is available where waste can be spread over. Bacteria decompose organic matter. The area of waste depends on the quantity of waste and rate of decomposition. The decomposition rate depends on composition of waste and atmospheric temperature.

Digestion of the sludge is normally done in the absence of air (anaerobic), the aerobic digestion of biological sludge has recently been adopted. Sludge can be treated in mechanical aeration tank and the extent of treatment possible is related to biodegradable fraction of biological sludge present.

Incineration. Incineration is generally applicable where sufficient land for disposal of raw waste is not available. The main advantage is reduction of waste containing organic material to inert material. It has high initial as well as operating cost.

Tank bottom sludges. Sludges from the bottom of crude oil storage tanks are having very high contents of wax, water and oil. A process has recently been developed to extract the useful micro-crystalline wax which has a very high melting point. The acidic sludges which are generated in this process is neutralised and used for grease making or incineration and the spent clay can be regenerated and reused.

Solvents and effluents

Refineries generally use solvents like toluene, methyl ethyl ketone, benzene for removing waxes and furfural for the removal of aromatics during the manufacture of lubricating oil. During the operation with these solvents starting from receiving in the refineries to final operation there is every possibility for the loss of these through spillage, solvent extraction and maintaining the proper ratio. The solvents lost by these processes ultimately accumulates in the effluent water system through surface drains with washout waters. Removal of these solvents contamination from the effluent water is a typical specialised job.

Solvent contaminated water streams from the units in the refineries gets diluted in the total influent water which undergoes the various treatment. The physical treatment removes free oils and suspended solids and chemical treatment removes emulsified oils, sulphides, etc., while the biological treatment removes the dissolved degradable hydrocarbons. So there is every possibility that the solvents remain unremoved in the final effluent water unless special treatment procedures are followed. On the contrary the chemicals like furfural are harmful for the biological treatment through trickling filter and activated sludge because the micro-organisms responsible for biodegradation gets damaged through furfural.

36.4 Control of effluents

36.4.1 Air pollution control

For the recovery of SO_2 and H_2S, etc., refineries have sulphur recovery unit (SRU) and monoethanol amine treating unit (MEA unit).

Sulphur recovery unit (SRU): This unit is designed to recover elemental sulphur from H_2S. The process involved in this recovery are 'Clause process' and ARI Process. The clause unit consists of thermal stage where hydrogen sulphide (H_2S) in the feed gas is partially burnt with air to sulphur dioxide (SO_2). The resulting mixture of H_2S and SO_2 reacts to form elemental sulphur and water vapour. There are two catalytic stages to achieve the recovery of 94% of sulphur in the feed gas.

The general reaction can be given as:

$$H_2S + 1/2\ O_2 \rightarrow H_2O + 1/n\ Sn$$

where, n is number of sulphur molecules.

Mono-ethanol amine treating unit: This unit can be a part of the sulphur recovery unit or it may be a separate unit. This unit is also designed to remove H_2S from the sour gases obtained from hydrofiner or other units, using an aqueous solution of Mono-ethanol Amine (MEA) as absorbing medium. The unit consists of two sections namely the absorption section and the regenerating sections. The sour gases are scrubbed of H_2S with MEA solution in the absorption section. The regeneration section strips off H_2S gas absorbed in MEA solution and this regenerated solution is recirculated to the absorber in the absorption section. The H_2S rich gas is then sent to sulphur recovery unit. Provisions are made available in the refineries to divert these gases to the flare also.

Flare stack: The function of the flare stack is to burn the accidental hydrocarbon release from different units. This systems has a knockout drum along with the water seal drum. The liquid hydrocarbons released from the units are collected in the knock out drum and is pumped back. Vapours that escape are burnt at the top of the flare stack.

Gas bottling plant: The LPG cylinders which are used for the domestic as well industrial purpose are filled in the gas bottling plant in the refineries. The general emission in this place are mercaptans and hydrocarbons.

The usual norms followed in a refinery to avoid pollution in these areas are good housekeeping and ventilation. Hydrocarbon detectors are also placed in this area to detect any leaks as they may lead to Vapour Cloud Explosion (VCE) if an ignition source is present.

The bottling area has all over sprinkler which can act as a water curtain in case of disastrous leak. All the filled cylinders are passed through a water bath for detecting leaks.

Efficient absorption of SO_2: The need of the hour is to devise efficient methods for SO_2 pollution abatement. So far many processes have been developed, particularly in the west and a few of them have been commercialised. In SO_2 removal processes, flue gases from stacks are subjected to physical and chemical treatments. Depending upon the nature of the techniques used, the processes may be broadly classified into wet and dry processes. The wet methods make use of a slurry or a solution of absorbent in water for absorption of SO_2, while the dry systems make use of dry particles of absorbents in fluidised bed, fixed bed or moving bed type reactors. In both these processes a further classification has been made depending on the reuse of the absorbent on regeneration or simply its throwing away. In the throwaway process cheap absorbent is used so that it can be disposed of as a less harmful waste. Limestone, because of its low cost and easy availability in each continent, has gained popularity as the absorbent of choice. In the regeneration/recovery process a separate step is introduced to convert the spent absorbent back into usable form. The sulphur recovered in this regeneration process adds to the economics of the process.

Dry absorption process: The conventional dry absorption processes have been popular because of their apparent advantage of stack-gas treatment at elevated temperatures thereby sustaining thermal buoyancy of the flue gases. So far only a few inorganic oxides like copper oxides, manganese oxide have found application in dry recovery process.

In the Shell Flue Gas Desulphurisation (SFGD) process, copper oxide deposited on alumina support is used as the absorbent.

$$CuO + SO_2 + 1/2\ O_2 \rightarrow CuSO_4$$

Regeneration is carried out using reducing gases like H_2, CO, or methane and results in sulphur rich gas from which either sulphur or H_2SO_4 can be produced. The need for a large amount of reducing gas for sulphur recovery has made the process uneconomical.

Another dry absorption process developed by Mitsubishi Heavy Industries, Japan, employs precipitated manganese dioxide for SO_2 capture.

$$MnO_2 + SO_2 \rightarrow MnSO_4$$

High cost of sorbent and consequent complex regeneration steps are the major drawbacks of the process.

Novel system for SO_2 absorption and regeneration: The need of the hour is an efficient solid system for SO_2 capture. Realising this necessity we have studied SO_2 capture by two solid systems, which incorporate the advantages of inert solids described above. In industrial practice solid pellets need to be prepared using inerts along with active sorbents and suitable binder followed by processing involving special techniques. To overcome this drawback we have picked up red mud which is a solid waste from Bayer's Alumina process. The red mud contains considerable amounts of iron oxide (41% Fe_2O_3) and rest are inert solid components, mainly consisting of Al_2O_3, TiO_2 and SiO_2. Thus the red mud serves as an easily available cheap sorbent for SO_2, with advantages of avoiding incorporation of inerts in active Fe_2O_3.

Iron oxide belongs to the class of regenerable solid sorbents. It captures SO_2 at 300–550°C.

The iron sulphate formed can be decomposed easily to iron oxide and a gas containing SO_2 in high concentration.

$$Fe_2(SO_4)_3 \xrightarrow[\text{Catalyst}]{650\text{–}700^\circ C} Fe_2O_3 + SO_2 + O_2$$

Initially 12 mm dia cylindrical pellets made up of pure iron oxide with porosity of 0.44 were subjected to SO_2 capture in a single pellet reactor. The maximum conversion reached was less than 10%. On the other hand with the same porosity pellets consisting of red mud, the conversion of Fe_2O_3 was more than 50% in the same time and temperature ranges. In order to examine the practical suitability of red mud in SO_2 capture and regeneration, five sets of SO_2 capture and regeneration runs were carried out—capture runs at 440°C and regeneration runs at 700°C. SO_2 capture and regeneration cycles were of 60 minutes and 45 minutes duration, respectively. The SO_2 capture rate, expressed in terms of fractional conversion of the initial oxide present. It is observed that SO_2 capture efficiency is quite attractive even after fifth cycle and regeneration efficiency is quite high in all the cycles. The fact that a good efficiency was maintained for 5 cycles and more, is further confirmed by the marginal decrease of effective diffusivity D_e from 2.56×10^{-2} cm^2/second for the first cycle to 2.35×10^{-2} cm^2/second for the fifth cycle. This is a sure sign of effective use of red mud for a few cycles more beyond the five cycles studied here.

Special behaviour of an efficient and cheap sorbent for SO_2: An interesting solid system for SO_2 capture discovered in the laboratory is the mineral red ochre. Here the active component is iron oxide ($\cong$ 14%) and rest are inerts like oxides of Si, Al and Ti. The mineral in the original form contains iron in the goethite structure (FeOOH). It can be calcined easily to change to Fe_2O_3 form, by removal of water and subsequent oxidation. The mineral red ochre was soft and quite porous in nature. It could be turned into cylindrical pellets easily using a lathe. Cylindrical pellets of red ochre of 12 mm dia were prepared by turning a lump of ore. It was calcined at 650°C to attain a constant weight and then subjected to SO_2 capture at low temperatures of 420–560°C. The conversion-time plots of these pellets showed a sigmoidal behaviour characterised by slow initiation accompanied by accelerated rate and terminating into slow rate, i.e., deceleration in rate. The porosity of the calcined product used for SO_2 capture was 0.44. The pellet prepared from powders of red ochre of similar porosity also showed similar sigmoidal behaviour and it was possible to attain quite high utilisation of solid (above 70%) in two hours reaction time.

36.5 Wastewater treatment

Removal of pollutants from oil refinery can be classified in three types of treatments:

Primary treatment	Gravity separation
Intermediate treatment	Neutralisation, chemical coagulation and sedimentation, dissolved air floatation.
Final treatment	Biological, physical, chemical Activated sludge aeration lagoon, trickling filter, cooling tower oxidation.

The classification can also be made on the type of treatment that is : Physical, Chemical and Biological.

36.5.1 Physical treatment

Wastewater may contain coarse, suspended and floating solids, grease, etc. These need to be removed before wastewater is subjected either to chemical or biological treatment. Common unit operations of physical treatments are bar screens, grinders, grit chambers, grease traps, preparation, flocculation, sedimentation, floatation, chemical precipitation, sludge pumping.

This treatment basically removes inert material which may hinder the subsequent treatments.

Bar screens: The purpose of bar screen is to remove large floating matter as well as to avoid clogging of filter media and obstructing aeration in activated sludge. These can be hand cleaned or mechanically cleaned. For the hand cleaned screen the velocity of approach should be limited to about 0.456 m/s.

Grit chambers: Grit chambers removes grit, i.e., sand, gravel, etc., which have specific gravities greater than the suspended organic matter. Purpose of providing grit is to protect mechanical equipments and avoid deposition in pipelines. Two types of grit chamber generally provided are horizontal flow grit chamber and aerated grit chamber.

Primary sedimentation: Purpose of this process is to reduce settleable suspended solid contents of wastewater. Here the wastewater is left undisturbed, the particles of higher specific gravity than wastewater will settle and those with lower specific gravity will float. About 50–65% removal of suspended solids and 20–40% of BOD_5 (at 20°C) can be achieved in a properly designed sedimentation tank. Common detention period is 90–150 minutes based on average rate of flow. Surface loading rates for untreated wastes are 20–40 m^3 pd/m^2 with a peak flow of 40 m^3 per day.

Oil separation: Oil can be present as free or emulsified form or both, in the wastewater. It is necessary to remove free oil before breaking the oil water emulsion. Attempts are made to float the free oil in suitably designed tank and then it is skimmed by a rotating arm or disc. The rise velocity (v) of the oil globules is governed by Stoke's law.

API separator is generally used for such type of oil separations. These separators are normally operated without using any chemicals or coagulant or aids. Oil and grease separated from these units as stop oil are recovered and reused in the refinery. Recently Titled Plate Separators (TPS) or corrugated plate Separators are used to separate, the free oil from the wastewater. TPS gives better performance over gravity separators as it reduces the rising distance of the oil droplet before reaching the surface where it coalesce to form oil layer which rises to top from where it is skimmed off.

The gravity separators are essentially rectangular chambers equipped with oil skimming and sludge scrapping device and design of the separators is based on removal of oil globules of 0.015 cm in diameter and above. The efficiency of oil removal is around 98%.

An emulsion is defined as a mixture of two immiscible liquids, one of the liquids being dispersed throughout the other in the shape of very fine droplets. Both the oil-in-water and water-in-oil emulsions are present in a refinery wastewater. The de-emulsifying agents which are normally adsorbed on the surface of the emulsified particles rendering the emulsion stable are soaps, sulphates, sulphonic and naphthenic acids, quartenary ammonium compounds, organic ethers and esters.

To remove emulsified oil dissolved air floatation (DAF) technique is used. This technique can be used alone or in combination with flocculation. This combination can remove 97% of the oil, 75% of the suspended solids and reduce both biochemical oxygen demand (BOD) and chemical oxygen demand (COD) by 80%. This reduction varies with the characteristics of the waste stream.

The basic mechanism by which an oil-in-water emulsion is broken involves neutralisation of the charges carried by the oil droplets by forming a floc that is initially precipitated with a charge opposite of that on the oil. Adsorption of the oil by a flocs which initially has a high adsorptive capacity for the oil and entrapment of the oil as the floc forms and grown around the oil droplets. The above objective can be achieved by adjusting the pH of the wastewater between 5 and 6, subsequently precipitating the emulsion stabilising agents like sulphonic and naphthenic acids as water insoluble calcium salts by the addition of hydrated lime until a pH of 7.5 to 8.5 reached.

Cationic polyelectrolytes are available for de-emulsification of oil. This has remarkable advantage over the conventional methods. Also polypropylene adsorbent can be utilised for de-emulsification. This adsorbent is in a continuous belt form which is oleophillic (attracts oil) and also hydrophobic (water repellent). This when passed in the wastewater adsorbs oil on its surface.

Various coagulating agents like ferrous sulphate, ferric sulphate (chlorinated copper as) calcium chloride, calcium carbonate and hydrated lime have been used to de-emulsify an oil separator effluent. One particular type of coagulant may be effective for the wastewater of a particular refinery but may, however be ineffective for the wastewater of another refinery. The type and behaviour of the emulsifying agents would dictate the application of a particular coagulant best suited for the purpose. The flocs that are formed by such treatment adsorbs most of the oil and depending on the initial oil concentration, an effluent may be produced with an oil content of 10 mg/L or even less.

36.5.2 Chemical methods

After removal of grit and floating matter, suspended and dissolved organic matter are removed. Important unit operations and processes involved in the chemical treatment of wastewater are discussed below:

Chemical coagulation, flocculation and sedimentation: Dissolved air floatation, both with and without flocculation, has been successfully used to obtain a low oil content in a refinery wastewater. Under favourable conditions it is possible to obtain an oil content of as low as 10 mg/L from a flotation cell using a coagulant and coagulant aid.

However, dissolved air flotation units require more skilled and regular supervision and power than gravity separators, hence their installations have been restricted to a few refineries.

Aluminium sulphate, lime, ferrous sulphate when used in the way they are used as in normal water clarification, on wastewaters, result in producing better effluent than plain sedimentation. With favourable conditions suspended matter can be reduced by about 90%, if the plant is operated skillfully. Interest in chemical treatment of wastewaters has been revived because there is:

1. Decrease in cost of chemicals.
2. Better understanding of mechanisms involved in coagulation.
3. Improvement in sludge handling processes.

Setting velocities of finely divided and colloidal particles are very low and hence require very long detention period. Also special efforts or mechanisms have to be provided so that these small particles aggregate together to settle under gravity at accelerated rates. Aggregation or building up of flocs is induced by addition of chemical coagulants to decrease the zeta potential which stabilises the colloids. To encourage collisions between destabilised particles, agitation is necessary. Also coagulants have to be used between optimum pH range to achieve optimum efficiencies.

Sulphide precipitation: Sodium hydroxide, when used to scrub cracked hydrocarbons for the removal of sulphur and phenol compounds, forms sodium sulphide and other products as per the following equation:

$$2\ NaOH + H_2S \rightarrow Na_2S + 2\ H_2O$$

$$NaOH + C_6H_5OH\ (\text{Phenol}) \rightarrow C_6H_5PNa\ (\text{Na-phenolate}) \rightarrow H_2O$$

$$NaOH + C_6H_4\ (CH_3)OH\ (\text{Cresol}) \rightarrow C_6H_4\ (CH_3)\ ONa\ (\text{Na-Cresolate}) + H_2O$$

If the spent caustic is derived from caustic scrubbing cracked gasoline which has been debutanised, then it would contain very little sodium sulphide and the sulphur compounds will be mostly mercaptans and thiophenols. In some refineries which discharge significant amount of spent caustic liquor containing very high sulphide concentration, chemical precipitation of sulphides by iron salts has been found to be effective and more economical than by neutralisation followed by steam or air stripping. Sulphide precipitation is particularly to wastewater containing sulphides and mercaptans as the latter is also precipitated as iron sulphides. By using chlorine with ferrous sulphate, chlorinated copperas is formed which becomes more effective in the precipitation of the sodium sulphide as per the following equation:

$$3FeSO_4.\ 7H_2O + 1.5\ Cl_2 \rightarrow Fe_2\ (SO_4)_3 + FeCl_3\ 21H_2O$$

$$Fe_2(SO_4)_3 + 3\ Na_2S \rightarrow Fe_2S_3 + 3Na_2SO_4$$

$$Fe_2S_3\ \text{hydrolysis} \rightarrow 2FeS + S$$

$$FeCl_3 + 3\ NaOH \rightarrow Fe\ (OH)_3 + 3NaCl$$

The above reactions are carried out optimally in alkaline pH range.

Filtration: This removes finely divided suspended material. This may or may not be preceded by chemical coagulation.

Air stripping: For removal of ammonia and hydrogen sulphide.

Ion-exchange: For the removal of phosphates, nitrogen and total dissolved solids.

Reverse osmosis: For removal of organic and inorganic substances.

Carbon adsorption: For reducing organic matter.

36.5.3 Biological methods

Biological treatment unit is primary meant for removal of pollutants like phenol, residual sulphide and BOD and also the non-recoverable oil present in the secondary effluent. Bacterial seeding and fertilisation of the oily wastewith appropriate bacterial species will accelerate biological degradation provided that dissolved oxygen and sufficient time are available. Approximately 66% of the hydrocarbon oxidising species are Pseudomonas species. Next in order of abundance are species of Mycobacteria, Proactinomyces, Zctinomyces, Yeast and molds.

The BOD of mineral oil is relatively high as about 3–4 mg of oxygen are required for the oxidation of 1 mg of various kinds of hydrocarbons. Depending upon their chemical composition, the oxygen demand of various kinds of oils ranges between 3.1 to 3.5 mg/mg of oil. These values may be compared with oxygen demand values of 1.07 for glucose, 1.18 for starch and cellulose, 1.5–1.8 for proteins and 2.5 to 2.9 for vegetable oils. It is found that the bacteria in biofilters and activated sludge to be upto 84% effective in removing oil from refinery waste. Grease and oils are particularly resistant to anaerobic digestion and when present in sludge they cause excessive scum accumulation in digesters, clog the pores of filters and deter the use of filters and deter the use of sludge fertilisers. When these substances are discharged in wastewaters or treated effluent they often cause surface films and shoreline deposits.

A knowledge of the quantity of grease and oil present in a waste is helpful in overcoming difficulties in plant operation, in determining plant efficiencies and in controlling the subsequent discharge of these materials to receiving stream. A knowledge of the amount of grease present in sludge can aid in the diagnosis of digestion and de-watering problems and indicate the suitability of a particular sludge for use as fertiliser. Depending on the type of oxidising agent used, the biological methods which employ micro-organisms, is grouped under aerobic (when free molecular oxygen is present) or anaerobic (absence of free oxygen) methods of treatment.

The rate of decomposition by microbes is importantly a function of the oil water interface. The greater the oil water interface the faster the microbial decomposition rate. It is reported that microbial disposal of oil appeared to be feasible with specifically adopted cultures (about 51 to 61% of crude oil was utilised by a species of Pseudomonas over a period of 21 days). Now-a-days a concept of engineering the species to utilise a particular waste has been developed. Also mixed cultures can be used to degrade a particular types of waste from oil refinery.

36.5.4 Aerobic method of treatment of industrial wastes

The methods of biological treatment used in industrial waste treatment are essentially the same as the those used in domestic sewage treatment.

The common methods used are:

1. Trickling filters.
2. Activated sludge.
3. Oxidation ditch.
4. Aerated lagoons.
5. Oxidation ponds.

Trickling filters

A trickling filter is a device for bringing wastewater into contact with biological growths. These filters decrease the oil and phenol contents of the refinery waste.

Activated sludge process

This technique is used for the large scale processes. The advantages of activated sludge plant over trickling filters includes:

1. Small area requirement.
2. Freedom from odour and fly nuisance.
3. Small head requirement.
4. Lower cost of construction.

Oxidation ponds

This is a simple method for the microbiological disposal of oily wastes. In 30 days retention period the sulphides decreased from 15 to 0 mg/L it and phenols dropped from 20 to 7 mg/L it and reduction of oil was from 150 to 0 mg/Lit. The retention period is particularly important for the effectiveness of oxidation ponds.

Aerated lagoons

It is generally an earthern basin having a depth of 2.4 to 3.0 metres of wastewater which is aerated either by diffused air or mechanical surface aerators. It acts as a settling cum aeration basin.

36.5.5 Anaerobic lagoons

Anaerobic lagoons for the treatment of wastes are designed on the principle of anaerobic digestion with the exception of not providing gas collection devices and are operated as open digesters. The main advantages of anaerobic lagoons are:

1. High organic loading rates can be applied on relatively small land area because of larger depth.
2. Low capital cost because only earthwork is needed.
3. Low running costs because of nutrient requirement is much less compared to aerobic treatment.
4. The problems of excess sludge disposal is also minimum.

36.6 Wet air oxidation of waste streams

Wet air oxidation is a liquid phase oxidation of waste at high temperature and pressure. Wet air oxidation offers an opportunity to destruct toxic pollutants in wastewater from chemical, petrochemical, refinery dyestuff and other industries. This water then can be further treated conventionally and recycled back into the process. The capital cost of wet air oxidation system depends on several factors such as capacity of the system, severity of the oxidation conditions to meet the treatment objectives and presence or absence of corrosive constituents in the waste. Hence, judicious choice needs to be made while selecting wet air oxidation in the overall water recycling scheme.

The search for new sources of water is never ending, particularly in developing countries, not only due to increasing population but also due to increasing demand of the industry. Because of the steepness of the demand curves and the tendency of the steepness to increase, engineers everywhere have sought ways to recycle the water as much as possible. Thus, the need to improve the wastewater treatment and enhance the removal of priority pollutants has focused attention on the application of new processes like wet air oxidation.

Once the toxic pollutants from the chemical and petrochemical industry wastewater are oxidised and made amenable to biological treatment, normal water recycling treatments such as chlorination/dechlorination, clariflocculation and/or membrane processes can be used, depending on the effluent characteristics.

Wet air oxidation involves oxidation of organics and some inorganics in the aqueous phase at high temperatures and pressures. The oxidation takes place in the temperature range of 100 to 350°C and pressures ranging from 10 to 300 atm. The dissolved oxygen in the water reacts with the oxidisable substance at these high temperatures, to produce a wide range of oxygenated products and generating heat. The water moderates the oxidation reaction by acting as a diluent and by supplying oxidant at low concentration. Higher pressure is required to keep the water in liquid phase, at the high operating temperatures. Few other names for this process are flameless combustion and oxidation, liquid phase oxidation and oxidative detoxification.

36.6.1 Process description

In this process, the wastewater which contains the oxidisable constituents is brought up to the system pressure using a high pressure pump. Compressed air or oxygen gas is introduced into the pressurised wastewater stream using a compressor. The mixture of air and wastewater is heated in process heat exchanger by heat exchange with the oxidised hot effluent leaving the reactor. An external source of heat is used to initiate the wet air oxidation process or to sustain the oxidation temperature if insufficient heat of reaction is released in the wet air oxidation reaction.

After heating, the mixture of gas and wastewater flows into the reactor, where it is detained for a period of time, sufficient to complete the desired degree of oxidation. The reactor is a vertical bubble column pressure vessel, which is sized to provide the desired hydraulic residence time. Wet air oxidation reactions are exothermic and raise the temperature.

The hot oxidised effluent that flows from the reactor is used to recover heat by pre-heating the incoming mixture. An optional water cooler may be used to further cool the oxidised effluent.

After cooling, the oxidised effluent passes through a Pressure Control Valve (PCV) and is directed into an atmospheric separator where the non-condensable gases separate from the oxidised liquid phase.

Advantages of WAO process

Energy efficiency: Wet air oxidation does 'not require auxiliary fuel, i.e., it proceeds autothermally, if the COD levels are above 15000 mg/L. This is because, in wet air oxidation, the only energy required is the difference between the enthalpies of incoming and outgoing streams. While in incineration, the sensible heat, heat of vapourisation of liquid, as well as heat required to heat the water vapours, combustion products and excess air to combustion products and excess air to combustion temperatures in the range of 800–1000°C need to be provided. Thus autogenous incineration requires COD levels as high as 300,000 to 400,000 mg/L.

Toxic wastes: Wet air oxidation easily detoxifies wastewaters which are too toxic for biotreatment, e.g., stream containing cyanides, phenols, etc., or having BOD/COD ratio less than 0.5. It can handle shock loads unlike biological processes.

No air pollution: Wet air oxidation does not cause any air pollution, as the gaseous products contain mainly spent air and carbon dioxide. In wet air oxidation, sulphur or nitrogen oxides are not formed.

Non-toxic products: The wet air oxidation effluent is readily biotreatable as it converts non-biodegradable, toxic. High molecular weight oxygenated compounds, predominantly carboxylic acids which are non-toxic and easily biodegradable in subsequent steps. With progressive oxidation, the BOD/COD ratio increases, indicating the reduction of toxicity to biotreatment.

Applications of wet air oxidation

Thermal sludge conditioning: Sewage sludge disposal poses a unique and difficult problem because of the high proportion of waste activated sludge from the secondary treatment of industrial as well as municipal wastewater. Sewage sludge is a complex mixture of waste solids forming a gelatinous mass which is nearly impossible to dewater without further treatment. The organic fraction of the sludge, largely of biological origin, consists of lipids, proteins and carbohydrates, all bound by physicochemical forces in predominantly water gel-like structure. The 'bound water' content of sewage sludge is 90–98%.

When sludge is heated under pressure at temperatures of 175–200°C in a wet air oxidation system, the gel like structure of the sludge is destroyed, liberating the bound water. Dewatering by filtration, centrifugation or drainage beds without chemical conditioning is then possible. The residue is sterile and suitable for land-fill or incineration. The mixing of air with sludge in WAO system provides oxidation compounds, thereby giving odour reduction. Thermally conditioned sludge cake possesses a heat value equivalent to low grade fuel - about 6600–8325 kcal/kg of volatile solids - and can be combusted without auxiliary fuel after start-up.

Activated carbon regeneration: Generally, the treated effluent from the primary or secondary treatment plant can not be recycled directly, due to the presence of organic contaminants. This water can be further purified by addition of chemicals or polyelectrolyte and powdered activated carbon is added. Chemical treatment followed by settling provides simple and effective precipitation of phosphorus and removal of suspended and colloidal matter. The resulting dissolving organic material is adsorbed onto powdered activated carbon resulting in high quality water which can be recycled.

The use of powdered carbon has been limited in the past due to difficulties in economically and efficiently regenerating the spent carbon. Wet air oxidation can be used here to regenerate the carbon, since only the adsorbed organics get selectively oxidised without destroying the carbon. In this process, the spent carbon is withdrawn from the wastewater contact system and concentrated by gravity thickening. Thickened carbon slurry is pressurised to system pressure, mixed with compressed air and heated to reaction temperature before introducing in WAO reactor where selective oxidation of organics takes place. Typically this regeneration is carried out at 200–245°C.

Materials of construction: Depending on the effluent nature, normal materials of construction like carbon steel, stainless steel, etc., can be used. However, if the effluent contains corrosion inducing constituents like chlorides in high concentration, then costlier materials of construction like titanium, alloy 20 are used. Thus the nature of inorganic constituents and the severity of oxidation conditions have a significant bearing on the capital cost involved. It is assumed that the wet oxidation unit would operate at maximum conditions of 280°C and 135 atm and would accomplish a 50 g/L COD reduction.

Operating cost: The major sources of energy consumption in wet air oxidation system are electrical energy consumption for air compression and high-pressure liquid pumping and pre-heating of feed wastewater. The energy required for pre-heating depends on the flowscheme chosen and the amount of oxidation occurring. The electrical power for air compression corresponds to about 0.136–0.167 kW per kg of oxygen consumed per second. Wet air oxidation systems consume approximately the amount of oxygen equivalent to the oxygen demand reduction for the treated wastewater. The electrical power for high pressure liquid pumping is generally about 10% of that for the air compression.

The energy required for pre-heating is inversely proportional to the amount of oxidation occurring. The heat required for pre-heating is less if more heat is released in oxidation. No external energy is required if the oxidation occurring corresponds to 15–20 g/L of oxygen demand reduction. If the oxidation corresponds to less than 15–20 g/L oxygen demand reduction, auxiliary energy addition would be necessary.

When a process heat exchanger is used, maximum auxiliary energy consumption is in the 139–167 MJ/m^3 range. When a process heat exchanger is not used, i.e., all pre-heating is provided by external means, the energy requirement for pre-heating can range from nil for highly concentrated wastes to about 1115 MJ/m^3 for very dilute wastewaters.

The operating costs are made on the basis that wet air oxidation unit would operate at maximum operating conditions of 280°C and 135 atm and would accomplish a 50 g/L COD reduction.

In conclusion, wet air oxidation offers a bright possibility of treating toxic/primary pollutants, to make the effluent amenable for recycling. However, the overall cost is significantly affected by the presence of corrosive constituents and hence judicious choice will have to be made to see the overall impact on the project cost involved in recycling.

36.7 Refinery wastewater treatment using heavy crudes as feedstock

Refinery profitability heavily depends upon the cost of the crudes processed and therefore, processing low cost crudes is a significant consideration. Crudes extracted from tar sands by Steam Assisted Gravity Drain (SAGD) or crudes extracted by steam flooding are extra heavy sour crudes and are relatively cheaper than the light low sulphur crudes. This section identifies wastewater process streams that are required to be treated in refineries and also identifies the possible sources from where they may be originating and suggests point source segregation for enhancing the overall collection and wastewater process efficiency and effectiveness. The section also highlights an innovative and cost effective process, developed and patented by Fluor. for the treatment of desalter mud wash waste generated in crude units.

The wastewater treatment of refinery waste streams such as oily, non oily, high Total Dissolved Solids (TDS) from ion exchange regenerate systems, spent caustic, ammonia and other streams typically consists of:

1. Selectively segregating the streams.
2. Pre-treating.
3. Combining those streams in a large equalisation tank.

Streams containing benzene and aromatics are segregated from their point source and after suitable intermediate treatment can also be combined in the equalisation tank. Processes downstream of the equalisation tank may include cooling and pH adjustment prior to proceeding to biological aerobic activated sludge processes to reduce biological oxygen demand and chemical oxygen demand levels. Additional processes include suspended solids reduction and also ammonia, hydrogen sulphide and benzene stripping to comply with outfall dis-charge, and/or to meet internal recycle to wash water makeup/cooling tower make up limits. In petroleum refineries, wastewater is generated by the refining process itself and also by the support systems (Fig. 36.2). The refinery wastewater list provided below is meant to be typical.

Utility and off sites wastewater:

1. Cooling tower blow down.
2. Boiler blow down.
3. Once through service water.
4. Pump seal flush water.
5. Water treatment ion exchange regeneration waste streams.
6. Filter backwash water.
7. Contaminated storm water.
8. Potentially contaminated storm water.

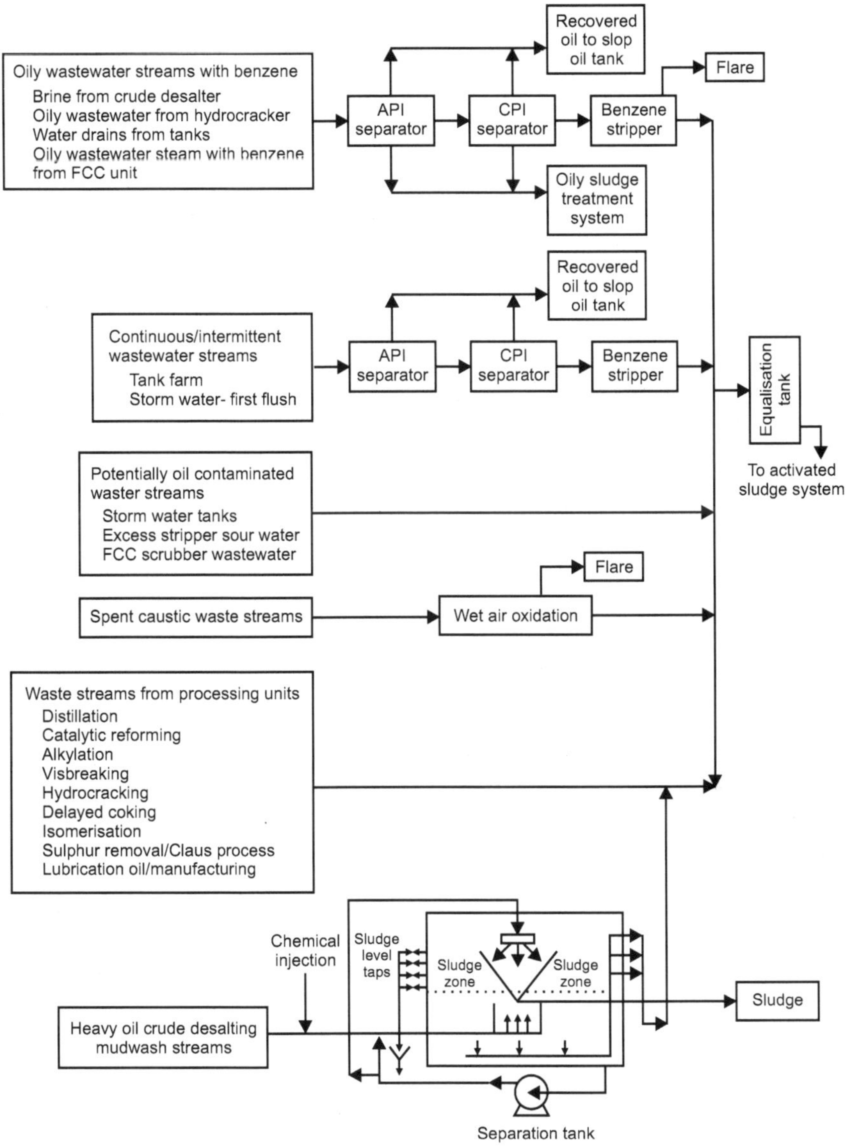

Figure 36.2: Wastewater stream collection and pre-treatment.

Process wastewater from:

1. Crude and vacuum units (stripping steam, desalting water).
2. Distillation (striping steam).
3. Catalytic reforming.

4. Fluid catalytic cracking (fluidisation, stripping and purge steam, process wash water).
5. Alkylation.
6. Thermal cracking/visbreaking.
7. Catalytic hydrocracking/hydrotreating.
8. Delayed coking.
9. Isomerisation.
10. Sulphur removal/Claus process.
11. Lubricating oil manufacturing.
12. Sweetening/Merox process.

36.7.1 Wastewater minimisation, classification, collection and treatment

It is prudent to identify wastewater streams which can be minimised by internal recycling at the point source with minimal treatment or without significant treatment. Waste streams normally are classified on the basis of their contents and attributes, such as, free and floating oil content, temperature, flow rate variations, benzene content, spent caustic content, dissolved solids content, ammonia, cyanides, hydrogen sulphide and other contents are indicated in the Fig. 36.2.

36.7.2 Refinery wastewater collection and preliminary pre-treatment

All streams containing free and/or floating oil are combined together and are pre-treated to remove oil before entering and mixed with other streams in a common equalisation tank. Streams containing hazardous material such as benzene, hydrogen sulphide and spent caustic are segregated at the point sources and pre-treated before being mixed with other streams. They are usually mixed in an equalisation tank for pH and temperature control, prior to undergoing advanced biological activated sludge treatment with subsequent discharge to an approved out-fall.

Waste streams with low dissolved solids content and without any contamination can be separately collected, treated and recycled back as a make-up water for internal plant use such as for heat exchangers and cooling towers, etc. In India, as elsewhere, refineries which process opportunity crude oils extracted by steam flooding will produce excessively high chemical oxygen demand (COD) wastewater streams from crude desalter units during the mud wash step. The COD from the steam flooded crude desalters could be in the range of 20,000 to 26,000 ppm. These high level COD wastewater streams will exert a heavy dissolved oxygen demand shock load on the downstream activated sludge reactor and will kill micro-organisms in the mixed liquor/ suspended solids stream section. These shock loads may be mitigated by spending significant capital in expanding the refinery's wastewater plant aeration system. Typical acceptable biological oxygen demand and COD levels are in the range of 500 ppm and 1000 ppm respectively. Aeration systems are designed to operate within these ranges. Fluor has developed an innovative cost effective process, as shown in the bottom stream in Fig. 36.2, to prevent such shock loads and maintain the performance integrity of existing wastewater treatment systems without spending significant capital. The process includes:

1. Segregating the desalter mud wash stream.
2. Chemically precipitating and removing up to 90% to 98%of the dissolved asphaltenes.
3. Reducing the Total Organic Carbon (TOC) and free and floating oil.

All of the above operations are achieved in a totally enclosed tank. High BOD/CODlevel wastewater stream from the desalter mud wash step is transferred into this tank at a controlled rate into which measured dosages of coagulant plus flocculent chemicals are injected. Selection and feed rates of the chemicals are established by laboratory jar tests, in order to produce low density floating precipitate in minimum time. The clean and clear water is separated from the precipitate (sludge), collected at the bottom of the tank and transferred to the equalisation tank. In this processing unit, approximately 20% to 25% of treated water is recycled back into the incoming mixture (of raw water and chemicals) to improve the solids contact effect of the stream (acting as catalyst) for a swift chemical reaction in the center well of the separation tank. Maintaining the correct rise rate in the center well, as observed in the jar test, is key factor in large size floc formation and its separation from the influent at the operating liquid level at the funnel bottom mounted inside the tank. If required, inert gas can be injected in the influent water which may help with the precipitate flotation and its separation in the center column.

Adjustable operating liquid water level inside the separation tank is maintained by an inverted seal type arrangement to trap the treated effluent with modulating valve arrangement (outside the tank) at the exit pipe. Low density precipitate will start accumulating at the operating liquid level inside the tank where the water will have to turn down to exit the tank.

Water settling velocity is controlled by tank cross-section to maximise the precipitate separation without having any pin floc carried downward with the treated exiting water. The sludge level is monitored by an instrument mounted at the tank top (not shown in the sketch), at a preset level, the effluent valve is automatically energised to slowly start closing. The water level in tank will start rising, which in turn will push the accumulated sludge upward in to the fixed funnel shaped container and be flushed off to an outside tank for disposal. Spray nozzles at the top of funnel, which are also actuated automatically for a specified interval, will ensure the funnel and drain pipe are completely empty and available for the next round of sludge disposal. When all the sludge have been transferred, the level sensing device will automatically open the effluent valve, thereby lowering the water level inside the tank back to the normal level and the sludge disposal cycle will repeat itself on a batch mode. This system is robust with high availability, automatic operation with essential manual over rides, with no moving parts inside the tank and the system requires minimum operator attention.

Further processing in an equalisation tank

Subsequent to the refinery wastewater streams' collection and preliminary pre-treatment as shown in process block flow diagrams Figs 36.2 and 36.3, the individual streams are collected and comingled in an equalisation tank. Adequate low shear mechanical mixing or inline educators are used to blend in the contents of all the incoming streams. Adequate retention times and internal baffles are necessary to prevent short circuiting of the hydraulic and organic load surges to the downstream processing units, especially the bio-reactor. Equalisation tank(s) can be operated on variable or fixed liquid levels depending on the flow rates from the processing units, (dry weather flow rate) versus wet (first flush + potentially contaminated) rain water flow rates which can vary substantially, depending upon the geographic location of the plant site. Equalisation tanks are also equipped with floating telescopic oil skimmers to remove free and floating oil which is subsequently transferred to the slop oil tank. As indicated in Fig. 36.3, flow from the equalisation tank is transferred at a controlled rate to the downstream Induced Gas Flotation (IGF) unit and or Dissolved Gas/Air Flotation (DGF/DAF) units to further remove free and floating oil to less than 5 ppm. These units are completely enclosed to prevent air emissions from the entrained volatile organic compounds in the waste stream. The equalised and oil free stream is cooled down to approximately 35°C for optimum micro-organism growth in the bioreactor process within the activated sludge system, wherein a mix regime of mixed liquor/suspended solids (MLSS) is maintained. Dissolved oxygen from atmospheric air is obtained by injecting through either coarse or fine bubble diffusers to promote and sustain micro-organism growth to stabilise the organic matter. A minimum concentration of 2.0 ppm of oxygen is required throughout the aeration basin to sustain an adequate food to mass ratio and adequate mixed liquor to suspended solids (MLSS) concentration for an extended period of time.

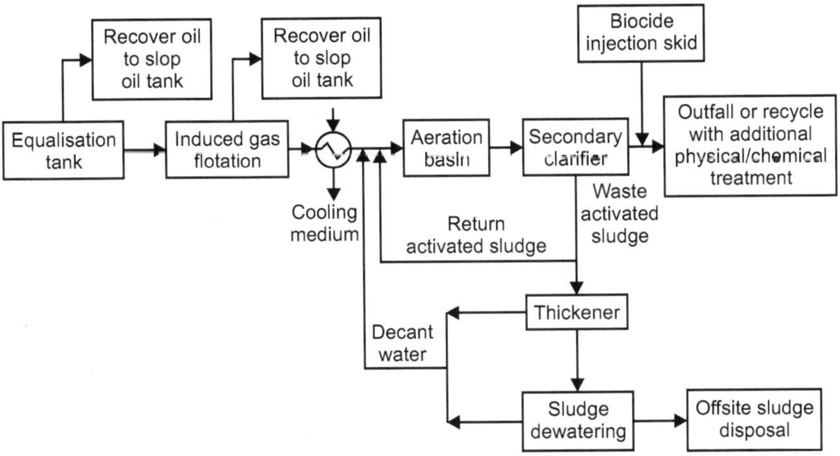

Figure 36.3: Activated sludge system.

Clarification of the treated mixture

The mixture of treated effluent and suspended solids from bioreactor flows by gravity to the secondary clarifier where MLSS settles to the bottom of the tank. The settled solids are mechanically scraped to the center of clarifier basin where clear water (containing less than 15 to 20 ppm of suspended solids) flows over the top of the clarifier to the effluent storage tank and chlorine contact chamber before discharge to the out fall. To maintain proper concentration of MLSS in the bioreactor, a certain amount of settled sludge from the clarifier bottom is returned back to the bio-reactor (called RAS) and remaining excess sludge (called WAS) is pumped to the sludge thickener. There can be several modifications to the conventional activated sludge process, such as Membrane Bio Reactor (MBR), Moving Bed Bio-Reactor (MBBR) and Integrated Fixed Film Activated Sludge (IFAS) each with their own specific pros and cons.

The regulatory mandated constraints on denitrification, low limits on heavy metals and total toxicity discharge, make it necessary to implement these modified processes either as an add on to existing plants or replace the conventional activated sludge process on new installations. More than 95% of the refineries and chemical plants use the tried and tested conservatively designed activated sludge process, some of these even take feed stock from unspecified sources producing further complex waste streams. Regardless of the bio-treatment used, a large amount of sludge is generated and recycled internally to provide sufficient Mean Cell Residence Time (MCRT), to cultivate the specific types of microbes which promote the degradation of ammonia, phenols and phosphates, etc. Excess sludge volume can be further reduced by using either aerobic or anaerobic digestors or sent directly to thickeners. Clear water from the thickener is decanted back to the aeration and or secondary clarifier and the thickened sludge is dewatered by centrifuges, belt filter presses or plate and frame type equipment to the desired concentrations for disposal to an approved offsite land fill facility.

Biochemically treated wastewater as indicated above is suitable for out fall discharge; however, if this water is required to be recycled either as a makeup water for cooling towers and or Boiler Feed Water (BFW), it will require additional treatment. Cooling tower make up streams will require removal of suspended solids, dissolved organics, reactive silica and this can be achieved by cold lime softening followed by passing the stream through multimedia sand or walnut shell media filters.

Free and floating oil is not expected in this stream; however, should there be oil in the stream its removal becomes necessary and this needs to be accomplished on a priority basis in the front end of the treatment plant. If this water is to be used for steam generation, its treatment will depend upon the required quantity and quality of steam. In oil fields, steam flooding to Enhance Oil Recovery (EOR), 70 to 80% quality stream is frequently generated using Once Through Steam Generators (OTSG). These boilers require elevated pH make up water, from which only scale prone hardness divalent and higher valence cations, need to be removed in lieu of all dissolved solids.

At OTSG high pressure, high temperature and elevated pH operating conditions, silica solubility increases substantially and remains soluble in the boiler water without laying any scale or deposits on the boiler tubes.

In most cases, 80% quality steam along with 20% high temperature boiler water is injected into the oil bearing formation. In some formations however, 100% quality steam is injected and this can be achieved by separating 20% hot boiler water from the gross 80% quality steam produced in the OTSG. High temperature boiler water which has been separated to produce 100%steam can be diverted to falling film evaporator/brine concentrator and crystalliser zero liquid discharge system yielding additional 95 to 98%of high quality condensate.

Much like in any modern refinery and other waste treatment operations, it is imperative that all contributory waste streams be properly segregated and identified of their contents from the originating point source. Suitable instrumentation and sufficient capacity storage tanks to intercept and divert off-specification streams are absolutely essential to prevent and mitigate downstream shock loads to the activated sludge process. Refineries with feed stocks procured from different sources, especially where steam flooding is used to extract heavy crude, will produce uniquely challenging waste streams with very high organic loads either coming from reservoir or the desalting step. As a consequence, there is a need for implementing the latest available state of the art technologies as described above to consistently and effectively treat and produce environmentally compatible effluents at the outfalls. The quality of treated water from these processes can be enhanced for recycle to reduce raw water use and inventory.

37

Pollution control in petrochemical industry

37.1 Introduction

Petrochemicals are chemicals made from petroleum (crude oil) and natural gas. Petrochemicals, also called petroleum distillates, are chemical products derived from petroleum. Some chemical compounds made from petroleum are also obtained from other fossil fuels, such as coal or natural gas or renewable sources such as corn or sugar cane.

Primary petrochemicals: Primary petrochemicals include: olefins (ethylene, propylene and butadiene) aromatics (benzene, toluene and xylenes) and methanol.

Petrochemical intermediates and derivatives: Some typical petrochemical intermediates are: (i) vinyl acetate for paint, paper and textile coatings vinyl chloride, (ii) polyvinyl chloride (PVC), (iii) resin manufacture and (iv) ethylene glycol for polyester - textile fibres.

Source of environmental pollution from the petrochemical industry are: (i) air emissions, (ii) combustion emissions, (iii) process emissions, (iv) fugitive emissions, (v) from storage and handling of petroleum liquids and (vi) secondary emissions: (a) combustion emissions are produced with onsite burning of fuels for energy production and transportation purposes, (b) fugitive emissions are released through leaking valves, pumps and other process devices, (c) process emissions are generated in the process units and released from process vents (d) storage and handling emissions are contributed from the storage and manipulation of natural gas and crude oil, as well as their intermediate and finished derivatives and (e) the water system of a production or processing site (tanks, ponds) are the main source of secondary emissions.

The principal pollutants from these air emission sources are: (i) nitrogen oxides, (ii) sulphur oxides, (iii) carbon monoxide, CO_2 and particulate, (iv) additional pollutants: (a) hydrogen sulphide, (b) volatile organic compounds, (c) methane and ethane, (d) benzene, (e) toluene, xylenes, etc.

37.2 Pollution control in petrochemical manufacture

37.2.1 Water pollution

Huge volume of water is required for various purposes in manufacture of petrochemicals. Water is used for cooling, process operations, fire fighting, drinking, housekeeping, steam raising, etc. Whole water after use reappears as effluents which requires treatment and disposal. Characteristics of effluents from various petrochemical plants are shown in Table 37.1.

Table 37.1: Characteristics of effluents from various petrochemical plants.

Plant	*Water pollutants*	*Mode of treatment*
DMT	Total dissolved solids, pH	Segregation and neutralisation, chlorine recovery
Aromatics	pH, BOD, COD, suspended solids	Neutralisation followed by biological treatment
Olefins	pH, COD, BOD, sulphides phenol, hydrocarbon, oil and spent caustic and cyanides	Oil separation, segregation of fluoride effluent, biological treatment
LAB	pH, COD, BOD, F, oil surfactants, sulphides	Oil separation segregation of fluoride effluent, biological treatment
ACN	pH, COD, BOD, cyanides, sulphates, ammonia	Segregation of cyanide effluents, biological treatment
EG	pH, COD, BOD	Biological treatment
LDPE, PP, PBR	pH, polymers, COD, surfactants, BOD	Segregation of polymer biological treatment
AF	pH	Segregation and neutralisation
ACR	pH, COD, BOD	Biological treatment
VC/PVC	pH, suspended soilds	Neutralisation coagulation

Segregation and treatment of wastewater

Following effluents are segregated and treated separately in the first stage and later on are mixed with process waste water, sanitary waste for further combined treatment: (i) acidic waste, (ii) spent caustic, (iii) fluoride, (iv) cyanides and (v) polymers.

Sulphide containing effluents and their treatment

Sulphides are undesirable in the stream because of their oxygen depletion capacity. Sulphides are also not desirable in biological treatment system hence sulphide effluents are separately treated in such a manner so that these effluents do not interfere with biological treatment.

Production of olefins generate effluent containing sulphides, pH, COD, BOD phenols, polymeric oils. Volumes vary 5–50 M^3/day. Table 37.2 shows solid waste disposal scheme for petrochemical plant. These effluents can be reused for sulphur soap manufacture, pulp and paper manufacture and processing of hides in tanneries.

Table 37.2: Solid waste disposal scheme for petrochemical plant.

Plant	*Type of waste*	*Treatment*
ACN	Toxic	Land disposal
	Cyanides,	Incineration
	Acrylonitrite	
	Ferrous sulphate, etc.	
Olefins	Oily waste	Reuse as fuel
LAB, EG	Contains various types of hydrocarbons	
Wastewater effluent treatment plant	Biological sludge from various process units	Anaerobic/Aerobic digestion, subsoil Application as fertiliser.
ACR, DMT	Bottom residues	Reuse as fuel incineration
AF, and	Polymer waste	Land fill up in small pits.
PBR, PP	Self igniting type	

Effluents can be oxidised and converted into thiosulphate (biodegradable), sulphate, (in presence of catalyst, i.e., ironoxide, copper oxide).

Sulphide can be removed by precipitation and addition of $FeCI_3$, ($FeSO_4$, + Cl_2). Solar evaporation is economical method of disposal on impervious-land.

Cyanide treatment: Cyanide is toxic to all forms of life hence it is undesirable in the effluents. Cyanide 0.1 mg/kg of body weight is lethal for animals in the form of hydro cyanic acid. Following methods are generally used for treatment of wastewater containing cyanides.

Hydrolysis and biological treatment: This treatment is most economical and highly reliable. Cyanide waste is hydrolysed at higher temperature under alkaline conditions thus most of the cyanide gets converted into ammonia and subsequently removed from the stream. It is possible to treat wastewater containing about 5000 to 10,000 ppm of cyanides and reduce to less than 1 ppm provided hydrolysis is followed by biological treatment. Biological media should have pseudomonas, fusarium as bacteria apart from other active aerobic bacteria.

Alkaline chlorination: This treatment is a conventional treatment for cyanide waste. Wastewater is made alkaline upto pH 9.0 and chlorination is carried out to oxidise cyanides. Generally excess chlorine is used for chlorination. But chlorine containing effluent are undesirable hence dechlorination is carried out by sodium sulphite. Hence this method is quite expensive.

Oxidation: Catalytic oxidation of cyanide waste is carried out in presence of cupric oxide as catalyst. Thus cyanide is converted into less toxic cyanates.

Oxidation by ozone: Cyanide waste when treated with ozone, cyanides are converted into cyanates which are less toxic. Both these methods are not much practicable.

Precipitation: Cyanides wastes are converted into less toxic ferricyanide compounds. Treatment is carried out by ferric salts. However this treatment offers another problem of solid waste disposal because there is again chance of break down of ferricyanides into cyanides.

Incineration: Most expensive method of cyanide treatment and gives rise to several air pollution problems which are difficult and expensive to handle.

Fluoride treatment

Fluorides are not desirable in water because of deleterous effects on health/ fluorides more than 1.5 ppm in potable water are harmful. Fluorides also interfere with biological activity in the biological treatment if effluent has more than 15 ppm of fluoride hence fluoride wastes are segregated and treated by following methods.

Precipitation: Soluble fluorides are converted into insoluble fluorides by addition of lime. Thus calcium is fluoride precipitate is then centrifuged and converted into cakes. Transported to disposal site and disposed of to about 95 to 99% removal of fluoride is achieved.

Precipitation: Calcium phosphate is also used to convert soluble fluoride into insoluble fluoride. This precipitate can be centrifuged in better form of cake and higher removal of fluoride is envisaged, i.e., 99.5%.

Activated silica: Fluorides are generally adsorbed by certain activated chemical like activated silica. Thus removal of fluoride depends upon type of activated silica and regeneration of silica is also a problem.

Activated alumina: Fluorides are also adsorbed on activated alumina but again regeneration of adsorbent is a problem and quite expensive.

Activated carbon, charcoal, saw dust: Economical methods for deflouridation on the large scale has been a subject of research and development. In this context above materials are used successfully and found to be quite satisfactory and suitable on large scale.

Recovery: Fluoride wastes are converted into aluminium fluoride which has a commercial value.

Common effluent treatment system

Most of the conventional waste treatment methods are modified for petrochemical waste treatment. Unit operations generally followed are:

1. Screen and grit chamber followed by surge tank.
2. Oil separator.
3. Clariflocculator.
4. Aeration tank.

5. Final clarifier.
6. Polishing tank.

Under ground gravity drainage system is used for transport of effluent. It is advisable to use screw pump instead of centrifugal pump for transfer of oily effluent from underground to above ground level so rest of all treatment is carried out by gravity.

Screen and grit chamber: Screen and baffles are provided to regulate the velocity/lft/second so that large settleable particles are removed and does not interfere later on with subsequent process.

Surge tank: Heterogenous waste is allowed to intermix over a period so that maximum homogenous waste flows to oil separator.

Oil separator: Oil floats on the surface of water. Separation is affected by gravity. Skimmers are used to drive oil globules up to slope where separation of oil is carried out. Separation of oil however depends on viscosity, specific gravity, temperature of oil and angle of inclination of slot pipe, presence of other suspended material, pH of the medium. Several demulsifying chemicals are also used to enhance oil separation. In some cases air floatation technique is used so that maximum oil appears on the surface and recovery becomes easier. Gravity separators are widely used and efficiency of oil recovery is appreciable. Foam is also used for recovery of oil. Foam is plunged on the oily surface and kept for a reasonable time and then squeezed to get oil.

Clariflocculator: In this system coagulating chemicals like alum, are added to flocculate small particles so that colloidal impurities are precipitated in the centre and removed subsequently and supernatant from suspended and colloidal impurities flows to aeration tank.

Aeration tank: Aeration tank contains active biomass which are being agitated by powerful aerators. Aeration is carried out for intense mixing (contact) of food (BOD) (waste material) and bacteria as well as transfer of oxygen from the atmosphere. Thus bacteria degrades waste material in presence of oxygen which brings about propogation of bacterial culture along with degradation of organic material (BOD) into carbon dioxide (stable end products) some of stabilised material is removed in the next stage and allowed to dry on sludge drying bed.

Final clarifier: Effluent from aeration tank contains mixed liquor suspended solids which are removed by gravity. Part of suspended solids are sent back to aeration tank, part is transferred to sludge drying bed for drying purpose and subsequent removal as solid wastes/fertiliser.

Polishing tank: Polishing tanks are only used if the effluent is to be recycled for some specific purpose, i.e., cooling towers, fire fighting. Generally very fine impurities/oily substance escaping from final clarifier are treated by filteration and activated carbon so that final effluent is crystal clear. Unit operations mentioned above ensure effluent quality conforming to 1S-2490 and compiled with statutory requirements of all state pollution control boards.

37.2.2 Air pollution control

Industry cannot afford to lose its valuable product by-product in the emissions. Hence, in most of the industries for economical consideration certain pollution control equipments are used to recover valuable products and avoid pollution of the atmosphere. Most of these equipments are part of the process. In aromatics plants wherever *p*-xylene is produced activated carbon bed is invariably used for scrubbing *p*-xylene from emissions. Thus effective recovery of *p*-xylene is carried out.

Fugitive emissions

Fugitive emissions emanate from a variety of non point sources in the process operations primarily in the form of leakage. Process losses or leakage from valves, steams, valve bonnet gasket, pipe flange gaskets, rotating equipment seals, etc. In addition leaks from seals, on internal or external floating tanks roofs, with the diurnal leakage from cone roof tanks vents, waste-water-treating units, lagoons, sludge treatment plants and land disposal areas. Following are typical fugitive emission process sources but do not include threaded fittings valves, flanges, relief valves, pump seals, compressor seals, drains, manual sample connections. Fugitive emission is affected by following variables. Fluid phase, pressure, temperature, unit type, batch/continuous process, fluid toxicity/ corrosiveness, service life, inspection frequency, valve packing/configuration, flange gasket, flange type, flange bolting procedure, pump mechanical seal, pump revolutions per minute, compressor seal type, equipment size.

Point source emissions

Generally these emissions are of continuous nature through stacks wherever fuel is used for heating, raising steam, generation of power.

Oxides of sulphur: Now a days most of the industries use fuel oil for combustion purpose. Fuel oil invariably contains variable amount of sulphur compounds.

Oxides of nitrogen: Oxides of nitrogen are emitted whenever nitrogen containing compounds are burnt. Oxides of nitrogen are also emitted where nitric acid rectification columns are used. From petrochemical industry burning of acrylonitrile, acetonitrile, polyacrylonitrile also gives rise to nitrogen oxides. Apart from this, oxide of nitrogen are also formed if combustion is carried out at higher temperature is the formation of oxides of nitrogen, because air contains

80% nitrogen which is also burnt out at higher temperature. Most of the power plant operations for raising power as well as steam are carried out at high temperature giving rise to substantial emissions of oxides of nitrogen. Power plant operation emits about 80 ppm of oxide of nitrogen which are using fuel oil.

Chlorine: Chlorine is highly reactive substance and is emitted from vinyl chloride polyvinyl chloride plant but generally this is controlled by vent scrubbers.

Particulate: Petrochemical plants are generally free from suspended particulate pollution except in the case of system failure.

Hydrochloric acid: Demineralisation plant in variably used hydrochloric acid which gives rise to HCI emissions which are controlled by blowers and suitable exhaust arrangement.

Hydrocarbons: Fugitive emissions and transfer of petroleum product result in release of hydrocarbons in the atmosphere. Even a trace of hydrocarbon reacts with oxides of nitrogen resulting in formation of secondary pollutants like (PAN and other eye irritating compounds).

Carbon monoxide: Carbon monoxide emissions take place whenever combustion process is not smoothly progressing. This happens only when plant operation is upset. Temperature in the petrochemical industries are always higher because of various heating operations. Inversion is observed in winter. Visibility reduction also takes place in winter in petrochemical complex and surrounding area.

37.2.3 Solid waste pollution in petrochemical industries

Solid wastes differ from other pollutants. Unlike air or water pollutants, solid wastes are yet to have seen any unpredicted growth of public concern. However, more and more concern on solid wastes too is anticipated in the coming years. Solid wastes differ from air and water pollutants in another sense too, viz., the solid wastes remain at their origin points till such a time a decision is taken by those concerned to collect and dispose them off.

Petroleum refining and petrochemical industries are feeding the blood into the veins of today's industrialisation. Considerable expenditure has been made in the area of treatment of air and water pollutants. But, solid waste treatment is yet to catch up to the required level.

Solid wastes in petrochemical industries can be in different forms: (i) toxic solids, (ii) non-toxic solids, (iii) actual solids, (iv) semi solids, (v) combustible solids, (vi) non-combustible solids, (vii) degradable solids, (viii) non-degradable solids, (ix) bio-degradable solids, (x) organic solids, (xi) inert solids, (xii) dry solids, (xiii) aqueous solids, (xiv) non-aqueous solids, etc.

37.2.4 Collection and disposal of wastes

The means of collection and methods of disposal of these differ from one to the other. Few are listed below:

1. Recycling.
2. Chemical conversion.
3. Incineration.
4. Pyrolysis.
5. Landfill.
6. Ocean dumping.
7. Lagooning.
8. Salvaging.
9. Biological treatment.
10. Reclamation.
11. Open dump burning.
12. Landforming.

Basic factors to be considered

These methods are applicable either alone or in combinations. Before reviewing the basic factors that are required to be considered in depth before attempting to design the relevant method of storing, collecting and disposing the solid wastes, a look at the various types of such wastes will be of great interest.

Some of them are: (i) lunch room wastes, (ii) catalysts, (iii) organic chemicals, (iv) wastewater treatment sludges, (v) filter cakes, (vi) viscous solids, (vii) ashes, (viii) fly-ashes, (ix) incinerator residue, (x) plastics, (xi) ferrous metals, (xii) non-ferrous metals, (xiii) water treatment sludges, (xiv) inorganic chemicals, (xv) polymers - pellets, dust, powder, (xvi) off-specification products, (xvii) contaminated products, etc.

Considering the wide varieties of types, forms, etc., as stated above, the major factors considered while designing the facilities are narrated here.

Physical characteristics: (i) volume, (ii) weight, (iii) density and (iv) handling problems.

Chemical characteristics: (i) combustibility, (ii) degradability, (iii) toxicity, (iv) potential value of materials salvaged for recycling, (v) availability of land for disposal, (vi) anticipated future effects on land due to the disposal and (vii) effect of method on environment.

Cost of process

Wasteless processing probabilities: Almost every petrochemical industry creates some or other type of solid waste that can be categorised in any of the above listed types, etc. For their processes, many petrochemical units need water treatment facilities. The chemicals that are required for satisfactory operation of these treatment units are of various types and nature. Flocculants, the precipitates they form, etc., too add to the complexity of the solid wastes that are generated. 'Plant trash' is the generic term coined to define miscellaneous wastes that may be found in the manufacturing facility. Trash is considered harmless if they consist only of non-combustibles. Striking examples of such a category are glass, metal scraps, waste wood, floor sweepings, etc. Trash can originate from boilers, incinerators, etc., too. The ash or residues left of fuel (solid) is trash. The incinerated material (such as sludge) may also leave trash. Power generation facility also is capable of inducing trash. The commonly identifiable scrap in any petrochemical unit is the metal chips, parts or portions of discorded, corroded or damaged vessels, tanks, pumps, etc., and are mostly ferrous in nature.

This ferrous scrap contains within it the capability of getting contaminated by the exposure to toxic substances generated elsewhere in the plant. Recent day technologies play around catalysts and they have become a must for any petrochemical unit. By definition, catalysts do not take part in the reactions they catalyse and hence is a commodity with that highly beneficial and economic factor of recoverability.

Over a period of usage, they too get 'Spent'. Spent catalysts may be solid, semi-solid, etc. Catalysts are generally possessing complexity in their configuration, nature and chemical characteristics. Spent catalysts of solid/ semi solid configuration impart challenges in selection of treatment/disposal methods. Production processes that encompass combinations of multiple complexities end up with equally complex solid pollutants. A wide variety of other organic/inorganic chemicals have every chance of entering the waste streams as a result of all these complexities.

The family of plastics, viz., polypropylene, polyester, polystyrene, polyvinyl chloride, polyurethane, etc., is present to a high degree in the solid wastegroup of petrochemical units. These polymers can be identified in any of the following forms: (i) pellets, (ii) chops, (iii) cleanouts, (iv) shreds, (v) dust, (vi) powder, (vii) fines, (viii) granules, (ix) emergency dumps, (x) off-specification products and (xi) contaminated products.

Not only physical but also chemical processes are made use of to treat waste water streams emanating from any petrochemical unit. Unit operations are available to effect the process via., biological routes. Methods like usage of trickling filters, aeration, stabilisation ponds and activation of sludge do work well but not without their own solid waste derivatives. Variance of production techniques or combinations of treatment processes alter the end waste characteristics by a huge variance.

37.2.5 Disposal and treatment

Chemical treatment

This is a factor often neglected or overlooked as a method of useful waste management in petrochemical industries. Processes that help in chemical conversion of wastes may be enumerated as: (i) stripping, (ii) distillation, (iii) leaching and (iv) extraction, etc.

Some physical characteristics of materials also are made use of in conversion processes of wastes. They are: (i) separation by magnets, (ii) flotation, (iii) hand picking, (iv) jigging and (v) electrical conductivity, etc.

Saleable products are possible outcomes of adoption of the right combination of selective methods in the correct sequence during waste treatment. Even if such a cleverly worked out strategy fails to derive fruitful products, they can render innocuous materials that are easier to dispose off or are not detrimental to the environment. Some possibilities in this direction, are enumerated below:

1. Organics are hydrolysable to product fuels.
2. Waste cellulose is acetylated to give cellulose acetate.

Recycling

This is a way to set up a wasteless closed circuit flow scheme. Here, a stream is returned back into the process. Recycling helps in more vigorous and effective conversions with savings in materials, energy and safety from objectionable emissions to surroundings. Usage of catalysts and solvents are acquired to the maximum extent possible in this process, when-performed at optimum conditions. This is more true and proves cheap especially when the cost of the reactants are high or when treatment of residual reactants before disposal, etc., is costly and/or hazardous. The cheaper reagent is used in excess and the costlier one is circulated repeatedly to get more yields of the products. Here, reactor design plays a vital role, however.

Incineration

This is a combustion process, conducted in controlled conditions, for burning solid wastes (in addition to the liquid and gas phase wastes) and hence is a valuable route for solid waste disposal. The common products of incineration are carbon dioxide, water, ash, etc.

These are non-combustibles. The wastes are converted to a less bulky, toxic or noxious material. However, this process has its limitations also, in the sense that, compounds containing the following elements must not be incinerated as they will form oxides: (i) sulphur, (ii) nitrogen, (iii) chlorine, (iv) fluorine, (v) bromine, (vi) iodine, (vii) mercury, (viii) arsenic, (ix) selenium, (x) lead and (xi) cadmium.

Incineration is considered useful where availability of land is scarce and methods other than land disposals are to be used.

Following variables, however, have a great effect on completion of proper incineration:

1. Residence time of waste in incinerator.
2. Flammability limits of wastes.
3. Temperature inside the incinerator.
4. Ignition temperature of the wastes.
5. Turbulence in the reaction zone.
6. Combustibility of wastes being incinerated.

Reduction of operating costs of incineration can be envisaged by heat recovery through steam generation. Incineration helps not only reduce one's investment in pollution control equipments but also gives out products which are a very small fraction of the original weight and volume of the input wastes for landfill. More of organic wastes find incineration as their funeral, compared to inorganics. This also constitutes a way out for harmful bacterial, viral and toxic constituents. Necessity of support fuels may be considered as the negative aspect of this process, in addition to the limitations of items that can be handled in this process. Several processes are in use for incineration of petrochemical solid wastes, such as: (i) fluidised bed reactors, (ii) dual chamber incinerators, (iii) rotary kilns, (iv) open chamber incinerators, (v) liquid burner furnaces, (vi) hearths, (vii) multiple hearth furnaces, (viii) catalytic chambers, (ix) molten salt incinerators and (x) pyrolysis units.

Factors that tend to limit the selection of incineration as the favoured method of treating solid wastes are given below:

Physical state of wastes: (i) solid, (ii) semi-solid, (iii) others, (iv) ease of handling, (v) reacting capacity: (a) within themselves at higher temperatures and (b) with the material of construction of the incinerator, (vi) combustion temperature, (vii) moisture content, (viii) combustion products, (ix) ash content, (x) combustibility and possibility of disintegration at elevated temperatures, etc.

The incombustibles are the ones that require a support fuel for sustaining combustion and thus incineration. These are generally identified to be due to the following factors: (i) moisture content, (ii) inorganic salts, (iii) halogens, (iv) nitrogen, (v) sulphur and (vi) phosphorus.

Combustibles are born for easy incineration. They do not need any support fuels and hence their incineration is cheaper, compared to the incineration process of non-combustibles. Functioning of combustibles, during incineration too depends on the following factors: (i) mixing with air, (ii) physical state, (a) solid and (b) others, (iii) spraying (for liquids only), etc.

Dual chamber incinerators have been designed to combust solid wastes at the rate of 5700 kg/day from petrochemical units. Here one of the two chambers is for vapourisation and the other for combustion of the vapour. The latter is operated at three times higher a temperature of the first, which operates at 400°C or so. Mixtures of organics with inorganics pose the toughest task to incineration. This is easy to understand by considering the fact that inorganics may contain water and that inorganic salts form oxides, carbonates, etc., unlike the organics which tend to decompose. Usage of equipments like scrubbers are necessitated. Hazardous wastes that are given the incineration treatment as a method of disposal include organohalogen compounds and chlorine, fluorine and bromine derivatives. Presence of halogens depresses the combustion temperatures. The organohalogen wastes are first hydrolysed to form respective acid gas that are water soluble and thus treatable through easy and well known routes like water absorption or alkaline absorption methods. Compounds that may pose explosion hazards (like organo nitrogen compounds) are treated using long residence, low temperature incineration methods. Incinerators can also handle organo phosphorus compounds and sulphur bearing compounds also, of course followed by few secondary treatment methods.

Pyrolysis: While normal incineration demands 50–100% excess air over stoichiometric value, pyrolysis is a zero-air process, as this functions on indirect heating principles only. Temperatures as high as 1650°C are envisaged in an air-free or air starved chamber in pyrolysis. Basically, this resembles processes like distillation or cracking. Pyrolysis wastes are distilled to combustible forms that gets discharged from the furnace. Pyrolysis poses comparatively less air pollution than incineration. This process is capable of handling large volumes. Burnable materials are degraded marginally in pyrolysis. The greatest plus point of this technique is the possibility of recovering chemicals or synthesis gas. Partial combustion of pyrolysis gas within the furnace provides the process heat. This results in using the unoxidised portions of the combustible gases as fuel elsewhere and recovering that heat. Only wastes having high calorific value are processed in this route. Problems of clinkering or fusion of ash generally encountered with the incineration process are absent here, mainly because of the application of abnormally high temperatures in pyrolysis. Waste polymers are conveniently converted to fuel oils by pyrolysis.

The pyrolysis unit can be a tubular, plug flow reactor and the process adapted is thermo cracking. It is proposed in many experiments that if operating conditions are maintained properly, even other polymers can be converted into various other useful end products. Few of these are: (i) additives for gasolene, (ii) fuel oils, (iii) gaseous fuels, (iv) stain removers, (v) solvents of inferior strengths, (vi) lighter fractions and (vii) feedstocks for few other petrochemical processes, etc.

Waste sludges of non-autogenous character are often combined with high calorific value materials to facilitate releasing of high energy levels required for pyrolysis. It is quite interesting to note here that few wastes such as spent filter media, shredded wood, torn paper wastes, unusable oils, etc., comprise the family of such 'high calorific value materials'.

Open dumping: Open dumping or open dump burning is an unacceptable method of solid waste disposal and is illegal in many countries. This demands availability of huge area of land which will become ultimately unusable.

Landfill: This simple technique involves spreading and compacting solid wastes into cells and covering them. This method of 'sanitary landfill' requires low level labour and is low in cost in addition to requiring low degree of expertise and ease of operation. Wastes that are allowed to be dumped must be inert or capable of degradation of microbial attack to harmless products. Production of inflammable gases during this degradation is possible in some cases.

Land forming: Land forming is the spreading of wastes over a vast area wastes in a thin layer and is applicable for semi-solids or solids mixed with liquids. On spreading, the liquid either evaporates or seeps down the earth, depending upon the type of liquid, thus leaving the solids for degradation. Predominantly paraffin wastes are given this treatment. To some extent, even aromatic sludges are land-formed. Heavy metal concentrates pose a limitation for this method. Materials having no contamination of the following metals or their salts can be disposed off using this process: (i) lead, (ii) nickel, (iii) manganese, (iv) copper, (v) chromium and (vi) arsenic.

Lagooning: A method generally found suitable for solids, semi-solids and liquid wastes generated by petrochemical industries, lagooning comprises dumping wastes into ponds or pits. Inspite of its advantages (such as easy operation, low labour requirement, low construction cost and low degree of expertise required), this process is inherent of a bad characteristic, viz., if being odouriferrous and unsightly. As degradation of wastes is limited in lagoons, this method is necessarily to be followed by a secondary treatment of concentrates the lagoon will produce.

Hazards can be faced in either of the following four forms: (i) corrosive, (ii) ignitive, (iii) reactive and (iv) toxic. At least in case of these wastes, waste management extends further to:

1. Minimise production of such wastes.
2. Investigate possibilities of using these in industries elsewhere as a desirable material.
3. Reprocessing to minimise end wastes.
4. Separate these from other common and non-hazardous wastes.

5. Treat to give a non-hazardous waste (however high that treatment may cost).
6. Dispose with care, security and caution in landfills.

Various products that produce hazardous waste streams are listed below in their descending order of quantity generated in major petrochemical units: (i) atrazines, (ii) lead alkyls, (iii) furfural, (iv) explosives, (v) perchloroethylenes, (vi) methyl methacrylates, (vii) epichlorohydrin, (viii) parathion, (ix) malathion, (x) chlorobenzene, (xi) vinyl chloride, (xii) trifluoroaniline, (xiii) ethanolamines, (xiv) toluene di isocyanate, (xv) maleic anhydride, (xvi) chloromethane, (xvii) acrylonitrile, (xviii) nitrobenzene, (xix) fluorocarbon, (xx) chlorotoluene, etc.

It may be hard to believe that many wastes do contain valuable substances that can be extracted from concentrated wastes economically. As pollution control facilities aim at the production of the least environmental effect processes within the constraints of cost, it is mandatory to look for indirect generation of pollutants while selecting such methods that may seem suitable and cost-effective at first sight. Means of avoiding the inadvertent creation of 'indirect pollutants' must also be derived for implementation.

Technologies do exist for tactful and efficient management of any waste. But they may be unaffordable. It is much more simple to assign monetary values to the long-term damage to humanity and mother earth or environment that may result from a poor management of such wastes, than to set aside the process of coining a suitable treatment and disposal method of neglecting the total issue as 'highly costly'.

Regulations that are not available from governments or agencies may be another possible reason of not caring for a proper waste management system. But then, there will be no government or agency tomorrow to pose any regulation whatsoever, as we ourselves have not cared to pose self-regulation in spite of our being aware of the fact that tomorrow's world is to be made worthy of living (if not beautified) for our off-springs.

Treatment of polymers

Polymers are generally in the form of fine powder. It is a valuable product. Because of economical consideration it has been always endeavour of enterpreneur to recover maximum product escaping alongwith effluent. Moreover most of polymers manufactured in the petrochemical industries has specific gravity very close to oils (hydrocarbons).

Hence polymers when comes in contact with oil, emulsified scum is formed which interferes with deoiling process as well as biological treatment. Polymers are removed by the following methods.

Vibro-screen: Screen having very fine mesh size and vibrating mechanisms are used to recover fine polymer powder. Frequent chocking of screen is common phenomenon. Screen requires quite a bit of maintenance.

Side hill screen: This type of screens are used in the paper industry for recovery of fine fibre. Generally screens are inclined at 30° to 45° and normally mesh size of screen is kept at 60, effluent containing polymers is screened through for recovery of polymers. Good recovery is possible and quite economical.

Gravity separators: Polymers being light floats on the surface of water, hence slot pipe arrangement is used to recover polymers from the surface of water and water under flows for further treatment.

37.3 Reducing hazardous operations in olefin plant

Light olefins are one of the most important material of petrochemical industry. They are mainly produced by thermal cracking of hydrocarbon feed stock in the cracking furnace of an olefin plant. The cracking furnaces are parallel tubular reactors with the pyrolysis reaction taking place inside the tubes at high temperature, low pressure and short residence time.

Besides the primary reactions to light olefins, secondary reactions are taking place, forming a wide variety of hydrocarbon molecules but also coke. This coke builds up a layer on the inner surface of the reactor tubes and has to be removed periodically. Typical cycle times are one to three months.

The coke removal, called decoking operation, is carried out with steam and air in a duration of one to two days. However, in combination with hydrocarbon service in an olefin plant, oxygen is a hazardous component. For plant safety, any air entry has to be strictly avoided. Therefore, the decoking off gas which still contains oxygen has to be routed separately from the normal process route. While the crack gas is routed in the transfer piping to the separation section for fractionation and purification, the decoking off gas is routed in the decoking piping to the atmosphere after removal of coke particulates. A changeover from the process route to the decoking route and vice versa has to be executed repeatedly.

With several parallel furnaces, the effluent switching from the process side to the atmosphere and vice-versa is a regular phenomenon in the olefin plant operation. With the risk that hydrocarbons get into contact with air or released to atmosphere it is at the same time considered as one of the most critical activities and has to be performed by experienced plant personnel.

37.3.1 Risks during effluent change over

Crack gas and decoking gas piping are separated from each other with a motor operated valve in each line. During change over from one line to the other, the furnace is still in operation with steam as process medium. The permanent steam flow is necessary for cooling of the convection tubes to avoid overheating. Reduction of firing is not an option as the radiant tubes have to be kept in hot condition to avoid coil ruptures or blockages caused by the inside coke. For these reasons process steam cannot be fully shut off during switch over. Consequently, the crack gas valve and the decoking gas valve are open for a certain period at the same time.

The possible risks normally assessed in an analysis together with all involved parties are as follows:

1. Process upsets, like a pressure disturbance in the effluent system.
2. Mal-operation in combination with/or mis-interpretation of an situation, like operating the wrong valve at the wrong time.
3. Material failures, like coil cracks.

During the change over procedure a complex operation of two (1 and 2 in the Fig. 37.1) valves is required to fulfil the process requirements. Situations occur where both valves are open at the same time. The risk of reverse flow of highly flammable crack gas from the effluent header, which could be released to atmosphere, is present. This increases the risk of creating explosive mixtures in the olefin plant regularly and frequently. For environmental and safety reasons, any release of crack gas into atmosphere has to be mandatorily avoided at all times and at all costs.

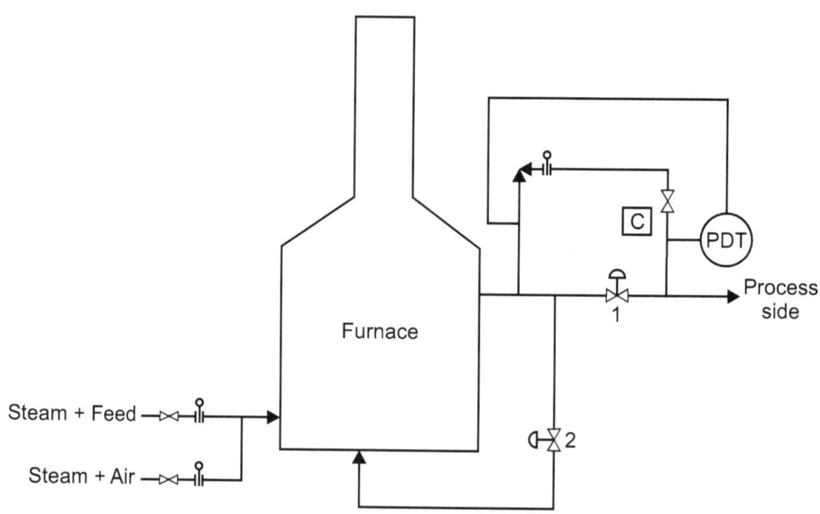

Figure 37.1: Process scheme of cracking furnaces of olefin plant.

Usually the reverse flow is avoided by building up a higher pressure in the furnace unit ensuring a positive flow towards the effluent header. During opening and closing of the valves (1 and 2), the operator in the field is carefully monitoring the pressure profile and the panel operator is checking the process conditions. Critical situations have to be handled by well trained and experienced field operator. However, the probability for human errors, plant upsets or sudden events, like a coil rupture, increases the safety risk during this change over operation. This risk can be reduced by automation of the operating procedure and introduction of an emergency logic.

Further, the preparations for the change over are usually quite time consuming. The maintenance intensive swapping of line spools is required to avoid the introduction of air, in form of decoking air, into the hydrocarbon processing separation units. A combination of key interlocking and advanced blinding systems can provide the same safety level and reduce the time for the preparations. Consequently, the time when the furnace coils are fully loaded with coke, transition state is minimised and coke palling can be avoided.

37.3.2 Automated versus manual operation

The automated switch over operation principally works in the same way as the manual one. A positive flow in the direction of the effluent header and not backwards is monitored by a differential pressure measurement and controlled by the opening/closing of the valves. However, through automation, there are clear advantages over the manual operation:

1. No probability for mal-operation by human errors.
2. Defined reaction during plant upsets like CGC suction pressure up and down or trip.
3. Included emergency functions, like emergency shut off or avoiding of blocked outlet.
4. Following a clear sequence.
5. Auto change over is faster and smoother operation, especially in upsets.
6. Change over still can be executed manually (operator has to be experienced and to act carefully).
7. Automatic or semiautomatic switch over possible with minimum of additional instrumentation and hence low costs for hardware upgrade.

Action during and prior to effluent change over operation from process side to atmosphere and vice versa

The decoke air line and the hydrocarbon line block valve will be closed and blinded. By using the local control panel of motor operated valve, the cracked gas valve will be closed using pressure differential and open the decoking gas valve in sequential opening. In this operation, the pressure difference across the process and atmosphere side will be maintained automatically by the pressure differential control. In the effluent line, the safety valve is ensuring against the high differential pressure during the changeover operation.

Thus, by implementing the above procedure and control philosophy in the design of the olefin plant, the risk involved in creating an explosive mixture in running plant while performing the effluent change over operation can be minimised to zero level.

38

Pollution control in chlor-alkali industry

38.1 Introduction

The chlor-alkali industry is one of the largest electrochemical technologies in the world. It is the second largest consumer of electricity (2400-bn kWh) among electrolytic industries.

Chlorine is produced electrolytically using three types of electrolytic cells. Chlorine is produced by the electrolysis of sodium chloride solution, called 'brine'. When sodium chloride is dissolved in water, it dissociates into sodium cations and chloride anions. The chloride ions are oxidised at the anode to form chlorine gas and water molecules are reduced at the cathode to form hydroxyl anions and hydrogen gas.

The sodium ions in the solution and the hydroxyl ions produced at the cathode constitute the components of sodium hydroxide.

The main difference in three chlorine process technologies lies in the manner by which the chlorine gas and the sodium hydroxide are prevented from mixing with each other to ensure generation of pure products. Thus, in diaphragm cells, brine from the anode compartment flows through the separator to the cathode compartment, the separator material being either asbestos or polymer modified asbestos composite deposited on a perforated cathode.

In membrane cells, on the other hand, an ion-exchange membrane is used as a separator. Anolyte-catholyte separation is achieved in the diaphragm and membrane cells using separators and ion-exchange membranes, respectively, whereas mercury cells contain no diaphragm or membrane and the mercury itself acts as a separator. The anode in all technologies is titanium metal coated with an electrocatalytic layer of mixed oxides.

All modern cells use these so-called Dimensionally Stable Anodes (DSA). Earlier cells used carbon based anodes. The cathode is typically steel in diaphragm cells, nickel in membrane cells and mercury in mercury cells. These cell technologies are shown in Figs 38.1 to 38.3.

38.2 Manufacturing process of chlor-alkali

38.2.1 Mercury cells

The mercury cell (Fig. 38.1) has steel bottoms with rubber-coated steel sides, as well as end boxes for brine and mercury feed and exit streams with a flexible

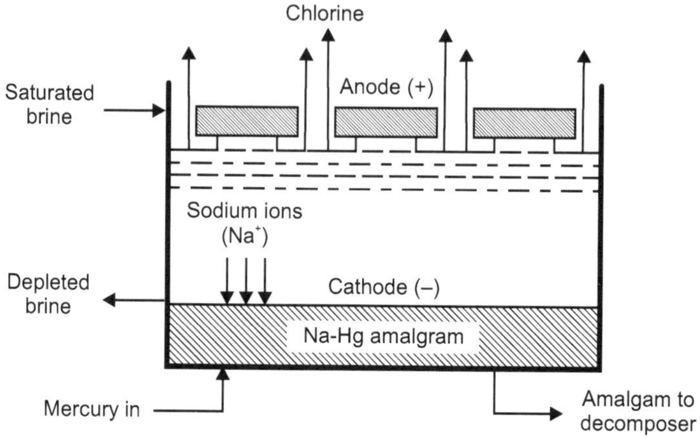

Figure 38.1: Schematic of a mercury cell.

rubber or rubber-coated steel cover. Adjustable metal anodes hang from the top and mercury (which forms the cathode of the cell) flows on the inclined bottom. The current flows from the steel bottom to the flowing mercury. Saturated brine fed from the end box is electrolysed at the anode to produce the chlorine gas, which flows from the top portion of the trough and then exits. The sodium ion generated reacts with the mercury to form sodium amalgam (an alloy of mercury and sodium) which flows out of the end box to a vertical cylindrical tank. About 0.25% to 0.5% sodium amalgam is produced in the cell. The sodium amalgam reacts with water in the decomposer, packed with graphite particles and produces caustic soda and hydrogen. Hydrogen, saturated with water vapour, exits from the top along with the mercury vapours. The caustic soda then flows out of the decomposer as 50% caustic. The unreacted brine flows out of the exit end box.

Some cells are designed with chlorine and anolyte outlets from the end box, which are separated in the depleted brine tank. The mercury from the decomposer is pumped back to the cell.

38.2.2 Diaphragm cells

The diaphragm cell (Fig. 38.2) is a rectangular box with metal anodes supported from the bottom with copper-base plates; which carries a positive current. The cathodes are metal screens or punch plates connected from one end to the other end of the rectangular tank. Asbestos, dispersed as a slurry in a bath, is vacuum deposited onto the cathodes, forming a diaphragm. Saturated brine enters the anode compartment and the chlorine gas liberated at the anode during electrolysis, exits from the anode compartment. It is saturated with water vapour at a partial pressure of water over the anolyte. The sodium ions are transported

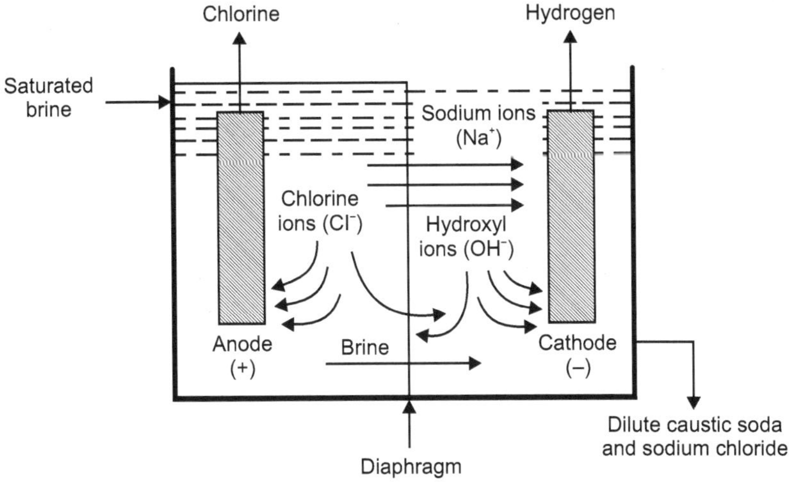

Figure 38.2: Schematic of a diaphragm cell.

from the anode compartment to the cathode compartment, by the flow of the solution and by electromigration, where they combine with the hydroxyl ions generated at the cathode during the formation of the hydrogen from the water molecules. The diaphragm resists the back migration of the hydroxyl ions, which would otherwise react with the chlorine in the anode compartment. In the cathode compartment, the concentration of the sodium hydroxide is ~12% and the salt concentration is ~14%. There is also some sodium chlorate formed in the anode compartment, dependent upon the pH of the anolyte.

38.2.3 Membrane cells

In a membrane cell (Fig. 38.3), an ion-exchange membrane separates the anode and cathode compartments. The separator is generally a bilayer membrane made of perfluorocarboxylic and perfluorosulphonic acid-based films, sandwiched between the anode and the cathode. The saturated brine is fed to the anode compartment where chlorine is liberated at the anode and the sodium ion migrates to the cathode compartment.

Unlike in the diaphragm cells, only the sodium ions and some water migrate through the membrane. The unreacted sodium chloride and other inert ions remain in the anolyte. About 30–32% caustic soda is fed to the cathode compartment, where sodium ions react with hydroxyl ions produced during the course of the hydrogen gas evolution from the water molecules. This forms caustic, which increases the concentration of caustic solution to ~35%. The hydrogen gas, saturated with water, exits from the catholyte compartment. Only part of the caustic soda product is withdrawn from the cathode compartment.

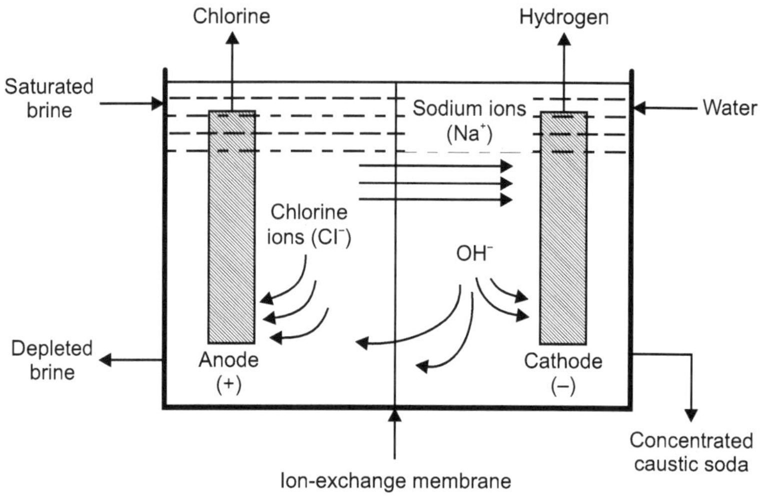

Figure 38.3: Schematic of a membrane cell.

The remaining caustic is diluted to ~32% and returned to the cathode compartment. Thus, all three basic cell technologies generate chlorine at the anode and hydrogen along with caustic soda in the cathode compartment. The distinguishing difference between the technologies lies in the manner by which the anolyte and the catholyte streams are prevented from mixing with each other. Separation is achieved in a diaphragm cell by a separator and in a membrane cell by an ion-exchange membrane. In mercury cells, the cathode itself acts as a separator by forming an alloy of sodium and mercury (sodium amalgam) which is subsequently reacted with water to form sodium hydroxide and hydrogen in a separate reactor.

38.2.4 Chlorine processing

The chlorine gas from the anode compartment contains moisture, by-product oxygen and some back-migrated hydrogen. In addition, if the brine is alkaline, it will contain carbon dioxide and some oxygen and nitrogen from the air leakage via the process or pipelines. Chlorine is first cooled to 60°F (16°C) and passed through demisters to remove the water droplets and the particulates of salt and sodium sulphate.

The cooled gas goes to sulphuric acid circulating towers, which are operated in series. Commonly, three towers are used for the removal of moisture. The dried chlorine then goes through demisters before it is compressed and liquefied at low temperatures.

The non-condensed gas, called snift gas, is used for producing hypochlorite or hydrochloric acid. If there is no market for hydrochloric acid, the snift gas is neutralised with caustic soda or lime (calcium hydroxide) to form hypochlorite. The hypochlorite is either sold as bleach or decomposed to form salt and oxygen.

38.2.5 Caustic soda processing

Caustic soda is marketed as 50%, 73% or anhydrous (dry) beads or flakes. The mercury cell can produce 50% and 73% caustic directly. The caustic from the decomposer is cooled and passed once or twice through an activated carbon filter to reduce the mercury levels in the caustic. After filtration, the mercury concentration is lowered to the parts-per-million (ppm) levels.

Even these low levels of mercury may be unacceptable to some customers, who then have to switch to using membrane grade caustic soda. The mercury cell caustic soda has a few ppm salt and <5 ppm sodium chlorate. The mercury cell caustic is the highest purity caustic that can be made electrolytically if trace concentrations of mercury are tolerable in the end use of caustic.

The membrane cell caustic is concentrated in a multiple effect falling film evaporator, which increases the caustic soda concentration to 50% with a high steam economy. Caustic soda from membrane cells generally has 30 ppm sodium chloride and 5–10 ppm sodium chlorate.

The catholyte from the diaphragm cells contains ~12% sodium hydroxide, ~14% sodium chloride, 0.25–0.3% sodium sulphate and 100–500 ppm sodium chlorate. The catholyte is evaporated in a multi-effect evaporator. Most of the salt from the catholyte will precipitate during the concentration of the caustic soda to 50% sodium hydroxide. The 50% caustic soda product will contain about 1% sodium chloride. The 50% caustic also has a high chlorate concentration (~0.1 %) compared to the caustic from membrane or mercury cells (~10 ppm). The salt, separated from the caustic during evaporation, is used to resaturate the brine fed to the cell. An additional single-effect evaporator is needed to produce 73% caustic soda.

Anhydrous (dry) caustic soda is produced in a rising film evaporator, operating at 725°F (385°C) and at a few inches of water vacuum.

38.3 Emission and consumption levels of the chlor-alkali industry

Emission and consumption levels of the chlor-alkali industry are quite specific to the cell technique used but also depend on the specifications of the products (O_2 or CO_2 content, for example), the purity of the incoming salt and the geographic location of the plant. The inputs are primarily salt and water as feedstock, acids and chemical precipitants used to remove impurities in the input brine or output chlorine/caustic soda, as well as refrigerants (CFCs,

HCFCs, HFCs, ammonia, etc.), for liquefying and purifying the chlorine gas produced. The chlor-alkali process needs huge amounts of electricity and hence electrical energy is a major input.

The main pollutant outputs which are common to all three electrolytic processes are emissions of chlorine and refrigerants to air, emissions of noise, emissions of free chlorine, chlorate, bromate, chloride, sulphate, heavy metals, sulphite, organic compounds and halogenated organic compounds to water, as well as the generation of spent acids from chlorine drying and sludges from brine purification.

The major pollutant in terms of environmental impact, originating from the mercury cell technique, is mercury. Due to the process characteristics, mercury can be released from the process to air, water, wastes and in products. The diaphragm cell and membrane cell techniques are more concerned with spent materials generation, such as asbestos waste, in the case of asbestos diaphragms.

38.4 Overview of emission and consumption levels of all cell plants

38.4.1 Water

For the production of the required demineralised water, several water sources can be used, such as groundwater, river water and drinking water. The techniques used to produce demineralised water include ion exchange, membrane filtration and evaporation.

Water is mainly added to the process for:

1. The preparation of the brine.
2. The preparation of some reagents used for the brine purification.
3. The production of caustic in the decomposer, in the case of the mercury cell technique.
4. The dilution of the produced caustic in the caustic circuit, in the case of the membrane cell technique.
5. The direct cooling of chlorine.
6. The dilution of caustic if necessary (e.g., for the chlorine absorption unit).

Water which is consumed leaves the process either via the products (hydrogen and caustic solution) or as waste water:

1. During electrolysis, water is consumed for the synthesis of hydrogen and caustic in a stoichiometric amount of approximately 0.51 $m^3/T\ Cl_2$ produced.

2. In addition, water is used as a solvent for the production of caustic. This amount depends on the caustic concentration and the salt concentration therein and cannot be reduced unless the caustic is further concentrated and the condensate recycled.

38.4.2 Energy

The energy consumption in chlor-alkali production originates from four main processes:

1. Energy to prepare and purify the raw materials, mainly the salt (sodium chloride or potassium chloride).
2. Electrical energy used for the electrolysis process itself.
3. Energy (steam) to obtain the caustic soda (or potash) at its commercial concentration.
4. Energy for auxiliary equipment such as heating devices, pumps, compressors, transformers, rectifiers and lighting.

Energy is used both as electricity and as heat (steam). About half of the electricity consumed is converted into the enthalpy of the products. The rest is converted into heat transferred to the air in the building and the products, which have to be cooled. The heat is partly recirculated through pre-heating of the brine. Surplus heat might also be used for heating surrounding buildings or for the production of steam which could be used for the concentration of caustic soda. Insulation of the salt dissolvers can be used to reduce the heat losses of the brine system.

Energy consumption depends on a number of factors such as:

1. The cell technique used.
2. The purity of the salt used as raw material.
3. The specific cell parameters, such as nominal current density, anode/cathode gap, adherence of developed gas bubbles on electrode structures, diaphragm/membrane type and thickness, catalytic electrode coatings.
4. The age of the diaphragm, the membrane and the catalytic electrode coatings.
5. Other technical characteristics of the installation such as the configuration of the electrolysers (monopolar or bipolar), the number of evaporative stages in the caustic concentration unit and the chlorine liquefaction conditions.
6. The actual current density.

The number of influencing factors explains why energy consumption data of chlor-alkali plants vary significantly, even when using the same cell technique.

38.5 Emissions and waste generation from all cell plants

The presence of a waste gas, waste water or solid waste stream depends on the specific layout of each plant. The major potential sources of emissions and waste for a membrane cell plant also occur in mercury and diaphragm cell plants (except for mercury which occurs in mercury cell plants and asbestos which occurs in diaphragm cell plants using asbestos). However, in diaphragm cell plants there is usually no brine purge and in mercury cell plants no condensate from caustic concentration. Moreover, diaphragm and mercury cell plants usually do not carry out secondary brine purification, while some mercury and membrane cell plants using vacuum salt do not carry out primary brine purification. Several of the waste water (e.g., condensates, spent sulphuric acid and bleach from the chlorine absorption unit) and waste gas streams (e.g., chlorine from treatment of depleted brine) may be partially or fully recycled within the process or elsewhere.

38.5.1 Emissions from the storage and handling of solids

Emissions from the storage and handling of solids include dust emissions to air originating from salt and ancillary materials, as well as emissions to water caused by the seepage of rainwater and discharge with run-off water. Percolation into the soil can be prevented by an impermeable layer (sometimes with hardened salt) under the salt.

38.6 Emissions and waste generation from the brine circuit

Emissions from the brine circuit to air include fugitive emissions of chlorine and carbon dioxide. Emissions to water occur via the brine purge. The emission levels depend on the purity of the incoming salt used, the cell technique used (mercury, diaphragm or membrane), the techniques used for brine purification and whether the brine is recirculated or wasted. Brine sludges which are generated during brine purification are one of the largest waste streams of the chlor-alkali industry.

38.6.1 Emissions to air

Emissions to air from the brine circuit include mainly fugitive emissions of carbon dioxide and chlorine. Carbon dioxide is emitted from the brine acidification tanks due to the decomposition of carbonate and hydrogen carbonate ions to water and carbon dioxide. The carbonate and hydrogen carbonate stem from the auxiliary chemicals used in the brine purification step. The carbon dioxide either escapes from the brine and is emitted to the atmosphere or is led to the chlorine absorption unit, depending on the presence of chlorine in the brine. The concentration of (hydrogen) carbonate is expected to be lower when the depleted brine is recirculated and resaturated with solid salts (membrane and mercury cells). CO_2 emissions due to brine acidification are approximately 1.2 kg per T of chlorine produced.

When the mercury cell technique is used, mercury emissions may occur from brine resaturators and salt dissolvers if there is no oxidising species.

Because chlorine is a hazardous gas, leakage from electrolytic cells is avoided. However, small amounts of chlorine might be emitted through leakages from the brine system and the cells, if these are under pressure. In some cases, chlorine detectors are placed in the cell room, giving an immediate indication of any significant leakage.

The chlorine concentration in the cell room can be below the odour detection level if the cells are operated under a slight vacuum. Odour detection levels are in the range of 0.2–3.5 ml/m^3 (ppmv), equivalent to 0.6–11 mg/m^3). Due to their diffuse nature, emissions from the cell room and the brine circuit are difficult to measure but may nevertheless contribute significantly to the overall chlorine emissions to air. One plant reports chlorine concentrations in the cell room of 0.08–0.1 ppmv and an average emission load of 3.1 g/T annual chlorine capacity, while another reports emission loads from the cell room ranging from 3.6 to 14 g/T annual chlorine capacity.

38.6.2 Emissions to water

Potential sources of waste water in chlor-alkali plants include:

1. Brine purge in mercury and membrane cell plants using a brine recirculation system.
2. Depleted brine in mercury and membrane cell plants using a once-through brine system.
3. Backwashing water from primary brine purification.
4. Backwashing water of ion-exchange resins from secondary brine purification, in the case of membrane cell plants.
5. Backwashing water of ion-exchange resins from water demineralisation.
6. Water from the condensers during chlorine cooling.
7. Diluted sulphuric acid from chlorine drying.
8. Water from the condensers during hydrogen cooling.
9. Water from the condensers during caustic soda evaporation in diaphragm and membrane cell plants.
10. Water from the recrystallisation of salt recovered during caustic evaporation in diaphragm cell plants.

11. Bleach produced by the chlorine absorption unit if it cannot be used or sold.
12. Water from cleaning of equipment.

Relevant pollutants include sulphate, chloride, free chlorine, chlorate, bromate, heavy metals, organic compounds, halogenated organic compounds and sulphite.

Sulphate emissions

Sulphate emissions may also be due to the neutralisation and discharge of spent sulphuric acid from chlorine drying. In addition, (hydrogen) sulphite is frequently used for complete dechlorination of the brine in the membrane cell technique, as well as for the treatment of waste water containing free chlorine. In both cases, (hydrogen) sulphite is converted to sulphate.

38.7 Emissions and waste generation from chlorine processing, storage and handling

Air emissions consist of fugitive emissions of carbon dioxide and chlorine from the processing steps, from the storage and handling of chlorine, as well as of channelled emissions from the chlorine absorption unit. Emissions of carbon tetrachloride during chlorine processing occur at a few installations that make use of this substance. Small amounts of chlorine dioxide could also be emitted from the chlorine absorption unit, as well as small amounts of refrigerants from the chlorine liquefaction system.

The condensed water formed after cooling is usually recycled as brine make-up, although some facilities combine this stream with other waste water streams prior to treatment. The remaining water vapour is removed by scrubbing the chlorine gas with concentrated sulphuric acid.

The spent sulphuric acid is used for other purposes, is recycled, or is sent to the site's waste water collection system.

38.7.1 Emissions to air

Carbon dioxide

Small amounts of carbon dioxide are released from the anode compartment and are collected and treated together with the chlorine. During chlorine liquefaction, carbon dioxide and other gases (N_2, O_2, H_2) remain in the gas phase and are finally purged from the system, together with a small amount of chlorine. This waste gas stream is usually processed in the chlorine absorption unit. Carbon dioxide that is not absorbed by the chlorine absorption unit is emitted into the atmosphere.

Chlorine

Because chlorine is a hazardous gas, extreme precaution is taken to prevent emissions of chlorine from the process and from handling and storage. Therefore, emissions of chlorine gas into the atmosphere are generally low. The sources of significant potential emissions are normally connected with the chlorine absorption unit.

Chlorine dioxide

Small amounts of chlorine dioxide originating from side reactions can be emitted from the chlorine absorption unit. When measuring chlorine in the exhaust from the chlorine absorption unit, chlorine dioxide is usually included in the analytical result.

Carbon tetrachloride

Carbon tetrachloride for the elimination of nitrogen tetrachloride or the recovery of chlorine from tail gas is circulated in a closed loop. However, emissions may occur through leakages and so the gas tightness of the system is of paramount importance to achieve low emission values.

Refrigerants

Refrigerants used for chlorine liquefaction are circulated in closed loops, but fugitive emissions may occur through leaks.

38.7.2 Emissions to water

Sulphuric acid mists coming from the drying towers are captured by candle filters and may be released as sulphates when the filters are washed for maintenance.

38.7.3 Generation of wastes

Spent sulphuric acid from chlorine drying

Concentrated sulphuric acid (92–98 wt-%) is used to dry chlorine. Often the spent sulphuric acid is returned to the supplier for refortification. The spent acid can also be used to control pH in process and waste water streams or to destroy surplus hypochlorite, or it can be sold to a user who can accept this quality of acid. In rare cases, the spent acid becomes waste.

Carbon tetrachloride

When carbon tetrachloride is used to absorb NCl_3 or to recover chlorine from tail gas, it needs to be replaced periodically because of the build-up of pollutants (bromine, $FeCl_3$ and chlorinated hydrocarbons, for example). The removed carbon tetrachloride is recycled, reclaimed or destroyed, in line with the provisions of the regulation on substances that deplete the ozone layer.

Bleach

During normal operation of a chlor-alkali plant, the bleach produced can be sold or used on site. When the bleach production unit needs to handle a large amount of chlorine in a short period of time (e.g., in the event of release of chlorine due to a malfunctioning of equipment), the bleach produced usually does not comply with the product specifications. In this case, the 'off-spec' bleach is either destroyed on site and discharged with the waste water, or is removed and processed elsewhere.

38.7.4 Generation of wastes from mercury cell

Wastes generated specific to mercury cell plants relate to those contaminated with mercury. Wastes contaminated with mercury include:

1. Solids from brine purification.
2. Solids from caustic filtration.
3. Solids from waste water treatment.
4. Solids from sludge (sewer, traps, channels).
5. Activated carbon from the treatment of gaseous streams.
6. Graphite from decomposer packing.
7. Residues from retorts.
8. Wastes from maintenance and renewal.

Use on site or off site: This technique consists in using the spent sulphuric acid from chlorine drying instead of neutralising and discharging it.

Concentrated sulphuric acid (92–98 wt-%) is used to dry chlorine. Up to 35 kg of acid is consumed per tonne of chlorine produced. The spent acid can be used to control the pH in the process and waste water streams or to destroy surplus hypochlorite if there is a respective demand on site. It can also be reconcentrated or sold as a by-product to any user who can use this quality of acid. Typical uses include the production of ammonium sulphate (fertilisers) and aluminium sulphate (water treatment, paper manufacturing).

Environmental benefits

The achieved environmental benefits of these techniques include:

1. Reduction of raw material and energy consumption, due to a replacement of fresh sulphuric acid.
2. Reduction of sulphate emissions and caustic consumption, if the spent acid would otherwise be neutralised and discharged.

38.7.5 Reduction of mercury emissions to air, water and hydrogen

Removal of mercury from process exhausts

Process exhausts refer to all gaseous streams by which mercury can be emitted to air, apart from cell room ventilation air and hydrogen. Some streams may be combined prior to treatment, while others require separate treatment units. For example, hydrogen streams are usually not mixed with streams containing significant amounts of air, in order to prevent the formation of explosive mixtures.

Mercury is mainly removed by:

1. Adsorption on iodised or sulphurised activated carbon.
2. Scrubbing with hypochlorite or chlorinated brine to form mercury chloride.
3. Adding chlorine to form dimercury dichloride (calomel), which is collected on a solid substrate, such as rock salt in a packed column.

Mercury concentrations in treated process exhaust are typically $\leq$ 30 μg/m^3.

Adsorption on iodised or sulphurised activated carbon

Many plants use adsorption on iodised or sulphurised activated carbon. Activated carbon filtration generates a waste contaminated with mercury that requires further treatment and/or disposal. Metallic mercury can be recovered from this waste by distillation.

Scrubbing with hypochlorite or chlorinated brine

Scrubbing is generally carried out in packed columns with a countercurrent flow of hypochlorite or chlorinated brine. Elemental mercury is oxidised and the resulting liquid can be recycled to the brine system. In the case of using chlorinated brine as scrubbing liquid, the waste gas requires further treatment in the chlorine absorption unit to remove chlorine, while in the case of using alkaline hypochlorite solution, the waste gas only needs to pass through a separator to remove entrained liquid droplets.

Residual mercury concentrations after treatment were reported to be 50–100 μg/m^3, corresponding to 0.006–0.01 g/T annual chlorine capacity.

38.7.6 Reduction of mercury emissions to water

Most of these streams can be recycled into the process, thereby achieving waste water volumes of 0.3–1.0 m^3/T of chlorine produced.

Various techniques exist to reduce the mercury levels in the residual waste water prior to its discharge.

38.7.7 Removal of mercury from caustic soda

The 50 wt-% caustic soda leaving the mercury cells usually contains metallic mercury in the range of 2.5–25 mg/L, depending on the type of denuder and temperature. For some applications very pure sodium or potassium hydroxide are needed.

The predominant technique to reduce mercury levels in caustic soda is activated carbon filtration in pre-coated candle or plate filters. The mercury is absorbed on a special carbon pre-coat and removed from the filter as a dry cake. This technique guarantees a maximum mercury content of 50 μg/L in 50 wt-% caustic, if the mercury content before filtration is in the range of 5–20 mg/L. The caustic filtration unit is sensitive to flowrate changes and low mercury levels are only achieved during undisturbed operation.

38.8 Treatment of wastes contaminated with mercury

Most of the mercury leaving a mercury cell chlor-alkali plant is contained in the various wastes from the process including wastes from mercury emission abatement. Additionally, there are solid wastes from maintenance, construction and demolition that may be contaminated with mercury.

The choice of the appropriate decontamination technique depends on the location of the mercury, its concentration and its chemical state for each category of waste. It is therefore essential to characterise each waste prior to treatment, preferably by chemical analysis. However, representative sampling may be difficult, because solid wastes are often inhomogeneous. This may result in estimates having to be made.

A waste management plan is usually drawn up and shared with the permitting authorities. Such a plan generally aims at reducing the quantity of waste contaminated with mercury sent for disposal and at recycling mercury back to the process.

Elements of such a plan include:

1. Detailed service routines.
2. Provisions for records to be kept.
3. Descriptions of the different types of wastes that are generated and techniques to reduce their generation.
4. Provisions for the quantification of the different types of waste.
5. Provisions for storage.
6. Provisions for waste treatment, including stabilisation and final disposal, both on site and off site.
7. Methods and frequencies for sampling and analysis before and after waste treatment.

38.8.1 Mechanical and physical treatment

Mechanical and physical treatment include washing with water (with or without pressure), ultrasonic vibration and vacuum cleaners with appropriate adsorption or condensation systems. Mechanical and physical treatments are suitable if significant quantities of metallic mercury are present. Other techniques include cryogenic treatment with nitrogen for rubber-coated steel or coated metallic materials and warm sandblasting in a fluidised bed for coated metallic materials. The resulting liquid streams usually require further treatment. All treatment techniques may result in mercury emissions to air.

38.8.2 Chemical treatment

Chemical treatment includes treatment with hypochlorite, chlorinated brine or hydrogen peroxide. The liquid streams may be recycled to the brine system or treated as waste water, with recovery and recycling of mercury.

38.8.3 Distillation/retorting

Distillation/retorting is used by many mercury cell chlor-alkali plants. Some of these plants have their own retorting unit, while others use the services of an external retorting company.

Distillation of mercury from solid waste can be carried out in special distillation units, if mercury is present in its elemental form. Some solid waste can therefore not be distilled (for example, when the mercury is present as a salt). The mercury is recovered as metallic mercury and can be used for future make-up to the cells. Distillation largely removes mercury from waste to residual levels of typically less than 50 mg/kg.

Process exhausts from the mercury distillation unit are usually further treated. After treatment, the residue may be cooled by ventilation air. This ventilation air contains a small amount of mercury. At the Solvin plant in Antwerp-Lillo (Belgium), the residual mercury concentration in ventilation air is less than 10 $\mu g/m^3$, corresponding to less than 0.005 g/T annual chlorine capacity. The solid residue from distillation is landfilled or stored underground.

38.8.4 Stabilisation and final disposal

In most cases, the solid residues contaminated with mercury that result from treatment require stabilisation before being disposed of. Stabilisation serves to transform mercury and its compounds into a less soluble form, to reduce the risk of leaching from the final disposal site. In many cases, mercury is treated with sulphur or sulphur compounds to transform it to mercury sulphide. This can also be combined with physical stabilisation, for example in cement-based materials. After stabilisation, the waste is landfilled or stored underground.

38.8.5 Treatment of asbestos-containing wastes

Asbestos-containing wastes are thermally treated, chemically treated and/or landfilled, depending on the legislation of the country. This section briefly describes a thermochemical treatment technique that is applied at the dow plant in stade (Germany). During this process, the dangerous fibre structure is destroyed and converted to 'sand type' silicate structures through thermal treatment. In order to achieve this, all spent asbestos is dewatered in a centrifuge and chemicals are added to generate a mixture with a significantly reduced melting point. This mixture is then pelletised and the pellets are fed to a rotary kiln for the vitrification process, which takes place at 1300°C. The resulting slag no longer contains asbestos fibres and can be used at landfill sites or to fill in underground cavities.

Monitoring: Asbestos fibres are usually measured in air by fibre counting, using phase-contrast microscopy, according to the method recommended by the world health organisation.

38.8.6 Emissions of noise from chlor-alkali plant

Emissions of noise originate from equipment commonly used in the chemical sector and are therefore not specific to the chlor-alkali manufacturing industry.

The major sources of noise emissions include.

1. Trucks for unloading and loading (e.g., salt).
2. Transformers and rectifiers.
3. Pumps (e.g., brine circuit, cooling water).
4. Motors.
5. Control valves.
6. Punched disks.
7. Fans (e.g., cell room ventilation, hydrogen emission).
8. Compressors (e.g., chlorine, hydrogen, refrigerant).
9. Centrifuges (e.g., caustic evaporation unit, in the case of diaphragm cell plants).
10. Cooling machines.
11. Transport of media in pipelines and other systems without optimal dimensions.
12. Transport on and near the site including railways.
13. Cleaning of process equipment (e.g., mechanical cleaning of cells).

39

Pollution control in paint industry

39.1 Introduction

Paint is defined a coloured substance which is spread over a surface and dries to leave a thin decorative or protective coating. Based on its use, paint products can be categorised into architectural coatings, product finishes and special-purpose coatings. Architectural coatings are products used to coat interior and exterior surfaces. Product finishes provide first coating on newly manufactured equipment and products. Special purpose coatings are products formulated to meet specific use of requirements such as extreme temperatures or heavy wear.

Based on carrier used in paint formulation, classification refers to the volatile solvent rather than to the combined solvent and binder. The volatiles typically water or solvent, evaporate after the paint has been applied to the substrate.

Major raw materials used to manufacture paint are binders (resins, drying oils), solvents, pigments, extenders, petroleum thinners, additives/chemicals, etc. The chemical composition of paints varies depending on the desired paint properties.

Binders: Binders form a continuous phase, hold the pigment in dry film and cause it to adhere to the surface to be coated. The majority of binders in paint films are composed of resins and drying oils, which are largely responsible for the protective and general mechanical properties of the film. Resins provide characteristics such as durability and flexibility. Alkydes, acrylics and vinyls are the three commonly used resins.

Solvents: Solvents are used to keep paints in liquid form so that they can be easily applied and evaporate completely. It is used to transfer the pigment mixture to a surface in a thin, uniform film and plays no role in film formation. Materials used as solvents are aromatic and aliphatic hydrocarbons, alcohols, ketones and esters. Water is the solvent in water-based and emulsion paints.

Pigments: Pigments provide the coating with colour, opacity and degree of durability to the paints. Pigments are either organic or inorganic. The major pigment used is titanium dioxide. This is white in colour and has a high refractive index (bends light and adds to hiding power).

Extenders: These are mainly used to modify the gloss level of paints as, the more pigment in a paint film, the flatter the sheen will be. Extenders are used to lower the sheen or gloss. These are also used to add body to paint and to increase its filling properties. The main extenders are calcium carbonate, talc, barytes and diatomaceous earth.

Additives/chemicals: These perform a special function or impart a certain property to the coating. Some of the additives include driers, thickeners, biocides, surfactants, dispersing agents, antifoams and catalysts, etc.

39.2 Waste streams in paint manufacturing

Major wastes that the paint industry must manage are the empty raw material packages, dust from air pollution control equipment, off-specification paint, spills and equipment cleaning wastes. Equipment cleaning wastes are a dominant waste stream.

Primary wastes specifically associated with paint manufacturing are:

1. Spills from accidental discharges.
2. Waste rinse water from equipment cleaning using water and/or caustic solutions.
3. Leftover raw materials during unloading of materials into mixing tanks.
4. Waste solvent from equipment cleaning using solvents.
5. Cotton waste from equipment cleaning.
6. Filter cartridges from undispersed pigments, etc.
7. Pigment dust from air during unloading of pigment.
8. Emission of volatile organic compounds from open processing equipments.

Wastes generated by the industry are usually managed in one of four ways: on-site reuse, on-site recycling, off-site recycling and off-site treatment/ disposal.

39.2.1 Wastewater

Water consumption during production process widely range depending up on the type of products manufactured at the time of study and the extent of water consumed in the cooling towers and floor washings. Thus, we could find no direct correlation between products manufactured and water consumed. Major portion of wastewater is generated from cleaning operations and its quality depends on the chemicals/solvents used for cleaning.

Wastewater from the paint manufacturing industries generally tends to be alkaline, contain some oil and grease and Biochemical Oxygen Demand (BOD), Chemical Oxygen Demand (COD) and suspended solids. Wastewaters may also contain small amounts of the solid products. The BOD and COD values give only a gross measure of organics in the wastes. Some of the organic and inorganic compounds used in the manufacturing operations are classified as hazardous.

Raw materials used have very less chances of entering the effluent stream, as majority of them being toxic are carefully used. These toxic powders, containing chromium, copper, lead, zinc, titanium, etc., cannot be traced as such in the effluent because during the manufacturing process they are mixed and blended with a variety of other substances and solvents. Phthalic anhydride is a common raw material widely used in Integrated paint industry for manufacture of resins. The components in the wastewater corresponds to the respective constituents in the wastewater, i.e., phenols in case of manufacturing phenolic resins, etc. These are at a high temperature and when condensed are traced in the condensate water.

They are in high concentrations in the condensate water but get diluted further when mixed with other effluent streams.

1. Wastewater from caustic cleaning: Caustic cleaning results in maximum production of wastewater in the paint industry. Caustic cleaning residuals are generated when equipment, tank and secondary piping are washed with water, caustic, or a blend of water and caustic. Wash water liquid residuals can be recycled. Typically the wash water is filtered prior to reuse which generates sludge. The sludge typically is disposed of at off-site landfills or sent off site for fuel blending. Wash water is disposed of predominately at off-site common wastewater treatment plants. Caustic wash water sludge requires proper offsite disposal at authorised facilities.
2. Wastewater from resin house: Water accumulates from condenser into separator. Water layer is disposed off in wastewater stream and the upper layer of solvents is reused. Water from separator is generally acidic.
3. Wastewater from water-based paints: Distemper is the main product in water based paints. Cleaning of hoppers, grinders and containers consume large amounts of water and contribute 400 to 700 mg/L of suspended solids, a large fraction of which is settleable.
4. Wastewater from solvent cleaning: Solvent cleaning residuals are generated when equipment, tanks and secondary piping are washed with a solvent or blend of solvents. Solvent cleaning wastes can be recycled back through paint production process. To extend the life of solvent wash water, many facilities send the solvent wash water through a reclamation process. The residues from reclamation process require proper disposal at authorised facilities.

39.3 Air emissions

The two major types of air emissions that occur in paint manufacturing process are volatile organic compounds and pigment dusts. Many paints are solvent-based and solvents are released during application, drying and (to a lesser degree) disposal. Solvents in paint add to volatile organic compounds (VOCs) in the atmosphere. Dust is generated during the assembly, pre-mix and blending steps and is a very fine powder comprising resins and pigments that were added into the batch. Generally, dust collectors are found over the tanks used during these steps to capture the emission and collect the dust in a bag house.

Emissions during fuel burning: (i) Flue gases are emitted from boilers for raising steam. Pollutants produced are minimal in case of Liquefied Petroleum Gas (LPG) and natural gases, somewhat higher for furnace oil, Low Sulphur Heavy Stock (LSHS) and light diesel oil (LDO) and maximum for coal, especially in form of particulates. Gases are generally discharged through 20–30 m high stacks, (ii) 'thermopacs' are often used for heating thermic fluids (e.g., oils) which in turn are used for heating in certain processes.

Such units are operated on furnace oils, LSHS or LDO.

Emissions during process:

1. Resin production:
   (a) Phthalic condenser-fumes from reaction kettles are passed through phthalic condensers. These condensers contain baffles to make the fumes follow a sinuous path, forcing the phthalic particles to deposit on the walls. The efficiency of the condenser is further enhanced by providing a water jacket for cooling the fumes.
   (b) Special chimney-here fine particles of phthalic are separated. Base of the chimney being conical helps the fumes to slow down to give better deposition of phthatlic particles. Weekly removal of phthalic particles should be done to maintain the trap efficiency.
2. Varnish production: Decomposition products of resin and oils are released as fumes. These fumes are scrubbed by jets of water spray.
3. Bituminous paint production: During bitumen melting, fumes are given out and scrubbed by water.
4. Air extraction: Air extractors and exhaust fans can be provided with suitably designed hoods at key positions to help improve ventilation in working areas. The air is dispersed through stacks. The flue gas emission from fuel sources should meet the standards prescribed in the emission regulations of the CPCB.

39.3.1 Volatile organic compounds

VOCs are released from several types of equipment and handling operations throughout paint manufacturing process and during cleanup operations. Emissions can be categorised according to the four manufacturing processes and cleaning operations.

Preassembly and premix: Majority of paint manufacturing facilities begin the process by thinning resins with solvents, alcohols, oils, and/or water. The equipment most often used in this premix and preassembly operation are drums and portable tanks in combination with mixers. Portable mix tanks, either alone or in combination with an agitator, are a common emissions source. Portable mix tanks are used to mix product and to keep the pigment in suspension. They are also used to transfer material from one manufacturing stage to the next. While they are being used for mixing, the tanks are often, but not always, covered with lids. If a cover is used on a mix tank during mixing, it will have a small opening through which the agitator shaft extends. In some cases, only a splash guard is used to cover the back half of the mix tank. If mix tanks are used for temporary storage, they are often covered with a solid lid. None of the lids seal with the mix tanks. Like the portable mix tanks, drums are often covered with non-sealing lids. If a cover is used on a drum during mixing, it will have a small opening for the agitator shaft.

Emissions occur during material loading when the tank or drum is uncovered or when the lid is open. VOCs may also be released through the agitator shaft opening and from around the edges of the lid during the mixing process. The quantity of material released varies with type of solvent, agitator mixing speed, material temperature and type of cover. More organics will be released with highly volatile solvents, increased agitator speed and warmer temperatures.

One source of emissions is scale systems, where solvent and resin raw materials are measured and transferred from storage tanks to the process tanks, between process tanks, or from process tanks to shipment containers. Emissions may occur during transfer and hose connecting and disconnecting. Another type of scale system consists of a floor scale, a drum, a drum dispenser and a receiving container.

Material is pumped out of the drum into the receiving container. Emissions occur during material transfer and free-fall into the receiving container.

In some cases, material is transferred by bucket and dip method. Here, emissions occur while the product is exposed to the air and while it is being scooped and transferred to the second container.

Another emission source is product filtering. As product flows through a filtering device, it is often exposed to the air, resulting in releases of VOCs.

Filling operations also result in VOC emissions. In one plant, portable mix tanks are mechanically lifted and tilted, allowing the finished product to gravity feed into containers for shipment. Some facilities allow product to gravity feed from processing equipment through filters into shipment containers. Emissions result from product freefall and material splashing.

Equipment cleaning: In addition to emissions from process operations, VOCs are also released from a variety of cleaning operations. Solvent-based materials are used to clean equipment in which solvent-based products are manufactured, while water-based supplies are used to clean after water-based production. Emissions occur during solvent addition and removal as well as during the cleaning process.

In many facilities, manufacturing equipment is cleaned manually on the production floor on an as-needed basis. In some cases, cleaning is performed after each batch and in other cases equipment is cleaned after a series of batches. The cleaning frequency depends on the number and size of batches processed, the size of equipment to be cleaned and the colour and type of product manufactured. The standard method of cleaning grinding equipment involves emptying the mill of product and then adding solvent to the vessel to capture remaining product residue. Wash solvent is normally drained from the tank and either disposed off as hazardous waste or recycled. Mix tanks and agitator blades may be cleaned with solvents, brushes, and/or rags.

Roller mills are often cleaned by hand using rags and solvent. Larger facilities may have areas designed specifically for cleaning operations. In these facilities, equipment cleaning may be more automated (i.e., automatic tank washers and spray guns), but emissions still occur during the process.

Equipment cleaning operations account for over 80% of paint industry's waste. Although solvents are not the only waste generated during cleaning processes, their contribution is high.

39.4 Solid and hazardous wastes

39.4.1 Solid wastes

Solid wastes generated from paint industry are as follows:

1. Off-specification paint, if not reused.
2. Retained product samples, is not reused.
3. Waste generated during effluent treatment.
4. Waste generated as scrap (cartons, drums, containers, etc.).
5. Paint skin.
6. Waste generated from spillage.

Small quantities of retained product samples or quality control samples are retained for quality reference. Most off-specification paint is reworked back into the process. Off specification paint which cannot be used or reworked back into manufacturing process or cannot be sold as a lesser grade product is usually stored in drums or tanks and sent offsite for disposal. Disposal options include incinerators, cement kilns, fuel blender or burned as a fuel.

Wastewater treatment sludge is generated from the on-site treatment of plant, equipment and tank cleaning washes and other miscellaneous wash water streams through physical and/or chemical treatment (e.g., sludge generated from the wastewater treatment of floor washings). Due to the size and nature of paint manufacturing facilities, wastewaters typically are not treated on-site. Facilities that treat their wastewaters on-site in tanks or at the on-site wastewater treatment facilities generate a sludge that usually is sent to an off-site landfill for disposal. Facilities also may send their waste off-site for incineration or to a fuel blender. The sludge generated is of two types – chemical and biological.

Chemical sludge comes from oil and grease removal in the oil and grease trap, coagulation and setting in primary clarifier. Biological sludge is the organic sludge arising from biological treatment which converts non-settleable organic matter to settleable biomass. Both the types are dewatered readily on sludge drying beds.

Solid wastes generated as scrap material can be classified into metallic and non-metallic. Metallic solids such as drums, containers, tins, etc., are sold as scrap. Non-metallic solids such as cartons, plastic bags, cotton waste, etc., are also sold.

Another solid waste is in the form of 'paint skin'. Sometimes after proper thinning and mixing, the batch is emptied in mild steel containers before packing. Paint sticking on the sides of containers dries and is removed as paint skin. This cannot be put to any use and is sold along with other solid wastes.

Powders and other raw materials spilt on the floor are often collected dry and disposed off by burial. However, landfilling of dewatered sludge from Effluent Treatment Plant (ETP) or burial of spilt powders can be done only in areas where the ground water table is low. In these areas there are remote chances of pollution of ground water by these solid wastes. In places where the ground water table is high, powders and spilt raw material may be better flushed down the drain with water and treated along with the wastewaters collected from other units. This will increase the load of suspended solids on ETP and its running cost marginally.

In areas where this waste is dumped close to drinking water sources, these wastes may enter the water body by runoff or as leachate. While disposing off solid waste by burial or by filling in low lying areas, due consideration should be given to the depth of groundwater table and nearness of fresh water sources to the site of disposal. Flow diagram Fig. 39.1 shows general sources of solid and liquid wastes generation and recovery of solvents.

Chemical and biological sludges

Both chemical and biological sludges can be dewatered readily on sludge drying beds. The dried sludges can be conveniently disposed off by filling in low-lying areas.

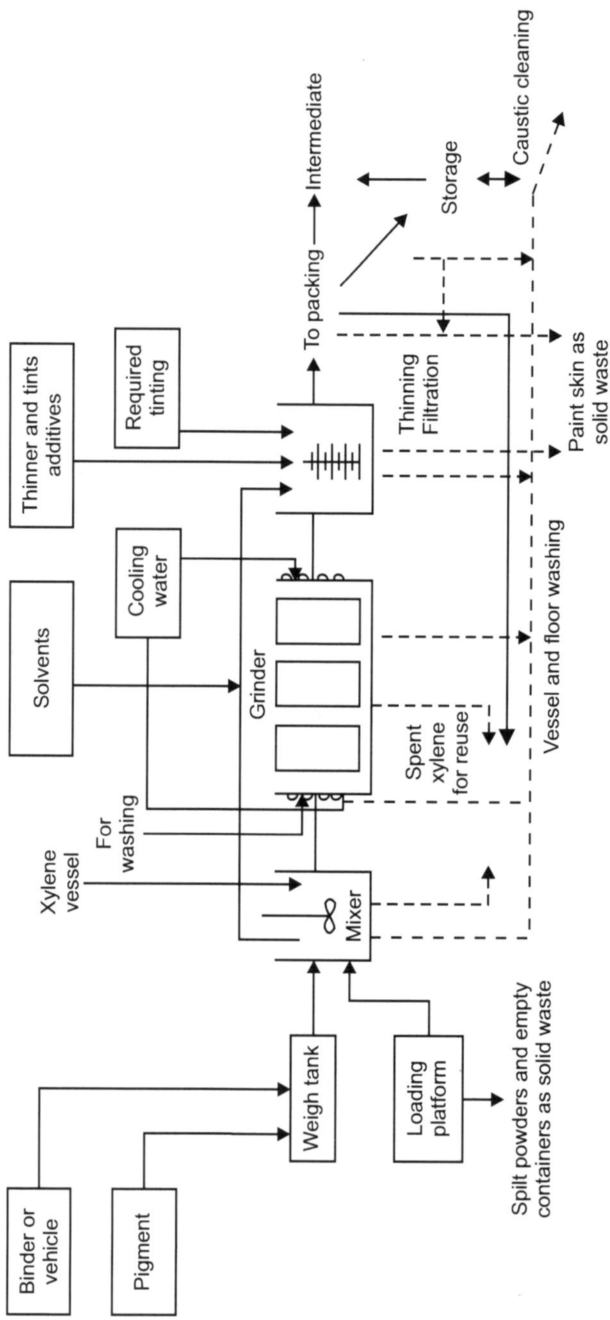

Figure 39.1: Flow diagram for manufacturing paints with general sources of solid and liquid waste generation and possible solvent recovery steps.

Metallic solids such as drums, containers, tins, etc., are sold as scrap. Non-metallic solids such as cartons, plastic bags, cotton waste, etc., are also sold.

Another solid waste is in the form of 'paint skin'. Sometimes after proper thinning and mixing, the batch is emptied in mild steel containers before packing. Paint on the sides of the containers dries and is removed as paint skin. This cannot be put to any use and is sold along with the other solid wastes.

Powders and other raw materials spilt on the floor are often collected dry and disposed off by burial. However, land filling of dewatered sludge from effluent treatment plant or burial of spilt powders can be done only in areas where the ground water table is low. In these areas there are remote chances of pollution of ground water by these solid wastes. In places where the ground water table is high, powders and spilt raw material may be better flushed down the drain with water and treated along with the wastewaters collected from other units. This will increase the suspended solids load on effluent treatment plant and its running cost marginally.

Also in areas where this waste is dumped close to drinking water sources, these wastes may enter the water body by runoff or as leachate. So while disposing solid waste by burial or by filling in low-lying areas, due consideration should be given to the depth of ground water table and nearness of fresh water sources to the site of disposal.

39.4.2 Hazardous wastes

Waste that might be hazardous during manufacturing process of paint includes the following:

1. Unusable liquid paints, stains, or inks.
2. Paint-thinner wastes of all types.
3. Paint spray-booth filters and arrestors.
4. Scrapings from paint booth walls and floors.
5. Paint-stripping waste.
6. Rags containing paint, ink, and/or solvent.
7. Sludge from distilling paint-thinner waste.
8. Blanket and fountain washes and other cleanup materials.

39.5 Waste minimisation technologies

Primary waste streams associated with paint manufacturing are listed in Table 39.1 along with recommended control methods. At the facility, the waste streams are equipment cleaning wastes; spills and off specification paint; leftover inorganic pigment in bags and packages; pigment dust from baghouses; filter cartridges and obsolete products. These waste minimisation methods as listed in Table 39.1 can be classified generally as source reduction, which can be achieved through material substitution, process or equipment modification, or better operating practices or as recycling.

Table 39.1: Waste minimisation methods for the paint manufacturing industry.

Waste stream	*Waste minimisation methods*
Equipment cleaning wastes (rinse water, solvent and sludge)	Use mechanical wipers on mix tanks.
	Use high pressure wash systems.
	Install Teflon liners on mix tanks.
	Use foams to clean lines.
	Reuse equipment cleaning wastes.
	Schedule production to minimise need for cleaning.
	Clean equipment immediately.
	Use countercurrent rinse methods.
	Use alternative cleaning agents.
	Increase spent rinse settling time.
	Use de-emulsifiers on spent rinses.
Spills and off specification paint	Increase use of automation.
	Use appropriate clean up methods.
	Recycle back into process.
	Implement better operating practices.
Air emissions, including VOCs and pigment dust	Modify bulk storage tanks.
	Use paste pigments.
	Install dedicated baghouse systems.
Filter cartridges	Improve pigment dispersion.
	Use bag or metal mesh filters.
Obsolete products/customer returns	Blend into new products.

39.5.1 Minimisation of wastewater

There is considerable variation in the nature of wastewater normally generated in paint manufacturing industry. Before evolving wastewater treatment scheme, the following aspects should be looked into:

1. Segregation of wastewaters based on characteristics and strength.
2. Reduction of volume and strength of wastewaters by adopting in-plant control measures

Treatability studies for various wastewater streams may be carried out to decide the best combination of treatment system.

Segregation of wastes is, however, recommended to reduce capital costs, improve the treatment efficiency and reduce chemical consumption. Effluents generated from paint industry can be segregated as caustic cleaning effluent, stiff paint effluent, effluent from remaining units and domestic waste

Caustic cleaning effluent is highly alkaline and needs neutralisation. Effluents from other units comprise thermopac burner cleanings, resin house waste, etc., which contain oil and has to be separated before mixing with the other effluents. Stiff paint effluent contains easily settleable solids which are settled in the primary clarifier and then wastewater is given further treatment.

Schemes for reducing the generation of wastewater at source (in the plant) should be practiced. This is to reduce the effluent load rather than finding methods to treat it. Unnecessary use of water not only adds to the quantity of effluent and cost of treating it, but also increases the wastage of heat, power and/or product in the effluent.

The cooling water is usually uncontaminated and thus should be collected and reused. It could be used for floor washing or discharged separately into the receiving water bodies, rather than mixing with polluted water and discharging into the treatment plant. Accidental spills and leakages should be reduced to a minimum through proper maintenance of equipment and training of personnel. In case of caustic cleaning, instead of washing away the caustic solution, it can be collected, stored and used for further cleaning. In case of stiff paints, water from first cleaning should be collected and used later as process water for a similar type of batch. Wastewater volume can also be reduced through reuse of rinse water for preparation of alkali solution. The above procedures can reduce the quantity of generation of caustic cleaning water significantly.

The dry powders and raw materials spilt on the floors during filling into the grinders and/or churners as well as while packing the finished products should be removed dry as far as possible and disposed of as dry solid wastes by burial instead of washing them with water. This will significantly reduce the pollution load of the combined wastewater, which in turn will reduce the size of the treatment plant required and its cost. Alternatively, the above material along with their containers may be sold to outside parties as scrap material.

39.5.2 Minimisation of VOCs

Methods to minimise VOC emissions include process and equipment modifications, improved operating practices and recycling. It is difficult, however, to determine the overall efficiency or impact of these VOC minimising methods on individual emission sources because many paint manufacturing facilities estimate total plant emissions rather than estimating or testing emissions by process or source (i.e., filling operations, grinding operations, cleaning processes).

Equipment or process modifications

Two stages which are amenable to equipment and process modifications are paint manufacturing and equipment cleaning.

Tank lids: Tank lids are the most common equipment modification used during the manufacturing process to control VOC emissions. Mix and blend tanks are a primary source of manufacturing VOC emissions because the solvent-containing materials spend a significant amount of time in this equipment. All open-top equipment may be covered during the manufacturing process to control these emissions.

The cover remains closed, except when production, sampling, maintenance, or inspection procedures require access. The cover is maintained in good condition, such that when in place, it maintains contact with the rim of the opening for at least 90 % of the circumference of the rim.

Many of the lids currently used in industry are flat and some are conical. Flat lids control emissions relatively well, but they do have some inherent flaws. The lids do not form a seal with the mix tank and the hinged door product adds chute does not always remain closed. Conical lids, a better engineering design, are considered as more efficient means of controlling emissions. However, they too have associated difficulties caused by added weight and bulky shape. The conical lids are more difficult to handle and damage more easily than the flat lids.

Lids may be constructed of plastic, wood, aluminium, or stainless steel. Plastic and wooden lids are normally one piece except for the center agitator shaft opening, while aluminium and stainless steel lids normally have hinged openings for product additions and sampling. Some facilities currently using aluminium lids question their safety. A study conducted in Germany indicates that having steel (e.g., carbon steel mix tank) scraping against aluminium containing silicon (e.g., mix tank cover) could be a potential source of sparks. A fire may break out if the sparks contact possible flammable vapours from solvent-containing paints. The control efficiency of covers on mix tanks ranges from 40 to 96 % depending on the method used to determine emissions. These values represent the ratio of emission reduction to uncontrolled emissions. They do not account for any subsequent venting to control devices. The value of 96 % arose from studies conducted with mix tanks in the polymeric coating industry. In this case, the demonstrator considered only evaporative losses during the mixing process. This method of emission determination fails to include the working losses that occur during filling and emptying a vessel containing a solvent saturated air space.

Modified milling equipment: In some cases paint manufacturers could reduce total VOC emissions by converting some of their older milling equipment to newer, more efficient closed-systems such as horizontal media mills. Although a wide range of products can be processed in the horizontal mills, some cannot be done. The mill base must be of a low viscosity to allow the grinding media to move with maximum velocity. The low viscosity requirement prevents some materials currently made in other types of milling equipment from being manufactured in horizontal mills. The viscosity of a product, along with other characteristics such as colour, gloss, type of raw materials and processing time, often determines the appropriate type of milling equipment.

39.5.3 Pollution control technologies

Wastewater treatment

Combined effluent from paint industries can be satisfactorily treated using the usual physicochemical and/or biological treatment methods. The treatment consists of coagulant addition and adjustment of pH to an optimum level for maximum precipitation. The precipitated material is removed by gravity separation, either on batch basis or in a continuous flow tank. It is understood that if wastewater contains a high settleable solids and the wastewater quantity is low, primary treatment followed by treatment in oxidation pond may result in reasonably high BOD removal but for a higher quantity of wastewater primary treatment should be followed by secondary treatment. On the other hand, removal of COD always demands tertiary treatment. The wastewater treatment removes oil and grease, suspended solids and toxic substances.

Primary treatment:

1. Oil and grease removal: Effluents from all units except stiff paint section and caustic cleaning waste are passed through an oil and grease removal device.
2. Equalisation-cum-neutralisation: Effluent from caustic cleaning operation is highly alkaline in nature and requires neutralisation prior to further treatment. An equalisation-cum- neutralisation tank is provided with an agitator. Effluent from stiff paint is mixed with the neutralised wastewater, dosed with a coagulant and sent to flash mixer. The effluent is then subjected to clariflocculation.
3. Clariflocculation: The effluent is clarified in clariflocculator and subjected to biological treatment. Sludge generated in this unit is carried to the sludge drying beds for dewatering.

Secondary treatment:

1. Extended aeration: Domestic wastewater from the factory premises is mixed with the supernatant from clariflocculator and is biologically treated by extended aeration process.

2. Secondary clarification: Mixed liquor from the aeration tank overflows to the - secondary clarifier. The settled sludge is recycled continuously through return sludge pumps to the aeration tank and excess sludge is discharged to sludge drying beds. Effluent from the secondary clarifier is fit for discharge to the environment.
3. Sludge drying: Sludge from oil and grease trap, clariflocculator and secondary clarifier is dewatered on sludge drying beds. Filtrate from these beds is returned to equalisation-cum-neutralisation tank

The above treatment process is expected to achieve 90 to 95% efficiency in removing pollutants and thus acceptable to the recipient environment.

Recycling techniques

One common recycling technique among paint manufacturers is using spent cleaning solvent in subsequent compatible batches. After a mill or tank has been emptied of product, solvent is added to the vessel to capture remaining product residue. The wash solvent is drained from the tank, staged and recycled into the next compatible product batch. Mills may be cleaned by replacing the residual heel of the exiting product with an equivalent amount of solvent which is compatible with both the preceding and the ensuing batches.

Another recycling technique which reduces total solvent consumption and VOC emissions is using countercurrent rinsing sequences. This method uses recycled 'dirty' solvent to initially clean the tank. Following this step, 'clean' recycled or virgin solvent is used to rinse away the 'dirty' solvent.

Recovery of paint from wastewater

Most off-specification (off-spec) paint is produced by small shops that deal in specialty paints. Since these paints cost more to produce and therefore sell at a premium price, most off-spec paint is reworked into a salable product. Since elimination of off-spec paint production has built-in economic incentives, the following techniques are widely used:

1. Unless the sludge from wet cleanup can be recycled into a marketable product, the use of dry cleanup methods should be maximised wherever possible.
2. By closing floor drains and discouraging employees from routinely (i.e., needlessly) washing down areas, some facilities have been able to achieve a large decrease in wastewater volume.
3. By employing volume-limiting hose nozzles, using recycled water for cleanups and actively involved supervision.

Bags and packages

Inorganic pigments, which may contain heavy metals and therefore be classified as hazardous, are usually shipped in bags. After emptying the bag, an ounce or two of pigment usually remains inside. Empty containers of liquid raw materials that constitute hazardous waste (e.g., solvents and resins) are typically cleaned or recycled to the original raw material manufacturer or to a local drum recycler.

The following are some of the waste reduction techniques for bags and packages:

1. When empty, the bags could be dissolved or mixed in with the paint. Such a method is commonly used for handling mercury compounds and other paint fungicides. This method cannot be used, however, when producing high quality, smooth finish paint since the presence of this material could affect the paint's film forming property or could increase the load on the filters which would increase filter waste.
2. Use of rinseable/recyclable drums with plastic liners instead of paper bags.
3. Segregation of hazardous and non-hazardous wastes.
4. Hazardous materials may be collected in plastic bags and stored in a special container to wait collection.

Air emissions

As mentioned in the previous sections, the two major types of air emissions that occur in the paint manufacturing process are volatile organic compounds and pigment dusts. Volatile organics may be emitted from the bulk storage of resins and solvents and from their use in open processing equipment such as mix tanks. Since most existing equipment is of open design, reducing or controlling organic emissions from process equipment could require substantial expenditures in retrofit costs. Following are some of the measures for bulk storage and pigment handling.

1. Use of pigments in paste form instead of dry powders: Pigments in paste form are supplied in drums which can be recycled and no dust would be generated when opened.
2. Dedicated baghouse system for pigment loading area: All the collected pesticide dust could be recycled.

Spills

Spills are due to accidental or inadvertent discharges usually occurring during transfer operations or equipment failures (leaks). Spilled paint and the resulting clean up wastes are usually discharged to the wastewater treatment system or are directly drummed for disposal. If the plant has floor drains, large quantities of water may be used to clean up water-based paint spills. Dry cleaning methods are employed for cleaning of solvent-containing spills or for water-based spills where floor drains are not available. Some of the better operating practices may include:

1. Dry cleanup methods can be maximised wherever possible until sludge from wet cleanup is recycled.
2. Employing volume-limiting hose nozzles, using recycled water for cleanups and actively involved supervision.

Filter cartidges

These are produced during the paint loading operation. These are designed to remove undispersed pigment from the paint during loading and are saturated with paint when removed. Hence, waste minimisation and economy both call for as small a cartridge as possible so as to reduce the amount of paint lost and the capital spent for the filters. If frequent filter plugging is a problem, then it should be first addressed from the standpoint of improving pigment dispersion and not from the standpoint of increasing filter area.

Viable alternatives to cartridge filters include bag filters and metal mesh filters. Metal mesh filters are available in very fine micron sizes and they can be cleaned and reused. Since it is very important to minimise all wastes, the issue of mesh filter cleaning waste reuse or recyling would need to be addressed before switching to these filters.

Obsolete products

Obsolete products and customer returns can be blended into new batches of paint. Obsolete products result from changes in customer demand, new superior products and expired shelf life. Marketing policies, such as discounting older paints, can reduce the amount of obsolete products requiring disposal.

Recovery of wastes

A large number of solvents are used in paint manufacturing and a majority of them are recovered and therefore not lost in the wastewater streams. In case of oil paints, solvents are added in grinders which are closed units; therefore, loss of solvents through evaporation is considerably reduced. High temperature is maintained in resin and varnish manufactures, resulting in evaporation of solvents added. These solvent vapours, along with the water vapours generated through chemical reactions are condensed and collected in a separator. The solvent layer is removed and reused in the next batch.

40

Pollution control in dyes and dye intermediates industry

40.1 Introduction

A dye is a coloured substance that has an affinity to the substrate to which it is being applied. Dyes are any substances synthetic, used to colour various materials and have wide industry applications ranging from textiles, leather, food and paper, etc.

Dyes and dye intermediates are synthesised in a reactor, filtered, dried and blended with other additives to produce the final product. In case of dyes, the synthesis step involves reaction such as sulphonation, oxidation, reduction, isolation, halogenation, amination, diazotisation and coupling followed by separation processes, which may include distillation, precipitation and crystallisation. The dye is then separated from the mixture and purified. On completion of the manufacture of actual colour, finishing operation including drying, grinding and standardisation are performed and these are important for maintaining consistent product quality. In case of dye intermediates, the process steps cover sulphonation, oxidation, reduction, isolation, filtrations and drying which are unit processes and operation.

Filtration: Filtration is most widely used separation process. Equipments used to carry out filtration are filter press, centrifuge and cake filters. At present, filter press and nutsche filter are most common equipments used by the industries.

Drying: Drying is the process of removing liquid from solids by evaporation. Different types of dryers used in the industry are hot air dryers – bin dryers, cabinet dryers – tray dryers, continuous conveyor dryers – belt dryers, pneumatic dryer, fluidised bed dryer, spray dryer.

To improve drying efficiency with energy conservation, following options are given like improving existing method of tray drying, flash drying, spray dryer, solar drying techniques.

Sulphonation: The introduction of sulphonic group in an aromatic nucleus is known as 'sulphonation'. This group is introduced to render intermediates soluble in water or to provide a route to other substituents, such as the hydroxyl group, which is obtained by subsequent alkaline fusion.

Sulphonation of the aromatic compounds is carried out by heating the compounds with the sulphonating reagent during which water is eliminated. The process in manufacturing of dye are shown in Fig. 40.1.

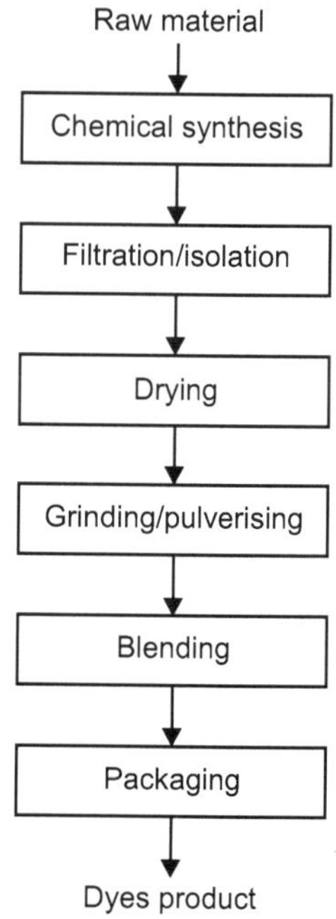

Figure 40.1: Operation sequence in dye manufacture.

The current technology uses following sulphonating agents because of their low cost, ease of availability and handling:

1. Sulphuric acid.
2. Oleum.
3. Sodium bisulphate.

Reduction: The most common reduction method in the manufacture of dye intermediates is the conversion of a nitro compound to the corresponding amines. The only reducing agent used in this sector is iron powder with acid. This technology is employed on a large scale because of its simplicity. The effective reducing agent is nascent hydrogen, which produced by the reaction of iron and water having a small HCl acid to promote a reaction. Using iron powder as reducing agent generates large amount of iron oxide sludge, which is hazardous and therefore required to be send to the landfill sites.

The dyes and dye intermediate industries sector are one of the most polluting industrial sectors. The waste generated from this sector is highly toxic/ hazardous, difficult to treat and very large in quantum. The causes of severe pollution generation from this sector is discussed below:

1. Multistage batch operations.
2. Poor process control of the unit operations/processes.

3. Poor housekeeping practices.
4. Use of obsolete technology.
5. Poor or no recovery of by-products.
6. Excessive use of reactants, solvents and utilities.
7. Poor quality control.
8. Poor maintenance and record keeping.
9. Unskilled/untrained workers.

40.2 Dyes intermediates

Dye intermediates are petroleum downstream products, which are further processed for any application. On processing they are transformed to finished dyes and pigments. The dye intermediates serve many industries like textiles, plastics, paints, printing inks and paper. Apart from this, dye intermediates also serve as an important raw materials for the acid, reactive and direct dyes. A major application of dye intermediates are found in hair dyes.

Two types of dye intermediates, that are broadly used in the hair dyes are paraphenylenediamine and paratoluenediamine.

The following sub-sections give brief descriptions of the various wastes generated by the sector.

40.2.1 Air emission

Mostly, in dye manufacturing Volatile Organic Compounds (VOCs), nitrogen oxides (NO_x) hydrogen chloride (HCL) and sulphur oxides (SO_x) are emitted. Moreover, air pollution is caused by emission of particulate for example dust during material transport, grinding or as a result from open dumping while drying material such as spray dryer. The other source of air pollution from this sector is the particulate matter emission from the drying and grinding operations. Flue gas from boilers is also a source of air pollution.

40.2.2 Liquid wastes

The environmental problems attributed to this industry arose first after industrialisation replaced traditional natural dyes by the synthetic dyes used today. Since synthetic dyes are designed to resist bleaching by ultraviolet light and chemicals to improve the quality of the end product, they are also persistent in the environment and some dyes can be biologically modified into carcinogenic compounds. The release of untreated wastewater poses a threat to the environment and the most serious problems are ground water and surface water pollution. Further, the discharge of coloured effluents into water bodies affects the sunlight penetration which in turn decreases both the photosynthetic activity and dissolved oxygen levels. The removal of dyes from wastewater is one of the major environmental challenges. Since the majority of dyes are recalcitrant to conventional biological methods the most common techniques used today are physicochemical in nature.

However, many of these methods suffer from limitations. They are usually not cost-efficient and most of them lead to generation of hazardous waste that needs to be dealt with separately.

The quantum of wastewater generated per T of dyestuff production is very low compared to that generated per T of dye intermediate production. The specific wastewater generation for major dye intermediates is about 15–20 m^3/T of product. The main sources of wastewater generation are:

1. Mother liquor or filtrate streams from filtration operations.
2. The wastewater streams from the washing of filter cake to remove either salt impurities or residual filtrate adhere from the cake.
3. Leakage and spillage.
4. Floor washing of the work area.

The typical characteristics of the wastewater from this sector are mentioned in the Table 40.1.

Table 40.1: Typical characteristics of wastewater from the production of dyes and dye intermediates.

Parameter	*Value*
Colour	Varying deep colours
pH	4–6
Chemical oxygen demand (COD)	50,000–1,00,000 mg/L
Total dissolved solids (TDS)	15,000–2,00,000 mg/L
Ratio of biochemical oxygen demand (BOD) to COD	< 0.2
Sulphates	61,000 to 73,000 mg/L
Sodium	40,000 to 47,000 mg/L
Chlorides	16,000 to 20,000 mg/L

The effluent discharged from this industrial sector is highly acidic and contains toxic compounds; many of them are carcinogenic and highly hazardous to human health and the environment. This is due to the presence of benzene, naphthalene and other nitro-aromatic based compounds in the wastewater, which are used as raw materials during the production of dye intermediates. Due to the excess use of acid and alkali quantity, the wastewater contains high concentration of inorganic salts that results in the high concentration of TDS. Due to this, treatment of effluent is very difficult and highly expensive.

40.2.3 Solid wastes

Major solid wastes such as sludge are generated from effluent treatment process and container residues. In conventional wastewater effluent treatment process sludge produced after it is treated, it becomes sludge cake that is mainly contains gypsum. Some industries sell it to cement manufacturers or dispose at landfill.

Large quantities of solid waste are generated from the dyes industry. In fact, solid waste items generated in the process of the production and industrial use of synthetic dyes, dye intermediates and pigments account for 26% of the entire volume of hazardous wastes sent to treatment, storage and disposal facilities. Disposing off such large volumes of hazardous waste entails a huge cost to various dye industries.

The sources of waste generation are both from the process itself as well as from the treatment of wastewater. For instance, a large quantum of gypsum sludge is generated from the primary treatment of acidic wastewater, where lime is used for neutralisation. The types of wastes from the process include gypsum sludge, iron sludge, residues from the filter press, tarry waste, waste dye powder and packaging material. These solid wastes contain toxic chemical compounds and thus proper disposal or treatment is required to minimise their serious impacts on ecology and environment.

40.2.4 Gaseous emissions

The major gaseous emissions from the dyes and dye intermediate industry are unrecovered gases generated from the process. Emissions generally observed as a by-product of unit processes and fugitive emissions contain gases like chlorine, sulphur dioxide, sulphur trioxide, nitrogen oxides and fumes of acid and organic solvents. If these gases are not recovered and marketed as by-products, they pose a serious pollution threat. Nevertheless, complete recovery is not possible and the unrecovered product is vented out through the stack.

40.3 Cleaner production

Conventional approaches to industrial pollution management tend to focus on end-of-pipe treatment of wastes. There is no attempt to reduce waste streams at the source and/or undertake resource recovery. However, this kind of approach leads to wastage of resources, associated costs of managing the pollution generated, environmental concerns and poor worker health and safety.

Instead, the concept of cleaner production provides a better solution. According to the researchers the capital cost of only effluent treatment comes to 8-10% of the total investment of the unit and recurring cost towards effluent treatment comes to around 5–6% of the sales price. The reduction of waste at source through cleaner production can be the only solution productive for this problem. Cleaner production is a preventive business strategy designed to conserve resources, mitigate risks to humans and the environment and promote greater overall efficiency through improved production techniques and technologies.

In addition to environmental, health and safety benefits, many cleaner production techniques provide opportunities to substantially reduce operating costs and improve product quality.

Cleaner production makes sound business sense to industries and service providers as it often results in cost reduction, improved productivity and enhanced competitiveness of wider markets.

Cleaner production is also known by other names such as pollution prevention, waste minimisation and green productivity, but for all intents and purposes, these terms mean one and the same thing.

40.3.1 Benefits of cleaner production

The benefits of cleaner production are discussed below:

1. Conservation of raw material and energy: Given the increasing cost of raw materials and the growing scarcity of good quality of water, no industry can afford to use these resources inefficiently. Cleaner production measures help in overcoming constraints posed by scarce, or increasingly costly, raw materials, chemicals, water and energy.
2. Lower costs: The ultimate goal of cleaner production is to minimise the generation of emissions and waste. In doing so, the amount of waste and emissions that need to be treated is reduced, as are associated costs.
3. Improved environment: Cleaner production minimises not just the amount but also the toxicity of waste generated. This leads to a decrease in the pollution load thus leading to a cleaner environment.
4. Better compliance with environmental regulations: By minimising waste generation, cleaner production makes it easier for an industrial unit to meet existing environmental regulations and standards, which may not have been the case prior to adopting cleaner production.
5. Enhanced working environment: Cleaner production not only improves the environment outside the industry but also improves it on the factory floor itself. It reduces the likelihood of accidents at the work place and increases worker productivity.
6. Improved quality: Through its focus on the process, cleaner production can lead to improvements in the quality of the product, which in turn leads to better market share or improved profitability.
7. Improved efficiency: Cleaner production leads to better efficiency of production, which means more output of product per unit input of raw

materials. An improvement in efficiency also leads to better financial performance for the industrial unit.

8. Market requirements: Increasing consumer awareness about environmental issues has brought about a need for the companies to demonstrate the environmental friendliness of their products and manufacturing processes, particularly in international markets. By adopting the cleaner production approach, many of these market requirements can be met and a company's ability to compete and get access such markets increases.
9. Public image: The environmental profile of a company is an increasingly important part of its overall reputation, particularly when the common man is aware about issues pertaining to the environment. Adopting cleaner production helps the concerned company build confidence with the public regarding its environmental responsibility.

Cleaner production assessments

Exploring the various possible benefits of cleaner production is done through what is known as the cleaner production Assessment (CPA). The CPA consists of a systematic and an objective review of manufacturing processes, products and services. It is designed to identify cleaner production opportunities for increasing productivity, profitability and sustainability of the enterprise, while reducing its environmental impacts and associated risks. The approach can be sub-divided into a number of tasks, as shown in Fig. 40.2.

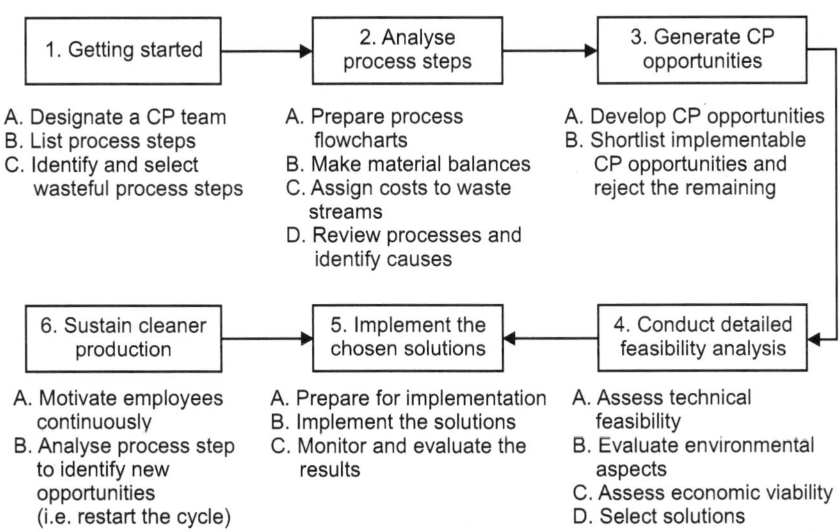

Figure 40.2: The cleaner production assessment (CPA) approach.

The predominant aspect of the methodology is the examination and re-evaluation of the production process. The activity starts with the basic aspect of team formation, followed by a process study, material balance, stoichiometric calculations and waste and energy audits to establish and characterise the sources as well as quantities of wastes generated. Investigating the factors that influence the volume and composition of the generated waste and emissions is also required.

Step 3 of the above figure is concerned with the generation of 'cleaner production opportunities'. What are these? These are the various possible ways to derive cleaner production benefits and are driven by cleaner production techniques.

The following section focuses on the types of cleaner production techniques and will provide examples for each within the context of the dyes and dye intermediate sector.

Lastly, as with other investments or technological innovation options, the implemented cleaner production measures should be taken up for performance evaluation, to ensure they perform as intended.

40.4 Cleaner production techniques

Cleaner production may be achieved by adopting one technique or a combination of various techniques, as shown in Fig. 40.3 and explained in detail in this section. In general, the techniques entail: (a) waste reduction at source, (b) recycling, (c) product modification and (d) energy conservation.

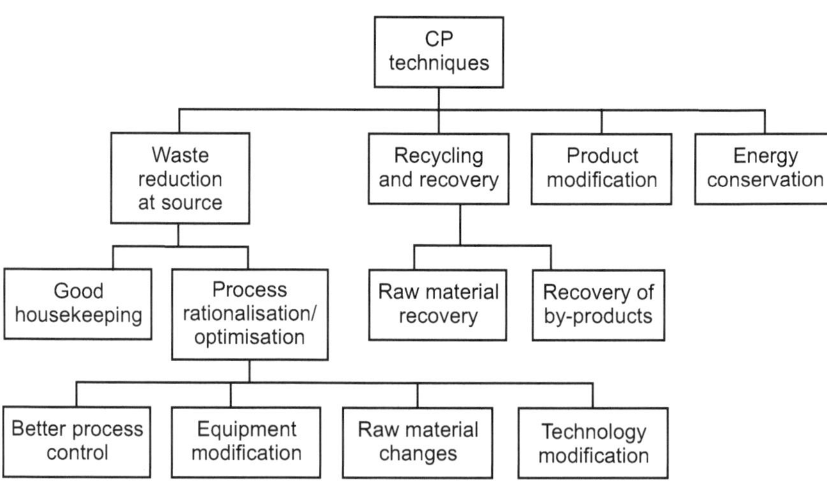

Figure 40.3: Cleaner production techniques.

40.4.1 Waste reduction at source

As the name suggests, reducing waste at the source itself is one of the basic tenets of cleaner production. This may be achieved through good housekeeping and process rationalisation/optimisation.

Good housekeeping

Good housekeeping entails improvements to work practices and methods and proper maintenance of equipment. These actions alone can produce significant benefits in terms of saving resources. Housekeeping related techniques are low cost and provide low-to-moderate benefits. Examples include training for proper material handling, minimising materials loss (e.g., stopping leakages), storage of materials in compatible groups, implementing a predictive and preventive maintenance system, using energy efficient devices, etc.

Typically, the following elements are covered under this technique.

Procedural measures	• Changes in routine but faulty operating practices • Maximisation of batch size to reduce cleaning waste • Dedication of equipment to a single product • Optimisation of raw material usage
Prevention of fugitive emissions, spillages and leakages	• Training for detecting and minimising material loss • Provisions for prevention of fugitive emissions, spillages and leakages • Spilled material management system • Reward schemes for attitudinal development of shop floor personnel
Waste stream segregation	• Preventing mixing of hazardous and non-hazardous wastes • Storage of materials in compatible groups • Segregation of different spent solvents • Isolation of liquid wastes from solid wastes
Improvements in material handling	• Training of employees for proper material handling practices • Providing appropriate tools for material handling

Process rationalisation/optimisation

Process rationalisation/optimisation involves slight reworking of the process sequence (for example the elimination of a redundant washing sequence), combining or modifying process operations to save on resources and time and improve the process efficiency.

In some cases, the changes may best be produced by piloting or demonstrating on a small scale. These techniques are typically low-to-medium cost and provide moderate-to-high benefits.

Process rationalisation in turn could happen in a number of ways – better process control, equipment modification, raw material changes and changes to technology (including technology upgradation). Given below are some examples of process rationalisation/optimisation which have been recorded in the dyes and dye intermediate industry.

Equipment modification

As the name suggests, modification of equipment can also serve as a cleaner production technique. These techniques are typically low to medium cost and can provide moderate to high benefits.

The following elements are covered under this technique:

Equipment improvement	• Replacing of equipment
	• Redesigning and retrofitting of equipments
Layout changes	• Rearranging equipment for better material movement

Raw material changes

Raw materials may be changed if better options exist in terms of costs, process efficiency and reduced health and safety related hazards. Such an approach may be necessary if the materials already in use are difficult to source, have become expensive, or have come under the purview of new environmental regulations. Before adopting raw material changes, it is essential to assess results through laboratory/bench-scale studies and pilots to ensure that the product quality is not changed and is acceptable to the market. Given below are some examples of raw material changes which have been recorded in the dyes and dye intermediate industry.

Technology modification

Modifying a technology or upgrading to a new technology can often reduce consumption, minimise wastes as well as increase the throughput of productivity. However, the uptake of such initiatives is largely dependent on the circumstances of each unit.

Given below are some measures of technology modification/upgradation possible in the dyes and dye intermediate industry.

The technology modification/upgradation technique may be capital intensive in some cases, but can lead to potentially high benefits. Due to the initial (i.e., capital) expense incurred, technology upgrades have been cited as an unfulfilled need for the dyes and dye intermediate sector.

Recovery and recycling

The recovery and recycling technique entails material recovery followed by on-site recycling and the recovery of by-products. The effectiveness of this technique depends on the segregation of the recoverable waste from other process wastes. This technique is typically low to medium cost and can provide moderate to high benefits.

Material recovery

It is possible that some waste streams or constituents thereof can be reused directly as raw material in the original production process, or substituted without reprocessing into another process. A number of physical and chemical techniques are available for such recovery. Most recovery systems generate some type of residue and the management and associated costs of these residues in turn determine the feasibility of material recovery in many cases. Wash water waste streams and solvents can be considered for possible recovery opportunities.

Recovery of by-products

In this industry in particular, by-products recovery from waste generated is a possibility worth exploring. Recovered by-products may have useful applications within the industry itself or outside it. Such recovered material not only leads to a minimisation of waste but cost savings as well.

Product modification

Modifying a product can cause impacts on both the 'upstream' as well as 'downstream' side of the product life cycle. Product redesign can, for instance, reduce the quantity or toxicity of materials in a product, or reduce the use of energy, water and other materials during use, or reduce packaging requirements, or increase the 'recyclability' of used components. This can lead to benefits such as reduced consumption of natural resources, increased productivity and reduced environmental risks. Often, this helps in both establishing as well as widening the market. At times, product modification may include modifying the packaging as well.

Product modification is however a major business strategy and may require feasibility studies and market surveys, especially if the supply-chain around the product is already established and is complex.

Energy conservation

This technique is an integral and yet often overlooked part of cleaner production. Industries are major consumers of energy thermal as well as electrical. With the rising costs of fuel (and therefore energy), it has become imperative to consider energy conservation as one way to help industry reduce processing costs. Moreover, with the added benefit of green house gas emission reductions and subsequent monetisation of the same, industries would be well-advised to consider this technique as part of their CPA. In the dyes and dye intermediate industry however, energy consumption plays a minor part as compared to raw material wastage and waste generation.

40.5 Environmental issues in dyes and dye intermediates industry and CP options

1. Gaseous emission such as SO_x, NO_x, HCl and NH_3 are generally scrubbed. So, properly designed scrubber with recovery reuse of scrubbed liquid is required.
2. Centrifuge is a highly energy intensive operation, which requires running time of around 12 hours a day. Optimum time required for removal of moisture is around 30–35 minutes. But because of the negligence of the worker, almost it runs for around 40–45 minutes or 1 hour consuming more energy. To overcome this, a simple timer can be installed on centrifuge. The digital timer will be connected to the centrifuge motor. Now after installing the timer one can set definite time for which they want to operate the centrifuge. As soon as the set time will be over, the motor will automatically stop running and thus saving the electrical energy.
3. The most commonly filtration system used are filter press and nutsche filter. Because the all products in dyes and dye intermediates are highly water soluble, there is high filtration loss.
   (a) The end product of filter press contains higher moisture content which increases the load on the dryer.
   (b) The technology may be suggested which either can reduce the moisture content of end product of filter press or some methods which can recover the filter loss which is otherwise inevitable.

   Use agitated nutsche filter which is a closed vessel designed to separate solid and liquid by filtration under pressure and vacuum, it offers an economical operation where maximum percentage of liquid in slurry is separated through mechanical means.
4. Use (i) improving existing management practice – efficient energy utilisation in drying, (ii) flash dryer, (iii) spray dryer, (iv) solar dryer as a technology up gradation option for drying.
5. Use sun drying: Use solar system for heating thermic fluid up to 180°C. One chamber with trays for solar drying of dye material can be used.

Material may be kept in trays and hot air is blown in the chamber. Using solar energy, hot air is generated and that can be blown by two blowers with sensor to control temperature. Benefit of the system could be: (i) reduction in time for drying, (ii) elimination of dust contamination (value addition to the material) reduction in labour cost.

6. Possibilities for adoption of cleaner process option for reducing the water consumption and effluent generation; better management practices for segregation and reuse/recycle of the treated effluent; effective utilisation of raw materials; improvement in efficiency of process. The effluent generated from manufacturing of some of the dyes and intemediates such as H-acid is not biodegradable, which requires process change.
7. Sulphonation with SO_3, chloro sulphonic acid and sulphamic acid can reduce effluent generation.
8. For reduction process by-product from iron sludge can be produced. Cleaner process technologies, e.g., catalytic hydrogenation, use of spent acid after nitration for acidification of fusion mass, which can eliminate generation of iron and gypsum sludge.
9. Areas which can be addressed with common facility: There are certain areas wherein technology up gradation is possible with collective approach i.e., common facility can be created like Common Effluent Treatment Plant (CETP) and secured land fill site. The areas are as under:
   (a) Common drying facility.
   (b) Gypsum recovery.
   (c) Spend sulphuric acid concentration and reuse.
10. On – site reuse and recovery. It is recycling or reuses a waste material each of originating process that is substituted input material or another process within the same production site.

40.5.1 Gaseous

1. Separation of entrained liquid from reactors.
2. Separation of dust from dryers, pulverises.
3. Condensation of organic vapours.
4. Adsorption of uncondensed organic.
5. Scrubbing of HCl by multiple scrubbers to get concentration HCl from 1st scrubber.
6. Scrubbing of lean HCl, Cl_2 with alkali.
7. Recovery of NaSH, Na_2SO_3 by scrubbing H_2S, SO_2 by NaOH.
8. Treatment of scrubber solution.
9. Incineration of toxic gases.

40.5.2 Liquid

1. Segregation of effluent streams and recycling wherever possible.
2. Recovery of sodium sulphate from aqueous effluent.
3. Recovery/reuse of H_2SO_4.
4. Removal of non-biodegradable sulphonic acids by extraction.
5. Removal of heavy metals by precipitation.
6. Biodegradable and biodegradable impurities.
7. Removal of colour by adsorption/oxidation/bleaching.

40.5.3 Solid

1. Recovery of salts such as sodium sulphite, sodium sulphate.
2. Effective washing of gypsum sludge and use in cement manufacture.
3. Recovery of mercury naphthalene from their sludge.
4. Manufacture of pigment from iron sludge.
5. Regeneration of spent carbon.
6. Incineration of organic residue.
7. Secured land filling of ash, Effluent Treatment Plant (ETP) sludge.

40.6 Pollution control in H-acid and G-acid

In this section we are discussing pollution from two important dye intermediates such as H-acid and G-acid.

H-acid (1-amino-8-naphthol-3,6-disulphonic acid)

It is in the form of a grey powder and is soluble in water, alcohol and ether. Having a molecular weight of 319.31, it is prepared by fusion of 1-naphthylamine-3,6,8-trisulphonic acid (koch acid) with 30% NaOH solution at 180°C. It is used as an azo dye intermediate. H-acid is a crystalline acid made from napthalene and used as a dye intermediate.

G-acid (2-naphthol-6,8-disulphonic acid)

It is in the form of white needles, soluble in water and has a molecular weight of 304.30. It is prepared by sulphonation of beta-naphthol with 94% H_2SO_4 at 95°C, yielding a mixture of G-acid (10%), R-acid (15%) and schaeffer acid (56%). Separation is effected by dilution with water, boiling to hydrolyse the sulphonic acid group and addition of metallic salts. It is used as an azo dye-intermediate. Liquid effluent is mainly the mother liquor from H-acid isolation and some washes. Additional mother liquor comes from Koch acid isolation in industries which are isolating it. This mother liquor, after H-acid and Koch acid isolation, has been subjected to HPLC analysis for estimating the organics present. The major organic components present were identified and characterised as H-acid monosodium salt, Omega and Koch acids. G-acid is an acid made by sulphonating beta-naphthol and used as an intermediate for manufacturing azo dyes.

40.6.1 Identification and characterisation of major source of emission

The gaseous emissions from H-acid/G-salt plants mainly constitute SO_x, CO_2, NO_x and NH_3 from the sulphonation, nitration and alkali-fusion processes.

40.6.2 Identification and characterisation of major source of solid wastes

There are two major sources of solid effluents from H-acid plant, viz., gypsum sludge from neutralisation and iron sludge from reduction. These solid effluents also have retained organics to the extent of 1.08% and 3.7% (w/w), respectively. It is imperative to leach them to levels of 0.05% and 0.1%, respectively, before their use as land filling agents. However, most industries do not conform to these limits for retained organics before dumping, thus giving rise to soil and water pollution.

40.6.3 Water consumption and wastewater generation

The present H-acid process followed by various manufacturers requires water of about 19 T in the process and 1 T for washing per 1 T production of H-acid. Out of this water, about 9 T is evaporated in the ‘concentration’ step. The steam generated can be used for further concentration, if multiple effect evaporation system is used. This reduces the steam requirement of the process. All the condensates can be recycled to the boiler, so that no make-up water is required for utility plants. About 11 T of water is in the liquid effluent. Most dye intermediates manufacturers give only primary treatment of neutralisation to the liquid effluent. Effluent after the primary treatment contains large quantities of inorganic and organic compounds and this water cannot be reused.

It is possible to reduce the requirement of water by modifying the process. The process of G-acid requires about 5 T of water – 4.2 T are required in the process and about 800 litres for scrubbing and washing. There are no effluents, except spent scrubber solution and washes. This is a small quantity and can be treated with effluents of other plants.

Recycling and reuse of wastes

1. Ammonia issuing out of alkali fusion reactors may be suitably scrubbed into water.

2. This resultant ammonia/ammoniacal solution may be utilised to neutralise SO_x and NO_x vapours.
3. The solid waste of gypsum may be given a thorough washing with alcoholic solvents to considerably reduce its organic content.
4. The treated gypsum may be used as a building material in place of cement.
5. The iron-sludge may be converted into value-added ferrous and ferric salts by reacting it with spent acid.
6. The iron-sludge may be converted into a valuable inorganic red pigment consisting mainly of iron oxide.
7. Alternatively, the process can be modified to eliminate solid wastes. This will involve isolation of nitromass, use of spent acid after isolation for neutralising fusion mass, catalytic hydrogenation of nitromass and recycle of catalyst. Methanol can be used as a solvent in hydrogenation and fusion. Methanol-water mixture can be recovered and reused.
8. The bio-degradability of the waste-water effluent can be considerably enhanced by separating the organics from it by extraction with a tertiary alkylamine dissolved in a water-immiscible solvent.
9. The naphthalenesulphonic acids thus obtained in a concentrated form can be further concentrated by drying techniques such as spray or vacuum drying.

40.6.4 Treatment of H-acid and G-acid effluents

The dyestuff industry is an easy target for conservationists because an effluent coloured with waste dye or dye intermediate, however small and innocuous the amount of colour may be, attracts attention. In addition to toxic gaseous, particulate and solid effluents, the dyestuff industry also generates considerable quantities of wastewater so that a treatment plant designed to cope adequately with them may have to be large and, therefore, expensive. However, such costs have to be absorbed by the large industries and economical methodologies to treat wastewater in terms of both running costs and capital expenditure have to be designed. As per government regulations, individual industrial units are responsible for the treatment of their effluents and ensuring that the effluent discharged conforms to the standards prescribed by the pollution control boards. Since it is not viable for a large number of units to have individual effluent treatment plants, an ideal alternative is to have a common treatment facility. The dye/dye intermediates manufacturing units must be preferably located in close proximity to the sea for the safe disposal of treated effluent, since it may not be techno-economically viable to degrade the organic chemicals to the extent fit enough for disposal in inland water. Alternatively, efforts should be made to recycle or recover the waste materials/organic chemicals in the waste streams to the extent possible and convert them to value-added products.

The liquid effluents treated to the required specifications may also be safely used for irrigation of lands, plants and fields for growing non-edible crops. In any case, it becomes the prime responsibility of the individual units to provide preliminary treatment to the waste liquid effluents in the form of (i) equalisation, (ii) neutralisation, (iii) flocculation and coagulation, (iv) sedimentation, (v) sludge collection and (vi) sludge disposal prior to the various terminal treatment processes, which may be executed in a common treatment plant provided by a third agency.

H-acid effluent

Effluents containing H-acid and other naphthalene sulphonic acids are toxic and non-biodegradable in nature. The wastes generated from H-acid industry fall under hazardous waste category and quantity generated far exceeds the regulatory limits. Improper disposal of H-acid wastes has led to pollution of drinking water sources and contamination of agricultural land, thereby directly affecting human health and the environment. Within the country, safe disposal of these wastes has not received adequate attention.

The toxic effluents generated in this process may be categorised into gaseous, solid and wastewater (liquid) effluents. Flow sheet of suggested liquid effluent treatment is given in Fig. 40.4.

Types of treatment

Gaseous effluents: These mainly consist of oxides of sulphur and nitrogen emanating from the neutraliser and ammonia gas evolving during the alkali fusion.

Solid effluents: Gypsum is the main solid waste formed by the neutralisation of nitro mass prior to the reduction of Koch acid. It is also precipitated during the neutralisation of wastewater streams. In addition, substantial amounts of iron oxide sludge is also generated during the reduction of the nitro mass to Koch acid.

Wastewater (liquid) effluents: Large volumes of mother liquor and washings are generated both during the Koch acid manufacturing stage and after the final isolation of H-acid. These wastewater effluents are highly acidic and contain significant amounts of dissolved naphthalene, nitronaphthalene, aminonaphthalene, hydroxyaminonaphthalene, di- and tri-sulphonic acids in addition to inorganic salts. These are potent enough to cause a very high degree of pollution.

G-acid effluent

SO_x vapour comes out in the sulphonation step. It is to be treated in the same way as mentioned in that for H-acid. There are no solid effluents. Mother

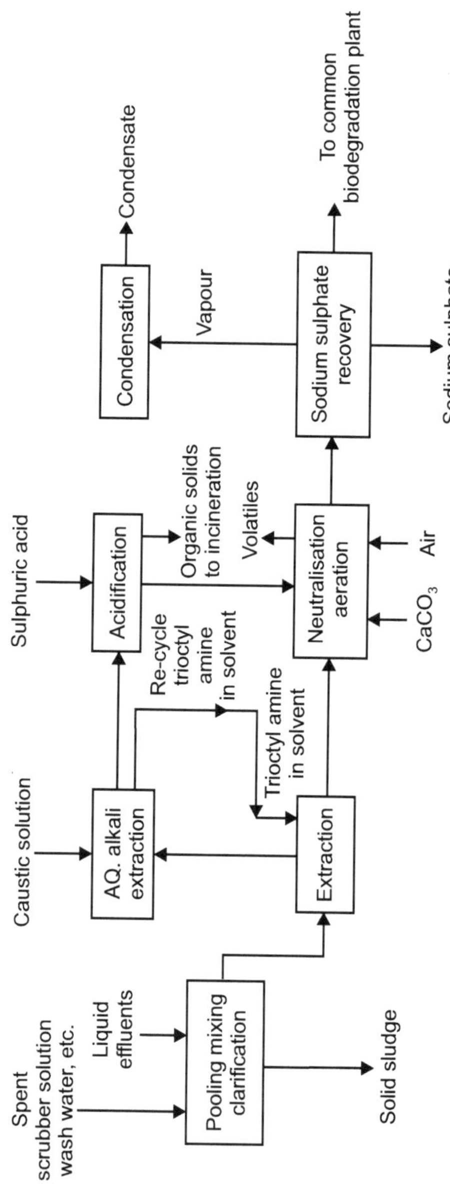

Figure 40.4: Flow sheet of suggested liquid effluent treatment of H-acid.

liquor coming out from G-acid isolation is sent to R-acid/Schaeffer's acid plants. Hence, there are no liquid effluents in G-salt plant.

Types of treatment

General effluent treatment practices followed for effluents of H-acid manufacturing plants.

Gaseous effluents. Suitable scrubbers are provided above the neutraliser to absorb the toxic nitrogen oxides and sulphur dioxide gases. The scrubbers are generally packed rubber or lead-lined cylindrical vessels using dilute aqueous alkali solutions as scrubbing agents. Similarly, ammonia can be scrubbed by water in suitably packed mild steel scrubbers.

Solid effluents. The gypsum formed during neutralisation is filtered out, thoroughly washed and the dried cake is transported to specific areas for land filling. The iron oxide sludge (after washing and drying) is mainly disposed of as a solid waste and utilised for land filling.

Wastewater effluents. The major thrust of effluent treatment is in this area since considerable volumes of deeply coloured, highly acidic mother liquors and washings having high salt content, toxic and excessive organics are encountered. Approximately 20 m^3 of waste liquid effluents are generated per T of H-acid manufactured. In general, these effluents are subjected to three modes of treatments viz., (i) primary treatment, (ii) secondary treatment (bio-treatment and others) and (iii) tertiary physicochemical treatment.

1. Primary treatment: This treatment makes the effluent suitable for bio-treatments and other secondary and tertiary treatments. The wastewater effluent from the manufacturing plants are taken into equalisation-cum-neutralisation tanks. More sophisticated neutralisation lagoons are provided with two stage automatic lime dosing system with the incorporation of stirrers, floating surface aerators, vibroaerators and pH control instruments. Generally, a residence time of 24 hours is provided which gives sufficient provision for normalisation of shock load due to wide variations in pH and organic load. Efficient mixing and agitation is ensured through vibro- and floating aerators.
   The neutralised effluent is then pumped into the neutral effluent holding cum settling tanks. The settled effluents from these tanks are filtered through a filter press (wooden) to remove suspended particles and the filtrate is pumped into hopper bottom settling tanks fitted with overhead clarifiers. They are generally made of mild steel. The settled sludge is removed from the bottom of the tanks.
   The normalised and neutralised clear effluent contains trace concentration of floating oil. To remove this floating oil/grease, the effluent is run into

the polishing tank made of brick masonary and provided with oil trapping arrangements.

The settled materials and solid wastes from settling tanks and filter presses are periodically removed into sludge drying beds (brick). The dried sludge is disposed of for land filling.

2. Secondary treatment: This generally consists of biological treatment of the effluent which has already been subjected to primary treatment. A specially designed 'activated sludge process' incorporates nutrient dosing, high rate aeration tank equipped with rotary surface aerators, static circular clarifier with bottom sludge raking, surface skimming devices, sludge recovery station for recirculation and removal of excess sludge for dewatering.
3. Tertiary physicochemical treatment: Bio-treated effluent is subjected to tertiary physicochemical treatment in a specially designed sludge blanket clarifier, followed by coagulant and flocculant dosing, treatment and desludging. The bio-assay of the treated effluent is determined.

40.6.5 Safe disposal of solid and liquid wastes

Generally, the solid and liquid effluents emanating from a chemical process are disposed of by such means as sub-surface storage, conversion of wastes to innocuous end products, storage in ponds or land spreading, discharge in river and discharge in ocean. The solid and liquid effluents that are generated in the H-acid/G-salt industries are gypsum and iron sludges, containing retained synthetic organics and liquid effluents, containing synthetic organics, inorganic metal salts and spent acids. The synthetic organics are non-biodegradable to a large extent and may fall into the toxic chemical category. In such a case, recovery and reuse is to be resorted to before adopting other ultimate disposal methods. This observation assumes great significance, particularly for improperly treated effluents from H-acid/G-salt industries, wherein disposition methods such as discharge in river or ocean can have a deleterious effect on aquatic and marine life.

Solid wastes such as gypsum and iron oxide sludges can be subjected to landfill disposal methods such as mixing with soil, evaporation and infiltration and/or shallow burial. Combinations of these methods can be involved in a disposal process. However, extreme care should be taken to first leach out retained organics from the sludges (% w/w) to prevent soil and water pollution. The landfills should be located in areas of low population density, low alternative land use value and low ground water contamination potential. The sites selected should have low rainfall and high evaporation rates.

Before disposal, the liquid effluents may be cooled to crystallise inorganic salts such as sodium sulphate. They may then be subjected to wet oxidative degradation and ozonolysis to reduce BOD and COD levels. The reduction in BOD and COD levels can also be achieved by separating the organics, viz., aminonaphthalene sulphonic acids by extraction with a tertiary alkylamine dissolved in a water immiscible solvent.

The organics thus obtained in a concentrated form by re-extraction with controlled aqueous alkali can be further concentrated by drying techniques such as spray drying or vacuum drying. These dried wastes may be used either for conversion to useful end products or may be subjected to incineration at high temperatures (~500°C). The principal products of such incinerations from a volume standpoint are CO_2, H_2O and ash along with oxides of sulphur and nitrogen (SO_x and NO_x). Ash can safely be landfilled. The formation of SO_x and NO_x may necessitate secondary treatment such as scrubbing, prior to atmospheric release. Liquid effluent with reduced BOD and COD levels is to be biotreated to bring these down below the specified limits.

This treated effluent may then be taken for discharge in river or ocean. Coupled with the observation that the mixing and diffusion of the treated effluent occurs rapidly in the wake of the discharge, it is expected there will be no effect on marine life from such a discharge. More care, however, needs to be exercised while discharging in river.

41

Pollution control in pesticides industry

41.1 Introduction

Pesticides are produced in two stages the manufacturing consists of chemical synthesis of active ingredients for crop protection, which is very often the synthesis of complex organic chemical compounds and subsequent formulation of these active ingredients (usually mixing and grinding process). Major chemical reactions involved in production of technical grade pesticides are: alkylation, carboxylation, acetylation, condensation, cyclisation, dehydration, halogenation, oxidation, sulphonation, nitration and amination. Main physical (mostly separation or purification) operations, which are usually called as 'unit operations' include: Liquid/liquid extraction, liquid/liquid separation, liquid/solid separation, gas/solid separation, distillation, crystallisation, gas absorption, drying, grinding and mixing (Fig. 41.1).

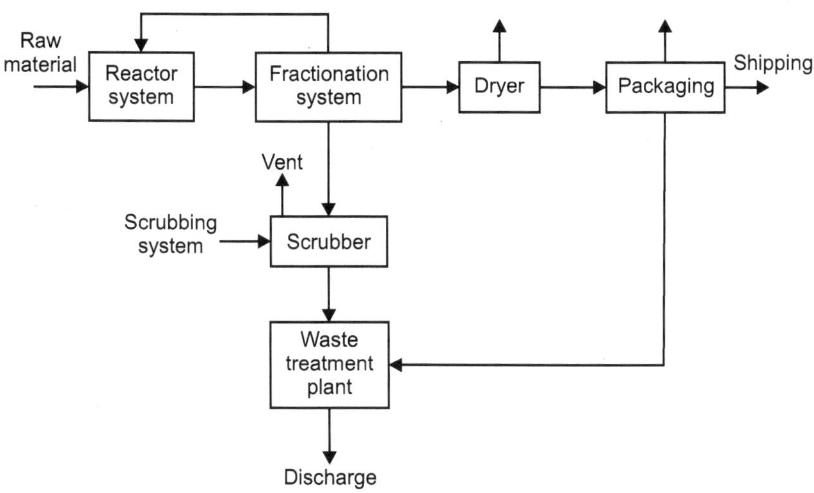

Figure 41.1: Flow diagram for manufacturing pesticides.

After a pesticide is manufactured in its relatively pure form (the technical grade material) the next step is formulation-processing a pesticide compound into liquids, granules, dusts and powders to improve its properties of storage, handling, application, effectiveness or safety. The technical grade material may be formulated by its manufacturer or sold to a formulator/packager (Fig. 41.2).

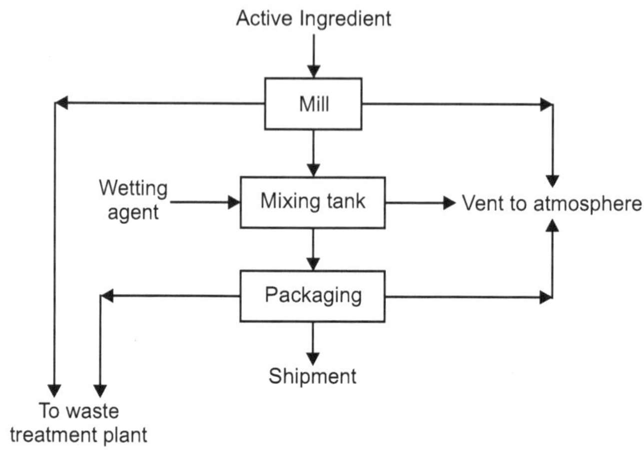

Figure 41.2: Typical unit operation of chemical synthesis.

Very often, during these operations, wastewater and solid waste is separated, whereas waste gas is directly released from the reaction itself.

41.2 Wastewater generation in pesticides industry

The entire manufacturing process for a particular product is a combination of various unit operations. The unit operations, where water is used and wastewater is generated, are to be identified so as to accurately identify the characteristics of the wastewater and the quantum of pollution load generated. The wastewater flow lines from process to treatment to final disposal are to be identified so that the entire status of the wastewater generation from the factory is depicted at one place.

The observations on wastewater generation and its characteristics are given below:

1. The possible sources of pollutants are raw material used, in pesticide synthesis in excess of their stoichiometric requirements, impurities in raw materials, solvent used as a carrier medium, solvent used as extraction medium, impurities in solvents, catalysts, manufacturing products, etc.
2. There is a great variation in quality and quantity of wastewater generated per unit of product. The biochemical oxygen demand (BOD) and chemical oxygen demand (COD) are widely varying in wastewater from product to product having no fixed correlation and has to be determined only on case to case basis. In general, BOD is in the range of 1000–5000 mg/L. The variation in the waste characteristics is hourly, daily and seasonal.
3. Some process streams are inorganic and do not need biological treatment.

4. Some streams are organic and/or inorganic and have very high dissolved solids (TDS). These cannot be treated biologically.
5. Details of various other pollutants in the wastewater are given below:
   (a) Volatile organics: Benzene, toluene, chlorobenzene, etc., which are used as raw materials.
   (b) Halomethanes: Methylchloride, chloroform, carbon tetrachloride, etc., which are used as raw materials and extraction solvents.
   (c) Cyanides: Raw materials favouring cyanide formation are cyanamides, cyanates, thiocyanates and cyanuric chloride.
   (d) Phenols: Phenols are compounds having hydroxyl group (OH) attached directly to an aromatic ring. These may be found throughout the pesticides industry as raw materials, impurities in raw materials or as by-products of reactions utilising related compounds, such as chloro-benzenes.
   (e) Heavy metals: These are used as catalysts or as raw materials which are incorporated into the active ingredient (technical) as in case of metallo-organic pesticides.
   (f) Pesticides: Traces of the finished product, i.e., the pesticides themselves are present.

In view of the wide variety of process technology options, the quantity of water used and wastewater generation are widely varying from product to product and industry to industry. It is to be seen that for monocrotophos, an organophosphorus pesticide, water use is varying from 0.5 to 120 kl per T of product. Similarly, wide variations are observed for other products. Hence, it is difficult to summarise a specific limit for water use and wastewater generation as is done for other industries such as distilleries, sugar, breweries, etc. However, the pesticides industries may realise that there is wide scope in reducing the quantities of water usage and wastewater generation which may effect in reduction in cost of treatment.

41.3 Wastewater treatment

Existing treatment facilities: The flow diagrams of wastewater treatment adopted in some of the industries is shown in Fig. 41.3. Pesticides industries generally use two schemes of pollution control which are given below:

1. The first scheme is evaporation system which further includes the following steps:
   (a) Segregation of production waste from utility and sanitation waste.
   (b) Evaporation of the segregated waste in suitably designed evaporation pans (solar evaporation or forced evaporation).

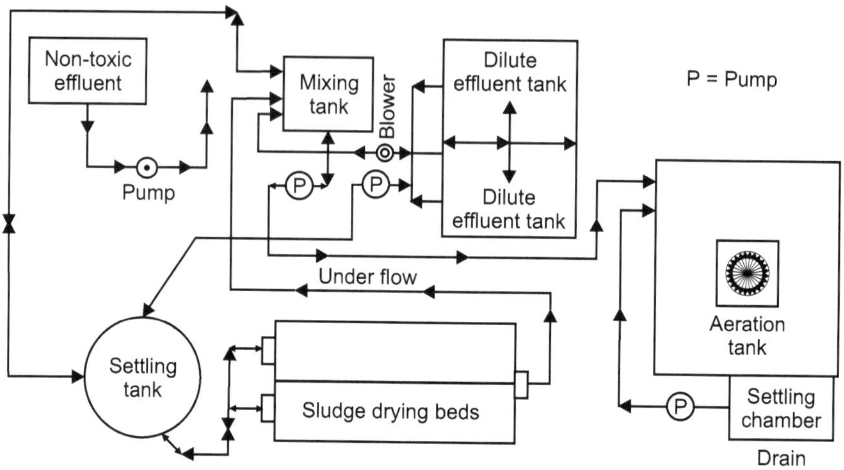

Figure 41.3: Wastewater treatment in an organophosphorous industry.

(c) Incineration of the concentrated waste.

Some pesticides industries, having effluents which are not easily biodegradable and also toxic, are adopting solar evaporation of effluents for the following reasons:

(a) Ease in management of effluents by solar evaporation.

(b) To overcome high cost involved in the construction and operation of a full-fledged effluent treatment plant.

(c) To avoid treatment of effluents to the levels of Minimal National Standards (MINAS) and to overcome effluent disposal problems.

2. The second scheme is detoxification followed by biological treatment (conventional activated sludge). In this system, the wastewater from various streams is detoxified, wherever it is toxic and is then combined and treated in an effluent treatment plant which includes primary and secondary treatment. The primary treatment includes oil and grease removal, equalisation, neutralisation, coagulation and settling. The secondary treatment includes aerobic biological treatment with aeration tanks, secondary clarifier and sludge drying beds.

Observations on existing treatment systems: The problems associated with the present wastewater treatment options adopted by the industries are given below.

41.3.1 Solar evaporation system

1. The solar evaporation system has limitations based on quantity of effluent and climatic conditions at the site. The effluent in the evaporation tanks

if not a leachate-proof holding arrangement, may cause pollution by virtue of percolation, overflow and evaporation of toxic volatile matter into the atmosphere. The rate of evaporation generally varies from 2.5 mm/sq.m/day in monsoon season to 12.0 mm/sq.m./day in summer. The predominant climatic conditions, i.e., rainfall and rate of evaporation in an area play a major role in limiting the quantity of wastewater that can be evaporated.

2. The rate of evaporation is low during rainy season and hence an industry will be left with excess quantity of wastewater, unless the evaporation system is well designed. This excess quantity needs to be managed through alternate arrangements or else it can be a potential source of pollution.
3. There are chances of contamination of groundwater where the groundwater table is high and the evaporation pond is permeable.
4. The toxic volatile gases from the effluents may cause air pollution during the process of evaporation and the pollutants such as oil and grease, organic solvents may reduce the rate of evaporation.

41.3.2 Biological treatment system

1. The characteristics and the quantity of wastewater in a pesticides industry are varying from product to product and also seasonally, daily and hourly, whereas the ASP (Activated Sludge Plant) type of biological treatment system needs a relatively uniform rate of organic loading and without shock loads due to toxicity, non-uniform flows, etc.
2. The organic, compounds through biodegradable, have complex organic chains due to which the operation of a biological system needs skilled staff.
3. There cannot be a single generalised treatment solution for the wastes from the pesticides industries due to the varying processes of manufacture stated earlier and the varying characteristics of waste.
4. The wastewaters, unless properly segregated, bear high dissolved solids thus affecting the biological treatment.
5. Wherever the biological treatment is affected, the microbial growth is limited and hence sludge return rate is to be maintained at a higher level in order to maintain food to micro-organism (F/M) ratio. But in the process sludge gets aged sooner and thus poses problem for biological treatment.
6. Under extreme climatic conditions, biological treatment does not work efficiently as the microbes find difficulty in surviving.

7. Activated sludge plants are generally adopted for aerobic treatment. These are high energy intensive and need skilled manpower for operation. Also, for high temperature locations, oxygen requirement is high and hence energy consumption is high. Thus, activated sludge plants have limitations as a technological option for aerobic treatment in India where temperature is high and skilled manpower is not provided for ETP operation.
8. In most cases it is observed that the treatability of the wastewater is not tested, operation of activated sludge plant not monitored and the continuous aeration not provided. This results in unsatisfactory performance. For optimal operation of activated sludge plants, skilled manpower, back-up technical infrastructure and uninterrupted power supply are needed.

41.4 Wastewater management

The conventionally adopted end-of-the-pipe treatment technology is not only ineffective but is also costly. Also, in view of the effluent disposal problems and the complexity of the pollution problem from the pesticides industries, it is required to manage the wastewater very effectively such that the waste is minimised, cost in treatment and risks due to pollution are lowered and the environment is protected. Various aspects for an effective wastewater management are discussed below.

Location: The environmental impact due to an industry not only depends on wastewater generations, emissions and solid waste release into environment, but also depends to a large extent on the quality of the environment at the location of the industry. Before the advent of pollution control legislation, industries were located considering largely economic factors, such as availability of raw material, labour, market, etc., which would maximise their profits. But now, with increased concern for environmental protection, a need has arisen to consider environmental factors for locating an industry.

The impact on the receiving environment from the pesticides industry which handles hazardous and toxic chemicals and emits obnoxious gases and wastewater, depends on:

1. Nearness to residential settlement and size of the settlement.
2. Climate, soil type, groundwater and geology at the site.
3. Adjointing landuses - agricultural, industrial, economically unproductive areas, etc.
4. Nearness to sensitive zone/areas.
5. Infrastructure availability for wastewater discharge and hazardous/solid waste disposal.
6. Existing quality of air, surface water, groundwater, etc., at the site.

Hence, it is very important to select a suitable site so that the impact on the receiving environment is the least.

Plant layout: Plant layout is to be well-designed for an effective drainage system, with ease in collection and disposal of wastewater and stormwater. It also helps in safety, makes various activities of storage, scarp yard, etc., well-defined and reduces misuse of areas. A well-laid out plant with well-defined areas, buffer zones and plantation, display boards and with colour codes for various process lines and storage tanks also helps in building a good industrial culture which has far-reaching results in improving process efficiency and preventing pollution besides safety.

41.5 Waste minimisation

Environmental audit: The conventional end-of-the-pipe treatment technology is not only ineffective but also costly, sometimes imposing a very heavy burden on the industry. The growing environmental pollution and the complexity of this problem with increasing risks from the regulatory controls need an effective management tool so as to prevent pollution by minimising losses, through recycle/recovery and to make pollution control programmes cost-effective and feasible. Environmental audit is an effective management tool in this direction for an industry. It also acts as 'liaison' with the government and the public in view of the mandatory requirement to submit an environmental statement to the government before the 30th day of september every year from 1993. Environmental audit comprises a systematic, documented, periodic and objective evaluation of how well the management systems are performing with the aim of:

1. Waste prevention and reduction.
2. Assessing compliance with regulatory requirements.
3. Facilitating control of environmental practices by a company's management.
4. Placing environmental information in the public domain.

The environmental audit helps in pollution control, improved production, safety and health and conservation of resources in an industry and as a whole helps in achieving sustainable development of the country. Environmental audit has far-reaching benefits to the industry, to the society and the nation at large. Some of the benefits of environmental audit are:

1. Helps in assessment of performance of process system and pollution control systems.
2. Serves as an eye-opener for waste minimisation.
3. Identifies areas of cost savings by reduction in losses and pollution load and by recycle and recovery of wastes.

4. Increases environmental awareness.
5. Helps in understanding 'technical capabilities and attitude' of the organisational setup.
6. Provides up-to-date environmental data-base for use in plant modification, emergencies, etc.
7. Ensures independent verification, identifies matters needing attention and provides timely warning to management on potential future problems.
9. Helps safeguard environment.
9. Assists in compliance with laws and regulations, company's policy and environment standards, etc.

The audit procedure broadly includes the following:

1. Pre-audit activities.
2. Activities at the site.
3. Post-audit activities.

41.5.1 Reduction of raw material losses

1. Keeping only appropriate inventory of raw materials to ensure minimum material handling losses, evaporation losses, etc.
2. Adopting mechanical handling of materials with proper monitoring facilities so as to dose only the predetermined quantities as per norms prescribed.
3. Having plant layout properly made so as to minimise transfer distance of materials between storage and process or between unit operations.
4. Preventing cross-contamination due to usage of same storage tanks for different materials depending on the batch product. Separate storages are to be provided.
5. Separating process lines for separate products or separate equipment for each unit operation to minimise losses due to residues in the equipment, which are usually washed out.
6. Providing storage tanks with proper dip arrangements for exhausts/vents and insulation provided so as to reduce evaporation losses.
7. Enclosing and covering material storage areas to keep them secured and reduce losses due to carry over by wind and rain.
8. Making enclosures to collect spills and overflows of materials at the material transfer and sampling points for reducing.
9. Having regular maintenance to check flange leaks, breaks/cracks, pump failures, etc.
10. Ensuring raw material purity.

11. Following good housekeeping practices.
12. Fixing norms for performance of various process operations so that the material usage is minimised and hence the material losses are minimised.

Recovery: There is a great potential in pesticides industry, for recovery recycle/reuse of useful products which otherwise go waste. These recoveries result in waste minimisation. Some of the recycle/reuse practices that can be easily adopted are:

1. The solvents which are widely used in the manufacture of pesticides, can be recovered and recycled/reused.
2. In most of the chemical reactions, there are more products formed than the required products. These unwanted products form wastes if left unrecovered. For example, methyl chloride is formed while manufacturing DDVP. These unwanted products such as, methyl chloride, HCl, monochlorobenzene, dichloroethane, trimethyl phosphite ester, sodium sulphite, sulphur dioxide, sodium thiosulphate, etc., can be recovered as valuable by-products by employing a suitable technique such as condensation, distillation, absorption, etc.
3. Condensate water can be collected and recycled/reused.
4. Floor washing, at times, can be collected, treated if necessary and recycled/reused.
5. Seal pot water, vacuum pump water and cooling water can be recycled and reused in operations, such as neutralisation, washing, etc.

41.6 Recycling and reuse of wastewater

Recycling and reuse of wastewater within the industry would help in minimisation of fresh water requirement and simultaneous reduction in wastewater volume for final treatment and discharge thereby reducing the related costs. From the investigation of material balance and unit operations of a particular process, it will be possible to characterise the waste and identify the possibilities of recycling or recovery of useful products. The process of identification of recoverable matter and recycling of wastewater can be made effective through 'environmental audit'.

Recycling in industries will be a major step in reducing regulatory risks and public litigation due to wastewater discharge outside causing impact on environment. The recycling options in an industry can be as follows:

1. Treating some or all process wastewater to make it suitable for plant process, cooling water make-up, floor wash, etc.
2. Recirculating the same water within an unit operation several times before it becomes unfit.

3. Sequential use of effluent from one process with treatment, if necessary, as input into another.
4. Usage of wastewater from a process to a lesser duty usage where an inferior quality of water will do. For example, it can be used for coal ash quenching in boiler house, for making lime solution used for neutralisation in the Effluent Treatment Plant (ETP), etc.

In-plant control: Good-housekeeping is the least expensive means to reduce the overall burden on treatment and disposal. Loss of raw material, solvent and product can be restricted by installing monitoring devices and the process managers held responsible for spills and poor housekeeping practices. Also the flow in various drains can be checked using flow indicators and flow meters and controlled accordingly. In some cases, wastewater generation can be substantially reduced by substitution of an organic solvent for water in the synthesis and separation operation of the process with subsequent solvent recovery. The steam jet ejectors and barometric condensers can be replaced in some cases by vacuum pumps and surface condenser systems. Barometric condenser produces a high volume of dilute waste stream and poses problems by contaminating the plant effluent. Methyl chloride gas, hydrogen chloride gas, hydrogen sulphate gas, etc., which are generated in the manufacturing process of pesticides, may be recovered or useful by-products produced instead of scrubbing and discharging to ETP. Waste segregation and treatment also form an important step in in-plant control of pollution.

Treatment methods: For arriving at the requirement of treatment, the sources of wastewater and its characteristics are to be identified. The amount of treatment to be provided depends on the standards to be achieved for recycle and reuse or disposal to water body, land, etc. The treatment technology that is generally available for removing various pollutants is briefed below.

Suspended solids are removed by screens, grit chambers, sedimentation, floatation, centrifugation and filtration. Effluents are neutralised using acid/alkali. Oil and grease is removed by gravity separation and air floatation. For removing heavy metals, pH is adjusted so as to precipitate these metals for removal by coagulation, settling and filtration. For the treatment of biodegradable organic effluents, biological treatment is adopted where in micro-organisms degrade the waste in presence of oxygen (aerobic treatment) or in absence of oxygen (anaerobic treatment). Aerobic treatment systems are—activated sludges process, extended aeration, aeration ponds, trickling filters and rotating horizontal tanks or vertical reactors (digesters). Oxidation of organic matter is also carried out chemically using chlorine, ozone, hydrogen peroxide, etc. For removing dissolved solids, ion exchange resin and reverse osmosis techniques are used. Adsorption by activated carbon, activated clay, etc., is used to remove specific pollutants. Highly toxic effluents which cannot be economically treated are incinerated. In case the effluents are small in quantity, a pre-treatment is given to detoxify the effluent and to remove volatile matter and are solar evaporated in lined tasks. More details on treatment methods for some of the pesticides are given below.

41.6.1 Waste segregation and pre-treatment

Based on the characteristics of wastewater from each of the unit operations, the streams should be segregated and preliminary treatment provided to the extent required and feasible. This will be economical and will make the final treatment effective. Various streams generated from a pesticides industry are as follows:

1. High acidic or alkaline stream.
2. Stream carrying oil and grease and floating organics.
3. Inorganic stream.
4. Organic stream.
   (a) Easily biodegradable.
   (b) Not easily biodegradable.
   (c) Toxic.
5. Cooling water bleed-off.
6. Boiler blow down.
7. Domestic wastewater from canteen, toilets, cloth-washings, etc.
8. Streams of highly variable characteristics.
9. Miscellaneous—floor wash, spills, leaks, drum/vessel washing, etc.

The pre-treatment that may be employed to various streams are 'neutralisation' for highly acidic/alkali stream, 'oil and grease trap' for oil and floating organics bearing streams, 'settling tank' for stream bearing high suspended solids and 'detoxification' for toxic streams. Separate equalisation for streams of highly variable characteristics can be effective in achieving overall treatment efficiency. The recoveries employed in various streams including that of inorganic/organic, form a part of pre-treatment and help in reducing pollution load to a considerable extent. The separation technologies may be employed for removal of organic and pesticidal chemicals from wastewater. The destruction and detoxification technologies, that prove very effective as pre-treatment or part of final treatment, in simplying the complex and stable organic structures of (chemical) pollutants and in removing toxicity are:

1. Chlorine dioxide oxidation.
2. Dye sensitised photo-oxidation.
3. Electro-chemical oxidation.

4. Flameless catalytic oxidation (low temperature vapour oxidation).
5. High energy radiation.
6. Hydrogen peroxide oxidation.
7. Incineration/combustion process.
8. Micro-biological and other metabolic systems.
9. Ozonation.
10. Ultraviolet light assisted ozonation.
11. Photo-decomposition - ultraviolet radiation.
12. Potassium permanganate oxidation.
13. Pyrolysis.
14. Wet catalytic oxidation.

Chemical oxidation

Oxidation by chemicals is an effective method for decomposition of many pesticides. The chemicals commonly used for oxidation are chlorine, chlorine dioxide, potassium permanganate, ozone, hydrogen peroxide and sodium peroxide. In general, unsaturated organic compounds are more susceptible to ozone oxidation than saturated compounds. Parathion is oxidised by chlorine and ozone to a more toxic product known as paraoxone. Lindane in aqueous solution is readily degraded by ozonation and only partially affected by potassium permanganate. Treatment of lindane with chlorination, peroxides and aeration has no measurable effect. Aldrin is readily attacked by chlorination, $KMNO_4$, ozone and aeration, but peroxides have no measurable effect. Dieldrin concentration in aqueous solution is decreased by aeration. The organo-nitrogen pesticides, during manufacture, emanate cyanide bearing effluent which can be converted to lesser toxic cyanates using chlorine in alkaline medium.

Coagulation: Suspended particulate matter, to some extent, can be removed using coagulant aids. DDT is easily removed by alum coagulation of doses typical of conventional water treatment. Efficiency of DDT removal is 98% for a 10 ppb load and 97% for a 25 ppb load. Lindane and parathion have low efficiency of removal by conventional alum coagulation and sand filtration. Aldrin, dieldrin, BHC and malathion have low efficiencies of removal by chemical coagulation.

Adsorption: The adsorbents that prove effective in removing some of the pesticides are activated carbon, saturated clay systems, humic acid, organic acid, bentonite, aluminium silicates and hydrous magnesium aluminium silicate. Activated carbon is effective in removal of pesticides, solvents, emulsifiers and odours. Aldrin, dieldrin, DDT, 2,4-D, BHC, captan and 2,4,5-T have partial adsorption by clays.

Photochemical degradation: UV-ozonation is remarkably effective for degradation of halogenated hydrocarbon. The effect of UV radiation is isomerisation in case of dieldrin, reduction of chlorinated compounds, replacement of aromatic halogens by hydroxyl in case of chlorinated herbicides and elimination in case of carbamates.

41.7 Effluent treatment plant

Treatment technology: The treatment of liquid wastes can be achieved employing physical, chemical and biological methods. The common physical methods are flow equalisation, oil and grease removal, sedimentation with or without coagulation, sludge removal, filtration and adsorption. Equalised and steady flow are very important for improved efficiency, reliability and control of treatment plant. Hence, flow equalisation is employed in the treatment of industrial wastewaters. Equalisation dampens flow variation and helps achieving a constant or nearly constant flow rate to ETP and prevents variation in quality (characteristics) due to blending of various streams. The equalisation tank needs proper mixing and aeration. Mixing is required to blend the contents of the tank and to prevent deposition of solids and aeration is needed to prevent the wastewater from becoming septic (anaerobic–devoid of oxygen). Proper sedimentation, with the removal of suspended solids including colloids from the wastewater, helps in reduction of BOD, COD, oil and other contaminants. Usage of a coagulant helps in agglomeration of colloids and subsequent sedimentation. The sedimentation system includes flash mixing. A rapid mix helps in rapid and uniform dispersion of the coagulant throughout the suspension and results in more efficient use of the coagulant. Flocculation is a gentle mixing that provides increased opportunities of contact between particles in suspension, thereby forming bigger flocs. Sedimentation, flash mixing and flocculation may be carried out in indiviudal units or in one combined unit such as 'clariflocculator'. The settled sludge is to be periodically removed to a filter press or sludge drying bed.

The chemical treatment includes neutralisation, oxidation, detoxification, etc. Neutralisation is required to maintain a pH of 6.5–8.5. It is required for:

1. Corrosion control.
2. Reaction control on many chemical treatment processes.
3. Protection of system organisms, if direct discharge is employed.
4. Protection of micro-organisms in biological treatment system.

Neutralisation is generally done using lime solution. Employing caustic soda entirely being costlier, lime may be used partly followed by automatic dosing of caustic. In manual control of pH, especially where lime is used, it is observed that either over-dosages or under-dosages are done. The quality of lime varies and so the dosage rate varies. The usage of lime or caustic leads to huge quantities of sludge and the usage of caustic leads to increased TDS. Hence, the usage of chemicals for neutralisation has to be optimised. On-line pH metres and automatic pH adjustment may be employed.

Highly alkaline conditions can detoxify certain pesticidal compounds. Caustic soda is often used for pH adjustment so as to bring about hydrolysis of some chlorinated and organophosphate manufacturing wastes.

Biological treatment includes aerobic treatment and anaerobic treatment. For aerobic treatment, adequate food, oxygen, nutrients and proper pH and temperature are required. It may be designed to operate in a pH range of 6 to 8. The biological population requires inorganic nutrients for cell synthesis, including nitrogen, phosphorous, iron, silica and others. If these nutrients are lacking in the effluent, they should be added.

Biodegradability is an important factor for determining the application of biological treatment to an effluent. Almost every organic compound can be broken down biologically but some, such as tertiary aliphatic compounds, benzene, trichlorophenols, pentachlorophenol, etc., pose problem and do not degrade easily. Some of the factors which affect biodegradability are:

1. Compounds in emulsified or chelated forms are not readily available to micro-organisms (e.g., DDT).
2. Large and complex structure of molecules often limits the rate at which the compound can be broken down (e.g., carbamates and carboxylic acid compounds).
3. Aliphatic (straight and cyclic) compounds are in general more degradable than aromatic compounds.
4. The substitution of elements other than carbon in the molecular chain often makes the compound more resistant. Esters and epoxides, salts, etc. are more resistant than the base pesticidal compound.
5. Halogen substitution to an aromatic compound renders it less degradable.

A detailed study of the waste properties and biodegradation tests are necessary before a biological treatment method is designed.

The biological methods used for pesticide containing wastewater include:

1. Trickling fillers.
2. Activated sludge plants.
3. Aerated lagoons.
4. Stabilisation ponds.

The treatment system to be adopted depends mainly on the standards to be achieved, reliability of the treatment system and its cost. The cost factor in an industry is expressed in terms of the percentage ratio of annual burden for

ETP (annual loan repayment on capital costs including civil, mechanical and electrical components plus running costs) to annual sales turnover. This percentage is acceptable, in general to a pesticides industry upto about 5%. Regarding reliability, physicochemical treatment systems with automation are more reliable than the manual handling systems and biological systems.

Treatment options: For the final treatment of wastewater from the pesticides industry, there are many options depending on the type of waste.

1. Highly toxic streams are generally concentrated and limited in quantities and hence may not be feasible to treat to the required standards economically. Hence this waste may be incinerated.
   The spills and leasks in process areas, may be wiped out using cotton, sawdust, etc., instead of using water and then this material incinerated. The incinerator should be designed for suitable detention time and temperature so as to destroy the waste and should be provided with suitable air pollution control equipment for the flue gas treatment.
2. The vessels/reactors in a formulation unit can be washed using the same solvent that is used in the formulation and the washings reused for this formulation.
3. The toxic wastewater, which is not easily biodegradable, may be treated physicochemically instead of treating biologically. This treatment includes detoxification, oil separation, equalisation, clariflocculation, oxidation with H_2O_2/NaOCl/$KMnO_4$, etc., neutralisation and clariflocculation.

Section VII

Polluction control in other important industries

42

Pollution control in pulp and paper industry

42.1 Introduction

Pulp and paper is made out of wood fibres originating from natural forest or pulpwood plantations. Pulp and paper production is highly energy intensive with 75–85% of the energy requirement being used as process heat and 15–25% as electrical power. The combination of these two energy requirements qualifies paper production for the use of cogeneration.

In general, the production process consists of 5 stages:

1. Raw material preparation.
2. Pulping.
3. Bleaching.
4. Chemical recovery.
5. Paper-making.

Most of the energy is used in form of heat within the pulping process (digester, evaporator and washing) when raw materials have to be cooked and mechanically or chemically treated for further use in the production chain. Furthermore, paper making requires considerable amounts of energy in form of both heat and electricity for forming, pressing and drying of the paper.

Energy consumption is also highly dependent on the type of raw material used in the production process. Energy consumption for pulping and digesting, for example, is lower if wastepaper is used instead of wood chips or agricultural residue.

Mechanical pulping: Mechanical pulping is the oldest form of pulping. The process employs mechanical energy to weaken and separate fibres from wood and waste paper feedstock via a grinding action. The advantage to mechanical pulping is that it produces much higher yields than chemical pulping processes (up to 95%). However, because this process does not dissolve lignin, the fibre strength and age resistance of the resulting pulp are low.

Chemical pulping: Chemical pulping is by far the most common method of producing wood pulp. Chemical pulping processes have low yields but generate pulp with strong and stable fibres for high quality products such as office paper. Chemical pulping separates the fibres in wood feedstock by dissolving the lignin bonds that hold these fibres together, often at elevated temperatures and pressures.

There are two primary forms of chemical pulping: (i) the kraft (or sulphate) pulping process and (ii) the sulphite pulping process.

Chemical recovery: The primary purpose of the chemical recovery process is to recover pulping chemicals from spent cooking liquor (i.e., black liquor) for reuse in subsequent pulping processes. Chemical recovery allows a mill to regenerate pulping chemicals at a rate of up to 98%, which significantly reduces the costs of purchased process chemicals. An added benefit is that chemical recovery allows a mill to generate a significant portion of its steam requirements by combusting the pulp residue contained in black liquor as part of the refining process.

The chemical recovery process for Kraft pulping consists of four key stages: (i) black liquor concentration, (ii) black liquor combustion (recovery boiler), (iii) recausticising and (iv) calcining (lime burning).

Bleaching: Raw pulp can range in colour from brown to crème due to the remaining lignin that was not removed during the pulping process. For paper products for which brightness and resistance to colour reversion are important, such as office and printing paper, the pulp must be whitened by a bleaching process prior to the paper making phase.

Bleaching can be defined as any process that chemically alters pulp to increase its brightness. The pulping process (i.e., chemical or mechanical) is a major driver of the type of bleaching that is required. Mechanical and semi-chemical pulping process will generate pulps with high lignin content, which requires a chemical-intensive bleaching process to decolourise the remaining lignin. The bleaching process for chemical pulps—which have low lignin content—focuses on the removal of remaining lignin from the pulp.

Mechanical pulp is often bleached using hydrogen peroxide and/or sodium hydrosulphite. Bleaching chemicals can be added into the mechanical pulping process, or added to the pulp in multi-stage reactions which occur in a series of post-pulping bleaching towers.

The number of bleaching reactions employed depends on the brightness requirements of the final paper product.

The bleaching of chemical pulp comprises multiple stages that alternate between washing the pulp and treating it with chemicals in bleaching towers.

The specific chemicals that are applied in bleaching processes for chemical pulp and the number of stages, vary by mill and depend on a number of factors including local environmental regulations, costs and desired pulp properties.

Pulp drying: In situations where pulping and papermaking operations are not located at the same facility, or when a temporary imbalance between pulp production and paper machine requirements exists, pulp is dried to reduce its moisture content. Flow diagram of paper mill is shown in Fig. 42.1.

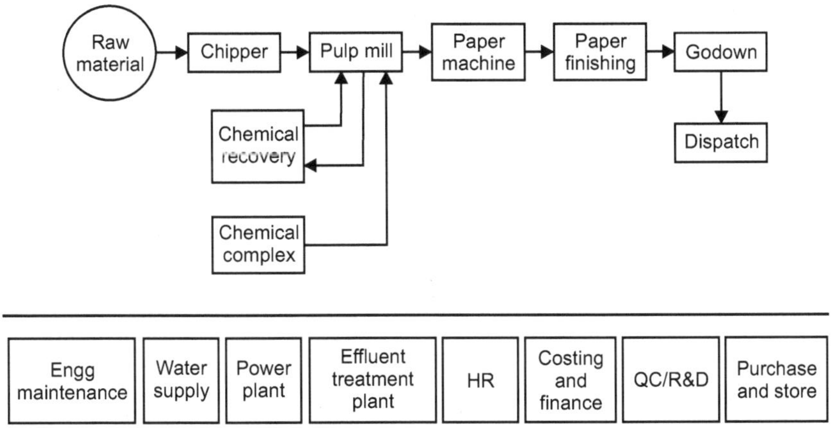

Figure 42.1: Flow diagram of paper mill.

Papermaking: The papermaking process can be divided into three basic stages: (i) stock preparation, (ii) 'wet end' processing where sheet formation occurs and (iii) 'dry end' processing where sheets are dried and finished.

42.2 Wastewater generation in pulp and paper industry

42.2.1 Sources of wastewater

In agricultural residue based units, the wastewater is released from the following sections:

1. Black liquor from cooking section.
2. Pulp wash water from pouchers.
3. Beater section specially when rags are used.
4. Bleaching section.
5. Thickener.
6. Paper machine.

Black liquor is the most polluting among the different streams. Black liquor is not segregated and it ends up in pulp washing wastewater. Thus pulp washing section contributes nearly 80% of the total pollution load. Further, flow and composition of pulp wash wastewater will be highly varying because the operations are carried out in batches. Only in a very few mills counter-current washing using two stages is adopted.

Paper machine wastewater is least polluting among all the sections but contributes appreciable amounts of suspended solids. In most of the mills, paper machine wastewater either as such or after fibre recovery, is used in pulp washing, beaters, etc.

42.3 Water pollution in pulp and paper industry

The main polluting constituents in pulp and paper mill wastewater are suspended solids, colour, foam, inorganics such as sodium carbonate, bicarbonates, chlorides and sulphates, toxic chemicals such as mercaptans and inorganic sulphides, mercury if caustic chlorine plant forms a part of pulp and paper mill, BOD and COD.

The wastewater when discharged untreated will damage the water course and the colour in water persists for a long distance since lignin is not readily biologically degraded. 'As paper industry being scattered all over the country, it may be said that no river is spared from pollution due to the discharge of these wastewaters'.

Another important pollutant (that is yet to be assessed) in pulp and paper industry is mercury. This is so because some of the mills make their own caustic and chlorine in mercury cells and it has been reported that about 0.25 kg of mercury is lost per T of caustic made. Further, some organic mercury compounds such as phenyl mercuric acetate (50–100 ppm of dry weight of pulp), methoxy ethyl mercury and pyridyl mercury salts are being used as slimicides on paper machines which finally find their way into the effluent. In Sweden it is reported that about 25 T of mercury salts are used per annum.

42.4 Characteristics of effluents

Waste paper and/or purchased pulp based mills the flow and characteristics of combined wastewater indicate that although there is variation in the flow of the combined wastewater, the minimum and maximum values respectively are 0.6 and 1.5 times the average flow. There is also variation in the characteristics of the wastewater and is mainly due to the different practices adopted for fibre recovery and extent of recycle of both recovered fibre and water. On an average for each tonne of paper made 107 ± 28.4 cubic meters of wastewater is generated containing 58 ± 10.7 kg of suspended solids, 20 ± 10.5 kg of BOD and 70 ± 15.2 kg of COD respectively.

The proportion of agricultural residue to waste paper used varies appreciably in all the agricultural residue and waste paper based mills. Even in the same mill, depending on the raw material availability, season, type of paper made, the proportions vary even within a week. Hence, wastewater flow and characteristics are found to vary. It is not possible to generalise the flow and pollution loads contributed per T of paper made from these mills.

Since most of the small paper mills are located in agricultural areas and some of them are using the combined wastewater for crop irrigation, the Sodium Absorption Ratio (SAR) and per cent sodium values in the combined wastewater are high in mills where unbleached paper is made using soda pulping. Leaving a few high values, the SAR ranged between 3.5 and 7.6. If the ratio of calcium to sodium can be increased in the combined wastewater from small paper mills, then it can be used for irrigation after suitable dilution to keep the total dissolved solids and BOD in the ranges recommended for irrigation waters.

BOD rate constant for black liquor and combined wastewater after settling compares with that of domestic wastewater. The non-biodegradable fraction of COD is high and is mainly due to the lignin in black liquor. In case of combined settled wastewater, 70% of the non-biodegradable COD is due to lignin and the remaining due to suspended solids, etc. Hence, it can be concluded that the COD due to lignin cannot be removed even after biological treatment. The only alternative method is to remove colour *vis-a-vis* lignin by chemical treatment.

42.4.1 Solid wastes generation

Solid wastes are generated from the following operations :

1. Raw material handling.
2. Rejects from screening and centri-cleaners.
3. Primary and secondary sludges from wastewater treatment system.
4. Coal or boiler ash from steam generation.

42.5 Treatment for the wastewater

42.5.1 Equalisation for pulp wash wastewater

Pulp washing section accounts for about 20–25% of total wastewater and contributes around 70–80% of pollution load from small paper mills using agricultural residues as raw materials. These wastewaters are discharged intermittently since, in most of the mills, washing of pulp is done in batches using pouchers. Normally two pouchers are employed and operated in series and thus generate two washes. The period of washing is more or less same in both but the quantity of water used varies appreciably and thus the first wash water is more concentrated than the second wash water.

The flow variation has been observed to be fairly wide since the minimum and maximum values, respectively, are 0.15 and 2.3 times the average flow. Discharge of these washes to the main wastewater stream (on an intermittent basis), as is being practised will alter the composition of the combined wastewater appreciably. This necessitates provision of equalisation to the pulp washes and discharge at a constant rate into the main sewer. Two alternatives can be considered for this purpose.

42.5.2 Primary clarifier and sludge drying

If secondary treatment like activated sludge, oxidation ditch or rotating biological disc is used. The excess secondary biological sludge should be added to the combined wastewater before primary settling such that the settled sludge can be filtered on drying bed. The combined primary and secondary sludges do not require biological stabilisation. The dried sludge can be disposed off by burning in an incinerator or dumping in open pits in a controlled manner.

42.5.3 Other effluent treatment methods

Sedimentation technology

Sedimentation technology is the simplest and most economical method of separating solid substances from the liquid phase. High efficiency is achieved in subsequent effluent treatment processes when the solid substances suspended in the effluents settle in a sedimentation tank as completely as possible and settled sludge is removed from the sedimentation tank. Sedimentation equipment with sets of lamella-shaped passages, are employed in the paper industry, especially for effluents with high fibre concentrations.

Biological treatment

Biological wastewater treatment is designed to degrade pollutants dissolved in effluents by the action of micro-organisms. The micro-organisms utilise these substances to live and reproduce. Pollutants are used as nutrients. Prerequisite for such degradation activity, however, is that the pollutants are soluble in water and non-toxic.

Anaerobic technology

Since the early 1980s, anaerobic treatment of industrial effluents has found widespread application in the pulp and paper industry. Several hundreds of installations are treating a large variety of different pulp and paper mill effluents. Anaerobic treatment is most commonly used for effluents originating from recycle paper mills, especially during production of containerboard. Moreover wastewater of mechanical pulping (peroxide bleached), semi-chemical pulping and sulphite and kraft evaporator condensates may be treated.

Aerobic technology

Aerobic micro-organisms require oxygen to support their metabolic activity. In effluent treatment, oxygen is supplied to the effluent in the form of air by special aeration equipment. Aerobic treatment allows fully biological degradation of paper mill effluents. Aerobically operated plants exhibit higher plant stability and are less sensitive to fluctuations in effluent and plant parameters.

Secondary clarification

Secondary clarification is intended to separate the biomass (activated sludge) formed in biological reactors and is therefore a key element in all processes employed in the final stage of a treatment plant. The quality of the separation process is just as crucial for the final effluent quality as is biological treatment itself.

Advanced and tertiary treatment

Tertiary and advanced wastewater treatment is used to remove specific wastewater constituents that cannot be removed by secondary treatment. Different treatment processes are necessary to remove nitrogen, phosphorus, additional suspended solids, refractory organics or dissolved solids. Sometimes it is referred to as tertiary treatment because advanced treatment usually follows high-rate secondary treatment. However, advanced treatment processes are sometimes combined with primary or secondary treatment (e.g., chemical addition to primary clarifiers or aeration basins to remove phosphorus) or used in place of secondary treatment (e.g., overland flow treatment of primary effluent). Advanced wastewater treatment in the pulp and paper industry is focused mainly on additional biological membrane reactors, ozone treatment and membrane filtration techniques such as micro-, ultra- or nanofiltration and reverse osmosis.

Filtration assisted crystallisation technology (FACT)

Filtration Assisted Crystallisation Technology (FACT) is a hybrid process, patented by swiss company, combining heterogeneous crystallisation and a simple filtration. The heterogeneous seeds should allow fast crystallisation and easy filtration.

Softening and controlled precipitation technologies

Especially, in the pulp and paper industry, the removal of scaling compound, called calcium carbonate, is a key point in the perspective of a reuse of the wastewater. According to crystallisation theory, precipitation is defined as a 'reactive crystallisation'. This definition is preferred as it emphasises the formation of the solid product *via* a chemical reaction.

Electrodialysis

Electrodialysis is an electrochemical membrane separation technique for ionic solutions that has been used in industry for several decades. It can be used in the separation and concentration of salts, acids and bases from aqueous solutions, the separation of monovalent ions from multivalent ions and the separation of ionic compounds from uncharged molecules.

Evapoconcentration

Evapoconcentration is a unit operation which consists in the separation of two liquid constitutes: the removal of solvent (generally water) as vapour from a solution or slurry. The aim of evapoconcentration is to obtain a very clean distillate which could be rejected in the nature, sent in a sewage treatment plant or reused.

Evapoconcentration has proven to be a very efficient and reliable technology for treatment of industrial wastewater mainly due to its flexibility in the treatment of very different types of wastes, with a very high load of pollutants.

Advanced flocculation

Chemical flocculation process is a crucial process as it promotes the aggregation of particles after being destabilised by a chemical agent.

In pulp and paper industry, flocculation is involved in different parts of the process: it is essential to form the paper sheet in the forming wire, determining retention, drainage rate and the formation and it is also used in the wastewater treatment to separate the colloidal material and in the sludge thickening.

Ozone/AOP technologies

Today Ozone and UV are well known and proven in the field of water and wastewater treatment ozone as multifunctional powerful oxidant and UV as best available technique for disinfection regarding treatment results, plant design and cost. The main goal of these combined processes is to enhance the oxidation potential.

Combined chemical (AOP) and biological oxidation processes are well-known for removing recalcitrant and anthropogenic substances from wastewater. Additionally the decolourisation with ozone has already been established as an application of polishing of biological treated effluents.

Membrane technologies (UF, NF, RO)

Membrane treatment in pulp and paper industry serves to optimise loop closure and therefore helps to reduce fresh water intake as well as wastewater treatment. Other purposes of membrane processes are: improved product quality because of lowered pollution of loop water, reuse of treated effluent in production, recovery of valuable substances, e.g., coating pigments and minimising environmental impact because of improved effluent quality. Full scale membrane filtration of pulp and paper effluents has been already installed in some mills. Nanofiltration (NF) treatment of total effluent is installed in a newsprint paper mill. The advantage of NF in the recovery of water for recirculation is mainly that the clean water can be used even in the most demanding places in the paper mill.

MBR processes

There is currently growing interest in the MBR (membrane bioreactor) process in municipal and industrial wastewater treatment. MBR is used in the paper industry as end-of-pipe technology as well as process integrated measure for the reduction of the concentration of detrimental substances in the water circuit. Especially in terms of effluent quality and economical aspects a MBR is a sustainable technology for the industrial wastewater treatment.

3FM technology

The purpose of water filtration is to remove particles and colloids which either disturb the industrial process, deteriorate the quality of the final product or support bacteria and viruses that are a danger for human health.

The 3FM® system (flexible fibre filter module) is a new high speed filtration device that can be substituted for conventional solid-liquid separation process such as coagulation, settling and sand filtration.

Ultimately, the solutions to the profound challenge ahead – the shifting of industry awareness and the transformation of the industry's lifeblood – requires continuous research and technology to be developed in the years to come.

42.6 Air pollution in pulp and paper industry

Particulates: Particulates arise predominantly chipper house, pulp washing, paper machine, caustic extraction. Chemical recovery power plant, coal handling and disposal sections have various sizes of particulate pollution. Good housekeeping, ventilation and appropriate exhaust and blowers controls this type of pollution to a considerable degree.

Apart from this certain mills have cyclones and electrostatic precipitator in the recovery sections which reduces air pollution to the desired level.

Large-sized particles are removed by setting as solid wastes and some times reused in the process, i.e., coal dust, fine fibre, etc.

Sulphur dioxide: Sulphur compounds are used for working wood material for manufacture of paper hence major pulping processes emit sulphur dioxide. Sulphur dioxide is emitted at the blow tank, evaporators, recovery furnace, lime kiln, sulphitation tower, etc. In addition sulphur dioxide emissions take place at the power boilers either under normal operation or during periods of auxiliary fuel operation.

It is estimated that about 20 to 25 Kg of sulphur bearing compounds are emitted per T of pulp.

Sulphur dioxide pollution is mostly controlled by scrubbing sulphur dioxide through the solution of alkali. Some items such as sulphur compounds are converted into sulphur and recovered in developed countries.

Carbon monoxide: Power plants are mostly based on coal. Coal-based power plants invariably emit carbon monoxide. However carbon monoxide emissions depend upon type of coal used and firing time required for the carbon monoxide. The pollution restricted to initial firing hence generally no control measures are adopted though, however this can be controlled by using better quality of coal. Methyl mercaptans, hydrogen sulphides, methyl sulphides, methyl disulphides are emitted in the recovery and evaporating and cooking sections. To a greater extent these can be controlled by process control and control of fugitive emissions. Wherever air pollution control equipments like scrubber or electrostatic precipitators are used most of these air pollutants are kept at a low level. However much depends on good housekeeping.

Chlorine: Chlorine, chlorine dioxide and various compounds of chlorine are used for bleaching pulp. During process of bleaching, chlorine always finds its way into the environment. The control equipments like exhaust and vacuum cleaners are used for considerable reduction of chlorine pollution. Most of the mills do not have any control equipment for this. However some mills have blowers which reduce chlorine level in the work room environment. Chlorine control is also achieved by scrubbing chlorine through alkaline solution.

42.7 Colour removal

Kraft pulp mills discharge dark brown coloured wastes. The colour is due to lignin and its derivatives. Only a small fraction of the total lignin of the wood processed comes out in the waste. Although lignin and its derivatives present in the waste are not toxic to human beings, animals and aquatic life, they create aesthetic pollution. Since they are not readily biodegradable when present in water, they also reduce the utility of the water and increase the cost of its treatment. The COD of 1 mg of lignin isolated from Kraft black liquor is 1.8 mg. A solution of 1 mg of lignin per litre at pH 7.0 exhibits nearly 10 units of colour on platinum - cobalt scale. The same solution has 1.3 Klett units when measured with 42 filter (420 nm) and 1 cm cell path. The colour intensity increases with increase in pH and is fairly constant between pH 7 and 8. In the manufacture of 1 T of bleached paper from bamboo, as much as 40–50 kg of lignin will be discharged into waste. Most of the lignin will be present in the wastes from pulp washing and caustic extraction stage bleaching. In case of straw, used in small mills, as much as 225 kg of lignin per T of paper will be discharged into the waste. This high value is because no chemical recovery is adopted in small mills. Studies on biological degradation of lignin revealed that:

1. There are only a limited number of micro-organisms capable of degrading lignin—higher organisms like fungi are more capable than bacteria.
2. Biological decomposition is extremely slow and incomplete.

3. There seems to be of not much scope for removal of lignin in the existing biological waste treatment methods.

Among the chemical methods employing precipitation or coagulation, massive lime treatment for colour removal has been found to be technically and economically feasible. This method besides achieving over 90% colour removal, also result in 20–40% reduction in BOD from the coloured pulp will wastes. In India, some mills are presently establishing the massive lime treatment for colour removal. Lime recovery is an essential part of this process. A modified lime treatment for colour removal using only 1000 mg/L CaO for the total unclarified kraft pulp and paper mill waste with lime recovery system has a great promise as this method also helps in the removal and disposal of the suspended solids in the combined waste. Among the coagulants only alum treatment is still being pursued with some promise. Activated carbon treatment for colour removal from raw wastewater may not prove economical.

However, activated carbon treatment preceded by massive lime treatment will yield effluents that can be reused in the mill. Such a combination will make activated carbon treatment an economic proposition. A microlime carbon sequence for removal of 90% colour and 40% TOC from clarified wastewater of unbleached kraft mill, has been shown to yield reuseable water at a relatively low cost compared to massive or minilime activated carbon treatment.

Soils as a medium for colour removal can be used successfully provided the right type of soil and adequate land are available. This method of disposal of coloured pulp mill wastes will be most suited for Indian conditions. The land requirement will be high in most cases. Colour removal in the waste from small mills, where black liquor is not recovered, will pose serious problems and prove uneconomical. In such cases, it is desirable to segregate the small volume of strong black liquor and store it in lagoons before discharge into water courses in the monsoon. The remaining coloured wastes can be used on land or treated by coagulation. This aspect has to be given a serious consideration before locating the mills.

42.8 Pollution prevention options

As mentioned earlier, the pollution prevention options available to the paper and pulp industry will be focused on the reduction of the wastewater generated. Among other pollution prevention options, process modification plays the most important role; it will be emphasised in this section.

42.8.1 Process modifications

Pulping process modifications

Pulp washer system efficiency is an important factor in reducing the amount of wastewater generated. The number of washers required depends on whether the pulping is carried out on a batch or continuous basis. Generally, continuous digesters employ diffusion washing as a final step. This final diffusion washer may be equivalent to one to three vacuum drum washers, the type commonly used in most mills. For batch processing, four or five stages of drum washers are common in a modern mill. A diffusion washer consists of a series of concentric baffles that move with the pulp as flows through the washer body and return to their original position at the end of a wash cycle. The washing solution is forced hydraulically through the pulp across the baffles, providing a wash cycle at least 50 times as long as a drum washer.

The pulp is kept at high consistency during the wash cycle and is not exposed to air, thus eliminating odour problems and the need to use defoamers. Therefore, diffusion washer systems should be employed to maximise overall washing efficiency.

The operating conditions of a drum washer are also important in maintaining high efficiency. The pulp mat that is formed should be uniform and the showers must spray evenly across the surface. The optimum temperature of the spray water is 60°C to 70°C. High water temperature decreases the liquor viscosity (the lower the viscosity, the higher the efficiency) and a temperature that is too high increases the vapour pressure, interfering with the washer vacuum. The wastewater volume may also be reduced by using high-pressure, low-volume showers.

There are alternatives to drum washers and diffusion washers. One example is a horizontal, enclosed belt washer. This washer is operated in a pressure hood under positive pressure. Another example is the Impco Compaction Baffle Filter. It is a slightly smaller unit than a drum washer and operates under pressure. Because it is fully enclosed, it is free of odour and foam problems. It operates at high feed consistency and the volume of the liquor (or wastewater) handled is reduced to approximately 25% of the volume required by a drum washer.

Another general pollution prevention option is to use closed cycle mill processes. An example of a closed cycle bleached Kraft pulp mill is shown in Fig. 42.2. This system is completely closed and water is added only to the bleached pulp decker or to the last dioxide stage washer of the bleach plant. The bleach plant is countercurrent and a major portion of the filtrate from this plant is recycled to the stock washers, after which it flows to the black liquor evaporators and then to the recovery furnace.

The evaporator condensate is steam stripped and is used as a major water source at various points in the pulp mill. A white liquor evaporator is used to separate NaCl, since the inlet stream to the white liquor evaporator contains a large amount of NaCl owing to the recycling of the bleach plant liquors to the recovery furnace.

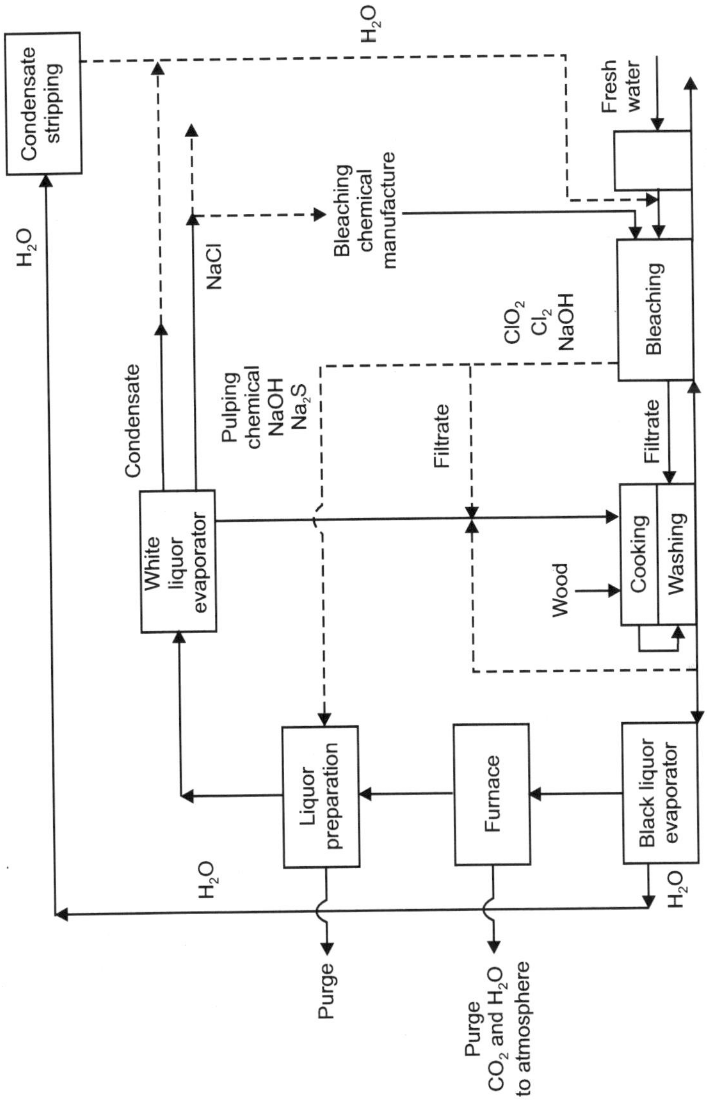

Figure 42.2: Closed-cycle mill.

Bleaching process modifications

The most commonly used chemicals for bleaching pulps are chlorine, sodium hypochlorite, calcium hypochlorite, chlorine dioxide, peroxide and oxygen. Bleaching is usually performed in a number of stages to preserve the strength of the pulp by avoiding extremely strong chemical treatments. All stages are typically carried out at greater than 10% consistency, with the exception of the chlorination stage that is usually carried out at 3% consistency. Therefore, a considerable amount of process water must be added and extracted before and after the chlorination stage. Thus, the chlorination stage plays an important role in wastewater pollution prevention efforts. Approaches to minimise the volume of wastewater are summarised in the following guidelines.

1. Replace the chlorination stage with an oxygen or ozone stage.
2. Recycle the chlorination stage process water.
3. Use water from the countercurrent washing system in the chlorination stage.
4. Perform high consistency gas phase chlorination.

Although using water directly from the countercurrent wash system produces the least volume of effluent, serious corrosion problems can occur if washers or piping in contact with chlorine dioxide filtrates are not made of corrosion-resistant materials.

Resources recovery and recycling

When paper-making stock is run on a paper machine wire, a certain amount of solids are passed through the wire together with most of the water used for suspending the fibre. The water removed from the stock on the wire is known as 'white water', which can be recycled to reduce the volume of effluent.

Typically, white water contains fibre debris, small fibres, soluble matter and a high percentage of non-fibrous suspended matter, such as pigments, starches and dyestuffs. The total amount of solid matter in the white water varies widely from mill to mill, depending on the quality of paper and the types of process employed. The richest white water is obtained from the trays under the first part of the wire. This can generally be returned directly to the mixing box, refiner, or other points as make-up water. A less rich white water is obtained from the pit under the wire. In some mills, the pit is divided into sections to collect water of different richness, with the richest white water used directly as make-up and any excess or less rich white water sent to savealls.

The design of a closed water system with a white water return should follow the general guidelines listed below:

1. Use rich white water from the wire pit and couch pit in as many applications as possible.

2. Make maximum use of saveall clarified clear water. This water should be used for all machine showers that can tolerate water of this quality (e.g., wire showers, wire knock-off showers, headbox showers and breast roll showers).
3. Consider better handling of the large-volume seal water used on water ring vacuum pumps.
4. Reduce the amount of fresh water used in felt showers.
5. Reduce the amount of fresh water used in gland seals and cooling water.

The large volume of seal water used on water ring vacuum pumps may be reduced by installing a separate recirculation system for the vacuum pumps. The use of saveall clarified water is not a practical option, unless the pH of the system is kept sufficiently low, because of pump corrosion problems. Use of vacuum pump seal water can be eliminated entirely by replacing the water ring pumps with high-speed exhausters or lobe-type pumps.

The volume of water used in felt showers can be reduced by the use of high-pressure, low-volume showers. If each low-pressure cleaning shower in a mill is replaced by a high-pressure intermittent cleaning shower, an approximate 90% reduction in water volume can result.

Many of the gland seals can be replaced by mechanical seals. The cooling water can be handled by a clear water sewer, in which all uncontaminated cooling water is collected and discharged untreated.

There are some potential operating problems that arise from recycling of white water. The higher level of suspended solids can cause plugging problems in wire showers, headbox showers, felt filling and lines. Suspended solids can also cause deposit formations that may be biological and/or non-biological in nature. Dissolved and colloidal matters present a more serious problem than suspended solids. After long periods of running under conditions of low fresh water use (or of high recycling), problems of scale and corrosion are observed. With very high degrees of recirculation, the concentration of dissolved matter increases to a point where the chemistry of the paper-making system may be affected and difficulties with foam and sizing can occur.

42.8.2 Good housekeeping and operating practices

To minimise sewer losses and spills from pulping processes, it is recommended that a segregated system be developed to treat each sewer as a separate stream. Sewer systems can be classified in four categories.

1. Low suspended solids.
2. High suspended solids.
3. Strong wastes.
4. Sanitary sewers.

Low suspended solids streams are cooling water, bleaching effluent and evaporator condensates that do not require clarification. High suspended solids streams are decker filtrate, paper machine effluent and wood yard effluent. Strong waste sewers collect floor drains, overflows, spills and strong condensates. The strong waste streams should be contained and reused if possible. Waste generation resulting from spills can be minimised by following the guidelines listed below:

1. Prevent spills by improving process controls.
2. Redirect overflow pipes to trenches leading to 'contaminated' surge vessels for ultimate use.
3. Minimise overflows or spills by installing level controls in process tanks and storage tanks.
4. Install reduntant key pumps and other equipment to avoid losses caused by equipment failure and routine maintenance.
5. Install monitoring systems to alert operating personnel of potential and actual spills so that corrective action can be initiated promptly.
6. Provide a storage lagoon before the biological treatment system to accept long-term shock loads.
7. Train personnel to prevent spills whenever possible and to take immediate corrective actions when they occur.

43

Pollution control in sugar industry

43.1 Introduction

Sugarcane is an energy crop and it is a renewable source. It is the nature's gift to mankind. Sugarcane synthesises energy from the sun and man extracts it from sugarcane in the form of sugar, heat energy, mechanical energy and electrical energy. The production of electrical energy from sugarcane fibre which is termed as cogeneration is gaining great importance day by day because of its renewable nature and attractive economics it offers to the sugar industry. Newer and better technologies are under constant development to harness this energy from sugarcane in the sugar factory. The success of cogeneration schemes depend upon maximising the energy output and minimising the energy expenditure in the factory operations. With the advent of cogeneration in sugar industry with high pressure technologies and export to the government grids, a lot of emphasis is being laid on reducing the power inputs in all the areas of sugar and power production in the factory operations in order to maximise the power export to the grid and increase the revenue generation to the factory.

The concept of cogeneration has enhanced the value of bagasse as a main source of raw material for generation of power and those sugar factories which do not have cogeneration systems look to bagasse as a major source of revenue earner for the factories and therefore implement all systems to save more bagasse by increasing efficiency of operations and reducing power consumption in all areas. This welcome awareness has been capitalised only by a few factories and majority of the factories nearly 10% of them are yet to adopt the changes and reap the benefits because of their own limitations.

The sugar industry is a sector of significant importance to the national economy. While sugar consumption has been growing over the years, the production has been cyclical. The sugar industry is a green industry and is largely self-sufficient in energy needs through use of bagasse for generating electricity and steam. In fact, the sugar industry generates surplus exportable energy through cogeneration and contributes in reducing the energy deficit that India is currently facing. The sugar industry is also the primary source of raw material for the alcohol industry in India.

Sugarcane is comprised of two main components - fibre and juice. Fibre is the most abundant substance in sugarcane, formed by the cellulose in it. Fibre acts as the skeleton of sugarcane, which comes out as bagasse during cane-crushing. Starch, proteins and other organic constituents of sugarcane remain in the cellwalls. The juice contains sucrose, some inorganic non-sucrose substances and water. Sucrose is a disaccharide formed by the chemical bonding of simple sugars, namely, dextrose and fructose, also known as mono-saccharides.

These monosaccharides occur also in free form in sugarcane. They are called reducing sugars and dissolve in a thick syrup during processing along with other chemical constituents, while sucrose crystallises to form the sugar. The inorganic chemicals such as silica, phosphates, chlorides, sulphates of potassium, sodium and magnesium are soluble in water and go into the juice. Sugar processing essentially consists of a series of liquid-solid separation to isolate the sucrose from the sugarcane. Flow diagram of sugar production process is shown in Fig. 43.1.

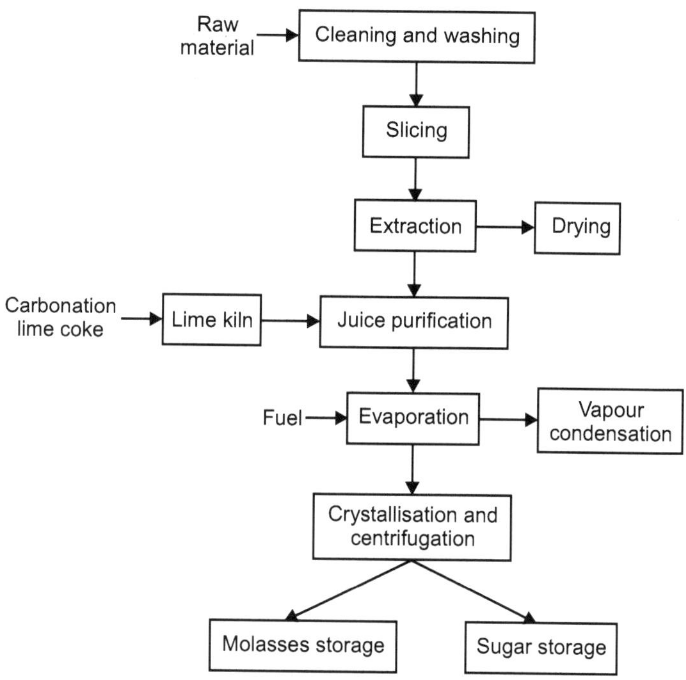

Figure 43.1: Flow diagram of sugar production process.

The manufacturing process in a typical sugar factory involves five basic processes:

1. Juice extraction: The juice extraction plant consists of cane handling, cane preparation and milling sections.
   (a) Cane handling: Cane brought to the mill, is mechanically unloaded by a grab type attachment. A truck tippler is sometimes provided to

unload cane, facilitating loading of sugar cane to cane carrier. Juice extraction from the prepared cane is done by two types of processes. About 95~97% of the sugar factories follow milling process and 3~5% of the sugar factories follow diffusion process.

(b) Milling: The prepared cane is passed through a milling tandem composed of four to six three-roller mills. The juice is extracted from the cane by squeesing under high pressure in these rollers. Extraction is maximised by leaching the disintegrated exposed cane with weak juice and make-up water in a counter current system. The fibrous matter or bagasse, which is left after milling, is used as a fuel for steam generation.

(c) Diffusion: Diffusion process is based on a systematic counter current washing of the cane or bagasse by means of imbibition water. Water is added at the discharge end of the conveyor and percolates through the bed of bagasse and the perforated slats of the conveyor. The water dissolves the sugar in the bagasse and the thin juice thus formed is collected in a hopper. This juice is moved forward one stage by pumping and the process is repeated until the juice reaches maximum concentration at the feed end of the diffuser. The diffuser may be conditioned either for single-flow or for parallel-flows juice circulation.

2. Juice clarification: The purification of juice involves: (a) juice heating (b) sulphitation (c) clarification and (d) filtration. The mixed juice from the mills is heated in raw juice heater(s). The heated juice is treated with chemicals to precipitate various dissolved impurities in the mixed juice. The precipitated impurities are separated to obtain clear sparkling juice in clarifiers. The clear juice is again heated to obtain a temperature of about 105°C.
3. Evaporation: The juice is concentrated from 15 Brix to around 60 Brix in a multiple-effect evaporator. The vapours are bled from the evaporators for juice heating in various heat exchangers and for boiling of massecuite (a mixture of molten liquid and crystals) in vacuum pans. This is the major steam consuming section of the plant.
4. Crystallisation: Crystallisation is an important unit operation, which in sugar industry is known as pan boiling. A major part of the crystallisation process is done in most of the sugar mills in batch type vacuum pans. The massecuite is then transferred to crystallisers where the process is completed by cooling the mass under stirred conditions.
5. Centrifuging: The massecuite from the vacuum pans is sent to the centrifuges, where the sugar crystals are separated from the molasses.

These centrifugal machines can be batch type or continuous type. There are separate centrifugal machines for massecuites. The molasses separated out from this section is a useful by-product, which is an excellent raw material for distilleries.

6. Drying, grading and packing: The moist crystals obtained from centrifugal machines normally contain about 15–20% surface moisture. They are dried in traditional dryers, graded according to crystal sizes and then packed in bags.

43.2 Cogeneration of bagasse

Cogeneration of bagasse is one of the most attractive and successful energy projects that have already been demonstrated in many sugarcane producing countries such as Mauritius, Reunion Island, India and Brazil. Combined heat and power from sugarcane in the form of power generation offers renewable energy options that promote sustainable development, take advantage of domestic resources, increase profitability and competitiveness in the industry and cost-effectively address climate mitigation and other environmental goals. According to World Alliance for Decentralised Energy (WADE) report on bagasse Cogeneration, bagasse-based cogeneration could deliver up to 25% of current power demand requirements in the world's main cane producing countries. The overall potential share in the world's major developing country producers exceeds 7%. There is abundant opportunity for the wider use of bagasse-based cogeneration in sugarcane-producing countries. It is especially great in the world's main cane producing countries like Brazil, India, Thailand, Pakistan, Mexico, Cuba, Colombia, Philippines and Vietnam. Yet this potential remains by and large unexploited.

Using bagasse to generate power represents an opportunity to generate significant revenue through the sale of electricity and carbon credits. Additionally, cogeneration of heat and power allows sugar producers to meet their internal energy requirements and drastically reduce their operational costs, in many cases by as much as 25%. Burning bagasse also removes a waste product through its use as a feedstock for the electrical generators and steam turbines.

Most sugarcane mills around the globe have achieved energy self-sufficiency for the manufacture of raw sugar and can also generate a small amount of exportable electricity. However, using traditional equipment such as low-pressure boilers and counter-pressure turbo alternators, the level and reliability of electricity production is not sufficient to change the energy balance and attract interest for export to the electric power grid. On the other hand, revamping the boiler house of sugar mills with high pressure boilers and condensing extraction steam turbine can substantially increase the level of exportable electricity. This experience has been witnessed in Mauritius, where, following major changes in the processing configurations, the exportable electricity from its sugar factory increased from around 30–40 kWh to around 100–140 kWh per T cane crushed. In Brazil, the world's largest cane producer, most of the sugar mills are upgrading their boiler configurations to 42 bars or even higher pressure of up to 67 bars.

43.2.1 Technology options

The prime technology for sugar mill cogeneration is the conventional steam-Rankine cycle design for conversion of fuel into electricity. A combination of stored and fresh bagasse is usually fed to a specially designed furnace to generate steam in a boiler at typical pressures and temperatures of usually more than 40 bars and 440°C respectively. The high pressure steam is then expanded either in a back pressure or single extraction back pressure or single extraction condensing or double extraction cum condensing type turbo generator operating at similar inlet steam conditions.

Due to high pressure and temperature, as well as extraction and condensing modes of the turbine, higher quantum of power gets generated in the turbine–generator set, over and above the power required for sugar process, other by-products and cogeneration plant auxiliaries.

The excess power generated in the turbine generator set is then stepped up to extra high voltage of 66/110/220 kV, depending on the nearby substation configuration and fed into the nearby utility grid. As the sugar industry operates seasonally, the boilers are normally designed for multi-fuel operations, so as to utilise mill bagasse, procured bagasse/biomass, coal and fossil fuel, so as to ensure year round operation of the power plant for export to the grid.

Latest trends

Modern power plants use higher pressures, up to 87 bars or more. The higher pressure normally generates more power with the same quantity of bagasse or biomass fuel. Thus, a higher pressure and temperature configuration is a key in increasing exportable surplus electricity. In general, 67 bars pressure and 495°C temperature configurations for sugar mill cogeneration plants are well-established in many sugar mills in India. Extra high pressure at 87 bars and 510°C, configuration comparable to those in Mauritius, is the current trend and there are about several projects commissioned and operating in India and Brazil. The average increase of power export from 40 bars to 60 bars to 80 bars stages is usually in the range of 7–10%. A promising alternative to steam turbines are gas turbines fuelled by gas produced by thermochemical conversion of biomass. The exhaust is used to raise steam in heat recovery systems used in any of the following ways; heating process needs in a cogeneration system, for injecting back into gas turbine to raise power output and efficiency in a

Steam Injected Gas Turbine cycle (STIG) or expanding through a steam turbine to boost power output and efficiency in a Gas Turbine/Steam Turbine Combined Cycle (GTCC). Gas turbines, unlike steam turbines, are characterised by lower unit capital costs at modest scale and the most efficient cycles are considerably more efficient than comparably sized steam turbines.

43.2.2 Energy potential of bagasse

Sugarcane is one of the most promising agricultural sources of biomass energy in the world. Sugarcane produces mainly two types of biomass, cane trash and bagasse. Cane trash is the field residue remaining after harvesting the cane stalk while bagasse is the fibrous residue left over after milling of the cane, with 45–50% moisture content and consisting of a mixture of hard fibre, with soft and smooth parenchymatous (pith) tissue with high hygroscopic property. Bagasse contains mainly cellulose, hemi cellulose, pentosans, lignin, sugars, wax and minerals. The quantity obtained varies from 22 to 36% on cane and is mainly due to the fibre portion in cane and the cleanliness of cane supplied, which, in turn, depends on harvesting practices.

The composition of bagasse depends on the variety and maturity of sugarcane as well as harvesting methods applied and efficiency of the sugar processing. Bagasse is usually combusted in furnaces to produce steam for power generation. Bagasse is also emerging as an attractive feedstock for bioethanol production. It is also utilised as the raw material for production of paper and as feedstock for cattle.

The value of bagasse as a fuel depends largely on its calorific value, which in turn is affected by its composition, especially with respect to its water content and to the calorific value of the sugarcane crop, which depends mainly on its sucrose content.

Moisture contents is the main determinant of calorific value, i.e., the lower the moisture content, the higher the calorific value. A good milling process will result in low moisture of 45% whereas 52% moisture would indicate poor milling efficiency.

Most mills produce bagasse of 48% moisture content and most boilers are designed to burn bagasse at around 50% moisture. Bagasse also contains approximately equal proportion of fibre (cellulose), the components of which are carbon, hydrogen and oxygen, some sucrose (1–2%) and ash originating from extraneous matter. Extraneous matter content is higher with mechanical harvesting and subsequently results in lower calorific value.

For every 100 T of sugarcane crushed, a sugar factory produces nearly 30 T of wet bagasse. Bagasse is often used as a primary fuel source for sugar mills; when burned in quantity, it produces sufficient heat and electrical energy to supply all the needs of a typical sugar mill, with energy to spare. The resulting

CO_2 emissions are equal to the amount of CO_2 that the sugarcane plant absorbed from the atmosphere during its growing phase, which makes the process of cogeneration greenhouse gas-neutral.

43.3 Environmental aspects of sugar industry

Environmental pollution has been recognised as one of the major problems of the modern world. The increasing demand for water and dwindling supply has made the treatment and reuse of industrial effluents an attractive option. The problem of environmental pollution on account of essential industrial growth is practical terms, the problem of disposal of industrial water, whether solid, liquid or gaseous. All the three types of wastes have the potentially of ultimately polluting water. Use of industrial effluent and sewage sludge on agricultural land has become a common practice all over the world as a result of which these toxic metals can be transferred and concentrated into plant tissues from the soil. These metals have damaging effects on plants themselves and may become a health hazard to man and animals. Above certain concentrations and over a narrow range, the heavy metals turn into toxins.

The most important effluent discharging industries are sugar mills, thermal power plants, paper mills, textiles, distilleries, fertiliser units, electroplating plants, tannery industries, sago factories, oil refineries, pesticide and herbicide industries. These industrial effluents containing heavy metals pose a serious threat to the ecosystem. Use of these industrial effluents and sewage sludge for agriculture have become a common practice in developing countries as a result of which these toxic metals get transferred and accumulated into plant tissues from soil.

Sugar industry is seasonal in nature and operates only for 120 to 200 days in a year. A significant large amount of waste is generated during the manufacture of sugar and contains a high amount of production load particularly in items of suspended solids, organic matters, effluent, sludge, press mud and bagasse.

43.4 Effluents from sugar mills

Sugar industry, the most important among agro based industries contributing significantly to rural economy in particular and national economy in general generates huge amount of waste during the manufacture of sugar. The most important crop from which sugar can be produced in commercial quantity are sugarcane. The sugar industries are playing an important role in the economic development of the country but the effluents released produce a high degree of organic pollution in both aquatic and terrestrial ecosystems. They alter the physicochemical characteristics of the receiving aquatic bodies and affect aquatic flora and fauna. Sugar factory effluent produces obnoxious odour and unpleasant colour when released into the environment without proper treatment. Farmers have been using these effluents for irrigation, found that the growth, yield and soil health were reduced. The life in effluent is highly diverse and consists of interacting population of micro-organisms and effluent fauna and their activities affect physical, chemical and biological characteristics of effluent. Some potential fungal strains such as *Penicillium pinophilum*, *Alternaria gaisen*, *Aspergillus flavus*, *Fusarium monolifome*, *A. niger* were isolated from sugarcane industrial effluent.

The sugar mill wastes contains a high amount of production load particularly, suspended solids, organic matters, press mud and several air pollutants. Wastewater from sugar mills with its high Biochemical Oxygen Demand (BOD), Chemical Oxygen Demand (COD) and Total Dissolved Solids (TDS) rapidly depletes available oxygen when discharged into water bodies and have adverse impact on aquatic life so much so rendering the receiving water unfit for drinking and domestic purposes, reducing crop yields if used for irrigation and exacerbating corrosion in water systems and pipe.

Sugar factory effluent has an obnoxious odour and unpleasant colour when released into the environment without proper treatment. Contaminants such as chloride, sulphate, phosphate, magnesium and nitrate are discharged with the effluent of various industries which create a nuisance due to physical appearance, odour and taste. Such harmful water is injurious to plants, animals and human beings. The effects of various industrial effluents on seed germination, growth and yield of crop plants have captivated the attention of many workers.

Diverse sugar industry effluents disposed in soil and water cause major pollution problems. The sugar industry plays an important role in the economic development of a country, but the effluents released produce a high degree of organic pollution in both aquatic and terrestrial ecosystems. The effluents also alter the physicochemical characteristics, flora and fauna of receiving aquatic bodies. In addition, sugar factory effluent discharged in the environment posses a serious health hazard to the rural and semi-urban populations that use stream and river water for agriculture and domestic purposes. Fish mortality and damage to paddy crops due to sugar industry aters entering agricultural land have been reported. Sugar factory effluent that has not been treated properly has an unpleasant odour when released into the environment. Farmers using these effluents for irrigation to reduce water demand have found that plant growth and crop yield were reduced and soil health was compromised. Because, sugar industry effluents are commonly used for irrigation and it is essential to determine how crops respond when exposed to industrial effluents.

The sugar mill effluent, as it leaves the premises, has a relatively clear appearance. However, after stagnating for sometime it turns black and starts emitting foul odour. In general, sugar mill effluents contain a significant concentration of suspended solids, dissolved solids, a high BOD, COD, considerable amount of chlorides, sulphates, nitrates, calcium, magnesium and sugar concentration. For mills that have an attached distillery, the numerous distillation stages produce a highly contaminated effluent with BOD and COD concentration of about 40,000–100,000 mg/L called sillage. In addition to that, various heavy metals were found in the effluent.

During earlier days treating effluents were achieved by ion exchange, chemical oxidation and chemical precipitation. For advanced purification, different physicochemical methods such as active carbon adsorption, ion exchange and reverse osmosis are used. Each of these methods has its own merits and demerits. Chemical precipitation and electrochemical treatments are ineffective, especially when metal ion concentration in aqueous solutions is lower than 50 mg/L. Moreover, such treatments produce large amount of sludge to be treated with great difficulties. Ion exchange, membrane technologies and activated carbon adsorption processes are extremely expensive.

Earlier several microbes were reported to take effective part in bioremediation of industrial wastes. The mechanism by which micro-organisms act includes biosorption, intracellular accumulation and enzyme-catalysed transformation. On the basis of energetic requirements, biosorption seems to be the most common mechanism. Furthermore, this is the only option where dead cells can be applied as bioremediation agent. Nevertheless, systems with living cells allow more effective bioremediation processes as they can self-replenish and remove metals via different mechanisms.

Physicochemical and biological characteristics of sugar mill effluent: Generally, the effluent is generated from mill house, wastewater from boiling house, wastewater from boiler blow down, condenser cooling water and soda and acid wastes. Sugar mills generate 0.16–0.76 m^3 of wastewater for every tonnes of cane crushed by them. The pollution standards stipulate that BOD of wastewater should be less than 30 mg/L for disposal into inland surface waters and less than 100 mg/L for disposal on land. Whereas, the combined sugar mill wastewater had a BOD of 1000 to 1500 mg/L. The untreated sugar mill effluent is toxic to plants when used for irrigation.

The trends in developing countries to use sugar mill effluent as fertiliser has gained much importance as it is considered as a source of organic matter and plant nutrients and serves as good fertiliser. Sugar mill effluent contains considerable amount of potentially harmful substances including soluble salts and heavy metals like Fe^{2+}, Cu^{2+}, Zn^{2+}, Mn^{2+}, Pb^{2+}. The long term use of this sugar mill effluent for irrigation which contaminates soil and crops to such an extent that it becomes toxic to plants and causes deterioration of soil. The physicochemical parameters of sugar mill effluent are, temperature (30°C), colour, turbidity (84.7 NTU), pH (8.1), EC (5530 dSm^{-1}), BOD (6856 mg/L),

COD (7432 mg/L), total dissolved solids (2516 mg/L), chloride (1894 mg/L), total alkalinity (254 mg/L), total hardness (342 mg/L), sulphate (540 mg/L), phosphate (224 mg/L), total acidity (45 mg/L), calcium (364 mg/L) and magnesium (151 mg/L) and according to the permissible level suggested by Bureau of Indian Standards (BIS) almost all the water quality parameters in the sugar effluents have been found to be very high and well above the permissible limits. The physicochemical characteristics of sugar industry effluent are high in TDS (422–608 mg/L), COD (1152–17680 mg/L) and BOD (380–650 mg/L) than ground water. There were some nutrients and heavy metals viz., P, N, K, Zn, Cu, Fe, Mn and Pb. The physicochemical properties such as silt, clay, electrical conductivity, water holding capacity, organic matter and total nitrogen contents, microbial population and cellulase activities were significantly higher in the samples collected from sugar industry wastes dump than in the non-dump sites. Analysis of sugar mill effluents and soil samples had shown high metal content than the permissible limits except Pb. Further, analysis of plant samples have indicated the maximum accumulation of iron (Fe) followed by Mn and Zn in root, shoot, leaves and seeds of wheat and mustard. The physicochemical analysis showed that the sugar mill effluent was acidic in nature and yellowish in colour. It was rich in total suspended and dissolved solids with large amount of biological oxygen demand and chemical oxygen demand. Higher amount of chloride, calcium, magnesium, sodium, potassium and iron were also present in the effluent. These effluents severely affected the plants and soil properties when used for irrigation.

The sugar mill effluent is typically associated with polluted problems related to high BOD, COD, toxicity, colour, suspended solids, etc., Sugar factory effluent when discharged into the environment, possess a serious health hazard to the rural and semi-urban populations where stream and river water was used for agriculture and domestic purposes. There are several reports of fish mortality and damage to the paddy crops in these areas due to wastewaters entering agricultural land. Farmers have been using these effluents for irrigation, found that the growth, yield and soil health were reduced. However, some potential fungal strains such as *Penicillium pinophilum*, *Alternaria gaisen*, *Aspergillus flavus*, *Fusarium monoliforme* and *Aspergillus niger* were isolated from sugarcane industrial effluent.

43.5 Bioremediation of various industrial effluent and toxic substances by microbial isolates

The term 'bioremediation' has been used to describe the process of using micro-organisms to degrade or remove hazardous pollutants from the environment. Bioremediation can also be defined as any process that uses micro-organisms, green plants or their enzymes to return to the natural environment altered by contaminants to its original condition. Heavy metal bioremediation involves removal of heavy metal from wastewater and soil through metabolically mediated or physicochemical pathways. This natural and environmental friendly technology is cost effective, aesthetically pleasant, soil organism- friendly, diversity enhancer, energy derivation from sunlight and more importantly, it is able to retain the fertility status of the soil even after the removal of heavy metals. Bioremediation, the use of micro-organisms to degrade environmental contaminants, is among these new technologies. Bioremediation has numerous applications, including cleanup of ground water, soils, lagoons, sludge and process-waste streams. Out of several methods that are used in the treatment of industrial effluents, biotreatment offers a cheaper and environmentally friendlier alternative for biological degradation of industrial effluents. Microbial treated industrial effluents contains an array of plant nutrients, trace elements, rich in silica which is essential for silicicolous plants like rice and sugarcane, for non lodging and also for improving resistance to pests and diseases. Industrial wastes for land application have been proven to be a cost effective. Biological removal of contaminants from aquatic effluents offers great potential when the contaminants are present in trace amounts. The term 'bioremediation' has been used to describe the process of using micro-organisms to degrade or remove hazardous pollutants from the environment.

Bioremediation can also be defined as any process that uses micro-organisms, green plants or their enzymes to return to the natural environment altered by contaminants to its original condition. Heavy metal bioremediation involves removal of heavy metal from wastewater and soil through metabolically mediated or physicochemical pathways. This natural and environmental friendly technology is cost effective, aesthetically pleasant, harmless to soil organism, diversity enhancer, energy derivation from sunlight and more importantly, it is able to retain the fertility status of the soil even after the removal of heavy metals. Bioremediation of toxic industrial effluents by micro-organisms serves as an effective method to substitute the conventional recovery and removal process. Fungal biomasses have huge capability of treating effluents discharged from various industries. White rot fungi are ubiquitous in nature and their adaptability to extreme conditions makes them good biodegraders. Their enzymatic degrading activity makes them effective decoloriser; they also remove toxic metals by biosorption ultimately rendering the effluents more ecofriendly.

43.6 Recycling of microbially remediated sugar mill effluent for agricultural use

The sugar industries consume huge quantity of water and throw back almost an equal quantity of effluent which contains highly toxic materials in dissolved or suspended form. If this water is properly used or it is purified to recycled, a part of water shortage will surely be solved. The term 'bioremediation' has been used to describe the process of using micro-organisms to degrade or remove hazardous pollutants from the environment. Bioremediation can also be defined as any process that uses micro-organisms, green plants or their enzymes to return to the natural environment altered by contaminants to its original condition. The beneficial use of treated effluents for agriculture is the major reuse application worldwide. Effluents of particular interest for land application include municipal wastewater, farm effluents, wastewater from the agricultural and forest industries such as effluents from animal and plant processing plants, food processing plants and paper mills. Currently, most countries still have no legal framework for wastewater reuse in agriculture. Many countries either refer to international guidelines or do not mention the option of effluent irrigation in agriculture in their resource management act at all. When considering effluent irrigation, the beneficial and harmful risks have to be assessed. Short term benefits associated with effluent irrigation, such as provision of nutrients for plant growth and development, improvement of soil structure and crop yield increase have been described in some detail. The major risk of effluent irrigation is the potential for infectious diseases in animals grazing sewage-irrigated pastures and people directly or indirectly exposed to the effluents. More recently, authorities are concerned about the potential long term tradeoffs associated with effluent irrigation, mainly the irreversible deterioration of groundwater quality.

Elevated concentrations of anthropogenic contaminants such as heavy metals, effluent-derived hormones, pharmaceuticals and other organic chemicals and soil - borne organic pollutants such as pesticides, have been identified in groundwater at effluent-irrigated sites.

Use of urban wastewater in agriculture is a centuries old practice, which has recently received renewed attention in many parts of the world, particularly in arid and semi-arid regions, due to the growing scarcity of water. Sewage or wastewater means untreated city effluent including industrial and domestic wastewater. High population growth rates and rapid industrialisation imparts immense pressure on existing water and land resources; and has resulted in the discharge of large volumes of sewage. Wastewater use is increasing in urban and peri-urban agriculture and even in distant downstream rural areas where such use derives significant economic activity and supports the livelihood of resource poor farmers but alters environment quality. In countries where treatment and safe effluent disposal facilities are limited or non-existent, sewage is used to irrigate fodders, ornamental and food crops including vegetables. Utility potential of industrial effluents for irrigation of crop-fields has been a controversial proposition due to the contradictory reports obtained on the effects of these effluents on crop plant responses. Most of our water resources are gradually becoming polluted by addition of huge amounts of sewage and industrial effluents. These effluents contain toxic materials with varying properties from simple nutrients to highly toxic substances. The discharge of industrial effluents with varying amounts of pollutants has altered the water quality. Sugar mill effluent at above 25% concentration significantly reduced the seed germination percentage of groundnut and paddy due to high concentration of solids in effluent. Effluent of sugar mill showed toxic effects on rice and blackgram. Rice remained more sensitive and susceptible to the toxic effects of industrial effluents but blackgram proved to be more tolerant.

Metribuzin present in sugar mill effluent is weakly absorbed in soil and therefore leaches easily to lower soil profiles and results in loss of activity. The lower concentration of sugar mill effluent showed promoting effect on seed germination, seedling growth, dry matter production and biochemical parameters whereas germination percentage and seedling growth were inhibited at 100% concentration. The sugar factory effluent drastically reduced the growth and germination of wheat, jowar, raddish and bhindi.

The seed germination and seedling growth of wheat, garden pea, black gram and mustard were significantly reduced with increase in concentration of mixed industrial (distillery and sugar) effluent. The sugar factory effluent was found to affect plant growth but bore water favoured seedling growth of greengram (*Vigna radiata*) and maize (*Zea mays*). All growth parameters of greengram (*Vigna radiata*), biochemical contents, enzyme activities and yield parameters were found to increase at 10% effluent concentration and decreased from 25% effluent concentration onwards. Increase in sugar mill effluent concentration resulted in corresponding decrease in germination rate of soyabeans (*Glycine max),* whereas 10% dilution of effluent enhanced seed germination and 100% effluent completely inhibit for both seed germination and seedling growth. The per cent root colonisation and the number of spores was less in sugar mill effluent polluted soil when compared to non polluted soil which resulted in decreased germination rate of sunflower (*Helianthus annus L.*)

Panoras and others studied the effects of effluent from reclaimed either by activated sludge with micro-organism or by stabilisation ponds on field-grown corn and reported that there is a potential risk of facing problems related to soil salinity and alkalinity if no consideration for soil reclamation is taken into account. Chandra and others studied the effect of sugar mill effluent on *Phaseolus aureus* and found that sugar mill effluent decreased the chlorophyll content. Similarly untreated sugar mill effluent decreases the seedling growth, chlorophyll and amylase contents in green gram.

Hulugalle and others reported that the treated sewage effluent using *Aspergillus niger* contained considerable amounts of nitrogen, phosphorus and high concentrations of sodium salts. Effluent water was moderately saline and compared with river water, had higher concentrations of Na, nitrate-N and K and lower concentrations of Ca and Mg. Irrigation with treated sewage effluent caused large increase in nitrate-N, small increase in exchangeable Mg, Na and K and small decrease in heavy metals.

Sompong meunchang and others stated that sugar mill by-products inoculated with N_2 fixing bacteria viz., *Azotobacter* sp., *Beijerinckia* sp. and *Azospirillum* sp. may be good soil amendment to promote tomato growth. Similarly, the higher concentration of the treated effluent using fungi (*Aspergillus niger*) had no effect on seed germination of *Triticum aestivum* and also the lower concentration of sugar factory effluent increased the seedling growth, chlorophyll and amylase contents in green gram seedlings.

Smith and others studied on *Phosphobacterium* treated distillery and sugar factory mixed effluent to investigate its effect on seed germination and seedling growth in wheat and black gram. Among the effluent analysed, the treated sugar mill effluent exhibited better performance in growth rate of wheat and black gram. Laine and Jurgensen found that mixing of *Bacillus megaterium*, along with nutrient addition degraded the heavy metals in the sugar mill contaminated soil and increased the germination percentage of Onion. Similarly, distillery contaminated soils were degraded using *Termitomyces clypeatus* and used in cultivation of olives, the length and number of new shoots, diameter of the main trunk, leaf area and dry weight of the plants were enhanced under the influence of treated soil.

43.7 Waste generation

43.7.1 Wastewater

The sugar mill operates for 3 to 6 months per year. A considerable amount of wastewater is released during crushing of sugar cane. These wastewaters are disposed into nearby water bodies and they are being used for irrigation. The discharge of this effluent into water bodies or on soil is causing a serious problem of water pollution resulting in severe damage to the flora and fauna and environmental degradation. Besides, the heavy metals are non-biodegradable and persist for longer periods in aquatic as well as terrestrial environments thus they can exert detrimental effect on human health and environment due to the toxicity of heavy metals.

Water pollution

1. Agrochemicals and sediments can pollute nearby water sources.
2. Most of the time it is observed that the sugar industries discharge their wastewater, i.e., spent wash during the rainy season in the nearby rivers

and other surface water bodies resulting in severe contamination of the surface water resources.

3. Siltation and eutrophication of nearby surface water bodies due to soil erosion due to flood irrigation and over irrigation.

Water scarcity

Sugarcane is a deep-rooted crop and can have a great impact on river flow as it reduces runoff from the catchment into rivers and draws heavily on ground water resources.

Mill house: Mill house wastewater is derived from continuous gland cooling and intermittent floor washing and contains high amounts of oils and grease and sugar from spills and leaks.

Boiler blow down: Boiler blow down is fairly clean water except that it contains high dissolved solids and phosphates.

Rotary filter: Filter cloth is periodically washed and constituents a source of wastewater.

Condensates: The vapours from the last effect evaporator and pan boiling are separately cooled in barometric condensers and the condensate goes to the pond. A part of the cooled water from the pond is recycled into the sugar mill, but a large portion is discharged as wastewater. If the mill operates without overloading, the evaporator and vacuum pan condensate is quite clean and the entire quantity can be reused. But many a times, overloading and poor operating conditions can lead to significant sugar losses in the condensates through entrainment and thus polluting the water.

Occasional spills and leaks: Leaks from pumps and pipes in the evaporators and centrifuge house, along with periodical floor washings, constitute another source of wastewater. Although the flow is intermittent and volume discharged is not large, it represents the most polluting fraction of sugar mill wastewater.

Condensate washings: Evaporators, juice heaters, pans, etc., are cleaned once in 20 days for removal of deposited scales. Caustic soda, sodium bicarbonate and hydrochloric acid are used for scale removal. Normally the caustic soda washings are stored and reused for cleaning operations. However, in India, most of the sugar mills discharge these chemicals into the drains. After the equipment is boiled with caustic soda and rinsed with fresh water, it is cleaned with dilute hydrochloric acid using an inhibitor. The wastewater is discharged into the drains, as the recovery of the chemicals may not prove to be economical. It is seen that the wastewater has small organic load but inorganic content may be high to pose a shock-load to wastewater treatment facility (occasional discharge, once in fortnight). It is suggested to have a holding tank and mix this wastewater gradually to the final effluent to avoid shock loading on the treatment plant.

Sulphur and lime houses: The washings of sulphur and lime house would contain a considerable amount of inorganic solids, which include carbonates and sulphates. The effluents from these two units when combined would give neutral pH value of waste. This wastewater does not contribute to organic pollution but can be characterised as inorganic wastewater.

Wastewater parameters

BOD: This is the measure of the oxygen consuming capabilities of organic matter. During decomposition, organic effluents exert a BOD that can deplete oxygen supply. BOD is generally measured and expressed in parts per million or milligrams per litre. The effluents from a raw sugar factory can vary between hundred to several thousands mg/L.

Dissolved oxygen: This is water quality constituent. It is measured and expressed as parts per million or mg/L.

Total suspended solids (TSS): Suspended solids when they settle form sludge on the stream, lakebed and they are most damaging to the life in water.

The different modes of disposal of wastes are:

1. Disposal into water bodies (River/Nallah, stream).
2. Evaporation in open pits.
3. Disposal into ocean.
4. Press mud for fertiliser.
5. Bagasse for paper and pulp and fibre.

43.7.2 Solid wastes

Bagasse: It is estimated that bagasse contributes to 33.3% residue of the total cane crushed. It has a calorific value of about 1920 kcal/kg and is mainly used as fuel in boilers for steam generation.

Press mud: It contains all non-sucrose impurities along with $CaCO_3$ precipitate and sulphate. Press mud from double sulphitation process contains valuable nutrients like nitrogen, phosphorous, potassium, etc., and therefore used as fertiliser. The press mud from double carbonation process is used for land filling and is not used as manure.

43.7.3 Effect of soils

1. Erosion is a significant issue in areas under sugar cane or beet cultivation, since erosion rates in tropical agro-ecosystems are usually greater than the rate of soil formation.
2. Cane harvesting can cause a significant removal of soil with the roots. Declining soil quality is associated with cane and beet production, due

to soil compaction, loss of organic matter, salinisation and acidification. Generally 10–30% of the total beet harvest weight is soil.

Impact on soil health

1. Heavy infield transport machinery is most commonly associated with soil compaction problems. Soil compaction decreases porosity and water infiltration rate, restricting the rooting ability of the crop.
2. Conventional tillage commonly promotes erosion by exposing soil aggregates to rainfall. Conventional tillage, i.e., deep ploughing, also drastically changes soil structure.
3. Acidification is also more prevalent in cane than beet growing areas, largely due to the use of inorganic nitrogenous fertilisers such as urea and ammonium sulphate.
4. Most of the farmers still use the flood irrigation pattern which results in huge wastage of water, electricity and results in salinisation of the soil which is another important cause of lower productivity.
5. Productivity of the sugarcane in the region is going down owing to excessive use of chemical fertilisers/herbicides.

43.7.4 Air pollutants

The bagasse, on burning, produces particulates, viz., unburnt fibres, carbon particles and gaseous pollutants like oxides of nitrogen, water vapour and other organic compounds. Of the particulate waste, the heavier particles slowly settle down in the surrounding area. Such dust fall leads to the problems of cleaning, reduction in property value, effect on vegetation, etc. The main gaseous pollutants are CO, which is altogether not measured by any unit and CO_2 is reported to be in the range of 12–14%.

43.8 Wastewater reduction and by-product recovery

The following areas are important to economise the usage of water.

43.8.1 Cooling water

1. Mainly used for condenser, bearing cooling, sulphur/lime houses and crystalliser for formation of crystal.
2. In condenser, water gets mixed with vapour. However, adjusting pH along with make-up water to keep dissolved solids in check can recycle it.
3. Evaporator cooling water contains entrained sugar and acid because of excess of SO_2 and can be recycled. Improvement in the designs of evaporator/pan boiler can reduce the loss. Losses will also be due to evaporation, splashing and percolations.

4. Keeping the temperature of incoming water between 30°C and 35°C can reduce losses due to evaporation. Splashing and percolation can be checked by proper maintenance.
5. Cooling water for bearings, power generation, etc., can be reused safely.

43.8.2 Process water

Sugar mill employs both hot and cold water for its various processes such as macreation, filter cake washing, lime preparation, dilution for lowering Brix, dilution in evaporators and pans, massecuite, magma making and crystal washing in centrifugals.

1. Water requirement before evaporator storage is about 1/5 to 1/4 of steam used while that used after evaporator requires approximately equal amounts, as for steam. Careful attention is required after evaporator stage to control water usage.
2. Hot water can be used in place of cold water to reduce the quantity of steam required.
3. It is preferable to use 18–20% maceration by equally adjusting it from the top and the bottom of bagasse bed feeding to the last mill at a pressure of 7–14 kg/cm^2 rather than merely pouring the same at 25 to 30% of cane (about 5–7% water can be saved).
4. If maceration is high enough, there will not be any need of dilution water for juice.
5. To reduce water quantity, light molasses can be used for magma making.

Washing water: Wash water may contain sugar and therefore requires treatment and should not be recycled. Periodic cleaning results in high BOD and it also contain caustic soda and weak acids. Returning it to the service water tank can reuse water.

Oil and grease: Providing suitable oil and grease traps can eliminate this.

Chemical reuse: The stored and settled supernatant can be reused with a little addition of fresh caustic soda for next cleaning operation.

Molasses handling: It is necessary to store molasses in RCC tanks or steel tanks above ground level. Otherwise, there is a possibility of ground water contamination. The high BOD of molasses may cause pollution problems due to mishandling.

Sugar industry wastewater treatment using adsorption: Most of the waste treatment plants have primary stage by addition of acid/alkali for neutralisation is causing side effects of increasing the amount of total dissolved and suspended solids. This effect hinders the effectiveness of subsequent treatment in lowering of COD, BOD, TDS, TSS amounts of oil and grease. Poisonous pesticides and organic substances difficult to degrade can also be easily removed by adsorption.

Different adsorbent like activated charcoal, bentonite, lignite, Mgo and fly ash can be used for adsorption of sugar industry wastewater. Use of the above adsorbents helps in 80% removal of TDS, TSS and oil and grease and in lowering of COD, BOD colour and smell.

Product recovery

The by-products available from sugar mills are bagasse, furnace ash, molasses and filter mud. The uses of these by-products are given below. If all the by-products are used for transformation into value added products, (resource recovery), it will minimise the pollution to large extent.

Bagasse: These are used for steam, power, charcoal, briquettes and methane and producer gas.

Molasses: These are used for fertiliser and cattle feed.

Filter mud: For fertiliser.

Boiler ash: For foundry material.

Safety

Sugar in boiler feed water causes water to foam, which will lead accidents. If not present in large quantity. It is decomposed by heat into products that are detrimental to the tubes and shells of boilers causing pitting and overheating.

If sugar is present in small amounts their traces will be eventually accumulated on the boiler tubes as a harmful and dangerous carbonaceous deposit. The break down of sugar also forms harmful organic acids. To prevent this lime is added to feed water to maintain pH = 8.0. A pronounced odour develops in the steam if boiler water contains sugar. Under such conditions the contaminated feed water is turned to sewer and the boilers are blown off. To prevent these hazards tests are conducted to determine amount of sugar traces in water. The most commonly used tests are Naphthol test and Aresenomolybdate test.

44

Pollution control in distillery industry

44.1 Introduction

Distillery wastewater, generally known as stillage, slops, vinnasse or dunder, constitutes a high value, high strength acidic waste that presents significant disposal or treatment problems. Intensive treatment of distillery wastewaters has become imperative because of strict water quality legislation and decrease in land availability. Production of ethanol from agricultural materials for liquid fuel supplement poses a serious threat to water quality in several regions of the world. A distillery is a premise where distillation takes place, especially distillation of alcohol. The distillery industry is based on sugar industry.

The distilling industry is that part of the food and beverage industries engaged in clarifying, flavouring, blending and ageing alcohol to make potable spirits (e.g., brandies, grain spirits, rum) and establishments which manufacture ethyl alcohol, whether they are or are not used in potable spirits.

Production of ethyl alcohol in distilleries based on cane sugar molasses constitutes a major industry in Asia. The aqueous distillery effluent stream known as spent wash is a dark brown highly organic effluent and is approximately 12–15 times by volume of the product alcohol. It is one of the most complex, troublesome and strongest organic industrial effluents, having extremely high COD and BOD values. Because of the high concentration of organic load, distillery spent wash is a potential source of renewable energy.

In recent years, membranes and membrane separation techniques have grown from a simple laboratory tool to an industrial process with considerable technical and commercial impact. Today membranes are used on a large scale to produce potable water from the sea by reverse osmosis, to clean industrial effluents (distillery wastewater) and to recover valuable constituents by electrodialysis. In many cases, membrane processes are faster, more efficient and more economical than conventional separation techniques. With membrane, the separation is usually at ambient temperature, thus allowing temperature sensitive solutions to be treated without the constituents being damaged or chemically altered.

There are two major types of wastes: (i) inorganic waste and (ii) organic waste. Organic wastewaters are potent sources of water pollution.

Various organic wastewaters that are known to cause serious problems may be attributed to distillery effluents, pulp and paper effluents, textile effluents and tannery effluents, among others. Among these types distillery wastewater is highly charged with organic matter, when dumped into water sources without treatment or with inappropriate treatment, causes serious pollution. Among the raw material sources for distillery, two very important raw materials are cane sugar molasses and beet sugar molasses. Molasses is a by-product of the extraction process and is heavily used as a raw material in many distilleries around the world. The discharge of wastewaters from wineries and distilleries is becoming increasingly restricted as pressures from environmental regulations increase and as awareness of the negative impacts of seasonal discharges of water containing high nutrient and organic loadings into water courses spreads. Raw stillage discharge has a highly deleterious effect on fish life. Stillage has been proposed for use as a fertiliser, food supplement, biomass production agent, animal feed and potash source.

The distillery wastewater has high amount of organic matter so without treatment pull down in water stream is not ethical. The first step is to characterise the distillery wastewater and membrane technology, then by these studies membrane techniques use for treatment of distillery wastewater.

Distillery wastewater which consist of composition of distillery wastewater, source, effect, zero discharge system of distillery then about membrane technologies which describe about membrane types, membrane modules, type of membrane techniques and also about how the membrane bioreactors are helpful for treatment of distillery wastewater.

44.2 Distillery wastewater

Among the raw material sources for distillery, two very important raw materials are cane sugar molasses and beet sugar molasses. Distillery wastewater (stillage) is the main by-product originating in distilleries and its volume is approximately 10 times that of ethanol produced. It is not surprising that the utilisation of the stillage raises serious problems and that many attempts have been made all over the world to solve them.

Distillery wastewater is usually comprised of a high volume of greatly acidic matter which presents many disposal and treatment problems. Waste streams generally contain high levels of both dissolved organic and inorganic materials. There has been increasing interest in the use of ethanol from biomass as a liquid fuel alternative.

Ethanol fermentation is examined in relation to distillery wastes. Reducing the volume of wastewater may be accomplished by fermenting higher strengths of molasses. To characterise distillery wastewater in detail so that proper insight may be gained in an attempt to treat the waste to reduce the pollution hazards. Oxygen consumption values can be used to quantify the amount of organic matter present in wastewater.

Waste streams of distillery wastewater generally contain high levels of both dissolved organic and inorganic materials. There has been increasing interest in the use of ethanol from biomass as a liquid fuel alternative. Ethanol fermentation is examined in relation to distillery wastes. Reducing the volume of wastes may be accomplished by fermenting higher strengths of molasses. There are various methods used in the treatment of distillery wastewater. Physical-chemical treatment of distillery wastewater has little success. Anaerobic digestion, anaerobic filters, lagoons, activated sludge and trickling filters have all been successfully applied to the treatment of distillery wastewater. This diverse spectrum of wastes requiring efficient treatment has focused the attention of researchers on membrane, ion-exchange and biological technologies. Membrane and membrane separation techniques with immobilised micro-organism or enzyme have very significant role in treatment of distillery wastewater.

44.2.1 Utilisation of stillage

As a result of difficulties of treating stillage and its high concentrations of dissolved organic and inorganic matter, a host of utilisation schemes have been proposed. These are discussed below.

Recycle

Recycle of stillage reduces the volume of waste to be treated and removes some of the organic and inorganic constituents. Where the distillery is associated with a sugar mill, there exists the possibility of using stillage as a portion of the cane wash water. Stillage can be recalculated as 10–20% of the molasses dilution water for a number of cycles. Beside reducing effluent quantities and fresh water requirements, this scheme decreases nutrient additions and steam consumption. Thus, the amount of stillage which can be recycled in this manner is upto 50%.

Land application

It has been observed that direct land application as an irrigation water and fertiliser have the following effects.

1. Increase in soil pH to neutrality because of high calcium and magnesium content.
2. Improvement in physical and chemical properties of the soil.
3. Increase in water retaining characteristics of the soil.
4. Increase in mineral salt retaining characteristics of the soil.
5. Production of ideal conditions for cane cultivation.
6. Restoration and maintenance of soil fertility.
7. Increase in soil micro-flora.

44.2.2 Measurement of water pollution

Water pollution may be analysed through several broad categories of methods:

Physical and chemical treatment methods

In general, physicochemical treatment of stillage has met with little success. Sedimentation was found to be unsatisfactory even with the addition of coagulants and other additives such as alum, lime, ferric chloride, other aluminium salts, iron, bentonite, activated charcoal and soap. Settling was also found to lead to anaerobic conditions and odour formation.

Some of the physical treatment methods are given below:

1. Sedimentation.
2. Screening.
3. Aeration.
4. Filtration.
5. Floatation.

Some of the chemical treatment methods are given below:

1. Chemical coagulation.
2. Chemical precipitation.
3. Chemical disinfection.
4. Chemical oxidation.
5. Advanced oxidation processes.
6. Ion exchange.
7. Chemical neutralisation, scale control and stabilisation.

44.3 Biological treatment methods

The characteristics of distillery wastewater have COD/BOD ratio of 1.8 to 1.9 indicating thereby that wastewaters are amenable to biological treatment. Further, BOD/N/P ratio of 100/2.4/0.3 to 100/2.75/0.8 and high BOD of the distillery effluent indicates that anaerobic treatment methods are more suitable to any aerobic biological treatment. Should the avenues of stillage utilisation and by-product recovery be closed, biological treatment of stillage offers the only means of disposal.

Additionally, the liquor remaining after by-product recovery is not generally suitable for discharge to a receiving water body and requires further treatment. The biological treatment is often the most affective. Biological treatment methods comprise anaerobic and aerobic systems and are discussed below.

44.3.1 Anaerobic lagooning

In this method, the effluent is retained in the lagoon for a period of about 90 days, which results in the reduction of BOD from 40,000 to 3000 ppm. This effluent has to be further diluted 6 to 7 times to bring down the BOD below 500 ppm to enable its disposal through surface irrigation. The main advantages of this method relate to the use of high organic loading rates on a relatively small land area because of larger depth, low capital cost as only earthwork is needed and low operational cost because the nutrient requirement is much less as compared to aerobic treatment. This method has the disadvantage that it leads to contamination of groundwater in the long run and also needs fairly impervious soil strata. It is not suited for urban locations, as lagoons emit a bad odour and breed mosquitoes and insects.

44.3.2 Solar drying of effluents

In this process, the effluent is dried in open shallow pans. Solar evaporation needs battery of trenches where from sediments are recovered for fertiliser use. This practice is not very popular for reasons of a vast requirement of land. A distillery of 6000 gallons capacity needs about 40 hectares of land for solar drying and it gives out an objectional smell. Besides, this method is operative during dry days only, thus limiting its application.

44.3.3 Activated sludge method

It is a biological oxidation treatment process. The basic characteristics of this system are three-fold: (i) mixed population of aerobic micro-organisms capable of degrading the noxious components of the waste, (ii) the required population must be able to grow in the environment of the aeration tank and (iii) the bacterial cells should agglomerate together in flocs which may settle down. Aeration period is usually 4–8 hours and in order to obtain a better effluent, longer periods can be used.

Under the microscope, it is found that the gelatinous portion of the sludge contains many filamentous and unicellular bacteria and algae, while over the mass, numerous protozoa and some bacteria crawl and feed. Some investigators have concluded that there must be a proper relationship between the bacteria and protozoa. This method takes less time and does not need much dilution of the effluent to give a reduction of 90–95% BOD, with a decrease in the degree of foul smell. Due to these advantages, it is employed in industrial and sewage waste treatment. Despite these advantages, there are some disadvantages. Activated sludge process needs a high capital cost and the time required to build up activated sludge and the quantum of electricity consumed may prove to be obstacles to flexible operation. The choice of a suitable organism is a matter of great importance in aerobic fermentation.

44.3.4 Anaerobic digestion and methane gas generation

A large amount of data on the anaerobic digestion of sewage for methane gas production has accumulated and considerable details have been worked out. Pilot plant studies have shown that methane gas can be increased by recirculating carbon dioxide. High capital cost is one of the many restricting factors discouraging plants to make use of this potentiality. Fermentation also leaves behind a sludge of BOD value around 500 ppm. The disposal of this sludge is not without its own problems.

44.3.5 Ammonification and nitrification of effluents

In this process, the effluent is first treated with ammonifying bacteria anaerobically. Pilot scale trials were carried out by the National Sugar Institute (NSI), Kanpur, in collaboration with Somaiya Organo Chemicals Ltd. The final BOD reached was 2500–3000 ppm, which needs further dilution before being discharged through irrigation. The main advantages of this process are a low capital cost and fast treatment of the waste without giving chance for seepage or foul odour; however, this method also suffers from the same problems as anaerobic lagooning.

44.3.6 Cultivation of torula yeast

A process has been patented in France where distillery slops have been used as a raw material for cultivation of torula yeast that can be used as a cattle feed. The yeast is developed under vigorous aeration and as a result, BOD is reduced by about 50%. Due to high ash content and other impurities in distillery waste, this method has not been found suitable in India. In Brazil, the average yield of dry cellular matter based on total soluble solids was found to be 18.8%. In this process, there is a need for a large amount of investment and power requirement is high; the residual water needs treatment to reduce pollution.

44.4 Effect of distillery wastewater on environment

In a distillery, sources of wastewater are stillage, fermenter and condenser cooling water and fermenter wastewater. The liquid residues during the industrial phase of the production of alcohol are liquor, sugar cane washing water and from the cleaning of the equipment, apart from other residual water. This extract is extremely polluting as it contains approximately 5% organic material and fertilisers such as potassium, phosphorus and nitrogen. The amount of water used in this process is large, generating a high level of liquid residues as shown in the Fig. 44.1.

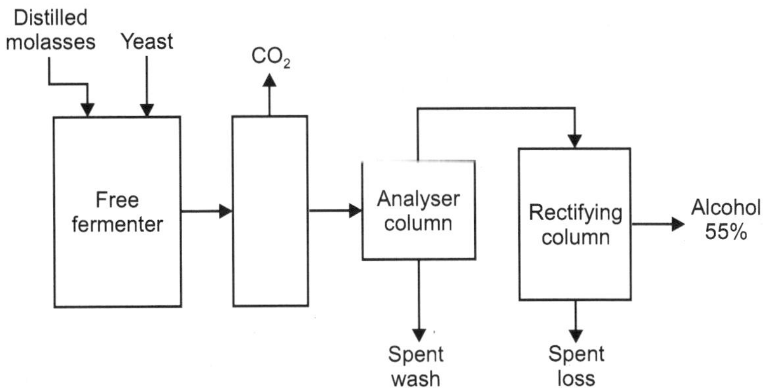

Figure 44.1: Schematic production of distillery wastewater.

The effluents from molasses based distilleries contain large amounts of dark brown coloured molasses spent wash (MSW). The molasses spent wash (MSW) is a potential water pollutant in two ways. First, the highly coloured nature of MSW can block out sunlight from rivers and streams, thus reducing oxygenation of the water by photosynthesis and hence becomes problem to aquatic life. Secondly, it has a high pollution load which would result in eutrophication of contaminated water sources. Due to the presence of putriciable organics like skatole, indole and other sulphur compounds, the MSW that is disposed in canals or rivers produces obnoxious smell.

In various countries there are a number of large scale distilleries integrated with sugar mills. The waste products from sugar mill comprise bagasse (residue from the sugarcane crushing), press mud (mud and dirt residue from juice clarification) and molasses (final residue from sugar crystallisation section). Bagasse is used in paper manufacturing and as fuel in boilers, molasses as raw material in distillery for alcohol production while press mud has no direct industrial application.

Ethanol manufacture from molasses generates large volumes of high strength wastewater that is of serious environmental problem. The effluent is characterised by extremely high chemical oxygen demand (COD) (80000 to 100000 mg/L) and biochemical oxygen demand (BOD) (40000 to 50000 mg/L), apart from low pH, strong odour and dark brown colour.

Other than high organic content, distillery wastewater also contains nutrients in thc form of nitrogen (1660 to 4200 mg/L), phosphorus (225 to 3038 mg/L) and potassium (9600 to 17475 mg/L) that can lead to eutrophication of water bodies. Further, its dark colour hinders photosynthesis by blocking sunlight and is therefore problem to aquatic life. Studies on water quality of a river contaminated with distillery effluent displayed high BOD values.

44.4.1 Waste characteristics of distillery effluents

The distillery wastes – liquid, solid and gas- are heavily polluting and noxious. The distilleries generate about 6–15 m^3 of wastewater (generally called spent wash) per m^3 of alcohol produced and a large amount of press mud. The spent wash poses a very serious pollution problem, as it has very high COD, BOD, inorganic load, low pH and dark brown colour. They cannot be discharged directly without reducing COD/BOD values below state pollution control norms. The treatment of distillery wastes is a priority area for environmental sustenance and its quality. The general treatment of spent wash involves its energy recovery through biomethanation. It is also used in the bio-composting of the press mud to produce good quality manure. It can be noted that the wastewater can generate biogas through biomethanation and biogas can be further utilised for the production of electricity and thermal energy through a technology integration comprising a boiler/steam turbine or 100% biogas engine. The distilleries can generate steam at desired pressure and temperature to meet their process requirements and produce electricity for captive use or for export. Several methods of waste treatment are in use; but a very few of them are able to meet the effluent quality standards.

The average characteristics of distillery spent wash total solids concentration of 7–9%, spent wash is a strong process wastewater and consists mainly of biodegradable organic constituents as indicated by the ratio of COD:BOD values of 2.2–2.5. The concentration of biodegradable matter, represented by high BOD values (40,000–50,000 mg/L) indicates a significant source of biochemical energy potential of spent wash. This stock of energy is available in the residual sugar present in spent wash and can be recovered as methane (bio-methane/biogas) by anaerobic fermentation reactions. The typical odour emanating from distilleries is a major nuisance. The colour of the spent wash interferes with its oxygenation and self purification.

44.4.2 Present status of treatment and disposal

Spent wash treatment is proposed by three different routes currently are as follows: (i) concentration followed by incineration, (ii) anaerobic digestion with biogas recovery followed by aerobic polishing and (iii) direct wet oxidation of stillage by air at high temperature with generation of steam followed by aerobic polishing.

44.4.3 Zero discharge of distillery wastewater

Worldwide environment regulatory authorities are setting for discharge of wastewaters from industries. In India for instance, distillery industry had been told to achieve zero discharge of spent wash as per the norms of central pollution control board as shown in the Fig. 44.2.

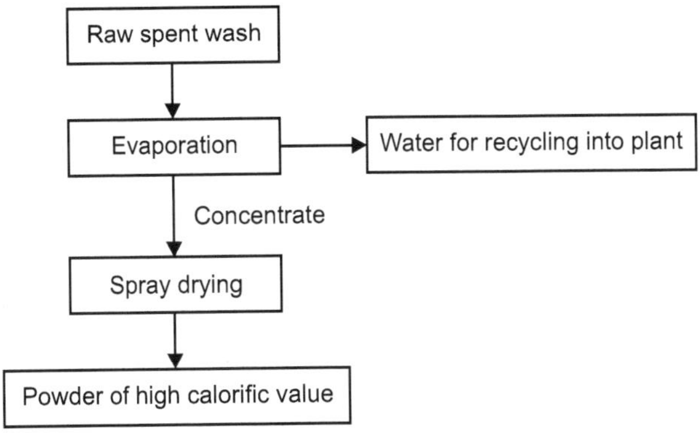

Figure 44.2: Zero discharge system for distilleries.

All methods of wastewater treatment such as lagooning, biodegradation wet air oxidation bio-methanation, membrane filtration evaporation composing were tried for last 25 years and found to be techno-economically nonfeasible. The present work proposes the process engineering approach based on experimental data on the same and similar fluid system. The process is experimented with achievements like zero discharge of wastewater, generation of distilled water to reuse in process, conservation of system energy, self refinance on utilise like steam, water and power. It further says that till 100% utilisation of spent wash is achieved, controlled and restricted discharge of treated effluent from lagoons during rainy season will be allowed by CPCB in such a way that the perceptible colouring of river water bodies does not occur.

44.5 Membrane technologies

Physical, chemical and biological treatment approaches have been employed for the treatment of distillery wastewater. The physical methods are sedimentation, screening, aeration, filtration (membrane technologies) and flotation. The chemical methods are chlorination, coagulation, adsorption and ion exchange. The biological methods are grouped into two types. Aerobic methods are activated sludge treatment, lagoons, trickling filtration and oxidation ponds and anaerobic methods are anaerobic digestion, septic tanks.

Municipal, agricultural and industrial liquid or solid wastes differ very much in their chemical, physical and biological characteristics. This diverse spectrum of wastes requiring efficient treatment has focused the attention of researchers on membrane, ion exchange and biological technologies. The most effective and ecological technological systems developed during the past 20 years are based on a combination of the chemical, physical and biological methods. Figure 44.3 explain all the fundamental of the membrane technology. Feed enters into stream and by membrane separated into concentrate and permeate.

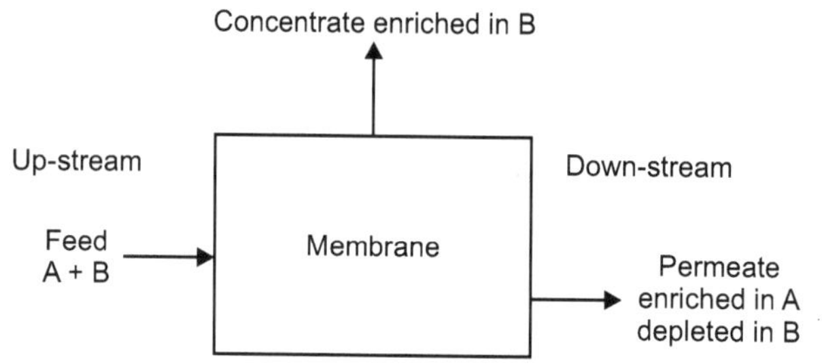

Figure 44.3: Process fundamental of membrane technology.

The fact behind the membrane technology, works without the addition of chemicals, with a relatively low energy use and easy and well-arranged process conductions. Membrane technology is a generic term for a number of different, very characteristic separation processes. These processes are of the same kind, because in each of them a membrane is used. Membranes are now competitive for conventional techniques. Membrane filtration can be used as an alternative for flocculation, sediment purification techniques, adsorption (active carbon and sand filter), extraction and distillation.

44.5.1 Growth of membrane technology

Membrane systems have been used in specialised applications for more than 30 years, largely for water treatment (distillery wastewater), including desalination of seawater and brackish water. With technical advances and corresponding cost reductions, membrane systems are now capable of decontaminating waters (including treated wastewaters) in single step processes at competitive costs. About two-thirds of the market will be for water and one-third for wastewater.

Membrane technologies are receiving special recognition as alternatives to conventional water treatment and as a means of polishing treated wastewater effluent for reuse applications. Membrane technologies are energy intensive. New membrane technologies feature the use of low pressure systems that significantly reduce energy use and operation and maintenance costs.

44.5.2 Components of a membrane system

Typical membrane systems consist of various steps which are describe below:

1. Pre-treatment.
2. Pumping.
3. Cartridge filtration.
4. Membranes.
5. Post-treatment.

The effluent collected from the distillery industry is highly acidic with a pH range of around 3. Hence, it is neutralised using sodium hydroxide. The neutralised solution has a lot of suspended solids, so the filtration is carried out to remove the suspended particles with a fine-pore thin cloth. This pre-filtrate is used as feed. Pre-treatment of alcohol-distillery wastes with ceramic membranes is performed prior to anaerobic digestion. Ceramic membranes of different pore size are chosen based on the particle size distribution in raw wastes. In some pre-treatment, chemical oxygen demand (COD) is reduced from 36000 to 18000 mg/L and suspended solids are almost completely removed. Mixed stillage exhibited higher fouling tendency than pure naked barley stillage.

Several cleaning methods are attempted to recover water flux. Although lumen flushing is effective, hydrogen peroxide proved to be the most effective cleaning agent.

The negative flux recovery after nitric acid cleaning could be explained by the ligand exchange theory. The performance of digester is greatly improved with membrane pre-treatment, especially in the case of naked barley based stillage. Pre-treatment may include the addition of chemicals to prevent organic materials or soluble salts from fouling the membrane.

Pumping is required to raise the pressure to the desired operating level and to maintain sufficient velocity across the membranes. A cartridge filter is nearly always provided by the membrane manufacturer, usually for the removal of particles. The filter provides protection against an upset in the pre-treatment step that could cause fouling of the membrane.

The membranes are the heart of the treatment system. They may be hydraulically connected in series or parallel configurations, depending upon the feedwater composition or desired water recovery. Post-treatment may include: (i) a degasifier to remove carbon dioxide and - hydrogen sulphide and (ii) the addition of lime or caustic to prevent corrosion of the subsequent piping or distribution system.

44.5.3 Design considerations

In addition to levels of constituent removal required factors to be considered in the design of membrane systems include membrane life, membrane fouling and disposal of concentrate. Typical membrane life is three to five years depending upon the type of service and type of membrane used. Membranes used in wastewater treatment typically have a life of four to five years. For distillery wastewater, the normal life of a membrane is three to five years and many have been in service for more than six years. For various membrane processes different types of considerations are taken into account, some are given below into Table 44.1.

Table 44.1: Various membrane processes.

Process	*Function*	*Driving force*
Reverse osmosis	Selective solvent transport	Pressure gradient
Electrodialysis	Selective ion transport	Electrical potential gradient
Ultrafiltration	Molecular size, shape and flexibility	Pressure gradient
Dialysis	Selective solute transport	Concentration gradient

Two types of membrane configuration used extensively for distillery wastewater treatment are hollow-fibre and spiral-wound. In a hollow-fibre element, fibres made of porous polymer material are bundled together and sealed in a pressure vessel. For some UF designs, feedwater enters through a perforated central tube and flows radially outward through the fibre bundle. Under pressure, water is forced through the hollow-fibre bores and exits through one or more ports.

For RO, feedwater enters from the outside surface of the fibre and product water is removed from the bores. Spiral-wound elements usually range from 2 to 10 inches in diameter and 10 to 60 inches in length. They consist of two flat membrane sheets separated by a thin, mesh-like porous support or spacer and are sealed on three sides like an envelope. The fourth side is fixed onto a perforated plastic centre tube that collects the product water. The membranes are rolled up around the tube in the form of a spiral. Feedwater is pumped through the layers and product water passes through the membranes and follows the spiral configuration to the central perforated tube. Water that does not penetrate the membrane exits the element as concentrate. Spiral wound elements are used for MF, UF and RO.

44.5.4 Treatment of distillery using wastewater membrane bioreactor

Membrane bioreactors (MBRs) are being increasingly recognised as an effective method for the treatment of industrial (distillery) wastewaters. MBRs offer the advantages of total solids retention at all biomass concentration, low sludge yield and better treated effluent quality. In addition, the high mixed liquor suspended solids (MLSS) concentration encourages the treatment of high strength wastewater. The widespread application of MBRs is however, limited by two reason high initial membrane cost and progressive membrane fouling, which leads to frequent membrane cleaning and eventual replacement, thus contributing to the high operating costs. There are very few investigations on distillery wastewater treatment in an MBR. The COD removal efficiency was 94.7%. Membrane coupled anaerobic bioreactor (MCAB) using 0.2 μm polypropylene and 0.14 μm zirconia skinned inorganic tubular membranes has also been investigated for the treatment of 40000 mg/L COD distillery wastewater at 55LC. High COD removal (90%) was observed in both the anaerobic MBRs as shown in the Fig. 44.4.

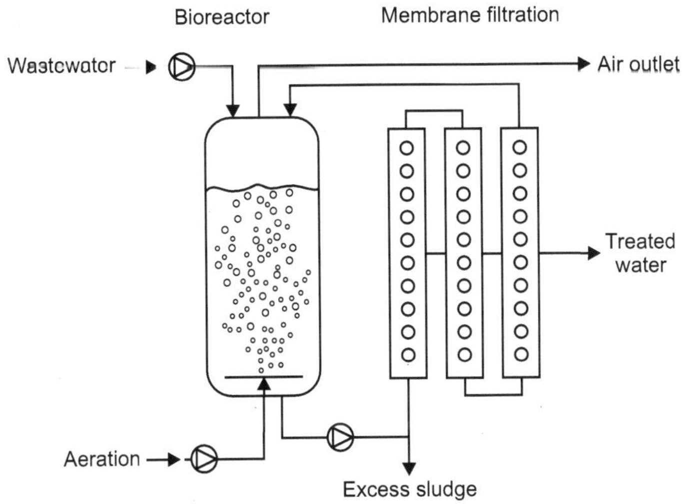

Figure 44.4: Process of wastewater treatment using membrane bioreactors.

Anaerobically treated spent wash from sugarcane molasses based distilleries has a high COD and requires further aerobic treatment. So the objective to investigate the optimum start up method and continuous operation of aerobic MBR using anaerobically treated spent wash as feed. The main objective behind using MBR was to provide long SRT (sludge retention time) so that the degradation of high molecular weight compounds could be achieved in the reactor. Nylon mesh was used instead of commercial microporous membranes to decrease the cost.

During MBRs study, the initial sludge acclimatisation phase where the focus was on biomass growth and sludge properties, followed by continuous operation that mainly deals with reactor operation and filtration performance.

44.6 Recovery of potash

Distillery effluents contain about 11.5% potash which can be recovered by multiple effect evaporators. The main problem in making use of this valuable plant nutrient lies in high capital investment as well as high energy needs involved in evaporation, drying and transportation.

44.6.1 Recovery of vitamin B_{12}

The synthesis of vitamin B_{12} was studied during treatment of distillery waste at NEERI, Nagpur. It has been found that it is worthwhile to recover the vitamin from the sludges, using plain water at boiling temperature. Addition of 1 mg/L cobalt increased vitamin synthesis by 50%.

A good deal of research work has been initiated at a number of institutes/ centres; namely, National Sugar Institute, Kanpur; National Environmental Engineering Research Institute, Nagpur, on anaerobic lagooning, anaerobic digestion in closed digesters with methane recovery, ammonification, drying, incineration, potash recovery and direct use for irrigation purposes; but the high cost involved in the treatment and recovery of by-products has been one of the main constraints. In order to tackle this problem, there is a need to adopt an integrated system (Fig. 44.5) which may help recover energy and water for inplant utilisation, water and potash for farmlands and feed for animal nutrition, consistent with minimum pollutional hazards.

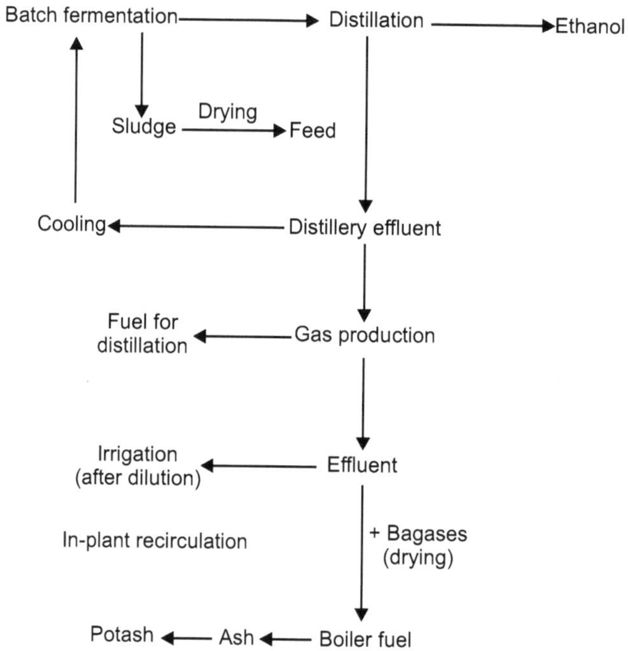

Figure 44.5: Recycling distillery effluents for fuel, ethanol, animal feed and potash.

As such, keeping in view the prospects and problems in the utilisation of distillery wastes, it is highly important to constitute a planning group to prepare a feasibility report and a coordinated programme. Efficient utilisation of effluents may not only help provide sanitary conditions all around the factory areas but also open up a vast potential to generate new resource materials.

45

Pollution control in rubber industry

45.1 Introduction

Natural rubber may be obtained from the latex of about 2000 species of plants containing rubber as its constituent. However, only a few species (*Heveabrasiliensis*, *Parthenium argentatum*, *castilla elastica*, *Ficus elastica*, etc.), are identified for commercial production of rubber. Natural rubber is obtained in two forms, namely field latex and field coagulm (or scrap rubber). Field latex and field coagulum are highly susceptible to degradation due to oxidation and contamination and hence are processed to obtain the following forms (products) that allow easy storage and marketing for utilisation by the rubber product manufacturing units.

Rubber manufacturing generally comprises the following operations: Raw materials handling, weighing and mixing, milling, extruding and calendering; component assembly and building, 'curing' or vulcanising, inspection and finishing, storage and dispatch (Fig. 45.1).

45.2 Natural rubber processing industry

45.2.1 Production process and pollution generation

Sieving: Latex collected from the field is first sieved through a mesh to remove impurities like bark shaving, leaves sand particles, etc.

Bulking: Bulking is the process of collecting latex from different fields in a common tank in order to have uniform properties.

Standardisation: The bulked latex is diluted to a standard Dry Rubber Content (DRC) of 12.5%. The dilution helps in settling denser and finer impurities at a faster rate and in obtaining a softer coagulums.

Preservation: Before coagulation, chemicals such as sodium bisulphite and paranitrophenol are added to the latex as preservative. Preservation is also accomplished by adding ammonia.

Latex concentration: The process of concentration involves removal of substantial quantity of serum from the latex and increase its rubber content to about 60%. This is accomplished either by creaming method or by centrifuging In creaming process, a creaming agent (tamorind seed powder) is added and allowed to remain undisturbed till the desired level of creaming (concentrated

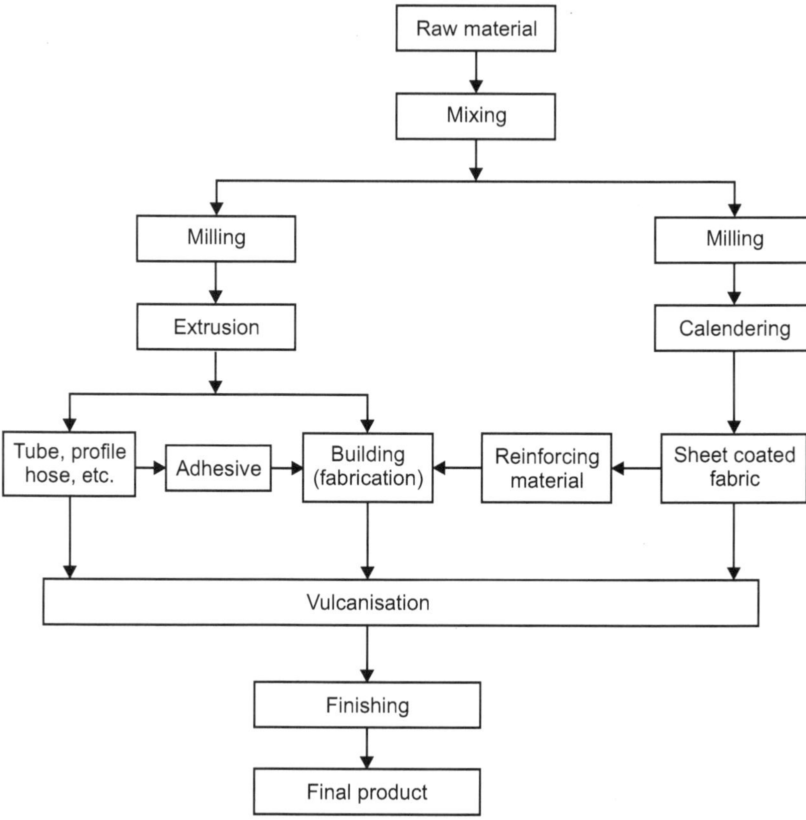

Figure 45.1: General process flow diagram for manufacturing of rubber products.

rubber) is obtained. In centrifuging, the latex is subjected to centrifugal force in a bowl rotating at speed around 6000 rpm whereby individual rubber particles tend to separate into a layer surrounding the axis of rotation.

Coaguluation: Coaguluation is the process of destabilisation by some means with a view to recover rubber from it. Acetic acid and formic acid are generally used for coagulation.

The above unit operations are most or less common for manufacturing of various products.

45.2.2 Pollution abatement

Pollution abatement involves: (i) in-plant control of waste and (ii) end-process treatment of wastewater. Few in-plant control measures are suggested to enable reduction in consumption of water, generation of pollutants and to increase the efficiency of the end-of-process wastewater treatment.

In-plant control measures

In the crepe and crumb rubber units, in which field coagulum is processed, higher than the required water quantity is generally used for soaking and also the soaking time allowed is not adequate. If the raw scrap rubber is properly soaked and primary, dirt removal is done by scrap-washer, the quantity of water consumed in milling can be reduced.

In the crumb units wastewater from final milling can be collected separately from the effluent of the other milling section and can be used either for soaking the scrap rubber or for the first milling process. This is compratively clean and amounts to reduction of about 25% of the total water consumption.

In centrifuge machine bowl washing is done at the interval of 3–4 hours to remove the sludge. About 0.5% rubber is lost during this washing. This washing can be done at two stages. The first washing which is more concentrated may be segregated and collected in a separate tank and coagulated for recovery of the rubber lost during washing. This will result in reduction of pollution load in the effluent. The possibility of diverting this waste stream into the skim coagulation tank can also be considered.

The quantity of acid used for coagulation of the latex specially skim latex kit after centrifuging operation is generally found to be higher than the actual requirement. The incomplete coagulation results in escape of rubber particles into the effluent alongwith skim serum. The excess acid not only causes acidic effluent but also redissolves the rubber protein and causes delay in coagulation. Hence, it is suggested that proper acid dosage and sufficient coagulation time should be provided so as to obtain more or less clear liquid after complete coagulation. The skim latex if demmoniated before coagulation, acid requirement can be reduced and the ammonia concentration in effluent may also be reduced. In the latex process units the segregated first washing of the coagulum may be diverted to the skim coagulum tank where after skim coagulum recovery, the effluent may join the other wastewater streams.

Extruding and calendering: The extruders force the rubber compound through a die into various forms, which are then cut to appropriate lengths. Strips of softened rubber compound are fed into multiple-roll milling machines (calenders) to form rubber sheeting, or to apply the rubber directly onto woven textile fabric, which can then be wound off onto a roll. During such manufacturing operations, fumes are often generated.

Component assembly and building: At this stage, solvents are frequently used, with the possibility of inhalation of solvent vapours or of direct effects of the solvent on the skin of the operator.

Curing or vulcanising: Heat is applied to the product, usually by use of steam, in a curing mold, press, or autoclave. Operators working in the area are exposed both to heat from the presses and to fumes from the heated rubber products. Chemical reactions take place throughout the manufacturing process and may give rise to new, more volatile chemicals.

Inspection and finishing: This involves the handling of cured rubber products, often while still hot. It usually involves direct and extensive skin-contact with the surface of the finished article (during inspection) and may also involve exposure to vulcanising fumes. Grinding, trimming, repair, painting and cleaning may also entail exposure to rubber dust, fumes and solvents.

Storage and dispatch: Large quantities of stored rubber goods may release considerable amounts of toxic substances, either as vapours or as constituents of the 'bloom' on the surface of finished goods.

Chemicals used in the rubber production process

A wide variety of natural or synthetic elastomers, fillers (e.g., carbon black, precipitated silica or silicates) and additives are used in compounding to create the necessary properties of the final rubber product. The actual chemicals used in this process have changed over time and vary extensively depending on the manufacturing sector (e.g., tyres, general rubber goods, retreading) and on the specific plant.

Compounding ingredients are classified as vulcanising agents (e.g., elemental sulphur, sulphur donors such as organic disulphides and higher sulphides, peroxides, urethane cross-linking agents), vulcanisation accelerators (e.g., sulphenamides, thiazoles, guanidines, thiurams, dithiocarbamates, dithiophosphates and miscellaneous accelerators such as zinc isopropyl xanthate and ethylene thiourea), vulcanisation activators (e.g., zinc oxide, magnesium oxide, lead oxide), retarders and inhibitors of vulcanisation (e.g., benzoic acid, salicylic acid, phthalic anhydride, *N*-nitrosodiphenylamine (NDPA), *N-(cyclohexylthio) phthalimide*), antidegradants; antioxidants (e.g., phenolics, phosphites, thioesters, amines, bound antioxidants such as quinone-diamines, miscellaneous antioxidants such as zinc and nickel salts of dithiocarbamates), antiozonants (e.g., *para*-phenylenediamines, triazine derivatives, waxes), anti-reversion agents (e.g., zinc carboxylates, thiophosphoryl derivatives, silane coupling agents, sulphenimide accelerator, hexamethylene-1,6-bisthiosulphate disodium dehydrate and 1,3-bis (citranimidomethyl) benzene), plasticisers and softeners (e.g., petroleum products such as petroleum waxes and mineral oils, coal-tar products such as coumarone resin, pine products, synthetic softeners and other products such as vegetable oils and fats) and miscellaneous ingredients (such as peptising agents, blowing agents, bonding agents and pigments).

End of process treatment

The rubber trap used for arresting suspended matters should have holding capacity of at least 12 hours with proper baffles to induce continuous up and down flow pattern. If designed properly, this can reduce suspended solids by 40 to 60%. The equalisation tank should have at least one day detention time. It is preferred to have two eqalisation tanks, each of one day detention time. For a latex processing unit, effluent from the equalisation tank to be sent for neutralisation and chemical treatment by alum and iron salt (about 200 mg/1). Combined wastewater of latex process units also need neutralisation by user of lime and settling of suspended solids by user of coagulants. The settlers/clarifier should have adequate detention time for removal of suspended solids. The sludge may be taken to sludge drying beds for dewatering.

After this, the effluent should be subjected to the biological treatment. If sufficient land area is not available then the effluent after primary settling may be subjected to an extended aeration activated sludge type biological treatment process.

Before going for biological treatment, it must be ensured that: (i) all the in-plant control measures are adopted and (ii) physicochemical/primary treatment, e.g., rubber trap equalisation neutralisation and clarification steps are incorporated. The above measures will reduce substantial quantity of pollutants particularly BOD and suspended solids. The primary treated affluent can be treated in a secondary/biological treatment unit. It is envisaged to render secondary treatment by adoption of extended aeration activated sludge process. The biological treated effluents should be settled in a secondary settling tank. If there is no constraint of land, the biological treatment could be anaerobic followed by aerobic pond system with the proper dimensions, holding capacity and adequate detention time (10 to 15 days) for anaerobic pond followed by 5 to 10 days for aerobic ponding system. The type of soil and proximity to the wastewater and ground water table condition should be taken into consideration before going for these treatment systems. Protective lining is recommended to eliminate any risk. In place of the anaerobic-aerobic system, an oxidation ditch of detention time of 2–3 days can also be considered as an alternative for treating the effluents of the crumb rubber unit.

45.2.3 Pollution control in rubber manufacturing

Basic polymer and ingredients used for making rubber products, either individually or after combinations make some carcinogenic gases and fumes during mixing, vulcanisation process and even during storage. Also, chemical dust comes in contact with the air, some gets contaminated in water and thus, directly or indirectly, affects living things.

However, waste disposal management is the big issue in the rubber manufacturing industry. The international standardisation forum is continuously reviewing the effect of the use of rubber chemicals in the industry. There is need for use of biodegradable materials in compounding, avoiding human beings direct contact with ingredients, developing techniques that will reduce contamination of particles, gases and solvents released during the different processes of rubber product manufacturing and hence avoiding them from polluting the environment.

Rubber products are manufactured in different stages. Basically, natural or synthetic rubber is used as the basic rubber for the manufacturing process.

The typical manufacturing processes in rubber industry are as follows:

1. Raw material storage, handling and weighing.
2. Milling of raw rubber, mixing rubber and rubber chemicals.
3. Intermediate processes, such as sheeting, extruding and calendering.
4. Component assembly, building and pre-shaping.
5. Vulcanisation or curing of rubber compound.
6. Product finishing, inspection and testing.
7. Packing, storage and dispatch.

In each of the above mentioned operational stages, there is the possibility of environmental pollution through dust, vapours, gases and fumes. They came in contact with air, water, soil and ultimately the environment. Many rubber product manufacturing units discharge their wastewater on land, rivulet and tunnel. Few examples, this pollution includes (but not limited to):oil mixed water, water from mixing mills, water with chemicals from testing labs, product wash water and water with different ingredients from latex plants. This pollutes drinking water resources, land for irrigation, air and ultimately the whole environment. However, no information is available regarding the long term effect on soil, ground water and also on crops. A detailed study needs to be carried out in this regard. There is a need to upgrade production processes so as to decrease the amount of pollution from rubber industry. In the latex industry in order to increase dry rubber content (DRC) of latex, it is essential to remove water and other impurities. Also, in washing the product water is needed. Although maximum care is taken, recycling of water is done, many times, this contaminated water is released into the open ground or tunnel.

This may contain ammonia solution, Tetramethyl Thiuram Disulphide (TMTD), zinc oxide and diamonium phosphate. Appropriate and low-cost methods need to be explored in dealing with wastewater from latex industry as well as in molded, extruded and other fabricated products, tyre and tube industry, reclaim rubber industry. Wastewater generation sources and the major water pollutants are given in Table 45.1.

Table 45.1: Wastewater generation sources and the major water pollutants.

Type of industry	*Wastewater sources*	*Contaminant*
Tyre and tube industry	Cooling water bleed and boiler blow down, floor wash down of different units (batch wise), leakages and spills contaminating cooling water, accidental overflow from antitack water tank, sanitary and other miscellaneous water	Oil and grease, rubber fines, antitack agents, etc.
Molded, extruded and fabricated products	Cooling water bleed and boiler blow down (batch wise), rinse and product wash water (batch wise), steam condensate with organic leach (batch wise), sanitary and other miscellaneous water	Oil and grease, rubber fines, antitack agents, powder as fillers, etc.
Latex based products	Cooling water bleed and boiler blow down, product wash water, wash down from latex storage, compounding and transfer areas, from wash and rinse waters, accidental overflow water, sanitary and other miscellaneous water, coagulation/coagulating agent tank wash/over flow	Latex spills, surfactants, coagulants, etc. Ball milling and washing containing flow chemicals including zinc, acetic/formic acid, etc.
Reclaimed rubber	Cooling water bleed and boiler blow down, steam condensate from auto-clave, area wash down of all processing areas, sanitary and other, miscellaneous water	Oil and greases, soluble and insoluble inorganics, rubber particles

During the manufacturing and storage of rubber products, various gases, vapours, fumes and aerosols are emitted due to the leaching out of chemicals and also during vulcanisation that occurs at high temperatures. These emissions consist of volatile components from the original compounding ingredients such as: plasticisers, antioxidants and residual monomers or polymer oligomers, as well as primary and secondary reaction products from the cross-linking systems. The formulae and vulcanisation conditions of individual substances as well as their origin have been identified in numerous studies. These substances are amines, amides, aliphatic/aromatic hydrocarbons, highly volatile sulphur compounds, aldehydes and ketones, phenols and esters. Fumes of solvents may also be present, depending on the manufacturing process, as these solvents are used for assembly and cleaning. Monitoring of the air quality at the workplace requires the implementation of a measurement method that is able to cope with the extremely complex situation in the rubber industry. From material inward stage and storage process, pollution starts.

Generally it is due to the shortage of storage area, humid climate, storage temperature, improper ventilation, storage in contact with other chemicals, inadequate knowledge of product and its maintenance, loading and unloading of material and the use of hooks and thus damaging the bags, etc., are some reasons for contamination as well as spillage of these ingredients. Sometimes they are very casually handled by operators, e.g., without packing, they are openly taken towards the weighing machine and mixing machine. Some of them get in contact with air and may spread unpleasant odour and dust. During the mixing process, fillers are dumped on polymer in the mixing mill. During that process, a lot of chemical dust gets spread into the air and the working area is affected, due to these dust particles.

During the vulcanisation process, fumes from heated or previously heated rubber create unpleasant odour which is a complex mixture of components and the associated hazard unknown. Epidemiological studies suggest that exposure to rubber fume may be a significant factor in the increased incidence of certain type of cancers and the cause of some other diseases. Airborne contamination from rubber fumes is either visible or non-visible. The invisible pollutants are gases and vapours of low molecular weight organic compounds (carbon disulphide and amines) or inorganic (hydrogen sulphide) compounds. The visible pollutants arise from aerosols. These aerosols evolved during the mixing, milling, blending of elastomers or combination of them with chemicals involved in the process. The aerosol fraction of rubber fumes is complex and with unknown composition. The concentration of airborne aerosols is very considerable at the workplace in the rubber industry and it should be monitored by approved trapping and detection methods. The major component of fumes is due to accelerators and their reactions with other ingredients of rubber compounds. For example, the major volatile component of tetramethyl thiuram disulphide (TMTD) and zinc dimethyl dithiocarbonate (ZDMC) accelerated cure, will include carbon disulphide, dimethyl amine and sometimes hydrogen sulphide. In case of the latex industry ammonia solution added to latex causes a strong smell. This can have adverse effect on workers health, especially the respiratory system.

45.2.4 Compounding ingredients and their effect

Generally, rubber compound is made from basic polymer, activators, accelerators, fillers, plasticisers, antidegradants, curatives, etc.

The followings are some few examples of the precursors for pollution:

Polymers

Raw or base rubber is the major part of rubber product as far as the weight of complete rubber product is concerned. Natural rubber and synthetic rubbers are used in the rubber industry. Within each type, based on the chemical structure, method of production and functional group present in the structure, there are many grades. Generally, each rubber has its own characteristic smell. We can experience the same in the storage area, during the rubber mastication process and vulcanisation process. It releases complex fumes and smell, which may either be individual or reactant products.

Rubbers are highly flammable. After coming into contact with heat and flame, it ignites spontaneously and generates considerable quantities of smoke while burning. Basic raw materials for synthetic polymer making are petro based products. It should be stored away from energy-generating sources such as boiler and oil storage room.

Fillers

In the rubber industry different types of fillers are used in order to improve the properties of rubber compound as well as for cost reduction. Few examples of such fillers are: carbon black, white fillers such as: silica, whiting, barytes, hard clay, soft clay, etc. These contribute (in weight) the main part of rubber compounds.

Carbon black: Approximately 65% of the carbon black is used in tyre production. The International Agency for Research on Cancer (IARC) conducted a comprehensive review of carbon black and concluded that there is limited evidence for carcinogenicity from carbon black, based on studies in humans. They concluded and classified carbon black as IARC classification 2B-possibly carcinogenic to humans and definitely carcinogenic to animals. Non-cancer respiratory effects in workers who work with carbon black have been reported include; cough, sputum production, bronchitis, chest radiographic opacities (e.g., Pneumoconiosis) and decrements in lung function. Short-term exposure to high concentrations of carbon black dust may produce discomfort to the upper respiratory tract through mechanical irritation. As carbon is the precursor for CO_2 emission and thus the warming of environment, there is the need to partially or completely replace carbon black with ecofriendly and biodegradable material or its use should be kept as minimum as possible.

Silica: In case of crystalline silica dust, dust particle can cause fibrosis (scar tissue formation in the lungs). Silicosis is the respiratory disease of the lungs that caused due to the inhalation of airborne crystalline silica dust. Individuals with silicosis are at increased risk of developing pulmonary tuberculosis, if exposed to person with tuberculosis.

Titanium dioxide (TiO_2): There is inadequate evidence of carcinogenicity of titanium dioxide in humans, but there is sufficient evidence through experimental work on animals, for the carcinogenicity of titanium dioxide. Titanium dioxide is possibly carcinogenic to humans (Group 2B).

Talc: In the case of talc, it is generally used as filler as well as dusting on rubber sheet in order to avoid sticking through adhesion of rubber sheet during storage. During dusting process, dust particles come in contact with the environment at working area.

Plasticisers

Plasticisers represent a special group of ingredients. They are used in the rubber industry as processing additives. They help in improving the processing of rubber as well as modify the physical and chemical properties of the rubber compound. The emission of carbon dioxide due to the consumption of fossil fuels is considerable in the rubber industry. In oils, based on aromatic content the important types are: aromatic oil, naphthenic oil and paraffinic oil. Also, there is one important group i.e., phthalates that are mostly used in polyvinyl chloride (PVC) and nitrile rubber. These are polar materials. They are released into the environment, either by spillage, in gaseous form, as odours, form complex structure with other ingredients and evaporate during the vulcanisation stage at high temperature. The followings are some of the examples.

Aromatic process oils: They contain polycyclic aromatic hydrocarbons (PAHs) or polycyclic aromatic (PCA) which are carcinogen and can cause mutation. Regulations (European Directive 2005/69/EC) imposed ban of process oils containing $\geq$ 10 mg/kg (ppm) of PAH since 2011. Aromatic process oils are labelled as carcinogenic under OSHA regulation.

Phthalates: Within the phthalates there are different types. Some of the commonly used are dioctyl phthalate (DOP), di-butyl phthalate (DBP), etc. It has been well-documented that endocrine disruptors, such as phthalates can be additive, therefore even very small amounts can interact with other chemicals to have cumulative, adverse 'cocktail' effects. Phthalates in pure form are usually clear liquids, some with faint sweet odours and some with faint yellow colour. With respect to health effects, phthalates are often classified as endocrine disruptors or hormonally-active agents (HAAs) because of their ability to interfere with the endocrine system in the body. In many countries there is ban on phthalate plasticisers.

Activators

In the rubber industry zinc oxide and stearic acid are used as activators. Generally, zinc oxide is used as a 5 phr (parts hundred rubbers). Recently the tendency has been to use between 2–4 phr and 3 is adequate in most purposes. The reduction is due to its high cost and high density. There is a more recent reason for the reduction of the zinc oxide quantity in rubber compound formulations. The U.S. Environmental Protection Energy (EPA) has classified zinc oxide as toxic chemical. The earliest symptoms of this metal oxides fume is metallic taste in mouth, accompanied by dryness and irritation of the throat in addition toflu-like symptoms, chills, fever, profuse sweating, headache and weakness. These symptoms usually occur within several hours after exposure to zinc oxide fumes and can persist for between 24–48 hours.

Accelerators

These types of ingredients are used to accelerate the rate of vulcanisation during the curing of rubber products. There are many types of accelerators, such as: amines, dithiocarbamates, guanidines, sulphenamides, thizoles, thioureas and thiurams, etc. It is known that rubber fume can contain *N*-nitrosamines. Many *N*-nitrosamines are carcinogenic. They are formed incidentally, when certain precursor compounds coexist under favourable reaction conditions. Most commonly, they are formed in the rubber industry by the reaction of an amine with a nitrosating agent. A number of accelerators are based on secondary amines. Few examples: thiurams, dithiocarbamate, sulphanamides-based accelerators, are likely to generate nitrosamines during curing. Nitrosating agents can include atmospheric oxides of nitrogen (NO_x), which are common pollutants from internal combustion engines, particularly diesel engines. The volatile *N*-nitrosamines are of most concern and they are the focus of the regulation. Measuring of *N*-nitrosamines emitted from elastomeric compounds is still an important issue that rubber industry has to deal with.

The chemical pathways of nitrosamine formation and degradation in the atmosphere (particularly in the aqueous phase) remain uncertain. Exposure to nitrosamine has been shown to form tumours in laboratory animals and have been linked in epidemiological studies to human cancers, including pancreatic cancer, childhood brain tumours. Nitrosamines are classified by a number of international organisations and regulatory authorities in accordance to their carcinogenicity. Under the International Agency for Research on Cancer (IARC), which is part of the World Health Organisation (WHO), *N*-nitrosodimethylamine (NDMA) and *N*-nitrosodiethylamine (NDEA) are classified as Group 2A substances (probably carcinogenic to humans). Earlier studies have shown that *N*-nitrosamines are evolved from heated elastomers. More recent studies by Ireland and others showed that *N*-nitrosamines can be extracted in water from elastomers.

Curing agent

Generally, rubber curing is done by sulphur and peroxide. There are also other chemicals used such as: metal oxides, resin curing agent, sulphur donors as a curing agent. They generate fumes and smell during storage as well as vulcanisation process. During sulphur vulcanisation with guanidine type accelerators (e.g., Diphenyl Guanidine-DPG, *N,N*' Di-O-Tolyguaniidine-

DOTG) and antioxidants based on phenylene diamine, cancerogenic primary aromatic amines (PAA) and isothiocyanates (ITC) are formed inevitable. Respiratory morbidity has been associated with exposure in the rubber industry. Chronic and acute bronchitis, decreased lung function and airway symptoms have been observed. Smith and others also surveyed other symptoms (e.g., headache and nausea) among workers doing vulcanisation. However, environmental conditions have improved during the last few decades in rubber industry. During peroxide curing, it generates fumes and characteristic smell. In case of dust inhalation, continuous exposure may cause light irritation to the eyes and skin, irritating to nasal mucous membranes and upper respiratory tract. Generally peroxides are heat sensitive. These are to be protected from heat, as they easily catch fire and burn rapidly. In peroxide cured rubber products, sometimes, there is need for post vulcanisation. During this process, a lot of fumes come in contact with air.

Antidegradants

This group of ingredients is used to protect the rubber article from degradation due to oxidation, ozone. These chemicals improve the dynamic properties of rubber products. According to their types, they are phenolic, phosphates, thioesters, amines and other multi functional types. Generally, they are volatile under severe temperature conditions, migrate towards surface of products and may give rise to staining and blooming. Antiozonant, e.g., *N*-phenyl-*N*'-(1,3-dimethylbutyl)-*p*-phenylenediamine (6PPD) partly decomposes and undergoes further reactions after migration in aqueous food simulants forming Primary aromatic amines (PAA), which is a critical hygiene concern.

These are the only few sources of pollution spreading. Besides this, there are many other sources such as: from scraps, cuttings, flash of rubber product, dust from rubber products grinding, different types of solvents that are used for cleaning and adhesive preparation, spillage from bags and oil drums, different types of dusting powders, surface finishing agents, wastewater and chemicals from testing labs, oil leakages from machineries, cuttings, rejected and scrapped products, e.g., tyre.

Suggestions and basic precautions: There is lot of scope for improvement. But some basic attention can minimise the issue at following levels:

1. Processing and storage: At the work area, during operation, masks, goggles, gloves and other safety wares should be used. The work area to be kept neat and clean. To prevent spreading of spilled chemicals, it is to be cleaned immediately. For collecting the material spread during storage and manufacturing process, dry vacuuming is the recommended method. Arrangement should be made for collecting fumes and dust. Fire extinguishers should be kept at working area at all times. Wastewater

treatment related care should be taken. The recycling of wastewater must be done.

2. Selection of material and process: Rubber compound ingredients that are degradable, environment friendly, creates minimum pollution are to be selected. While designing compound formulation, the pollution and health hazard nature of the individual ingredient, should be taken into account. Material safety data sheet should be referred to, for precaution. Product making process should be properly designed where human contact with process-generating pollutants, will be minimum, e.g., the use of machines having closed chambers from which contaminated air, fumes can be collected by suitable collecting arrangement should be employed.
3. Scrap: By using rubber as an auxiliary fuel in cement plants and heating plants, the fuel value of the waste can be utilised. However, large quantities of waste rubber still end-up in landfill. Recycling methods that are applied, to some extent, include: the reuse of scrap rubber in the form of finely ground powder in various rubber materials and the mixing of rubber waste into asphalt. Chemical recovery methods are also used to a limited extent. Examples of these methods include pyrolysis, devulcanisation and the production of reclaimed rubber. The problem, however, is that vulcanisation is an irreversible process.

45.3 Emissions from rubber industry

45.3.1 Air emissions

The manufacture of rubber products exposes workers to a large variety of chemicals as discussed above.

Volatile Organic Carbon (VOC) emissions are also an environmental concern in the rubber product manufacturing process. VOCs are added to rubber compounds in order to aid in mixing, promote elasticity, produce tack or stickiness and extend or replace a portion of the raw material. VOCs are also used as solvents to degrease equipment and tools and as a type of adhesive during building and fabrication. Typically releases of solvents occur either when the spent solvent solutions are disposed or when degreasing solvents are allowed to volatilise. In some facilities, mold release compounds, sprayed in the cavities of compression molds, produce significant fugitive emission. However, solvents are becoming less of an issue as water, silicon and nonsolvent based release compounds are now common.

Emission sources

All possible sources of air emissions from rubber product manufacturing processes are shown in Fig. 45.2.

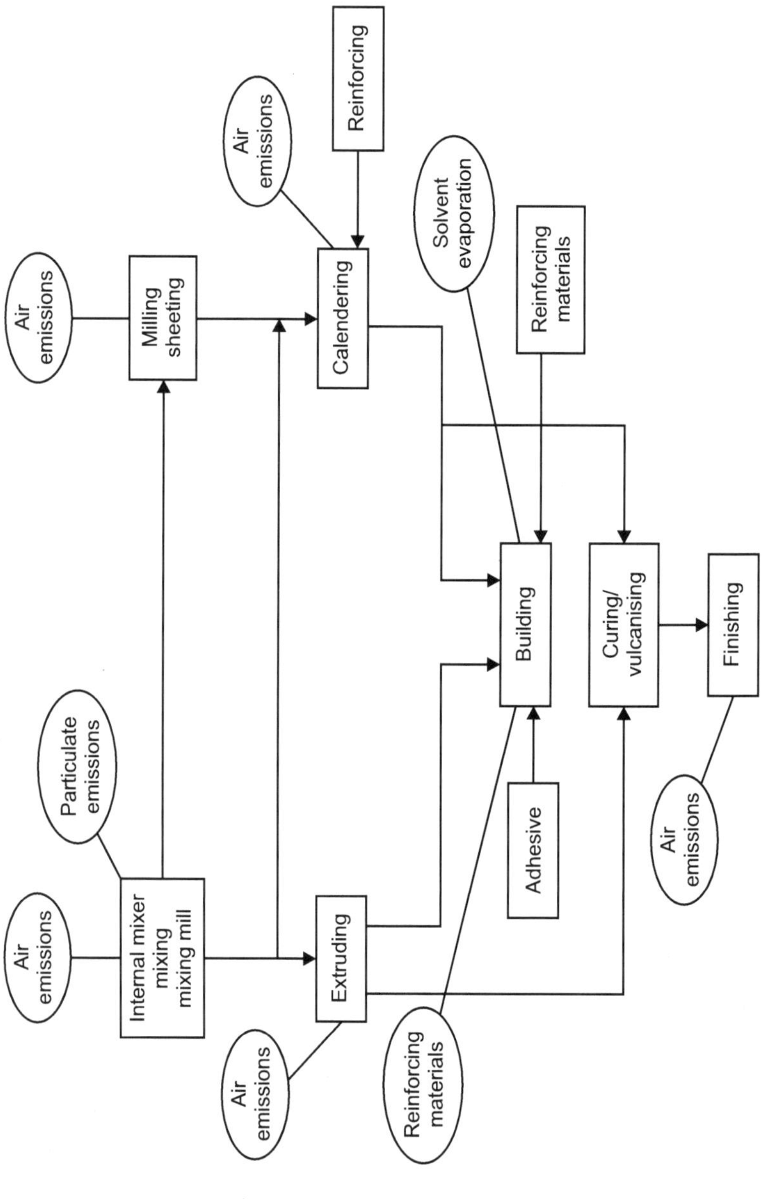

Figure 45.2: Rubber products manufacturing processes showing air emission sources.

45.3.2 Noise pollution

Process operations involving machineries are a source of noise in the in-plant environment. The sources of noise in the rubber products manufacturing industry are attributed to:

1. Machine operation.
2. Generator operation.
3. Boiler operation.
4. Mechanical workshop.
5. Cutting, grinding and finishing operations, etc.

Industrial clusters may also contribute to ambient noise levels.

45.4 Control of air emission and noise mitigation

45.4.1 Air pollution control

Prevention of emission is source controls. Transmission or path control interferes with transfer. Worker or operated oriented controls reduce exposure and uptake by the receiver. Source control is generally accepted as being most effective. Putting control in the transmission or exposure/uptake stages, without controlling the source, means that someone else could be unexpectedly exposed from the same source. Thus the classification according to pollution based process provides a classification of the solutions in terms of where the intrusion takes place.

The following approach should be adopted to control occupational hazards from emissions can take place at any of three locations:

1. At the source of the hazard (source emission control).
2. Along the path from the source to the worker (transmission control).
3. At the worker (reception exposure control).

45.4.2 Source control

Source control can be achieved by eliminating the use of materials causing air emissions by reducing the amount of material used (if possible) by substituting them with less polluting ones, or by changing their form. Some of measures at source control are described below:

Elimination

Elimination generally means avoidance of hazardous substances so that workers are no longer exposed to contaminated air. Elimination can be encouraged by local and national legislation.

Substitution

If elimination is impossible, substitution of less hazardous materials by nature and form is potentially the best way to reduce pollution risk. Elimination or substitution of harmful materials is perhaps the most effective means of engineering control of hazards in rubber manufacturing.

For example, beta-naphthylamine was removed from use because it is known to be carcinogen. Another known carcinogen, benzene, was once widely used as a solvent throughout the rubber industry; it is being steadily replaced by toluene.

Reducing toxic chemical usage by reformulation

Reformulation of compound master batch through trail may help to reduce use of toxic chemical without compromising the quality of product.

Form

It may be possible to effectively eliminate or decrease hazardous exposure by changing the form in which a substance is used, such as:

1. Some dusty materials (fine powders) can be palletised or used in liquid suspension.
2. This control method is particularly useful in the compounding and mixing area of a rubber plant.
3. Powdered raw materials can be pre-packed according to batch intake.
4. Powdered materials and chemicals in rubber industry can be incorporated in a rubber pre-mix for addition to the process, which can minimise the possibility of exposure.
5. Organic solvent can be substituted by water based solvents.

Process control

Significant reductions in fume levels may be achieved by avoiding compound temperatures in excess of process requirements. Fuming products should be cooled as soon as they have finished processing using water, air or by passing cured items over chilled surfaces.

45.4.3 Solid waste generations and disposal

Rubber product industries have most of waste with salvage value and face no problem for its disposal. Cured and off-specification rubber wastes serve as raw materials for inferior quality products in industry. Solid waste like discarded materials other than rubber such as sludge from wastewater treatment plant is disposed as land filling.

45.4.4 Noise control

The noise levels can be mitigated by adopting: (i) engineering measures and (ii) administrative measures.

Engineering control measures involve

1. Structural and mechanical modifications.
2. Proper maintenance of the machines.
3. Provide silencers on compressed air exhausts.
4. Replace old gear drive motors with new belt drive ones.
5. Plug air leaks.
6. Reducing the transmission path of noise from the source to the receiver.
7. Providing proper acoustic enclosures, vibration insulator paddings for the noise generating machines, generators, etc.

Administrative control measures involve

1. Ear mufflers/plugs for workers working in the noisy area.
2. Regulating the length of noise exposures by proper shift scheduling, job rotation and/or by restricting the operation of the noise source.
3. Education of workers.

45.5 Wastewater generation in tyre industry

45.5.1 Water consumption

Tyre and tube industry is basically dry process and water is mainly required for cooling of machinery used for rubber processing at different stages of final product. There is considerable heat generated by compounding, extruding, calendering and molding processes of solid raw materials involved in the manufacturing of tyre and tube and it must be dissipated and controlled to ensure the quality of final product. The other main area for water consumption is boiler for steam generation used in curing/vulcanisation process. The water uses in all possible areas of tyre and tube industry are listed as below:

1. Steam generation (boiler).
2. Non-contact cooling water of machinery.
3. Contact cooling of tread tyre.
4. Make-up water for anti tack solutions and water based spray of green tyres.
5. Washing of floor and machinery.
6. Domestic use in toilets, canteen and gardening, etc.

The industrial water consumption varies widely mainly due to different cooling water systems adopted by tyre and tube industry. The consumption variations are mainly due to the use of once- through cooling water in certain plants as compared to re-circulation cooling in others.

45.5.2 Wastewater generation sources

The wastewater from the process areas includes water and steam leakages, overflows, runoff from oil-storage areas soapstone solution spillages and wash down and runoff from process or storage areas.

Water leakages occur at various water-cooled machinery units including mills, Banburies, extruders and tread cooling tanks. In addition, water can escape from the hydraulic water system used in Banbury and Press areas. Water and steam leakages occur in the process area due to broken seals, failing bladder bags and overflows from the collection pumps. Oil and solid matter, which have collected on the floor area, scavenged by these various water streams and are carried, untreated to the drainage system. Oil on the floor spaces is lubricating oil, which is dripped or leaked from oil seals of mills, pumps and like equipment, from open gears, from gearboxes and from the hydraulic water system.

In mixing and compounding area, soapstone solution is used for coating rubber sheets to prevent them from sticking together during storage. Spills in the soapstone area are common and may create a wastewater problem. Wastewater problem in the tread processing area arise from the spillage of the solvent base cements, from oil and water leakages from the various mills and from accidental overflows from the cooling water system. The cooling water overflow would not normally be a problem since the rubber tread is relatively inert and therefore does not contaminate the water. However, it does serve as a wash-down agent for an area contaminated with the cements and oils.

Wastewater problems in tyre cords and belts arise due to the latex dipping operation in addition to problems with oil and water leaks and spillages, which are similar to those of the tread process.

In the final operation of grinding and painting involved in tyre manufacturing process, relatively small grinding particles and runoff due to over spraying of the paint may find their way to local drainage system creating wastewater problem. In general wastewater problems arising from compounding, extrusion, molding and curing operations in tube manufacturing are very similar to that of the tyre manufacturing.

Another component of wastewater arises in tyre and tube industry from utility services such as:

1. Once through cooling waters.
2. Boiler blow down.

3. Cooling tower/pond blow down in case of recirculating system.
4. Wastewater treatment.

The third component of wastewater is generated from toilet and canteen as domestic wastewater stream.

45.6 Control and treatment technology

1. The first approach adopted by some plants is to combine domestic and industrial wastewater and to treat the entire plant effluent. Generally, the reasons supporting this approach are as follows.
   (a) In-plant drains for domestic and industrial wastewaters are usually combined, thus making combined treatment more attractive.
   (b) Industrial flows in case of re-circulated cooling water management particularly for low production line of industries is highly intermittent and of low volume as compared to continuous domestic wastewater generation.
2. The second approach generally adopted by the plants of lower production lines is to use recirculating cooling water pond as a sink and diluting pond of intermittent and low volume industrial wastewater generated from process areas and utility services. Such industries claim to have zero industrial discharge and no treatment to industrial wastewater is given.
3. Once through cooling water as used by some plants is mixed with low volume and intermittent wastewater generated from other sources. Pollution load is diluted to such an extent requiring no further treatment as claimed by some of the industries.
4. The fourth approach adopted by very few plants particularly of high production lines is to control and treatment of a segregated and undiluted wastewater. This approach has been followed in plants having partially or wholly segregated industrial drains. This would, of course, include any plant using re-circulated cooling water. Treatment units adopted by such plant generally include oil segregation followed by settling tank.
5. The fifth in-plant control technology covers segregation and measures for handling, reuse, modification of processing and disposal of various types of wastewaters including spills and leakage, wash downs, control of runoff and house keeping practices. End-of-pipe treatment technology covers the primary treatment given at end of pipe after combining wastewater generated from different sources.

45.7 Reclaimed rubber

Reclaimed rubber is the product resulting from the treatment of ground scrap tyres, tubes and miscellaneous waste rubber articles with heat and chemical agents where the rubber compound is converted to its plastic state by devulcanisation (depolymerisatioin) process. It is then compounded and vulcanised like regular natural and synthetic rubber and in most cases is blended with new rubber to achieve balance in processing and physical properties. Reclaimed rubber manufacturing process involves the following main steps, which are shown in Fig. 45.3.

1. Debeading: Beads in scrap tyres are removed manually for its further processing. In most of the existing plants, debeaded tyres are being received.
2. Size reduction: Debeaded scrap tyre is reduced by mechanical chopping or cracking on a very heavy cracker mill to a suitable size for the devulcanisation step being used. The cracked ground stock is conveyed to a vibrating screen of a given mesh size. The oversized material is returned to the crackers for further grinding. The stock, which passes through the screen, is conveyed for the storage to use in the depolymerisation process. Some reclaiming plants use a series of screens, air separators and sizing equipment to remove/ reduce fibre content from ground rubber scrap.
3. Depolymerisation (devulcanisation): Rubber scrap separation and size reduction is followed by appropriate depolymerisation process. Three basic techniques are used to produce reclaimed rubber are used to produce reclaimed rubber as discussed below.

45.7.1 Wet digester process

The digester process consists of placing the ground scrap, water and reclaimed agents into a steam-jacketed agitator-equipped autoclave (digester). The batch is then cooked for 5–24 hours at 370–405°F. Reclaiming agents used include petroleum and coal tar-based oils and resins as well as various chemical softeners such as phenol alkyl sulphides and disulphides, thiols (mercaptans) and amino compounds.

The reclaiming agents generally function by catalysing the oxidative breakdown of the polymer chain and oxidative disruption of sulphur cross-links. Sometimes, defibreing agents such as caustic soda or chlorides of zinc and calcium and plasticising oils are added to the digester to complete the charge. At the end of the digestion period, the contents of the digester are screened, frequently washed, dewatered and dried in a hot-oven prior to further processing.

Chemical defibreing and the subsequent washing process create an effluent problem. Hence, this process is not opted by reclaim industries these days.

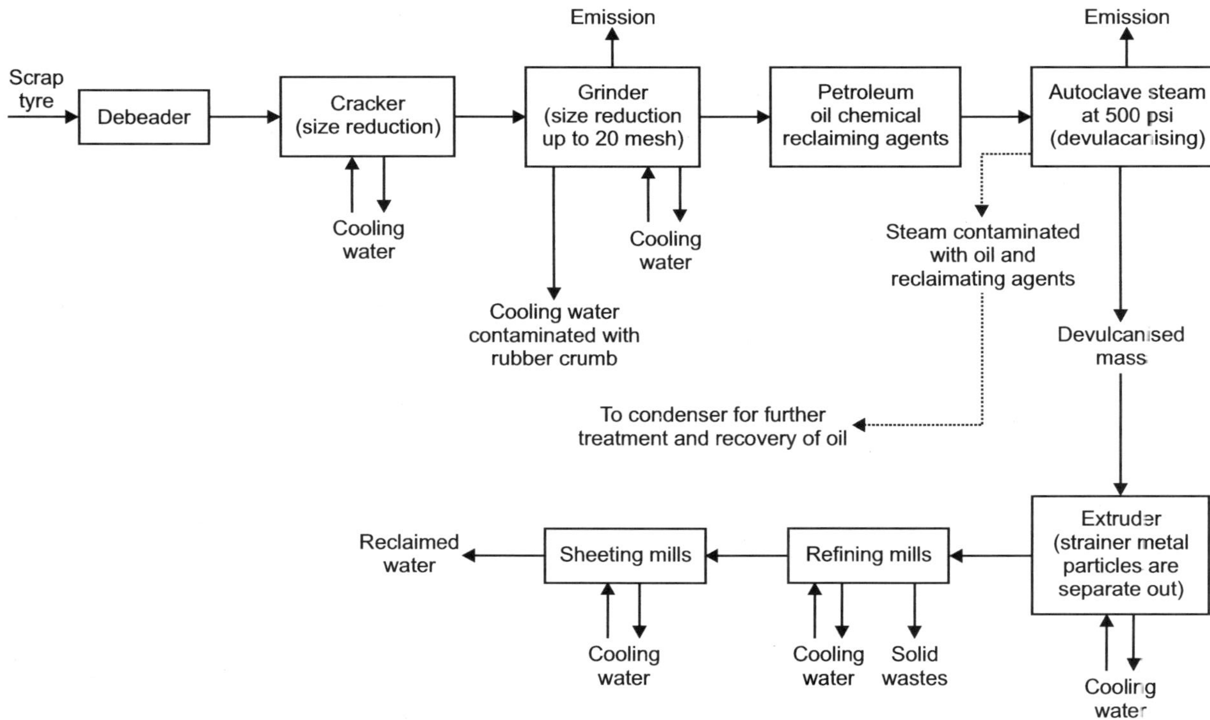

Figure 45.3: Process flow diagram for the production of reclaimed rubber.

45.7.2 Pan (heater) process

The finely ground scrap which is usually free from fibre is blended with the correct amount of reclaiming agents and placed in a single shell pressure vessel (autoclave) into which live steam is passed. Depolymerisation is carried out at about 365° F for 2–18 hours. After this treatment, the heater is vented, the pressure vessel discharged and the crumb of rubber sent on further processing. Since the condensate from this operation is highly contaminated with oils, resins, etc., it cannot be returned to the boilers and therefore, must be treated as a waste. This process is relatively inexpensive because the equipment is simple and the washing and drying steps are eliminated. This process is mainly used in India with some variation, known as dry digester process.

45.7.3 Mechanical reclaiming process

The mechanical reclaiming process, unlike the other two preceding processes, is continuous. The fine ground, fabric free rubber scrap is fed continuously into a high temperature, high shear machine.

The discharged reclaimed rubber needs no drying and is ready for further processing. This process is not common in India.

Final processing

The final stage of the reclaiming operations involves straining to remove foreign matter before going to refining and sheeting mill. The strainer is an extruder, which contains a wire mesh screen held between two strong perforated steel plates in the head of the machine. The reclaim is then given a preliminary refining on a short two-roll mill having a high-fraction ratio between the roll surfaces. The reclaim is then sheeted for the final thickness on a sheeting mill. Sheeted reclaim is cooled, dusted with talc to prevent sticking with each other, tested, packed and finally dispatched to the customer.

To sum up, in rubber industry for spreading pollution, different chemicals contribute. Also at different stages it occurs. The described data gives few examples of the chemicals from particular group, e.g., polymer, fillers, plasticisers, activators, accelerator, curing agent and antidegradants. During different stages of manufacturing, there is possibility of spreading pollution and same can be avoided and minimised by proper precautions. To avoid and minimise the pollution, there is need of proper storage, processing techniques, proper knowledge of chemicals, manpower training, proper selection of material and developing the new techniques and methods for scrap disposal and reuse of scrap.

46

Pollution control in plastic industry

46.1 Introduction

Plastic polymers and products, are extremely diverse, both in terms of chemical composition, properties and possible applications. Several hazardous substances may be released during the life cycle of a plastic product; and considering the large and growing global consumption of plastic products and their persistence in the environment, there is a need for assessing the hazards and risks of this large material group.

46.2 Lifecycle and ecological impact of plastics

The lifecycle of plastics involves three stages: manufacturing in the first stage, usage in the second and recycling and/or disposal in the third. The utilisation of plastics ranges from toys to aircrafts, from hosepipes to dolls, from soft drink bottles to refrigerators, from gramophone records to television sets.

46.2.1 Manufacturing of plastics

The starting point in the production of plastics is the heating of hydrocarbons or the cracking process. The process involves the conversion of natural gas or crude oil components into monomers like ethene, propene, butene and styrene in the presence of a catalyst.

Toxic gases and chemicals are emitted into the air or discharged into the water in the process of producing plastics, which eventually generate negative environmental and human health effects. The kind of emissions or effluents generated and their toxicity varies depending on the type of plastic being made. Nevertheless, the toxic chemicals that are most frequently released during the production of plastics include trichloroethane, acetone, methylene chloride, methyl ethyl ketone, styrene, toluene, benzene. Other major emissions include sulphur oxides, nitrous oxides, methanol, ethylene oxide and volatile organic compounds.

Benzene is believed to cause cancer, styrene has been ranked in the U.S., as extremely toxic, sulphur oxides are known to harm the respiratory system, nitrous oxides adversely affect the nervous system and child behavioural development and ethylene oxides harm the male and female reproductive capacity.

46.3 Recycling of plastics

The kind of recycling practiced in India is quite different from what is practiced in the rest of the world, in that state-of-the-art technologies are not employed here. The entire process of recycling is done on the basis of experience. The starting point is the sorting of plastic waste. This is done on the basis of colour, transparency, hardness, density and opacity of the scrap. The sorted waste is then sent to the granulators. The technology employed is mechanical with the traditional grinding and extrusion to obtain granules.

The final stage is reprocessing. The reprocessing sector can be divided into the granulators and the converters. The granulators make granules from the plastic scrap and sell these granules to the converters. The converters use these to make plastic products. A majority of the units in the informal sector are the granulators that utilise their storage shed in the houses to carry out the grinding. Such units are often located in slums and function with stolen power and single machine extruding units. Scrap storage is done in the backyards and washing is done in open drums. These activities are often termed as backyard recycling. Conversion units are small industrial units that process the granules into finished products. The technologies used in these industries are also old and local. The rate of recycling in India is extremely high. About 40% of the total plastics manufactured are sorted, collected and recycled as opposed to only 10–15% in developed countries. Of the types of plastics recycled in India, PVC (polyvinyl chloride) accounts for 45%, LDPE (low density polyethylene) for 25%, HDPE (high density polyethylene) for 20%, PP (polypropylene) for 7.6% and other polymers such as PS (polystyrene) for 2.4%. According to manufacturers, almost all these types of waste can be recycled up to four or five times. However, the quality of the recyclate deteriorates as additives and virgin material are added to give it strength.

The recycling of plastics is not always green. Recycling usually results in the down cycling of plastics into lower-quality products that have higher and more leachable levels of toxic additives. In addition, the incomplete combustion of polyethylene during recycling releases carbon monoxide. But a recycling plant generates the largest amounts of effluents during washing and cleaning. During recycling, the plastic scrap is cleaned to remove the dirt and foreign matter adhering to it. It is usually soap solution that is used for this purpose and it is reused several times before it is finally disposed of into open drains. This way wastewater is generated. The quantity and the characteristics of wastewater generated cannot be generalised and depends to a large extent on the contents of the plastic scrap. Nevertheless, this wastewater has high pollution load in terms of BOD, COD and TSS. This water needs treatment before proper disposal into the drains. As of today, recycling units in the country release the wastewater into open drains without prior treatment.

Recycling means making new products out of the waste materials. All types of plastics cannot be recycled. If we recycle the ones that can be, the environment will be saved to some extent. Plastic recycling involves the process of recovering scrap plastic and this waste plastic is then reprocessed to form new materials that may be different from their original state. Compared to other materials like glass and metal, recycling of plastic is expensive and complex. This is due to the high molecular weight of the large polymer chains that build the plastic material. Heating plastic doesn't dissolve the polymer chains and hence a tedious and complex process is essential. Different types of plastic cannot be mixed together because they phase separate. Such a resulting melting product cannot be recycled to make another plastic product. While making plastic products many fillers and other additives are used. These fillers cannot be separated from the plastic using inexpensive techniques. This makes the process more complex.

46.3.1 Recycling methods

Mechanical recycling

Mechanical recycling of plastics refers to processes which involve the melting, shredding or granulation of waste plastics. Plastics must be sorted prior to mechanical recycling. Mostly, sorting is done manually. Recently, technology is being introduced to sort plastics automatically, using various techniques such as X-ray fluorescence, infrared and near infrared spectroscopy, electrostatics and flotation. Following sorting, the plastic is either melted down directly and molded into a new shape, or melted down after being shredded into flakes and than processed into granules called regranulate.

Chemical or feedstock recycling

Feedstock recycling describes a range of plastic recovery techniques to make plastics, which break down polymers into their constituent monomers, which in turn can be used again in refineries or petrochemical and chemical production. A range of feedstock recycling technologies is currently being explored. These include: (i) pyrolysis, (ii) hydrogenation, (iii) gasification and (iv) thermal cracking. Feedstock recycling has a greater flexibility over composition and is more tolerant to impurities than mechanical recycling, although it is capital intensive and requires very large quantities of used plastic for reprocessing to be economically viable.

Strategies for reduction of environmental impact of plastics

Reduce the use: This can be done by source reduction. Retailers and consumers can select products that use little or no packaging. Select packaging materials that are recycled into new packaging–such as glass and paper. If people refuse plastic as a packaging material, the industry will decrease production for that purpose and the associated problems such as energy use, pollution and adverse health effects will diminish.

Reuse containers: Since refillable plastic containers can be reused for many times, container reuse can lead to a substantial reduction in the demand for disposable plastic and reduced use of materials and energy, with the consequent reduced environmental impacts. Container designers will take into account the fate of the container beyond the point of sale and consider the service the container provides.

Require producers to take back resins: Get plastic manufacturers directly involved with plastic disposal and closing the material loop, which can stimulate them to consider the product's life cycle from cradle to grave. Make reprocessing easier by limiting the number of container types and shapes, using only one type of resin in each container, making collapsible containers, eliminating pigments, using water-dispersible adhesives for labels and phasing out associated metals such as aluminium seals. Container and resin makers can help develop the reprocessing infrastructure by taking back plastic from consumers.

Legislatively require recycled content: Requiring that all containers be composed of a percentage of post-consumer material reduces the amount of virgin material consumed.

Standardise labelling and inform the public: Standardised labels for 'recycled,' 'recyclable,' and 'made of plastic type X' must be developed for easy identification.

46.3.2 Advantages of recycling plastics

Recycling plastic has many advantages:

1. Use of non-renewable fossil fuels is reduced by recycling as manufacturing new plastic materials require more of these fuels.
2. Consumption of energy is also reduced as already prepared plastic is recycled for new use.
3. Amount of plastic that reach the landfill sites are greatly reduced. This will eliminate land pollution to some extent.
4. Carbon emissions are reduced as manufacturing units emit more carbon.

Cutting back on global warming, preventing air and water pollution and saving our energy resources, are the reasons why we should recycle. Factories and industries that produce goods that have become an indispensable part of our lives, add an appalling amount of waste to the environment and recycling seems the only way to save our planet a ton of cans will reduce 70% of CO_2 production in landfills.

Reduce global warming

Carbon dioxide is a greenhouse gas that is a major contributor to global warming. Manufacturing certain products releases a lot more carbon dioxide than recycling them would. Aluminium is the best example of this. Manufacturing new aluminium goods produces 95% more carbon dioxide than recycling old aluminium objects. For each ton of paper that is recycled, 15 trees are saved from being felled! We are aware of the fact that trees use up carbon dioxide from atmosphere to make food through the process of photosynthesis. It is estimated that one tree absorbs almost 125 Kg of carbon dioxide each year. Here are some more points that would explain why recycling is encouraged these days.

Prevent air pollution

We are already aware of the amount of gaseous waste that industries spew out every day. Factories and industries manufacturing items made from plastic and metals release large amounts of toxic gases. With continuous increase in population, demand for these goods also keeps rising. If we do not recycle these objects then we have to set up new industries to manufacture these goods. This means more air pollution.

To solve the problem of scarcity of landfills

Lot of our garbage is disposed of in landfills. But the rate at which we are producing trash, is overwhelming. We are already running out of areas that could be used as landfills and soon there might not be any left either. This crisis is severe in the cities and suburbs. There, the problem is so pressing that landfills are slowly creeping closer to crowded human settlements. This is making human beings vulnerable to diseases.

Prevent water pollution

This problem is related in part to landfills. The waste that is disposed in landfills is not treated. Contaminants from the products in the landfills seep down to lower levels of the soil where they reach groundwater and pollute it. Other than polluting groundwater, scarcity of landfills and the overwhelming amount of waste that we produce has not left us with any choice but to dump waste in seas and oceans. The devastating effect of release of industrial waste in these water bodies coupled with civic waste that are being dumped in the seas and oceans, on marine ecology, is well-known to us. Environmental preservation is one of the most important benefits of recycling.

Save energy

Another reason why you should recycle is that it saves energy. Recycling most of the products that we use involves much lesser amount of energy than making them a new. For example, recycling paper saves almost 65% energy than making new sheets of paper. If instead of making new products, 1 kg of steel is recycled then it saves enough energy to light a 100 watt bulb for 24 hours straight!

Recycling facts of plastics

1. There are reasons enough for us to recycle plastic. One of them is that plastics are a major source of environmental pollution and are much easy to be recycled that generally thought. It has been found that Americans alone throw away 2.5 million plastic bottles every hour?
2. On an average, one person uses up two pine trees worth of paper each year.
3. To produce one ton of paper, 390 gallons of oil is used up.
4. For every ton of glass recycled, one ton of resources used to make it is saved.
5. With increase in use of electronic goods, people are also resorting to recycling them. These form a considerable portion of trash that is dumped in landfills. Needless to say that the amount of plastic and other electric components that are used to manufacture these products, are making them a potent source of toxins being released to the environment.

After learning about the various environmental changes that are the direct result of human activity, one should help create awareness about reducing pollution and saving energy. Buying recycled products is an important part of the recycling procedure which helps complete the loop. By doing so, you can take part in the 'Save Earth' campaign and can make it a success. Your contribution is very important for the success of the mission. It can help save the resources and can help keep the earth green, pollution-free and beautiful. You can use 'wet garbage' to produce manure for your garden. Thus, you can save a lot of energy and money. By using wind energy or solar energy, you can save other important resources on the earth. You can easily avoid unnecessary use of plastic bags and bottles. Last but not the least, recycling has helped create millions of new jobs. So, it is beneficial for all. The government has already set up units for the same. However, besides sending all the trash to these units, we ourselves could contribute to the process by reusing used products and checking the amount of waste that we generate.

Basic steps that are involved in the recycling

The basic steps that are involved in the recycling of plastic are:

Step 1: Collecting plastic waste from households as well as industrial wastes as well.

Step 2: Sorting the plastic waste in different categories such as PET bottles, bags, containers, etc.

Step 3: The Plastic is cut into tiny pieces.

Step 4: The tiny pieces are thoroughly washed for any dirt or unwanted particles on them.

Step 5: The washed pieces are melted and poured into small containers for reuse.

Factors affecting recycling of plastics

Recycling and reutilisation of waste plastics have several advantages. Recycling and reutilisation of waste plastics lead to a reduction of the use of virgin materials and of the use of energy, thus also a reduction of carbon dioxide emissions. Economically, in some cases, plastics recycling may be profitable. However, a number of factors can complicate the practice of plastics recycling, such as the collection of the plastics waste, separation of different types of plastics, cleaning of the waste and possible pollution of the plastics. A further complicating factor is the low-value nature of most of the products that can be manufactured from recycled plastics. Reusing plastic is preferable to recycling as it uses less energy and fewer resources.

Advantages of reuse and recycle of plastics

1. Conservation of non-renewable fossil fuels–plastic production uses 8% of the world's oil production, 4% as feedstock and 4% during manufacture.
2. Reduced consumption of energy.
3. Reduced amounts of solid waste going to landfill.
4. Reduced emissions of carbon-dioxide (CO_2), nitrogen-oxides (NO_x) and sulphur-dioxide (SO_2).

Plastic bags and their nuisances

Plastic bags are popular with consumers and retailers as they are a functional, lightweight, strong, cheap and hygienic way to transport food and other products. Most of these go to landfill and garbage heaps after they are used and some are recycled. Once littered, plastic bags can find their way on to our streets, parks and into our waterways. Although plastic bags make up only a small percentage of all litter, the impact of these bags is nevertheless significant. Plastic bags create visual pollution problems and can have harmful effects on aquatic and terrestrial animals. Plastic bags are particularly noticeable components of the litter stream due to their size and can take a long time to fully break down. Many carrier bags end up as unsightly litter in trees, streets, parks and gardens which, besides being ugly, can kill birds, small mammals and other creatures. Bags that make it to the ocean may be eaten by sea turtles and marine mammals, who mistake them for jellyfish, with disastrous consequences. In developed countries billion bags are thrown away every year, most of which are used only once before disposal. The biggest problem with plastic bags is that they do not readily break down in the environment. It has been found that, the average plastic carrier bag is used for five minutes, but takes 500 years to decompose.

Plastic bags are made from ethylene, a gas that is produced as a by-product of oil, gas and coal production. Ethylene is made into polymers (chains of ethylene molecules) called polyethylene. This substance, also known as polyethylene or polythene, is made into pellets which are used by plastic manufacturers to produce a range of items, including plastic bags.

Plastics can be broadly split into two groups, those that consist of long strands (thermo softening) and those that also contain short cross-links (thermosetting). Thermo softening plastic will deform when heated and can be remolded into new shapes. Thermosetting plastics are much stronger, but once they have been formed into a shape, they will hold that shape indefinitely and if heated they will merely burn.

Recycle and reuse of plastic bags can mitigate environmental problems.

Issues relating to plastics

There are about 50 different groups of plastics, with hundreds of different varieties. All types of plastic are recyclable. To make sorting and thus recycling easier, the American Society of Plastics Industry developed a standard marking code to help consumers identify and sort the main types of plastic. Before recycling, plastics are sorted according to their resin identification code.

46.3.3 Disposal of plastics

The final stage in the life cycle of plastics is disposal. In India, there are three common ways of getting rid off plastics - by dumping them in landfills, by burning them in incinerators or by littering them. In the case of littering, plastic wastes fail to reach landfills or incinerators. It is the improper way of disposing plastics and is identified as the cause of manifold ecological problems. Incineration is a process in which plastic and other wastes are burnt and the energy produced, as a result, is tapped. In Sweden, 95% of the heat generated from incineration is used for district central heating thereby covering roughly 10% of the country's total need. Policy makers in India too advocate it as a sound option. Several big cities like Mumbai and Chennai have entered into agreements for constructing waste to energy plants. In Chennai, for instance, a 14.85 MW waste to energy plant will be set up by 2020 in which 6000 T per day of municipal solid waste would be converted to electricity. Incineration of plastic wastes also significantly reduces the volume of waste requiring disposal. It is believed that the volume reduction brought about by incineration ranges from 80 to 95%. It is also a suitable option for disposing waste that cannot be recycled further or is non-recyclable.

The pollution that occurs in the disposal stage is largely during incineration and when plastic wastes fail to reach landfills or incinerators. Given the limited recyclability of plastics, a large amount of plastic wastes is burnt in incinerators. Even in the villages in India, plastic and other portions of the waste stream are frequently burned in 'back-yard' fires. But the burning of these chlorine-containing substances releases toxic heavy metals and emits noxious gases like dioxins and furans. The latter two are two of the most toxic and poisonous substances on earth and can cause a variety of health problems including damage to the reproductive and immune system, respiratory difficulties and cancer. In fact, dioxin has been shown to have hormonal activity and is an endocrine disruptor.

It has been observed that due to an inefficient and faulty waste collection and transit system, a large amount of plastic waste fails to reach landfills or incinerators. Instead they are left behind to find their way into the soil, the sewage system and the water bodies. They choke the gutters and drains and during the monsoons flood streets causing severe health problems. When plastic wastes get dispersed in urban fringes or in rural zones, they clog the soil preventing the free flow of water through it and depleting its fertility. It is also said that when plastics reach the rivers, seas and oceans, they pose a serious threat to marine animals like sea turtles, seabirds and fish. If marine animals mistaking them to be authentic food consume plastic objects and pellets, they can clog their intestines leading to death out of starvation or malnutrition.

46.4 Plastics in the environment

Consumption, of course, leads to waste and in the case of plastics, large volumes of extremely persistent waste are created. The problems with plastic waste are many.

46.4.1 Plastic waste management

The largest plastic waste fraction goes to landfill. For Europe the average is around 50%. Landfill space is scarce in many countries which leads to emissions from transportation over long distance and increasing the landfill area may require valuable land. Emissions from landfills may also contaminate ground and surface water. Alternatives to landfilling are mechanical recycling, energy recovery or chemical recovery. The heterogeneity of plastic products and types obstructs recycling, or makes plastic recycling difficult in many cases. Recycling is possible for some fractions, but in order to get a similar, or not a too low grade product, a homogenous fraction is required, which requires labour intensive sorting. Plastic waste for recycling is often transported over long distances, for instance exported from the industrial to the developing countries, particularly in Asia. Energy recovery by incineration of plastic waste in Europe has a larger share (30 %), than recycling (20%).

Combustion of plastics may cause emissions of hazardous substances and contributes to global warming (since most plastics are fossil based). Chemical recovery is under development and means that chemicals, e.g., raw materials such as monomers and gases, are recovered/converted from the plastic material. This can, for instance, be done by controlled thermal degradation such as thermolysis, which is a non-catalytic cracking process.

Some solutions to plastic waste and recycling problems have resulted in new uses for plastic waste and sometimes even more chemically complex products are created and further dispersion in the environment may follow.

46.4.2 Plastic waste in the marine and terrestrial environment

A very large fraction of plastic waste ends up as litter in the terrestrial and marine environment. The marine environment is especially exposed and has been most widely studied. Large and increasing amounts of plastic products, debris, fragments and even microparticles are found in the open ocean, on the surface, in the deep ocean and ocean bed, in coast lines, in sediments and in organisms. The sources are not solely from dumping at sea, but a large quantity is likely to come from land sources, e.g., littering on land (especially coastal areas), plastics blown from landfill sites and sewage effluents, as well as spillage during transport and accidents. This plastic waste can be transported far by currents, for instance the great oceanic gyres. It can also accumulate in the centre of the gyres. The plastic products and debris are fragmented into smaller pieces, even to sizes below 5 mm, called microplastics.

However, the extent of this plastic pollution, especially by microplastics, is much more prevalent than previously thought, both in terms of greater quantities and smaller particles. For instance, in surface trawl studies (net mesh size 330 m) plastics were found at 55 of 76 locations in the Kuroshio Current area (North Pacific), in quantities ranging 03.52 $\times 10^6$ pieces/km^2, with a mean abundance of 1.74 $\times 10^5$ pieces/km^2 and near the central pressure cell of the North Pacific subtropical high a mean abundance of 3.34 $\times 10^5$ plastic pieces/km^2 were found.

Effects from entanglement or ingestion of plastic waste, including suffocation or blocking of digestive tract causing death, have been well documented for birds, turtles and marine mammals.

46.4.3 Plastic degradation

The persistence of plastic waste is another problem. Most plastic polymer types are resistant to biodegradation, i.e., degradation by microorganisms and the two most abundant ones, polyethylene and polypropylene, are extremely resistant to biodegradation. In a polyethylene polymer only 0.1% of the carbon will be transformed into CO_2 per year by biodegradation under optimal laboratory exposure conditions, according to Andrady. There are a few biodegradable plastics which today only have a minor, but growing, share in the plastic market. Not all of them, however, are completely biodegradable in the natural environment. Non-biodegradable polymers can be degraded by heat, oxidation, light, ionic radiation, hydrolysis and mechanical shear and by pollutants such as carbon monoxide, sulphur dioxide, nitrogen oxide and ozone. This causes the polymer to get brittle, to fragment into small pieces and to release degradation products. Different degradation mechanisms exist and which of them that will dominate depends on the polymer type. Chain scission involves breaking the chemical bonds in the polymer molecule and is often random, but for some polymers it proceeds at the polymer end chains and the initial monomers are broken off, a process called depolymerisation. Another degradation mechanism is chain stripping in which the side atoms/groups attached to polymer main chain are released; usually these are volatile molecules. Cross-linking can also occur for some polymers and bonds are then created between the polymer chains. Complete conversion of the plastic product and its breakdown products, into carbon dioxide, water and inorganic molecules by photo- and biodegradation, is for most plastic types extremely slow. In the marine environment degradation is especially slow, since degradation mainly is likely to occur by solar radiation and slow thermal oxidation. This means that the time frame for complete degradation is very long and could, in some situations, be several hundred years.

46.4.4 Plastic composition and hazardous chemicals

Plastic products are made from plastic polymers to which additives are added to enable processing and/or to give certain desired properties for a specific application. The polymers are made by polymerising monomers into macro-molecular chains. These monomers are almost exclusively derived from non-renewable crude oil.

Other substances (besides monomers) are often needed for polymerisation to occur, for instance initiators, catalysts and depending on manufacturing process, solvents may also be used. The resulting plastic polymer can be blended with different additives, for instance plasticisers, flame retardants, heat stabilisers, antioxidants, light stabilisers, lubricants, acid scavengers, antimicrobial agents, anti-static agents, pigments, blowing agents and fillers and is finally processed into a plastic product. There are many different plastic polymers and several thousand different additives, which results in an extremely large variation in chemical composition of plastic products.

Plastic polymers are not particularly reactive and their large size limit transport across biological membranes. They are, therefore, not considered as toxic. In the polymeric material, however, non-polymeric components such as residual monomers, oligomers, low molecular weight fragments, catalyst remnants, polymerisation solvents and a wide range of additives can be present. Several of these are hazardous to human health and the environment, for instance carcinogenic, mutagenic, toxic for reproduction, sensitising and hazardous to the aquatic environment with long lasting effects. Since the non-polymeric compounds usually are of low molecular weight and are either weakly bound or not bound at all to the polymeric macro-molecules, they, or their degradation products, can be emitted from the plastic product to air, water or other contact media (e.g., food).

Many additives are hazardous for human health and the environment. Some are especially hazardous, for instance brominated flame retardants used to retard ignition and prevent fire from spreading; some phthalate plasticisers mainly used to make PVC flexible; and lead heat stabilisers used to prevent degradation of PVC during processing.

Among the phthalate plasticisers the most hazardous ones, i.e., BBP, DEHP and DBP, are classified as toxic for reproduction.

46.4.5 Release and fate of chemicals from plastics

Release of chemicals associated with plastic products may occur in all phases of the life cycle, i.e., during production, use and end-of-life (Fig. 46.1). The environmental fate of the polymer and of the substances released during the life cycle, including the degradation products, as well as the bioaccumulation potential, will affect the exposure for humans and the environment. The physical chemical properties of these substances, e.g., boiling point, vapour pressure, water solubility and octanol-water partition coefficient, can be used to predict the environmental fate and bioaccumulation potential of these substances. Since plastic products are composed of many different chemicals and the main part of these broken down into something completely different, this complicates the predictions.

The size and type of emissions from plastic products are controlled by many factors. The content of non-polymeric substances controls what can be released, while other factors control the potential of release into a surrounding medium, i.e., the migration potential. Migration is generally favoured if the polymer matrix is permeable; if the size of gaps between polymer molecules is larger than the size of migrant; if the migrant is small, has a similar solubility

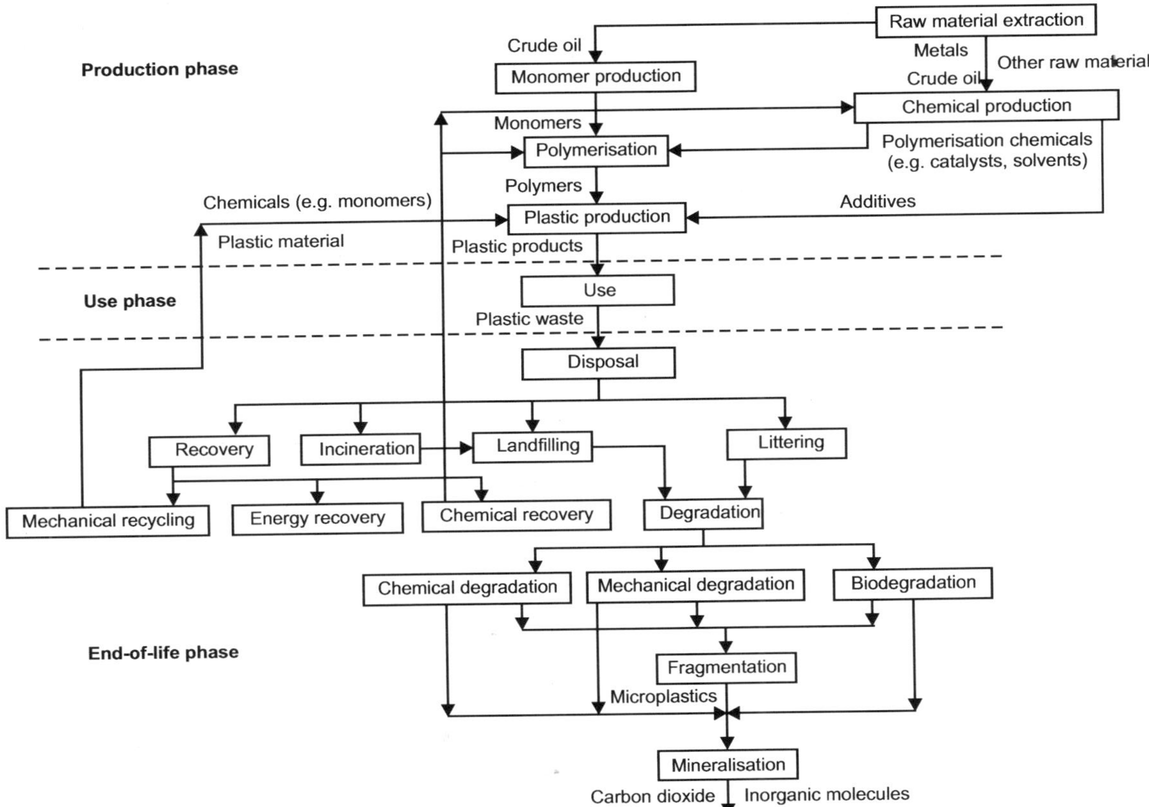

Figure 46.1: The life cycle of a plastic product.

parameter as the polymer and is volatile; if the temperature is high and if the surrounding medium is water for water soluble migrants, fat containing for hydrophobic migrant and acidic for metals.

46.4.6 Exposure of employees in plastic industry

For workers in the plastics industry the main route of exposure to toxic substances is by inhalation and absorption through the lungs, which according to Lokensgard and Richardson accounts for nearly 90% of the toxic symptoms observed in the plastics industry. This is quite expected since many of the hazardous chemicals used in plastic production are volatile organic compounds (VOCs). The VOCs are mainly emitted during the production phase, but also during the use and the end of life phase. This causes indoor air pollution which may be harmful for human health. VOCs also contribute to elevated ground-level ozone levels which may damage vegetation, can irritate the respiratory system, aggravate asthma and lung diseases, cause permanent lung damage and affect the immune system.

46.5 Pollution waste management

Plastic products have become an integral part in our daily life as a basic need. Its broad range of application in films, wrapping materials, shopping and garbage bags, fluid containers, clothing, toys, household and industrial products and building materials. It is a fact that plastics will never degrade and remains on landscape for several years. Mostly, plastics are recyclable but recycled products are more hazardous to the environment than the virgin products. The recycling of a virgin plastic material can be done 2–3 time only, because after every recycling, the plastic material is deteriorated due to thermal pressure.

46.5.1 Utilisation of plastic waste in road construction

Polymer blended bitumen roads

The process of road laying using waste plastics is designed and the technique is being implemented successfully for the construction of flexible roads at various places in India.

46.5.2 Utilisation of plastic waste in cement kiln

Plastic waste generated from different cities and towns is a part of Municipal Solid Waste (MSW). It is a matter of concern that disposal of plastic waste is causing many problems such as leaching impact on land and ground water, choking of drains, making land infertile, indiscriminate burning causing environmental hazards, etc. This waste plastic can be used as a fuel in cement industry.

46.5.3 Converting plastic waste as a alternative fuel and raw material (AFR)

Co-processing refers to the use of waste materials in industry process such as cement, lime or steel production and power stations or any other large combustion plants. Co-processing indicate substitution of primary fuel and raw material by waste, recovering industry and material from waste. Waste material such as plastic waste used for co-processing are referred to as alternative fuels and raw material (AFR). Co-processing of plastic waste offers advantages for cement industry as well as for the municipal authorities responsible for waste management. On other hand, cement producers can save fossil fuel and raw material consumption, contributing the more eco-efficient production. In addition, one of the advantage recovery method used in existing facility, eliminating the need to invest on other plastic waste practices and to secure land filling.

46.6 Biodegradable and compostable plastics

46.6.1 Compostable plastics

The plastics that undergoes degradation by biological process during composting to yield CO_2, water, inorganic compounds and biomass at rate consistent with other known compostable material and leave no visible, distinguishable or toxic residue. Biodegradable plastics made with bio-based polymers have been available for many years. Their high cost, however, has meant they have never replaced traditional non-degradable plastics in the mass market.

Types of biodegradable plastics

There are several degradable plastic are reported such as: biodegradable, compostable, hydro-biodegradable, photo-degradable and biodegradable. Biodegradable plastic products, starch-based products including thermoplastic starch, starch and synthetic aliphatic polyester blends and starch, naturally produced polyesters, renewable resource polyesters such as PLA, synthetic aliphatic polyesters, aliphatic-aromatic (AAC) polyesters, hydro-biodegradable polyester such as modified PET, water soluble polymer such as polyvinyl alcohol and ethylene vinyl alcohol, photo-degradable plastics, controlled degradation additive master batches.

46.6.2 Specifications for compostable plastics (ISO 17088 : 2008)

The Indian standards for compostable plastic is identical with ISO 17088 : 2008 Specification for compostable plastics issued by the International

Organisation for Standardisation (ISO) was adopted by the Bureau of Indian Standards. The purpose of this specification is to establish standards for identifying and labelling plastic products and materials that will compost satisfactorily in well-managed composting facilities where the typical conditions of composting can be consistently obtained (i.e., a long thermophilic phase, aerobic conditions, sufficient water content, a suitable carbon/nitrogen ratio, etc.). Products meeting the requirements outlined below are appropriate for labelling as 'compostable', 'compostable in municipal and commercial facilities' or 'biodegradable during composting'.

46.6.3 Disintegration during composting

A plastic product is considered to have demonstrated satisfactory disintegration if, after 845 days in a controlled composting test, no more than 10% of its original dry mass remains after sieving through a 2,0 mm sieve. The test shall be carried out in accordance with ISO 16929, ISO 20200, ISO 14855-1 or ASTM D 5338 under thermophilic composting conditions without the CO_2-trapping equipment.

46.6.4 Ultimate aerobic biodegradation

A plastic product is considered to have demonstrated a satisfactory rate and level of biodegradation if, when tested in accordance with ISO 14855-1 ISO 14855-2 or ASTM D 5338, it achieves the ratio of conversion to carbon dioxide.

The ultimate aerobic biodegradability shall be determined for the whole material and for each organic constituent which is present in the material at a concentration of more than 1% (by dry mass).

Constituents which are present at concentrations of less than 1% do not need to demonstrate biodegradability. However, the sum of such constituents shall not exceed 5%.

For all polymers, 90% of the organic carbon (relative to a positive-control reference material) shall have been converted to carbon dioxide by the end of the test period both the positive control and the test sample shall be composted for the same length of time and the results compared at the same point in time after the activity of both as reached a plateau. The positive control used shall be microcrystalline cellulose. As an alternative, 90% (in absolute terms) of the organic carbon shall have been converted to carbon dioxide by the end of the test period.

Note: Although the biodegradation test includes the conversion of the polymers into biomass and humic substances in additions to carbon dioxide, no recognised standard test methods or specifications exist for the quantifications of these conversion products. When such tests and specifications become available, this International Standard may be revised.

46.6.5 Plasma pyrolysis technology (PPT)

Plasma Pyrolysis is a state of the art technology, which integrates the thermo-chemical properties of plasma with the pyrolysis process. The intense and versatile heat generation capabilities of plasma pyrolysis technology enable it to dispose of all types of plastic waste including polymeric, biomedical and hazardous waste in a safe and reliable manner. Pyrolysis is the thermal disintegration of carbonaceous material in oxygen-starved atmosphere. When optimised, the most likely compounds formed are methane, carbon monoxide, hydrogen carbon dioxide and water molecules.

In Plasma Pyrolysis, firstly the plastics waste is fed into the primary chamber at 850°C through a feeder. The waste material dissociates into carbon monoxide, hydrogen, methane, higher hydrocarbons, etc. Induced draft fan drains the pyrolysis gases as well as plastics waste into the secondary chamber where these gases are combusted in the presence of excess air. The inflammable gases are ignited with high voltage spark. The secondary chamber temperature is maintained at 1050°C. The hydrocarbon, CO and hydrogen are combusted into safe carbon dioxide and water. The process conditions are maintained such that it eliminates the possibility of formation of toxic dioxins and furans molecules (in case of chlorinated waste). The conversion of organic waste into non toxic gases (CO_2, H_2O) is more than 99%. The extreme conditions of plasma kill stable bacteria such as *bacillus* stereo-thermophilus and *bacillus* subtilis immediately. Segregation of the waste is not necessary, as the very high temperatures ensure treatment of all types of waste without discrimination.

46.6.6 Recycling of plastics through environmentally sound manner

The main goal for developing green recycling of waste plastic was to design an extruder, which would have 'zero significant adverse environmental impact'. This has been achieved by assigning right motor of minimum capacity, selecting optimum L/D ratio, heat sealing and right temperature for the processes and trapping all the emission in pollution control gadget and treating the pollutants to produce by-products. The extrusion and palletisation processes have been redesigned to make the pollution from the process to a minimum level and as a result to enhance the efficiency of the process. The details of process are shown in Fig. 46.2.

46.7 Polystyrene and its impact on environment

Polystyrene is made from petroleum, a non-sustainable, non-renewable, heavily polluting and fast-disappearing commodity. The environmental impacts of polystyrene production in the categories of energy consumption, greenhouse

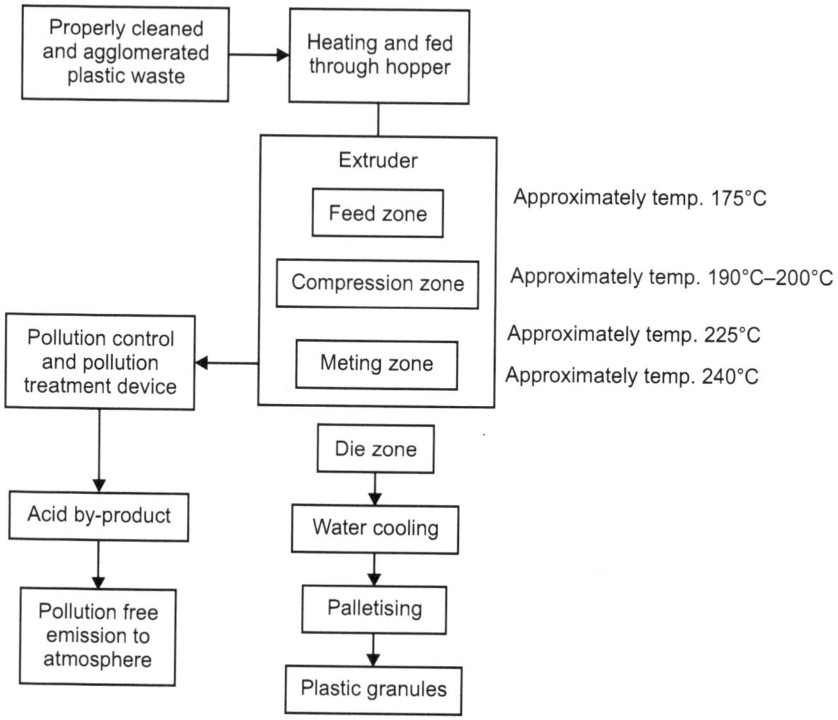

Figure 46.2: Flow chart of the green recycling process.

gas effect and total environmental effect ranks second highest, behind aluminium. Polystyrene is a principle component of urban litter and marine debris. It is detrimental to wildlife that ingests it.

Polystyrene is not biodegradable. It takes several decades to hundreds of years for polystyrene to deteriorate in the environment or in a landfill.

Polystyrene takes up more space in landfills than paper and will eventually re-enter the environment when water or mechanical forces breach landfills.

Food service polystyrene packaging is not recycled because it is not economically viable to store, transport, degrease and wash it before recycling.

Polystyrene contains the toxic substances styrene and benzene, suspected carcinogens and neurotoxins that are hazardous to humans. Polystyrene food containers leach the toxin styrene when they come into contact with warm food or drink, alcohol, oils and acidic foods causing human contamination and posing a health risk to people.

Polystyrene foam, commonly known by the name 'styrofoam' is formed by adding a blowing agent to polystyrene, a petroleum-based plastic material. Polystyrene foam is light-weight (about 95% air), with good insulation properties and is used in all types of products from cups that keep beverages hot or cold to materials that protect items during shipping.

There are two widely used types of expanded polystyrene: EPS (expanded polystyrene) and XPS (extruded polystyrene). They are made from the same plastic (polystyrene), but the manufacturing processes are different. EPS begins as small plastic 'beads' that are expanded and fused together, forming products such as protective packaging for electrical appliances. XPS begins as a continuous mass of molten material, which is then shaped into products such as take-out lunch boxes and trays for fresh cuts of meat at the supermarket.

XPS has little recycling ability since it is often contaminated by food and grease that render it non-recyclable. On the other hand, in most countries, including Hong Kong, EPS waste that is not contaminated is recyclable despite difficulties involved with storage, transportation, relatively low scrap value and negative environmental impacts.

In the past, polystyrene foam was banned by cities around the world due in part to the ozone-depleting CFC gases used as blowing agents. More recent bans have been enacted because of the litter and marine debris impacts of polystyrene foam food packaging as well as overall impacts on human and environmental health. As a result of the impacts on marine pollution and adverse effect to marine wildlife, several coastal cities across the United States, including California coastal cities such as San Francisco, Malibu, Oakland and Huntington Beach have banned the use of polystyrene food packaging altogether.

46.7.1 Environmental impacts of polystyrene

Polystyrene foam presents unique management issues because of its lightweight nature, floatability and prevalence to be blown from disposal sites even when disposed of properly. The lightweight and buoyant polystyrene travels easily through gutters and storm drains, eventually reaching the ocean. Plastic from urban runoff is the largest source of marine debris worldwide. Pollution of waterways and waterfront negatively affects tourism and quality of life. When polystyrene travels down waterways and storm drains into the ocean, it breaks down into smaller, nonbiodegradable pieces that are ingested by marine life and other wildlife thus harming or killing them. In one Californian study, at least 162 marine species including most seabirds were reported to have eaten plastics and other marine litter.

The process of manufacturing polystyrene pollutes the air and creates large amounts of liquid and solid waste. A 2009 EPA report on solid waste named the polystyrene manufacturing process as the fifth largest creator of hazardous waste in the United States. In the product manufacturing process as well as the use and disposal of the products, energy consumption, greenhouse gas effect and total environmental effect, polystyrene's environmental impacts were second highest, behind aluminium, according to the California Integrated Waste Management Board. Polystyrene recycling, if and when it is carried out properly, is not 'closed loop'. For example, collected polystyrene protective packaging is not remanufactured into protective packaging, but into other products, such as packing filler and cafeteria trays. This means that more resources will have to be used and more pollution created, to produce more polystyrene protective packaging.

Human health impacts of polystyrene

There are potential health impacts from polystyrene foam food packaging associated with its production and with the leaching of some of its chemical components into food and drink. The general public is not typically warned of these public hazards. Styrene, a component of polystyrene, is a known hazardous substance that medical evidence and the U.S. Food and Drug Administration suggests leaches from polystyrene containers into food and drink. Polystyrene food containers leach Styrene when they come into contact with warm food or drink, alcohol, oils and acidic foods causing human contamination and posing a direct health risk to people. It is for this reason that it is not recommended that plastic take-out containers ever be heated in a microwave. Styrene is a suspected carcinogen and neurotoxin which potentially threatens human health. It has been detected in the fat tissue of every man, woman and child tested by the EPA in a 2009 study.

Several adverse health effects have been attributed to styrene. Some studies have concluded that at exposure levels above 100 parts per million, humans experience acute mucous membrane irritation with the eyes, nose and throat particularly affected. Increased levels of styrene concentration may cause fatigue, irritation and decrease in concentration ability. Styrene has been linked to increased levels of chromosomal damage, abnormal pulmonary function and cancer in workers at polystyrene and styrene plants.

Another known toxin, benzene, is a chemical component of polystyrene foam. It is a known carcinogen and enters the human body either through the skin or the respiratory system.

46.8 Biopolymers

Biopolymers are polymers which are naturally found in nature. Like polymers biopolymers are chain-like molecules made up of repeating chemical blocks and can be very long in length.

Biopolymers can be classified in three groups, depending on the nature of the repeating unit they are made of: (i) polysaccharrides are made of sugars, (ii) proteins of amino acids and (iii) nucleic acids of nucleotides.

The following substances are example-biopolymers for each group: cellulose (found in plants), myoglobin (muscle tissues) and DNA (genetic material of a given organism).

46.8.1 Biodegradation of polysaccharides

A large number of polysaccharides such as guar gum, xanthan gum, carboxymethyl cellulose, starch, amylose and amylopectin have been investigated for their effectiveness in turbulent drag reduction, flocculation and shear stability. The solutions of these polysaccharides biodegrade. However, when grafted with polyacrylamide branches, these graft copolymers become highly efficient drag reducers and flocculants. They have robustness of rigid polysaccharide chains and efficiency of flexible polyacrylamide in causing drag reduction. In case of flocculant, the dangling polyacrylamide chains on rigid polysaccharide chains, increase the approachability to contaminants hence they are very effective flocculants. When polysaccharides are grafted with polyacrylamide chain, their regularity is diminished. This decreases their biodegradability which is observed in all cases. The branched polysaccharides are less susceptible to biodegradation. Grafted and ungrafted amylopectin are clear examples of this. In case of drag reducers and flocculants, the grafting reduces biodegradation and thus increases their self-life which is a desirable attribute for their applications in these phenomena.

Biopolymers are complex molecules with biological activity

In contrast to synthetic polymers which have a simpler and more random structure, biopolymers are complex molecular assemblies that adopt precise and defined 3D shapes and structures. This feature is essential because this is what makes biopolymers active molecules *in vivo*. Their defined shape and structure are indeed keys to their function. For example, haemoglobin would not be able to carry oxygen in the blood if it was not folded in a quaternary structure.

Biopolymers have a low environmental footprint

Unlike synthetic polymers which feedstock can be derived from petrochemicals or chemical processes, biopolymers are produced from renewable resources such as plant and/or living organisms. They can be degraded by natural processes, micro-organisms and enzymes down to elemental entities that can be resorbed in the environment. Biopolymers thus offer the possibility to create a sustainable industry and reduce CO_2 emissions.

46.8.2 Polymer degradation

The degradation of polymers is possible by physical, chemical, mechanical and biological means. The chemical and physical degradation of polymers implies that micro organism, macro organisms or enzyme are not present and ageing is dependent on physical, chemical and or mechanical influences. The degradation refers to any chemical change in polymer structure. The biodegradation involves living organism or their secretions and can be defined as a molecular phenomenon in which biological systems mediate at least one step of the overall transformation of the molecule into gases products and biomass. Polymers may be degraded as a result of exposure to various agents such as heat, light, atmosphere, hydrolysis and biodegradation.

The degradation may take place by exposure to one agent or combination of them such as photo-degradation can be defined as breakdown of polymeric molecules initiated by sunlight. The photo-oxidative degradation involves sunlight and oxygen. Photo-oxidation decreases the life time of most materials used in outdoor applications. For biomedical materials, the body fluids and compound dissolved in these fluids decrease the lifetime of the polymers. For clinical applications, it is often to distinguish clearly between 'biodegradable polymers' and 'bioabsorbable polymers'. Biodegradable polymers are polymers which are decomposed in the living body but whose degradation products remain in tissues long-term. On the other hand bioabsorbable can be defined as polymers which degrade after implantation into non-toxic product which are then eliminated from body or metabolised therein.

The bioabsorption of polymeric materials can be divided into two distinct stages : decomposition and absorption. The former stage involves bond breaking in the main chain, a decrease in molecular weight and production of oligomer and monomers. The main reaction for this process is hydrolysis.

Polymers that are decomposed by enzyme-specific reactions are called enzymatically degradable polymers while polymers which are decomposed by contact with water or serum are non-enzymaticaly degradable polymers. Hydrolysis affects a number of synthetic polymers as well as such as polyesters, polyanhydrides, polyamides (proteins) and polyurethanes.

The degradation can be artificial using phosphate buffers simulating body fluids or medicated by different enzyme (hydrolases). The biodegradation attack by living organism can take place on biopolymers as well as synthetic polymers having hydrolysable backbone chains like aliphatic polyesters.

Factors affecting degradability

The degradation times of polymers vary considerably from a couple of days upto several years depending upon the type of bond in the repeat unit and the environment. It also depends on several other factors.

1. Amorphous regions are more susceptible to hydrolysis reactions and microbial attacks than crystalline.
2. Hydrophobic polymer is often less degradable than a hydrophilic one.

Hydrophobicity substantially decreases the hydrolyse rate and microbial attack.

3. The relationship of surface/volume dictates the degradation rate-small polymer granule degrade faster than the corresponding more compact film.
4. Linear polymers are generally more subject to degradation than are the branched polymers in biodegradation test.
5. The most important factor that influences controlled degradation is degradation in presence of oxygen in some form.
6. Hydroxyl, hydroxyperoxide and carbonyl groups formed during photodegradation may lead to increased hydrophilicity which leads to biodegradation. Ester and anhydride groups containing polymers that represent a substantial portion of macromolecular unit result in highly oxygen containing units and are easily hydrolysed.
7. Polymers that are hydrolytically unstable and water permeable give structures that we subject to rapid degradation, e.g., aliphatic polyesters. Some polyamides and polyurethanes might also be hydrolytically unstable but they usually have a lower water penetration and permeability and subsequent degradation is slow.
8. Some polymers are highly resistant to hydrolytic degradation. They include polyacrylates and silicones. Both are essentially inert to hydrolysis but at the same time water or liquid can penetrate them and cause some swelling or they might be subject to cracking in cycling experiment.

46.8.3 Biodegradable polymers

The use of biodegradable polymers in single use disposable packaging in concert with composting infrastructure is an ecologically sound approach to management of the resultant waste and its diversion from land fill. Composting biodegradable materials alongwith other organic waste represent biologically recycling of the polymeric carbon by way of carbon cycle.

Different approaches have been followed to make classical polymers such as PE and PP (bio) degradable by physical incorporation of starch as biodegradable element. Three approaches have been followed. The granular starch has been used as filler in polyolefins. Gelatinised starch plastics composites and grafted starch itself have been proposed plastics materials. In the first case, the incorporation of starch is limited to 6–12% while in the second case it may be upto 40%. The phase transfer agents along with other ingredients are needed to produce the films. The graft copolymers of starch with PMMA and PS have shown potential for production of extrusion shaped products and films. However, the mechanical properties of these materials are not upto the mark. The constituent plastics biodegradation is also slow. The current trend is to use 100% biodegradable materials. Polymers such as poly(caprolactone), poly(lactic acid) and poly(hydroxy butyrate-CO-hydroxy valerate) have been introduced into the market place.

The last one has mechanical properties comparable to those of polypropylene. However, these materials are costly. Hence these materials can be blended with starch to reduce their cost and enhance their flexibility. It has been shown that although the addition of 0–50% starch granules to P(HB-CO-19% HV) results in an increased biodegradation rate, it also results in reduced tensile strength and flexibility because of formation of composite rather than blend. This lacuna can be partly removed by using thermoplastic starch in place of granular starch. Mixing of thermoplastic starch with poly(hydroxy butyrate-CO-12% hydroxy valerate) resulted in blends. Nevertheless, the blends are still brittle despite the presence of flexible starch. Improved properties of the incompatible blend can also be obtained by optimising the blend morphology during processing. There is considerable scope to enhance the compatibility of these blends by incorporating grafted starch in the blends. Such blends will have less brittleness and higher mechanical strength.

46.8.4 Environmental implications of recycling and polymer degradation

1. In recycling process various steps such as collection, sorting and separation, washing and purification processing and marketing are involved. The individuals handling urban solid waste are under threats by the pathogenic micro-organisms that may be present in faceal material in raw municipal waste. These micro-organisms cause diarrhoea and dysentery. Waste collection and recovery activities pose environmental and health risks, ways should sought to minimise these dangers as far as possible.
2. Most of the polymers under the influence of light, heat or mechanical presence during recycling decompose and release benign degradation products but some polymers such as PVC and PS when decompose on heating emanate HCl and styrene which are toxic. Polyethylene and polypropylene are inert materials but ethylene and propylene may also cause problems.
3. Environmental effects of plastics also differ according to the type and quantity of additives that have been used. Some flame retardant may pollute environment for example bromine emissions. Pigments or colourants may contain heavy metals that are highly toxic to humans such as Cr, Cu, Co, Se, Pb and Cd are often used to produce brightly coloured plastics. The additives used as heat stabilisers frequently contain heavy metals such as Ba, Sn, Pb and Cd and sometimes in combination. These

metals are serious environmental pollutants and have deleterious effect on human health. They also interfere with processes in plant, animal tissues and in the soil when present at or above some specific concentrations.

4. Taking into consideration the process of plastic recycling, the most important problem caused by the polluting substances is air pollution either within the reprocessing units or in the open air. During extrusion process several substances such as additives may be released. In polymers like PVC along with additives, vinyl chloride and HCl are also emanated.
5. During incineration unless the combustion is complete burning plastics within the waste may release considerable quantities of polluting substances. With complete combustion 90% of plastic materials is converted to CO_2 and H_2O. Incomplete combustion of PE, PP, PS and PVC can cause further problems as CO and smoke may be produced. PVC when burnt emanates HCl. During waste incineration chlorodioxins are formed which are highly toxic.
6. During the recycling, combustion of plastics should be avoided at all cost. In reprocessing machines, PVC may decompose and give off fumes of HCl which may damage the machine and discolour and spoil finished product. The processing temperatures of PVC range between 150–210°C. Heat decomposition may occur at temperatures as low as 180°C. These difficulties can be overcome by keeping the moulding period short and adding a heat stabiliser.

Precautions during recycling

It is, therefore, advisable that following precautions should be undertaken during recycling.

1. All waste should be thoroughly cleaned and sorted before it is reprocessed. This will help to minimise odours and will improve the quality of the end product.
2. Do not allow plastics to burn; the gases released may be toxic and pose a health risk to the workers.
3. Avoid the use of PVC for recycling.
4. Some recycled plastics should not be used in products that will come in contact with food. Waste plastics from unknown sources should not be used for toys, kitchen utensils and drinking water pipes.
5. Plastic waste from hospitals must be approached with caution as they may contain dangerous or unhygienic materials.

47

Pollution control in drug and pharmaceutical industry

47.1 Introduction

Like many other industries, the pharmaceutical industry produces a wide variety of products. This industry uses both inorganic and organic materials as raw materials, the latter being either of synthetic or of vegetable and animal origin.

Some pharmaceutical plants do not discharge liquid waste at all, some discharge very small but concentrated liquid waste, while others discharge highly alkaline and toxic liquid wastes. Therefore, it is very difficult to make any generalisation in regard to the characterisation of the pharmaceutical plant wastes. The drug manufacturing and processing industry can be categorised into five areas:

1. Fermentation.
2. Chemical synthesis.
3. Natural extraction.
4. Formulation and packaging.
5. Research and development.

47.2 Pharmaceutical industry

47.2.1 Fermentation

Batch fermentation processes are typically used for producing steroids and antibiotics. The fermentation process consists of three steps:

1. Inoculum and seed preparation.
2. Fermentation.
3. Product recovery or isolation.

Inoculum preparation begins with a population of a microbial strain and a few cells from this culture are grown into a dense suspension through a series of test tubes and flasks. The cells are then transferred to a seed tank for further growth. The seed tank is essentially a fermenter designed for the maximum growth. The cells from the seed tank are charged to a sterilised fermenter and sterilised nutrients are added to begin fermentation. During fermentation, the fermenter contents are agitated and aerated by a sparger maintaining an optimum air flow rate and temperature. After the cells are matured, the fermentation broth is filtered to remove the solid remains of the inoculum micro-organisms (i.e., mycelia). The product is then recovered from the resulting filter beer through various recovery processes, for example, solvent extraction, precipitation, adsorption, or others. An example of a fermentation process is illustrated in Fig. 47.1.

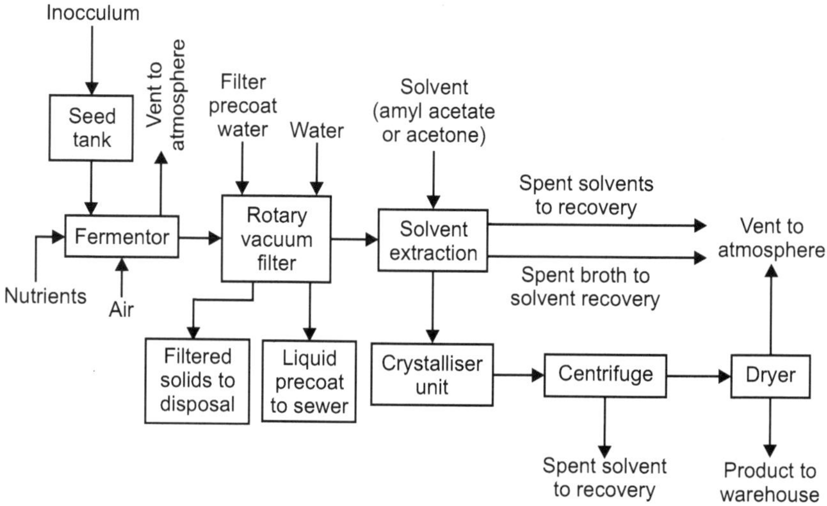

Figure 47.1: Fermentation process.

Chemical synthesis

The majority of drugs today are produced by chemical synthesis. Typically, one or more batch reactors (depending on the number of chemical steps) are utilised in series to make the final product. The product recovery (or isolation) steps may involve separation and/or purification. The intermediates can also be isolated in order to avoid carrying impurities to the next chemical reaction (impurities may interfere with the next reaction, reducing yield). The types of chemical reactions and the isolation processes used are quite diverse, depending on the chemical structure of the desired final product.

An example of a chemical synthesis is illustrated through a process flow diagram of product X (Fig. 47.2). Potassium permanganate (Product X precursor) and water are mixed in a 3000-gallon reactor and manganese dioxide is formed.

Manganese dioxide is removed from the solution by a rotary drum filter coated with a filter aid (celite). The filter cake, containing manganese dioxide and celite, is deposited into trash bins for disposal. The filtrate is neutralised with sulphuric acid and sent to an evaporator to remove water.

The product solution is then sent to an 800-gallon Pfaudler reactor where a final pH adjustment is made with additional sulphuric acid. The mixture is

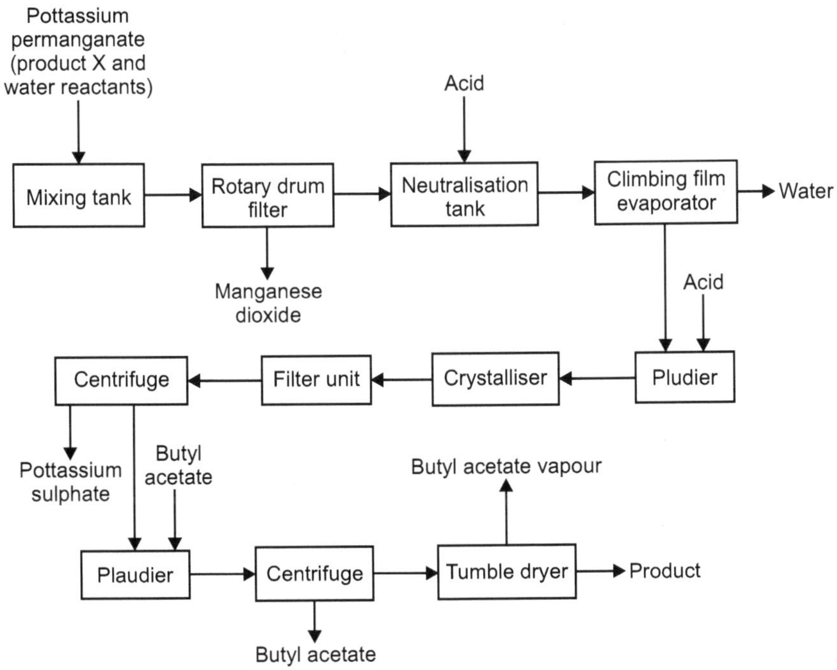

Figure 47.2: Chemical synthesis.

cooled, potassium sulphate (product of neutralisation) crystallises and the potassium sulphate crystals are filtred by centrifuge.

Butyl acetate is then added to the filtrate and the resulting mixture is azeotropically dried (by azeotropic distillation) until all the water is removed. The mixture is again filtered to remove additional potassium sulphate salt, which comes out of solution as more water is removed. When the filtrate is cooled, the product crystallises and the product crystals are isolated by filtering (centrifuge). The product is then dried in a tumble dryer and packaged.

Formulation

The active ingredient produced by either fermentation, chemical synthesis, or natural extraction is then formulated to tablets, capsules, liquids, creams and ointments. The active ingredient is blended to the right dosage by adding fillers or binders. Fillers or binders used for tablet formulation are starch and sugar. The blend is then compressed via wet granulation, direct compress, or slugging.

In wet granulation, the active ingredient and filler in powder form are blended and wetted with a binder solution to form coarse granules. These granules are dried, mixed with lubricants (e.g., magnesium stearate) and compressed into tablets. Direct compression is done by using a dye that holds a measured amount of material and a punch that compresses the tablet. Slugging is used for drugs that are not sufficiently stable for wet granular formulation or cannot be compressed directly. The slugging process blends and compresses relatively large size tablets (20 to 30 grams) which can then be ground and screened to a desired mesh size and recompressed into tablets.

After tablets, capsules are the most widely-used oral dosage form solid drugs. Hard gelatin capsules are prepared by depositing a gelatin film on the capsule (using a solution of gelatin and cooling). The capsules are dried, trimmed and the separate sections of the capsule are filled and joined.

Liquids are prepared for injection and oral use in aqueous solutions, syrups, suspensions and tinctures. The liquids are prepared by mixing the active ingredient with a solvent and preservatives (to prevent mold and bacterial growth) and filtering at high pressure. Suspensions and emulsions are prepared by using colloid mills and homogenisers, respectively. Although liquid solutions prepared for oral or topical use do not require sterilisation, liquid solutions for ophthalmic use must be sterilised. The liquid dosages that are injected into the fluid systems of the body (e.g., intradermal, hypodermic, intramuscular, intravenous, etc.), also need to be sterilised by dry or moist heat under pressure. Products that are not stable at high temperatures can be sterilised by using bacteria-retaining filters.

Creams and ointments are prepared for topical use, usually by blending the active ingredient with petroleum derivative petrolatum and passing the mixture through a colloid or roller mill.

47.3 Waste from drug industry

47.3.1 Fermentation

Fermentation processes generate large volumes of spent fermentation medium. Filtration processes, which remove the solid remains of micro-organisms (mycelia), yield large quantities of solids as spent filter cake. The filter cake contains mycelia, filter aids and small amounts of residual product. After product recovery, the spent filtrate is discharged as wastewater. Other sources of wastewater are equipment-cleaning operations and fermenter vent gas scrubbing. Wastewaters from fermentation processes are typically high in BOD (biological oxygen demand), COD (chemical oxygen demand) and TSS (total suspended solids), with a pH range of 4 to 8.

47.3.2 Chemical synthesis

Wastes from chemical synthesis are complex and diverse as a result of the variety of reactions and operations employed. Chemical synthesis processes generate acids, bases, solvents, cyanides and metal wastes among others. Solid wastes are generated as waste filter cake, which may include various inorganic salts (e.g., sodium sulphate, ammonium sulphate, etc.), organic by-products and metal complexes. Typically, spent solvents are recycled on-site within the process (via extraction and distillation). Still bottom wastes are generated as a result of recovery and recycling operations. Chemical synthesis processes also generate aqueous waste streams, which include mother liquors from crystallisation operations, aqueous wash layers from extractions, contaminated scrubber water and equipment-cleaning wastes. Wastewaters from chemical synthesis processes typically have high BOD, COD and TSS, with the pH ranging from 1 to 11.

47.3.3 Natural extraction

Wastes resulting from natural extraction processes include raw materials such as leaves and roots, spent solvents and wastewater. Wastewaters from natural extraction processes are typically low in BOD, COD and TSS levels, with the pH ranging from 6 to 8.

Formulation

Wastes generated in formulation processes include wastes resulting from cleaning, sterilisation, spills and rejected products. During blending and/ or tableting operations, dusts are generated, which are typically recycled back into the formulation process. During tablet coating and capsule manufacturing, there are air emissions from solvent handling. The primary source of wastewater is equipment cleanouts. Wastewaters typically contain inorganic salts, sugars, syrups and residual amounts of the active ingredient. Wastewaters are usually low in BOD, COD and TSS, with neutral pH. A general flow diagram for the treatment of wastewater from pharmaceutical industries is shown in Fig. 47.3.

47.4 Management of wastes

Good housekeeping is the least expensive means to reduce the overall burden on treatment and disposal. Loss of materials, solvent and product can be restricted by installing monitoring devices so that the process managers can be held responsible for spills and poor housekeeping practices. By adopting suitable process modifications, substantial reduction in wastewater generation can be achieved. Substitution of organic solvent for water in synthesis and subsequent solvent recovery can reduce the quantity of wastewater generated. Barometric condenser systems can be a major source of contamination in plant effluent producing large volume of dilute waste stream. Steam jet ejectors and barometric condenser systems can be replaced with vacuum and surface condenser systems.

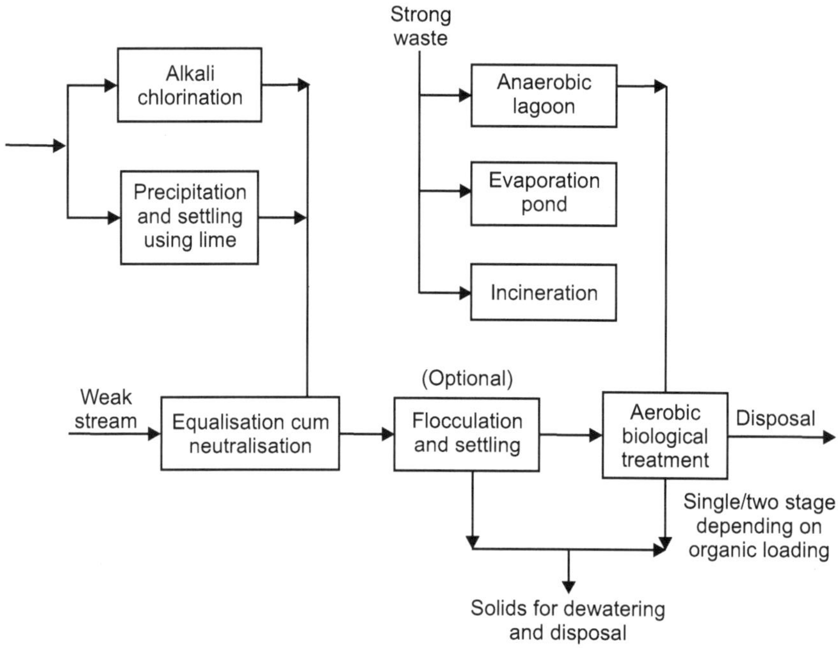

Figure 47.3: A general flow diagram for the treatment of wastewater from pharmaceutical industries.

Dust collector devices and vacuum type packing can reduce powder flying around in the tablet and packing sections.

47.4.1 Gaseous/particulate waste

Gaseous effluents like hydrogen chloride, hydrogen cyanide, sulphur dioxide, chlorine and solvent vapours are emitted from various processing units manufacturing basic drugs. Particulate emissions are predominant in the material handling section and in formulation units where the products are in dry powder form.

Various physicochemical processes are employed for controlling the emission of gases as fumes before discharging through stack into the atmosphere. Adsorption is a control process by which the gas molecules attach themselves on solid surfaces. Activated carbon, alumina, bauxite, silica gel and molecular sieves are common adsorbents. Adsorption is carried out either in fixed bed or moving bed. Thermal regeneration is generally employed for the regeneration of the adsorbants.

Absorption is a process in which a soluble gas is transferred from a gas stream into a liquid. The liquid phase dispersed system include packed column, wetted wall column, spray column, venturi scrubber or flooded disc scrubber.

The physical and chemical properties of the particulate as well as the size play an important role in the design of particulate control equipment. Various types of equipment for the particulate control are gravity settling chambers, cyclones, scrubbers, electrostatic preciptators and fabric filters.

47.4.2 Solid wastes management

Treatment of solid wastes depends upon the nature and composition of the waste. The present disposal practice mainly consists of hauling solid wastes to off-premises land disposal. Little attempt has been made to salvage solid wastes for reuse or making by-products. Reclamation possibilities include biological treatment of extracted botanical drugs to produce useful by-products and regeneration and reuse of filter materials. The reported solid wastes recycling consists of selling mycellium and biological sludge as high protein animal or plant feed supplements. Toxic solid wastes from processes are to be incinerated inside factory premises. Radioactive wastes generated from research activities are to be sealed into special containers and disposed of in some designated area. Solid wastes such as card-board boxes, packaging materials, glass and metal containers from damaged or defective batches warrants investigation for salvaging with necessary precautions. Specific suggestions for minimising solid wastes disposal problems are as follows:

Recovery of solids like spent catalysts, unreacted materials, side products etc., can be separated to reduce pollution load to the treatment plant. These can be separated, settled and reused. Some of these possibilities are:

1. Mycellium from antibiotic fermentation is non-toxic and a rich source of protein and vitamins. This can be dried and used as animal feed. It has very high BOD and as such necessary to isolate from wastewater to be treated in the effluent treatment plant.
2. Volatiles in the by-products may be condensed and reused.
3. Exhaust gases in the fermentation process can be recovered and used as potential fuel source. For example, methane is a good fuel which is generated in anaerobic fermentation process.
4. Tar produced in some organic reactions like Friedel Craft's acrylation and alkylation can be used as a cheap fuel source.
5. Volatile raw materials and solvents can be separated by rectification and these can be recycled into the process.

47.5 Pollution prevention options

Pollution prevention options in the drug manufacturing and processing industry are source reduction through process modifications, recycling and resource recovery and good housekeeping and operating practices. Process modifications include product reformulation, material substitution and process step modifications. The following sections describe some of the pollution prevention options available to the drug manufacturing and processing industry.

47.5.1 Process modifications

Product reformulation involves the substitution of one or more raw materials, which may reduce the volume and/or toxicity of wastes generated. However, product reformulation is difficult for the pharmaceutical industry because of the time required for product redevelopment and testing. Testing is required to ensure that the reformulation has the same medicinal effect as the original drug. The time and the testing required for FDA approval of the reformulated drug can be long and drawn out. Therefore, product reformulation is usually not an economic or practical option. Material substitution is the replacement of a hazardous material in manufacturing or processing with a non-hazardous material. Material substitution may also reduce the volume and toxicity of the wastes generated, but without changing the product itself. For example, organic solvents used during tablet-coating processes can be substituted with water-based solvents; reducing the volatile organic compound emissions into the atmosphere. Another example is the use of less hazardous organic solvents (e.g., ethyl acetate, isopropyl acetate, toluene, etc.), in the place of more toxic (more rigidly controlled by CPCB) solvents, such as methylene chloride and benzene. In fact, many drug companies have specific guidelines prohibiting the use of certain solvents (e.g., methlyene chloride and benzene) even at the research and development stage. Another approach in material substitution includes the use of aqueous cleaning solutions (e.g., soap and water) in place of solvent-based solutions. The current trend in the industry is to replace chlorinated solvents with non-chlorinated solvents.

Generation of wastes can be reduced through process modification or modernisation. Process modifications involve changes in operating conditions and/or equipment. Process modernisation includes installing updated control systems or increasing control levels. Another form of modernisation is increasing the degree of process automation.

Incomplete chemical reactions can result in the increased formation of by-products, thus increasing the amount of wastes generated. Possible causes of incomplete reaction are inadequate feed flow control and/or reactor temperature control. Installing updated control systems can improve reactor/reaction efficiency and reduce by-product formation. An increased degree of automation can also reduce wastes generated by preventing operator errors. Other process considerations for waste minimisation are:

1. Install internal recycling systems for solvents and cooling water.
2. Select new or improved catalysts to improve reaction conversion.

3. Consider using a continuous process in place of batch process.
4. Modify tank and vessel dimensions to minimise wetting loss (e.g., use of conical vessels).
5. Optimise crystallisation conditions to reduce the amount of product loss to the mother liquor and cake wash.
6. Optimise crystallisation conditions to obtain the desired crystal size, thus minimising the need for recrystallisation (crystal size too small to filter) or milling (crystal size too large).
7. Consider new reactor-filter-dryer systems that can handle more than one operation (e.g., Rosenmund filter-dryer).
8. Utilise available computer programmes/expert systems that can evaluate the entire process for efficiency and characterise the waste streams, thus determining the major source of waste generation and optimising the process.

One example of pollution prevention via process modification is the use of supported liquid membranes (SLMs) for liquid solution separation during reaction, work-up, and/or isolation sequences. The SLMs are thin layers of pure liquid or liquid solutions immobilised in microporous inert supports. They have the extraordinary capacity of removing solutes from a feed solution to a strip solution. For example, an aqueous solution containing a solute is fed into an SLM and the solute is extracted into a contained or supported organic solvent, which in turn, is stripped to the aqueous strip solution. The organic solvent is continuously renewed and this specific feature allows significant reduction in solvent usage.

Other advantages of SLMs are the high separation factor per stage and low capital/operating costs. The major disadvantage is the instability of the SLMs, which prevents long-term operation. In an attempt to overcome SLM's shortcomings, the hollow fibre contained liquid membrane (HFCLM) technique was developed. HFCLM uses microporous hollow fibres to contain liquid membranes in a permeator (Fig. 47.4). In theory, the solvent contained in the membrane can be used indefinitely, thus reducing the amount of solvent used for extractions. However, depending on the type of process, the solvent may have to be replaced as the performance starts to degrade because of the accumulation of impurities. Although significant waste reduction can be achieved through process modifications, there are problems associated with this option. Extensive process changes can be very costly, owing to the installation of new equipment and downtime. New processes must also be tested and certified by the appropriate agency to ensure that the resulting product is acceptable. An important parameter to consider is the impurity profile of the drug manufactured by the new process, because an identical or better

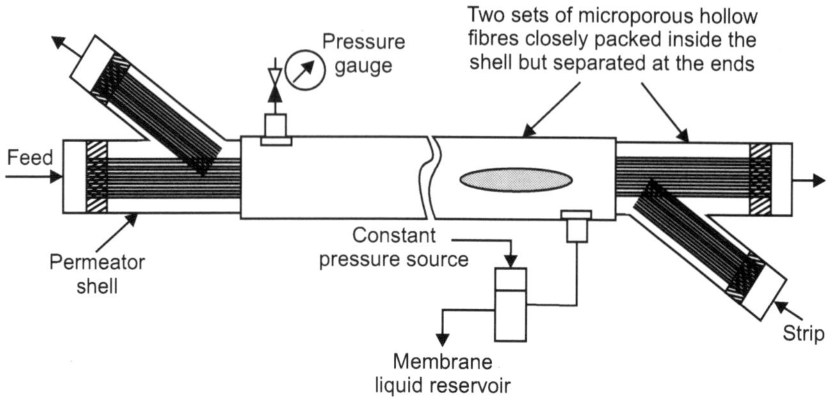

Figure 47.4: Hollow fibre contained liquid membranes.

impurity profile (i.e., fewer impurities) is essential for the validation of clinical testing. In addition, the existing management may be reluctant to change a working (successful) process. Such resistance is particularly likely against adopting continuous processes in place of existing batch processes.

Equipment cleaning efficiency can be improved significantly by employing high-pressure spray nozzles (as a preliminary cleaning method) to clean reactor walls, rather than conventional solvent or caustic solution boil-outs. The use of high-pressure nozzles can also reduce solvent usage appreciably.

Resource recovery and recycling

Recycling and resource recovery include direct reuse of the waste generated-reclaimed for a separate use or direct reuse within the process via separation of impurities. However, the high degree of quality control required in the pharmaceutical industry limits the options available for reuse within the process. Fermentation wastes may be used as supplements to animal feeds, soil conditioners and fertilisers. For example, mycelium filter cake and other by-products of the production of antibiotics by fermentation are widely sold as protein supplements to animal feed, as long as the antibiotic activity in the filter cake does not exceed 2 grams per tonne of cake and less than 3 pounds of cake is used per tonne of feed.

Inorganic salts such as ammonium sulphate and sodium sulphate (typical by-products of some chemical synthesis processes) can be reclaimed. Ammonium sulphate can be used as raw material for fertiliser and sodium sulphate can be dried for sale to the glass industry. Recycling is a particularly good option for spent solvents. Spent solvents can be used for equipment cleaning. They can also be purified for reuse in the process for reaction mediums, extraction mediums and coating mediums.

The following steps can improve solvent waste recyclability:

1. Segregate chlorinated solvents from non-chlorinated solvents, aliphatic from aromatic solvents and aqueous from flammable solvents.
2. Minimise solids concentration in solvent wastes.
3. Label all solvent wastes properly, including the composition and generation source.

47.5.2 Good housekeeping and operating practices

Good housekeeping and operating practices can reduce the amount of wastes generated and material losses. Such practices may be categorised into three components: (i) plant management, (ii) raw materials handling and (iii) waste management. Good operating and housekeeping practices in plant management include the following:

1. Management's awareness of environmental responsibilities.
2. Closer supervision of plant personnel to increase production efficiency and reduce waste generation.
3. Effective employee training for safe equipment operation, material handling and spill clean-up.
4. Improved documentation of process steps, including start-up, shutdown, emergency, special and normal operating procedures.
5. Effective scheduling to reduce waste generation.

Raw material handling procedures should be updated to improve materials tracking and inventory control. Utilisation of a computer to monitor inventories can reduce over-stocking. Proper labelling and use of material safety data sheets (MSDS) facilitate material handling. Spill and leak prevention programmes are also important for pollution prevention. A spill and leak prevention programme should include starting with properly designed storage tanks and process vessels (including level indicators and alarms, installing secondary containment, setting up procedures that prevent bypass of the interlock systems and keeping a regular maintenance schedule that includes checking the integrity of containers and testing alarms). In addition, a proper maintenance programme can prevent the generation of (hazardous) waste by equipment failure. Preventive maintenance programmes should include cleaning, making minor adjustments, lubrication, testing, calibrating and replacing of minor parts. Once wastes are generated, it is essential that waste tracking and inventory control are established. There should also be frequent environmental and waste audits. The waste creams can be segregated (hazardous from non-hazardous, liquid from solid, etc.), to reduce waste volume, to simplify disposal and to facilitate recycling and recovery.

48

Pollution control in textile industry

48.1 Introduction

The textile industry is not a single entity but encompasses a range of industrial units which use a variety of natural and synthetic fibres to produce various fabrics. All over the world cotton has retained its dominant position not only because of its easy domestic availability but also due to the climatic conditions in the country which dictate the need for cotton based wear. The growth of the textile industry in any region is greatly influenced by the climatic conditions and availability of raw materials.

The textile industries are very complex in nature as far as varieties of products, process and raw materials are concerned. During production, the cloth has to pass through various processes and chemical operations like sizing, desizing, scouring, mercerising, bleaching, dying, printing and finishing. In a textile industry, a number of dyes chemicals and auxiliary chemicals are used to impart desired quality in the fabrics. The wastewater of the industry is highly alkaline in nature and contains high concentration or BOD, COD, TDS and alkalinity. It can cause environmental problems unless it is properly treated before disposal. The industry also generates air pollution. Processing of fibres prior to and during spinning and weaving generates dust, lint, etc., which degrades working environment in the industry. Dust may cause respiratory diseases in workers. A chronic lung disease, byssinosis is commonly observed among workers exposed to cotton, flax and hemp dust. Besides this, there are a number of process operations including spinning weaving that produce noise to the tune of 90 dB (A).

48.2 Textile manufacturing processes

In general, the entire textiles manufacturing process can be described in five different stages (Fig. 48.1) of production, as given below:

1. Preparation of fibre natural (e.g., wool, cotton) and manmade, i.e., cellulosic (e.g., rayon, acetate) and synthetic (e.g., polyester, nylon).
2. Conversion of fibre into yarn (spinning).
3. Manufacturing of textile from yarn (weaving and knitting).
4. Colouring and finishing of textiles.
5. Garmenting by cutting and stitching.

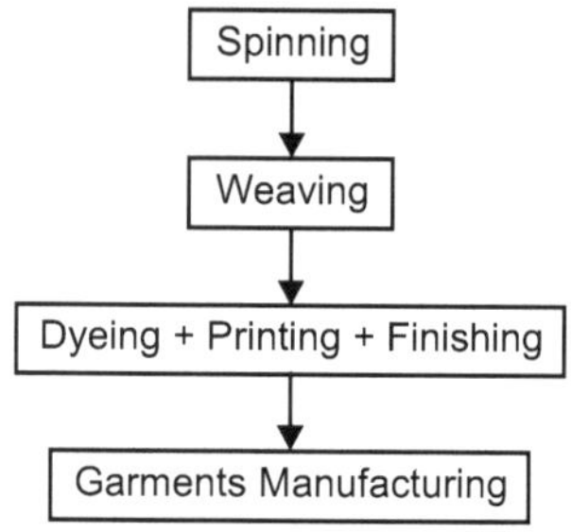

Figure 48.1: Process flow diagram of textile manufacturing.

The following sections describe each of these stages of production.

48.2.1 Preparation of fibre

Natural fibres

The natural fibres must be opened, blended, carded and/or combed and drafted before spinning. The main steps used for processing wool and cotton are briefed below. Although the equipment used for cotton is designed somewhat differently from that used for wool, the machinery operates in essentially the same fashion.

Man-made fibres

The manmade fibres (both synthetic and cellulosic) are manufactured by processes that simulate or resemble the manufacturing of silk (i.e., forcing a liquid through a small opening where the liquid solidifies to form a continuous filament). The main methods of fibre manufacturing are: (i) melt spining and (ii) dry-spinning and (iii) wet spinning. After the spinning process, the filaments are drawn to increase the orientation of the macromolecules and thereby the tensile strength of the yarns.

48.2.2 Spinning - conversion of fibre into yarn

The formation of spun yarn is done in spinning mills. Before spinning preparatory processes take place, the tasks of the processes are opening of the fibre bales, mixing of the fibres, cleaning, arrangement, paralleling of the fibres, drafting and twining of the fibres to a yarn. Ring spinning is the most important technology (80% of worldwide yarn production). The open end technique is mostly used non-conventional spinning technology (Fig. 48.2).

The natural fibres as well as the man-made staple fibres are produced into yarns with different types of spinning systems. The kind of system used depends on the fibre length, fibre thickness and the end use of the product.

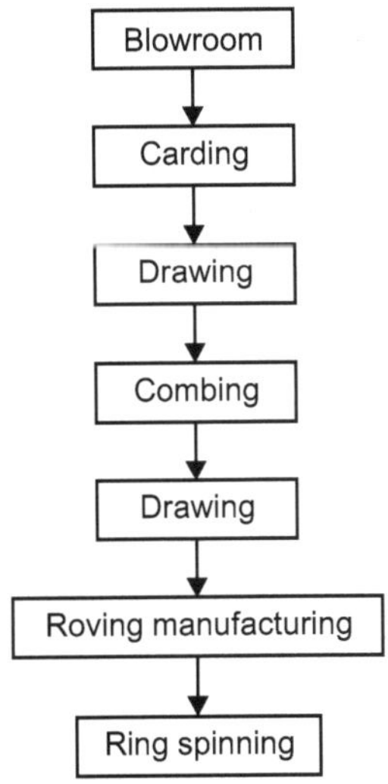

Figure 48.2: Flow diagram of spinning.

48.3 Textile industry and its impact on environment

Any industrial activity causes pollution in one form or the other and so is the textile industry. This industry covers a wide spectrum of manufacturing activities and is diverse in terms of raw materials and techniques employed, chemicals used and the final products. The impact of textile production on the environmental aspects such as air, water, land and human body and the social aspects such as child labour and poor unhygienic working conditions must be considered.

48.3.1 Industry's attitude towards environmental problems

Textile processing industry is characterised not only by the large volume of water required for various unit operations, but also by the variety of chemicals used for various processes. There is a long sequence of wet processing stages requiring input of water, chemical and energy and generating wastes at each stage. The other feature of this industry, which is a backbone of fashion garment, is large variation in demand of type, pattern and colour combination of fabric resulting into significant fluctuation in waste generation volume and load. Textile processing generates many waste streams, including liquid, gaseous and solid wastes, some of which may be hazardous. The nature of the waste generated depends on the type of textile facility, the processes and technologies being operated and the types of fibres and chemicals used.

The textile industry is a significant contributor to many national economies, encompassing both small and large-scale operations worldwide. In terms of its output or production and employment, the textile industry is one of the largest industries in the world.

The textile manufacturing process is characterised by the high consumption of resources like water, fuel and a variety of chemicals in a long process sequence that generates a significant amount of waste. The common practices of low process efficiency result in substantial wastage of resources and a severe damage to the environment.

The main environmental problems associated with textile industry are typically those associated with water body pollution caused by the discharge of untreated effluents. Other environmental issues of equal importance are air emission, notably Volatile Organic Compounds (VOC)'s and excessive noise or odour as well as workspace safety.

48.3.2 Pollution aspects of textile industry

Air pollution

The textile industry is plagued by air pollution problems which must be resolved. In particular, smoke and odour arising in the process require abatement. Air pollution is the introduction of chemicals, particulate, or biological material that causes harm or discomfort to humans or other living organisms, or damages the natural environment into the atmosphere. Main sources of air pollution are boilers(s), thermo pack and diesel generator(s) which generate gaseous pollutants such as Suspended Particulate Matter (SPM), sulphur dioxide gas, oxide of nitrogen gas, etc.

The major air pollution problem in the textile industry occurs during the finishing stages, where various processes are employed for coating the fabrics. Coating materials include lubricating oils, plasticisers, paints and water repellent chemicals-essentially, organic (usually hydrocarbon) compounds such as oils, waxes or solvents. After the coatings are applied, the coated fabrics are cured by heating in ovens, dryers, tenter frames, etc. A frequent result is the vapourisation of the organic compounds into high molecular weight Volatile Organic (usually hydrocarbon) Compounds (VOCs). In terms of actual emissions, the industry must also deal with larger particles, principally lint.

Most processes performed in textile mills produce atmospheric emissions. Gaseous emissions have been identified as the second greatest pollution problem (after effluent quality) for the textile industry. Speculation concerning the amounts and types of air pollutants emitted from textile operations has been widespread but, generally, air emission data for textile manufacturing operations are not readily available. Air emissions include dust, oil mists, acid vapours, odours and boiler exhausts. Cleaning and production changes result in sludges from tanks and spent process chemicals, which may contain toxic organics and metals. Summary of wastes generated during textile manufacturing is shown in Table 48.1.

Table 48.1: Summary of the wastes generated during textiles manufacturing.

Process	*Source*	*Pollutants*
Energy production coating, drying and curing	Emissions from boiler Emission from high temperature ovens	Particulates, nitrous oxides (NO_x) sulphur dioxide (SO_2) volatile organic components (VOCs)
Cotton handling activities	Emissions from preparation, carding, combing and fabrics manufacturing	Particulates
Sizing	Emission from using sizing compound (gums, PVA)	Nitrogen oxides, sulphur oxide, carbon monoxide.
Bleaching	Emission from using chlorine compound	Chlorine, chlorine dioxide
Dyeing	Disperse dyeing using carriers: sulphur dyeing, aniline dyeing	Carriers H_2S Aniline vapours
Printing	Emission	Hydrocarbons, ammonia
Finishing	Resin finishing. Heat setting of synthetic fabrics	Formaldehyde Carriers - low molecular weight Polymers - lubricating oils
Chemical storage	Emissions from storage tanks for commodity and chemicals	Volatile organic components (VOCs)
Wastewater treatment	Emissions from treatment tanks and vessels	Volatile organic components, toxic emissions

Air pollution is the introduction of chemicals, particulate, or biological material that causes harm or discomfort to humans or other living organisms, or damages the natural environment into the atmosphere.

Air emissions can be classified according to the nature of their sources:

Point sources: (i) boilers, (ii) ovens and (iii) storage tanks.

Diffusive: (i) solvent-based, (ii) wastewater treatment (iii) warehouses and (iv) spills.

Textile mills usually generate nitrogen and sulphur oxides from boilers. Other significant sources of air emissions in textile operations include resin finishing and drying operations, printing, dyeing, fabric preparation and wastewater treatment plants. Hydrocarbons are emitted from drying ovens and from mineral oils in high-temperature drying/curing. These processes can emit formaldehyde, acids, softeners and other volatile compounds. Effect of pollutants on human being is shown in Table 48.2.

Table 48.2: Effect of pollutants on human being.

Pollutants	*Effects on human being*	*Sources*
Sulphur dioxide	Irritates respiratory system and causes bronchitis	Boiler flue gas, rayon plant, etc.
Aldehydes	Irritates all parts of respiratory system	Polyester plant
Chlorine	Causes lung irritation and also irritation in eyes	Processing house
Carbon dioxide	Deprives body cells of oxygen and cause unconsciousness by CO combining with haemoglobin	Boiler house

Residues from fibre preparation sometimes emit pollutants during the heat setting process. Carriers and solvents may be emitted during dyeing operations depending on the types of dyeing processes used and from wastewater treatment plant operations. Carriers used in batch dyeing of disperse dyes may lead to volatilisation of aqueous chemical emulsions during heat setting, drying, or curing stages. Acetic acid and formaldehyde are two major emissions of concern in textiles.

Effect of dust:

1. Formation of dust film on every surface.
2. The particles fall into moving parts of machinery.
3. Dirty appearance of product.

Pollution control in spinning mill:

1. Maintaining m/c parts.
2. Maintaining hardness of rubber coats.
3. Maintaining relative humidity%.
4. Using overhead pneumatic cleaners.
5. Cyclone filters, cloth filters.

Pollution control in wet processing unit:

1. Electrostatic precipitator.
2. Scrubber.
3. Oxidiser.

General control measures

Height of chimneys: Chimneys height should not be less than 30 meters and release the pollutants not in the vicinity of living organism.

Graves and inertial separator: These are working on gravitational and inertial concepts of collecting, filtering, etc., of the particulate matter, e.g., settling chambers, dynamic separator and wet cyclones and multiple cyclones.

Filters: Woven or sintered metal beds of fibres, metal turning, fibrous mats and aggregate bed filter, paper filters and fabric filters are used for the filtration of particulate matter like dust, lint and fumes.

Reducing air pollution

Some steps to reduce the emissions to air include:

1. Decreasing emissions of organic solvents by changing to water-based products.
2. Using scrubbers to collect particulate matter.
3. Optimising boiler operations to reduce the emissions of nitrous and sulphur oxides.
4. Pre-screening chemicals using the material safety data sheets to ensure that chemicals are not toxic.
5. Identifying sources of air pollution and quantifying emissions.
6. Designing and manufacturing products that do not produce toxic or hazardous air pollutants.
7. Avoiding fugitive air emissions from chemical spills through improved work.

Water pollution

The textile industry uses high volumes of water throughout its operations, from the washing of fibres to bleaching, dyeing and washing of finished products. On average, approximately 200 litres of water are required to produce 1 kg of textiles. The large volumes of wastewater generated also contain a wide variety of chemicals, used throughout processing. These can cause damage if not properly treated before being discharged into the environment. Of all the steps involved in textile processing, wet processing creates the highest volume of wastewater.

The aquatic toxicity of textile industry wastewater varies considerably among production facilities. The sources of aquatic toxicity can include salt, surfactants, ionic metals and their metal complexes, toxic organic chemicals, biocides and toxic anions. Most textile dyes have low aquatic toxicity. On the other hand, surfactants and related compounds, such as detergents, emulsifiers and dispersants are used in almost each textile process and can be an important contributor to effluent aquatic toxicity, BOD and foaming.

Solid waste pollution

The primary residual wastes generated from the textile industry are non-hazardous. These include scraps of fabric and yarn, off-specification yarn and fabric and packaging waste. There are also wastes associated with the storage and production of yarns and textiles, such as chemical storage drums, cardboard reels for storing fabric and cones used to hold yarn for dyeing and knitting. Cutting room waste generates a high volume of fabric scraps, which can often be reduced by increasing fabric utilisation efficiency in cutting and sewing.

Thus, cleaner production is an attractive approach to tackle environmental problems associated with industrial production and poor material efficiency. Since the cleaner production approach has been successfully implemented in some areas in the textile sector, it shows that significant financial saving and environmental improvements can be made by relatively low-cost and straightforward interventions. This improves the quality of products and minimises the cost of production, enabling the branch to compete in the global market. Moreover, cleaner production also improves the company's public image by highlighting the steps it has taken to protect the environment.

48.3.3 Waste stream

The waste stream generated in these processes is essentially the water based effluent. The process of singeing causes air pollution. Solid wastes and hazardous wastes are also generated in the processes such as yarn production. The nature of the waste generated depends on the type of the process used and the technology involved and types of raw materials and chemicals used. The textile industry consumes a vast quantity of water and generates an equally vast quantity of wastewater. The level of pollution of the wastewater can be appreciated from the data presented in Table 48.3.

Table 48.3: Characteristics of water generated from textile processing units, vis-a-vis the standards.

Property (mg/L)	*Standard*	*Cotton*	*Synthetic*	*Wool scouring*	*Wool dyeing, finishing*
pH	5.5–9.0	8–12	7–9	3–10	5–10
BOD (mg/L/5 day)	30–350	150–750	150–200	5000–8000	500–600
COD (mg/L/day)	250	200–2400	400–650	10000–20000	1700–2400
Alkalinity(as $CaCO_3$)	–	180–7300	550–630	80–100	240–300
Phenols	–	0.030–1.00	–	–	–
Oils and grease	10–20	4.5–30	–	2000–2500	400–500
Suspended solids	100–600	35–1750	50–150	5000–6000	500–700
Dissolve solids	–	2100–7700	1060–1080	10000–13000	800–1000

Table 48.3 shows distinctly that the wastewater generated is not conforming to the prescribed standards and thus it is highly polluted and dangerous.

48.3.4 Some facts regarding environmental issues

Some facts, we should all know regarding the environmental issues in this industry, are:

1. Treatment is essential before discharge of effluents.
2. Small units cannot afford sophisticated technologies such as reverse osmosis.
3. ETPs (effluent treatment plants) are installed in a few units only.
4. Honesty of functioning of ETP is a matter of concern (total dissolved solids, chemical oxygen demand, biochemical oxygen demand, are generally not within the prescribed limits.
5. ETP maintenance, replacement of corroded parts, etc., is not well done regularly.
6. Low priority for environmental technologies by the entrepreneurs.
7. Enforcement of legislation is not strict.
8. Govt's attention and support is also not acting as sufficient incentive for adopting environmental technologies.
9. Awareness and education have to be increased.

48.3.5 Wastes generated

Wastes generated in textile industry may be classified into four types:

1. Hard to treat wastes.
2. Dispersible wastes.
3. Hazardous or toxic wastes.
4. High volume wastes.

Hard to treat waste

Textile wastes include colour, metals, phenol, toxic organic compounds and phosphates. The first two, i.e., colour and metals originated primarily from dyeing and printing operations, although metals are sometimes found in other processes also. Phosphates are used primarily in preparatory processes and dyeing. In addition to these three, (i.e., colour, metals and phosphates) other hard to treat wastes include non-biodegradable organic materials, such as certain surfactants, solvents, etc. These can resist treatment pass through standard activated sludge system and produce aquatic toxicity in effluent from treatment plants. The primary problem in that case is toxic, hence these wastes is also classified in the hazardous and toxic category.

Dispersible wastes

The wastes generated from the following processes and operations in the textile industry are classified as highly dispersible:

1. Waste stream from continuous operation (e.g., finishing, dyeing, printing and preparatory).
2. Print paste (especially from the screen, squeeze and drum cleaning).
3. Lint.
4. Waste from coating operations (especially foam).
5. Waste solvents from machine cleaning.
6. Still bottoms from solvent recovery (especially dry cleaning operation).
7. Batch dumps of unused processing (especially finishing mixes).

A prominent source of dispersible wastes in textile wet processing is the following.

Pastes: Pastes usually come from printing and include oil/water pastes and acrylic polymers. Both types tend to gel and form lumps in drains. This waste can be difficult to sample with automated equipment because of their tendency to stop up sampler liners, pumps and filters.

Lint: Lint originates from many textile operations, particularly preparatory, dyeing and washing operations. Usually removing lint is fairly easy using primary control measures such as filters, which can be placed in the circulation line of dyeing machines and other equipments. The filters must be maintained and cleaned out on a regular basis to ensure proper operation. The collected lint usually can be dried and then used as - landfill or incinerated. Higher quality lint can be marketed for various other uses.

Solvent: Due to lack of proper handling and disposal of solvents, we generate solvent based wastes. Once the solvent is mixed with water, it becomes difficult to separate. Also, solvent reclamation and reuse is much easier before solvents are mixed with other wastes. In general, it is advisable to avoid dumping solvents down the drain with other wastes.

Wastes from continuous processes: Dumps of dye and chemical mixes sometimes occur in continuous textile operations such as slashing, preparatory, continuous dyeing, printing, coating and finishing.

Hazardous or toxic wastes

Hazardous or toxic wastes are generally a subgroup of hard to treat waste. Their impact on the environment is very significant. Hazardous or toxic wastes include:

1. Metal.
2. Chlorinated solvents.

3. Non-degradable surfactants.
4. Other non-degradable or volatile organic materials.

These wastes originate often from non-process operations, such as machine cleaning. Appropriate reduction strategies include conservation, substitution, process modification and maintenance/housekeeping.

High volume waste

Large volume wastes are sometimes a problem for the textile processing units. The most common large volume wastes include:

1. Hydraulic loading (high volume of wastewater).
2. Wash water for preparation and continuous dyeing processes and alkaline waste from preparatory processes.
3. Batch dyeing waste, especially containing large amounts of salt, acid or alkali.

These can sometimes be reduced by appropriate recycling process and/or by process and equipment modifications. However, there are many cases where little or nothing can be achieved by the above approaches.

Other wastes

In addition to size and resize chemicals removed from the textile, sizing and desizing operations generate additional wastes that deserve attention, including.

1. Packaging material for size.
2. Dumps of the unused portion of size mixes.
3. Machine cleaning and maintenance.
4. Fibre lint and yarn waste.

48.3.6 Solution to environmental problems

Solutions to environmental problems are already found by appropriate technological approaches.

1. Adoption of technologies of processing giving minimum or zero environmental pollution.
2. Effluent treatment plants are the most widely accepted approaches for the wet processing of textiles. Environmental monitoring and testing is an essential step for the successful implementation of any of the above mentioned two approaches.

Environmental technologies related to textile processing industry are grouped in following categories.

Solid waste separation: Scrubbers, cyclone separators, electrostatic precipitators, air pollutant treatments, etc.

Wastewater treatment: Total dissolved solids removal, heavy metals removal, colour removal, BOD and COD reduction. Small package treatment plants, common effluent treatment plants, toxicity reduction, chemical recovery, primary, secondary and tertiary wastewater treatment processes.

Sludge from textile industry

Solid waste management is a very acute problem for industrial establishments. In such circumstances, management of sludge in textile industry has now become a burning issue due to its surplus volume and undesired characteristics.

The sludge can be utilised for manufacture of bricks. It thus serves two important purposes, namely - efficient management of textile sludge and production of useful by-product, which fetches cash to the industry and a low cost construction material to the public. The bricks, made up of this sludge, were tested for water absorption, compressive strength, presence of soluble salts and drop test and the results were compared with those of ordinary bricks.

Textile sludge has a variable composition and normally contains high organic matter, N, P, K and micronutrients contents. Additionally, dyes, heavy metals and pathogenic micro-organisms may be presented and composting is often required for utilisation and transformation of these wastes in a soil amendment. During the composting, the plant nutrients present in the wastes are converted by microbial action into soluble forms, available to plants. Composting can also decrease or eliminate the toxicity of textile sludge.

Vermicomposting of solid textile mill sludge

Vermicomposting is commonly used for the management of organic wastes. It was investigated that the potential of an epigeic earthworm, *Eisenia foetida*, to transform Solid Textile Mill Sludge (STMS) spiked with anaerobically digested Biogas Plant Slurry (BPS) into vermicompost to evaluate the feasibility of vermicomposting. The growth and reproduction of *Eisenia foetida* was monitored in a range of different feed mixtures for 15 weeks in laboratory under controlled experimental conditions. *Eisenia foetida* did not survive in fresh STMS. But worms grew and reproduced in STMS spiked with BPS feed mixtures. A greater percentage of STMS in feed mixture affected biomass gain and cocoon production by earthworms. The maximum growth was recorded in 100% BPS. The net weight gain by *Eisenia foetida* in 100% BPS was two-four-fold higher than STMS-containing feed mixtures.

Vermicomposting resulted in pH shift toward acidic, significant reduction in C:N ratio and increase in nitrogen, phosphorus and potassium contents.

Microbial activity measured as dehydrogenase activity increased with time up to day 75 but decreased on day 90, indicating the exhaustion of feed and decrease in microbial activity.

Textile sludge management by anaerobic technology

Composite samples of textile mill sludge obtained from a factory were characterised for their pollution characteristics, pH, solids concentration, oxygen demand, nitrogen, phosphorus, total bacteria counts, etc. The analysis revealed that the sludge has high pollution potentials and therefore needed treatment before disposal to the environment.

The ratio of chemical oxygen demand, (COD) to that of biochemical oxygen demand, (BOD) was 3.08; meaning the sludge has high substrate biodegradability.

Samples were subjected to mesophilic anaerobic treatment at the temperature of 35°C. The method achieved solids reduction of 61% total solids, 68% settleable solids and 51% volatile solids and a total bacteria reduction of 99.99%. The reduction in BOD and COD were 89% each. Nitrate and phosphate were found to reduce substantially thereby preventing eutrophication due to undesirable nutrients.

The anaerobic treatment was found to have an additional benefit of producing biogas (methane and carbon (IV) oxide) which if harnessed may be used as fuel. Figure 48.3 shows the various treatment operations in textile mills and formation of sludge.

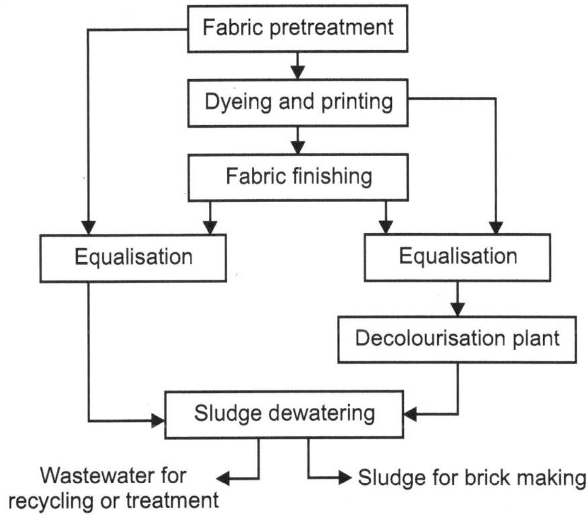

Figure 48.3: Sludge from textile mills and its treatment.

Anaerobic digestion has been reported to achieve a reasonable reduction in the organic content of sludge. The main products being methane gas and carbon (IV) oxide.

Efficiency of activated sludge in the treatment of textile sludge: The efficiency of activated sludge in the treatment of textile finishing wastewater can be improved through acclimation of bacteria to the various textile chemicals. The negative effect of thiosulphate and sulphide on the activated sludge biodegradation of surface active substances in viscose fibre wastewater was reported by Smith and others.

Biochemical Oxygen Demand (BOD), Chemical Oxygen Demand (COD) and dehydrogenase activity were used to monitor the biochemical oxidation of concentrations up to 1200 mg/L of *N*-methylpyrrolidone.

Activated sludge in the treatment of textile sludge: Biological treatment with activated sludge was found to be more efficient and reliable in terms of organic matter removal than chemical coagulation for the treatment of textile-processing industry waste. Smith and others estimated the economic effectiveness of preventing environmental harm in connection with biological treatment of wastewater from acetate fibre manufacturing.

Energy efficient bricks from sludge

The bricks are obtained by molding clay in rectangular blocks of uniform size and then by drying and burning these blocks. Burnt clay bricks have good resistance to moisture, insects and erosion and create a good room environment. They are medium in cost and have medium to high compressive strength. In brick making the major input is fuel followed by labour. Bricks manufactured from dried sludge collected from textile wastewater treatment plant were investigated. Results of tests indicated that the sludge proportion and the firing temperature were the two key factors determining the brick quality. Results showed that the brick weight loss on ignition was mainly attributed to the organic matter content in the sludge being burnt off during the firing process. With up to 6.66% sludge added to the bricks, the strength measured at temperatures 500°C met the requirements of the National Standards. This study showed that the pulverised sludge could be used as a brick material in reducing the firing temperature for the production of energy efficient bricks. The bonding strength can be further enhanced by controlling operating conditions.

48.4 Efficient methods for the removal of chromium from textile effluents

Chromium is a metal that exists in several oxidation or valence states, ranging from Cr(-II) to Cr(+VI). Many effluents from leather industries, textile, dye industries, cement industries give away these toxic heavy metals that disturb our eco system. This section includes the methods to remove this hexavalent chromium. Chemical precipitation, adsorption and biosorption, reverse osmosis, ion exchange, electrodialysis and photocatalysis.

Chromium compounds are very stable in trivalent state and occure naturally in this state in ores such as ferrochromite or chromite ore. Chromium III is an essential nutrient for maintaining blood glucose levels. Hexavalent chromium is the second most stable compound that rarely occurs in nature. It is generally toxic and man made. It is used in many industrial application for its anti corrosive properties and it is used in electroplating. The sources of chromium are aircraft painting, leather tanning, textile manufacturing, dyeing and cement industry. But it is hazardous to health. It is erosive to stomach, causes hemorrhage. Direct eye contact with chromic acid or chromate dusts can cause permanent eye damage. Hexavalent chromium can irritate eye, nose and throat. It can also cause ulcers, allergic reaction, skin rashes and is carcinogenic. It even causes death.

48.4.1 Methods of removal of chromium

Chemical precipitation

Formation of seperable solid substance from a solution, either by converting the substance into an insoluble form or by changing the composition of the solvent to diminish the solubility of the substance in it. The removal of chromium can be accomplished by the addition of ferrous sulphate and lime. Ferrous ion first reduces hexavalent chromium to trivalent chromium by simultaneous oxidation of ferrous ion to ferric. The resulting forms can be precipitated as hydroxides by lime. Chromium is precipitated as hydroxide:

$$Cr^{3+} + (OH)^{-} \rightarrow Cr(OH)^{3}$$

The process requires addition of other chemicals, which finally leads to the generation of a high water content sludge, the disposal of which is cost intensive. Its efficiency is affected by low pH and the presence of other salts (ions) and is ineffective in removal of the metal ions at low concentration.

Adsorption and biosorption

Adsorption is defined as accumulation of liquid or gas phase on the surface of a solid phase. The material that absorbs is called adsorbent and the substance getting adsorbed is called adsorbate. Adsorption may be physical adsorption or chemical adsorption or a combination of both (Fig. 48.4). Sometimes the adsorbent materials have negatively charged ligands that can form complexes with metal ion via electrostatic interactions. If we have to remove soluble

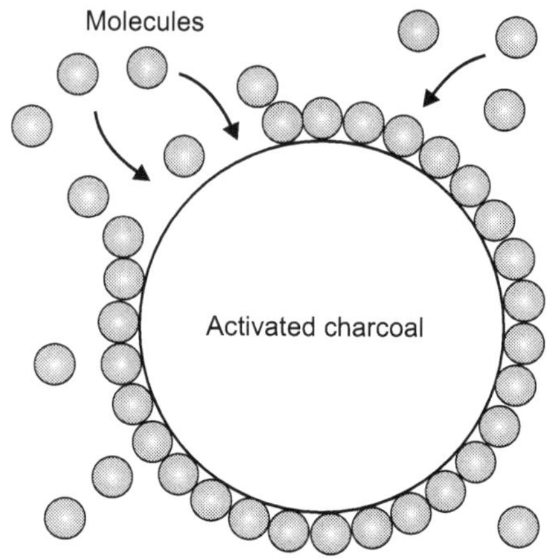

Figure 48.4: Adsorption of chromium by using activated charcoal.

material from the solution phase, but the material is neither volatile nor biodegradable, we often employ adsorption processes. Biosorption is a property of certain types of inactive, dead, microbial biomass to bind and concentrate heavy metals from even very dilute aqueous solutions. It is particularly the cell wall structure of certain algae, fungai and bacteria which was found responsible for this phenomenon. The advantages of biosorption process are that these are cost effective, technically feasible and eco-friendly. This method suffers from low adsorption capacity and less intensity of biosorption.

Reverse osmosis

Reverse osmosis is filteration process that is often used for water. It works by using pressure to force a solution through a membrane, retaining the solute on one side and allowing the pure solvent to pass to the other side. This is the reverse of the normal osmosis process, which is the natural movement of solvent from an area of low solute concentration, through a membrane, to an area of high solute concentration when no external pressure is applied. A semi permeable membrane, like the membrane of a cell wall or a bladder, is selective about what it allows to pass through and what it prevents from passing. These membranes in general pass water very easily because of its small molecular size; but also prevent many other contaminants from passing by trapping them. Water will typically be present on both sides of the membrane, with each side having a different concentration of dissolved minerals. Since the water with the less concentrated solution seeks to dilute the more concentrated solution, water will pass through the membrane from the lower concentration side to the greater concentration side. Eventually, osmotic pressure will counter the diffusion process exactly and an equilibrium will form. The semi permeable membrane can be fabricated by a variety of materials in a way to support a high transmembrane pressure. Generally membranes made up of polyamide are used for the treatment of chromium containing effluent. The reverse osmosis technique has been successfully used in the treatment of electroplating rinse waters, not only to meet effluent discharge standards, but also to recover concentrated metal salt solutions for reuse. Its main demerits are high priced equipment and/or expensive monitoring system, high energy requirement, sludge generation.

Ion exchange

It is based on the exchange of cations or anions on synthetic resins with essential characteristics of its regeneration after the elution of ions. Resins are classified based on the type of functional group they contain:

Cation exchangers: Strongly acidic-functional groups derived from strong acids, e.g., R–SO_3H (sulphonic). Weakly acidic functional groups derived from weak acids, e.g., R–COOH (carboxylic).

Anionic exchangers: Strongly basic-functional groups derived from quaternary ammonia compounds, R-N-OH. Weakly basic- functional groups derived from primary and secondary amines, R–NH_3OH or R–R′–NH_2OH.

About 100% removal of Cr (VI) was achieved in the studies. Its advantages over other processes are the recovery of the metal's value, high selectivity, less sludge volume produced and the ability to meet strict discharge specifications. Its limitations are high operating costs compared to other treatment systems. There can be incomplete removal of the chromium from the salt solution.

Electrodialysis (ED)

Electrodialysis is an electro membrane process in which ions are transported through ion permeable membranes from one solution to another under the influence of a potential gradient. The electrical charges on the ions allow them to be driven through the membranes fabricated from ion exchange polymers. Applying a voltage between two end electrodes generates the potential field required for this. Since the membranes used in electro dialysis have the ability to selectively transport ions having positive or negative charges.

The ion permeable membranes used in eletrodialysis are essentially sheets of ion exchange resins. They usually contain other polymers to improve mechanical strength and flexibility. The resin components of a cation exchange membrane would have negatively charged groups chemically attached to the polymer chains. Polymer chain forms anion permeable membranes, which are selective to transport of negative ions, because the fixed –NR^{3+} groups repel positive ions. The recovery percentage of chromium is quite good, the chromium concentration is not high enough to be cycled to the tanning process. Other problems with electrodialysis are high capital and operating costs involved and the requirement of highly trained human resources. The fouling and scaling of membranes is another drawback which can be controlled to an extent by employing flushing step.

Photo catalysis

Photo catalysis over a semiconductor oxide such as TiO_2 is initiated by the absorption of a photon with energy equal to, or greater than the band gap of the semiconductor producing electron hole pairs.

$$TiO_2 \rightarrow e^-_{cb}\ (TiO_2) + h^+_{vb}\ (TiO_2)$$

Oxidation of water by the hole produces the hydroxyl radical. Similarly O_2 radical are also formed. OH radicals rapidly attack pollutants in solution. The oxidation pathway is not yet fully understood. But OH radical can be formed in two different manners.

$$TiO_2(h_{vb}^+) + H_2O_{ads} \rightarrow TiO_2 + HO_{ads} + H^+$$

$$TiO_2(_{hvb}^+) + HO_{ads} \rightarrow TiO_2 + HO_{ads}$$

Cr(VI) will be reduced to Cr(III) and precipitated out. Photo catalysis has large capability for the removal of trace metals. The drawback of this method is that of being slow compared with traditional methods but it has the advantage not leaving toxic by product or sludge to be disposed.

Result and discussion

500 ml of effluent water is taken for each analysis, the methods are considered in the order for the removal of maximum chromium percentage. The first method, the chemical precipitation shows that it removes 69% of chromium and the second method, adsorption and biosorption to remove 72% of chromium. The further study shows from that reverse osmosis method the chromium removal of 76% is the highest chromium removal method among ion exchange, electrodialysis and photocatalysis method which is discussed in (Table 48.4).

Table 48.4: Removal of chromium using various method in %.

Method	*% present*	*% removed*
Chemical precipitation	80	69
Adsorption and biosorption	80	72
Reverse osmosis	80	76
Ion exchange	80	75
Electrodialysis	80	74
Photo catalysis	80	68

To sum up each method has its own merits and limitation. The versatility, simplicity, cost effectiveness and technical feasibility are a few factors that must be considered while selecting a particular method. High cost and technical complication are the problem associated with reverse osmosis and electrodialysis. Ion exchange is also comparatively costly whereas chemical precipitation leads to sludge generation and involve high capital costs. Photo catalysis process is still in the developmental stage. Adsorption and biosorption are found to be technically uncomplicated as well as economical but the desorption studies on the adsorbents need to be carried out before going for large scale applications.

48.5 Zero liquid discharge in textile industry

In today's business environment, companies are beginning to understand the importance of sustainable operations. With the increasing focus on water usage and water discharge, companies are looking for ways to conserve water consumption and reduce discharge to the environment.

A method that not only helps reduce water usage, but also concentrates on eliminating water discharge is known as Zero Liquid Discharge (ZLD). ZLD processes purify and recycle plant wastewater, changing liquid waste into disposable dry solids and delivering effluent water back into the plant process stream to be reused. ZLD processes can incorporate many different unit operations to eliminate water discharge including cold lime softening, evaporation and crystallisation and evaporation ponds. These technologies can deliver valuable financial returns where water conservation and strict permitting regulations have significantly increased the cost of industrial use.

U.S. water's integrated zero liquid discharge (iZLD) approach combines our chemical and equipment knowledge to design a water re-use system that meets your specific plant needs and goals on both the mechanical and chemical side. A successful iZLD design must take into account both chemical treatment needs and advanced water treatment equipment technologies in regards to your: (i) plant design, (ii) plant operating conditions, (iii) plant personnel resources, (iv) available water conditions and (v) budget realities.

Thus, zero liquid discharge is a process that is beneficial to industrial and municipal organisations as well as the environment because it saves money and no effluent or discharge, is left over. ZLD systems employ the most advanced wastewater treatment technologies to purify and recycle virtually all of the wastewater produced.

48.6 Treatment and disposal of fibre wastes

Recycling and reuse are the most desirable methods of pollution control. In the man made fibre industry water is a basic raw material although it does not form part of the product and with the increasing scarcity of the resource, conservation of it becomes increasingly important.

Much work is being done on solvent dyeing in the absence of water and the use of solvents for cleaning is becoming more common. In the production of the oldest of the man made fibres viscose rayon, control of the process is in many plants achieved by batchwise production of the viscose solution. This in turn leads to waste which can be avoided by continuous manufacture and the difficulties of process control are overcome by more sophisticated instrumentation and the application of improved techniques.

In both instances of course there are other major benefits but the control of waste is a significant factor in the selection of a process route and the plant to be installed.

Almost all processes in the industry have however irreducible wastes which economically are not worth preventing. Often conventional methods require modification and with new products it is possible that quite novel effluents may arise.

Treatment, wherever possible, is designed to lead to recovery. If a chemical is to be removed it is often economical to consider processes to recover it. It may well be that recovery is not immediately viable but in the long term growing scarcity may make it so.

Considering those wastes which need treatment, they can be categorised as below: (i) gaseous effluents, (ii) liquid effluents and (iii) solid wastes.

48.6.1 Gaseous effluents

Gaseous effluents in this industry are in general not a health risk, potentially toxic gases are discharged, after treatment, through tall chimneys with considerable dilution. They can, however, present a considerable nuisance. There is a dilemma too in that good working conditions may require large volumes of air to be extracted from the work rooms. This dilutes the gas to be treated and so increases the size of the treatment plant.

Odours can arise from many textile processes, when the operation is a small one, for example stentering of fabrics, much can be done with simple scrubbers or small activated carbon units. A major problem arises however from the production of cellulose fibre or film from viscose solutions. Carbon disulphide needs to be used at a rate of about 30% by weight of finished product. Since the product contains no sulphur all this material will appear in some form as effluent. During the process, some of the disulphide reacts to become hydrogen sulphide and very small quantities react to form more or less complex sulphur compounds. These latter substances have very low odour thresholds and are believed to be the main cause of the viscose 'reek' for which the industry was at one time notorious.

In a typical factory manufacturing viscose rayon staple the exhaust air would be extracted at a rate of 17000 m^3/min. containing about 600 ppm CS_2 and 300 ppm H_2S, together with detectable quantities of organic sulphur compounds, e.g., thioformaldehyde, carbonyl sulphide, furfural mercaptan, etc. From this whole stream it is necessary to remove as much hydrogen sulphide and related chemicals as possible. It is, however, possible to separate an enriched stream of carbon disulphide of about 8000 m^3/min.

The method of removal of H_2S and related compounds is restricted by the high air flow and comparatively low concentration. The mass emission is such that dispersion through tall chimneys alone may in some units be sufficient. Absorption onto treated activated carbon with subsequent sulphur regain is feasible but not economical. Oxidation by combustion is uneconomical unless part of the stream can be used as the source of air for factory boiler plants though this is clearly limited by the demand of the furnaces and the great care is required to ensure that combustion is complete. Incomplete combustion can produce intermediate organic substances even more odourous than the original contaminants. Oxidation in this way also produces SO_2 which is undesirable although it does not give an odour nuisance.

The methods of choice are therefore limited to chemical oxidation in scrubbers. There are various methods of chemical removal of H_2S. The eventual choice must be governed by the efficiency and cost of the operation. The final choice can only be properly made after extensive pilot plant studies and when the process route has been selected, operating conditions optimised by further pilot studies. The method well-known 'Ferrox' process. The chemistry of the process is superficially simple, the sulphur gases are absorbed in alkaline liquor in the presence of ferric hydroxide acting as a catalyst. Air is blown into the liquor and the sodium hydrogen sulphide converted back to hydroxide with sulphur being liberated.

The hydroxide restores the sodium carbonate/bicarbonate balance and the process is theoretically self supporting. In its simplest form the scrubber is fabricated as a long horizontal chamber, the liquor being sprayed through jets mounted in the side walls. The choice of a horizontal chamber in place of the more common vertical type can be made because of the high pumping cost for some 800 litres/sec of circulating liquor. Spray design and mounting are critical, it is advantageous to have a heavy curtain of liquor broken up by the impinging gases rather than a finely divided liquor spray. The chamber floor is designed to collect and return the liquor to adjacent regeneration (aeration) tanks. After scrubbing the purified air stream passes through eliminator plates, any entrained liquor being released and returned to the circulating tank. The catalyst liquor is prepared by reacting ferrous sulphate and sodium carbonate giving ferrous hydroxide, oxidised to ferric hydroxide before use.

Regeneration is performed by oxidation with atmospheric air the liquor being stirred whilst diffused air is introduced below the stirrer. Alternatively a mixer-aerator of the helical type may be used. The flow of liquor through the tank carries the released sulphur into a trough where it is diluted before going forward as effluent and thus sulphur recovery is possible.

Carbon disulphide (CS_2) does not present so great an odour problem though its removal from air streams is desirable. In some instances it is possible to absorb this material onto beds of activated carbon and subsequently recover it for reuse. Both static and fluidised beds are in use for this purpose. In certain installations it is possible to obtain good removal efficiencies by condensation of rich streams of CS_2. Thus, the only complete oxidation of the substance or its removal are sure remedies for odour complaint. As with other methods of effluent treatment it is important that the system chosen does not itself produce a further problem. Thus it remains common practice to treat sulphur gases in a scrubber using caustic soda to absorb them. The treatment of the resulting sodium hydrogen sulphide is however a major problem.

48.6.2 Liquid effluents

The textile industry in the past has tended to operate in small manufacturing units and liquid effluents arise from many sources, for example during manufacture of the polymer and the dyeing and processing of the finished fabric. The liquid effluents from dyeing and processing are usually rather stronger (i.e., have a higher oxygen demand) than normal domestic wastes but are usually present in comparatively small volume. If, therefore, they do not contain substances which cause difficulties in the public sewer systems or interfere with the biological plant at the treatment works it is normally advantageous to treat them together with domestic wastes at the waste treatment works of the relevant water authority.

This is because the trade waste is normally sterile and the admixture with other wastes which already contain bacteria and nutrients (phosphorus and nitrogen) improves markedly the treatability of the effluent.

If the effluent contains substances which have to be removed before discharge is permitted, such treatment can be difficult and expensive.

The man made fibres industry is developing and subject to frequent innovation. New fashions as well as new materials bring with them chemicals that present new problems in treatment. For example, dealing with sizing compounds, before the resultant effluent can be put to public sewer. The sizing process is typical of the changing pattern of the industry. Sizing is the operation of coating threads, usually warps to facilitate weaving. Over the last decade the traditional size materials, starch, gelatins, gums, etc., have been largely replaced by synthetic polymers such as carboxymethyl cellulose and polyvinyl alcohol. This has led to greatly increased weaving efficiencies when processing man made fibres.

The type of size is crucial since the whole quantity applied is completely washed from the cloth subsequent to weaving and prior to finishing. Polyvinyl Alcohol (PVA) has a very low biological oxygen element (BOD) but a high chemical oxygen demand (COD) and of course the total organic carbon (TOC) is a direct measure of the concentration of chemical present.

Treatment charges at wastewater treatment plants are likely to be based on COD or TOC in the future and since large amounts of size are used to increase productivity, the charges will be proportionately high. Removal of PVA by chemical means is difficult and costly, conventional agents such as lime, alum and ferric chloride being ineffective.

This has led to the investigations of ultrafiltration. Preliminary trials have shown this technique not only successfully removes polyvinyl alcohol (PVA) but it is feasible if there are no other agents present, to recycle the hot permeate and to hopefully reuse the concentrated PVA in sizing.

The system under consideration employs porous carbon tubes to support the semi-permeable membrane which is incorporated in them. The ultrafiltration loops are fitted with two 1000 modules and a main circulation pump capable of transferring 150 L/sec. The loops are inter-connected forming stages for optimum economic operation.

The operating temperature of the last stage is 100°C at a working pressure of 6.9 bar. The flux rates depend upon concentration levels, approximately 0.16 L/sec at 3% and 0·075 L/sec. at 6% per 1000 tube module.

In most liquid wastes from man made fibre production and the subsequent dyeing and finishing the prime constituents are usually soluble organic material. The treatment of these effluents will therefore inevitably include a biological oxidation step, but whereas the design parameters for works dealing with domestic effluent are well known, they are not usually known for textile wastes. The dilemma again occurs, that whereas flexibility is required to accommodate possible changes in effluent composition, certain discharges are quite specific in their treatment requirements. In order to provide maximum flexibility and maintain special conditions if they are required, an activated sludge plant can be constructed for the treatment of dyeing and finishing wastes incorporating two quite different aeration systems. In adjoining aeration chambers two types of aerators are installed. One uses surface aerators, in the other air is supplied to a series of helical plastic distributors. The aeration chambers may be used independently or in series and the facility exists for using them as either extended aeration systems or as conventional high rate chambers. It is also possible to use a modification of the contact stabilisation process by introducing sludge back into the aeration tanks at an appropriate location.

The manufacture of viscose rayon has a particular problem of liquid effluent pollution. It is necessary to incorporate into the process quantities of zinc salts which ultimately appear in the factory effluents. The dilution is such that removal by chemical means is very costly. Simple precipitation with lime creates a difficult solid waste disposal problem and the metal cannot readily be recovered from the sludge. Similarly, if a more expensive alkali are used such as caustic soda the gelatinous sludge of zinc hydroxide cannot be thickened successfully and the recovery of zinc from it is uneconomical. New technologies have been investigated in some detail. The electrolytic winning of zinc from dilute streams has been shown to be feasible on a laboratory scale using the newly developed cell design involving semi-permeable membranes and a rotating cathode. The product is however not sufficiently pure for immediate reuse. The energy requirement of the process too is very high and the practical use of the process is doubtful. Ion exchange processes for zinc removal have been in use for some years, but simple base-exchange resins are limited in their effectiveness by the ratio of sodium/zinc ions in the effluent. They do not function efficiently as zinc exchangers, if this ratio exceeds 12 or 15:1 and viscose factory effluents normally exceed this. The use of the newer chelating resins has, therefore, been studied. In neutral solutions these preferentially absorb zinc ions in the presence of high ionic concentration rations of sodium. At the present time however resin life is apparently too short to proceed to the design of a plant based on this system. Finally the removal of zinc from effluents by extraction in the liquid phase with organic reagents has been attempted on a factory scale, so far unsuccessfully. Operating in a liquid/solid state is clearly more attractive if the capacity of the resin is large enough to permit plants of reasonable size to be constructed.

48.6.3 Solid wastes

The chemical or biological treatment of liquid wastes inevitably lead to the production of a solid waste. Problems of removing surplus water to reduce the difficulties of final disposal are in general common to all industries. Again no generalisation can be made, no two biological sludges are exactly similar and certainly chemical sludges vary widely in their composition. The approach is therefore often empirical, laboratory tests such as capillary suction times, sludge volume indices, specific filterability indices, etc., can give indications. The relative merits of rotary vacuum filtration, centrifuges, conventional filtration or the newly developed filter belt presses are best compared with trials if possible on site. The disposal of solid wastes has become a major national problem, In general the man madefibre industry does not generate unusually difficult or toxic solid wastes, nor are the amounts involved as large as in many other manufacturing processes.

In common with many industries, however toxic substances, e.g., wet fibre containing residual CS_2 are produced in small quantities. Incineration in such cases is not possible and other wastes can be dangerous or give rise to toxic fumes if burnt. Where possible, waste of this nature is disposed of by controlling landfill at special sites. Some wastes, e.g., lactam wastes from the manufacture of nylon polymer have very high oxygen demands if leached out by rainwater. If these are disposed of by land-fill great care is needed in the selection of sites and treatment of the wastes.

It will be seen that over the whole spectrum of pollution and throughout the industry the disposal of wastes is dealt with in a pragmatic way. The industry is very diverse both in its products and in the size of the manufacturing units. Also it is a developing industry and the nature of the pollution can change very significantly and unpredictably. The special problems of odour control, the control of dyestuffs and new auxiliary compounds and zinc removal, etc., require considerable research effort before completely satisfactory techniques can be developed.

In addition to the technical matters referred to briefly in this section there is wide scope for managerial control including good housekeeping, adequate separation of heavily contaminated process streams, a properly designed procedure for preventing as far as possible accidental discharges, etc.

48.7 Pollution aspects in jute mills

48.7.1 Air pollution

Boiler

The major source of air pollution from a jute mill is the emission from the stack of the furnace used for the boiler. Coal is the principal fuel burnt in the furnaces used for raising steam. Jute waste/jute dust mixed with mineral oil collected during floor sweepings are also used as auxiliary fuel. The composition of the fuel and the performance efficiency of the furnace primarily determine the composition of stack emission.

Diesel generator set (DG)

The other source of air pollution in jute processing industry is the DG set. Power shut downs are quite common in the industrial belt where the mills are situated.

Miscellaneous sources

Fugitive emission: The third source of air pollution from jute mill is dust blown up in the ambient air by air currents. Dust and jute fibres present on floor, machine tops, beams and rafters, they settle down on any resting place they find. Unless these are regularly removed and good house-keeping is observed, a strong wind is likely to lift them up and blow the dust into the atmosphere. Jute fibres are seen outside the mill shed as well. They too can be blown up by wind movement. Air pollution caused by such casual wind movement is of a minor nature. Good house-keeping, indoor and outdoor of the mill sheds, will minimise if not eliminate the pollution on this account.

In certain mills local exhaust ventilation is employed for control of dust and air contaminants in the working environment inside the mill shed. Exhaust air containing the dust and fumes is discharged outside the mill building. Unless the dust and the gaseous pollutants are separated from the exhaust air, the discharge of the exhaust air will add to the pollution of the ambient air. Suitable collectors, such as bag filters cyclones, etc., need to be incorporated in the local exhaust system.

Dust fall: The dust discharged in the ambient air is carried by wind over long distances. Dust particles will settle down eventually. The distance and direction these dust particles are conveyed depend on several factors amongst which the meteorological data—such as humidity, precipitation, velocity and direction of wind movement are the major ones.

The following measures are recommended:

1. Use of coal for:
   (a) Low sulphur content, (less than 0.4%).
   (b) Low ash content, (less than 40%).
2. Jute waste/jute fines which are used as an auxiliary fuel in the furnace be sieved properly before use to remove inorganic dust.
3. Improve furnace performance by proper maintenance.
4. Improve the performance of the DG sets.

48.7.2 Water pollution

Wastewater characterisation

Liquid effluents originate from manufacturing processes and utilities. The individual waste streams from various sources have distinct characteristics.

The flow, type and concentration of pollutants vary widely depending upon process and water use. In this section, the factors contributing to the variations in waste flow and its characteristics in respect of different sources are analysed to develop a waste profile for the industry. The development of waste profile involves logical interpretation and analysis of 'the data from literature, questionnaire survey and indepth study of operating service.

Water use pattern

Water is consumed for the following purposes in a jute mill:

1. Jute processing.
2. Production of steam in boilers.
3. Cooling system, gardening and fire fighting.
4. Plant domestic use.

It is observed from the manufacturing process that, in jute processing, water is required mostly in make-up and steam generation in different unit operations mainly in hatching, fixing and finishing (calendering and damping).

In jute processing industries, the water consumption for utility section like boiler make-up, cooling, pond make-up, fire fighting and gardening is not properly assessed. However, it should not exceed 1.0 cubic metre/T.

Pattern in water consumption: It is observed that variation of water consumption in mills are so wide that there is no direct relationship between production and water consumption.

Wastewater generation

Industrial waste: Liquid wastes are generated from the following sections in conventional jute processing: (i) batching, (ii) sizing and beaming, (iii) finishing and (iv) sanitary block, canteen, etc.

Characterisation of wastewater

It is understood that the two major streams coming out from jute processing industries are from sizing and batching operations. It is, therefore, necessary to study the waste streams of batching and sizing, alongwith the combined streams. Samples were examined for 26 parameters covering the expected physico-chemical characteristics in waste streams and the combined one in jute processing industry and are summarised. Thus, it is seen that batching and sizing effluent does not differ much except in case of oil and grease.

It is noticed that in sizing effluent, the pH is low (6.8 to 6.9) and fluoride is present. Combined waste reflects that BOD-COD ratio and carbon to nitrogen ratio favours biodegradability and biological treatment. The wastewater too contain no toxic element to be eliminated. But the sodium content is high and thus needs to be considered if these wastewater is to be used for irrigation. In summary, the wastewater has low organic content and no toxic element, hence it is biodegradable.

However, before using it for irrigation, sodium content and conductivity values need to be assessed properly with the combination of soil characteristics and crop-choice.

48.8 Carbon footprint in textile industry

The term 'carbon footprint' has become a topic of hot discussion all over the world. Carbon foot print can be described as the extent of damage caused to the environment due to some actions. It is the measure of severity of our activities on the environment, especially on the climate change. Many of the activities in our everyday life produce emissions, through the burning of fossil fuels for electricity, heating, etc. These activities have carbon footprint, producing large amount of greenhouse gases, causing a disastrous effect on the environment.

Greenhouse gases and global warming: Greenhouse gases are produced by human activities, which result in global warming. Carbon dioxide is a major gas that accounts for almost 80% of the emissions. Burning of fossil fuels, oil, natural gas and petrol releases carbon dioxide, methane, nitrous oxide, sulphur hexafluoride, perfluorocarbons, etc. are a few other greenhouse gases originating from industrial processes. These gases accumulate and absorb infrared radiation from the atmosphere, affecting the balance between energy received from the sun and the energy that escapes.

The green house gas emission is caused by the production and consumption of fuels, manufactured goods, materials, wood, roads and services. For simplicity of reporting, it is often expressed in terms of the amount of carbon dioxide, or its equivalent of other GHGs emitted. Just as walking on the sand leaves a footprint, burning fuel leaves carbon dioxide in the air, which is called a 'carbon footprint'. Thus the carbon footprint basically relates to the amount of carbon released into the air based on the fuel consumption.

The carbon footprint is assessed in 2 layers:

1. Primary footprint - monitors carbon emission directly through energy consumption - burning fossil fuels for electricity, heating and transportation, etc.
2. Secondary footprint- relates to indirect carbon emissions (life cycle of products and sustainability).

48.8.1 Reducing of carbon foot print

Thus, the most effective way to decrease a carbon footprint is to either decrease the amount of energy needed for production or to decrease the dependence on carbon emitting fuels.

Machinery/equipment related

1. Use of low and ultra low liquor ratio machines–to reduce consumption of water during pre-treatment, dyeing and post dyeing wash-off sequence. Simultaneously reducing the energy required for water heating at various processing steps and effective load on the effluent treatment.

2. Pre-heating of process water by solar panels to reduce consumption of other non renewable energy sources (fossil fuels, wood, husk, etc.).
3. Adequate insulation of dyeing, drying and stenter machines and appropriate heat recovery systems to avoid undesired energy loss.
4. Recycle and reuse of process water and alkali by installing adequate filtration process.

Chemicals and dyes

1. Use of enzymes – biodegradable and non corroding for desizing, scouring, bleach neutralising, biosoftening and post dyeing wash off. Suppliers and formulators of enzymes are offering specialised products for combined processes to reduce number of processing steps.
2. Cationisation of cotton for salt-free dyeing with reactive and direct dyes.
3. High fixation reactive dyeing with reduced salt for exhaustion.
4. Digital ink jet printing.
5. Low temperature curing pigment printing.

48.8.2 Creating new green paradigm

To create new green paradigm the textile and apparel industry needs to adopt 3R concept, i.e., reduce, reuse and recycle.

Reduce

Low carbon foot print processes cut costs by reducing waste of raw materials and energy. Water and energy usage reductions by the textile dyeing and finishing sector can help reduce global carbon dioxide emissions. By saving energy and water, the textile industry cannot only save a lot of money, but also help to slow down climate change. Substituting organic fibres for conventionally grown fibres as it uses less energy, no petrochemical-based fertilisers and pesticides for production, emits fewer GHG and supports organic farming (which has myriad environmental, social and health benefits). Other 'greener' alternatives include organic wool, linen, bamboo, hemp, abaca, soyabean fibre, biopolymers and polyester recycled from used clothing.

Reuse

Effluents of chemically treated textiles are discharged in water. Treatment of wastewater obtained from chemically treated textiles is a must. Use of chrome mordant dyeing and limiting the emission of copper, chromium and nickel into water reduces impurities in dyes and pigments. Using dyeing carriers with high chlorine content should be evaded. During the process of bleaching, alternative agents that are less or not hazardous can be used.

Households currently throw out 1.17 million tonnes of textiles each year, most of it clothing, which could be recycled or reused. It's therefore important, whether as designers, retailers or consumers, that we begin to tackle some of these issues that have been highlighted today.

Recycle

The textile and apparel industry should more utilise recycled fibres. The environmental impact of recycling worn-out polyester or cotton waste into new polyester or cotton fibre respectively, for instance, is significantly lower than making that same fibre a new. A wide range of innovative, sustainable clothing can be made from recycled textiles.

We should take care of the ways to combat 'fast-fashion' and to reduce its negative environmental impact as the issues of textile recycling, cheap clothing or 'throwaway fashion' affects us all.

1. Eco circle environmentally friendly closed-loop recycling system chemically converts used polyester products into new polyester raw materials.
2. It is biodegradable and recyclable and the production process is more sustainable and includes recovery of most chemicals.

48.8.3 Creating a carbon free environment

Role of trees in offsetting emissions: Trees are the gift of nature to filter our air. They absorb carbon-dioxide and release oxygen. Apart from filtering the air, sustainably managed forests aid multiple environmental and socio-economic functions which will be crucial at the global level in creating a sustainable development.

Treating textile effluents: Effluents of chemically treated textiles are discharged in water. Treatment of wastewater obtained from chemically treated textiles is a must. Use of chrome mordant dyeing and limiting the emission of copper, chromium and nickel into water reduces impurities in dyes and pigments. Using dyeing carriers with high chlorine content should be evaded. During the process of bleaching, alternative agents that are less or not hazardous can be used.

Adopting nature's way of life: Natural fibres are a gift to mankind. Using these fibres makes recycling easy and quicker and is also environmentally friendly. Fibres from linen, bamboo and rice straw are a few to name.

Organic cotton: Using organic cotton is integral, as organic soil scrub the atmosphere to global warming gases by trapping the carbon dioxide and converting the same into soil material. In the global warming arsenal, organic farming is a vital tool.

48.8.4 Educate consumers to change attitude

Consumer education about the huge carbon footprint mainstream textiles have and to help inspire consumers to change their habits is a must. It also assists in changing consumer attitudes.

Their inclination towards 'organic fabrics' not simply fabric made from organic fibres, eco-friendly fibres, not cotton or synthetics; minimising purchase of fabrics that are blends of natural and synthetic fibres (i.e., cotton and polyester), or blends of two or more different synthetic fibres (polyester and acrylic), because there is no hope of recycling these fabrics right now.

48.8.5 Low-carbon manufacturing programmes

Low-carbon manufacturing programmes and carbon accounting in factories, carbon footprint calculation projects, benchmarking energy consumption across the textile and apparel supply chain are few of the strategic measures required to reduce carbon footprint of textile and apparel industry.

49

Pollution control in tannery industry

49.1 Introduction

Tanning is the chemical process that converts animal hides and skin into leather and related products. The transformation of hides into leather is usually done by means of tanning agents and the process generates highly turbid, coloured and foul smelling wastewater. The major components of the effluent include sulphide, chromium, volatile organic compounds, large quantities of solid waste, suspended solids like animal hair and trimmings. For every kilogram of hides processed, 30 litres of effluent is generated and the total quantity of effluent discharged by Indian industries is about 50,000 m^3/day. The various components present in the effluent affect human beings, agriculture and livestock besides causing severe ailments to the tannery workers such as eye diseases, skin irritations, kidney failure and gastrointestinal problems.

Tannery waste material also varies considerably in volume and concentration due to continuous operation and intermittent discharge. The composition of tannery wastewater is shown in Table 49.1.

Table 49.1: Composition of tannery wastewater.

BOD (mg/L)	210–4300
COD (mg/L)	180–27000
Total suspended solids (mg/L)	925–36000
Total chromium (mg/L)	3–350
Sulphides (mg/L)	1–500
Chlorides (mg/L)	1500–28000
Total phenolic compounds (mg/L)	0.4–100
Ammonium nitrogen (mg/L)	17–380
Kjehdahl nitrogen (mg/L)	90–630
Fats and oils (mg/L)	49–620
pH	1–13

Sulphide is one of the major components of the tannery effluent. It causes an irritating, rotten-egg smell above 1 ppm (1.4 mg m^3) and at concentrations above 10 ppm, the toxicological exposure limits are exceeded. It is highly toxic to human beings. It can cause headaches, nausea and affect central nervous system even at low levels of exposure. It causes death within 30 min at concentrations of only 800–1000 mg/L and instant death at higher concentrations. The upper concentration limit of sulphide in water intended for human consumption is 250 mg/L. The corrosive properties of sulphide are apparent in the damage done to concrete walls of reactors, sewer systems and steel pipelines. It also inhibits the methanogenesis process. Soluble sulphide ranging from 50–100 mg/L can be tolerated in anaerobic treatment with little or no acclimation. Sulphide has high oxygen demand of 2 mols O_2/mol sulphide and causes depletion of oxygen in water.

Chromium salts used during the tanning process generate two forms of chrome; hexavalent chromium and trivalent chromium. Hexavalent chromium is highly toxic to living organisms even at low concentration causing carcinogenic effect. Trivalent chromium may be present in the waste or can be produced from the hexavalent chromium by chemical treatment. Soluble trivalent chromium causes toxicity in anaerobic digestion due to the accumulation of the metal in the intracellular fraction of biomass. Several components in the effluent contain nitrogen as part of their chemical structure, which can lead to development of anaerobic conditions harmful to the aquatic life.

The environmental protection regulations stipulate that industries are not allowed to emit sulphide and chromium in the wastewater. Thus, removal of sulphide and chromium from the wastewater is very important. A number of researchers worked on the removal of sulphide and chromium from the wastewater streams, but little has been reported on the sulphide removal from the tannery wastewater.

49.2 Manufacturing of leather

The manufacturing of leather can be divided into two parts; beam house operations and tanning process. In beam house operations, the removal of dirt and blood by washing is the first step after which the hides are then soaked in water for softening and removal of salts. After the removal of salts, fatty tissue is removed by fleshing. Liming is done to swell the hides for the better penetration of tanning agents and hair removal. Chemical dissolution of the hair and epidermis with an alkaline medium of sulphide and lime takes place. During liming, a high concentration of sodium sulphide, lime and organic matter is delivered to the effluent. Hides are then neutralised with acid ammonium salts and treated with enzymes to remove the hair remnants and to degrade proteins. This results in a major part of the ammonium load in the effluent. Pickling is usually done to prepare the hides for tanning. The pH value of hides is adjusted by addition of acids (mainly sulphuric acid). Salts are added to prevent the hides from swelling.

Tanning is the reaction of the collagen fibres in the hides with tannins, chromium, alum or other chemical agents. Alums, syntans, formaldehyde, glutaraldehyde and heavy oils are used as tanning agents. During the tanning process, about 300 kg chemicals are added per ton of hides. Based on the tanning agents, tanning operations are further divided into vegetable tanning and chrome tanning. Vegetable tanning is usually done in series of vats by using natural organic substances.

Chrome tanning is done at a higher pH using chromium salts. After tanning, tanned leather is piled down, wrung and graded for thickness and quality, split into flesh and grain layers and shaved to desired thickness. In chrome tanning, retanning, dyeing and fatliquoring are the additional steps as compared to the vegetable tanning. Fatliquoring is the process of introducing oil into the skin before the leather is dried to replace the natural oils lost in beam house and tan yard processes. After drying, a number of finishing operations like buffing, plating and embossing are carried out to make the leather softer and aesthetic.

49.3 Sources of wastes from tannery

Of the wet salted hide purchased by the tannery only 30–35% is potentially convertible into leather and of this amount, only some 30% is eventually converted into high quality leather with further 10% yielding lower grade material. The rest of the hide together with excess processing chemicals and the large volumes of water employed form the solid and liquid residues.

49.3.1 Gaseous wastes

Tanneries discharge into the atmosphere odorous gases, smoke and dust. The main source of smells in a tannery are the compounds containing nitrogen and sulphur. The end products of anaerobic decomposition or putrefaction of proteins include indole, skatole, mercaptans and miscellaneous aldehydes, all of which are odorous.

The other odour producing compounds in tanneries include sulphide, fatty acids like butyric acid, valeric acid and caproic acids, solvents, lacquers, formalin and some of the chemicals used in finishing operations. Smells in tanneries intensify from unhygienic practices in skin and hide processing and delayed disposal of liquid and solid wastes. In many tanneries it is the foul odour which emanate from the putrescible solid and liquid wastes which account for much of the smell traditionally associated with the tanneries.

Immediately after the hide or skin is removed from the animals, decay starts unless it is cured properly. The operations like soaking, liming, deliming, fleshing, etc., are the most disagreable steps in leather manufacture. They involve use of bad smelling materials and production of putrescible organic matter like soak pit sludge, lime sludge, green fleshings, limed fleshings. During soaking, the removal of curing salt and dehydration of the skin introduces the possibility of bacterial growth and protein putrefaction. Many unhairing systems in practical use are based on balance between sodium sulphide, sulphohydrate, dimethylamine sulphate and sodium hydrosulphide. Some of these sulphides have potential to liberate hydrogen sulphide when mixed with pickling liquors. Hydrogen sulphide is an extremely bad smelling gas which can be detected in concentrations as low as 0.1 ppm. Hydrogen sulphide and organic sulphides cause odour nuisance when present in the air at concentrations of 10 to 150 times smaller than the lowest concentration of sulphur dioxide detectable by smell. There is a possibility of hydrogen sulphide generation during deliming from the sulphides sticking to hide if the limed hides are not properly washed. Indiscriminate disposal of liquid and solid wastes generated during tanning operations may also give rise to considerable smell. Solid wastes produced during beam house operations are highly putrefactive.

The composite liquid waste let out from a tannery contains large amounts of putrescible organic matter and is highly polluting in nature. Although the relatively high pH of the composite wastes tends temporarily to inhibit putrefaction, the wastewaters are inherently putrescible and eventually become highly offensive. Sludge produced during primary or biological treatment of tannery effluents can also give rise to smells.

49.3.2 Liquid wastes

In the development of the tanning industry, water plays a vital role as the industry consumes large quantities of water. Approximately 30–40 litres of water is used for processing one kg of raw hide/skin into finishing leather. Most of the Indian tanneries which are located near the river banks or natural water bodies draw surface water. Ground water from their own open wells/tubewells existing within their premises is also used by some tanneries. Most of the traditional tanneries do not have overhead water tanks for proper distribution system. Water is being pumped directly to the process and in a few tanneries, it is stored in open cement lined pits and ground level tanks.

In general, the quantity of water usage and nature of wastewater discharge varies from process to process and tannery to tannery and from time to time. Most of the discharges are intermittent.

The average water usage and wastewater discharge per kg of hide/skin for different process are as follows: (i) raw to E.I – 25–30 L/kg of raw weight, (ii) raw to wet blue – 25–30 L/kg of raw weight, (iii) raw to finish – 30–40 L/kg of raw weight, (iv) E.I. to finish – 40–50 L/kg of E.I. weight and (v) wet blue to finish – 20–25 L/kg of wet blue weight.

Most of the tanneries neither have proper drainage system for collection of the wastewater nor any effluent treatment system. The wastewater is discharged from various sectional operations intermittently and it takes its own course to the nearby low-lying area neighbouring land, pond, street, roadside, etc.

Characteristics of wastewater

Characteristics of the effluents vary from tannery to tannery and in any one tannery with respect to time. The wastewater from beamhouse process, viz. soaking, liming, deliming, etc., are highly alkaline, containing decomposing organic matter, hair, lime, sulphide and organic nitrogen with high BOD and COD. The wastewater from tanyard process, viz. pickling, chrome tanning are acidic and coloured. Vegetable tan wastewater contain high organic matter. The chrome tanning wastes contain high amounts of chromium mostly in the trivalent form.

Characteristics of the composite wastewater

The characteristics of the composite wastewater is governed by the following factors:

1. Intermittent discharge of wastewater from different sectional operations.
2. Wide variation in the volume and quality of wastewater from section to section.
3. Partial operations in one tannery and balance operation in another tannery.

49.3.3 Solid wastes

Solid wastes generated in a tannery can be classified under four heads, viz.

1. Non-proteinous wastes.
2. Non-collagenous protein wastes.
3. Untanned collagen.
4. Tanned collagen.

Some of these solid wastes or by-products find consumers in the gelatine, glue, textile and artificial leather industries. It is estimated that 33% of the original protein material leaves the tannery in the form of effluent sludge, fleshing, trimmings, shavings, etc., of which the greater part would appear as solid wastes. Solid wastes may be produced in a tannery in varying quantities from the following processes.

The salted, 'green' or dried hides are trimmed to remove irregular matter which will produce raw trimmings. Salted hides are then dusted. This operation will produce salt dust containing impurities like blood, dung, etc. Following soaking, the skins or hides are again trimmed. The method of (using lime sulphide, enzymes, or dimethylamine) will determine whether the hair will be recovered as solid waste or pulped into the effluent. Lime sludge is produced in the lime pits which is either discharged into the effluent drains or removed as solid wastes. After liming, hides and skins are fleshed, producing a waste of flesh with a certain amount of hide pieces. The dehaired and fleshed pelts are then passed through deliming, pickling and tanning operations. The tanning material may be vegetable, chrome or other agents. In case of vegetable tanning, solid wastes generated are spent tan liquor sludge. After tanning, the leather undergoes a number of machine-shop operations, such as shaving, splitting, buffing, trimming, etc., resulting in tanned solid wastes such as split pieces, trimmings and buffing dust. Sludge is obtained as solid waste from equalisation and settling of wastewaters from different sections of tannery and the biological treatment of tannery wastewaters. All these solid wastes have to be either utilised or disposed off safely without affecting the environment.

49.4 Wastewater treatment in tannery

Various physicochemical techniques used for wastewater treatment can be applied to tannery wastewater (to the entire process or to individual step in the process) but these processes are expensive. Biological treatment of wastewater is more favourable and cost effective as compared to other physicochemical methods. Various micro-organisms are capable of reducing the content of pollutants significantly by utilising them as energy and nutrient source in the presence or absence of oxygen.

49.4.1 Aerobic treatment

Aerobic micro-organisms use organic carbon in the effluent and convert it to biomass and carbon dioxide. A large amount of sludge is generated along with high energy consumption in the process.

Aerobic treatment of tannery wastewater reduces chemical oxygen demand (COD) by 60–80% and biological oxygen demand (BOD) reduction is 95%, when combined with physicochemical pre-treatment.

In a combined biochemical oxidation and chemical ozonation step, chemical oxygen demand (COD), Total Kheldal Nitrogen (TKN) and total suspended solids (TSS) removals of 96%, 92% and 98%, respectively were obtained. Ozonation step was integrated with sequencing batch biofilm reactor. Ozonation partially oxidises the refractory compounds present in tannery wastewater and increases their biodegrability. Sludge production was 0.1 kg VSS/ kg COD removed, which is lower than the value used in conventional biological systems. Aerobic treatment followed by chemical ozonation and again aerobic treatment further increases the biodegrability of refractory compounds.

A combination of electrochemical and biological treatment can also be used to eliminate ammonia and avoid implementation of biological nitrification.

Respirometry combined with sequencing batch reactor is an effective method for the removal of COD in tannery effluent. At 12 h sequencing batch reactor cycle with a loading rate of 1.9–2.1 kg/m^3 day, removals of COD, TKN and NH_3-N were 80–82%, 78–80% and 83-99%, respectively. The removal efficiencies were much higher than conventional aerobic systems.

49.4.2 Anaerobic treatment

Anaerobic treatment of wastewater converts the organic pollutants into a small amount of sludge and large amount of biogas (methane and carbon dioxide).

The sulphide present in wastewater inhibits the anaerobic treatment. Table 49.2 shows the effect of sulphide formation in anaerobic reactors.

Table 49.2: Effect of sulphide formation in anaerobic reactors.

Disadvantages	*Advantages*
Reduced COD removal efficiency	Removal of sulphate, sulphite and thiosulphate from the waste stream
Corrosion	Heavy metal removal
Accumulation of inert material in the sludge (e.g., metal sulphides)	Precipitated metal sulphides
Less methane formation	
Poor biogas quality + need for H_2S removal from the biogas	
Malodor	
Potential toxicity	
Disturb the aerobic activated sludge post treatment system (activated sludge bulking; excessive growth of phototrophs)	

Methanogenic bacteria are inhibited by sulphide, whereas acidifying and sulphate reducing bacteria do not inhibit. Three inhibiting effects of sulphide or sulphide reduction are known: direct toxicity of sulphide, substrate competition between sulphate reducing and methanogenic bacteria and precipitation of trace elements by sulphide. The extent of these effects depends on the experimental system used.

In a continuous fixed film reactor, the efficiency of degradation was improved by 15% at a hydraulic retention time of 1.9 days when the concentration of undissociated sulphide was reduced from 100 to 30 mg/L. In two stage anaerobic treatment of tannery wastewater, 30% of the sulphate was reduced independent of the sulphate influent concentration in the first stage. With high concentration of sulphate in influent, percentage of desulphurisation decreased in the second stage.

Pre-treatment of wastewater for reducing the tannin, chromium and sulphide levels gives better results in COD removal efficiency. In anaerobic up flow contact filter, COD removal efficiency was in the range of 79–95% after pre-treatment compared to 60–86% for untreated wastewater. The biogas production was in the range of 95–198 mL/hr after pre-treatment compared to 98–200 mL/hr for untreated wastewater. In batch process, 60 mg/L of sulphide,

60 mg/L chromium and 400 mg/L of tannin inhibited microbial growth whereas in continuous process, toxicity occurs at higher range.

In up flow fixed biofilm reactor, 60–75% COD removal and 0.36 m^3/kg COD removed methane yield has been obtained. It has been found that porous polyurethane foam material is more suitable than Raschig rings as a micro carrier in fixed film reactor.

Sulphur recovery unit integrated with UASB up flow anaerobic sludge blanket reactor for the treatment of tannery wastewater resulted in stripper efficiency of 65 to 95% in terms of sulphide removal. Sulphur recovery unit consisted of a stripper column, absorber column, regeneration unit and sulphur separator.

Up flow anaerobic sludge blanket reactor with activated sludge reactor without recirculation of the sludge gave 96% COD removal at 8 days of hydraulic retention time and 71 g/L of total dissolved solids.

In up flow anaerobic sludge blanket reactor, average COD removal efficiency is 65% at organic loading rate in the range of 0.2–7 kg/m^3d. The average gas production is 0.3 m^3/kg of COD removed.

A combined aerobic and anaerobic treatment of tannery with 900 mg/L dissolved organic carbon corresponding to discharge of 23 kg/T of raw hides gave 85% removal efficiency of the dissolved organic carbon.

49.5 Sulphide removal methods in tannery

Use of sodium sulphide and sodium hydrosulphide in tannery for dehairing the hides result in the sulphide content varying from 10–5000 mg/L. Sulphate generated during the process also gets converted into sulphide during anaerobic treatment. In the absence of dissolved oxygen and nitrate, sulphate reducing bacteria converts sulphate into sulphide In order to remove sulphide from wastewater streams, a number of physicochemical methods like direct air stripping, chemical precipitation and oxidation are in common use today. Many of the metals such as iron, zinc, copper, etc., can be used to precipitate the sulphide into insoluble metal sulphide.

Oxidation processes used for sulphide removal are aeration (catalysed and uncatalysed), chlorination, ozonation, potassium permanganate treatment and hydrogen peroxide treatment. During the sulphide oxidation by aeration, there is some loss of sulphide directly in the atmosphere. Sulphide consumes the oxygen in the aerator and reduces the effectiveness of the equipment. In chlorination, chlorine reacts with certain metals and organic matter in the water to form hazardous chlorinated organic chemicals. Catalytic chemical oxidation of the sulphide with air removes the sulphide quantitatively but it is a time consuming and expensive process.

Following reaction takes place, when potassium permanganate reacts with hydrogen sulphide:

$$3\ H_2S + 4\ KMnO_4 \rightarrow 2\ K_2SO_4 + S + 3\ MnO + MnO_2 + 3\ H_2O$$

Hydrogen peroxide oxidises the hydrogen sulphide to elemental sulphur via following reaction:

$$8\ H_2S + 8\ H_2O_2 \rightarrow S_8 + 16\ H_2O$$

In all these processes, apart from elemental sulphur, sulphate and thiosulphate are generated, which are difficult to separate.

For the removal of hydrogen sulphide from the gas streams, various methods are reported in the literature. In commercially used Claus process, hydrogen sulphide is oxidised with air to produce sulphur dioxide. The mixture of H_2S/SO_2 is then passed over bauxite catalyst to yield elemental sulphur and water. The relatively high energy requirements or the high chemical and disposal costs besides environmental problems constitute important drawbacks of these methods. Therefore, alternative techniques for the hydrogen sulphide removal are required. Partial biological oxidation of sulphide to sulphur is a cheap alternative, which also allows sulphur reclamation, since sulphur is non-soluble and thus, it can be removed from the wastewater. The partial oxidation of H_2S to elemental sulphur instead of sulphate has several advantages. Elemental sulphur is a non-toxic, non-corrosive solid containing more sulphur per unit mass (3–8 times more valuable than H_2SO_4). Moreover, the elemental sulphur generated during the process can be used as a feedstock for the chemical, fertiliser and materials manufacturing industries.

Several micro-organisms have been studied for application in biotechnological hydrogen sulphide removal processes. The following (biological) overall reaction occurs in an aerobic sulphide removal system:

$$2\ HS^- + O_2 \rightarrow 2\ S^\circ + 2\ OH^-$$

$$2\ S^\circ + 3\ O_2 \rightarrow 2\ SO_4^{2-} + 2\ OH^+$$

Aerobic micro-organisms or chemotrophes used for the oxidation of hydrogen sulphide are the species of *Thiobacillus*, *Pseudomonas*, *Beggiatoa* and *Thiothrix* which have been thoroughly studied by various researchers. These micro-organisms use inorganic carbon as a carbon source and chemical energy from the oxidation of reduced inorganic compounds.

The simpler nutritional requirements and higher sulphide tolerance of chemotrophic organisms favoured their application in biological sulphide oxidation. Various *Thiobacillus* species are widely used in conversion of hydrogen sulphide to sulphur on the laboratory scale, i e. *T. thiooxidans* and *T. ferroxidans* grow at low pH 1–6 while *T. denitrificans*, *T. thioparus* and *T. novellas* can grow at pH 6–8. Sublette and Sylvester focused on the use of *Thiobacillus denitrificans* for the oxidation of sulphide to sulphate.

Buisman and others used a mixed culture of *Thiobacilli* for the aerobic oxidation of sulphide to elemental sulphur. Smith and others used three phase fluidised bed bioreactor with *Thiobacillus* sp. IW immobilised on activated carbon and showed 94% hydrogen sulphide removal efficiency in the concentration range of 100–200 ppm with flow rate of 1–2 L/m.

A biofilter system immobilised with *Thiobacillus* species showed a 95% removal efficiency of hydrogen sulphide at a gas flow rate up to 93 L/h with an inlet concentration of 60 ppm, but efficiency was reduced to 78% for a gas flow rate of 180 L/h. Visser and others found *Thiobacillus* sp. W5 as a dominant organism in the process. Collin and others showed nearly 100% removal efficiency of hydrogen sulphide with sulphide loading rate up to 19 kg/m^{3}d in a reverse fluidised loop reactor inoculated with *T. denitrificans*. The efficiency declined to 90% with accumulation of 15–18% thiosulphate for sulphide loading rate of 30 kg/m^{3}d.

Anaerobic micro-organisms or phototrophs such as Chlorobium and Chromatium can be used for the conversion of hydrogen sulphide into sulphur. Phototrophs use carbon dioxide as a carbon source and light as an energy source. Borkenstein and Fischer obtained 98.7% sulphide removal efficiency with 60.4% sulphur recovery by using purple sulphur bacteria *Allchromatium vinosum* strain 21 D. Major disadvantages in using photosynthetic bacteria on a large scale lie in their anaerobic nature and their requirement for strong light source. Also, phototrophic bacteria generally store the produced sulphur internally, making a separation of cells and sulphur impossible.

Kleinjan and others suggested that only those micro-organisms should be used for the sulphide removal that can store sulphur extracellularly for easy separation of sulphur. Thus, tannery wastewater is difficult to treat because of complex characteristics like high BOD, COD, suspended solids, sulphide and chromium. The main source of sulphide in tannery effluent is beam house operations. Anaerobic treatment of tannery wastewater gives better results but formation of sulphide in anaerobic reactors restricts its application.

Various phototrophes and chemotrophes have been used for sulphide removal but requirement of light source is the major problem in case of phototrophes. Chemotrophes require careful control of oxygen to produce sulphur instead of sulphate but still some sulphate formation is there.

49.6 Impact of pollutants on environment

49.6.1 Impact of gaseous pollutants on environment

The problems associated with gaseous pollutants from a tannery include noxious smell, hydrogen sulphide and dust. The effect of noxious smells on people is primarily a nuisance effect. The loss of property values near poorly operated tanneries is partly a consequence of offensive odours. Dust problems normally arise in tanneries from leather buffing operations. Leather dust of finer sizes are reported to be harmful to human health and comfort. Hydrogen sulphide which is liberated during some of the tanning operations is an irritant gas and has a very bad smell. The maximum allowable concentration for 8 hours exposure in working areas is 20 ppm.

49.6.2 Impact of liquid wastes on environment

Tannery effluents contain vegetable tannins and non-tannins which exert oxygen demand. They also contain high amounts of protein, especially when a hair pulping unhairing system is used. These proteins are biologically degradable and exert high BOD. About 79–80% of the total dissolved volatile solids in the settled and filtered composite tannery wastes is composed of organic matter present in the form of proteins, fatty acids, ether solubles and tannins and 20–21% composed of other organic compounds.

Chlorides, trivalent chromium, nitrogen, phosphorous, sulphate, ammonium salts, lime, etc., are the inorganic pollutants present in significant quantities. These pollutants are of more permanent nature, unless they are in suspended state or precipitated from solution and settled. Using normal tanning techniques chromium will be present in the trivalent form which will be precipitated in the mixed effluent. The discharge of untreated tannery effluents effects streams, groundwater, land and sewer in which they are discharged.

The discharge of untreated wastewaters in water courses may affect the physical, chemical and biological characteristics of the water and deplete dissolved oxygen from the water bodies. The high oxygen demand of tannery wastes is due to proteins, fatty matter and tannins. High pH, excessive alkalinity, suspended matter, sulphides are injurious to fish and other aquatic life in streams. Sulphides present in tannery wastewaters can cause unpleasant odour problems, react with iron and other metals causing black precipitate, render the water unfit for industrial uses and affect fish and other aquatic life. The sulphide toxicity to fish increase as the pH value is lowered. Nitrogen and phosphorus from tannery effluents encourage uncontrolled growth of algae and other aquatic plants in water bodies. High amounts of chloride present in tannery wastewaters can make the receiving water less suitable for drinking, industrial and agricultural purposes.

Suspended solids both inorganic and organic present in tannery wastewaters may settle in a stream and affect fisheries by covering the bottom of the stream thereby destroying bottom fauna necessary for fish as food or reduce the spawning ground of fisheries. In general the toxic effects of chromium salts particularly hexavalent chromium salts towards aquatic life varies with species, temperature, valence of chromium and the complex synergistic and antagonistic effects due to other factors such as hardness. The effluent from vegetable tanning is coloured and contain some amounts of non-biodegradable matter. When these wastewaters are discharged into streams it is reported that the colour attributed persisted for a long time. The discharge of untreated tannery wastes into water course may also increase the turbidity of water thereby reducing light penetration and impairing photosynthetic activity of aquatic plants.

Groundwater has been found to be affected where wastewater from tanneries are ponded or lagooned spread out on land or discharged into dry river beds. The groundwater is reported to be rendered unfit for drinking and irrigation where the tanneries are concentrated together. When the tannery waste gains access to cultivable lands or when the lands are irrigated with such waste, fertility of the soil is reported to have been affected. It may change the characteristics of the soil and interfere with the intake of water by plants. When tannery wastewaters are applied to land, the soil productivity decreases.

The problems associated with discharge of tannery wastes into municipal sewers, include incrustation of sewers, sewer clogging and other forms of interference with sewage treatment. Tannery effluents are known to cause deposition of calcium carbonate with the consequent choking of the receiving sewer. Lime is converted to calcium carbonate by the carbon dioxide produced by decomposition of the organic matter present in the effluent and the hair and fleshings help to form a binder with this calcium carbonate which firmly adhere and build up gradually on the surface of sewer. Concrete sewers are likely to suffer damage when they are made to carry sewage containing a high concentration of hydrogen sulphide due to admixture with tannery wastes. If the tannery waste concentration in domestic sewage is higher, the interference with waste treatment operation could be due to:

1. Excessive alkalinity or pH.
2. Hair and fleshings which form scum.
3. Lime sludge and adhering deposits.
4. Higher concentration of sulphides and chromium.
5. Excessive loads of organic matter.

49.6.3 Impact of solid wastes on environment

Generation of solid wastes generated during various tanning operations are not properly utilised or disposed they are likely to cause a number of problems. Salt dust if stored in heaps outside the tanneries is likely to be washed away during rains and cause groundwater pollution. Hair and lime sludge if discharged along with the effluents are likely to choke the drains. Raw and green fleshings, limed fleshings, splits and trimmings putrify easily and give rise to noxious smells. In many tanneries, it is the foul odour which emanate from some of these putrescible solid wastes which account for much of the smell traditionally associated with tannery wastes. Vegetable tan bark and vegetable tan sludge if discharged into effluent stream can cause problems in effluent treatment. Vegetable and chrome tanned shavings and splits do not easily decompose. If they are not utilised, problems of disposal are encountered. Primary and secondary sludges obtained during the treatment of tannery wastes are also putrescible. The methods adopted for disposal of solid wastes from tanneries vary considerably from area to area and country to country. In general where tanning industries have long been established there usually has been a simultaneous growth in industries which can use these wastes. This pattern is specially helpful to the tannery industry in the case of obnoxious rag and limed fleshings, trimmings and splits which are lifted by glue, gelatine and fat rendering units.

49.7 Pollution abatement and control in tannery

49.7.1 Control of gaseous pollution

Source control is the most effective means of abating odour. Good sanitation practices are usually cheaper than control measures. Source control of odours can be carried out by following methods given below:

1. Drawing the odorous air from the working atmosphere by exhaust fans and diluting with relatively clean air.
2. Removal of causative impurities from the tannery.
3. Masking the odour with objectionable additives.
4. Removal of odour bearing dusts by cyclone separators.
5. Sorption of odorous gases through a granular sorbents like active carbon.

Chemical air scrubber can be used to eliminate the odour nuisance in the tannery wastewater treatment plant. Odour masking chemicals can be added. Masking agents can also be used to control odours outdoors in such places as waste lagoons. The smell caused by putrefaction of solid and liquid wastes generated during tannery operations can be reduced by quickly disposing them off without allowing for putrefaction. Aerobic biological methods of effluent treatment and dewatering of sludge produced by methods like vacuum filtration, drying on sand drying bed will considerably reduce the smell caused by them.

49.7.2 Control of water pollution

Tannery effluents if disposed off without any treatment either on land or in inland surface waters, may create severe problems leading to damage of the environment. Tannin, trivalent chromium, proteinous matter, sulphides and high BOD/COD of the wastewater call for proper treatment.

Tannery wastes in general, can be treated to get the desired end results in the following main stages: (i) in-plant measures, (ii) primary treatment, (iii) chemical treatment and secondary treatment.

In-plant control measures

In-plant measures include reduction of water consumption in the tannery, process modifications to reduce pollutional load of the waste being generated, segregation of soak liquor and lastly, recovery of the by-products and reuse of various process liquors.

Reduction of water usage in a tannery. To reduce the volume of the effluent, the water usage in tanneries can be considerably reduced by:

1. Better housekeeping.
2. Alteration of processes and low float systems to use less water.
3. Separation of cleaner fractions of the waste for direct reuse without treatment.
4. Recycle after complete or partial treatment.

49.8 Waste management in tanneries

49.8.1 In-house measures

Before designing treatment system the following aspects should be looked into:

1. Segregation of wastewater.
2. Reduction of quantity and strength of wastewaters by adopting in-process and in-plant control measures.
3. Treatability studies of various wastewaters to decide the best combination of treatment system.

49.8.2 Segregation of wastewater

In few countries the salted hides and skins are first dusted to remove as much of dry salt as possible and then only soaked in water. The salt-laden effluents from the soaking process are collected separately in solar evaporation pans and allowed to evaporate. For operational reasons, the solar evaporation pans are divided into four or more compartments.

The dried salt in the pans is manually scraped, collected and dumped along with the salt dustings in nearby places. It has been suggested that the final disposal of the salt residue should only be discharged into the sea at selected places as per the pollution control norms. There is scope for segregation of wastes in some additional cases also as follows:

1. Liming effluent carrying sulphides can be segregated and treated separately by aeration with diffused air in the presence of a catalyst like magneous sulphate to reduce sulphide content before mixing with other wastewater.

2. The waste chrome liquor can be segregated for chromium recovery which will enable the sludge from the other wastes to be used for composting, land fill, etc.

Wastewater treatment combinations and alternatives: The primary and secondary treatment combinations for various types of tannery wastewater depends upon the quality of wastewater, level of treatment required depending upon the final mode of disposal, etc.

49.9 Waste control measures

49.9.1 Good housekeeping and water conservation

As in all waste control programmes, good housekeeping is the first step to prevent wastage of water and materials in tannery. Economical use and reuse of water are necessary to reduce the volume of the wastewater.

The water usage in tanneries can be considerably reduced by:

1. Better housekeeping.
2. Alteration of processes and low float systems to use less water.
3. Segregation of cleaner fractions of the wastewater for direct reuse without treatment.
4. Recycle after complete or partial treatment.

The adoption of batch washing as an alternative to continuous rinsing using the lattice door can reduce water consumption. It has been observed that paddles may have some advantages over drums for certain types of production from the point of view of water usage.

Direct reuse is possible in a tannery since some wash waters are relatively clean, such as from washing after bating, pickling, neutralising and dyeing which can be used for less important tasks such as washing after soaking and liming and floor washing.

49.9.2 Recovery and reuse

Segregation of chrome waste stream, chrome recovery and reuse is proved to be one of the viable procedures. The chrome recovery and reuse in medium and large size tanneries should be practised because it makes good economical sense and prevents pollution due to chromium.

The general methods that may be employed are given below:

1. The spent chrome liquor is captured, filtered, precipitated, reacidified, strengthened and reused as the tanning liquor as depicted in Fig. 49.1. The tannery wastewater is difficult to treat because of complex characteristics like high BOD, COD, suspended solids, sulphide and

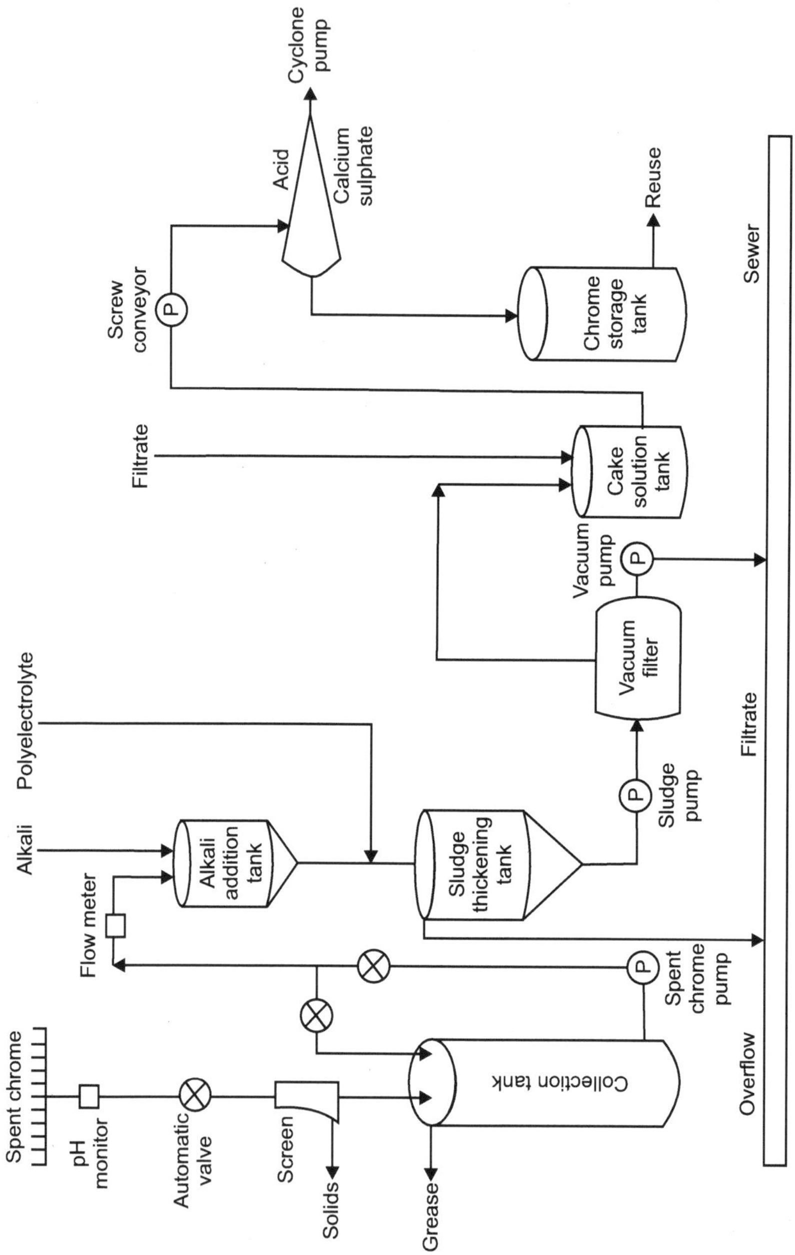

Figure 49.1: Chrome recovery system

chromium. The main source of sulphide in tannery effluent is beam house operations. Anaerobic treatment of tannery wastewater gives better results but formation of sulphide in anaerobic reactors restricts its application. Various phototrophes and chemotrophes have been used for sulphide removal but requirement of light source is the major problem in case of phototrophes. Chemotrophes require careful control of oxygen to produce sulphur instead of sulphate but still some sulphate formation is there.

2. The chrome tanning of hides is carried out by employing Basic Chrome Sulphate (BCS) as the tanning agent. The quantity of BCS employed is 60–80 kg per 1000 kg hide processed. This BCS quantity on average contains 9.04 kg chromium as metal. The chronology of the chrome recovery is summarised in the Fig. 49.2.

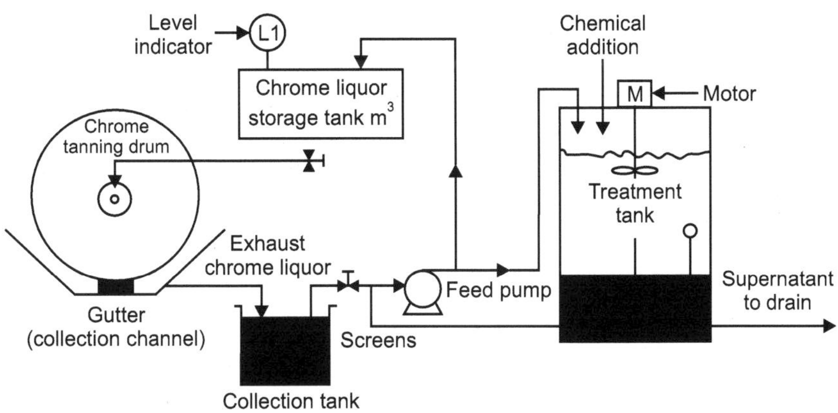

Figure 49.2: Process flow diagram of a typical chrome recovery plant.

It is established that in case of conventional process, even after the chrome tanning is complete wherein the chrome absorption by the hide is only 60%, the spent tan liquor still contains a significant concentration (upto 40%) of chromium which can be recovered by the mechanism (upto 40%) described below:

Step 1: Collection of spent tan liquor.

Step 2: Maintenance of pH of spent liquor at 8 to 8.2 by adding MgO (0.5% of spent liquor) to optimise the precipitation of chrome (in sludge).

Step 3: The mixture is stirred and allowed to settle down for few hours. The chrome in the form of chrome hydroxide gets settled.

Step 4: The supernatant containing insignificant quantity of chromium concentration is liable to be discarded.

Step 5: The chrome sludge is dissolved in 98% concentration sulphuric acid (quantity of which is 0.1% of spent liquor). This mixture is worth of blending with freshly made chrome tan liquor.

Step 6: It is generally believed that recovery of chromium in a properly designed and operated recovery plant could be somewhere around 98% of the chromium in the spent tanned liquor.

50

Nanotechnology for industrial pollution prevention

50.1 Introduction

Environmental nanotechnology is considered to play a key role in the shaping of current environmental engineering and science. The nanoscale has stimulated the development and use of noval and cost effective technologies for remediation, pollution detection, pollution monitoring and remediation of pollutants. The nanostructured materials are used as biosensors for monitoring and detection of different compounds. The use of nanoparticals may have advantage over conventional method due the much larger surface area of nanoparticles on a mass basis. The unique structure and electronic properties of some nanoparticles are adsorbent of pollutants.

Many nano materials have properties of adsorbents, depending on size. Scientific and technical methods to mitigate environmental pollution rely on many approaches and vary for the cases of soil, water and air purification. The remediation approach chosen depends on the complexity and nature of the contaminated media and economic costs of the treatment. Environmental remediation plays a pivotal role in the decision to employ a particular technology.

50.2 Nanotechnology and pollution control

Pollution results from resource production and consumption, which in their current state are very wasteful. Nanofabrication holds much potential for effective pollution control, but it currently faces many problems that prevent it from mass commercialisation particularly its high cost. Nanotechnology plays a vital role in air and water pollution control.

50.2.1 Air pollution

Air pollution can be remediated using nanotechnology in several ways. One is through the use of nano-catalysts with increased surface area for gaseous reactions. Catalysts work by speeding up chemical reactions that transform harmful vapours from cars and industrial plants into harmless gases. Catalysts currently in use include a nanofibre catalyst made of manganese oxide that removes volatile organic compounds from industrial smokestacks. Other methods are still in development. Another approach uses nanostructured membranes that have pores small enough to separate methane or carbon dioxide from exhaust. John Zhu of the University of Queensland is researching carbon nanotubes (CNT) for trapping greenhouse gas emissions caused by coal mining and power generation. CNT can trap gases up to a hundred times faster than other methods, allowing integration into large-scale industrial plants and power stations. This new technology both processes and separates large volumes of gas effectively, unlike conventional membranes that can only do one or the other effectively.

50.2.2 Water pollution

As with air pollution, harmful pollutants in water can be converted into harmless chemicals through chemical reactions. Trichloroethene, a dangerous pollutant commonly found in industrial wastewater, can be catalysed and treated by nanoparticles. Studies have shown that these 'materials should be highly suitable as hydrodehalogenation and reduction catalysts for the remediation of various organic and inorganic groundwater contaminants'.

Nanotechnology eases the water cleansing process because inserting nanoparticles into underground water sources is cheaper and more efficient than pumping water for treatment. The deionisation method of using nano-sized fibres as electrodes is not only cheaper, but also more energy efficient. Traditional water filtering systems use semipermeable membranes for electrodialysis or reverse osmosis. Decreasing the pore size of the membrane to the nanometer range would increase the selectivity of the molecules allowed to pass through. Membranes that can even filter out viruses are now available.

Also widely used in separation, purification and decontamination processes are ion exchange resins, which are organic polymer substrate with nano-sized pores on the surface where ions are trapped and exchanged for other ions. Ion exchange resins are mostly used for water softening and water purification. In water, poisonous elements like heavy metals are replaced by sodium or potassium. However, ion exchange resins are easily damaged or contaminated by iron, organic matter, bacteria and chlorine. Recent developments of nanowires made of potassium manganese oxide can clean up oil and other organic pollutants while making oil recovery possible. These nanowires form a mesh that absorbs up to twenty times its weight in hydrophobic liquids while rejecting water with its water repelling coating. Since the potassium manganese oxide is very stable even at high temperatures, the oil can be boiled off the nanowires and both the oil and the nanowires can then be reused.

50.2.3 Removal of pollutant by adsorbant

The nanoparticles have been investigated as adsorbant for removal of organic and inorganic contaminants. The nano size metal oxides and natural nanosized clays have been investigated for the removal of metals and inorganic ions. Besides oxidised and hydrooxylated CNTs are good absorbers for metals such as Cu, Ni, Cd and Pb. Pristine multiwalled CNTs has been found to be stronger adsorber materials for organometallic compounds. CNTs has also been found as a powerful adsorbant for a wide variety of organic compounds from aquatic environment, which include: dioxin, polynuclear aromatic hydrocarbons (PAHs), DDT and its metabolites, PBDEs, chlorobenzenes and chlorophenoles, trihalomethanes, bisphenol A and nonyphenol, phthalate easters, dyes, pesticides and herbicides such as sulphuron derivatives, atrazine and dicamba. Nanoporous polymers which has cross linked and copolymerised with fuctionalised CNTs have been demonstrated for a high sorption capacity for a variety of organic compounds such as *p*-nitrophenol and trichloroethylene.

50.2.4 Degradation of pollutants using semiconductors-TiO_2 and ZnO

The semiconductor-TiO_2 nanoparticles have been extensively studied for oxidative transformation of organic and inorganic contaminants. They are now used in a variety of products such as self-cleaning glass, disinfectant tiles and filters for air purification. TiO_2 electrodes have the capacity to determine the chemical oxygen demand of water and are used as sensors for monitoring contaminated water. TiO_2 nanoparticles can be immobilised on different supports which are used for the solar detoxification of water and air. These engineered nanoparticles are known for their interaction with organic, inorganic and biological contaminants such as heavy metals, organochlorine pesticide, arsenic and phosphates in water, induced by ultravilot light. TiO_2 leads to pollutant degradation through two everyday chemical reactions: reduction and oxidation. Once excited by UV and TiO_2 electron-hole pairs develop. These electrons have sufficient oxidising potential to oxidise pollutants in wastewater. Interestingly, the combination of UV and TiO_2 generates bactericidal activity, which attacks several types of bacteria. This approach thus provides a comprehensive treatment procedure since chemical species and pathogens can be removed from wastewater simultaneously.

Other nanoparticles with semiconducting properties, such as ZnO, ZnS, F_2O_3 and CdS can be used for photocatalysis oxidation. TiO_2 is biologically and chemically inert and has demonstrated great resistance to corrosion along with the capacity to be used repetitively without substantial loss of catalytic activity and it is therefore inexpensive to use. In light of these properties, TiO_2, is potentially more attractive for environmental applications than other oxidative nanoparticles. However, TiO_2 requires ultraviolet light and consequently, is effective only for treatment of transparent wastewater. Several research groups have attempted to overcome this limitation by extending the nanoparticle excitation into visible light with the doping of TiO_2 with transition metal ions or sensitising dye such as Ru (II) polypyridyl complex or investigated an approach involving a tube reactor production based on hollow glass tubes. The tubes are extremely coated with TiO_2 and UV-Light passes through the hollow tubes. Preliminary results suggested that combining ultrasound processes such as sonolysis and photocatalysis improves pollutant oxidation and could be an effective approach. Several pilot projects are under way to evaluate the potential of solar reactors to generate the energy required for TiO_2 excitation and detoxification of polluted water. If these attempts are successful sunlight could play an economic and ecologically part in the treatment of wastewater. Unlike TiO_2, semiconductor nanoparticles such ZnO are able to emit strongly in the visible region and the visible emissions of ZnO are usually very sensitive to hole scavengers such as phenols or iodide ions. ZnO particles thus seem good candidates for use as sensors for chemical compounds. It has been claimed that sensor system based on ZnO could reach a detection sensitivity of 1 ppm. Furthermore, ZnO has the ability to induce contaminant degradation under ultraviolet light. Consequently, use of ZnO can be considered as a promising way to simultaneously sense and destroy toxic chemicals. It has been reported that nanostructured ZnO films could simultaneously detect and degrade organic compounds in water. Such a catalyst system is useful to induce contaminant degradation where the system senses a targeted molecule, thus avoiding destruction of inoffensive molecules present in the environment.

50.2.5 Degradation of pollutants using iron nanoparticles

Laboratory research has established that nanoscale metallic iron is very effective in destroying a wide variety of common contaminants such as chlorinated methanes, brominated methanes, trihalomethanes, chlorinated etenes, chlorinated benzenes, other polychlorinated hydrocarbons, pesticides and dyes. The basis for the reaction is the corrosion of zerovalent iron in the environment. Contaminants such as tetrachloroethene can readly accept the electros from iron oxidation and be reduced to ethane.

However, nanoscale zerovalent iron (nZVI) can reduce not only organic contaminants but also the inorganic anions nitrate, which is reduced to ammonia perchlorate (plus chlorate or chlorite) and then reduced to chloride, selenate, arsenate, arsinite and chromate. nZVI is also efficient in removing dissolved metals from solution, e.g., Pb and Ni28. The reaction rates for nZVI are atleast 25–30 times faster and also sorption capacity is much highercompared with granular iron. The metals are either reduced to zerovalent metals or lover oxidation states, e.g., Cr (III), or are surface complexed with the iron oxides that are formed during the reaction. Some metal can increase the dechlorination rate of organics and also lead to more benign products, whereas other metals decrease the reactivity. The most of the research using nZVI has been devoted to groundwater and soil remediation.

50.2.6 Nanotech for hazardous waste clean-ups

Nanoscale materials can make a huge difference in the cleanup of hazardous waste. There are two reasons for the optimism: firstly, the size of nano materials lets them penetrate otherwise impossible- to-reach ground water or soil and secondly, their engineered coatings allow them to stay suspended in groundwater, a major asset in cleanups. If practically feasible, nanomaterials could slash clean-up prices by avoiding the extraordinary costs and risks of hauling waste away for burning or burial. Most nanoremediation research projects undertaken by the department of defence are focused on cleaning up ground water contaminated by chlorinated solvents like trichloroethylene. Research shows that results have been promising in most demonstrations with most of the contaminant being destroyed – a finding that has been replicated by researchers elsewhere. However, there are complications. For one, iron also reacts with non-targeted materials, making it to degrade too quickly, before remediation is comparticles clump after their release, making it difficult for them to travel beyond where they are injected. Scientists are trying coatings that might enable nanoparticles to travel. They are also trying to learn how to make the nanoiron react only with target contaminants as well as to get it to self-destruct after it has done its job. Such 'smarter' nanomaterials will ultimately make the technology won't be a silver bullet for clean ups because most contaminated sites are fouled by more than a single contaminant and a phased approach will be needed.

50.2.7 Nanoparticles in wastewater

The environmental management of nanoparticle waste could greatly improve following a new study into substances found in wastewater. Researchers from several scientific institutions around the world collaborated on the study have focused on the behaviour of potentially hazardous nanoparticles in sewage treatment plants, particularly silica-shelled nanoparticles. During the course of the studies, it was discovered that coating these nanoparticles with a commercial surfactant enabled their separation from the water together with other waste particles. The scientists say that if the nanoparticles were allowed to 'settle out' in this way, then they could be stopped from passing on to the subsequent stage of the waste treatment process, which would help prevent industrial accidents.

50.3 Remediation technologies to clean up

50.3.1 Environmental pollutants

Environmental cleanup has promoted the development of highly efficient photocatalysts that can participate in detoxification reactions. Environmental remediation by photocatalysts comes with several advantages: direct conversion of pollutants to non toxic by products without the necessity for any other associated disposal steps; use of oxygen as oxidant and elimination of expensive oxidising chemicals; potential for using free and abundant solar energy; self regeneration and recycling of photocatalyst, etc. A significant amount of research on semiconductor catalysed photooxidation of organic chemicals has been carried out during the past 25 years. The ability to catalyse the destruction of a wide variety of organic chemicals and complete oxidation of organics to CO_2 and dilute mineral acids in many cases, lack of inherent toxicity and resistance to photodegradation at low cost render this process highly suitable for environmental remediation. Environmental protection and pollution issues are frequently discussed worldwide as topics that need to be addressed sooner rather than later. Nanotechnology can strive to provide and fundamentally restructure the technologies currently used in environmental detection, sensing, remediation and pollution removal.

Some nanotechnology applications are near commercialisation: nanosensors and nanoscale coatings to replace thicker, more wasteful polymer coating that prevent corrosion, nanosensors for dectection of aquatic toxins, nanoscale biopolymers for improved decontamination and recycling of heavy metals, nanostructured metals that break down hazardous organics at room temperature, smart particles for environmental monitoring and purification, nanoparticles as novel photocatalyst for environmental cleanup.

50.4 Pollution prevention/reduction

Reduction of waste in manufacturing processes; reduction in the use of harmful chemicals; reduction in the emission of 'greenhouse'-effect gases during fuel combustions; use of biodegradable plastics: these are only few of the many approaches that can be taken to reduce the pollution of the environment. Nanotechnology is already actively involved in this sector, either as a technology to produce advanced materials that pollute less, or as a method to increase the efficiency of certain industrial processes.

50.4.1 Materials

Materials that are more environment-friendly fabricated using nanotechnology include biodegradable plastics made of polymers that have a molecular structure optimal for degradation; non-toxic nanocrystalline composite materials to replace lithium-graphite electrodes in rechargeable batteries; and self-cleaning glasses, such as Activ™ glass, a commercial product available worldwide from Pilkington. The glass is composed of a special coating made of nano-crystals of TiO_2 which, once exposed to daylight, reacts in two ways. First, it breaks down any organic dirt deposits on the glass and second, when exposed to water, it allows rain to 'sheet' down the glass easily and washes the loosened dirt away. Another area where nanotechnology is making a contribution is the development of fertilisers and wood treatment products that are more stable and leach less into the environment. For instance, researchers at the Michigan State University have incorporated biocides for wood treatment inside polymeric nanoparticles.

Their small size allows them to efficiently travel inside the very fine, sieve-like structure of wood. At the same time, the biocide, being safely trapped inside a 'nanoshell', is protected from leach and random degradative processes.

50.4.2 Nanocatalysis

A catalyst is a substance that increases a chemical reaction rate without being consumed or chemically altered. One of the most important properties of a catalyst is its 'active surface' where the reaction takes place. The 'active surface' increases when the size of the catalysts is decreased. The higher is the catalysts active surface, the greater is the reaction efficiency. Also, research has shown that the spatial organisation of the active sites in a catalyst is important as well. Both properties (nanoparticle size and molecular structure/distribution) can be controlled using nanotechnology. In the environmental field, nanocatalysis is being investigated for desulphurising fuels, with the aim of developing 'clean' fuels containing very low sulphur products (produced in the fuel during its refining process and responsible for generating sulphuric acid upon fuel combustion).

A commercial example is Oxonica's Envirox fuel which uses nano-sized cerium oxide as a catalyst to enhance efficiency. This enhanced-fuel has been tested in 2003 and 2004 in 1000 buses in the U.K. (another 500 buses were tracked as control). It was found that the test buses used 5% less fuel then the controls and that the fuel savings more then paid for the additive.

Nanoscale catalysts are also promising for improving air quality and for treating particularly challenging contaminants in water that must be reduced to a very low level.

50.4.3 Green manufacturing

Manufacturing processes are always accompanied by the production of diverse waste products, many of which pose a threat to the environment and thus need to be removed and treated. Ideally, manufacturing processes should be designed to minimise material usage and waste production, while ensuring the use of the least amount of energy possible. 'Green manufacturing' is a generic name to broadly cover methods and technologies that are directed towards achieving this goal. It includes the development of new chemical and industrial procedures (for instance water-based rather the solvent-based processes); reduction in the use of unsafe compounds (such as metals); development of 'green' chemicals that are more environment-compatible and efficient use of energy. In terms of its application to the reduction of manufacturing waste, nanotechnology can contribute in two ways: by directing the manufacturing to be more controlled and efficient and by using nanomaterials (such as catalysts) that can raise the manufacturing efficiency while reducing or eliminating the use of toxic materials. Overall, nanotechnology has the potential of making industrial processes more efficient in terms of energy usage and material usage, while minimising the production of toxic wastes. The application of 'green nanotechnology' to manufacturing includes bottom-up, atomic-level synthesis for developing improved catalysts; inserting information into molecules to build new materials (such as DNA) through highly specific synthetic routes; scaling down material usage during chemical reaction by using nanoscale reactors and improving manufacturing to require less energy and less toxic materials. An example of 'green nanotechnology' is the development of aqueous-based microemulsions to be used in alternative to volatile organic compounds (VOCs) in the cleaning industry. These toxic and potentially carcinogenic compounds, such as chloroform, hexane, percholoroethylene, are conventionally used in the cleaning industry (like the textile industry) as well as in the oil extraction industry.

Microemulsions contain nano-sized aggregates that can be used as 'receptors' for extracting specific molecules at a nanoscale level. Researchers from the University of Oklahoma have synthesised microemulsions having water attractive and water-repellent 'linkers' inserted between the head and tail parts of a surfactant molecule. The result is a surfactant that has a very low interfacial tension with a wide range of oils. When tested for cleaning textiles from motor oil residues, as well as for extracting edible oil from oilseeds, the microemulsions were found to be very competitive with conventionally used VOCs, both in terms of extraction yield and simplicity of the process.

50.4.4 Environment sensing

Protection of the human health and of the environment requires the rapid, sensitive detection of pollutants and pathogens with molecular precision. Accurate sensors are needed for *in situ* detection, as miniaturised portable devices and as remote sensors, for the real-time monitoring of large areas in the field. Generally speaking, a sensor is a device built to detect a specific biological or chemical compound, usually producing a digital electronic signal upon detection. Sensors are now used for the identification of toxic chemical compounds at ultra low levels (ppm and ppb) in industrial products, chemical substances, water, air and soil samples or in biological systems. Nanotechnology can improve current sensing technology in various ways. First, by using nanomaterials with specific chemical and biological properties, the sensor selectivity can be improved, thus allowing isolating a specific chemical or biological compound with little interference. Hence, the accuracy of the sensors is improved. As with other nanoengineered products discussed in this section, the high surface-to-volume ratio of nanomaterials increases the surface area available for detection, which in turn has a positive effect on the limit of detection of the sensor, therefore improving the sensitivity of the device. Scaling down using nanomaterials allows packing more detection sites in the same device, thus allowing the detection of multiple analytes. This scaling-down capability, together with the high specificity of the detection sites obtainable using nanotechnology, will allow the fabrication of super-small 'multiplex' sensors, this way lowering the cost of the analysis and reduce the number of devices needed to perform the analysis with an economic benefit. Advancements in the field of nanoelectronics will also allow the fabrication of nanosensors capable of continuous, real time monitoring.

Research in the field of nanosensors includes various areas, like synthesising new nanomaterials with specific detection sites able to recognise a certain pollutant; developing new detection methods, to increase the limit of detection of the sensors while ensuring a 'readable' electrical signal; and miniaturising the size of the sensor elements while integrating these with larger parts of the device. An example of how nanoscience can be applied to the sensing technology which discusses the operational principle of a heavy-metal nanosensor developed for monitoring heavy metals in drinking water. The sensor is made of an array of electrode pairs fabricated on a silicon chip and separated by few nanometres. When the electrodes are exposed to a solution of water containing metal ions, these deposit inside the nano-gap in between the electrodes. Once the deposited metal bridges the gap a 'jump' in conductance between the electrodes is registered. The size of the gap, being only few nanometres, allows the detection of a very low concentration of metal ions. This type of sensor is called 'nanocontact sensor'.

Some nanomaterials in the form of nanowires or nanotubes offer outstanding opportunities as sensor elements in chemical and biological sensors. Individual single-walled carbon nanotubes (SWNTs) have been demonstrated to exhibit a faster response and a substantially higher sensitivity for instance towards gaseous molecules (such as NO_2 and NH_3) than that of existing solid-state sensors. In this case, direct binding of the gaseous molecule to the surface of the SWNT is the mechanism involved in sensing, upon which the electrical resistance of the SWNT dramatically increases or decreases. Moreover, this sensitivity was registered at room temperature, whereas conventional solid-state sensors operate at very high (200 to 600°C) temperatures in order to achieve enhanced chemical reactivity between molecules and the sensor material.

Although SWNTs are promising candidates as nanosensors, they also have some limitations that could limit their development. First, existing synthetic methods produce a mixture of metallic and semiconducting NTs, only the latter being useful as sensors. Second, in order to be able to sense a variety of chemical and biological species, the surface of NTs needs to be modified to have specific functionalities to bind those species. Flexible methods to modify the surface of NTs to bind a large variety of analytes are not well established yet. Conversely, nanowires of semiconductors such as Si don't have these limitations: they are always semiconductors and there is established knowledge for the chemical modification of their surface. Boron-doped silicon nanowires (SiNWs) have been used for the sensitive real-time electrical detection of proteins, antibodies the metabolic indicator calcium and glucose in water. The small size and the capability of these semiconductor nanowires to detect in real-time a wide range of analytes could be used for developing sensors for detecting pathogens, chemical and biological agents in water, air and food.

50.5 Nanotechnology for pollution prevention in textile industry

Nanotechnology is growing by leaps and bounds and it has been introduced in many fields including the textile industries. There is a considerable potential for profitable applications of nanotechnology in cotton and other textile industries. Its application can economically extend the properties and values of textile processing and products. By deploying nanotechnology, ultra strong, durable and specific function oriented fabrics can be efficiently produced for a number of applications including medical, military and industrial apparels, etc. Nanotechnology overcomes the limitations of applying conventional methods to impart certain properties to textile materials. There is no doubt that in the next few years, nanotechnology will penetrate into every area of textile industry.

Azo dyes and Cr (VI) most often coexists in industrial wastewater. These contaminants enter the environment when released through wastewater in textile industry and exert many carcinogenic and mutagenic effects on plants, animals and human health. Therefore, treatment of wastewater and its reuse is a practice related not only to a number of benefits with regard to eater balance and management but also to a number of question marks.

Immediate research must be launched towards the direction so as to protect human health and environmental ecosystems. Micro-organism (bacteria, fungi) is the potential source to remove these contaminants from soil and wastewater because they are cost effective and environmentally friendly and immediate contaminants at a faster rate.

However, their immobilisation can accelerate this treatment more efficiently. In this regard, surface modified nanomaterial and biochar have achieved much attention to remove heavy metals and azo dyes simultaneously.

Nanomicrobiological techniques has been proposed or are under active development for treatment of polluted soil and wastewater, but many techniques are still at experimental or pilot stage. In conclusion, there is much recent interest in the use of engineered nanomaterial combined with biotechnology as in *in situ*. In general, the attention in this field is not enough and additional studies should be conducted to increase the knowledge in this field. In addition, health impacts and environmental fate of these nanomaterials need to be addressed before their widespread application.

50.6 Nano-silica-silver composite material for removal of textile effluents and dyes

The effluents of textile industries containing different type of dyes are often discharged untreated into water bodies. Some 72 toxic chemicals have been identified in water solely from textile dyeing, 30 of which cannot be removed.

This causes a serious environmental threat to aquatic and human life. Moreover, water treatment plants are very prone to fouling due to micro-organism growth in the contaminated water, resulting in higher energy consumption and operating cost.

The development of sustainable, robust, energy-efficient and cost-effective water purification technologies is a challenging task. Conventional practices adopted for water purification-which can be classified into physical, chemical and biological methods-suffer from certain limitations such as high cost, low adsorption capacity, generation of toxic sludge, etc.

These technologies-which include coagulation, flocculation, reverse osmosis, membrane separation, oxidation and ozonation, adsorption-are expensive or inadequate to remove the dye. Adsorption with activated carbons, which is a cheap and effective method, has been demonstrated to remove dye from waste. But this approach is not suitable for industrial wastewater treatment because activated carbon can only be used once and then it is commonly disposed of in landfills. Moreover, removal of pathogens from treated water requires additional processes like chlorination, ozonation, etc., which increases the cost of treatment.

50.6.1 Nano-silica-silver composite material

A possible solution to tackle this problem has been demonstrated by scientists in India. They developed nanotechnology-based water purification using nano-silica-silver composite material as antifouling, antimicrobial and dye adsorptive material. Using this process, pathogenic bacteria and dye present in contaminated water can be treated simultaneously without using any chemicals, high-temperature, pressure or electricity.

'Researchers synthesised a nanosilica supported silver nanocomposite material through ecofriendly protein mediated reduction of nano-silica bound silver ions.' 'The proteins, extracted from *Rhizopus oryzae* - a zygomycetes fungy, served both as a reducing and a protecting agent for the silver nanoparticles and prevented their oxidation under environmental conditions. The result is a low-cost, highly effective nanomaterial for sustainable water purification.'

The simple, low-temperature bio-synthesis fabrication process–it does not require any elaborate or expensive equipment–works without any chemicals for reduction of silver ions and subsequent production of silver nanoparticles, thus minimising the environmental load of toxic chemicals during the fabrication of this nanocomposite material. The coating of proteins on the nanoparticles' surface prevents the leaching of silver ions – which in itself could be a source of water contamination – and provides long stability of the nanocomposite.

Researchers found that the synthesised nanocomposite demonstrated very high dye removal capacities and exhibited antimicrobial and antifouling properties. The nanocomposite removes the dyes at wide pH, temperature and dye concentration in solution. Moreover, the nanocomposite kills the micro-organisms frequently present in the contaminated water.

It was found that the silver nanocomposite very efficiently removes dyes and pathogenic micro-organisms from water bodies in a single-step operation. In addition, the nanocomposite material could be regenerated after treatment of dye bearing wastewater and the regenerated nanocomposite could be stored and reused for several more cycles.'

Nanosilica has a very high surface to volume ratio and contains a large number of surface hydroxyl groups, which provide electrostatic binding energy for dye molecules on its surface. Combining this with the antibacterial activity of the silver nanoparticles on the surface of nano-silica results in a synergistic effect of the nanocomposite, which is responsible for high removal of dyes and micro-organisms from contaminated water.

The nanocomposite also prevented attachment of floating micro-organisms and inhibited the formation of biofilms on its surface. This makes it possible to use it for prolonged times in contaminated water.

Thus, the long term antibacterial, antifouling and high dye adsorption properties of our functional nanomaterial are exceptionally promising for the development of high-efficiency and low cost water purification technologies.

References

Arceivala, S.J., *Wastewater Treatment and Disposal,* Marcel Dekker Inc., New York.

Bradley, S.R., *Industrial Pollution*, Academic Press, London.

Bryan, H.K, *Handbook of Industrial Pollution,* Academic Press, London.

Chaphekar, S.B., and Shetye, R.P., *Effects of Industrial Pollutants on Plants,* Final Report of the U.G.C. Sponsored Research Project, Inst. of Science, Bombay.

Chapman, S.B., *Air Pollution Sources,* Blackwell Scientific Publications, London.

Connell, D.W. and Miller, G.J., *Chemistry and Ecotoxicology of Pollution,* John Wiley & Sons, New York.

Cox, C.R., *Operation of Control of Water Treatment Process.* No. 49, WHO, Geneva.

Creighton, K.E., *Atmospheric Dispersion of Pollution*, Academic Press, London.

Dix, H.M., *Environmental Pollution,* John Wiley & Sons, New York.

Ellis, K.V. and Warn, A. E., *Surface Water Pollution and Its Control*, Macmillan Education, UK.

Evison, L., *Fundamentals of Air Pollution*, Reinhold Publishing Corporation, New York.

Horald, G., *Air Pollution and Its Control*, Prentice-Hall, London.

James, A., *Refinery Operations*, John Wiley & Sons, New York.

Jencks, W.P., *Chemical Analysis of Pollutants*, John Wiley & Sons, New York.

Lawrence, K., Wang, Nazih K. Shammas and Yung-Tse Hung, *Advances in Hazardous Industrial Waste Treatment*, CRC Press, New York.

Ledbetter, J.O., *Air Pollution: Analysis,* Marcel Dekker, New York.

Liptak, B.G., *Water Pollution Environmental Engineers Handbook,* Chilton Book Company, Radnor, Pennsylvania.

Mansfield, T.A., *Effects of Air Pollutants on Plants,* Cambridge University Press, Cambridge.

Martin, B.H., *Pollution Prevention in Process Industries*, Academic Press, London.

Masters, G.M., *Introduction to Environmental Science and Technology,* John Wiley & Sons, New York.

McCaull, J. and Crossland, J., *Water Pollution,* Harcourt Brace Jovanovich, New York.

Metham, R.A., *Atmospheric Pollution and its History, Origin and Prevention*, Pergamon Press, New York.

Munn, R.F., *Environment Impact Assessment,* John Wiley & Sons, New York.

Nancy, J. Sell, *Industrial Pollution Control: Issues and Techniques*, Wiley-Blackwell.

Nemerow, N.L., *Industrial Water Pollution: Origins, Characteristics and Treatment*, Addison-Wesley Publishing Company, Philippines.

Nyer, E., *Chemistry of Air Pollution*, Van Nostrand Reinhold, New York.

Painter, D.E., *Air Pollution Technology,* Reston Publishing Co., Reston, Virginia.

Reid, G., *Industrial Pollution*, Reinhold Publishing Corporation, New York.

Rudd, R.L. and Herman, S.G., *Environmental Toxicology of Pesticides,* Academy Press, New York.

Sax, N.I., *Industrial Pollution,* Van Nostrand Reinhold, New York.

Smith, C.N., *Handbook of Air Pollution*, McGraw-Hill, Tokyo.

Snell, I.D. and Snell, C.T., *Handbook of Industrial Pollution,* D.Van Nostrand, New York.

Teal, J.M., *Air Pollution and Its Fundamentals*, Pergamon Press, New York.

Wesley Eckenfelder Jr., *Industrial Water Pollution Control*, McGraw-Hill. New York.

William, Heumann, L., *Industrial Air Pollution Control Systems*, McGraw-Hill Professional. New York.

Wolfrom, M., *Environmental Contaminants*, Reston Publishing Co., Reston, Virginia.

Index

F

P